Cohen–Macaulay Rings

Already published

Cohen–Macaulay Rings
Revised edition

Winfried Bruns
University of Osnabrück

Jürgen Herzog
University of Essen

CAMBRIDGE
UNIVERSITY PRESS

CAMBRIDGE UNIVERSITY PRESS
Cambridge, New York, Melbourne, Madrid, Cape Town, Singapore, São Paulo

Cambridge University Press
The Edinburgh Building, Cambridge CB2 2RU, UK

Published in the United States of America by Cambridge University Press, New York

www.cambridge.org
Information on this title: www.cambridge.org/9780521566742

First published 1993
Reprinted 1996
First paperback edition with revisions 1998

A catalogue record for this publication is available from the British Library

ISBN-13 978-0-521-56674-2 paperback
ISBN-10 0-521-56674-6 paperback

Transferred to digital printing 2005

For our wives,

Ulrike and Maja

Contents

II   Classes of Cohen–Macaulay rings   205

III   Characteristic p methods   321

Preface to the revised edition

The main change in the revised edition is the new Chapter 10 on tight closure. This theory was created by Mel Hochster and Craig Huneke about ten years ago and is still strongly expanding. We treat the basic ideas, F-regular rings, and F-rational rings, including Smith's theorem by which F-rationality implies pseudo-rationality. Among the numerous applications of tight closure we have selected the Briançon–Skoda theorem and the theorem of Hochster and Huneke saying that equicharacteristic direct summands of regular rings are Cohen–Macaulay. To cover these applications, Section 8.4, which develops the technique of reduction to characteristic p, had to be rewritten. The title of Part III, no longer appropriate, has been changed.

Another noteworthy addition are the theorems of Gotzmann in the new Section 4.3. We believe that Chapter 4 now treats all the basic theorems on Hilbert functions. Moreover, this chapter has been slightly reorganized.

The new Section 5.5 contains a proof of Hochster's formula for the Betti numbers of a Stanley–Reisner ring since the free resolutions of such rings have recently received much attention. In the first edition the formula was used without proof.

We are grateful to all the readers of the first edition who have suggested corrections and improvements. Our special thanks go to L. Avramov, A. Conca, S. Iyengar, R. Y. Sharp, B. Ulrich, and K.-i. Watanabe.

Osnabrück and Essen,
October 1997

WINFRIED BRUNS
JÜRGEN HERZOG

Preface to the first edition

The notion of a Cohen–Macaulay ring marks the cross-roads of two powerful lines of research in present-day commutative algebra. While its main development belongs to the homological theory of commutative rings, it finds surprising and fruitful applications in the realm of algebraic combinatorics. Consequently this book is an introduction to the homological and combinatorial aspects of commutative algebra.

We have tried to keep the text self-contained. However, it has not proved possible, and would perhaps not have been appropriate, to develop commutative ring theory from scratch. Instead we assume the reader has acquired some fluency in the language of rings, ideals, and modules by working through an introductory text like Atiyah and Macdonald [15] or Sharp [344]. Nevertheless, to ease the access for the non-expert, the essentials of dimension theory have been collected in an appendix.

As exemplified by Matsumura's standard textbook [270], it is natural to have the notions of grade and depth follow dimension theory, and so Chapter 1 opens with the introduction of regular sequences on which their definition is based. From the very beginning we stress their connection with homological and linear algebra, and in particular with the Koszul complex.

Chapter 2 introduces Cohen–Macaulay rings and modules, our main subjects. Next we study regular local rings. They form the most special class of Cohen–Macaulay rings; their theory culminates in the Auslander–Buchsbaum–Serre and Auslander–Buchsbaum–Nagata theorems. Unlike the Cohen–Macaulay property in general, regularity has a very clear geometric interpretation: it is the algebraic counterpart of the notion of a non-singular point. Similarly the third class of rings introduced in Chapter 2, that of complete intersections, is of geometric significance.

In Chapter 3 a new homological aspect determines the development of the theory, namely the existence of injective resolutions. It leads us to the study of Gorenstein rings which in several respects are distinguished by their duality properties. When a Cohen–Macaulay local ring is not Gorenstein, then (almost always) it has at least a canonical module which, so to speak, acts as its natural partner in duality theorems, a decisive fact for many combinatorial applications. We then introduce local cohomology and prove Grothendieck's vanishing and local duality theorems.

Chapter 4 contains the combinatorial theory of commutative rings which mainly consists in the study of the Hilbert function of a graded module and the numerical invariants derived from it. A central point is Macaulay's theorem describing all possible Hilbert functions of homogeneous rings by a numerical condition. The intimate connection between homological and combinatorial data is displayed by several theorems, among them Stanley's characterization of Gorenstein domains. In the second part of this chapter the method of associated rings and modules is developed and used for assigning numerical invariants to modules over local rings.

Chapters 1–4 form the first part of the book. We consider this material as basic. The second part consists of Chapters 5–7 each of which is devoted to a special class of rings.

Chapter 5 contains the theory of Stanley–Reisner rings of simplicial complexes. Its main goal is the proof of Stanley's upper bound theorem for simplicial spheres. The transformation of this topological notion into an algebraic condition is through Hochster's theorem which relates simplicial homology and local cohomology. Furthermore we study the Gorenstein property for simplicial complexes and their canonical modules.

In Chapter 6 we investigate normal semigroup rings. The combinatorial object represented by a normal semigroup ring is the set of lattice points within a convex cone. According to a theorem of Hochster, normal semigroup rings are Cohen–Macaulay. Again the crucial point is the interplay between cellular homology on the geometric side and local cohomology on the algebraic. The fact that the ring of invariants of a linear torus action on a polynomial ring is a normal semigroup ring leads us naturally to the study of invariant rings, in particular those of finite groups. The chapter closes with the Hochster–Roberts theorem by which a ring of invariants of a linearly reductive group is Cohen–Macaulay.

Chapter 7 is devoted to determinantal rings. They are discussed in the framework of Hodge algebras and algebras with straightening laws. We establish the straightening laws of Hodge and of Doubilet, Rota, and Stein, prove that determinantal rings are Cohen–Macaulay, compute their canonical module, and determine the Gorenstein rings among them. In view of the extensive treatment available in [61], we have restricted this chapter to the absolutely essential.

The third part of the book is constituted by Chapters 8 and 9. They owe their existence to the fact that a Noetherian local ring is in general not Cohen–Macaulay. But Hochster has shown that such a ring possesses a (not necessarily finite) Cohen–Macaulay module, at least when it contains a field. The construction of these 'big' Cohen–Macaulay modules in Chapter 8 is a paradigm of characteristic p methods in commutative algebra, and we hope that it will prepare the reader for the more recent developments in this area which are centered around the notion of tight closure introduced by Hochster and Huneke [190].

In Chapter 9 we deduce the consequences of the existence of big Cohen–Macaulay modules, for example the intersection theorems of Peskine and Szpiro and Roberts, the Evans–Griffith syzygy theorem, and bounds for the Bass numbers of a module.

Chapters 8 and 9 are completely independent of Chapters 4–7, and the reader who is only interested in the homological theory may proceed from the end of Section 3.5 directly to Chapter 8.

It is only to be expected that the basic notions of homological algebra are ubiquitous in our book. But most of the time we will only use the long exact sequences for Ext and Tor, and the behaviour of these functors under flat extensions. Where we go beyond that, we have inserted a reference to Rotman [318]. One may regard it as paradoxical that we freely use the Ext functors while Chapter 3 contains a complete treatment of injective modules. However, their theory has several peculiar aspects so that we thought such a treatment would be welcomed by many readers.

The book contains numerous exercises. Some of them will be used in the main text. For these we have provided hints or even references to the literature, unless their solutions are completely straightforward. A reference of type A.n points to a result in the appendix.

Parts of this book were planned while we were guests of the Mathematisches Forschungsinstitut Oberwolfach. We thank the Forschungsinstitut for its generous hospitality.

We are grateful to all our friends, colleagues, and students, among them L. Avramov, C. Baeţica, M. Barile, A. Conca, H.-B. Foxby, C. Huneke, D. Popescu, P. Schenzel, and W. Vasconcelos who helped us by providing valuable information and by pointing out mistakes in preliminary versions. Our sincere thanks go to H. Matsumura and R. Sharp for their support in the early stages of this project.

We are deeply indebted to our friend Udo Vetter for reading a large part of the manuscript and for his unfailing criticism.

Vechta and Essen,   WINFRIED BRUNS
February 1993   JÜRGEN HERZOG

Part I

Basic concepts

1 Regular sequences and depth

After dimension, depth is the most fundamental numerical invariant of a Noetherian local ring R or a finite R-module M. While depth is defined in terms of regular sequences, it can be measured by the (non-)vanishing of certain Ext modules. This connection opens commutative algebra to the application of homological methods. Depth is connected with projective dimension and several notions of linear algebra over Noetherian rings.

Equally important is the description of depth (and its global relative grade) in terms of the Koszul complex which, in a sense, holds an intermediate position between arithmetic and homological algebra.

This introductory chapter also contains a section on graded rings and modules. These allow a decomposition of their elements into homogeneous components and therefore have a more accessible structure than rings and modules in general.

1.1 Regular sequences

Let M be a module over a ring R. We say that $x \in R$ is an *M-regular element* if $xz = 0$ for $z \in M$ implies $z = 0$, in other words, if x is not a zero-divisor on M. Regular sequences are composed of successively regular elements:

Definition 1.1.1. A sequence $x = x_1, \ldots, x_n$ of elements of R is called an *M-regular sequence* or simply an *M-sequence* if the following conditions are satisfied: (i) x_i is an $M/(x_1, \ldots, x_{i-1})M$-regular element for $i = 1, \ldots, n$, and (ii) $M/xM \neq 0$.

In this situation we shall sometimes say that M is an *x-regular module*. A *regular sequence* is an R-sequence.

A *weak M-sequence* is only required to satisfy condition (i).

Very often R will be a local ring with maximal ideal $\mathfrak{m}$, and $M \neq 0$ a finite R-module. If $x \subset \mathfrak{m}$, then condition (ii) is satisfied automatically because of Nakayama's lemma.

The classical example of a regular sequence is the sequence $X_1, \ldots, X_n$ of indeterminates in a polynomial ring $R = S[X_1, \ldots, X_n]$. Conversely we shall see below that an M-sequence behaves to some extent like a sequence of indeterminates; this will be made precise in 1.1.8.

The next proposition contains a condition under which a regular sequence stays regular when the module or the ring is extended.

3

Proposition 1.1.2. *Let R be a ring, M an R-module, and $x \subset R$ a weak M-sequence. Suppose $\varphi : R \to S$ is a ring homomorphism, and N an R-flat S-module. Then $x \subset R$ and $\varphi(x) \subset S$ are weak $(M \otimes_R N)$-sequences. If $x(M \otimes_R N) \neq M \otimes_R N$, then x and $\varphi(x)$ are $(M \otimes_R N)$-sequences.*

PROOF. Multiplication by x_i is the same operation on $M \otimes N$ as multiplication by $\varphi(x_i)$; so it suffices to consider x. The homothety $x_1 : M \to M$ is injective, and $x_1 \otimes N$ is injective too, because N is flat. Now $x_1 \otimes N$ is just multiplication by x_1 on $M \otimes N$. So x_1 is an $(M \otimes N)$-regular element. Next we have $(M \otimes N)/x_1(M \otimes N) \cong (M/x_1 M) \otimes N$; an inductive argument will therefore complete the proof.   □

The most important special cases of 1.1.2 are given in the following corollary. In its part (b) we use $\hat{M}$ to denote the $\mathfrak{m}$-adic completion of a module M over a local ring $(R, \mathfrak{m}, k)$ (by this notation we indicate that R has maximal ideal $\mathfrak{m}$ and residue class field $k = R/\mathfrak{m}$).

Corollary 1.1.3. *Let R be a Noetherian ring, M a finite R-module, and x an M-sequence.*
(a) *Suppose that a prime ideal $\mathfrak{p} \in \operatorname{Supp} M$ contains x. Then x (as a sequence in $R_\mathfrak{p}$) is an $M_\mathfrak{p}$-sequence.*
(b) *Suppose that R is local with maximal ideal $\mathfrak{m}$. Then x (as a sequence in $\hat{R}$) is an $\hat{M}$-sequence.*

PROOF. Both the extensions $R \to R_\mathfrak{p}$ and $R \to \hat{R}$ are flat. (a) By hypothesis $M_\mathfrak{p} \neq 0$, and Nakayama's lemma implies $M_\mathfrak{p} \neq \mathfrak{p} M_\mathfrak{p}$. *A fortiori* we have $x M_\mathfrak{p} \neq M_\mathfrak{p}$. (b) It suffices to note that $\hat{M} = M \otimes \hat{R}$ is a finite $\hat{R}$-module.
                                                                            □

The interplay between regular sequences and homological invariants is a major theme of this book, and numerous arguments will be based on the next proposition.

Proposition 1.1.4. *Let R be a ring, M an R-module, and x a weak M-sequence. Then an exact sequence*

$$N_2 \xrightarrow{\varphi_2} N_1 \xrightarrow{\varphi_1} N_0 \xrightarrow{\varphi_0} M \longrightarrow 0$$

of R-modules induces an exact sequence

$$N_2/xN_2 \longrightarrow N_1/xN_1 \longrightarrow N_0/xN_0 \longrightarrow M/xM \longrightarrow 0.$$

PROOF. By induction it is enough to consider the case in which x consists of a single M-regular element x. We obtain the induced sequence if we tensor the original one by $R/(x)$. Since tensor product is a right exact functor, we only need to verify exactness at N_1/xN_1. Let $^-$ denote residue classes modulo x. If $\bar{\varphi}_1(\bar{y}) = 0$, then $\varphi_1(y) = xz$ for some $z \in N_0$ and

$x\varphi_0(z) = 0$. By hypothesis we have $\varphi_0(z) = 0$; hence there is $y' \in N_1$ with $z = \varphi_1(y')$. It follows that $\varphi_1(y - xy') = 0$. So $y - xy' \in \varphi_2(N_2)$, and $\bar{y} \in \bar{\varphi}_2(\bar{N}_2)$ as desired. □

If we want to preserve the exactness of a longer sequence, then we need a stronger hypothesis.

Proposition 1.1.5. *Let R be a ring and*

$$N_\bullet : \cdots \longrightarrow N_m \xrightarrow{\varphi_m} N_{m-1} \longrightarrow \cdots \longrightarrow N_0 \xrightarrow{\varphi_0} N_{-1} \longrightarrow 0$$

an exact complex of R-modules. If x is weakly N_i-regular for all i, then $N_\bullet \otimes R/(x)$ is exact again.

PROOF. Once more one uses induction on the length of the sequence x. So it is enough to treat the case $x = x$. Since x is regular on N_i, it is regular on $\operatorname{Im} \varphi_{i+1}$ too. Therefore we can apply 1.1.4 to each exact sequence $N_{i+3} \to N_{i+2} \to N_{i+1} \to \operatorname{Im} \varphi_{i+1} \to 0$. □

Easy examples show that a permutation of a regular sequence need not be a regular sequence; see 1.1.13. Nevertheless there are natural conditions under which regular sequences can be permuted.

Let x_1, x_2 be an M-sequence, and denote the kernel of the multiplication by x_2 on M by K. Suppose that $z \in K$. Then we must have $z \in x_1 M$, $z = x_1 z'$, and $x_1(x_2 z') = 0$, whence $x_2 z' = 0$ and $z' \in K$, too. This shows $K = x_1 K$ so that $K = 0$ if Nakayama's lemma is applicable. Somewhat surprisingly, x_1 is always regular on $M/x_2 M$; the reader may check this easily.

Proposition 1.1.6. *Let R be a Noetherian local ring, M a finite R-module, and $x = x_1, \dots, x_n$ an M-sequence. Then every permutation of x is an M-sequence.*

PROOF. Every permutation is a product of transpositions of adjacent elements. Therefore it is enough to show that $x_1, \dots, x_{i+1}, x_i, \dots, x_n$ is an M-sequence. The hypothesis of the proposition is satisfied for $\bar{M} = M/(x_1, \dots, x_{i-1})M$ and the $\bar{M}$-sequence $x_i, \dots, x_n$. So it suffices to treat the case $i = 1$ and to show that x_2, x_1 is an M-sequence. In view of the discussion above we only need to appeal to Nakayama's lemma. □

Quasi-regular sequences. Let R be a ring, M an R-module, and $X = X_1, \dots, X_n$ be indeterminates over R. Then we write $M[X]$ for $M \otimes R[X]$ and call its elements *polynomials with coefficients in M*. If $x = x_1, \dots, x_n$ is a sequence of elements of R, then the substitution $X_i \mapsto x_i$ induces an R-algebra homomorphism $R[X] \to R$ and also an R-module homomorphism $M[X] \to M$. We write $F(x)$ for the image of $F \in M[X]$ under this map. (Since the monomials form a basis of the free R-module $R[X]$, we may speak of the coefficients and the degree of an element of $M[X]$.)

Theorem 1.1.7 (Rees). *Let R be a ring, M an R-module, $x = x_1, \dots, x_n$ an M-sequence, and $I = (x_1, \dots, x_n)$. Let $X = X_1, \dots, X_n$ be indeterminates over R. If $F \in M[X]$ is homogeneous of (total) degree d and $F(x) \in I^{d+1}M$, then the coefficients of F are in IM.*

PROOF. We use induction on n. The case $n = 1$ is easy. Let $n > 1$ and suppose that the theorem holds for regular sequences of length at most $n - 1$. We must first prove an auxiliary result which is an interesting fact in itself: *let $J = (x_1, \dots, x_{n-1})$; then x_n is regular on $M/J^j M$ for all $j \geq 1$.*

In fact, suppose that $x_n y \in J^j M$ for some $j > 1$. Arguing by induction we have $y \in J^{j-1}M$; so $y = G(x_1, \dots, x_{n-1})$ where $G \in M[X_1, \dots, X_{n-1}]$ is homogeneous of degree $j - 1$. Set $G' = x_n G$. Then the theorem applied to $G' \in M[X_1, \dots, X_{n-1}]$ yields that the coefficients of G' are in JM. Since x_n is regular modulo JM, it follows that the coefficients of G are in JM too, and therefore $y \in J^j M$.

The proof of the theorem for sequences of length n requires induction on d. The case $d = 0$ is trivial. Assume that $d > 0$. First we reduce to the case in which $F(x) = 0$. Since $F(x) \in I^{d+1}M$, one has $F(x) = G(x)$ with G homogeneous of degree $d+1$. Then $G = \sum_{i=1}^{n} X_i G_i$ with G_i homogeneous of degree d. Set $G_i' = x_i G_i$ and $G' = \sum_{i=1}^{n} G_i'$. So $F - G'$ is homogeneous of degree d, and $(F - G')(x) = 0$. Furthermore, $F - G'$ has coefficients in IM if and only if this holds for F.

Thus assume that $F(x) = 0$. Then we write $F = G + X_n H$ with $G \in M[X_1, \dots, X_{n-1}]$. The auxiliary claim above implies that $H(x) \in J^d M \subset I^d M$. By induction on d the coefficients of H are in IM. On the other hand $H(x) = H'(x_1, \dots, x_{n-1})$ with $H' \in M[X_1, \dots, X_{n-1}]$ homogeneous of degree d. As

$$(G + x_n H')(x_1, \dots, x_{n-1}) = F(x) = 0,$$

it follows by induction on n that $G + x_n H'$ has coefficients in JM. Since $x_n H'$ has its coefficients in IM, the coefficients of G must be in IM too.   □

Let I be an ideal in R. One defines the *associated graded ring of R with respect to I* by

$$\mathrm{gr}_I(R) = \bigoplus_{i=0}^{\infty} I^i/I^{i+1}.$$

The multiplication in $\mathrm{gr}_I(R)$ is induced by the multiplication $I^i \times I^j \to I^{i+j}$, and $\mathrm{gr}_I(R)$ is a graded ring with $(\mathrm{gr}_I(R))_0 = R/I$. If M is an R-module, one similarly constructs the *associated graded module*

$$\mathrm{gr}_I(M) = \bigoplus_{i=0}^{\infty} I^i M/I^{i+1}M.$$

It is straightforward to verify that $\mathrm{gr}_I(M)$ is a graded $\mathrm{gr}_I(R)$-module. (Graded rings and modules will be discussed in Section 1.5. The reader not familiar with the basic terminology may wish to consult 1.5.) Let I be generated by $x_1, \ldots, x_n$. Then one has a natural surjection $R[X] = R[X_1, \ldots, X_n] \to \mathrm{gr}_I(R)$ which is induced by the natural homomorphism $R \to R/I$ and the substitution $X_i \mapsto \bar{x}_i \in I/I^2$. Similarly there is an epimorphism $\psi : M[X] \to \mathrm{gr}_I(M)$. One first defines ψ on the homogeneous components by assigning to a homogeneous polynomial $F \in M[X]$ of degree d the residue class of $F(x)$ in $I^d M/I^{d+1}M$; then ψ is extended additively. As the reader may check, ψ is an epimorphism of graded $R[X]$-modules. Obviously $IM[X] \subset \mathrm{Ker}\,\psi$; via the identification $M[X]/IM[X] \cong (M/IM)[X]$, we therefore get an induced epimorphism $\varphi : (M/IM)[X] \to \mathrm{gr}_I(M)$. The kernel of ψ is generated by the homogeneous polynomials $F \in M[X]$ of degree d, $d \in \mathbb{N}$, such that $F(x) \in I^{d+1}M$. So we obtain as a reformulation of 1.1.7

Theorem 1.1.8. *Let R be a ring, M an R-module, $x = x_1, \ldots, x_n$ an M-sequence, and $I = (x)$. Then the map $(M/IM)[X_1, \ldots, X_n] \to \mathrm{gr}_I(M)$ induced by the substitution $X_i \mapsto \bar{x}_i \in I/I^2$ is an isomorphism.*

This theorem says very precisely to what extent a regular sequence resembles a sequence of indeterminates: the residue classes $\bar{x}_i \in I/I^2$ operate on $\mathrm{gr}_I(M)$ exactly like indeterminates. Since a regular sequence may lose regularity under a permutation, whereas 1.1.8 is independent of the order in which x is given, it is not possible to reverse 1.1.8; see however 1.1.15. Later on it will be useful to have a name for sequences x satisfying the conclusion of 1.1.8; we call them *M-quasi-regular* if, in addition, $xM \neq M$.

Exercises

1.1.9. Let $0 \to U \to M \to N \to 0$ be an exact sequence of R-modules, and x a sequence which is weakly U-regular and (weakly) N-regular. Prove that x is (weakly) M-regular too.

1.1.10. (a) Let $x_1, \ldots, x_i, \ldots, x_n$ and $x_1, \ldots, x_i', \ldots, x_n$ be (weakly) M-regular. Show that $x_1, \ldots, x_i x_i', \ldots, x_n$ is (weakly) M-regular. (Hint: In the essential case $i = 1$ one finds an exact sequence as in 1.1.9 with $M/x_1 x_1' M$ as the middle term.)
(b) Prove that $x_1^{e_1}, \ldots, x_n^{e_n}$ is (weakly) M-regular for all $e_i \geq 1$.

1.1.11. Prove that the converse of 1.1.2 holds if, in the situation of 1.1.2, N is faithfully flat over R.

1.1.12. (a) Prove that if x is a weak M-sequence, then $\mathrm{Tor}_1^R(M, R/(x)) = 0$.
(b) Prove that if, in addition, x is a weak R-sequence, then $\mathrm{Tor}_i^R(M, R/(x)) = 0$ for all $i \geq 1$.

1.1.13. Let $R = K[X, Y, Z]$, k a field. Show that X, $Y(1 - X)$, $Z(1 - X)$ is an R-sequence, but $Y(1 - X)$, $Z(1 - X)$, X is not.

1.1.14. Prove that $x_1, \ldots, x_n$ is M-quasi-regular if and only if $\bar{x}_1, \ldots, \bar{x}_n \in I/I^2$ is a $\mathrm{gr}_I(M)$-regular sequence where $I = (x_1, \ldots, x_n)$.

1.1.15. Suppose that x is M-quasi-regular, and let $I = (x_1, \ldots, x_n)$. Prove
(a) if $x_1 z \in I^i M$ for $z \in M$, then $z \in I^{i-1} M$,
(b) $x_2, \ldots, x_n$ is $(M/x_1 M)$-quasi-regular,
(c) if R is Noetherian local and M is finite, then x is an M-sequence.

1.2 Grade and depth

Let R be a Noetherian ring and M an R-module. If $x = x_1, \ldots, x_n$ is an M-sequence, then the sequence $(x_1) \subset (x_1, x_2) \subset \cdots \subset (x_1, \ldots, x_n)$ ascends strictly for obvious reasons. Therefore an M-sequence can be extended to a maximal such sequence: an M-sequence x (contained in an ideal I) is *maximal* (in I), if $x_1, \ldots, x_{n+1}$ is not an M-sequence for any $x_{n+1} \in R$ ($x_{n+1} \in I$). We will prove that all maximal M-sequences in an ideal I with $IM \neq M$ have the same length if M is finite. This allows us to introduce the fundamental notions of grade and depth.

In connection with regular sequences, finite modules over Noetherian rings are distinguished for two reasons: first, every zero-divisor of M is contained in an associated prime ideal, and, second, the number of these prime ideals is finite. Both facts together imply the following proposition that is 'among the most useful in the theory of commutative rings' (Kaplansky [231], p. 56).

Proposition 1.2.1. *Let R be a Noetherian ring, and M a finite R-module. If an ideal $I \subset R$ consists of zero-divisors of M, then $I \subset \mathfrak{p}$ for some $\mathfrak{p} \in \mathrm{Ass}\, M$.*

PROOF. If $I \not\subset \mathfrak{p}$ for all $\mathfrak{p} \in \mathrm{Ass}\, M$, then there exists $a \in I$ with $a \notin \mathfrak{p}$ for all $\mathfrak{p} \in \mathrm{Ass}\, M$. This follows immediately from 1.2.2. □

The following lemma, which we have just used in its simplest form, is the standard argument of 'prime avoidance'.

Lemma 1.2.2. *Let R be a ring, $\mathfrak{p}_1, \ldots, \mathfrak{p}_m$ prime ideals, M an R-module, and $x_1, \ldots, x_n \in M$. Set $N = \sum_{i=1}^{n} Rx_i$. If $N_{\mathfrak{p}_j} \not\subset \mathfrak{p}_j M_{\mathfrak{p}_j}$ for $j = 1, \ldots, m$, then there exist $a_2, \ldots, a_n \in R$ such that $x_1 + \sum_{i=2}^{n} a_i x_i \notin \mathfrak{p}_j M_{\mathfrak{p}_j}$ for $j = 1, \ldots, m$.*

PROOF. We use induction on m, and so suppose that there are $a_2', \ldots, a_n' \in R$ for which $x_1' = x_1 + \sum_{i=2}^{n} a_i' x_i \notin \mathfrak{p}_j M_{\mathfrak{p}_j}$ for $j = 1, \ldots, m-1$. Moreover, it is no restriction to assume that the $\mathfrak{p}_i$ are pairwise distinct and that $\mathfrak{p}_m$ is a minimal member of $\mathfrak{p}_1, \ldots, \mathfrak{p}_m$. So there exists $r \in (\bigcap_{j=1}^{m-1} \mathfrak{p}_j) \setminus \mathfrak{p}_m$. Put $x_i' = rx_i$ for $i = 2, \ldots, n$ and $N' = \sum_{i=1}^{n} Rx_i'$. Since $r \notin \mathfrak{p}_m$ we have $N'_{\mathfrak{p}_m} = N_{\mathfrak{p}_m}$. On the other hand, as $r \in \mathfrak{p}_j$ for $j = 1, \ldots, m-1$, it follows that $x_1' + x_i' \notin \mathfrak{p}_j M_{\mathfrak{p}_j}$ for $i = 2, \ldots, n$ and $j = 1, \ldots, m-1$. If $x_1' \notin \mathfrak{p}_m M_{\mathfrak{p}_m}$, then x_1' is the element desired; otherwise $x_1' + x_i' \notin \mathfrak{p}_m M_{\mathfrak{p}_m}$ for some $i \in \{2, \ldots, n\}$, and we choose $x_1' + x_i'$. □

Note that if $M = R$ and $N = I \subset R$, then the condition $N_{\mathfrak{p}_j} \not\subset \mathfrak{p}_j M_{\mathfrak{p}_j}$ simplifies to $I \not\subset \mathfrak{p}_j$.

Suppose that an ideal I is contained in $\mathfrak{p} \in \operatorname{Ass} M$. By definition, there exists $z \in M$ with $\mathfrak{p} = \operatorname{Ann} z$. Hence the assignment $1 \mapsto z$ induces a monomorphism $\varphi' : R/\mathfrak{p} \to M$, and thus a non-zero homomorphism $\varphi : R/I \to M$. This simple observation allows us to describe in homological terms that a certain ideal consists of zero-divisors:

Proposition 1.2.3. *Let R be a ring, and M, N R-modules. Set $I = \operatorname{Ann} N$.*
(a) *If I contains an M-regular element, then $\operatorname{Hom}_R(N, M) = 0$.*
(b) *Conversely, if R is Noetherian, and M, N are finite, $\operatorname{Hom}_R(N, M) = 0$ implies that I contains an M-regular element.*

PROOF. (a) is evident. (b) Assume that I consists of zero-divisors of M, and apply 1.2.1 to find a $\mathfrak{p} \in \operatorname{Ass} M$ such that $I \subset \mathfrak{p}$. By hypothesis, $\mathfrak{p} \in \operatorname{Supp} N$; so $N_{\mathfrak{p}} \otimes k(\mathfrak{p}) \neq 0$ by Nakayama's lemma, and since $N_{\mathfrak{p}} \otimes k(\mathfrak{p})$ is just a direct sum of copies of $k(\mathfrak{p})$, one has an epimorphism $N_{\mathfrak{p}} \to k(\mathfrak{p})$. (By $k(\mathfrak{p})$ we denote the residue class field $R_{\mathfrak{p}}/\mathfrak{p}R_{\mathfrak{p}}$ of $R_{\mathfrak{p}}$.) Note that $\mathfrak{p}R_{\mathfrak{p}} \in \operatorname{Ass} M_{\mathfrak{p}}$. Hence the observation above yields a non-zero $\varphi' \in \operatorname{Hom}_{R_{\mathfrak{p}}}(N_{\mathfrak{p}}, M_{\mathfrak{p}})$. Since $\operatorname{Hom}_{R_{\mathfrak{p}}}(N_{\mathfrak{p}}, M_{\mathfrak{p}}) \cong \operatorname{Hom}_R(N, M)_{\mathfrak{p}}$, it follows that $\operatorname{Hom}_R(N, M) \neq 0$. (See [318], Theorem 3.84 for the isomorphism just applied.) $\square$

Lemma 1.2.4. *Let R be a ring, M, N be R-modules, and $x = x_1, \ldots, x_n$ a weak M-sequence in $\operatorname{Ann} N$. Then*

$$\operatorname{Hom}_R(N, M/xM) \cong \operatorname{Ext}_R^n(N, M).$$

PROOF. We use induction on n, starting from the vacuous case $n = 0$. Let $n \geq 1$, and set $x' = x_1, \ldots, x_{n-1}$. Then the induction hypothesis implies that $\operatorname{Ext}_R^{n-1}(N, M) \cong \operatorname{Hom}_R(N, M/x'M)$. As x_n is $(M/x'M)$-regular, $\operatorname{Ext}_R^{n-1}(N, M) = 0$ by 1.2.3. Therefore the exact sequence

$$0 \longrightarrow M \overset{x_1}{\longrightarrow} M \longrightarrow M/x_1 M \longrightarrow 0$$

yields an exact sequence

$$0 \longrightarrow \operatorname{Ext}_R^{n-1}(N, M/xM) \overset{\psi}{\longrightarrow} \operatorname{Ext}_R^n(N, M) \overset{\varphi}{\longrightarrow} \operatorname{Ext}_R^n(N, M).$$

The map φ is multiplication by x_1 inherited from M, but multiplication by x_1 on N also induces φ; see [318], Theorem 7.16. Since $x_1 \in \operatorname{Ann} N$, one has $\varphi = 0$. Hence ψ is an isomorphism, and a second application of the induction hypothesis yields the assertion. $\square$

Let R be Noetherian, I an ideal, M a finite R-module with $M \neq IM$, and $x = x_1, \ldots, x_n$ a maximal M-sequence in I. From 1.2.3 and 1.2.4

we have, since I contains an $(M/(x_1,\ldots,x_{i-1})M)$-regular element for $i = 1,\ldots,n$,

$$\operatorname{Ext}_R^{i-1}(R/I, M) \cong \operatorname{Hom}_R\left(R/I, M/(x_1,\ldots,x_{i-1})M\right) = 0.$$

On the other hand, since $IM \neq M$ and x is a maximal M-sequence in I, then I must consist of zero-divisors of M/xM, whence

$$\operatorname{Ext}_R^n(R/I, M) \cong \operatorname{Hom}_R(R/I, M/xM) \neq 0.$$

We have therefore proved

Theorem 1.2.5 (Rees). *Let R be a Noetherian ring, M a finite R-module, and I an ideal such that $IM \neq M$. Then all maximal M-sequences in I have the same length n given by*

$$n = \min\{i : \operatorname{Ext}_R^i(R/I, M) \neq 0\}.$$

Definition 1.2.6. Let R be a Noetherian ring, M a finite R-module, and I an ideal such that $IM \neq M$. Then the common length of the maximal M-sequences in I is called the *grade of I on M*, denoted by

$$\operatorname{grade}(I, M).$$

We complement this definition by setting $\operatorname{grade}(I, M) = \infty$ if $IM = M$. This is consistent with 1.2.5:

$$\operatorname{grade}(I, M) = \infty \quad \Longleftrightarrow \quad \operatorname{Ext}_R^i(R/I, M) = 0 \text{ for all } i.$$

For, if $IM = M$, then $\operatorname{Supp} M \cap \operatorname{Supp} R/I = \emptyset$ by Nakayama's lemma, hence

(1) $$\operatorname{Supp} \operatorname{Ext}_R^i(R/I, M) \subset \operatorname{Supp} M \cap \operatorname{Supp} R/I = \emptyset;$$

conversely, if $\operatorname{Ext}_R^i(R/I, M) = 0$ for all i, then 1.2.5 gives $IM = M$.

The inclusion in (1) results from the natural isomorphism

$$\operatorname{Ext}_{R_\mathfrak{p}}^i(N_\mathfrak{p}, M_\mathfrak{p}) \cong \operatorname{Ext}_R^i(N, M)_\mathfrak{p}$$

which holds if R is Noetherian, N a finite R-module, M an arbitrary R-module, and $\mathfrak{p} \in \operatorname{Spec} R$; see [318], Theorem 9.50.

A special situation will occur so often that it merits a special notation:

Definition 1.2.7. Let $(R, \mathfrak{m}, k)$ be a Noetherian local ring, and M a finite R-module. Then the grade of $\mathfrak{m}$ on M is called the *depth of M*, denoted

$$\operatorname{depth} M.$$

Because of its importance we repeat the most often used special case of 1.2.5:

Theorem 1.2.8. *Let $(R, \mathfrak{m}, k)$ be a Noetherian local ring, and M a finite non-zero R-module. Then* $\operatorname{depth} M = \min\{i : \operatorname{Ext}_R^i(k, M) \neq 0\}$.

Some formulas for grade. We now study the behaviour of $\operatorname{grade}(I, M)$ along exact sequences.

Proposition 1.2.9. *Let R be a Noetherian ring, $I \subset R$ an ideal, and $0 \to U \to M \to N \to 0$ an exact sequence of finite R-modules. Then*

$$\operatorname{grade}(I, M) \geq \min\{\operatorname{grade}(I, U), \operatorname{grade}(I, N)\},$$
$$\operatorname{grade}(I, U) \geq \min\{\operatorname{grade}(I, M), \operatorname{grade}(I, N) + 1\},$$
$$\operatorname{grade}(I, N) \geq \min\{\operatorname{grade}(I, U) - 1, \operatorname{grade}(I, M)\}.$$

PROOF. The given exact sequence induces a long exact sequence

$$\cdots \to \operatorname{Ext}_R^{i-1}(R/I, N) \to \operatorname{Ext}_R^i(R/I, U) \to \operatorname{Ext}_R^i(R/I, M)$$
$$\to \operatorname{Ext}_R^i(R/I, N) \to \operatorname{Ext}_R^{i+1}(R/I, U) \to \cdots$$

One observes that $\operatorname{Ext}_R^i(R/I, M) = 0$ if $\operatorname{Ext}_R^i(R/I, U)$ and $\operatorname{Ext}_R^i(R/I, N)$ both vanish. Therefore the first inequality follows from 1.2.5 and our discussion of the case $\operatorname{grade}(I, \_) = \infty$. Completely analogous arguments show the second and the third inequality. $\square$

The next proposition collects some formulas which are useful in the computation of grades. (In the sequel $V(I)$ denotes the set of prime ideals containing I.)

Proposition 1.2.10. *Let R be a Noetherian ring, I, J ideals of R, and M a finite R-module. Then*

(a) $\operatorname{grade}(I, M) = \inf\{\operatorname{depth} M_\mathfrak{p} : \mathfrak{p} \in V(I)\}$,

(b) $\operatorname{grade}(I, M) = \operatorname{grade}(\operatorname{Rad} I, M)$,

(c) $\operatorname{grade}(I \cap J, M) = \min\{\operatorname{grade}(I, M), \operatorname{grade}(J, M)\}$,

(d) *if $x = x_1, \ldots, x_n$ is an M-sequence in I, then* $\operatorname{grade}(I/(x), M/xM) = \operatorname{grade}(I, M/xM) = \operatorname{grade}(I, M) - n$,

(e) *if N is a finite R-module with $\operatorname{Supp} N = V(I)$, then*

$$\operatorname{grade}(I, M) = \inf\{i : \operatorname{Ext}_R^i(N, M) \neq 0\}.$$

PROOF. (a) It is evident from the definition that $\operatorname{grade}(I, M) \leq \operatorname{grade}(\mathfrak{p}, M)$ for $\mathfrak{p} \in V(I)$, and it follows from 1.1.3 that $\operatorname{grade}(\mathfrak{p}, M) \leq \operatorname{depth} M_\mathfrak{p}$. Furthermore, if $\operatorname{grade}(I, M) = \infty$, then $\operatorname{Supp} M \cap V(I) = \emptyset$ so that $\operatorname{depth} M_\mathfrak{p} = \infty$ for all $\mathfrak{p} \in V(I)$. Thus suppose $IM \neq M$ and choose a maximal M-sequence x in I. By 1.2.1 there exists $\mathfrak{p} \in \operatorname{Ass}(M/xM)$ with $I \subset \mathfrak{p}$. Since $\mathfrak{p}R_\mathfrak{p} \in \operatorname{Ass}(M/xM)_\mathfrak{p}$ and $(M/xM)_\mathfrak{p} \cong M_\mathfrak{p}/xM_\mathfrak{p}$, the ideal

$\mathfrak{p}R_\mathfrak{p}$ consists of zero-divisors of $M_\mathfrak{p}/xM_\mathfrak{p}$, and x (as a sequence in $R_\mathfrak{p}$) is a maximal $M_\mathfrak{p}$-sequence.

(b) and (c) follow easily from (a).

(d) Set $\bar{R} = R/(x)$, $\bar{I} = I/(x)$, and $\bar{M} = M/xM$. Elementary arguments show that $IM = M \Longleftrightarrow I\bar{M} = \bar{M} \Longleftrightarrow \bar{I}\bar{M} = \bar{M}$. Furthermore $y_1, \ldots, y_n \in I$ form an $\bar{M}$-sequence if and only if $\bar{y}_1, \ldots, \bar{y}_n \in \bar{I}$ form such a sequence. This proves the first equation. The second equation results from 1.2.5.

(e) The hypothesis entails that $\operatorname{Rad}\operatorname{Ann} N = \operatorname{Rad} I$. By (b) we may therefore assume that $I = \operatorname{Ann} N$. Now one repeats the proof of 1.2.5 (and the discussion of the case $IM = M$) with R/I replaced by N. $\square$

The name 'grade' was originally used by Rees [303] for a different, though related invariant:

Definition 1.2.11. Let R be a Noetherian ring and $M \neq 0$ a finite R-module. Then the *grade of M* is given by

$$\operatorname{grade} M = \min\{i : \operatorname{Ext}_R^i(M, R) \neq 0\}.$$

For systematic reasons the grade of the zero-module is infinity.

It follows directly from 1.2.10(e) that $\operatorname{grade} M = \operatorname{grade}(\operatorname{Ann} M, R)$. It is customary to set

$$\operatorname{grade} I = \operatorname{grade} R/I = \operatorname{grade}(I, R),$$

for an ideal $I \subset R$, and we follow this convention. (Of course, $\operatorname{grade} I$ has two different meanings now, but we will never use it to denote the grade of the module I.)

Depth and dimension. Let $(R, \mathfrak{m})$ be Noetherian local and M a finite R-module. All the minimal elements of $\operatorname{Supp} M$ belong to $\operatorname{Ass} M$. Therefore, if $x \in \mathfrak{m}$ is an M-regular element, then $x \notin \mathfrak{p}$ for all minimal elements of $\operatorname{Supp} M$, and induction yields $\dim M/xM = \dim M - n$ if $x = x_1, \ldots, x_n$ is an M-sequence. (Note that $\dim M/xM \geq \dim M - n$ is automatic; see A.4.) We have proved:

Proposition 1.2.12. *Let $(R, \mathfrak{m})$ be a Noetherian local ring and $M \neq 0$ a finite R-module. Then every M-sequence is part of a system of parameters of M. In particular* $\operatorname{depth} M \leq \dim M$.

The inequality in 1.2.12 can be somewhat refined:

Proposition 1.2.13. *With the notation of 1.2.12 one has* $\operatorname{depth} M \leq \dim R/\mathfrak{p}$ *for all* $\mathfrak{p} \in \operatorname{Ass} M$.

PROOF. We use induction on $\operatorname{depth} M$. There is nothing to prove for $\operatorname{depth} M = 0$. If $\operatorname{depth} M > 0$, then there exists an M-regular $x \in \mathfrak{m}$. For

$\mathfrak{p} \in \text{Ass}\, M$ we choose $z \in M$ such that Rz is maximal among the cyclic submodules of M annihilated by $\mathfrak{p}$. If $z \in xM$, then $z = xy$ with $y \in M$, and $\mathfrak{p}y = 0$ since x is M-regular; moreover, Rz is a proper submodule of Ry, contrary to the choice of z. Therefore $\mathfrak{p}$ consists of zero-divisors of M/xM, and is contained in some $\mathfrak{q} \in \text{Ass}(M/xM)$. As $x \notin \mathfrak{p}$, we have $\mathfrak{p} \notin \text{Supp}(M/xM)$, and thus $\mathfrak{p} \neq \mathfrak{q}$. Now $\text{depth}(M/xM) = \text{depth}\, M - 1$ by 1.2.10, whence, by induction,

$$\dim R/\mathfrak{p} > \dim R/\mathfrak{q} \geq \text{depth}(M/xM) = \text{depth}\, M - 1. \qquad \square$$

A global variant of 1.2.12 says that height bounds grade.

Proposition 1.2.14. *Let R be a Noetherian ring and $I \subset R$ an ideal. Then* $\text{grade}\, I \leq \text{height}\, I$.

PROOF. Since $\text{grade}\, I = \inf\{\text{depth}\, R_{\mathfrak{p}} : \mathfrak{p} \in V(I)\}$ by 1.2.10, and $\text{height}\, I = \inf\{\dim R_{\mathfrak{p}} : \mathfrak{p} \in V(I)\}$, the assertion follows from 1.2.12. $\qquad \square$

Depth, type, and flat extensions. Finally we investigate how depth behaves under flat local extensions. As a by-product we obtain a result on the behaviour of the type of a module under such extensions. This is an invariant which refines the information given by the depth:

Definition 1.2.15. Let $(R, \mathfrak{m}, k)$ be a Noetherian local ring, and M a finite non-zero R-module of depth t. The number $r(M) = \dim_k \text{Ext}_R^t(k, M)$ is called the *type of M*.

Proposition 1.2.16. *Let $\varphi : (R, \mathfrak{m}, k) \to (S, \mathfrak{n}, l)$ be a homomorphism of Noetherian local rings. Suppose M is a finite R-module, and N is a finite S-module which is flat over R. Then*
(a) $\text{depth}_S M \otimes_R N = \text{depth}_R M + \text{depth}_S N/\mathfrak{m}N$,
(b) $r_S(M \otimes_R N) = r_R(M) \cdot r_S(N/\mathfrak{m}N)$.

The proof of the proposition is by reduction to the case of depth 0. We collect the essential arguments in a lemma.

Lemma 1.2.17. *Under the hypotheses of 1.2.16 the following hold:*
(a) $\dim_l \text{Hom}_S(l, M \otimes N) = \dim_k \text{Hom}_R(k, M) \cdot \dim_l \text{Hom}_S(l, N/\mathfrak{m}N)$,
(b) *if y is an $(N/\mathfrak{m}N)$-sequence in S, then y is an $(M \otimes_R N)$-sequence, and N/yN is flat over R.*

PROOF. (a) Set $T = S/\mathfrak{m}S$. There is a natural isomorphism

$$(2) \qquad \text{Hom}_S\big(l, \text{Hom}_S(T, M \otimes N)\big) \cong \text{Hom}_S(l, M \otimes N),$$

since the modules on both sides can be identified with the submodule $U = \{z \in M \otimes N : \mathfrak{n}z = 0\}$ of $M \otimes N$. As N is flat over R, we have a natural isomorphism

$$\text{Hom}_S(T, M \otimes N) = \text{Hom}_S(k \otimes S, M \otimes N) \cong \text{Hom}_R(k, M) \otimes N.$$

(see [318], 3.82 and 3.83). Now $\operatorname{Hom}_R(k, M) \cong k^s$ for some $s \geq 0$, and so $\operatorname{Hom}_R(k, M) \otimes N \cong (N/\mathfrak{m}N)^s$. In conjunction with (2), this yields the equation asserted.

(b) One has a natural isomorphism $(M \otimes N)/J(M \otimes N) \cong M \otimes (N/JN)$ for an arbitrary ideal $J \subset S$. Therefore we may use induction on the length n of y, and only the case $n = 1$, $y = y$ needs justification.

By Krull's intersection theorem one has $\bigcap_{i=0}^{\infty} \mathfrak{m}^i(M \otimes N) = 0$. Suppose that $yz = 0$ for some $z \in M \otimes N$. If $z \neq 0$, then there exists i such that $z \in \mathfrak{m}^i(M \otimes N) \setminus \mathfrak{m}^{i+1}(M \otimes N)$, and y would be a zero-divisor on $\mathfrak{m}^i(M \otimes N)/\mathfrak{m}^{i+1}(M \otimes N)$. However, consider the embedding $\mathfrak{m}^i M \to M$. Since N is flat, the induced map $\mathfrak{m}^i M \otimes N \to M \otimes N$ is also injective, and its image is $\mathfrak{m}^i(M \otimes N)$. The same reasoning for $\mathfrak{m}^{i+1}$ and flatness again then yield an isomorphism

$$\mathfrak{m}^i(M \otimes N)/\mathfrak{m}^{i+1}(M \otimes N) \cong (\mathfrak{m}^i M/\mathfrak{m}^{i+1} M) \otimes N \cong k^t \otimes N \cong (N/\mathfrak{m}N)^t$$

for some $t \geq 0$. Since y is regular on $N/\mathfrak{m}N$, it must be regular on $\mathfrak{m}^i(M \otimes N)/\mathfrak{m}^{i+1}(M \otimes N)$.

In order to test flatness of N/yN it suffices to consider exact sequences

$$0 \longrightarrow M_1 \longrightarrow M_2 \longrightarrow M_3 \longrightarrow 0$$

of finite R-modules ([318], Theorem 3.53). By hypothesis

$$0 \longrightarrow M_1 \otimes N \longrightarrow M_2 \otimes N \longrightarrow M_3 \otimes N \longrightarrow 0$$

is also exact. As has been shown previously, y is regular on $M_3 \otimes N$, and $(M_3 \otimes N)/y(M_3 \otimes N) \cong M_3 \otimes N/yN$. Therefore 1.1.4 yields the exactness of

$$0 \longrightarrow M_1 \otimes N/yN \longrightarrow M_2 \otimes N/yN \longrightarrow M_3 \otimes N/yN \longrightarrow 0. \qquad \square$$

PROOF OF 1.2.16. Let $x = x_1, \ldots, x_m$ be a maximal M-sequence, and $y = y_1, \ldots, y_n$ a maximal $(N/\mathfrak{m}N)$-sequence. First, $\varphi(x) = \varphi(x_1), \ldots, \varphi(x_m)$ is an $(M \otimes N)$-sequence; see 1.1.2. Second, by 1.2.17, y is an $(\bar{M} \otimes N)$-sequence where $\bar{M} = M/xM$. Since $\bar{M} \otimes N \cong (M \otimes N)/\varphi(x)(M \otimes N)$, it follows that $\varphi(x), y$ is an $M \otimes N$-sequence.

Set $N' = N/yN$. Then $N'/\mathfrak{m}N' \cong (N/\mathfrak{m}N)/y(N/\mathfrak{m}N)$, and

$$(M \otimes N)/(\varphi(x), y)(M \otimes N) \cong \bar{M} \otimes N'.$$

An application of 1.2.4 therefore gives the isomorphisms

$$\operatorname{Hom}_R(k, \bar{M}) \cong \operatorname{Ext}_R^m(k, M), \quad \operatorname{Hom}_S(l, N'/\mathfrak{m}N') \cong \operatorname{Ext}_S^n(l, N/\mathfrak{m}N),$$
$$\operatorname{Hom}_S(l, \bar{M} \otimes N') \cong \operatorname{Ext}_S^{m+n}(l, M \otimes N).$$

Part (a) of 1.2.17 implies that $\dim_l \operatorname{Ext}_S^{m+n}(l, M \otimes N)$ has the dimension required for (b), and in particular is non-zero. Together with the fact that $\varphi(x), y$ is an $(M \otimes N)$-sequence this proves $\operatorname{depth}(M \otimes N) = m + n$. $\qquad \square$

The type of a module of depth 0 is the dimension of its socle:

Definition 1.2.18. Let M be a module over a local ring $(R, \mathfrak{m}, k)$. Then

$$\operatorname{Soc} M = (0 : \mathfrak{m})_M \cong \operatorname{Hom}_R(k, M)$$

is called *the socle of M*.

For ease of reference we formulate the following lemma which was already verified in the proof of 1.2.16.

Lemma 1.2.19. *Let* $(R, \mathfrak{m}, k)$ *be a Noetherian local ring, M a finite R-module and x a maximal M-sequence. Then* $r(M) = \dim_k \operatorname{Soc}(M/xM)$.

Exercises

1.2.20. Let k be a field and $R = k[[X]][Y]$. Deduce that X, Y and $1 - XY$ are maximal R-sequences. (This example shows that the condition $IM \neq M$ in 1.2.5 is relevant.)

1.2.21. Let R be a Noetherian ring, $I \subset R$ an ideal, $I = (x_1, \ldots, x_n)$, and M a finite R-module with $IM \neq M$. Set $g = \operatorname{grade}(I, M)$. Prove
(a) I can be generated by elements $y_1, \ldots, y_n$ such that $y_{i_1}, \ldots, y_{i_h}$ form an M-sequence for all $i_1, \ldots, i_h$ with $1 \leq i_1 < \cdots < i_h \leq n$, $h \leq g$,
(b) if $y_1, \ldots, y_n$ satisfies (a), then, in fact, every permutation of $y_{i_1}, \ldots, y_{i_h}$ is an M-sequence.
Hint: It is possible to choose $y_i = x_i + \sum_{j \neq i} a_j x_j$. Use the discussion above 1.1.6 for (b).

1.2.22. Let R be a Noetherian ring, $I \subset R$ an ideal, and M a finite R-module with $IM \neq M$. Set $\bar{R} = R/\operatorname{Ann} M$.
(a) Prove that $\operatorname{grade}(I, M) \leq \operatorname{height} I\bar{R}$.
(b) Give an example where $\operatorname{grade}(I, M) > \operatorname{height} I$.
(c) Show that if $I = (x_1, \ldots, x_n)$, then $\operatorname{grade}(I, M) \leq n$.

1.2.23. Let R be a Noetherian local ring, and $I \subset R$ an ideal. Show $\operatorname{grade} I \geq \operatorname{depth} R - \dim R/I$. (Hint: Use 1.2.13.)

1.2.24. Let R be a Noetherian ring, M a finite R-module, and I an ideal of R. Show that $\operatorname{grade}(I, M) \geq 2$ if and only if the natural homomorphism $M \to \operatorname{Hom}_R(I, M)$ is an isomorphism.

1.2.25. Let $\varphi : (R, \mathfrak{m}) \to (S, \mathfrak{n})$ be a homomorphism of local rings, and N an R-flat S-module such that $N/\mathfrak{m}N$ has finite length over S. Show that for every finite length R-module M, $\ell_S(M \otimes N) = \ell_R(M) \cdot \ell_S(N/\mathfrak{m}N)$. (The symbol ℓ denotes length). Hint: use induction on $\ell(M)$.

1.2.26. Let $\varphi : (R, \mathfrak{m}) \to (S, \mathfrak{n})$ be a homomorphism of Noetherian local rings, and M an S-module which is finite as an R-module.
(a) Suppose $\mathfrak{p} \in \operatorname{Ass}_S M$, and let $x \in M$ with $\operatorname{Ann}_S x = \mathfrak{p}$. Prove that φ induces an embedding $R/(\mathfrak{p} \cap R) \to S/\mathfrak{p} \cong Sx$ which makes $S/\mathfrak{p}$ a finite $R/(\mathfrak{p} \cap R)$-module. Conclude that $\mathfrak{p} \cap R \neq \mathfrak{m}$, if $\mathfrak{p} \neq \mathfrak{n}$.
(b) Show that $\operatorname{depth}_R M = \operatorname{depth}_S M$.
(c) Suppose in addition that φ is surjective. Prove $r_R(M) = r_S(M)$.

1.2.27. Let R be a Noetherian ring, M a finite R-module, and N an arbitrary R-module. Deduce that $\operatorname{Ass} \operatorname{Hom}_R(M,N) = \operatorname{Supp} M \cap \operatorname{Ass} N$.

1.3 Depth and projective dimension

Let R be a ring, and M an R-module; M has an augmented projective resolution

$$P_\bullet: \cdots \longrightarrow P_n \xrightarrow{\varphi_n} P_{n-1} \longrightarrow \cdots \longrightarrow P_1 \xrightarrow{\varphi_1} P_0 \xrightarrow{\varphi_0} M \longrightarrow 0.$$

(By definition a projective resolution is non-augmented, i.e. M is replaced by 0; for the most part it is clear from the context whether one uses a non-augmented resolution or an augmented one, so that one need not mention the attribute 'augmented' explicitly.) Set $M_0 = M$ and $M_i = \operatorname{Ker} \varphi_{i-1}$ for $i \geq 1$. The modules M_i depend obviously on $P_\bullet$. However, M determines M_i up to projective equivalence ([318], Theorem 9.4), and therefore it is justified to call M_i the i-th *syzygy of* M. The *projective dimension of* M, abbreviated $\operatorname{proj dim} M$, is infinity if none of the modules M_i is projective. Otherwise $\operatorname{proj dim} M$ is the least integer n for which M_n is projective; replacing P_n by M_n one gets a projective resolution of M of length n:

$$0 \longrightarrow M_n \longrightarrow P_{n-1} \longrightarrow \cdots \longrightarrow P_0 \longrightarrow M \longrightarrow 0.$$

For a finite module M over a Noetherian local ring $(R, \mathfrak{m}, k)$ there is a very natural condition which, if satisfied by $P_\bullet$, determines $P_\bullet$ uniquely. It is a consequence of Nakayama's lemma that $x_1, \ldots, x_m \in M$ form a minimal system of generators of M if and only if the residue classes $\bar{x}_1, \ldots, \bar{x}_m \in M/\mathfrak{m}M \cong M \otimes k$ are a k-basis of $M \otimes k$. Therefore $m = \dim_k M \otimes k$, and

$$\mu(M) = \dim_k M \otimes k$$

is the *minimal number of generators of* M. Set $\beta_0 = \mu(M)$. We choose a minimal system $x_1, \ldots, x_{\beta_0}$ of generators of M and specify an epimorphism $\varphi_0 : R^{\beta_0} \to M$ by $\varphi_0(e_i) = x_i$ where $e_1, \ldots, e_{\beta_0}$ is the canonical basis of R^{β_0}. Next we set $\beta_1 = \mu(\operatorname{Ker} \varphi_0)$ and define similarly an epimorphism $R^{\beta_1} \to \operatorname{Ker} \varphi_0$. Proceeding in this manner we construct a *minimal free resolution*

$$F_\bullet: \cdots \longrightarrow R^{\beta_n} \xrightarrow{\varphi_n} R^{\beta_{n-1}} \longrightarrow \cdots \longrightarrow R^{\beta_1} \xrightarrow{\varphi_1} R^{\beta_0} \xrightarrow{\varphi_0} M \longrightarrow 0.$$

It is left as an exercise for the reader to prove that $F_\bullet$ is determined by M up to an isomorphism of complexes. The number $\beta_i(M) = \beta_i$ is called the i-th *Betti number of* M.

Proposition 1.3.1. *Let $(R, \mathfrak{m}, k)$ be a Noetherian local ring, M a finite R-module, and*

$$F_\bullet: \cdots \longrightarrow F_n \xrightarrow{\varphi_n} F_{n-1} \longrightarrow \cdots \longrightarrow F_1 \xrightarrow{\varphi_1} F_0 \longrightarrow 0$$

a free resolution of M. Then the following are equivalent:
(a) $F_{\bullet}$ *is minimal;*
(b) $\varphi_i(F_i) \subset \mathfrak{m}F_{i-1}$ *for all* $i \geq 1$;
(c) $\operatorname{rank} F_i = \dim_k \operatorname{Tor}_i^R(M,k)$ *for all* $i \geq 0$,
(d) $\operatorname{rank} F_i = \dim_k \operatorname{Ext}_R^i(M,k)$ *for all* $i \geq 0$.

PROOF. The equivalence of (a) and (b) follows easily from Nakayama's lemma. Since $\operatorname{Tor}_i^R(M,k) = H_i(F_{\bullet} \otimes k)$, (c) holds if and only if $\varphi_i \otimes k = 0$ for all $i \geq 0$. The latter condition is evidently equivalent to (b). To relate (b) to (d) one uses that $\operatorname{Ext}_R^i(M,k) = H^i(\operatorname{Hom}_R(F_{\bullet},k))$. $\quad\square$

Corollary 1.3.2. *Let* $(R,\mathfrak{m},k)$ *be a Noetherian local ring, and* M *a finite* R-module. Then $\beta_i(M) = \dim_k \operatorname{Tor}_i^R(M,k)$ *for all* i *and*

$$\operatorname{proj dim} M = \sup\{i : \operatorname{Tor}_i^R(M,k) \neq 0\}.$$

The following theorem, the 'Auslander–Buchsbaum formula', is not only of theoretical importance, but also an effective instrument for the computation of the depth of a module.

Theorem 1.3.3 (Auslander–Buchsbaum). *Let* $(R,\mathfrak{m})$ *be a Noetherian local ring, and* $M \neq 0$ *a finite* R-module. If $\operatorname{proj dim} M < \infty$, then

$$\operatorname{proj dim} M + \operatorname{depth} M = \operatorname{depth} R.$$

The proof is by induction on $\operatorname{depth} R$. We isolate the main arguments in two lemmas, the first of which, in view of a later application, is more general than needed presently.

Lemma 1.3.4. *Let* $(R,\mathfrak{m},k)$ *be a local ring, and* $\varphi : F \to G$ *a homomorphism of finite* R-modules. Suppose that F is free, and let M be an R-module with $\mathfrak{m} \in \operatorname{Ass} M$. Suppose that $\varphi \otimes M$ is injective. Then
(a) $\varphi \otimes k$ *is injective;*
(b) *if* G *is a free* R-module, then φ *is injective, and* $\varphi(F)$ *is a free direct summand of* G.

PROOF. Since $\mathfrak{m} \in \operatorname{Ass} M$, there exists an embedding $\iota : k \to M$. As F is a free R-module, the map $F \otimes \iota$ is also injective. Furthermore we have a commutative diagram

$$
\begin{array}{ccc}
F \otimes k & \xrightarrow{\ F \otimes \iota\ } & F \otimes M \\
{\scriptstyle \varphi \otimes k}\downarrow & & \downarrow{\scriptstyle \varphi \otimes M} \\
G \otimes k & \longrightarrow & G \otimes M
\end{array}
$$

If $\varphi \otimes M$ is injective, then $\varphi \otimes k$ is injective too. This proves (a).

For (b) one notes that its conclusion is equivalent to the injectivity of $\varphi \otimes k$. This is an easy consequence of Nakayama's lemma. $\quad\square$

Lemma 1.3.5. *Let* $(R, \mathfrak{m})$ *be a Noetherian local ring, and* M *a finite* R-*module. If* $x \in \mathfrak{m}$ *is* R-*regular and* M-*regular, then*

$$\operatorname{proj dim}_R M = \operatorname{proj dim}_{R/(x)} M/xM.$$

PROOF. Choose an augmented minimal free resolution $F_{\bullet}$ of M. Then $F_{\bullet} \otimes R/(x)$ is exact by 1.1.5, and therefore it is a minimal free resolution of M/xM over $R/(x)$. Now apply 1.3.2.   □

PROOF OF 1.3.3. Let depth $R = 0$ first. By hypothesis M has a (minimal) free resolution

$$F_{\bullet} : 0 \longrightarrow F_n \overset{\varphi_n}{\longrightarrow} F_{n-1} \longrightarrow \cdots \longrightarrow F_1 \longrightarrow F_0 \longrightarrow M \longrightarrow 0$$

with $n = \operatorname{proj dim} M$. Since depth $R = 0$, the maximal ideal $\mathfrak{m}$ is in Ass R. If $n \geq 1$, i.e. if φ_n is really present, then, as shown in 1.3.4, φ_n maps F_n isomorphically onto a free direct summand of F_{n-1}, in contradiction to $\operatorname{proj dim} M = n$. Therefore $n = 0$, and furthermore depth $M = $ depth $R = 0$ since M is a free R-module.

Let now depth $R > 0$. Suppose first that depth $M = 0$. Then 1.2.9 yields depth $M_1 = 1$ for a first syzygy M_1 of M. Since $\operatorname{proj dim} M_1 = \operatorname{proj dim} M - 1$, it is enough to prove the desired formula for M_1. Thus we may assume depth $M > 0$. Then $\mathfrak{m} \notin \operatorname{Ass} R$ and $\mathfrak{m} \notin \operatorname{Ass} M$. So $\mathfrak{m}$ contains an element x which is both R-regular and M-regular. The formulas for the passage from M to M/xM in 1.2.10 and 1.3.5 yield

$$\operatorname{depth}_{R/(x)} R/(x) = \operatorname{depth} R - 1, \quad \operatorname{depth}_{R/(x)} M/xM = \operatorname{depth}_R M - 1,$$

$$\operatorname{proj dim}_{R/(x)} M/xM = \operatorname{proj dim} M.$$

Therefore induction completes the proof.   □

Exercises

1.3.6. Let R be a Noetherian local ring, M a finite R-module, and x an M-sequence of length n. Show $\operatorname{proj dim}(M/xM) = \operatorname{proj dim} M + n$.

1.3.7. Let R be a Noetherian local ring, and N an n-th syzygy of a finite R-module in a finite free resolution. Prove that depth $N \geq \min(n, \operatorname{depth} R)$.

1.4 Some linear algebra

In this section we collect several notions and results which may be classified as 'linear algebra': torsion-free and reflexive modules, the rank of a module, the acyclicity criterion of Buchsbaum and Eisenbud, and perfect modules.

Torsion-free and reflexive modules. Let R be a ring, and M an R-module. If the natural map $M \to M \otimes Q$, where Q is the total ring of fractions of R, is injective, then M is *torsion-free*; it is a *torsion module* if $M \otimes Q = 0$. The *dual of* M is the module $\operatorname{Hom}_R(M, R)$, which we usually denote by M^*; the *bidual* then is M^{**}, and analogous conventions apply to homomorphisms. The bilinear map $M \times M^* \to R$, $(x, \varphi) \mapsto \varphi(x)$, induces a natural homomorphism $h: M \to M^{**}$. We say that M is *torsionless* if h is injective, and that M is *reflexive* if h is bijective. Some relations between the notions just introduced are given in the exercises. Here we note a useful criterion:

Proposition 1.4.1. *Let R be a Noetherian ring, and M a finite R-module. Then:*
(a) *M is torsionless if and only if*
   (i) *$M_{\mathfrak{p}}$ is torsionless for all $\mathfrak{p} \in \operatorname{Ass} R$, and*
   (ii) *depth $M_{\mathfrak{p}} \geq 1$ for $\mathfrak{p} \in \operatorname{Spec} R$ with depth $R_{\mathfrak{p}} \geq 1$;*
(b) *M is reflexive if and only if*
   (i) *$M_{\mathfrak{p}}$ is reflexive for all $\mathfrak{p}$ with depth $R_{\mathfrak{p}} \leq 1$, and*
   (ii) *depth $M_{\mathfrak{p}} \geq 2$ for $\mathfrak{p} \in \operatorname{Spec} R$ with depth $R_{\mathfrak{p}} \geq 2$.*

PROOF. Consider the natural map $h: M \to M^{**}$ and set $U = \operatorname{Ker} h$, $C = \operatorname{Coker} h$. Note that the construction of h commutes with localization in the situation considered. Therefore the necessity of conditions (i) in (a) and (b) is obvious. Next Exercise 1.4.19 implies

$$\operatorname{depth} M_{\mathfrak{p}}^{**} \geq \min(2, \operatorname{depth} R_{\mathfrak{p}})$$

for all $\mathfrak{p} \in \operatorname{Spec} R$. That (b)(ii) is necessary for reflexivity follows directly from this inequality. If M is torsionless, then $M_{\mathfrak{p}}$ is isomorphic to a submodule of $M_{\mathfrak{p}}^{**}$, and we get depth $M_{\mathfrak{p}} \geq \min(1, \operatorname{depth} R_{\mathfrak{p}})$ for all $\mathfrak{p} \in \operatorname{Spec} R$. So (a)(ii) is necessary for M to be torsionless.

As to the sufficiency of (a)(i) and (ii), note that $U_{\mathfrak{p}} = 0$ for all $\mathfrak{p} \in \operatorname{Ass} R$ by (i), and, by (ii), depth $U_{\mathfrak{p}} \geq 1$ if depth $R_{\mathfrak{p}} \geq 1$. It follows that $\operatorname{Ass} U = \emptyset$, hence $U = 0$.

For the sufficiency of (b)(i) and (ii) we may now use that (a) gives us an exact sequence $0 \to M \to M^{**} \to C \to 0$. If depth $R_{\mathfrak{p}} \leq 1$, then $C_{\mathfrak{p}} = 0$ by (i). If depth $R_{\mathfrak{p}} \geq 2$, then depth $M_{\mathfrak{p}} \geq 2$ by (ii), and depth $M_{\mathfrak{p}}^{**} \geq 2$ by the inequality above. Therefore depth $C_{\mathfrak{p}} \geq 1$, and it follows that $\operatorname{Ass} C = \emptyset$. $\qquad\square$

Rank. The dimension of a finite dimensional vector space over a field is given either by the minimal number of generators or by the maximal number of linearly independent elements. The second aspect of 'dimension' is generalized in the notion of 'rank':

Definition 1.4.2. Let R be a ring, M an R-module, and Q be the total ring of fractions of R. Then M has *rank r* if $M \otimes Q$ is a free Q-module of rank r. If $\varphi: M \to N$ is a homomorphism of R-modules, then φ has *rank r* if $\operatorname{Im}\varphi$ has rank r.

Proposition 1.4.3. *Let R be a Noetherian ring, and M an R-module with a finite free presentation $F_1 \xrightarrow{\varphi} F_0 \longrightarrow M \longrightarrow 0$. Then the following are equivalent:*
(a) *M has rank r;*
(b) *M has a free submodule N of rank r such that M/N is a torsion module;*
(c) *for all prime ideals $\mathfrak{p} \in \operatorname{Ass} R$ the $R_\mathfrak{p}$-module $M_\mathfrak{p}$ is free of rank r;*
(d) *rank $\varphi = \operatorname{rank} F_0 - r$.*

PROOF. (a) $\Rightarrow$ (b): A free basis $x_1, \dots, x_r$ of $M \otimes Q$ can be formed from elements $x_i \in M$ (multiply by a suitable common denominator). Now take $N = \sum Rx_i$.

(b) $\Rightarrow$ (a): This is trivial.

(a) $\Rightarrow$ (c): $M_\mathfrak{p}$ is a localization of $M \otimes Q$.

(c) $\Rightarrow$ (a): Q is a semi-local ring. Its localizations with respect to its maximal ideals are just the localizations of R with respect to the maximal elements of $\operatorname{Ass} R$. By hypothesis M is therefore a projective module over Q, and moreover the localizations with respect to the maximal ideals of Q have the same rank r. Such a module is free; see Lemma 1.4.4 below.

(c) $\iff$ (d): In view of the equivalence of (a) and (c) we can replace (d) by the condition that $(\operatorname{Im}\varphi)_\mathfrak{p}$ is free and $\operatorname{rank}(\operatorname{Im}\varphi)_\mathfrak{p} = \operatorname{rank} F_0 - r$ for all $\mathfrak{p} \in \operatorname{Ass} R$. Now consider the exact sequence

$$0 \longrightarrow (\operatorname{Im}\varphi)_\mathfrak{p} \longrightarrow (F_0)_\mathfrak{p} \longrightarrow M_\mathfrak{p} \longrightarrow 0.$$

If $M_\mathfrak{p}$ is free, then $(\operatorname{Im}\varphi)_\mathfrak{p}$ must be free. Since $\mathfrak{p} \in \operatorname{Ass} R$, the converse is also true; see 1.3.4. $\qquad\square$

Lemma 1.4.4. *Let R be a semi-local ring, and M a finite projective R-module. Then M is free if the localizations $M_\mathfrak{m}$ have the same rank r for all maximal ideals $\mathfrak{m}$ of R.*

PROOF. We use induction on r. The case $r = 0$ is trivial. Suppose that $r > 0$. Then 1.2.2 (with $N = M$ and $\mathfrak{p}_1, \dots, \mathfrak{p}_m$ denoting the maximal ideals of R) yields an element $x \in M$ such that $x \notin \mathfrak{m} M_\mathfrak{m}$ for all maximal ideals of M. Thus x is a member of a minimal system of generators of $M_\mathfrak{m}$. Since every such system is a basis of the free module $M_\mathfrak{m}$, one concludes that $(M/Rx)_\mathfrak{m}$ is free of rank $r - 1$. By the induction hypothesis M/Rx is free of rank $r - 1$. Therefore $M \cong Rx \oplus M/Rx$. In particular Rx is a projective R-module. But Rx is also free: the natural epimorphism $\varphi: R \to Rx$ yields an isomorphism $\varphi_\mathfrak{m}: R_\mathfrak{m} \to (Rx)_\mathfrak{m}$ for every maximal ideal $\mathfrak{m}$. Since $(\operatorname{Ker}\varphi)_\mathfrak{m} = \operatorname{Ker}\varphi_\mathfrak{m}$ it follows that $\operatorname{Ker}\varphi = 0$. $\qquad\square$

Rank is additive along exact sequences.

Proposition 1.4.5. *Let R be a Noetherian ring, and $0 \to U \to M \to N \to 0$ an exact sequence of finite R-modules. If two of U, M, N have a rank, then so does the third, and* rank $M =$ rank $U +$ rank N.

PROOF. In view of 1.4.3 we may assume that R is local and of depth 0. Then two of U, M, N are free. If U and N are free, then so is M. Thus M is always free (after the reduction to depth 0), and the result follows from the equivalence of 1.4.3(a) and (d). □

Corollary 1.4.6. *Let R be a Noetherian ring, and M an R-module with a finite free resolution $F.: 0 \to F_s \to F_{s-1} \to \cdots \to F_1 \to F_0$. Then* rank $M = \sum_{j=0}^{s}(-1)^j$ rank F_j.

PROOF. Observe 1.4.5 and use induction on s. □

Corollary 1.4.7. *Let R be a Noetherian ring, and $I \neq 0$ an ideal with a finite free resolution. Then I contains an R-regular element.*

PROOF. By 1.4.6 I has a rank, and that rank $I +$ rank $R/I =$ rank $R = 1$ follows immediately from 1.4.5. Since I is torsion-free and non-zero, the only possibility is rank $I = 1$, whence rank $R/I = 0$. Thus R/I is annihilated by an R-regular element. □

Ideals of minors and Fitting invariants. Let U be an $m \times n$ matrix over R where $m, n \geq 0$. For $t = 1, \ldots, \min(m, n)$ we then denote by $I_t(U)$ the ideal generated by the t-minors of U (the determinants of $t \times t$ submatrices). For systematic reasons one sets $I_t(U) = R$ for $t \leq 0$ and $I_t(U) = 0$ for $t > \min(m, n)$. If $\varphi : F \to G$ is a homomorphism of finite free R-modules, then φ is given by a matrix U with respect to bases of F and G. It is an elementary exercise to verify that the ideals $I_t(U)$ only depend on φ. Therefore we may put $I_t(\varphi) = I_t(U)$. It is just as easy to show that $I_t(\varphi)$ is already determined by the submodule $\operatorname{Im} \varphi$ of G. As proved by Fitting in 1936, these ideals are even invariants of $\operatorname{Coker} \varphi$ (when counted properly), and therefore called the *Fitting invariants* of $\operatorname{Coker} \varphi$: let

$$F_1 \xrightarrow{\varphi} F_0 \longrightarrow M \longrightarrow 0 \quad \text{and} \quad G_1 \xrightarrow{\psi} G_0 \longrightarrow M \longrightarrow 0$$

be finite free presentations of the R-module M, and $n = $ rank F_0, $p = $ rank G_0; then $I_{n-u}(\varphi) = I_{p-u}(\psi)$ for all $u \geq 0$. (The proof is left as an exercise for the reader.) This justifies the term *u-th Fitting invariant of M* for $I_{n-u}(\varphi)$.

It is an important property of the ideals $I_t(\varphi)$ that their formation commutes with ring extensions: if S is an R-algebra, then $I_t(\varphi \otimes S) = I_t(\varphi)S$. (Simply consider φ as given by a matrix.)

The ideals $I_t(\varphi)$ determine the minimal number $\mu(M_{\mathfrak{p}})$ of generators of a localization in the same way that they control the vector space dimension of M if R is a field.

Lemma 1.4.8. *Let R be a ring, M an R-module with a finite free presentation $F_1 \xrightarrow{\varphi} F_0 \longrightarrow M \longrightarrow 0$, and $\mathfrak{p}$ a prime ideal. Then the following are equivalent:*
(a) $I_t(\varphi) \not\subseteq \mathfrak{p}$;
(b) $(\operatorname{Im} \varphi)_{\mathfrak{p}}$ *contains a (free) direct summand of* $(F_0)_{\mathfrak{p}}$ *of rank t;*
(c) $\mu(M_{\mathfrak{p}}) \leq \operatorname{rank} F_0 - t$.

PROOF. It is no restriction to assume that $R = R_{\mathfrak{p}}$. Nakayama's lemma entails that $\mu(M) = \mu(M/\mathfrak{p}M)$. Similarly it implies that $\operatorname{Im} \varphi$ contains a (free) direct summand of F_0 of rank t if and only if there are elements $x_1, \ldots, x_t \in \operatorname{Im} \varphi$ which are linearly independent modulo $\mathfrak{p}F_0$. (Note that every direct summand of a finite free module over a local ring is free itself – again an application of Nakayama's lemma.) After these observations we may replace R by the field $R/\mathfrak{p}$. For vector spaces over fields the equivalence of (a), (b) and (c) is an elementary fact. $\qquad\square$

Lemma 1.4.9. *With the notation of* 1.4.8, *the following are equivalent:*
(a) $I_t(\varphi) \not\subseteq \mathfrak{p}$ *and* $I_{t+1}(\varphi)_{\mathfrak{p}} = 0$;
(b) $(\operatorname{Im} \varphi)_{\mathfrak{p}}$ *is a free direct summand of* $(F_0)_{\mathfrak{p}}$ *of rank t;*
(c) $M_{\mathfrak{p}}$ *is free and* $\operatorname{rank} M_{\mathfrak{p}} = \operatorname{rank} F_0 - t$.

PROOF. We may assume that $R = R_{\mathfrak{p}}$. Then each of (b) and (c) is equivalent to the split exactness of the sequence $0 \to \operatorname{Im} \varphi \to F_0 \to M \to 0$.

If (a) holds, then, with respect to suitable bases of F_1 and F_0, the matrix of φ has the form
$$\begin{pmatrix} \operatorname{id}_t & 0 \\ 0 & 0 \end{pmatrix}$$

where id_t is the $t \times t$ identity matrix. This implies (b). The converse is seen similarly. $\qquad\square$

Let M be a finite module over a Noetherian ring R. Then M is a projective module (of rank r) if and only if $M_{\mathfrak{p}}$ is a free $R_{\mathfrak{p}}$-module (of rank r) for all $\mathfrak{p} \in \operatorname{Spec} R$. Combining this fact with 1.4.9 we obtain the global version of 1.4.9:

Proposition 1.4.10. *Let R be a Noetherian ring, and M a finite R-module with a finite free presentation $F_1 \xrightarrow{\varphi} F_0 \longrightarrow M \longrightarrow 0$. Then the following are equivalent:*
(a) $I_r(\varphi) = R$ *and* $I_{r+1}(\varphi) = 0$;
(b) M *is projective and* $\operatorname{rank} M = \operatorname{rank} F_0 - r$.

The rank of a homomorphism $\varphi : F \to G$ is determined by the ideal $I_t(\varphi)$, just as in elementary linear algebra:

Proposition 1.4.11. *Let R be a Noetherian ring, and let $\varphi : F \to G$ be a homomorphism of finite free R-modules. Then* $\operatorname{rank} \varphi = r$ *if and only if* $\operatorname{grade} I_r(\varphi) \geq 1$ *and* $I_{r+1}(\varphi) = 0$.

The easy proof is left as an exercise for the reader.

The Buchsbaum–Eisenbud acyclicity criterion. Let R be a ring. A complex

$$G_\bullet : \cdots \longrightarrow G_m \xrightarrow{\psi_m} G_{m-1} \longrightarrow \cdots \longrightarrow G_1 \xrightarrow{\psi_1} G_0 \longrightarrow 0$$

of R-modules is called *acyclic* if $H_i(G_\bullet) = 0$ for all $i > 0$, and *split acyclic* if it is acyclic and $\psi_{i+1}(G_{i+1})$ is a direct summand of G_i for $i \geq 0$.

Let R be a Noetherian ring, and

$$F_\bullet : 0 \longrightarrow F_s \xrightarrow{\varphi_s} F_{s-1} \longrightarrow \cdots \longrightarrow F_1 \xrightarrow{\varphi_1} F_0 \longrightarrow 0$$

a complex of finite free R-modules. We want to develop a criterion for $F_\bullet$ to be acyclic. This criterion will involve ideals generated by certain minors of the homomorphisms φ_i. A first relation between the ideals $I_t(\varphi)$ and the acyclicity of complexes is given in the next proposition.

Proposition 1.4.12. *Let R be a ring, M an R-module,*

$$F_\bullet : 0 \longrightarrow F_s \xrightarrow{\varphi_s} F_{s-1} \longrightarrow \cdots \longrightarrow F_1 \xrightarrow{\varphi_1} F_0 \longrightarrow 0$$

be a complex of finite free R-modules, and $\mathfrak{p} \subset R$ be a prime ideal. Set $r_i = \sum_{j=i}^{s} (-1)^{j-i} \operatorname{rank} F_j$ for $i = 1, \ldots, s$. Then the following are equivalent:

(a) *$F_\bullet \otimes R_\mathfrak{p}$ is split acyclic;*

(b) *$I_{r_i}(\varphi_i) \not\subset \mathfrak{p}$ for $i = 1, \ldots, s$.*

Furthermore, $I_t(\varphi_i)_\mathfrak{p} = 0$ for all $i = 1, \ldots, s$ and $t > r_i$, if one of these conditions holds.

If $\mathfrak{p} \in \operatorname{Ass} M$, then (a) and (b) are equivalent to

(c) *$F_\bullet \otimes M_\mathfrak{p}$ is acyclic.*

PROOF. We may suppose that $R = R_\mathfrak{p}$.

(a) $\Rightarrow$ (b): If $F_\bullet$ is split acyclic, then $F_\bullet \otimes R/\mathfrak{p}$ is a (split) acyclic complex of vector spaces over $R/\mathfrak{p}$; so we can refer to elementary linear algebra.

(b) $\Rightarrow$ (a): We again use induction, and may assume that $\operatorname{Coker} \varphi_2$ is a free R-module of rank r_1. According to 1.4.8, $\operatorname{Im} \varphi_1$ contains a free direct summand U of F_0 of rank r_1. So we get an induced epimorphism $\operatorname{Coker} \varphi_2 \to U$ of free R-modules, both of which have rank r_1. Such a map must be an isomorphism. One easily concludes that $\operatorname{Im} \varphi_1 = U$. Hence $F_\bullet$ is split acyclic.

That $I_t(\varphi_i) = 0$ for $t > r_i$, follows most easily from (a) in conjunction with 1.4.9.

(c) ⇒ (a): Let $F'_\bullet$ be the truncation $0 \to F_s \to \cdots \to F_1 \to 0$. Then $F'_\bullet \otimes M$ is acyclic; arguing inductively, we may therefore suppose that $F'_\bullet$ is split acyclic. Then $F'_1 = \operatorname{Coker}\varphi_2$ is free, and the induced map $F'_1 \otimes M \to F_0 \otimes M$ is injective by hypothesis. By virtue of 1.3.4, F'_1 is mapped isomorphically onto a free direct summand of F_0.

(a) ⇒ (c): This is evident.   □

We have completed our preparations for the following important and extremely useful acyclicity criterion.

Theorem 1.4.13 (Buchsbaum–Eisenbud). *Let R be a Noetherian ring, and*

$$F_\bullet : 0 \longrightarrow F_s \xrightarrow{\varphi_s} F_{s-1} \longrightarrow \cdots \longrightarrow F_1 \xrightarrow{\varphi_1} F_0 \longrightarrow 0$$

a complex of finite free R-modules. Set $r_i = \sum_{j=i}^s (-1)^{j-i} \operatorname{rank} F_j$. Then the following are equivalent:

(a) *$F_\bullet$ is acyclic;*

(b) *$\operatorname{grade} I_{r_i}(\varphi_i) \geq i$ for $i = 1, \ldots, s$.*

Before we prove the theorem the reader should note that $r_i = \operatorname{rank}\varphi_i \geq 0$ when $F_\bullet$ is acyclic; this is just a restatement of 1.4.6. Conversely, it is not necessary to require that $r_i \geq 0$ for the implication (b) ⇒ (a); if $r_i < 0$, then $r_{i+1} > \operatorname{rank} F_i$, and $I_{r_{i+1}}(\varphi_{i+1}) = 0$ in contradiction with (b). In the situation of 1.4.13 we call r_i the *expected rank of* φ_i.

PROOF. (a) ⇒ (b): By what has just been said and 1.4.11, we see that $\operatorname{grade} I_{r_i}(\varphi_i) \geq 1$ for $i = 1, \ldots, s$. In particular there is an R-regular element x contained in the product of the ideals $I_{r_i}(\varphi_i)$. If x is a unit, then $I_{r_i}(\varphi_i) = R$ for all i, and we are done. Otherwise we use induction. Let ‾ denote residue classes modulo x. It follows immediately from 1.1.5 that the induced complex $0 \to \bar{F}_s \to \bar{F}_{s-1} \to \cdots \to \bar{F}_2 \to \bar{F}_1 \to 0$ is acyclic. Furthermore $I_{r_i}(\varphi_i)^- = I_{r_i}(\bar{\varphi}_i)$, and $\operatorname{grade} I_{r_i}(\bar{\varphi}_i) \geq i - 1$ by induction. Then $\operatorname{grade} I_{r_i}(\varphi_i) \geq i$ for $i = 2, \ldots, s$.

The reader may have noticed that this implication follows immediately from the Auslander–Buchsbaum formula 1.3.3. In view of the generalization 9.1.6 an independent proof is useful, however.

(b) ⇒ (a): Using induction again we may assume that $F'_\bullet : 0 \to F_s \to F_{s-1} \to \cdots \to F_2 \to F_1 \to 0$ is acyclic. We set $M_i = \operatorname{Coker}\varphi_{i+1}$ for $i = 1, \ldots, s$, and show by descending induction that $\operatorname{depth}(M_i)_\mathfrak{p} \geq \min\{i, \operatorname{depth} R_\mathfrak{p}\}$ for all $\mathfrak{p} \in \operatorname{Spec} R$ and $i = 1, \ldots, s$.

As $M_s = F_s$, this is trivial for $i = s$. Let $i < s$ and consider the exact sequence

$$0 \longrightarrow M_{i+1} \longrightarrow F_i \longrightarrow M_i \longrightarrow 0.$$

If $\operatorname{depth} R_\mathfrak{p} \geq i + 1$, then $\operatorname{depth}(M_{i+1})_\mathfrak{p} \geq i + 1$, and we get $\operatorname{depth}(M_i)_\mathfrak{p} \geq i$ from 1.2.9. If $\operatorname{depth} R_\mathfrak{p} \leq i$, then $I_{r_{i+1}}(\varphi_{i+1}) \not\subseteq \mathfrak{p}$ by hypothesis; on the other hand rank M_{i+1} = rank φ_{i+1} = r_{i+1}, and therefore $I_t(\varphi_{i+1}) = 0$ for $t > r_{i+1}$. So 1.4.9 yields that $(M_i)_\mathfrak{p}$ is free, hence depth$(M_i)_\mathfrak{p}$ = depth $R_\mathfrak{p}$.

We still have to show that the induced map $\varphi_1' : M_1 \to F_0$ is injective. Let $N = \operatorname{Ker}\varphi_1'$. In order to get $N = 0$, we derive that $\operatorname{Ass} N = \emptyset$. If depth $R_\mathfrak{p} \geq 1$, then depth$(M_1)_\mathfrak{p} \geq 1$ as seen above; therefore $\mathfrak{p} \notin \operatorname{Ass} M_1 \supset \operatorname{Ass} N$. If depth $R_\mathfrak{p} = 0$, then $I_{r_i}(\varphi_i) \not\subset \mathfrak{p}$ for $i = 1,\ldots,s$, and $F_{\bullet} \otimes R_\mathfrak{p}$ is even split acyclic by 1.4.12. It follows that $N_\mathfrak{p} = 0$ since $N_\mathfrak{p} \cong H_1(F_{\bullet} \otimes R_\mathfrak{p})$. $\qquad\square$

Often one only needs the following consequence of 1.4.13.

Corollary 1.4.14. *Let R be a Noetherian ring, and $F_{\bullet}$ be a complex as in 1.4.13. If $F_{\bullet} \otimes R_\mathfrak{p}$ is acyclic for all prime ideals $\mathfrak{p}$ with depth $R_\mathfrak{p} < s$, then $F_{\bullet}$ is acyclic.*

PROOF. Let $\mathfrak{p}$ be a prime ideal with depth $R_\mathfrak{p} < i \leq s$. The implication (a) $\Rightarrow$ (b) of the theorem applied to $F_{\bullet} \otimes R_\mathfrak{p}$ yields grade $I_{r_i}(\varphi_i)_\mathfrak{p} \geq i$, which is only possible if $I_{r_i}(\varphi_i) \not\subset \mathfrak{p}$. This shows grade $I_{r_i}(\varphi_i) \geq i$, and the acyclicity of $F_{\bullet}$ follows from the implication (b) $\Rightarrow$ (a) of the theorem. $\qquad\square$

Theorem 1.4.13 is the most important case of the acyclicity criterion of Buchsbaum and Eisenbud. Its general form will be discussed in Chapter 9.

Perfect modules. Let R be a Noetherian ring, and M a finite R-module. Since one can compute $\operatorname{Ext}_R^i(M, R)$ from a projective resolution of M, one obviously has grade $M \leq \operatorname{proj dim} M$. Modules for which equality is attained have especially good properties.

Definition 1.4.15. Let R be a Noetherian ring. A non-zero finite R-module M is *perfect* if proj dim M = grade M. An ideal I is called *perfect* if R/I is a perfect module.

Perfect modules are 'grade unmixed':

Proposition 1.4.16. *Let R be a Noetherian ring, and M a perfect R-module. For a prime ideal $\mathfrak{p} \in \operatorname{Supp} M$ the following are equivalent:*
(a) $\mathfrak{p} \in \operatorname{Ass} M$;
(b) depth $R_\mathfrak{p}$ = grade M.
Furthermore grade $\mathfrak{p}$ = grade M *for all prime ideals $\mathfrak{p} \in \operatorname{Ass} M$.*

PROOF. For all finite R-modules M and $\mathfrak{p} \in \operatorname{Supp} M$ one has the inequalities

$$\operatorname{grade} M \leq \operatorname{grade} M_\mathfrak{p} \leq \operatorname{proj dim} M_\mathfrak{p} \leq \operatorname{proj dim} M,$$

and moreover proj dim $M_\mathfrak{p}$ + depth $M_\mathfrak{p}$ = depth $R_\mathfrak{p}$ by the Auslander–Buchsbaum formula 1.3.3. If M is perfect, then the inequalities become equations, and depth $M_{\mathfrak{p}} = 0$ if and only if depth $R_{\mathfrak{p}} = \text{grade } M$. This shows the equivalence of (a) and (b).

If $\mathfrak{p} \in \text{Ass } M$, then $\mathfrak{p} \supset \text{Ann } M$, and so grade $\mathfrak{p} \geq \text{grade } M$. For perfect M the converse results from (b) and the inequality grade $\mathfrak{p} \leq \text{depth } R_{\mathfrak{p}}$.
□

It follows easily from 1.3.6 that an ideal generated by a regular sequence in a Noetherian ring R is perfect. Some more examples are described in the following celebrated theorem:

Theorem 1.4.17 (Hilbert–Burch). *Let R be a Noetherian ring, and I an ideal with a free resolution*

$$F_{\bullet} : 0 \longrightarrow R^n \xrightarrow{\varphi} R^{n+1} \longrightarrow I \longrightarrow 0.$$

Then there exists an R-regular element a such that $I = aI_n(\varphi)$. If I is projective, then $I = (a)$, and if $\text{proj dim } I = 1$, then $I_n(\varphi)$ is perfect of grade 2.

Conversely, if $\varphi : R^n \to R^{n+1}$ is an R-linear map with grade $I_n(\varphi) \geq 2$, then $I = I_n(\varphi)$ has the free resolution $F_{\bullet}$.

PROOF. First we prove the converse part. Let $\varphi : R^n \to R^{n+1}$ be a map with grade $I_n(\varphi) \geq 2$. Then φ is given by an $(n + 1) \times n$ matrix U. Let δ_i denote the i-minor of U with the i-th row deleted, and consider the homomorphism $\pi : R^{n+1} \to R$ which sends e_i to $(-1)^i \delta_i$. Laplace expansion shows that we have a complex

$$0 \longrightarrow R^n \xrightarrow{\varphi} R^{n+1} \xrightarrow{\pi} I \longrightarrow 0,$$

which in fact is exact by 1.4.13.

Suppose now that an ideal I with free resolution $F_{\bullet}$ is given. Then 1.4.13 yields grade $I_n(\varphi) \geq 2$, and we can apply the first part of the proof to obtain $I \cong \text{Coker } \varphi \cong I_n(\varphi)$; equivalently, there exists an injective linear map $\alpha : I_n(\varphi) \to R$ with $I = \text{Im } \alpha$. According to 1.2.24, α is just multiplication by some $a \in R$. Because of 1.4.7 (or 1.4.13) a cannot be a zero-divisor.

If I is projective, then $I_n(\varphi) = R$ by 1.4.10, and thus $I = (a)$. If $\text{proj dim } I = 1$, then $\text{proj dim}(R/I_n(\varphi)) = \text{proj dim } R/I = 2$, and $R/I_n(\varphi)$ is perfect of grade 2.
□

Exercises

1.4.18. Let R be a ring, and M a finite torsion-free module. Prove that if M has a rank, then M is isomorphic to a submodule of a finite free R-module of the same rank.

1.4.19. Let R be a Noetherian ring, I an ideal, and M, N finite modules. Prove $\text{grade}(I, \text{Hom}_R(M, N)) \geq \min(2, \text{grade}(I, N))$.

1.4.20. Let R be a Noetherian ring, and M a finite R-module. Prove
(a) if M is torsionless, then it is torsion-free,
(b) M is torsionless if and only if it is a submodule of a finite free module,
(c) if M is reflexive, then it is a second syzygy, i.e. there is an exact sequence $0 \to M \to F_1 \to F_0$ with F_i finite and free.

1.4.21. Let R be a Noetherian ring, and M a finite R-module. Suppose $\varphi : G \to F$ is a homomorphism of finite free R-modules with $M = \operatorname{Coker} \varphi$. Then $D(M) = \operatorname{Coker} \varphi^*$ is the *transpose* of M. (It is unique up to projective equivalence.) Show that $\operatorname{Ker} h = \operatorname{Ext}^1_R(D(M), R)$ and $\operatorname{Coker} h = \operatorname{Ext}^2_R(D(M), R)$ where $h : M \to M^{**}$ is the natural homomorphism.

1.4.22. Let R be a Noetherian ring, and M a finite R-module such that M^* has finite projective dimension. Prove
(a) if $\operatorname{depth} M_\mathfrak{p} \geq \min(1, \operatorname{depth} R_\mathfrak{p})$ for all $\mathfrak{p} \in \operatorname{Spec} R$, then M is torsionless,
(b) if $\operatorname{depth} M_\mathfrak{p} \geq \min(2, \operatorname{depth} R_\mathfrak{p})$ for all $\mathfrak{p} \in \operatorname{Spec} R$, then M is reflexive.
Hint: $\operatorname{proj dim} M^* < \infty \Rightarrow \operatorname{proj dim} D(M) < \infty$.

1.4.23. Let R be a Noetherian ring, and M a finite R-module. Show that M has a rank if and only if M^* has a rank (and both ranks coincide). Hint: It is enough to consider the case $R = R_\mathfrak{p}$, $\operatorname{depth} R_\mathfrak{p} = 0$. Apply 1.4.22.

1.4.24. Let R be a Noetherian local ring, and $0 \to L_s \to L_{s-1} \to \cdots \to L_1 \to L_0 \to 0$ a complex of finite R-modules. Suppose that the following hold for $i > 0$: (i) $\operatorname{depth} L_i \geq i$, and (ii) $\operatorname{depth} H_i(L_\bullet) = 0$ or $H_i(L_\bullet) = 0$. Show that $L_\bullet$ is acyclic. (This is Peskine and Szpiro's 'lemme d'acyclicité' [297].)
Hint: Set $C_i = \operatorname{Coker}(L_{i+1} \to L_i)$, and show by descending induction that $\operatorname{depth} C_i \geq i$ and $H_i(L_\bullet) = 0$ for $i > 0$.

1.4.25. Let R be a Noetherian ring, I an ideal of finite projective dimension, and M a finite R/I-module. Prove the following inequality of Avramov and Foxby [29]:

$$\operatorname{grade}_{R/I} M + \operatorname{grade}_R R/I \leq \operatorname{grade}_R M \leq \operatorname{grade}_{R/I} M + \operatorname{proj dim}_R R/I;$$

if I is perfect, then equality is attained. (Use the Auslander–Buchsbaum formula.)

1.4.26. Let R be a Noetherian ring, and M a perfect R-module of grade n. Suppose $P_\bullet$ is a projective resolution of M of length n and set $M' = \operatorname{Ext}^n_R(M, R)$. Prove
(a) $P_\bullet^*$ is acyclic and resolves M',
(b) M' is perfect of grade n, and $M'' = M$,
(c) $\operatorname{Ass} M' = \operatorname{Ass} M$.

1.4.27. Let R be a Noetherian ring, x an R-sequence of length n, and $I = (x)$. Show that R/I^m is perfect of grade n for all $m \geq 1$. (Theorem 1.1.8 is useful.)

1.5 Graded rings and modules

In this section we investigate rings and modules which, like a polynomial ring, admit a decomposition of their elements into homogeneous components.

Definition 1.5.1. A *graded ring* is a ring R together with a decomposition $R = \bigoplus_{i \in \mathbb{Z}} R_i$ (as a $\mathbb{Z}$-module) such that $R_i R_j \subset R_{i+j}$ for all $i, j \in \mathbb{Z}$.

A *graded R-module* is an R-module M together with a decomposition $M = \bigoplus_{i \in \mathbb{Z}} M_i$ (as a $\mathbb{Z}$-module) such that $R_i M_j \subset M_{i+j}$ for all $i, j \in \mathbb{Z}$. One calls M_i the i-th *homogeneous* (or *graded*) *component of M.*

The elements $x \in M_i$ are called *homogeneous* (*of degree i*); those of R_i are also called *i-forms*. According to this definition the zero element is homogeneous of arbitrary degree. The degree of x is denoted by $\deg x$. An arbitrary element $x \in M$ has a unique presentation $x = \sum_i x_i$ as a sum of homogeneous elements $x_i \in M_i$. The elements x_i are called the *homogeneous components* of x.

Note that R_0 is a ring with $1 \in R_0$, that all summands M_i are R_0-modules, and that $M = \bigoplus_{i \in \mathbb{Z}} M_i$ is a direct sum decomposition of M as an R_0-module.

Definition 1.5.2. Let R be a graded ring. The *category of graded R-modules*, denoted $\mathcal{M}_0(R)$, has as objects the graded R-modules. A *morphism* $\varphi \colon M \to N$ in $\mathcal{M}_0(R)$ is an R-module homomorphism satisfying $\varphi(M_i) \subset N_i$ for all $i \in \mathbb{Z}$. An R-module homomorphism which is a morphism in $\mathcal{M}_0(R)$ will be called *homogeneous*.

Let M be a graded R-module and N a submodule of M. N is called a *graded submodule* if it is a graded module such that the inclusion map is a morphism in $\mathcal{M}_0(R)$. This is equivalent to the condition $N_i = N \cap M_i$ for all $i \in \mathbb{Z}$. In other words, N is a graded submodule of M if and only if N is generated by the homogeneous elements of M which belong to N. In particular, if $x \in N$, then all homogeneous components of x belong to N. Furthermore, M/N is graded in a natural way. If φ is a morphism in $\mathcal{M}_0(R)$, then $\operatorname{Ker} \varphi$ and $\operatorname{Im} \varphi$ are graded.

A (not necessarily commutative) R-algebra A is *graded* if, in addition to the definition, $A_i A_j \subset A_{i+j}$.

The graded submodules of R are called *graded ideals*. Let I be an arbitrary ideal of R. Then the graded ideal I^* is defined to be the ideal generated by all homogeneous elements $a \in I$. It is clear that I^* is the largest graded ideal contained in I, and that R/I^* inherits a natural structure as a graded ring.

Examples 1.5.3. (a) Let S be a ring, and $R = S[X_1, \ldots, X_n]$ a polynomial ring over S. Then for every choice of integers $d_1, \ldots, d_n$ there exists a unique grading on R such that $\deg X_i = d_i$ and $\deg a = 0$ for all $a \in S$: the m-th graded component is the S-module generated by all monomials $X_1^{e_1} \cdots X_n^{e_n}$ such that $\sum e_i d_i = m$. If one chooses $d_i = 1$ for all i, then one obtains the grading of the polynomial ring corresponding to the total degree of a monomial. Unless indicated otherwise we will always consider R to be graded in this way.

(b) Every ring R has the *trivial grading* given by $R_0 = R$ and $R_i = 0$ for $i \neq 0$. A typical example of a graded module over R is a complex

$$C_\bullet : \cdots \xrightarrow{\partial} C_n \xrightarrow{\partial} C_{n-1} \xrightarrow{\partial} \cdots$$

of R-modules. Such a complex may be equivalently described as a graded module $C_\bullet = \bigoplus_{i=-\infty}^{\infty} C_i$ together with an R-endomorphism ∂ such that $\partial^2 = 0$ and $\partial(C_i) \subset C_{i-1}$ for all i. (In the terminology to be introduced below, ∂ is a homogeneous endomorphism of degree -1.)

The most important graded rings arise in algebraic geometry as the coordinate rings of projective varieties. They have the form $R = k[X_1, \ldots, X_n]/I$ where k is a field and I is an ideal generated by homogeneous polynomials (in the usual sense). Then R is generated as a k-algebra by elements of degree 1, namely the residue classes of the indeterminates. Graded rings R which as R_0-algebras are generated by 1-forms will be called *homogeneous R_0-algebras*. More generally, if R is a graded R_0-algebra generated by elements of positive degree, then we say that R is a *positively graded R_0-algebra*.

We want to clarify which graded rings are Noetherian. Let us first consider positively graded rings.

Proposition 1.5.4. *Let R be a positively graded R_0-algebra, and $x_1, \ldots, x_n$ homogeneous elements of positive degree. Then the following are equivalent:*
(a) $x_1, \ldots, x_n$ generate the ideal $\mathfrak{m} = \bigoplus_{i=1}^{\infty} R_i$;
(b) $x_1, \ldots, x_n$ generate R as an R_0-algebra.
In particular R is Noetherian if and only if R_0 is Noetherian and R is a finitely generated R_0-algebra.

PROOF. For the implication (a) $\Rightarrow$ (b) it is enough to write every homogeneous element $y \in R$ as a polynomial in $x_1, \ldots, x_n$ with coefficients in R_0, and this is very easy by induction on $\deg y$. The rest is evident. $\square$

The last assertion of 1.5.4 holds for graded rings in general.

Theorem 1.5.5. *Let R be a graded ring. Then the following are equivalent:*
(a) every graded ideal of R is finitely generated;
(b) R is a Noetherian ring;
(c) R_0 is Noetherian, and R is a finitely generated R_0-algebra;
(d) R_0 is Noetherian, and both $S_1 = \bigoplus_{i=0}^{\infty} R_i$ and $S_2 = \bigoplus_{i=0}^{\infty} R_{-i}$ are finitely generated R_0-algebras.

PROOF. The implications (d) $\Rightarrow$ (c) $\Rightarrow$ (b) $\Rightarrow$ (a) are obvious. For (a) $\Rightarrow$ (d) we first note that R_0 is a direct summand of R as an R_0-module. It follows that $IR \cap R_0 = I$ for every ideal I of R_0, and thus (a) implies that R_0 is Noetherian. (Extend an ascending chain of ideals of R_0 to R, and contract the extension back to R_0.) A similar argument shows that R_i is a finite R_0-module for every $i \in \mathbb{Z}$.

Let $\mathfrak{m} = \bigoplus_{i=1}^{\infty} R_i$. We claim that $\mathfrak{m}$ is a finitely generated ideal of S_1. By hypothesis $\mathfrak{m}R$ has a finite system of generators $x_1, \ldots, x_m$, which may certainly be chosen to be homogeneous of positive degrees d_i. Let d be the maximum of $d_1, \ldots, d_m$. Then a homogeneous element $y \in \mathfrak{m}$ with $\deg y \geq d$ can be written as a linear combination of $x_1, \ldots, x_m$ with coefficients from S_1. Thus $x_1, \ldots, x_m$ together with a finite set of homogeneous elements spanning $R_1, \ldots, R_{d-1}$ over R_0 generate $\mathfrak{m}$ as an ideal of S_1. According to 1.5.4, S_1 is a finitely generated R_0-algebra, and the claim for S_2 follows by symmetry. $\square$

Very often we shall derive properties of a graded ring or module from its localizations with respect to graded prime ideals. The following lemma is basic for such arguments.

Lemma 1.5.6. *Let R be a graded ring.*
(a) *For every prime ideal $\mathfrak{p}$ the ideal $\mathfrak{p}^*$ is a prime ideal.*
(b) *Let M be a graded R-module.*
  (i) *If $\mathfrak{p} \in \operatorname{Supp} M$, then $\mathfrak{p}^* \in \operatorname{Supp} M$.*
  (ii) *If $\mathfrak{p} \in \operatorname{Ass} M$, then $\mathfrak{p}$ is graded; furthermore $\mathfrak{p}$ is the annihilator of a homogeneous element.*

PROOF. (a) Let $a, b \in R$ such that $ab \in \mathfrak{p}^*$. We write $a = \sum_i a_i$, $a_i \in R_i$, and $b = \sum_j b_j$, $b_j \in R_j$. Assume that $a \notin \mathfrak{p}^*$ and $b \notin \mathfrak{p}^*$. Then there exist integers p, q such that $a_p \notin \mathfrak{p}^*$, but $a_i \in \mathfrak{p}^*$ for $i < p$, and $b_q \notin \mathfrak{p}^*$, but $b_j \in \mathfrak{p}^*$ for $j < q$. The $(p+q)$-th homogeneous component of ab is $\sum_{i+j=p+q} a_i b_j$. Thus $\sum_{i+j=p+q} a_i b_j \in \mathfrak{p}^*$, since $\mathfrak{p}^*$ is graded. All summands of this sum, except possibly $a_p b_q$, belong to $\mathfrak{p}^*$, and so it follows that $a_p b_q \in \mathfrak{p}^*$ as well. Since $\mathfrak{p}^* \subset \mathfrak{p}$, and since $\mathfrak{p}$ is a prime ideal we conclude that $a_p \in \mathfrak{p}$ or $b_q \in \mathfrak{p}$. But a_p and b_q are homogeneous, and so $a_p \in \mathfrak{p}^*$ or $b_q \in \mathfrak{p}^*$, a contradiction.

(b) For (i) assume $\mathfrak{p}^* \notin \operatorname{Supp} M$; then $M_{\mathfrak{p}^*} = 0$. Let $x \in M$ be a homogeneous element. Then there exists an element $a \in R \setminus \mathfrak{p}^*$ such that $ax = 0$. It follows that $a_i x = 0$ for all homogeneous components a_i of a. Since $a \in R \setminus \mathfrak{p}^*$, there exists an integer i such that $a_i \notin \mathfrak{p}^*$. Since a_i is homogeneous, we even have $a_i \notin \mathfrak{p}$. Hence $x/1 = 0$ in $M_{\mathfrak{p}}$. This holds true for all homogeneous elements of M. Thus we conclude that $M_{\mathfrak{p}} = 0$, a contradiction.

For (ii) we choose an element $x \in M$ with $\mathfrak{p} = \operatorname{Ann} x$. Let $x = x_m + \cdots + x_n$ be its decomposition as a sum of homogeneous elements x_i of degree i. Similarly we decompose an element $a = a_p + \cdots + a_q$ of $\mathfrak{p}$. Since $ax = 0$, we have equations $\sum_{i+j=r} a_i x_j = 0$ for $r = m+p, \ldots, n+q$. It follows that $a_p x_m = 0$, and, by induction, $a_p^i x_{m+i-1} = 0$ for all $i \geq 1$.

Thus a_p^{n-m+1} annihilates x. As $\mathfrak{p}$ is a prime ideal, we have $a_p \in \mathfrak{p}$. Iterating this procedure we see that each homogeneous component of a belongs to $\mathfrak{p}$.

In order to prove the second assertion in (ii) one can now use the fact that $\mathfrak{p}$ is generated by homogeneous elements. It follows easily that $\mathfrak{p}$ annihilates all the homogeneous components of x. Set $\mathfrak{a}_i = \mathrm{Ann}\, x_i$; then, as just seen, $\mathfrak{p} \subset \mathfrak{a}_i$. On the other hand $\bigcap_{i=m}^{n} \mathfrak{a}_i \subset \mathfrak{p}$. Since $\mathfrak{p}$ is a prime ideal, there exists j with $\mathfrak{a}_j \subset \mathfrak{p}$, and therefore $\mathfrak{a}_j = \mathfrak{p}$. □

Let $\mathfrak{p}$ be a prime ideal of R, and let S be the set of homogeneous elements of R not belonging to $\mathfrak{p}$. The set S is multiplicatively closed, and we put $M_{(\mathfrak{p})} = M_S$ for any graded R-module M. For $x/a \in M_{(\mathfrak{p})}$, x homogeneous, we set $\deg x/a = \deg x - \deg a$. We further define a grading on $M_{(\mathfrak{p})}$ by setting

$$(M_{(\mathfrak{p})})_i = \{x/a \in M_{(\mathfrak{p})} : x \text{ homogeneous}, \deg x/a = i\}.$$

It is easy to see that $R_{(\mathfrak{p})}$ is a graded ring and that $M_{(\mathfrak{p})}$ is a graded $R_{(\mathfrak{p})}$-module; $M_{(\mathfrak{p})}$ is called the *homogeneous localization* of M. The extension ideal $\mathfrak{p}^* R_{(\mathfrak{p})}$ is a graded prime ideal in $R_{(\mathfrak{p})}$, and the factor ring $R_{(\mathfrak{p})}/\mathfrak{p}^* R_{(\mathfrak{p})}$ has the property that every non-zero homogeneous element is invertible.

Lemma 1.5.7. *Let R be a graded ring. The following conditions are equivalent:*
(a) *every non-zero homogeneous element is invertible;*
(b) *$R_0 = k$ is a field, and either $R = k$ or $R = k[t, t^{-1}]$ for some homogeneous element $t \in R$ of positive degree which is transcendental over k.*

PROOF. (a) $\Rightarrow$ (b): $R_0 = k$ is a field. If $R = R_0$, then R is a field. Otherwise $R \neq R_0$, and there exist non-zero homogeneous elements of positive degree. Let t be an element of least positive degree, say $\deg t = d$. As t is invertible there exists a homomorphism $\varphi : k[T, T^{-1}] \to R$ of graded rings where φ maps k identically to R_0 and where $\varphi(T) = t$. (The grading on $k[T, T^{-1}]$ is of course defined by setting $\deg T = d$.)

We claim that φ is an isomorphism. Let $f \in \mathrm{Ker}\, \varphi$, $f = \sum_{i \in \mathbb{Z}} a_i T^i$, $a_i \in k$; then $0 = \varphi(f) = \sum_{i \in \mathbb{Z}} a_i t^i$, and so $a_i t^i = 0$ for all i. As t is invertible, we get $a_i = (a_i t^i) \cdot t^{-i} = 0$ for all i, which implies that $f = 0$. Hence φ is injective. In order to show that φ is surjective, we pick a non-zero homogeneous element $a \in R$ of degree i. If $i = 0$, then $a \in \mathrm{Im}\, \varphi$. Thus we may assume that $i \neq 0$. Write $i = jd + r$ with $0 \leq r < d$. The element at^{-j} has degree r. As d was the least positive degree, we conclude that $r = 0$. Thus $a = bt^j$ for some $b \in R_0$, and hence $a = \varphi(bT^j) \in \mathrm{Im}\, \varphi$.

(b) $\Rightarrow$ (a) is trivial. □

The following theorem contains the dimension theory of graded rings and modules:

Theorem 1.5.8. *Let R be a Noetherian graded ring, M a finite graded R-module and $\mathfrak{p} \in \operatorname{Supp} M$.*
(a) *If $\mathfrak{p}$ is graded, then there exists a chain $\mathfrak{p}_0 \subset \cdots \subset \mathfrak{p}_d = \mathfrak{p}$, $d = \dim M_\mathfrak{p}$, of graded prime ideals $\mathfrak{p}_i \in \operatorname{Supp} M$.*
(b) *If $\mathfrak{p}$ is not graded, then $\dim M_\mathfrak{p} = \dim M_{\mathfrak{p}^*} + 1$.*

PROOF. A very special case of (b) is the following: if $\mathfrak{p}$ is not graded, then $\operatorname{height} \mathfrak{p}/\mathfrak{p}^* = 1$. In order to prove this equation we may replace R by $R/\mathfrak{p}^*$ and assume that $\mathfrak{p}^* = 0$. Then $\mathfrak{p}$ does not contain a non-zero homogeneous element. Therefore it is harmless to invert all these elements. This yields the homogeneous localization $R_{(0)}$. Since $\mathfrak{p}R_{(0)}$ is a non-zero prime ideal, $R_{(0)}$ has the form $k[t, t^{-1}]$ by 1.5.7, whence $\operatorname{height} \mathfrak{p} = \operatorname{height} \mathfrak{p}R_{(0)} = 1$.

Now let $\mathfrak{p} \in \operatorname{Supp} M$ be an arbitrary prime ideal, and $d = \dim M_\mathfrak{p}$. Both claims will be proved once we show that there exists a chain $\mathfrak{p}_0 \subset \cdots \subset \mathfrak{p}_d = \mathfrak{p}$ of prime ideals in $\operatorname{Supp} M$ such that $\mathfrak{p}_0, \ldots, \mathfrak{p}_{d-1}$ are graded. Note that in the case of (b) it follows that $\mathfrak{p}_{d-1} \subset \mathfrak{p}^*$, and therefore $\mathfrak{p}_{d-1} = \mathfrak{p}^*$ since there is no prime ideal properly between $\mathfrak{p}$ and $\mathfrak{p}_{d-1}$.

Let $\mathfrak{p}_0 \subset \cdots \subset \mathfrak{p}_d = \mathfrak{p}$ be a chain of prime ideals in $\operatorname{Supp} M$. Then $\mathfrak{p}_0$ is minimal in $\operatorname{Supp} M$, and therefore graded by 1.5.6. In the case $d = 1$ we are already done. Arguing inductively we may therefore suppose that $\mathfrak{p}_0, \ldots, \mathfrak{p}_{d-2}$ are graded.

If $\mathfrak{p}$ is not graded, we replace $\mathfrak{p}_{d-1}$ by $\mathfrak{p}^*$, which is properly contained in $\mathfrak{p}$, and properly contains $\mathfrak{p}_{d-2}$ because $\operatorname{height} \mathfrak{p}/\mathfrak{p}^* = 1$, as was proved above.

If $\mathfrak{p}$ is graded, then it contains a homogeneous element $a \notin \mathfrak{p}_{d-2}$, and we replace $\mathfrak{p}_{d-1}$ by a minimal prime $\mathfrak{q}$ of $\mathfrak{p}_{d-2} + (a)$ contained in $\mathfrak{p}$. Since $\operatorname{height} \mathfrak{p}/\mathfrak{q} = 1$, it is impossible for $\mathfrak{q}$ to equal $\mathfrak{p}$; furthermore $\mathfrak{q}$ is a minimal element of $\operatorname{Supp} R/(\mathfrak{p}_{d-2} + (a))$, and thus graded by 1.5.6. $\square$

Our next goal is an equation similar to 1.5.8 for the depth of a graded module. We shall need the result that the ordinary Ext-groups $\operatorname{Ext}_R^i(M, N)$ of graded R-modules admit a natural grading, provided R is Noetherian and M is finite.

If M is a graded R-module and i is an integer, then $M(i)$ denotes the graded R-module with grading given by $M(i)_n = M_{i+n}$.

The category $\mathcal{M}_0(R)$ has enough projectives. In fact, each module $M \in \mathcal{M}_0(R)$ is a homomorphic image (in $\mathcal{M}_0(R)$) of a module of the form $\bigoplus R(i)$. So every graded module has a graded free resolution. When we speak of a natural grading of modules appearing as the values of derived functors, then it is of course important that the standard argument of homological algebra ([318], Theorem 6.11), which guarantees that derived functors are well defined, can be made 'graded'. That $\mathcal{M}_0(R)$ has enough injectives will be shown in 3.6.2.

It is not hard to see that the tensor product $M \otimes N$ of graded R-modules is a graded R-module; its homogeneous component $(M \otimes N)_n$ is generated (as a $\mathbb{Z}$-module) by the products $x \otimes y$ with $x \in M$, $y \in N$ homogeneous such that $\deg x + \deg y = n$; see 1.5.19. Together with the fact that each graded module has a graded free resolution this implies that the modules $\text{Tor}_i^R(M, N)$ admit a natural grading.

Let M, N be graded R-modules. In general, the set of morphisms $\varphi: M \to N$ in $\mathcal{M}_0(R)$ is not a submodule of $\text{Hom}_R(M, N)$. Thus for the construction of a reasonable graded Ext functor one must consider a larger class of maps. An R-module homomorphism $\varphi: M \to N$ is called *homogeneous of degree i* if $\varphi(M_n) \subset N_{n+i}$ for all i. (A homogeneous homomorphism whose degree is not explicitly specified has degree 0.) Note that φ may be considered as a morphism $\varphi: M(-i) \to N$ in $\mathcal{M}_0(R)$. Denote by $\text{Hom}_i(M, N)$ the group of homogeneous homomorphisms of degree i. The $\mathbb{Z}$-submodules $\text{Hom}_i(M, N)$ of $\text{Hom}_R(M, N)$ form a direct sum, and it is obvious that $^*\text{Hom}_R(M, N) = \bigoplus_{i \in \mathbb{Z}} \text{Hom}_i(M, N)$ is a graded R-submodule of $\text{Hom}_R(M, N)$. In general $^*\text{Hom}_R(M, N) \neq \text{Hom}_R(M, N)$, but equality holds if M is finite: see Exercise 1.5.19.

For any $N \in \mathcal{M}_0(R)$ we define $^*\text{Ext}_R^i(M, N)$ as the i-th right derived functor of $^*\text{Hom}_R(\_, N)$ in $\mathcal{M}_0(R)$. Thus, if $P_\bullet$ is a projective resolution of M in $\mathcal{M}_0(R)$, then

$$^*\text{Ext}_R^i(M, N) \cong H^i(^*\text{Hom}_R(P_\bullet, N))$$

for all $i \geq 0$. It is immediate from this definition and the above remarks that $^*\text{Ext}_R^i(M, N) = \text{Ext}_R^i(M, N)$ for Noetherian R and finite M. Nevertheless we shall use the notation $^*\text{Ext}_R^i(M, N)$ to emphasize that these modules are graded.

Theorem 1.5.9. *Let R be a Noetherian graded ring, M a finite graded R-module, and $\mathfrak{p} \in \text{Supp} \, M$ a non-graded prime ideal. Then*

$$\text{depth} \, M_\mathfrak{p} = \text{depth} \, M_{\mathfrak{p}^*} + 1 \quad \text{and} \quad r(M_\mathfrak{p}) = r(M_{\mathfrak{p}^*}).$$

PROOF. In order to compute the depths and types of $M_\mathfrak{p}$ and $M_{\mathfrak{p}^*}$ we may consider both modules as modules over the homogeneous localization $R_{(\mathfrak{p})}$ of R with respect to $\mathfrak{p}$. Thus we may assume that $R/\mathfrak{p}^* \cong k[t, t^{-1}]$ where k is a field and t is an element of positive degree which is transcendental over k. It follows that $\mathfrak{p} = aR + \mathfrak{p}^*$ for some $a \in R \setminus \mathfrak{p}^*$. Hence we have an exact sequence

$$0 \longrightarrow R/\mathfrak{p}^* \overset{a}{\longrightarrow} R/\mathfrak{p}^* \longrightarrow R/\mathfrak{p} \longrightarrow 0,$$

which yields the long exact sequence

$$\cdots \longrightarrow {}^*\text{Ext}_R^i(R/\mathfrak{p}^*, M) \overset{a}{\longrightarrow} {}^*\text{Ext}_R^i(R/\mathfrak{p}^*, M) \longrightarrow \text{Ext}_R^{i+1}(R/\mathfrak{p}, M) \longrightarrow \cdots$$

The ${}^*\mathrm{Ext}^i_R(R/\mathfrak{p}^*, M)$ are graded $R/\mathfrak{p}^*(= k[t, t^{-1}])$-modules. Since every graded $k[t, t^{-1}]$-module is free (Exercise 1.5.20) and $a \notin \mathfrak{p}^*$, the map

$$ {}^*\mathrm{Ext}^i_R(R/\mathfrak{p}^*, M) \xrightarrow{\ a\ } {}^*\mathrm{Ext}^i_R(R/\mathfrak{p}^*, M) $$

is injective. Therefore

$$ \mathrm{Ext}^{i+1}_R(R/\mathfrak{p}, M) \cong {}^*\mathrm{Ext}^i_R(R/\mathfrak{p}^*, M)/a \cdot {}^*\mathrm{Ext}^i_R(R/\mathfrak{p}^*, M). $$

The equation $\mathfrak{p} = aR + \mathfrak{p}^*$ implies that $\mathrm{Ext}^{i+1}_R(R/\mathfrak{p}, M)$ is a free $(R/\mathfrak{p})$-module of the same rank as the free $(R/\mathfrak{p}^*)$-module ${}^*\mathrm{Ext}^i_R(R/\mathfrak{p}^*, M)$. Hence

$$ \dim_{k(\mathfrak{p})} \mathrm{Ext}^{i+1}_{R_{\mathfrak{p}}}(k(\mathfrak{p}), M_{\mathfrak{p}}) = \mathrm{rank}_{R/\mathfrak{p}} \mathrm{Ext}^{i+1}_R(R/\mathfrak{p}, M)) $$

$$ = \mathrm{rank}_{R/\mathfrak{p}^*} \, {}^*\mathrm{Ext}^i_R(R/\mathfrak{p}^*, M) = \dim_{k(\mathfrak{p}^*)} \, {}^*\mathrm{Ext}^i_{R_{\mathfrak{p}^*}}(k(\mathfrak{p}^*), M_{\mathfrak{p}^*}). $$

This equation in particular entails the assertion of the theorem.   □

What makes the proof of 1.5.9 more difficult than one might expect at first sight is illustrated by the following example. Let k be a field and $S = k[X, Y]$ be graded such that $\deg X = 0$ and $\deg Y = 1$. The residue class ring $R = S/(XY)$ is graded, and (x, y) is a graded maximal ideal of grade 1. Nevertheless every homogeneous element of (x, y) is a zero-divisor, in fact contained in a minimal prime ideal. However, as we shall see in 1.5.11, under suitable hypotheses there exist homogeneous regular sequences. First we prove a graded version of prime avoidance:

Lemma 1.5.10. *Let R be a graded ring and I an ideal generated by elements of positive degree. Let $\mathfrak{p}_1, \ldots, \mathfrak{p}_n$ be prime ideals such that $I \not\subset \mathfrak{p}_i$ for $i = 1, \ldots, n$. Then there exists a homogeneous element $x \in I$, $x \notin \mathfrak{p}_1 \cup \cdots \cup \mathfrak{p}_n$.*

PROOF. Let $S = \bigoplus_{j=0}^\infty R_j$. Since I is generated by elements of positive degree, one has $I \cap S \not\subset \mathfrak{p}_i \cap S$ for $i = 1, \ldots, n$. Therefore we may assume that R is positively graded. Furthermore it is harmless to replace $\mathfrak{p}_i$ by $\mathfrak{p}_i^*$ for all i.

Using induction on n, we may suppose that $\mathfrak{p}_n$ is a minimal element of $\{\mathfrak{p}_1, \ldots, \mathfrak{p}_n\}$ and that there is a homogeneous $x' \in I$ with $x' \notin \mathfrak{p}_1 \cup \cdots \cup \mathfrak{p}_{n-1}$. If $x' \notin \mathfrak{p}_n$, then we are done. Otherwise there exists a homogeneous $r \in (\bigcap_{i=1}^{n-1} \mathfrak{p}_i) \setminus \mathfrak{p}_n$. We choose a homogeneous $y \in I \setminus \mathfrak{p}_n$. Then $\deg x' > 0$ and $\deg ry > 0$ so that $(x')^u + (ry)^v$ is homogeneous for suitable exponents u, v. Furthermore, $(x')^u + (ry)^v \notin \mathfrak{p}_1 \cup \cdots \cup \mathfrak{p}_n$.   □

Proposition 1.5.11. *Let R be a Noetherian graded ring, and let I be an ideal in R generated by homogeneous elements of positive degree. Set $h = \mathrm{height}\, I$ and $g = \mathrm{grade}(I, M)$ where M is a finite R-module. Then there exist sequences $\mathbf{x} = x_1, \ldots, x_h$ and $\mathbf{y} = y_1, \ldots, y_g$ of homogeneous elements of I such that $\mathrm{height}(x_1, \ldots, x_i) = i$ for $i = 1, \ldots, h$ and $\mathbf{y}$ is an M-sequence.*

PROOF. It is enough to find x_1 and y_1 because we may use induction on n after having replaced all objects by their reductions modulo x_1 or y_1. But the choice of x_1 or y_1 only requires the avoidance of finitely many prime ideals none of which contains I. □

Often one needs a stronger version of 1.5.11.

Proposition 1.5.12. *In addition to the hypotheses of* 1.5.11 *assume that R_0 is a local ring with an infinite residue class field and that I is generated by elements of degree 1. Then the sequences $x = x_1, \ldots, x_h$ and $y = y_1, \ldots, y_g$ can be composed of elements of degree 1.*

PROOF. We choose a system $z_1, \ldots, z_n$ of degree 1 elements generating I. If height $I > 0$ and $\mathfrak{p}$ is a minimal prime ideal of R, then $I_1 \not\subset \mathfrak{p}$. Therefore $I_1 \cap \mathfrak{p}$ is a proper R_0-submodule of I_1. As k_0 is infinite, it is impossible for I_1 to be the union of the finitely many proper submodules obtained in this manner. (Modulo the maximal ideal $\mathfrak{m}_0$ of R_0 this turns into an elementary fact of linear algebra.) So I_1 has an element x_1 which is not in any minimal prime ideal of R. In order to construct $x_2, \ldots, x_h$ one proceeds by induction. The construction of y is similar. □

°Local rings. In the following definition we introduce the graded counterparts of local rings.

Definition 1.5.13. Let R be a graded ring. A graded ideal $\mathfrak{m}$ of R is called *°maximal*, if every graded ideal that properly contains $\mathfrak{m}$ equals R. The ring R is called *°local*, if it has a unique *°maximal ideal $\mathfrak{m}$. A *°local ring with *°maximal ideal $\mathfrak{m}$ will be denoted by $(R, \mathfrak{m})$.

Let $(R, \mathfrak{m})$ be a *°local ring. All non-zero homogeneous elements of the graded ring $R/\mathfrak{m}$ are invertible, and so $R/\mathfrak{m}$ is either a field, or else $R/\mathfrak{m} \cong k[t, t^{-1}]$, where k is a field and t is a homogeneous element of positive degree which is transcendental over k; see 1.5.7. In the first case $\mathfrak{m}$ is a maximal ideal, and in the second $\mathfrak{m}$ is a prime ideal with $\dim R/\mathfrak{m} = 1$. Note that R_0 is a local ring with maximal ideal $\mathfrak{m}_0 = \mathfrak{m} \cap R_0$, and that all homogeneous elements $a \in R \setminus \mathfrak{m}$ are units. We define the *°dimension of R as the height of $\mathfrak{m}$ and denote it by *°$\dim R$. According to 1.5.8, *°$\dim R$ equals the supremum of all numbers h for which there exists a chain of graded prime ideals $\mathfrak{p}_0 \subset \mathfrak{p}_1 \subset \cdots \subset \mathfrak{p}_h$ in R. If $x_1, \ldots, x_n$, $n = $ *°$\dim R$, are homogeneous elements such that $(x_1, \ldots, x_n)$ is $\mathfrak{m}$-primary, then $x_1, \ldots, x_n$ is called a *homogeneous system of parameters.*

Examples 1.5.14. (a) Let $\mathfrak{p}$ be a graded prime ideal. Then $R_{(\mathfrak{p})}$ is a *°local ring.

(b) Let R be a positively graded ring for which R_0 is a local ring with maximal ideal $\mathfrak{m}_0$. Then R is a *°local ring with *°maximal ideal

$\mathfrak{m} = \mathfrak{m}_0 \oplus \bigoplus_{n>0} R_n$. In particular a positively graded algebra over a field is *local.

(c) Let $(S, \mathfrak{n})$ be a local ring and t an indeterminate over S. Then $R = S[t, t^{-1}]$ is in a natural way a *local ring with *maximal ideal $\mathfrak{n}S[t, t^{-1}]$, and one has $\dim S = {}^*\dim R = \dim R - 1$.

With respect to its finite graded modules M, a *local ring $(R, \mathfrak{m})$ behaves like a local ring, as we shall now see.

Let $g_1, \ldots, g_n$ be a homogeneous minimal system of generators of M. Let $F_0 = \bigoplus_{i=1}^n R(-\deg g_i)$, the i-th summand being generated by an element e_i satisfying $\deg e_i = \deg g_i$. The R-module F_0 is free of rank n, and the assignment $e_i \mapsto g_i$ induces a surjective morphism φ_0 of graded modules. Of course $\operatorname{Ker} \varphi_0$ is a graded submodule of F_0. Suppose that $\operatorname{Ker} \varphi_0 \not\subset \mathfrak{m}F_0$. Then there exists a homogeneous element $u \in \operatorname{Ker} \varphi_0$, $u \notin \mathfrak{m}F_0$, and one of the coefficients a_i in $u = \sum a_i e_i$ is not in $\mathfrak{m}$; call it a_j. But all the a_i are homogeneous, and so a_j is a unit by hypothesis on $(R, \mathfrak{m})$. It follows that the given system of generators is not minimal, which is a contradiction. Localizing with respect to $\mathfrak{m}$ we see that $n = \mu(M_\mathfrak{m})$. In particular all homogeneous minimal systems of generators have the same number of elements. Furthermore, iterating the construction of F_0 and φ_0, one obtains an (augmented) free resolution of M which for the reasons given is called a *minimal graded free resolution* of M. It is easy to show that such a resolution is unique up to an isomorphism in $\mathcal{M}_0$.

Proposition 1.5.15. *Let $(R, \mathfrak{m})$ be a Noetherian *local ring, M a finite graded R-module, and I a graded ideal. Then*
(a) *every minimal homogeneous system of generators of M has exactly $\mu(M_\mathfrak{m})$ elements,*
(b) *if $F_\bullet$ is a minimal graded free resolution of M, then $F_\bullet \otimes R_\mathfrak{m}$ is a minimal free resolution of $M_\mathfrak{m}$,*
(c) *the functor $\_ \otimes R_\mathfrak{m}$ is faithfully exact on the category $\mathcal{M}_0$,*
(d) *M is projective if and only if it is free,*
(e) *one has*

$$\operatorname{proj\,dim} M = \operatorname{proj\,dim} M_\mathfrak{m}, \qquad \operatorname{grade}(\mathfrak{m}, M) = \operatorname{depth} M_\mathfrak{m},$$
$$\operatorname{grade} M = \operatorname{grade} M_\mathfrak{m}, \qquad \operatorname{grade}(I, M) = \operatorname{grade}(I_\mathfrak{m}, M_\mathfrak{m}).$$

PROOF. (a) and (b) were shown above, and (a) implies that $\_ \otimes R_\mathfrak{m}$ is faithful, which proves (c) because localization is always exact. Part (d) follows from (b) since the first Betti number $\beta_1(M_\mathfrak{m}) = 0$ if M is projective, and therefore $F_1 = 0$ in a minimal graded free resolution of M. The first equation in (e) is also a consequence of (b), whereas the remaining ones result from (c) and the fact that the modules $\operatorname{Ext}_R^i(R/I, M)$ and $\operatorname{Ext}_R^i(M, R)$ are graded. (One must of course use the description of grade by 1.2.5).   □

It is customary to collect the terms with the same 'shift' in each free module of a graded free resolution and to write it in the form

$$\cdots \longrightarrow \bigoplus_j R(-j)^{\beta_{ij}} \longrightarrow \cdots \longrightarrow \bigoplus_j R(-j)^{\beta_{0j}} \longrightarrow M \longrightarrow 0.$$

Though a minimal graded free resolution is uniquely determined, this is not true for the numbers β_{ij} if one only requires that $(R, \mathfrak{m})$ is *local. We need a slightly stronger hypothesis which is satisfied for example by all positively graded algebras over local rings:

Proposition 1.5.16. *Let* $(R, \mathfrak{m})$ *be a Noetherian *local ring such that* $\mathfrak{m}$ *is a maximal ideal of* R *(in the ordinary sense). Then for every finite graded R-module M the numbers* β_{ij} *in a minimal graded free resolution of M are uniquely determined by M.*

PROOF. Let $F = \bigoplus_j R(-j)^{\beta_j}$. Then $\beta_j = \dim_{R/\mathfrak{m}}(F \otimes R/\mathfrak{m})_j$ since one has $R(-j) \otimes R/\mathfrak{m} \cong R/\mathfrak{m}(-j)$. $\qquad\square$

In the situation of 1.5.16 not only is the cardinality of a minimal homogeneous system of generators unique, but also their degrees are fixed (up to a permutation).

Graded Noether normalization. The existence of Noether normalizations of affine algebras is stated in A.14. Here we want to prove its graded variant.

Theorem 1.5.17. *Let k be a field and R a positively graded affine k-algebra. Set* $n = \dim R$.
(a) *The following are equivalent for homogeneous elements* $x_1, \ldots, x_n$:
  (i) $x_1, \ldots, x_n$ *is a homogeneous system of parameters;*
  (ii) R *is an integral extension of* $k[x_1, \ldots, x_n]$;
  (iii) R *is a finite* $k[x_1, \ldots, x_n]$*-module.*
(b) *There exist homogeneous elements* $x_1, \ldots, x_n$ *satisfying one, and therefore all, of the conditions in* (a). *Moreover, such elements are algebraically independent over k.*
(c) *If R is a homogeneous k-algebra and k is infinite, then such* $x_1, \ldots, x_n$ *can be chosen to be of degree 1.*

PROOF. We set $S = k[x_1, \ldots, x_n]$ and $I = (x_1, \ldots, x_n)$.

The existence of $x_1, \ldots, x_n$ as claimed in (b) or (c) follows immediately from 1.5.11 and 1.5.12 if one observes that the *maximal ideal $\mathfrak{m}$ of R has height n. The algebraic independence of $x_1, \ldots, x_n$ results from (a)(ii) in conjunction with A.8 because $\dim S = n$ if and only if $x_1, \ldots, x_n$ are algebraically independent.

The equivalence of (a)(ii) and (iii) is a general fact: R is a finitely generated S-algebra. That (a)(ii) $\Rightarrow$ (i) follows from A.8 which entails that $\dim R/I = \dim(S/I \cap S) = 0$. Thus I is $\mathfrak{m}$-primary.

There remains the proof of (a)(i) $\Rightarrow$ (iii). We choose a system $y_1, \ldots, y_m$ of homogeneous elements of positive degree generating R over k. If (i) holds, then I is $\mathfrak{m}$-primary, and there exists an e such that $z = y_1^{e_1} \cdots y_m^{e_m} \in I$ whenever $\deg z \geq e$ (deg is the degree in R). Let M be the S-submodule of R generated by those monomials z with $\deg z < e$. We claim that $R = M$. In fact, every $r \in R$ is a k-linear combination of monomials $y_1^{e_1} \cdots y_m^{e_m}$, and thus it is enough that $s = y_1^{e_1} \cdots y_m^{e_m} \in M$ for all $e_i \in \mathbb{N}$. If $\deg s < e$, then $s \in M$ for trivial reasons. So assume $\deg s \geq e$. Then $s \in I$, and $s = \sum_{i=1}^{n} f_i x_i$ with elements $f_i \in R$. Since s and the x_i are homogeneous of positive degree, the f_i can be chosen homogeneous of degree $< \deg s$. Now write f_i as a k-linear combination of monomials in $y_1, \ldots, y_m$, and apply an inductive argument. $\square$

Dehomogenization. In concluding we want to study the relation between a graded ring R and a residue class ring $A = R/(x-1)$ where x is a non-nilpotent homogeneous element of degree 1. One calls A the *dehomogenization of R with respect to x.* The relationship between R and A is much closer than between a ring and a residue class ring in general. A typical example for R and A arises in algebraic geometry: R is the homogeneous coordinate ring of a projective variety, and A is the coordinate ring of the affine open subvariety complementary to the hyperplane given by the vanishing of x.

Let $\pi: R \to A$ be the natural homomorphism, and $S = R_x$. Then π factors in a natural way through a homomorphism $\psi: S \to A$. Since x is homogeneous, the grading of R induces a grading on S.

Proposition 1.5.18. (a) *The homomorphism $S_0[X, X^{-1}] \to S$ which is the identity on S_0 and sends X to x is an isomorphism.*
(b) *The restriction of ψ to S_0 is an isomorphism $S_0 \cong A$.*

PROOF. (a) This is a general fact: if T is a graded ring which has a unit x of degree 1, then $T \cong T_0[X, X^{-1}]$.

(b) The kernel of ψ is the ideal $(x-1)S$, and therefore ψ induces an isomorphism $A \cong S/(x-1)S \cong S_0$. $\square$

It follows easily that several properties transfer from R to A. For example, it is immediate that if R is reduced or an integral domain, then so is A. Also see Exercises 1.5.26, 2.2.34, and 2.2.35.

Exercises

1.5.19. Let R be a graded ring. All the modules in this problem are supposed to be graded.
(a) Prove that $\bigoplus M_i$ has a unique grading for which the natural embeddings $M_j \to \bigoplus M_i$ are morphisms in $\mathscr{M}_0$, that is, $\bigoplus M_i$ is the direct sum in $\mathscr{M}_0$.

(b) The direct product $\prod M_i$ lacks this property of the direct sum; nevertheless, prove there exists a direct product in $\mathcal{M}_0$: let ${}^*\!\prod M_i$ be the submodule of $\prod M_i$ generated by the elements (x_i) such that all the x_i are homogeneous of degree n, $n \in \mathbb{Z}$.

(c) What can be said about direct and inverse limits in $\mathcal{M}_0$?

(d) Prove that the tensor product $M \otimes N$ is a graded R-module with $(M \otimes N)_u$ generated over $\mathbb{Z}$ by the tensor products $x \otimes y$ of homogeneous elements with $\deg x + \deg y = u$. (Choose a presentation $G \to F \to M \to 0$ in $\mathcal{M}_0$ with $F = \bigoplus R(\alpha_i)$ and $G = \bigoplus R(\beta_j)$.)

(e) Show the functor ${}^*\!\mathrm{Hom}_R(\_, N)$ is left exact, and one has ${}^*\!\mathrm{Hom}_R(\bigoplus M_i, N) \cong {}^*\!\prod {}^*\!\mathrm{Hom}_R(M_i, N)$.

(f) Verify that ${}^*\!\mathrm{Hom}_R(M, N) = \mathrm{Hom}_R(M, N)$ if M is finite. In general, however, ${}^*\!\mathrm{Hom}_R(M, N)$ is a proper submodule of $\mathrm{Hom}_R(M, N)$.

(g) Prove ${}^*\!\mathrm{Hom}_R(M(-i), N(-j)) \cong {}^*\!\mathrm{Hom}_R(M, N)(i - j)$.

1.5.20. Let $R = k[t, t^{-1}]$ be a graded ring where $R_0 = k$ is a field, and $t \in R$ is a homogeneous element of positive degree which is transcendental over k. Show that every graded R-module is free.

1.5.21. Let k be a field, $S = k[X_1, X_2, Y_1, Y_2]$ the polynomial ring with the grading determined by $\deg X_i = 0$ and $\deg Y_i = 1$, and $R = S/(X_1 Y_1 - X_2 Y_2)$. Prove that the grade of the ideal $I = (x_1, x_2, y_1, y_2)$ in R is 3, but I does not even contain a homogeneous R-sequence of length 2.

1.5.22. Let k be a field. We consider the polynomial ring $R = k[X_1, \ldots, X_n]$ as a graded k-algebra with $\deg X_i = a_i$ for $i = 1, \ldots, n$. Show that R is ${}^*\!$local if and only if all a_i are positive or all a_i are negative.

1.5.23. Let $(R, \mathfrak{m})$ be a Noetherian ${}^*\!$local ring, and M a finite graded R-module. Show that every permutation of a homogeneous M-sequence is an M-sequence.

1.5.24. Prove the following variants of Nakayama's lemma.

(a) Let $(R, \mathfrak{m})$ be a ${}^*\!$local ring, M a finite graded R-module, and N a graded submodule. If $M = N + \mathfrak{m}M$, then $M = N$.

(b) Let R be a graded ring for which $(R_0, \mathfrak{m}_0)$ is local. Suppose that M is a graded R-module such that M_i is finite over R_0 for all i. If one has $M = N + \mathfrak{m}_0 M$ for a graded submodule N of M, then $M = N$.

1.5.25. Let R be a Noetherian positively graded ring, and M a finite graded R-module. Prove $\dim M = \sup\{\dim M_\mathfrak{p} : \mathfrak{p} \in \mathrm{Supp}\, M \text{ graded}\}$. Hint: consider an ideal $\mathfrak{m}$ which is maximal among the graded members of $\mathrm{Supp}\, M$ and use 1.5.7.

1.5.26. Let R be a graded ring, x a non-nilpotent element of degree 1, $A = R/(x-1)$, and $\pi: R \to A$ the natural homomorphism. As in 1.5.18 we set $S = R_x$, and identify A and S_0.

(a) One has $\pi(I) = IS \cap A$ for every homogeneous ideal $I \subset R$ and $J = \pi(JS \cap R)$ for every ideal J of A. (One calls $\pi(I)$ the *dehomogenization* of I, and $JS \cap R$ the *homogenization* of J.)

(b) The homomorphism π induces a bijective correspondence between the set of homogeneous ideals of R modulo which x is regular and the set of all ideals of A.

(c) This correspondence preserves inclusions and intersections, and the properties of being a prime, primary, or radical ideal.

1.6   The Koszul complex

We introduce the Koszul complex $K_.(x)$ of a sequence $x = x_1,\ldots,x_n$ of elements of a ring R. Under suitable hypotheses one can determine grade(I, M) from the homology of $K_.(x) \otimes M$ where I is the ideal generated by x. This fact and its 'universal' properties make the Koszul complex an indispensable tool.

Moreover, the Koszul complex is the paradigm of a complex with an algebra structure. In order to emphasize this fact we introduce more generally the Koszul complex of a linear form. A review of exterior algebra has been included for the reader's convenience.

Review of exterior algebra.   The following is an excerpt from Bourbaki [48], Ch. III, which we recommend as a source for exterior algebra. We hope that the details included will enable the reader to carry out the calculations on which the theory is based. When one has to check whether a map is well defined, it is usually the best strategy to exploit the universal properties of the objects under consideration.

Let R be a ring, and M an R-module. We consider R as a graded ring by giving it the trivial grading. Let $M^{\otimes i}$ denote the i-th tensor power of M, i.e. the tensor product $M \otimes \cdots \otimes M$ of i factors M for $i > 0$, and R for $i = 0$. The tensor powers form a graded R-module

$$\bigotimes M = \bigoplus_{i=0}^{\infty} M^{\otimes i}.$$

The assignment

$$((x_1,\ldots,x_m),(y_1,\ldots,y_n)) \longmapsto x_1 \otimes \cdots \otimes x_m \otimes y_1 \otimes \cdots \otimes y_n$$

induces an R-bilinear map $M^{\otimes m} \times M^{\otimes n} \to M^{\otimes(m+n)}$, and its additive extension to $\bigotimes M \times \bigotimes M$ gives $\bigotimes M$ the structure of a graded associative R-algebra. Henceforth 'R-algebra' always means 'associative R-algebra'. (Obviously $\bigotimes M$ is not commutative in general.) The tensor algebra is characterized by a universal property: given an R-linear map $\varphi : M \to A$ where A is an R-algebra, there exists a unique R-algebra homomorphism $\psi : \bigotimes M \to A$ extending φ; here we identify M and $M^{\otimes 1}$.

The exterior algebra $\bigwedge M$ is the residue class algebra

$$\bigwedge M = (\bigotimes M)/\mathfrak{J}$$

where $\mathfrak{J}$ is the two-sided ideal generated by the elements $x \otimes x$, $x \in M$. Since $\mathfrak{J}$ is generated by homogeneous elements, $\bigwedge M$ inherits the structure of a graded R-algebra. The product in $\bigwedge M$ is denoted $x \wedge y$. In general $\bigwedge M$ is not commutative; it is however *alternating*: one has

$$x \wedge y = (-1)^{(\deg x)(\deg y)} y \wedge x \qquad \text{for homogeneous } x, y \in \bigwedge M, \qquad \text{and}$$

$$x \wedge x = 0 \qquad \qquad \text{for homogeneous } x, \ \deg x \text{ odd.}$$

Let $x_1, \ldots, x_n$ be elements of M, and π a permutation of $\{1, \ldots, n\}$. Then

$$x_{\pi(1)} \wedge \cdots \wedge x_{\pi(n)} = \sigma(\pi) x_1 \wedge \cdots \wedge x_n;$$

here $\sigma(\pi)$ is the sign of π. Furthermore $x_1 \wedge \cdots \wedge x_n = 0$ if $x_i = x_j$ for some indices $i \neq j$. For a subset I of $\{1, \ldots, n\}$ we set

$$x_I = x_{i_1} \wedge \cdots \wedge x_{i_m} \qquad \text{when} \quad I = \{i_1, \ldots, i_m\} \text{ with } i_1 < \cdots < i_m.$$

For subsets $J, K \subset \{1, \ldots, n\}$ with $J \cap K = \emptyset$ let $\sigma(J, K) = (-1)^i$ where i is the number of elements $(j, k) \in J \times K$ with $j > k$; if $J \cap K \neq \emptyset$, let $\sigma(J, K) = 0$. Then

$$x_J \wedge x_K = \sigma(J, K) x_{J \cup K}.$$

Useful identities satisfied by σ are given in Exercise 1.6.23. It is clear that the notation x_I can be extended to the more general case in which $(x_g)_{g \in G}$ is a family of elements of M indexed by a linearly ordered set G and I is a finite subset of G.

The i-th graded component of $\bigwedge M$ is denoted by $\bigwedge^i M$ and is called the i-th *exterior power* of M. From the definition of $\bigwedge M$ it follows easily that one has natural isomorphisms $\bigwedge^0 M \cong R$, $\bigwedge^1 M \cong M$; so we may identify R and $\bigwedge^0 M$, M and $\bigwedge^1 M$.

Let $(x_g)_{g \in G}$ be a system of generators of M. Then $\bigwedge^j M$ is generated by the exterior products x_I with $I \subset G$ and $|I| = j$. In particular, if M is generated by $x_1, \ldots, x_n$, then $\bigwedge^i M = 0$ for all $i > n$.

The exterior algebra is characterized by a universal property which it inherits from that of the tensor algebra: given an R-linear map $\varphi : M \to E$ from M to an R-algebra E such that $\varphi(x)^2 = 0$ for all $x \in M$, there exists a unique R-algebra homomorphism $\psi : \bigwedge M \to E$ extending φ. It follows immediately that for every R-linear map $\varphi : M \to N$ there exists a unique R-algebra homomorphism $\bigwedge \varphi$ for which the diagram

$$
\begin{array}{ccc}
M & \xrightarrow{\;\varphi\;} & N \\
{\scriptstyle \text{nat}} \downarrow & & \downarrow {\scriptstyle \text{nat}} \\
\bigwedge M & \xrightarrow{\;\bigwedge \varphi\;} & \bigwedge N
\end{array}
$$

commutes; $\bigwedge \varphi$ is homogeneous of degree 0, and one has

$$\bigwedge \varphi(x_1 \wedge \cdots \wedge x_n) = \varphi(x_1) \wedge \cdots \wedge \varphi(x_n)$$

for all $x_1, \ldots, x_n \in M$. If φ is surjective, then $\bigwedge \varphi$ is also surjective, and $\operatorname{Ker} \bigwedge \varphi$ is the ideal generated by $\operatorname{Ker} \varphi$. (This is neither obvious nor indeed true in general; for example, if φ is injective, $\bigwedge \varphi$ need not be injective.) The map $\bigwedge^i M \to \bigwedge^i N$ induced by $\bigwedge \varphi$ is denoted by $\bigwedge^i \varphi$. Suppose that φ is surjective; then $\bigwedge^i \varphi$ is also surjective, and from the description of $\operatorname{Ker} \bigwedge \varphi$ just mentioned (and the alternating property of $\bigwedge M$) it follows easily that the sequence

$$\overset{i-1}{\bigwedge} M \otimes \operatorname{Ker} \varphi \longrightarrow \overset{i}{\bigwedge} M \xrightarrow{\bigwedge^i \varphi} \overset{i}{\bigwedge} N \longrightarrow 0$$

is exact where the map on the left hand side is induced by the exterior multiplication $\bigwedge^{i-1} M \times \operatorname{Ker} \varphi \to \bigwedge^i M$.

The exterior powers $\bigwedge^i M$ are also characterized by a universal property: for every alternating i-linear map $\alpha \colon M^i \to N$, N an R-module, there exists a unique R-linear map $\lambda \colon \bigwedge^i M \to N$ such that

$$\alpha(x_1, \dots, x_i) = \lambda(x_1 \wedge \cdots \wedge x_i)$$

for all $x_1, \dots, x_i \in M$.

An important property of the exterior algebra is that it commutes with base extensions: if $R \to S$ is a homomorphism of commutative rings, then one has a natural isomorphism

$$\left(\bigwedge M \right) \otimes_R S \cong \bigwedge (M \otimes_R S)$$

of graded S-algebras.

Let M_1, M_2 be R-modules. On $(\bigwedge M_1) \otimes (\bigwedge M_2)$ one defines a multiplication by setting

$$(x \otimes y)(x' \otimes y') = (-1)^{(\deg y)(\deg x')}(x \wedge x') \otimes (y \wedge y')$$

for all homogeneous elements $x, x' \in M_1$, $y, y' \in M_2$. It is straightforward to verify that $(\bigwedge M_1) \otimes (\bigwedge M_2)$ is an alternating graded R-algebra under this multiplication. Its degree 1 component is $(M_1 \otimes R) \oplus (R \otimes M_2) \cong M_1 \oplus M_2$. By the universal property of the exterior algebra the natural map $M_1 \oplus M_2 \to (\bigwedge M_1) \otimes (\bigwedge M_2)$ extends to an R-algebra homomorphism $\Phi \colon \bigwedge(M_1 \oplus M_2) \to (\bigwedge M_1) \otimes (\bigwedge M_2)$.

One gets an inverse $\Psi \colon (\bigwedge M_1) \otimes (\bigwedge M_2) \to \bigwedge(M_1 \oplus M_2)$ to Φ by setting

$$\Psi(x \otimes y) = \Psi_1(x) \wedge \Psi_2(y)$$

where $\Psi_i \colon \bigwedge M_i \to \bigwedge(M_1 \oplus M_2)$ is the extension of the natural embedding $M_i \to M_1 \oplus M_2$. The compositions $\Phi \circ \Psi$ and $\Psi \circ \Phi$ are the identities on $(\bigwedge M_1) \otimes (\bigwedge M_2)$ and $\bigwedge(M_1 \oplus M_2)$. Therefore we have an isomorphism

$$\left(\bigwedge M_1 \right) \otimes \left(\bigwedge M_2 \right) \cong \bigwedge(M_1 \oplus M_2)$$

of alternating graded R-algebras.

In what follows, the most important case for M is that of a finite free R-module F. Suppose $e_1, \dots, e_n$ is a basis of F. The elements

$$e_I, \qquad I \subset \{1, \dots, n\}, \quad |I| = i,$$

form a basis of $\bigwedge^i F$; this non-trivial fact amounts to the existence of determinants. In particular $\bigwedge^i F$ is free of rank $\binom{n}{i}$. A multiplication table of $\bigwedge F$ with respect to this basis in given by

$$e_I \wedge e_J = \sigma(I, J)e_{I \cup J}.$$

Suppose R is a graded ring, and $M = \bigoplus_{i \in \mathbb{Z}} M_i$ is a graded R-module. Then one can endow $\bigwedge M$ with a unique grading such that $M \subset \bigwedge M$ has the given grading, and $\bigwedge M$ is a graded algebra over R. We restrict ourselves to the case $M = F = \bigoplus_{i=1}^{n} R(-a_i)$. Let $e_1, \ldots, e_n$ be the basis of F corresponding to this decomposition. Then one assigns to e_I the degree $\sum_{i \in I} a_i$, and verifies easily that the induced grading on $\bigwedge F$ makes $\bigwedge F$ a graded (in fact, a bigraded) R-algebra.

Basic properties of the Koszul complex. Let R be a ring, L an R-module, and $f : L \to R$ an R-linear map. The assignment

$$(x_1, \ldots, x_n) \mapsto \sum_{i=1}^{n} (-1)^{i+1} f(x_i) x_1 \wedge \cdots \wedge \widehat{x}_i \wedge \cdots \wedge x_n$$

defines an alternating n-linear map $L^n \to \bigwedge^{n-1} L$. (By $\widehat{x}_i$ we indicate that x_i is to be omitted from the exterior product.) By the universal property of the n-th exterior power there exists an R-linear map $d_f^{(n)} : \bigwedge^n L \to \bigwedge^{n-1} L$ with

$$d_f^{(n)}(x_1 \wedge \cdots \wedge x_n) = \sum_{i=1}^{n} (-1)^{i+1} f(x_i) x_1 \wedge \cdots \wedge \widehat{x}_i \wedge \cdots \wedge x_n$$

for all $x_1, \ldots, x_n \in L$. The collection of the maps $d_f^{(n)}$ defines a graded R-homomorphism

$$d_f : \bigwedge L \to \bigwedge L$$

of degree -1. By a straightforward calculation one verifies the following identities:

$$d_f \circ d_f = 0 \quad \text{and} \quad d_f(x \wedge y) = d_f(x) \wedge y + (-1)^{\deg x} x \wedge d_f(y)$$

for all homogeneous $x \in \bigwedge L$. To say that $d_f \circ d_f = 0$ is to say that

$$\cdots \longrightarrow \bigwedge^n L \xrightarrow{d_f} \bigwedge^{n-1} L \longrightarrow \cdots \longrightarrow \bigwedge^2 L \xrightarrow{d_f} L \xrightarrow{f} R \longrightarrow 0$$

is a complex. The second equation expresses that d_f is an *antiderivation* (of degree -1).

Definition 1.6.1. The complex above is the *Koszul complex of f*, denoted by $K_\bullet(f)$. More generally, if M is an R-module, then $K_\bullet(f,M)$ is the complex $K_\bullet(f) \otimes_R M$, called the *Koszul complex of f with coefficients in M*; its differential is denoted by $d_{f,M}$.

Proposition 1.6.2. *Let R be a ring, L an R-module, and $f : L \to R$ an R-linear map.*
(a) *The Koszul complex $K_\bullet(f)$ carries the structure of an associative graded alternating algebra, namely that of $\bigwedge L$.*
(b) *Its differential d_f is an antiderivation of degree -1.*
(c) *For every R-module M the complex $K_\bullet(f,M)$ is a $K_\bullet(f)$-module in a natural way.*
(d) *One has $d_{f,M}(x.y) = d_f(x).y + (-1)^{\deg x} x . d_{f,M}(y)$ for all homogeneous elements x of $K_\bullet(f)$ and all elements $y \in K_\bullet(f,M)$.*

PROOF. (a) and (b) are part of the discussion preceding the proposition.
   (c) is obvious: if A is an R-algebra, then $A \otimes_R M$ is an A-module for every R-module M.
   (d) It is enough to verify the equation for elements $y = w \otimes z$ with $w \in K_\bullet(f)$, $z \in M$. Then $d_{f,M}(x.w \otimes z) = d_{f,M}((x \wedge w) \otimes z) = d_f(x \wedge w) \otimes z$, and the rest follows from the fact that d_f is an antiderivation. $\square$

For a subset S of $K_\bullet(f)$ and a subset U of $K_\bullet(f,M)$ let $S.U$ denote the R-submodule of $K_\bullet(f,M)$ generated by the products $s.u$, $s \in S$, $u \in U$.
   Set

$$Z_\bullet(f) = \operatorname{Ker} d_f, \qquad Z_\bullet(f,M) = \operatorname{Ker} d_{f,M},$$
$$B_\bullet(f) = \operatorname{Im} d_f, \qquad B_\bullet(f,M) = \operatorname{Im} d_{f,M}.$$

Definition 1.6.3. The homology $H_\bullet(f) = Z_\bullet(f)/B_\bullet(f)$ is the *Koszul homology* of f. For every R-module M the homology $Z_\bullet(f,M)/B_\bullet(f,M)$ is denoted by $H_\bullet(f,M)$ and called the *Koszul homology of f with coefficients in M*.

From 1.6.2(d) one easily derives the following relations:

$$Z_\bullet(f).Z_\bullet(f,M) \subset Z_\bullet(f,M), \quad Z_\bullet(f).B_\bullet(f,M) \subset B_\bullet(f,M),$$
$$B_\bullet(f).Z_\bullet(f,M) \subset B_\bullet(f,M).$$

We have a natural isomorphism $K_\bullet(f) \cong K_\bullet(f,R)$. So the first relation entails that $Z_\bullet(f)$ is a graded R-subalgebra of $K_\bullet(f)$, and the second and third show that $B_\bullet(f)$ is a two-sided ideal in $Z_\bullet(f)$.

Proposition 1.6.4. *Let R be a ring, L an R-module, and $f : L \to R$ an R-linear map.*
(a) *The Koszul homology $H_\bullet(f)$ carries the structure of an associative graded alternating R-algebra.*
(b) *For every R-module M the homology $H_\bullet(f,M)$ is an $H_\bullet(f)$-module in a natural way.*

PROOF. (a) That $H_*(f)$ is an R-algebra follows from the discussion preceding the proposition. The asserted properties are inherited by quotients of graded R-subalgebras of $K_*(f)$ modulo graded ideals.

(b) The first of the relations above shows that $Z_*(f, M)$ is a $Z_*(f)$-module; the second says that $B_*(f, M)$ is a $Z_*(f)$-submodule, and the third implies that $Z_*(f, M)/B_*(f, M)$ is annihilated by $B_*(f)$. $\square$

It results immediately from 1.6.4 that $H_*(f, M)$ is an R/I-module where $I = \operatorname{Im} f$. This will be stated in 1.6.5 where it follows from a somewhat stronger statement.

It is useful also to introduce the *Koszul cohomology* (*with coefficients in* M): we set

$$K^*(f) = \operatorname{Hom}_R(K_*(f), R), \qquad K^*(f, M) = \operatorname{Hom}_R(K_*(f), M),$$
$$H^*(f) = H^*(K^*(f)), \qquad H^*(f, M) = H^*(K^*(f, M)).$$

Let $I = \operatorname{Im} f \subset R$; then, by construction, $H_0(f) = R/I$ and $H_0(f, M) = M/IM$.

Proposition 1.6.5. *Let R be a ring, L an R-module, and $f : L \to R$ an R-linear map. Set $I = \operatorname{Im} f$.*
(a) *For every $a \in I$ multiplication by a on $K_*(f)$, $K_*(f, M)$, $K^*(f)$, $K^*(f, M)$ is null-homotopic.*
(b) *In particular I annihilates $H_*(f)$, $H_*(f, M)$, $H^*(f)$, $H^*(f, M)$.*
(c) *If $I = R$, then the complexes $K_*(f)$, $K_*(f, M)$, $K^*(f)$, $K^*(f, M)$ are null-homotopic. In particular their (co)homology vanishes.*

PROOF. We choose $x \in L$ with $a = f(x)$. Let ϑ_a denote the multiplication by a on $K_*(f)$, and λ_x the left multiplication by x on $K_*(f)$. Then $\vartheta_a = d_f \circ \lambda_x + \lambda_x \circ d_f$ as is easily verified.

Thus multiplication by a is null-homotopic on $K_*(f)$. Of course $\vartheta_a \otimes M$ and $\operatorname{Hom}_R(\vartheta_a, M)$ are the multiplications by a on $K_*(f, M)$ and $K^*(f, M)$, and the rest of (a) follows immediately. Part (b) is a general fact: if φ is a null-homotopic complex homomorphism, then the map induced by φ on homology is zero. For (c) we choose $a = 1$, and apply (a) and (b). $\square$

Let L_1 and L_2 be R-modules, and $f_1 : L_1 \to R$, $f_2 : L_2 \to R$ be R-linear maps. Then f_1 and f_2 induce a linear form $f : L_1 \oplus L_2 \to R$ by $f(x_1 \oplus x_2) = f_1(x_1) + f_2(x_2)$.

Proposition 1.6.6. *With the notation just introduced, one has an isomorphism of complexes $K_*(f_1) \otimes_R K_*(f_2) \cong K_*(f)$.*

PROOF. The graded R-algebras underlying $K_*(f_1) \otimes K_*(f_2)$ and $K_*(f)$, namely $(\bigwedge L_1) \otimes (\bigwedge L_2)$ and $\bigwedge L$, are isomorphic, as noted above. We may identify them. The differential d_f is an antiderivation on $\bigwedge L$ which on the degree 1 graded component $L = L_1 \oplus L_2$ coincides with $d_{f_1} \otimes d_{f_2}$.

An antiderivation on the exterior algebra $\bigwedge L$ is uniquely determined by its values on L. Therefore it is enough to check that $d_{f_1} \otimes d_{f_2}$ is an antiderivation too. The straightforward verification of this fact is left as an exercise for the reader. One has of course to remember the definition of tensor product of complexes: the n-th graded component of $K_\bullet(f_1) \otimes K_\bullet(f_2)$ is $\bigoplus_{i=0}^n \bigwedge^i L_1 \otimes \bigwedge^{n-i} L_2$, and

$$d_{f_1} \otimes d_{f_2}(x \otimes y) = d_{f_1}(x) \otimes y + (-1)^i x \otimes d_{f_2}(y)$$

for $x \otimes y \in \bigwedge^i L_1 \otimes \bigwedge^{n-i} L_2$. $\qquad\qquad\qquad\qquad\qquad\quad$ $\square$

The Koszul complex 'commutes' with ring extensions, and so does Koszul homology if the extension is flat:

Proposition 1.6.7. *Let R be a ring, L an R-module, and $f : L \to R$ an R-linear map. Suppose $\varphi : R \to S$ is a ring homomorphism.*
(a) *Then one has a natural isomorphism $K_\bullet(f) \otimes_R S \cong K_\bullet(f \otimes S)$.*
(b) *Moreover, if φ is flat, then $H_\bullet(f, M) \otimes S \cong H_\bullet(f \otimes S, M \otimes S)$ for every R-module M.*

PROOF. There is a natural isomorphism $(\bigwedge L) \otimes S \cong \bigwedge(L \otimes S)$, and $d_f \otimes S$ and $d_{f \otimes S}$ are antiderivations which coincide in degree 1. So we can use the same argument as in the previous demonstration. This proves (a), and (b) follows immediately since $H_\bullet(C_\bullet \otimes S) = H_\bullet(C_\bullet) \otimes S$ for an arbitrary complex $C_\bullet$ over R if S is R-flat. $\qquad\qquad\qquad\qquad$ $\square$

Suppose L and L' are R-modules with linear forms $f : L \to R$ and $f' : L' \to R$. Every R-homomorphism $\varphi : L \to L'$ extends to a homomorphism $\bigwedge \varphi : \bigwedge L \to \bigwedge L'$ of R-algebras, as discussed above. If $f = f' \circ \varphi$, then $\bigwedge \varphi$ is a homomorphism of Koszul complexes:

Proposition 1.6.8. *With the notation just introduced, if $f = f' \circ \varphi$, then $\bigwedge \varphi : K_\bullet(f) \to K_\bullet(f')$ is a complex homomorphism.*

The Koszul complex of a sequence. Let L be a finite free R-module with basis $e_1, \ldots, e_n$. Then a linear form f on L is uniquely determined by the values $x_i = f(e_i)$, $i = 1, \ldots, n$. Conversely, given a sequence $x = x_1, \ldots, x_n$, there exists a linear form f on L with $f(e_i) = x_i$. We set

$$K_\bullet(x) = K_\bullet(f),$$

and the rest of the notation is to be modified accordingly. Henceforth we shall only consider Koszul complexes $K_\bullet(x)$. Since f is just the 'direct sum' of the linear forms $f_i : R \to R$, $f_i(1) = x_i$, 1.6.6 specializes to the isomorphism

$$K_\bullet(x) \cong K_\bullet(x') \otimes K_\bullet(x_n) \cong K_\bullet(x_1) \otimes \cdots \otimes K_\bullet(x_n)$$

where $x' = x_1, \ldots, x_{n-1}$. Furthermore one should note that, by 1.6.8, $K_\bullet(x)$ is essentially invariant under a permutation of x.

We set $I = (x)$. Let $F_\bullet$ be a free resolution of R/I. As $H_0(x) = R/I$, there exists a complex homomorphism $\varphi : K_\bullet(x) \to F_\bullet$ lifting the identity on R/I; note that φ is unique up to homotopy.

Proposition 1.6.9. *Let R be a ring, $x = x_1, \ldots, x_n$ a sequence in R, and $I = (x)$. For all i there exist natural homomorphisms*

$$H_i(x, M) \to \mathrm{Tor}_i^R(R/I, M) \quad and \quad \mathrm{Ext}_R^i(R/I, M) \to H^i(x, M).$$

PROOF. The map φ introduced above yields complex homomorphisms $\varphi \otimes M : K_\bullet(f, M) \to F_\bullet \otimes M$ and $\mathrm{Hom}_R(\varphi, M) : \mathrm{Hom}_R(F_\bullet, M) \to K^\bullet(f, M)$. □

Let L be a finite free R-module with basis $e_1, \ldots, e_n$. Then $e_1 \wedge \cdots \wedge e_n$ is a basis of $\bigwedge^n L$, and there exists a unique R-isomorphism $\omega_n : \bigwedge^n L \to R$ with $\omega_n(e_1 \wedge \cdots \wedge e_n) = 1$. (An isomorphism $\bigwedge^n L \cong R$ is usually called an *orientation* on L.) We define $\omega_i : \bigwedge^i L \to (\bigwedge^{n-i} L)^*$ by setting

$$(\omega_i(x))(y) = \omega_n(x \wedge y) \quad \text{for} \quad x \in \overset{i}{\bigwedge} L, \; y \in \overset{n-i}{\bigwedge} L.$$

(This causes no ambiguity for $i = n$ if we identify R and R^* under the natural isomorphism.) It follows immediately that

$$(\omega_i(e_I))(e_J) = \begin{cases} 0 & \text{for } I \cap J \neq \emptyset, \\ \sigma(I, J) & \text{for } I \cap J = \emptyset. \end{cases}$$

In this formula I and J are multi-indices as introduced above. It shows that ω_i is an isomorphism. If we denote the dual basis of (e_I) by (e_I^*), the formula says that

$$\omega_i(e_I) = \sigma(I, \bar{I}) e_{\bar{I}}^*$$

where $\bar{I} = \{1, \ldots, n\} \setminus I$. Thus ω_i is an isomorphism. We consider the diagram

$$K_\bullet(x) : 0 \longrightarrow \overset{n}{\bigwedge} L \overset{d}{\longrightarrow} \overset{n-1}{\bigwedge} L \overset{d}{\longrightarrow} \cdots \overset{d}{\longrightarrow} L \overset{d}{\longrightarrow} R \longrightarrow 0$$

$$\downarrow \omega_n \qquad \downarrow \omega_{n-1} \qquad\qquad \downarrow \omega_1 \qquad \downarrow \omega_0$$

$$K^\bullet(x) : 0 \longrightarrow R \overset{d^*}{\longrightarrow} L^* \overset{d^*}{\longrightarrow} \cdots \overset{d^*}{\longrightarrow} (\overset{n-1}{\bigwedge} L)^* \overset{d^*}{\longrightarrow} (\overset{n}{\bigwedge} L)^* \longrightarrow 0$$

with $d = d_x$ and $d^* = (d_x)^*$.

Proposition 1.6.10. Let $x = x_1, \ldots, x_n$ be a sequence in a ring R.
(a) With the notation just introduced, one has $\omega_{i-1} \circ d_i = (-1)^{i-1} d^*_{n-i+1} \circ \omega_i$
for all i.
(b) The complexes $K_{\bullet}(x)$ and $K^{\bullet}(x) = (K_{\bullet}(x))^*$ are isomorphic (we say that
$K_{\bullet}(x)$ is self-dual).
(c) More generally, for every R-module M the complexes $K_{\bullet}(x, M)$ and
$K^{\bullet}(x, M)$ are isomorphic, and
(d) $H_i(x, M) \cong H^{n-i}(x, M)$ for $i = 0, \ldots, n$.

PROOF. The verification of (a) is left as an exercise for the reader (1.6.23 is
helpful). We observed above that ω_i is an isomorphism so that the maps
$\tau_i = (-1)^{i(i-1)/2} \omega_i$ define an isomorphism $K_{\bullet}(x) \cong K^{\bullet}(x) = (K_{\bullet}(x))^*$.

For (c) we note that there is a natural homomorphism $N^* \otimes M \to$
$\operatorname{Hom}_R(N, M)$ for all R-modules N, M. If N is finite and free, this
homomorphism is an isomorphism, and it induces an isomorphism
$K^{\bullet}(x) \otimes M \cong \operatorname{Hom}_R(K_{\bullet}(x), M)$. Now one uses (b). Part (d) is a trivial consequence of (c). □

The reader may have noticed that for a formally correct formulation
of 1.6.10(b) one would first have to convert the cochain complex $K^{\bullet}(x)$
into a chain complex $C_{\bullet}$ (by setting $C_i = K^{-i}(x)$) and then state that
$K_{\bullet}(x) \cong C_{\bullet}(-n)$. A similar manipulation would be necessary for (c).
The Koszul complex is an exact functor:

Proposition 1.6.11. Let R be a ring, $x = x_1, \ldots, x_n$ a sequence in R, and
$0 \to U \to M \to N \to 0$ an exact sequence of R-modules. Then the induced
sequence

$$0 \longrightarrow K_{\bullet}(x, U) \longrightarrow K_{\bullet}(x, M) \longrightarrow K_{\bullet}(x, N) \longrightarrow 0$$

is an exact sequence of complexes. In particular one has a long exact
sequence

$$\cdots \longrightarrow H_i(x, U) \longrightarrow H_i(x, M) \longrightarrow H_i(x, N) \longrightarrow H_{i-1}(x, U) \longrightarrow \cdots$$

of homology modules.

PROOF. The components of $K_{\bullet}(x)$ are free, hence flat R-modules. □

In place of an R-module M one can more generally consider a complex
$C_{\bullet}$, and then define the Koszul homology of $C_{\bullet}$ to be the homology of
$K_{\bullet}(x) \otimes C_{\bullet}$ etc. We consider this construction only for the special case in
which $x = x$:

Proposition 1.6.12. Let R be a ring, and $x \in R$.
(a) For every complex $C_{\bullet}$ of R-modules one has an exact sequence

$$0 \longrightarrow C_{\bullet} \longrightarrow C_{\bullet} \otimes K_{\bullet}(x) \longrightarrow C_{\bullet}(-1) \longrightarrow 0.$$

(b) *The induced long exact sequence of homology is*

$$\cdots \longrightarrow H_i(C_{\bullet}) \longrightarrow H_i(C_{\bullet} \otimes K_{\bullet}(x)) \longrightarrow H_{i-1}(C_{\bullet}) \xrightarrow{\pm x} H_{i-1}(C_{\bullet}) \longrightarrow \cdots$$

(c) *Moreover, if x is $C_{\bullet}$-regular, then there is an isomorphism*

$$H_{\bullet}(C_{\bullet} \otimes K_{\bullet}(x)) \cong H_{\bullet}(C_{\bullet}/xC_{\bullet}).$$

(According to our convention for graded modules $C_{\bullet}(-1)$ is just the complex $C_{\bullet}$ with all degrees increased by 1.)

PROOF. The complex $K_{\bullet}(x)$ is simply $0 \longrightarrow R \xrightarrow{x} R \longrightarrow 0$. The i-th graded component of $K_{\bullet}(x) \otimes C_{\bullet}$ is therefore $(R \otimes C_i) \oplus (R \otimes C_{i-1}) = C_i \oplus C_{i-1}$. So we have in each degree an exact sequence

$$0 \longrightarrow C_i \xrightarrow{\iota} C_i \oplus C_{i-1} \xrightarrow{\pi} C_{i-1} \longrightarrow 0,$$

where ι and π are the natural embedding and projection. If ∂ is the differential of $C_{\bullet}$, then the differential $d: C_i \oplus C_{i-1} \to C_{i-1} \oplus C_{i-2}$ is given by the matrix

$$\begin{pmatrix} \partial & (-1)^{i-1}x \\ 0 & \partial \end{pmatrix}$$

according to the definition of tensor products of complexes. Now (a) is obvious.

For (b) one looks up the definition of connecting homomorphism. It is defined by the following chain of assignments starting from $z \in C_{i-1}$ with $\partial(z) = 0$:

$$z \xrightarrow{\pi^{-1}} (0, z) \xrightarrow{d} ((-1)^{i-1}xz, 0) \xrightarrow{\iota^{-1}} (-1)^{i-1}xz.$$

So the connecting homomorphism $H_i(C_{\bullet}(-1)) = H_{i-1}(C_{\bullet}) \to H_{i-1}(C_{\bullet})$ is multiplication by $(-1)^{i-1}x$.

(c) The natural maps $C_i \oplus C_{i-1} \to C_i \to C_i/xC_i$ constitute a complex homomorphism $C_{\bullet} \otimes K_{\bullet}(x) \to C_{\bullet}/xC_{\bullet}$. We claim that the associated map of homology is an isomorphism. In fact, let $z \in C_i$ such that $\partial(z) \in xC_{i-1}$. Then there exists $z' \in C_{i-1}$ with $\partial(z) = xz'$, and $d(z, (-1)^i z') = (0, (-1)^i \partial(z'))$. Next one has $x\partial(z') = \partial(\partial(z)) = 0$; so $\partial(z') = 0$ since multiplication by x is injective on $C_{\bullet}$: $(z, (-1)^i z')$ is a cycle mapped to the cycle $\bar{z} \in C_i/xC_i$. That the map of homology is injective can be verified similarly. $\square$

Corollary 1.6.13. *Let R be a ring, $x = x_1, \ldots, x_n$ a sequence in R, and M an R-module.*

(a) *Set $x' = x_1, \ldots, x_{n-1}$. Then one has an exact sequence*

$$\cdots \xrightarrow{\pm x_n} H_i(x', M) \to H_i(x, M) \to H_{i-1}(x', M) \xrightarrow{\pm x_n} H_{i-1}(x', M) \to \cdots$$

(b) *Let $p \leq n$, $x' = x_1, \ldots, x_p$, and $x'' = x_{p+1}, \ldots, x_n$. If x' is weakly M-regular, then one has an isomorphism $H_{\bullet}(x, M) \cong H_{\bullet}(x'', M/x'M)$.*

PROOF. Part (a) is a special case of 1.6.12(b) when we take $C_\bullet = K_\bullet(x', M)$ and use the isomorphisms

$$K_\bullet(x', M) \otimes K_\bullet(x_n) \cong K_\bullet(x') \otimes M \otimes K_\bullet(x_n) \cong K_\bullet(x, M).$$

For part (b) it is enough to do the case $p = 1$ from which the general case follows by induction. Next we may permute x to the sequence $x_2, \ldots, x_n, x_1$, and then the assertion follows from 1.6.12(c). $\qquad\square$

It is an immediate consequence of 1.6.13 that $H_i(x, M) = 0$ for $i = n - p + 1, \ldots, n$ if (the first) p elements of $x_1, \ldots, x_n$ form an M-sequence. As we shall see in 1.6.16, there is a somewhat stronger vanishing theorem.

Corollary 1.6.14. *Let R be a ring, x a sequence in R, and M an R-module.*
(a) *If x is an M-sequence, then $K_\bullet(x, M)$ is acyclic.*
(b) *If x is an R-sequence, then $K_\bullet(x)$ is a free resolution of $R/(x)$.*

Remark 1.6.15. Let R be a graded ring and $x = x_1, \ldots, x_n$ a sequence of homogeneous elements. Then x induces a linear form of degree 0 on $F = \bigoplus_{i=1}^n R(-\deg x_i)$. The Koszul complex $K_\bullet(x)$ is a graded complex with a differential of degree 0 if we give $\bigwedge F$ the grading discussed above. In particular one has $\bigwedge^n F \cong R(-\sum_{i=1}^n \deg x_i)$.

The Koszul complex and grade. The main importance of the Koszul complex stems from the fact that $H_\bullet(x, M)$ measures grade(I, M) if M is a finite module over a Noetherian ring R and $I = (x)$. This will be made precise in 1.6.17. The finiteness assumption just stated will be necessary to establish the existence of an M-sequence in I from the vanishing of certain homology modules $H_i(x, M)$. The converse holds without such an assumption:

Theorem 1.6.16. *Let R be a ring, $x = x_1, \ldots, x_n$ a sequence in R, and M an R-module. If $I = (x)$ contains a weak M-sequence $y = y_1, \ldots, y_m$, then*

$$H_{n+1-i}(x, M) = 0 \quad for \quad i = 1, \ldots, m, \quad and$$

$$H_{n-m}(x, M) \cong \operatorname{Hom}_R(R/I, M/yM) \cong \operatorname{Ext}_R^m(R/I, M).$$

PROOF. The last isomorphism is given by Lemma 1.2.4. The remaining claims are proved by induction on m. For $m = 0$ we must show that

$$H_n(x, M) \cong \operatorname{Hom}_R(R/I, M).$$

In fact, by 1.6.10 one has $H_n(x, M) \cong H^0(x, M)$, and the latter is naturally isomorphic with $\operatorname{Hom}_R(R/I, M)$, as follows from the exactness of $R^n \to R \to R/I \to 0$ and the left exactness of $\operatorname{Hom}_R(\_, M)$. Explicitly, if we identify $\bigwedge^n R^n \otimes M$ and $R \otimes M \cong M$ via an orientation ω_n of R^n, then $H_n(x, M)$ is just the submodule $\{y \in M : Iy = 0\} \cong \operatorname{Hom}_R(R/I, M)$ of M.

Let $m \geq 1$. Then we set $\bar{M} = M/y_1 M$. The exact sequence

$$0 \longrightarrow M \xrightarrow{y_1} M \longrightarrow \bar{M} \longrightarrow 0$$

induces an exact sequence

$$\cdots \longrightarrow H_i(x, M) \xrightarrow{y_1} H_i(x, M) \longrightarrow H_i(x, \bar{M}) \longrightarrow H_{i-1}(x, M) \xrightarrow{y_1} \cdots;$$

see 1.6.11. Since, by 1.6.5, y_1 annihilates $H_i(x, M)$ for all i, this exact sequence breaks up into exact sequences

$$0 \longrightarrow H_i(x, M) \longrightarrow H_i(x, \bar{M}) \longrightarrow H_{i-1}(x, M) \longrightarrow 0.$$

It only remains to apply the induction hypothesis. $\square$

Theorem 1.6.17. *Let R be a Noetherian ring, and M a finite R-module. Suppose I is an ideal in R generated by $x = x_1, \ldots, x_n$.*
(a) *All the modules $H_i(x, M)$, $i = 0, \ldots, n$, vanish if and only if $M = IM$.*
(b) *Suppose that $H_i(x, M) \neq 0$ for some i, and let*

$$h = \max\{i : H_i(x, M) \neq 0\}.$$

Then every maximal M-sequence in I has length $g = n - h$; in other words, $\mathrm{grade}(I, M) = n - h$.

PROOF. (a) The implication '$\Rightarrow$' is trivial: $M = IM \iff H_0(x, M) \cong M/IM = 0$. For the converse choose a prime ideal $\mathfrak{p}$. By 1.6.7 and the flatness of localization one has $(H_i(x, M))_\mathfrak{p} \cong H_i(x, M_\mathfrak{p})$ where x is considered a sequence in $R_\mathfrak{p}$ on the right hand side. If $I \not\subset \mathfrak{p}$, then $H_i(x, M_\mathfrak{p}) = 0$ by 1.6.5. If $I \subset \mathfrak{p}$, then $M_\mathfrak{p} = 0$ by Nakayama's lemma, and again we have $H_i(x, M_\mathfrak{p}) = 0$.

(b) We give two proofs. (A third proof for the case $M = R$ is indicated in Exercise 1.6.30.)

(i) By part (a) we have $M \neq IM$. Let y be a maximal M-sequence in I; then y has length $g = \mathrm{grade}(I, M)$. It follows immediately from 1.6.16 and 1.2.5 that $H_i(x, M) = 0$ for $i = n - g + 1, \ldots, n$ and $H_{n-g}(x, M) \cong \mathrm{Ext}_R^g(R/I, M) \neq 0$.

(ii) Let y be a maximal M-sequence in I, and suppose that y has length g. Then $H_i(x, M) = 0$ for $i = n - g + 1, \ldots, n$ by 1.6.16, and furthermore $H_{n-g}(x, M) \cong \mathrm{Hom}_R(R/I, M/yM)$. Since I consists of zero-divisors of M/yM, this module is non-zero; see 1.2.3. $\square$

The second proof just given is independent of the 'homological' Lemma 1.2.4, and shows again that all maximal M-sequences in I have the same length. Therefore one could build the theory of grade upon 1.6.17.

Corollary 1.6.14 can be reversed for local rings. We need the following lemma:

Lemma 1.6.18. *Let* $(R, \mathfrak{m})$ *be a Noetherian local ring,* M *a finite R-module, and* $x = x_1, \ldots, x_n$ *a sequence in* $\mathfrak{m}$. *Set* $x' = x_1, \ldots, x_{n-1}$. *If* $H_i(x, M) = 0$, *then* $H_i(x', M) = 0$.

PROOF. By 1.6.6 we have $K_\bullet(x) \cong K_\bullet(x') \otimes K_\bullet(x_n)$. So 1.6.13 gives us an exact sequence

$$H_i(x', M) \xrightarrow{\pm x_n} H_i(x', M) \longrightarrow H_i(x, M).$$

These modules are finite. If $H_i(x, M) = 0$, then multiplication by x_n on $H_i(x', M)$ is surjective, whence $H_i(x', M) = 0$ by Nakayama's lemma. $\square$

Corollary 1.6.19. *Let* $(R, \mathfrak{m})$ *be a Noetherian local ring,* $M \neq 0$ *a finite R-module, and* $I \subset \mathfrak{m}$ *an ideal generated by* $x = x_1, \ldots, x_n$. *Then the following are equivalent:*
(a) grade$(I, M) = n$;
(b) $H_i(x, M) = 0$ for $i > 0$;
(c) $H_1(x, M) = 0$;
(d) x is an M-sequence.

PROOF. The equivalence of (a) and (b) follows from 1.6.17, and (b) $\Rightarrow$ (c) and (d) $\Rightarrow$ (a) are trivial. The proof of (c) $\Rightarrow$ (d) is an easy induction based on 1.6.18 and 1.6.13. $\square$

We saw in 1.1.6 that under the hypotheses of 1.6.19 every permutation of an M-sequence is again an M-sequence. Since, by 1.6.8, the Koszul complexes of x and every permutation of x are isomorphic, 1.6.19 yields another proof of 1.1.6.

Remark 1.6.20. For an arbitrary ring R and an arbitrary module M it follows from $H_1(x, M) = 0$ that x is M-quasi-regular, provided $xM \neq M$; see [48], Ch. X, §9, Théorème 1.

The Koszul complex as an invariant. Let R be a Noetherian local ring, I an ideal, and $x = x_1, \ldots, x_n$ and $y = y_1, \ldots, y_n$ minimal systems of generators of I. Then any $n \times n$ matrix $A = (a_{pq})$ such that

$$x_i = \sum_{j=1}^{n} a_{ji} y_j, \qquad i = 1, \ldots, n,$$

is invertible since the residue classes of x and y are bases of $I/\mathfrak{m}I$ over $R/\mathfrak{m}$. If f and f' are the linear forms on R^n defined by x and y respectively, then there exists an R-automorphism φ of R^n (defined by A) such that $f = f' \circ \varphi$, and it follows from 1.6.8 that the Koszul complexes $K_\bullet(x)$ and $K_\bullet(y)$ are isomorphic. This obviously fails if x and y have different lengths. Nevertheless the Koszul complexes $K_{\bullet}(x)$ and $K_{\bullet}(y)$ are closely related. The following proposition shows how to compare each of them to $K_{\bullet}(x, y)$.

Proposition 1.6.21. *Let R be a ring, $x = x_1, \ldots, x_n$ a sequence in R, and $x' = x_1, \ldots, x_n, x_{n+1}, \ldots, x_m$ with $x_{n+1}, \ldots, x_m \in (x)$. Then*

$$K_{\bullet}(x') \cong K_{\bullet}(x) \otimes \bigwedge R^{m-n}$$

as graded R-algebras; here $\bigwedge R^{m-n}$ is considered a complex with zero differential. In particular, for every R-module M one has

$$H_{\bullet}(x', M) \cong H_{\bullet}(x, M) \otimes \bigwedge R^{m-n}.$$

PROOF. Since $\bigwedge R^{k+1} \cong \bigwedge R^k \otimes \bigwedge R$ it suffices to treat the case $m = n + 1$. Write $x_{n+1} = \sum_{j=1}^{n} a_j x_j$. Let f be the linear form on R^{n+1} defined by x' and f' the linear form defined by $x'' = x, 0$. The assignment $e_i \mapsto e_i$ for $i = 1, \ldots, n$ and $e_{n+1} \mapsto \sum_{j=1}^{n} a_j e_j + e_{n+1}$ induces an automorphism φ of R^{n+1} such that $f = f' \circ \varphi$. As above one concludes that $K_{\bullet}(x') \cong K_{\bullet}(x'')$; in other words, there is no restriction in assuming that $x_{n+1} = 0$.

In the special situation we have reached, the first claim is a trivial consequence of 1.6.6. The second claim is easily verified. □

Corollary 1.6.22. *Let R be a ring, I a finitely generated ideal, and M an R-module. Suppose $x = x_1, \ldots, x_m$ and $y = y_1, \ldots, y_n$ are systems of generators of I, and let $g \in \mathbb{N}$. Then $H_i(x, M) = 0$ for $i = m - g + 1, \ldots, m$ if and only if $H_j(y, M) = 0$ for $j = n - g + 1, \ldots, n$.*

The corollary follows easily from 1.6.21. Note that for a finite module M over a Noetherian ring R it just restates part of 1.6.17. However, when we define the grade of a finitely generated ideal with respect to an arbitrary module in Chapter 9, 1.6.22 will be an essential result.

Exercises

1.6.23. Let I, J, I_1, I_2, I_3 be finite subsets of $\mathbb{N}$. Suppose $I = \{i_1, \ldots, i_p\}$, $J = \{i_{p+1}, \ldots, i_{p+q}\}$, the elements given in ascending order.
(a) Suppose $I \cap J = \emptyset$, and let π be the permutation of $I \cup J$ given by $\pi(j_k) = i_k$ where $I \cup J = \{j_1, \ldots, j_{p+q}\}$ is given in ascending order. Prove $\sigma(I, J) = \sigma(\pi) = (-1)^{pq} \sigma(J, I)$.
(b) Deduce that $\sigma(I_1, I_2)\sigma(I_1 \cup I_2, I_3) = \sigma(I_1, I_2 \cup I_3)\sigma(I_2, I_3)$.

1.6.24. Let R be a local ring, and M a finite R-module.
(a) Show $\mu(\bigwedge^i M) = \binom{\mu(M)}{i}$ for all $i \in \mathbb{N}$.
(b) Let $1 \leq i \leq \mu(M)$. Prove that M is free if and only if $\bigwedge^i M$ is free.

1.6.25. (a) Let R be a ring, and M an R-module of rank r. Prove rank $\bigwedge^i M = \binom{r}{i}$ for all $i \in \mathbb{N}$.

(b) Show the analogue for a homomorphism $\varphi : F \to G$ of finite free modules over a Noetherian ring.

Hint for (b): One may assume that R is local and of depth 0. Then $\operatorname{Im} \varphi$ is a free direct summand of G.

1.6.26. Let R be a Noetherian local ring, F a finite free R-module, $U \subset F$ a submodule of rank r and ι the natural embedding. Show that $\bigwedge^j \iota$ is injective if and only if $\bigwedge^j U$ is torsion-free. In particular $\bigwedge^j U$ is non-zero, but $\bigwedge^j \iota$ is not injective for rank $U < j \leq \mu(U)$.

1.6.27. Let R be a ring, and M an R-module. For $f_1, \ldots, f_p \in M^*$ let $\varphi(f_1, \ldots, f_p)$ be the restriction of $d_{f_1} \circ \ldots \circ d_{f_p}$ to $\bigwedge^p M$. Show that φ induces an R-linear map $\Phi : \bigwedge^p(M^*) \to (\bigwedge^p M)^*$. Prove that Φ is an isomorphism, if M is finite and free.

1.6.28. Let L' be an R-module, $x \in L'$, and ρ the right multiplication by x on $\bigwedge L'$. Prove

$$\widetilde{K}^*(x) : 0 \longrightarrow R \overset{\rho}{\longrightarrow} L' \overset{\rho}{\longrightarrow} \overset{2}{\bigwedge} L' \overset{\rho}{\longrightarrow} \cdots$$

is a complex.

Suppose that $L' = (R^n)^*$ and $f \in L'$. Then the complexes $\widetilde{K}^*(f)$ and $K^*(f)$ are isomorphic. (Since $K^*(f) \cong K_{\bullet}(f)$ by 1.6.10, one can introduce the Koszul complex via $\widetilde{K}^*(f)$ if one is satisfied with having it only for linear forms on finite free modules.)

1.6.29. Let R be a Noetherian ring, $x = (x_1, \ldots, x_n)$ an element of R^n, $M = R^n/Rx$, and I the ideal generated by $x_1, \ldots, x_n$. Prove that $\operatorname{grade} I \geq k$ if and only if

$$0 \longrightarrow R \overset{\rho}{\longrightarrow} R^n \overset{\rho}{\longrightarrow} \overset{2}{\bigwedge} R^n \overset{\rho}{\longrightarrow} \cdots \overset{\rho}{\longrightarrow} \overset{k}{\bigwedge} R^n \longrightarrow 0$$

is a free resolution of $\bigwedge^k M$. (The map ρ is right multiplication by x as in 1.6.28; one always has $\bigwedge^k M \cong (\bigwedge^k R^n)/\rho(\bigwedge^{k-1} R^n)$.)

1.6.30. Let $x = x_1, \ldots, x_n$ be a sequence in R, and denote by φ_i the differential $\bigwedge^i R^n \to \bigwedge^{i-1} R^n$ in the Koszul complex of x. Let $r_i = \binom{n-1}{i-1}$ be the expected rank of φ_i.

(a) Show that $\operatorname{Rad} I_{r_i}(\varphi_i) = \operatorname{Rad}(x)$.

(b) Derive 1.6.17 for $M = R$ from the Buchsbaum–Eisenbud acyclicity criterion.

1.6.31. Let R be a Noetherian ring, and $M \neq 0$ a finite R-module. Let I be an ideal, $x = x_1, \ldots, x_n$ a system of generators of I, and $g = \operatorname{grade}(I, M)$. Show $H_i(x, M) = 0$ for $i = n - g + 1, \ldots, n$, and $H_i(x, M) \neq 0$ for $i = 0, \ldots, n - g$. (This property is called the *rigidity* of the Koszul complex.) Hint: Reduce to the local case and use 1.6.18 for an inductive argument.

1.6.32. Let $(R, \mathfrak{m})$ be a Noetherian local ring, $I \subset \mathfrak{m}$ an ideal, $x \in \mathfrak{m}$, and M a finite R-module. Prove $\operatorname{grade}(I + (x), M) \leq \operatorname{grade}(I, M) + 1$.

1.6.33. Let $(R, \mathfrak{m})$ be a *local ring, and $\mathfrak{p} \neq \mathfrak{m}$ be a prime ideal such that $\mathfrak{p}^* = \mathfrak{m}$. Choose $a \in \mathfrak{p}$ with $\mathfrak{p} = \mathfrak{m} + (a)$ (see the proof of 1.5.9). Then a is R-regular (why?). If $F_{\bullet}$ is a graded minimal free resolution of $R/\mathfrak{m}$, show that $(F_{\bullet} \otimes K_{\bullet}(a))_{\mathfrak{p}}$ is a minimal free resolution of $R_{\mathfrak{p}}/\mathfrak{p}R_{\mathfrak{p}}$.

Notes

After the foundations of homological algebra had been laid by Cartan and Eilenberg [67], it invaded commutative ring theory through the epochal work of Auslander and Buchsbaum [17], [18], [19], Rees [303], and Serre [332]. These works cover the contents of Sections 1.1–1.3, and much more, to be developed in Chapters 2–4. Previously commutative algebra had been *ideal* theory (under which title Krull (in German) and Northcott published influential monographs); now *modules* were considered the objects that give structure to a ring. An intermediate position was taken by Gröbner's rather 'modern' treatise [141], but it introduced modules only as 'Vektormoduln', i.e. submodules of free modules over polynomial rings.

Proposition 1.2.16 and several theorems in Chapters 2 and 3 resemble a very successful method in topology, namely to relate the properties of the total space of a fibration to those of the base and the fibre. The algebraic analogue of this principle was studied systematically by Grothendieck [142] (which, by the way, contains various results on regular sequences not reproduced by us).

Torsion-freeness, reflexivity, and their 'higher' analogues are treated in the monograph [16] of Auslander and Bridger; see Bruns and Vetter [61] for a compact presentation. The definition of rank is taken from Scheja and Storch [323].

The very useful acyclicity criterion of Buchsbaum and Eisenbud appeared in [63]. It is closely related to Peskine and Szpiro's equally important 'lemme d'acyclicité' [297] which we reproduced in Exercise 1.4.24.

The notion of perfect ideal or module appeared in Rees [303]. It is an abstract version of Gröbner's [141] which in turn goes back to Macaulay [262]. A special form of the Hilbert–Burch theorem was proved by Hilbert [171] (and had been previously conjectured by Meyer [274]) whereas Burch [66] provided the first 'modern' version. The theorem has been re-proved several times; we have essentially reproduced the version of Buchsbaum and Eisenbud [64] who generalized the theorem to a factorization theorem for the ideals $I_{r_i}(\varphi_i)$ appearing in their acyclicity criterion.

Because of their importance for algebraic geometry, graded rings have been a standard topic in commutative algebra. Their enumerative theory will be developed in Chapter 4. Rees [305] ascribes Theorem 1.5.5 to Samuel. Theorem 1.5.8 is due to Matijevic [267], and 1.5.9 was given by Matijevic and Roberts [268]. (The proof of 1.5.9 has been drawn from Fossum and Foxby [109] and Goto and Watanabe [134].) These theorems are part of a programme aiming at characterizations of graded rings which only use localizations with respect to graded prime ideals.

We shall reproduce the pertinent results in the exercises of Chapters 2 and 3.

The Koszul complex [240] appeared for the first time in Hilbert [171]: after having proved his syzygy theorem (see 2.2.14) Hilbert determined the free resolution of the $k[X_1, \ldots, X_n]$-module k. That the Koszul complex is an utterly useful construction even when it is not acyclic seems to have been recognized by Auslander and Buchsbaum [19] and Serre [334]. Auslander and Buchsbaum established the main results of Section 1.6 whereas Serre found the connection with multiplicity theory; see Chapter 4.

2 Cohen–Macaulay rings

In this chapter we introduce the class of Cohen–Macaulay rings and two subclasses, the regular rings and the complete intersections. The definition of Cohen–Macaulay ring is sufficiently general to allow a wealth of examples in algebraic geometry, invariant theory, and combinatorics. On the other hand it is sufficiently strict to admit a rich theory: in the words of Hochster, 'life is really worth living' in a Cohen–Macaulay ring ([183], p. 887). The notion of Cohen–Macaulay ring is a workhorse of commutative algebra.

Regular local rings are abstract versions of polynomial or power series rings over a field. The fascination of their theory stems from a unique interplay of homological algebra and arithmetic. Complete intersections arise as residue class rings of regular rings modulo regular sequences, and, in a sense, are the best singular rings. Their exploration is dominated by methods related to the Koszul complex.

2.1 Cohen–Macaulay rings and modules

Let R be a Noetherian local ring, and M a finite module. If the 'algebraic' invariant depth M equals the 'geometric' invariant dim M, then M is called a Cohen–Macaulay module:

Definition 2.1.1. Let R be a Noetherian local ring. A finite R-module $M \neq 0$ is a *Cohen–Macaulay module* if depth $M = \dim M$. If R itself is a Cohen–Macaulay module, then it is called a *Cohen–Macaulay ring*. A *maximal* Cohen–Macaulay module is a Cohen–Macaulay module M such that $\dim M = \dim R$.

In general, if R is an arbitrary Noetherian ring, then M is a *Cohen–Macaulay module* if $M_{\mathfrak{m}}$ is a Cohen–Macaulay module for all maximal ideals $\mathfrak{m} \in \operatorname{Supp} M$. (So we consider the zero module to be Cohen–Macaulay.) However, for M to be a *maximal* Cohen–Macaulay module we require that $M_{\mathfrak{m}}$ is such an $R_{\mathfrak{m}}$-module for each maximal ideal $\mathfrak{m}$ of R. As in the local case, R is a *Cohen–Macaulay ring* if it is a Cohen–Macaulay module.

If I is an ideal contained in $\operatorname{Ann} M$, then it is irrelevant for the Cohen–Macaulay property whether we consider M as an R-module or an R/I-module. In particular, if R is local and M a Cohen–Macaulay module, then M is a maximal Cohen–Macaulay module over $R/\operatorname{Ann} M$.

The next theorem exhibits the fact that for a Cohen–Macaulay module the grade of an *arbitrary* ideal is given by its 'codimension'.

Theorem 2.1.2. *Let $(R, \mathfrak{m})$ be a Noetherian local ring, and $M \neq 0$ a Cohen–Macaulay R-module. Then*

(a) $\dim R/\mathfrak{p} = \operatorname{depth} M$ *for all $\mathfrak{p} \in \operatorname{Ass} M$,*

(b) $\operatorname{grade}(I, M) = \dim M - \dim M/IM$ *for all ideals $I \subset \mathfrak{m}$,*

(c) $x = x_1, \ldots, x_r$ *is an M-sequence if and only if $\dim M/xM = \dim M - r$,*

(d) x *is an M-sequence if and only if it is part of a system of parameters of M.*

PROOF. (a) We saw $\operatorname{depth} M \leq \dim R/\mathfrak{p}$ in 1.2.13, and $\dim R/\mathfrak{p} \leq \dim M$ holds since $\operatorname{Ass} M \subset \operatorname{Supp} M$.

(b) If $\operatorname{grade}(I, M) = 0$, then there exists $\mathfrak{p} \in \operatorname{Ass} M$ with $I \subset \mathfrak{p}$; therefore $\dim M/IM = \dim M$ follows from (a). If $\operatorname{grade}(I, M) > 0$, then we choose $x \in I$ regular on M. One has $\operatorname{grade}(I, M/xM) = \operatorname{grade}(I, M) - 1$, $\operatorname{depth} M/xM = \operatorname{depth} M - 1$, and $\dim M/xM = \dim M - 1$ so that induction completes the argument.

(c) It suffices now to quote 1.6.19.

(d) This is just a reformulation of (c). $\square$

The Cohen–Macaulay property is stable under specialization and localization:

Theorem 2.1.3. *Let R be a Noetherian ring, and M a finite R-module.*

(a) *Suppose x is an M-sequence. If M is a Cohen–Macaulay module, then M/xM is Cohen–Macaulay (over R or $R/(x)$). The converse holds if R is local.*

(b) *Suppose that M is Cohen–Macaulay. Then for every multiplicatively closed set S in R the localized module M_S is also Cohen–Macaulay. In particular, $M_\mathfrak{p}$ is Cohen–Macaulay for every $\mathfrak{p} \in \operatorname{Spec} R$. If $M_\mathfrak{p} \neq 0$, then $\operatorname{depth} M_\mathfrak{p} = \operatorname{grade}(\mathfrak{p}, M)$; if in addition R is local, then $\dim M = \dim M_\mathfrak{p} + \dim M/\mathfrak{p}M$.*

PROOF. (a) By the definition of Cohen–Macaulay module one may evidently assume that R is local. Let n be the length of x. Then $\dim M/xM = \dim M - n$ by 1.2.12 and $\operatorname{depth} M/xM = \operatorname{depth} M - n$ by 1.2.10.

(b) Let $\mathfrak{q}$ be a maximal ideal of R_S. The ideal $\mathfrak{q}$ is the extension of a prime ideal $\mathfrak{p}$ in R, and so $(R_S)_\mathfrak{q} \cong R_\mathfrak{p}$. Let $\mathfrak{m}$ be a maximal ideal of R containing $\mathfrak{p}$. Then $R_\mathfrak{p}$ is a localization of the Cohen–Macaulay local ring $R_\mathfrak{m}$. So we may again assume that R is local.

There is nothing to prove if $M_\mathfrak{p} = 0$. When $M_\mathfrak{p} \neq 0$, we use induction on $\operatorname{depth} M_\mathfrak{p}$. If $\operatorname{depth} M_\mathfrak{p} = 0$, then $\mathfrak{p} \in \operatorname{Ass} M$, and $\mathfrak{p}$ is a minimal prime of Supp M by 2.1.2; therefore dim $M_\mathfrak{p} = 0$. The same argument shows that $\mathfrak{p}$ cannot be contained in any $\mathfrak{q} \in \text{Ass } M$ if depth $M_\mathfrak{p} > 0$. So $\mathfrak{p}$ contains an M-regular element x, and the induction hypothesis applies to M/xM. It follows easily that $M_\mathfrak{p}$ is Cohen–Macaulay and that depth $M_\mathfrak{p} = \text{grade}(\mathfrak{p}, M)$. The second equation results from that and 2.1.2. $\qquad\square$

Corollary 2.1.4. *Let R be a Cohen–Macaulay ring, and $I \neq R$ an ideal. Then* grade I = height I, *and if R is local,* height I + dim R/I = dim R.

PROOF. One has height I = $\min\{\dim R_\mathfrak{p} : \mathfrak{p} \in V(I)\}$ and furthermore grade I = $\min\{\text{depth } R_\mathfrak{p} : \mathfrak{p} \in V(I)\}$. Theorem 2.1.3 yields dim $R_\mathfrak{p}$ = depth $R_\mathfrak{p}$ for all $\mathfrak{p} \in \text{Spec } R$. This proves the first equation, and the second follows from that and 2.1.2. $\qquad\square$

Let k be a field. We shall see in the next section that every finite module over a polynomial ring $k[X_1, \ldots, X_n]$ or a power series ring $k[[X_1, \ldots, X_n]]$ has finite projective dimension. Furthermore these rings are Cohen–Macaulay as will be shown below. This explains why the following theorem is a very effective Cohen–Macaulay criterion.

Theorem 2.1.5. *Let R be a Cohen–Macaulay ring, and M a finite R-module of finite projective dimension.*
(a) *If M is perfect, then it is a Cohen–Macaulay module.*
(b) *The converse holds when R is local.*

PROOF. Let M be perfect and $\mathfrak{p} \in \text{Supp } M$. Then $M_\mathfrak{p}$ is a perfect module as shown in the proof of 1.4.16. So we may assume that R is local. The Auslander–Buchsbaum formula gives proj dim M = dim R − depth M, and 2.1.4 yields grade M = dim R − dim M. Thus depth M = dim M if and only if proj dim M = grade M. $\qquad\square$

One says that an ideal I is *unmixed* if I has no embedded prime divisors or, in modern language, if the associated prime ideals of R/I are the minimal prime ideals of I. Macaulay showed in 1916 that an ideal $I = (x_1, \ldots, x_n)$ of height n in a polynomial ring over a field is unmixed, and for regular local rings this was proved by Cohen in 1946. (An n-generated ideal of height n is said to be *of the principal class*.) These facts and the following theorem explain the nomenclature 'Cohen–Macaulay'.

Theorem 2.1.6. *A Noetherian ring R is Cohen–Macaulay if and only if every ideal I generated by* height I *elements is unmixed.*

PROOF. '$\Rightarrow$': Suppose $I = (x)$, $x = x_1, \ldots, x_n$, and let $\mathfrak{p}, \mathfrak{q} \in \text{Ass } R/I$, $\mathfrak{p} \subset \mathfrak{q}$. Then there is a maximal ideal $\mathfrak{m}$ with $\mathfrak{q} \subset \mathfrak{m}$, and $\mathfrak{p}R_\mathfrak{m}, \mathfrak{q}R_\mathfrak{m} \in \text{Ass}(R_\mathfrak{m}/I_\mathfrak{m})$. If height I = n, then $\dim(R_\mathfrak{m}/I_\mathfrak{m})$ = dim $R_\mathfrak{m} - n$, and x is an $R_\mathfrak{m}$-sequence by 2.1.2. Therefore $R_\mathfrak{m}/I_\mathfrak{m}$ is Cohen–Macaulay, whence $\mathfrak{p}R_\mathfrak{m} = \mathfrak{q}R_\mathfrak{m}$ (again by 2.1.2), and so $\mathfrak{p} = \mathfrak{q}$.

'$\Leftarrow$': Let $J \subset R$ be an arbitrary ideal, say height $J = n$. Then there exist $x_1, \ldots, x_n \in J$ with height$(x_1, \ldots, x_i) = i$ for all $i = 0, \ldots, n$ (see A.2). It is impossible for x_{i+1} to be contained in a minimal prime ideal of $(x_1, \ldots, x_i)$. By hypothesis it therefore is an $(R/(x_1, \ldots, x_i))$-regular element. So $x_1, \ldots, x_n$ is an R-sequence. We have shown that grade J = height J for every proper ideal J of R. Then R is certainly Cohen–Macaulay. $\square$

Flat extensions of Cohen–Macaulay rings and modules. The behaviour of depth under flat local extensions was studied in Section 1.2. That makes it easy to prove an analogous theorem for the Cohen–Macaulay property.

Theorem 2.1.7. *Let* $\varphi : (R, \mathfrak{m}) \to (S, \mathfrak{n})$ *be a homomorphism of Noetherian local rings. Suppose M is a finite R-module and N is an R-flat finite S-module. Then $M \otimes_R N$ is a Cohen–Macaulay S-module if and only if M is Cohen–Macaulay over R and $N/\mathfrak{m}N$ is Cohen–Macaulay over S.*

In fact, according to 1.2.16 we have depth$_S M \otimes N =$ depth$_R M +$ depth$_S N/\mathfrak{m}N$. Since depth is bounded above by dimension, the theorem follows from the analogous equation for dimension; see A.11.

Corollary 2.1.8.   *Let $(R, \mathfrak{m})$ be a Noetherian local ring, M a finite R-module, and $\hat{M}$ its $\mathfrak{m}$-adic completion.*
(a) *Then* dim$_R M =$ dim$_{\hat{R}} \hat{M}$ *and* depth$_R M =$ depth$_{\hat{R}} \hat{M}$.
(b) *M is Cohen–Macaulay if and only if $\hat{M}$ is Cohen–Macaulay.*

PROOF. The extension $R \to \hat{R}$ is local and flat, and $\hat{M} = M \otimes_R \hat{R}$ since M is finite. $\square$

One can of course use more direct arguments in order to prove the previous corollary. Similarly there is a more 'elementary' approach to the following theorem; see for example [231].

Theorem 2.1.9. *Let R be a Noetherian ring, M a finite R-module, and $S = R[X_1, \ldots, X_n]$ or $S = R[\![X_1, \ldots, X_n]\!]$. Then $M \otimes S$ is a Cohen–Macaulay S-module if and only if M is a Cohen–Macaulay module.*

PROOF. Since the indeterminates can be adjoined successively, we may assume $n = 1$, $X = X_1$. The 'only if' part is easy: in both cases X is $(M \otimes S)$-regular, and $R \cong S/(X)$, $M \cong (M \otimes S)/X(M \otimes S)$. (That X is $(M \otimes S)$-regular is evident for $S = R[X]$; the reader should find a justification for $S = R[\![X]\!]$.)

Conversely, let $\mathfrak{m}$ be a maximal ideal of S, and set $\mathfrak{p} = \mathfrak{m} \cap R$. As outlined below A.12 the fibre $S_\mathfrak{m}/\mathfrak{p}S_\mathfrak{m}$ is a discrete valuation ring, and thus Cohen–Macaulay. Now we invoke 2.1.7 and complete the proof. $\square$

For polynomial extensions the proof of 2.1.9 shows that a stronger local version of 2.1.9 is valid: for $q \in \operatorname{Spec} R[X_1, \ldots, X_n]$ the localization $R[X_1, \ldots, X_n]_q$ is Cohen–Macaulay if and only if R_p is Cohen–Macaulay for $p = q \cap R$. Similarly, there is a local version of the following theorem:

Theorem 2.1.10. *Let k be a field, R a Noetherian k-algebra, and K an extension field of k. Suppose that R is a finitely generated k-algebra, or that K is finitely generated as an extension field of k. Then R is a Cohen–Macaulay ring if and only if $R \otimes_k K$ is.*

PROOF. If R is a finitely generated k-algebra, then $R \otimes_k K$ is a finitely generated K-algebra, and therefore Noetherian. Suppose that K is a finitely generated extension field. Then K is a finite algebraic extension of a finite purely transcendental extension K' of k. Since K' is the field of fractions of a polynomial ring $k[T_1, \ldots, T_n]$, we find again that $R \otimes_k K'$ is Noetherian, whence $R \otimes_k K = (R \otimes_k K') \otimes_{K'} K$ is also Noetherian.

Evidently $R \otimes K$ is a faithfully flat R-algebra. Therefore, given a prime ideal p of R, there exists $q \in \operatorname{Spec} R \otimes K$ such that $p = R \cap q$. The fibre of the extension $R_p \to (R \otimes K)_q$ is a localization of $k(p) \otimes K$. In conjunction with 2.1.7 this argument reduces the theorem to the assertion that $L \otimes_k K$ is Cohen–Macaulay for extension fields L and K of k, provided one of them is finitely generated. This follows from the next proposition. $\quad\square$

Proposition 2.1.11. *Let k be a field, R a k-algebra, and K a finitely generated extension field of k. Then $R \otimes_k K$ is isomorphic to a ring*

$$R[X_1, \ldots, X_n]_S / (f_1, \ldots, f_m)$$

where S is a multiplicatively closed subset of $R[X_1, \ldots, X_n]$, and $f_1, \ldots, f_m$ is a $R[X_1, \ldots, X_n]_S$-sequence.

PROOF. The extension $k \subset K$ decomposes into a series of cyclic extensions $k = K_0 \subset \cdots \subset K_t = K$. We use induction on t. Suppose that $T = R \otimes_k K_i = R[X_1, \ldots, X_n]_S / (f_1, \ldots, f_m)$.

If $K_{i+1} = K_i[Y]/(g)$ with a monic irreducible polynomial g, then

$$R \otimes_k K_{i+1} \cong T \otimes_{K_i} K_{i+1} \cong T[Y]/(g).$$

Since T is a flat K_i-algebra, g is not a zero-divisor of

$$T[Y] = R[X_1, \ldots, X_m, Y]_S / (f_1, \ldots, f_m).$$

If $K_{i+1} = K_i(Y)$, then $R \otimes_k K_{i+1} = R[X_1, \ldots, X_n, Y]_{S'} / (f_1, \ldots, f_m)$ where S' is generated by the image of S and the image of $K_i[Y] \setminus \{0\}$. $\quad\square$

Chain conditions in Cohen–Macaulay rings. Cohen–Macaulay rings were introduced as those rings for which depth equals dimension. Corollary 2.1.4 and the next theorem show that dimension theory itself is simpler for Cohen–Macaulay rings than for general Noetherian rings. One says a Noetherian ring R is *catenary* if every saturated chain joining prime ideals $\mathfrak{p}$ and $\mathfrak{q}$, $\mathfrak{p} \subset \mathfrak{q}$, has (maximal) length height $\mathfrak{q}/\mathfrak{p}$; R is *universally catenary* if all the polynomial rings $R[X_1, \ldots, X_n]$ are catenary. It is easy to see that R is universally catenary if and only if every finitely generated R-algebra is (universally) catenary.

Theorem 2.1.12. *A Cohen–Macaulay ring R is universally catenary.*

PROOF. 'Universally' may be dropped because of 2.1.9. So let $\mathfrak{p} \subset \mathfrak{q}$ be prime ideals of R. The localization $R_\mathfrak{q}$ is Cohen–Macaulay, and 2.1.4 applied to $R_\mathfrak{q}$ yields

$$\text{height}\, \mathfrak{q} = \dim R_\mathfrak{q} = \text{height}\, \mathfrak{p}R_\mathfrak{q} + \dim(R_\mathfrak{q}/\mathfrak{p}R_\mathfrak{q}) = \text{height}\, \mathfrak{p} + \text{height}\, \mathfrak{q}/\mathfrak{p}.$$

It is an easy exercise to show that R is catenary if this equation holds for all prime ideals $\mathfrak{p} \subset \mathfrak{q}$. $\qquad\Box$

Corollary 2.1.13. *A Noetherian complete local ring R is universally catenary.*

PROOF. Cohen's structure theorem (see A.21) tells us that R is a residue class ring of a formal power series ring $A = k[\![X_1, \ldots, X_n]\!]$ where k is a field or a discrete valuation ring. By 2.1.9, A is Cohen–Macaulay and therefore universally catenary; R inherits this property as a residue class ring of A. $\qquad\Box$

Remark 2.1.14. For the sake of clarity 2.1.4 and 2.1.12 were kept more special than necessary. If R has a Cohen–Macaulay module M with $\text{Supp}\, M = \text{Spec}\, R$, then we need only replace grade I by grade(I, M) in 2.1.4 to obtain an equally valid result. It follows that a local Noetherian domain which has a maximal Cohen–Macaulay module is universally catenary. One of Nagata's famous counter-examples is a non-catenary such domain ([284], Example 2, p. 203).

However, to be universally catenary is not the only necessary condition for R to have a Cohen–Macaulay module M with $\text{Supp}\, M = \text{Spec}\, R$; it must also satisfy Grothendieck's condition (CMU). This condition requires that for every prime ideal $\mathfrak{p}$ of R the spectrum of $R/\mathfrak{p}$ contains a non-empty open subset U such that $(R/\mathfrak{p})_\mathfrak{q}$ is Cohen–Macaulay for all $\mathfrak{q} \in U$; see [142], IV, 6.11. A local ring violating (CMU) was constructed by Ferrand and Raynaud [106].

A Noetherian complete local ring is universally catenary since it is a residue class ring of a Cohen–Macaulay ring, and for the same reason it satisfies (CMU). It is an open question whether every Noetherian complete local ring has a maximal Cohen–Macaulay module.

We shall see in Chapter 9 that the existence of maximal Cohen–Macaulay modules implies a wealth of homological theorems. Fortunately, it will not be essential that these Cohen–Macaulay modules M

are really finite; we 'only' need every system of parameters of the ring to be an M-sequence. In Chapter 8 such modules will be shown to exist for local rings containing a field.

For a prime ideal $\mathfrak{p}$ in a Cohen–Macaulay local ring R the residue class ring $R/\mathfrak{p}$ is not Cohen–Macaulay in general; it is however unmixed in the sense of Nagata [284]:

Theorem 2.1.15. *Let R be a Cohen–Macaulay local ring, and $\mathfrak{p}$ a prime ideal. Then* $\dim \hat{R}/\mathfrak{q} = \dim R/\mathfrak{p}$ *for all* $\mathfrak{q} \in \mathrm{Ass}(\hat{R}/\mathfrak{p}\hat{R})$. *In particular* $\mathfrak{p}\hat{R}$ *is an unmixed ideal.*

PROOF. If $\mathfrak{q} \cap R \neq \mathfrak{p}$, then $\mathfrak{q}$ would contain an $(\hat{R}/\mathfrak{p}\hat{R})$-regular element. Therefore $\mathfrak{q} \cap R = \mathfrak{p}$, and we have a flat local ring extension $R_{\mathfrak{p}} \to \hat{R}_{\mathfrak{q}}$. Applying 1.2.16 and since $\hat{R}_{\mathfrak{q}}$ and $R_{\mathfrak{p}}$ are Cohen–Macaulay we get

$$\dim \hat{R}_{\mathfrak{q}} = \operatorname{depth} \hat{R}_{\mathfrak{q}} = \operatorname{depth} R_{\mathfrak{p}} + \operatorname{depth}(\hat{R}_{\mathfrak{q}}/\mathfrak{p}\hat{R}_{\mathfrak{q}}) = \dim R_{\mathfrak{p}}.$$

In view of 2.1.4 this equation is equivalent to the theorem. □

Serre's condition (S_n). Sometimes one only needs a ring or a module to be Cohen–Macaulay in low 'codimension'. A finite module over a Noetherian ring R satisfies *Serre's condition (S_n)* if

$$\operatorname{depth} M_{\mathfrak{p}} \geq \min(n, \dim M_{\mathfrak{p}})$$

for all $\mathfrak{p} \in \operatorname{Spec} R$. The theorems of this section need some modification when Cohen–Macaulay is replaced by (S_n). As an example we treat the (S_n) analogue of 2.1.7.

Proposition 2.1.16. *Let $\varphi : R \to S$ be a flat homomorphism of Noetherian rings.*
(a) *Let $\mathfrak{q} \in \operatorname{Spec} S$ and $\mathfrak{p} = \mathfrak{q} \cap R$. If $S_{\mathfrak{q}}$ satisfies (S_n), then so does $R_{\mathfrak{p}}$.*
(b) *Suppose R and all the fibres $k(\mathfrak{p}) \otimes S$ with $\mathfrak{p} \in \operatorname{Spec} R$ satisfy (S_n). Then S satisfies (S_n).*

PROOF. (a) Replacing R by $R_{\mathfrak{p}}$ and S by $S_{\mathfrak{q}}$ we may assume that φ is a flat homomorphism of local rings. For $\mathfrak{p} \in \operatorname{Spec} R$ we now choose a minimal prime $\mathfrak{q}$ of $\mathfrak{p}S$. Then $\dim(S_{\mathfrak{q}}/\mathfrak{p}S_{\mathfrak{q}}) = 0$, and according to 1.2.16 we have

$$\operatorname{depth} R_{\mathfrak{p}} = \operatorname{depth} S_{\mathfrak{q}} \geq \min(n, \dim S_{\mathfrak{q}}) = \min(n, \dim R_{\mathfrak{p}}).$$

(b) For $\mathfrak{q} \in \operatorname{Spec} S$ and $\mathfrak{p} = R \cap \mathfrak{q}$ one similarly deduces

$$\begin{aligned}
\operatorname{depth} S_{\mathfrak{q}} &= \operatorname{depth} R_{\mathfrak{p}} + \operatorname{depth}(S_{\mathfrak{q}}/\mathfrak{p}S_{\mathfrak{q}}) \\
&\geq \min(n, \dim R_{\mathfrak{p}}) + \min(n, \dim(S_{\mathfrak{q}}/\mathfrak{p}S_{\mathfrak{q}})) \\
&\geq \min(n, \dim R_{\mathfrak{p}} + \dim(S_{\mathfrak{q}}/\mathfrak{p}S_{\mathfrak{q}})) \\
&= \min(n, \dim S_{\mathfrak{q}}).
\end{aligned}$$

□

Exercises

2.1.17. Let k be a field, and $R = k[X_1, \ldots, X_n]$. If $\mathfrak{p}$ is a prime ideal in R with height $\mathfrak{p} \in \{0, 1, n - 1, n\}$, show that $R/\mathfrak{p}$ is Cohen–Macaulay.

2.1.18. Let k be a field. Show
(a) the subalgebra $S = k[U^4, U^3V, UV^3, V^4]$ of $k[U, V]$ is not Cohen–Macaulay,
(b) for each m with $2 \leq m \leq n - 2$ there exists a prime ideal of height m in $R = k[X_1, \ldots, X_n]$ for which $R/\mathfrak{p}$ is not Cohen–Macaulay.

2.1.19. Let k be a field, and S the subalgebra of $k[X_1, \ldots, X_n]$ generated by the monomials of degrees 2 and 3. Show S is an n-dimensional domain; the maximal ideal $(X_1, \ldots, X_n) \cap S$ has height n and grade 1.

2.1.20. Prove (a) a one dimensional reduced Noetherian ring is Cohen–Macaulay, (b) a one dimensional Noetherian local ring has a maximal Cohen–Macaulay module.

2.1.21. Characterize (S_n) by an unmixedness property.

2.1.22. Prove that a module M satisfies (S_n) if and only if $M_\mathfrak{p}$ is Cohen–Macaulay for all prime ideals $\mathfrak{p}$ with depth $M_\mathfrak{p} < n$.

2.1.23. Let $R \to S$ be a faithfully flat homomorphism of Noetherian rings. Show the following are equivalent:
(a) S is Cohen–Macaulay;
(b) R and all the fibres $S_\mathfrak{q}/\mathfrak{p}S_\mathfrak{q}$ are Cohen–Macaulay where $\mathfrak{q} \in \operatorname{Spec} S$ and $\mathfrak{p} = \mathfrak{q} \cap R$.
Hint: use A.10.

2.1.24. Prove the analogues of 2.1.8(b), 2.1.9, and 2.1.10 for (S_n). For the passages from R to $\hat{R}$ and to $R[[X_1, \ldots, X_n]]$ assume that R is a residue class ring of a Cohen–Macaulay ring.

2.1.25. Prove the converse of 2.1.5(a) under the hypothesis that $\operatorname{Supp} M$ is connected. (The crucial point is to show that the function $\mathfrak{p} \mapsto \operatorname{proj} \dim M_\mathfrak{p}$ is locally constant on $\operatorname{Supp} M$ if M is locally perfect.)

2.1.26. Let R be a Cohen–Macaulay local ring of dimension d and M a finite R-module. Deduce that the d-th syzygy of M in an arbitrary finite free resolution is either 0 or a maximal Cohen–Macaulay module.

2.1.27. Let R be a Noetherian graded ring, and M a finite graded R-module. Show:
(a) For $\mathfrak{p} \in \operatorname{Spec} R$ the localization $M_\mathfrak{p}$ is Cohen–Macaulay if and only if $M_{\mathfrak{p}^*}$ is. (This follows easily from the results of Section 1.5.)
(b) The following are equivalent:
  (i) M is Cohen–Macaulay;
  (ii) $M_\mathfrak{p}$ is Cohen–Macaulay for all graded prime ideals $\mathfrak{p}$;
  (iii) $M_{(\mathfrak{p})}$ is Cohen–Macaulay for all graded prime ideals $\mathfrak{p}$.
(c) Suppose in addition that $(R, \mathfrak{m})$ is *local. Then M is Cohen–Macaulay if and only if $M_\mathfrak{m}$ is.

2.1.28. Let $(R, \mathfrak{m})$ be a Noetherian *local ring, and $x \in \mathfrak{m}$ a homogeneous R-regular element. Then R is Cohen–Macaulay if and only if so is $R/(x)$.

2.1.29. Let the Noetherian ring R be a free $\mathbb{Z}$-module such that $R \otimes K$ is Cohen–Macaulay for some field K of characteristic $p > 0$. Show that $R \otimes L$ is Cohen–Macaulay for every field L of characteristic 0.
Hint: reduce the problem to the case in which $K = \mathbb{Z}/(p)$, $L = \mathbb{Q}$ and use $R \otimes \mathbb{Z}_\mathfrak{p}$, $\mathfrak{p} = (p)$, as a 'bridge'.
This is the first and easiest example of *reduction to characteristic p*.

2.2 Regular rings and normal rings

The most distinguished of all Noetherian local rings are those whose maximal ideal can be generated by a system of parameters:

Definition 2.2.1. A Noetherian local ring $(R, \mathfrak{m})$ is *regular* if it has a system of parameters generating $\mathfrak{m}$; such a system of parameters is called a *regular system of parameters*.

Evidently, when $\dim R = 0$, then R is regular if and only if it is a field, and when $\dim R = 1$, R is regular if and only if it is a discrete valuation ring. Other examples of regular local rings are $k[[X_1, \ldots, X_n]]$ where k is a field, and $k[X_1, \ldots, X_n]_\mathfrak{m}$, $\mathfrak{m} = (X_1, \ldots, X_n)$.

We may rephrase the definition above as follows: R is regular if and only if $\mu(\mathfrak{m}) = \dim R$. In fact, $\mu(\mathfrak{m}) \geq \dim R$ by Krull's principal ideal theorem, and a system of generators of $\mathfrak{m}$ has $\dim R$ elements exactly when it is a system of parameters.

Proposition 2.2.2. *A Noetherian local ring $(R, \mathfrak{m})$ is regular if and only if its $\mathfrak{m}$-adic completion $\hat{R}$ is regular.*

PROOF. The maximal ideal of $\hat{R}$ is $\mathfrak{m}\hat{R}$, and we have natural isomorphisms $R/\mathfrak{m} \cong \hat{R}/\mathfrak{m}\hat{R}$, $\mathfrak{m}/\mathfrak{m}^2 \cong (\mathfrak{m}\hat{R})/(\mathfrak{m}\hat{R})^2$. Therefore $\mu(\mathfrak{m}) = \mu(\mathfrak{m}\hat{R})$. Furthermore $\dim R = \dim \hat{R}$, and by definition R is regular if and only if $\dim R = \mu(\mathfrak{m})$. □

It is easily proved that regular local rings are integral domains.

Proposition 2.2.3. *Let $(R, \mathfrak{m})$ be a regular local ring. Then R is an integral domain.*

PROOF. We use induction on $\dim R$. When $\dim R = 0$, R is a field. So suppose $\dim R > 0$, and let $\mathfrak{p}_1, \ldots, \mathfrak{p}_m$ be the minimal prime ideals of R. There exists an element $x \in \mathfrak{m}$ which is not contained in any of the ideals $\mathfrak{m}^2, \mathfrak{p}_1, \ldots, \mathfrak{p}_m$. (This follows easily from 1.2.2 with $M = N = \mathfrak{m}$.) Since x is part of a minimal system of generators of $\mathfrak{m}$, it is part of a regular system of parameters, and thus $R/(x)$ is regular (use that $\dim R/(x) = \dim R - 1$). As $\dim R/(x) < \dim R$ we may assume that $R/(x)$ is a domain. Thus (x) is a prime ideal, and therefore contains a minimal prime ideal of R, say $\mathfrak{p}_1$. Every $y \in \mathfrak{p}_1$ has the form $y = xz$, and since $x \notin \mathfrak{p}_1$, z is an element of $\mathfrak{p}_1$. It follows that $\mathfrak{p}_1 = x\mathfrak{p}_1$, which, by Nakayama's lemma, implies $\mathfrak{p}_1 = 0$, as required. □

Using the previous proposition, one can say precisely which residue class rings of a regular local ring are also regular:

Proposition 2.2.4. *Let R be a regular local ring, and $I \subset R$ an ideal. Then R/I is regular if and only if I is generated by a subset of a regular system of parameters.*

PROOF. The 'if' part is trivial. So suppose that R/I is regular. Then $\mu(\mathfrak{m}/I) = \dim R/I$; set $m = \dim R - \dim R/I$. By Nakayama's lemma I contains elements $x_1, \ldots, x_m$ which are part of a minimal system of generators of $\mathfrak{m}$. Then $R/(x_1, \ldots, x_m)$ is regular of dimension $\dim R - m = \dim R/I$. Since I and $(x_1, \ldots, x_m)$ are prime ideals, one must have $I = (x_1, \ldots, x_m)$. □

The next proposition gives useful characterizations of regularity.

Proposition 2.2.5. *Let $(R, \mathfrak{m}, k)$ be a Noetherian local ring, and $x_1, \ldots, x_n$ a minimal system of generators of $\mathfrak{m}$. Then the following are equivalent:*
(a) *R is regular;*
(b) *$x_1, \ldots, x_n$ is an R-sequence;*
(c) *the substitution $X_i \mapsto \bar{x}_i \in \mathfrak{m}/\mathfrak{m}^2$ yields an isomorphism $k[X_1, \ldots, X_n] \cong \mathrm{gr}_{\mathfrak{m}}(R)$.*

PROOF. (a) $\Rightarrow$ (b): Since $x_1, \ldots, x_n$ is a minimal system of generators of $\mathfrak{m}$, it is a regular system of parameters, and $R/(x_1, \ldots, x_i)$ is also regular for each i. Therefore $R/(x_1, \ldots, x_i)$ is a domain, and x_{i+1} is regular on $R/(x_1, \ldots, x_i)$.

(b) $\Rightarrow$ (a): An R-sequence is part of a system of parameters by 1.2.12.

(b) $\Longleftrightarrow$ (c): This follows from 1.1.8 and its converse 1.1.15. □

Corollary 2.2.6. *A regular local ring is Cohen–Macaulay.*

The Auslander–Buchsbaum–Serre theorem. Whereas the characterizations of regular local rings in 2.2.5 are rather close to the definition, this can hardly be said of the following theorem. Together with 2.2.19 below, it is considered to be the most important achievement of the use of homological algebra in the theory of commutative rings.

Theorem 2.2.7 (Auslander–Buchsbaum–Serre). *Let $(R, \mathfrak{m}, k)$ be a Noetherian local ring. Then the following are equivalent:*
(a) *R is regular;*
(b) *proj dim $M < \infty$ for every finite R-module M;*
(c) *proj dim $k < \infty$.*

PROOF. (a) $\Rightarrow$ (b): Let $d = \dim R$, and N a d-th syzygy module of M. Since R is Cohen–Macaulay, N is a maximal Cohen–Macaulay module or 0 by Exercise 2.1.26. If $N = 0$, we are done; otherwise every regular system of parameters x is a (maximal) N-sequence. Lemma 1.3.5 gives

$\operatorname{proj dim}_R N = \operatorname{proj dim}_{R/(x)}(N/xN) = \operatorname{proj dim}_k(N/\mathfrak{m}N) = 0$. So N is free, and $\operatorname{proj dim} M \leq d$.

(b) $\Rightarrow$ (c): This is trivial.

(c) $\Rightarrow$ (a): This is a special case of the following theorem. □

Theorem 2.2.8 (Ferrand, Vasconcelos). *Let $(R, \mathfrak{m})$ be a Noetherian local ring, and $I \neq 0$ a proper ideal with $\operatorname{proj dim} I < \infty$. If I/I^2 is a free R/I-module, then I is generated by a regular sequence.*

PROOF. Since I has a finite free resolution, it contains an R-regular element x by 1.4.7. It is no restriction to assume $x \notin \mathfrak{m}I$: if $x \in \mathfrak{m}I$, then we choose some $y \in I \setminus \mathfrak{m}I$; $Ry + Rx$ is not contained in any $\mathfrak{p} \in \operatorname{Ass} R$, and by 1.2.2 there is $a \in R$ for which $y + ax$ has the same property. This proves the theorem when $\mu(I) = 1$, and when $\mu(I) > 1$, we use induction, passing from R to $R/(x)$ and from I to $I/(x)$.

Of course, we must first verify that $\operatorname{proj dim}_{R/(x)} I/(x) < \infty$. Since $x \notin \mathfrak{m}I$, the residue class of x in I/I^2 is part of a basis $\bar{x}, \bar{x}_2, \ldots, \bar{x}_m$ of this free module. Set $J = (x_2, \ldots, x_m)$; we claim: $J \cap (x) \subset xI$. In fact, if $z = ax = a_2 x_2 + \cdots + a_m x_m$, then $a \in I$ because $\bar{x}, \bar{x}_2, \ldots, \bar{x}_m$ are linearly independent modulo I. Therefore we get a composition of maps

$$I/(x) = (J + (x))/(x) \cong J/J \cap (x) \longrightarrow I/xI \longrightarrow I/(x),$$

in which the residue class of x_i is sent to itself, and which therefore is the identity on $I/(x)$. So $I/(x)$ is a direct summand of I/xI; as x is I-regular, the latter has finite projective dimension over $R/(x)$ by 1.3.5.

Finally we need that $\bar{I}/\bar{I}^2$ is a free R/I-module where $\bar{I} = I/(x)$. But this is a very easy consequence of the linear independence of $x, x_2, \ldots, x_m$ modulo I. □

The proof of 2.2.7 can be varied: the Koszul complex of a regular system of parameters resolves k by 2.2.5 and 1.6.14, whence (a) $\Rightarrow$ (c). Moreover, the implication (c) $\Rightarrow$ (b) follows from 1.3.2: $\operatorname{proj dim} k < \infty \Rightarrow \operatorname{Tor}_i^R(M, k) = 0$ for $i \gg 0$, and this in turn gives $\operatorname{proj dim} M < \infty$. While this reasoning uses a truly homological argument, namely the fact that Tor can be computed from a free resolution of either module, the proof above merely exploits the existence of minimal free resolutions. Serre's original argument for (c) $\Rightarrow$ (a) will be indicated in Exercise 2.3.22.

Corollary 2.2.9. *Let R be a regular local ring, and $\mathfrak{p}$ a prime ideal in R. Then $R_\mathfrak{p}$ is regular.*

PROOF. By 2.2.7(a) $\Rightarrow$ (b) we have $\operatorname{proj dim} R/\mathfrak{p} < \infty$. It follows that $\operatorname{proj dim}_{R_\mathfrak{p}}(R/\mathfrak{p})_\mathfrak{p} = \operatorname{proj dim}_{R_\mathfrak{p}}(R_\mathfrak{p}/\mathfrak{p}R_\mathfrak{p}) < \infty$, whence $R_\mathfrak{p}$ is regular by 2.2.7(c) $\Rightarrow$ (a). □

Over a regular local ring the Cohen–Macaulay property is equivalent to perfection (see Section 1.4 for this notion):

Corollary 2.2.10. *A finite module M over a regular local ring is Cohen–Macaulay if and only if it is perfect.*

The corollary is an immediate consequence of 2.1.5 and 2.2.7.

Let $(R, \mathfrak{m})$ be a Noetherian local ring. By 2.1.8, R is Cohen–Macaulay if and only if its $\mathfrak{m}$-adic completion $\hat{R}$ is Cohen–Macaulay. Furthermore, if R contains a field or is a domain, then it is a finite module over a regular local subring (see A.22). Thus the following proposition may almost be considered a new description of the Cohen–Macaulay property for rings.

Proposition 2.2.11. *Let R be a Noetherian local ring and S a regular local subring such that R is a finite S-module. Then R is Cohen–Macaulay if and only if it is a free S-module.*

PROOF. By 2.2.7 one has $\operatorname{proj\,dim}_S R < \infty$; therefore R is S-free if and only if $\operatorname{depth}_S R = \dim S$. Choose a (regular) system of parameters x in S. Then x is also a system of parameters of R, and therefore R is Cohen–Macaulay $\Longleftrightarrow$ x is an R-sequence $\Longleftrightarrow$ $\operatorname{depth}_S R = \dim S$. (One could also use Exercise 1.2.26.) □

Flat extensions of regular rings. The behaviour of regularity under flat local extensions is described by the following theorem.

Theorem 2.2.12. *Let $\varphi : (R, \mathfrak{m}, k) \to (S, \mathfrak{n}, l)$ be a flat homomorphism of Noetherian local rings.*
(a) *If S is regular, then so is R.*
(b) *If R and S/$\mathfrak{m}S$ are regular, then so is S.*

PROOF. (a) Let $F_\bullet$ be a minimal free resolution of the R-module k. Then $F_\bullet \otimes S$ is a free resolution of $k \otimes S \cong S/\mathfrak{m}S$ because of flatness, and even a minimal one since $\varphi(\mathfrak{m}) \subset \mathfrak{n}$. Thus $\operatorname{proj\,dim}_R k = \operatorname{proj\,dim}_S S/\mathfrak{m}S < \infty$, and R is regular by 2.2.7.

(b) Let $m = \dim R$, $n = \dim S/\mathfrak{m}S$, and choose minimal systems of generators $x_1, \ldots, x_m$ of $\mathfrak{m}$ and $y_1, \ldots, y_n$ of $\mathfrak{n}/\mathfrak{m}S$. Then $\varphi(x_1), \ldots, \varphi(x_m)$, $y_1, \ldots, y_n$ generate $\mathfrak{n}$, and S must be regular because $\dim S = \dim R + \dim S/\mathfrak{m}S$; see A.11. □

An easy example shows that $S/\mathfrak{m}S$ need not be regular in the situation of 2.2.12(a): Let k be a field, and choose $S = k[\![X, Y]\!]/(Y - X^2)$ and $R = k[\![y]\!] \subset S$. Then R and S are regular, and S is a free R-module generated by 1 and x, but $S/yS \cong k[\![X]\!]/(X^2)$ is not regular. (The reader should imagine the geometry of this example.)

In order to formulate a theorem relating the regularity of R and that of $R[X_1, \ldots, X_n]$ we must first agree on calling a Noetherian ring R *regular* if its localizations $R_\mathfrak{m}$ with respect to maximal ideals $\mathfrak{m}$ are regular.

Theorem 2.2.13. *A Noetherian ring R is regular if and only if $R[X_1, \ldots, X_n]$ is regular. The same holds for R and $R[[X_1, \ldots, X_n]]$.*

PROOF. We may assume that $n = 1$ and set $X = X_1$. Suppose that $R[X]$ is regular, and, given a maximal ideal $\mathfrak{m}$ of R, choose $\mathfrak{n} = (\mathfrak{m}, X) \subset R[X]$. Then $X \notin \mathfrak{n}^2$, equivalently $X \notin (\mathfrak{n}R[X]_\mathfrak{n})^2$, and it follows immediately from the definition of regularity that $R_\mathfrak{m}$ is regular. The same argument shows that regularity descends from $R[[X]]$ to R. Of course, what has just been shown can also be derived from 2.2.12(a), and the reader is invited to use 2.2.12(b) in proving that regularity ascends from R to $R[X]$ and $R[[X]]$. (Compare the proof of 2.1.9.) □

In particular, a polynomial ring $k[X_1, \ldots, X_n]$ over a field k is regular.

Corollary 2.2.14. *Let k be a field, and $R = k[X_1, \ldots, X_n]$.*
(a) *(Hilbert's syzygy theorem) Every finite graded R-module M has a finite graded free resolution of length $\leq n$.*
(b) *Moreover, $\operatorname{proj\,dim} M \leq n$ for every finite R-module M.*
(c) *In fact, every finite R-module has a finite free resolution of length $\leq n$.*

PROOF. (a) Set $\mathfrak{m} = (X_1, \ldots, X_n)$ and consider a minimal graded free resolution $F_{\boldsymbol{.}}$ of M. Such a resolution exists, and furthermore $F_{\boldsymbol{.}} \otimes R_\mathfrak{m}$ is a minimal free resolution of $M_\mathfrak{m}$; see 1.5.15. Now $R_\mathfrak{m}$ is a regular local ring. Therefore $F_{\boldsymbol{.}} \otimes R_\mathfrak{m}$ has length at most n, and the same holds true for $F_{\boldsymbol{.}}$.

(b) Consider an arbitrary maximal ideal $\mathfrak{n}$ of R. Then $\dim R_\mathfrak{n} \leq \dim R = n$, and $R_\mathfrak{n}$ is a regular local ring. Hence $\operatorname{proj\,dim}_{R_\mathfrak{n}} M_\mathfrak{n} \leq n$. Taking the supremum over all maximal ideals, we get $\operatorname{proj\,dim} M \leq n$. (In fact, let N be the n-th syzygy of M in a resolution by finite projective R-modules. Then $N_\mathfrak{n}$ is a finite free $R_\mathfrak{n}$-module for all $\mathfrak{n}$, and therefore N is projective.)

(c) By the theorem of Quillen and Suslin ([318], Theorem 4.59) every finite projective R-module is free. □

In 2.2.14(a) and 2.2.15 below it is not essential that $\deg X_i = 1$ for all i. One may replace the standard grading of R by any grading which makes R a *local ring, for example by a grading such that $\deg X_i > 0$ for all i.

Corollary 2.2.15. *Let k be a field, $R = k[X_1, \ldots, X_n]$, $\mathfrak{m} = (X_1, \ldots, X_n)$, and M a finite graded R-module. Then the following are equivalent:*
(a) *M is Cohen–Macaulay;*
(a') *M is perfect;*
(b) *$M_\mathfrak{m}$ is Cohen–Macaulay;*
(b') *$M_\mathfrak{m}$ is perfect.*

PROOF. The implications (a') $\Rightarrow$ (a) $\Rightarrow$ (b) $\Rightarrow$ (b') follow from 2.1.5 and 2.2.10. The remaining implication (b') $\Rightarrow$ (a') is an immediate consequence of the equations $\operatorname{proj\,dim} M = \operatorname{proj\,dim} M_\mathfrak{m}$ and $\operatorname{grade} M = \operatorname{grade} M_\mathfrak{m}$ proved in 1.5.15. □

Remark 2.2.16. Let R be a Noetherian k-algebra where k is a field, and K an extension field of k. If R is finitely generated as a k-algebra or K is a finitely generated extension field, then $R \otimes_k K$ is a Noetherian ring as shown in the proof of 2.1.10. Since $R \otimes_k K$ is a flat R-algebra, it follows readily from 2.2.12 that R is regular if $R \otimes_k K$ is regular. We saw in the proof of 2.1.10 that the fibres of the extension $R \to R \otimes_k K$ are of the form $(L \otimes_k K)_\mathfrak{p}$ where L is an extension field of k and $\mathfrak{p} \in \operatorname{Spec}(L \otimes_k K)$. If $L \otimes_k K$ is regular for every extension field L of k (provided one of K, L is finitely generated), then one obtains from 2.2.12 that $R \otimes_k K$ is regular if R is regular.

The fields K satisfying the condition just formulated are the *separable* extensions of k. We refer the reader to [270], §26 for a discussion of separability, and to [142], IV, 6.7.4.1 for the theorem concerning the regularity of $L \otimes_k K$.

Factoriality of regular local rings. Our next goal is to show that a regular local ring is a factorial domain (a UFD in other terminology). We need two elementary lemmas whose proofs are left as an exercise for the reader.

Lemma 2.2.17. *A Noetherian domain R is factorial if and only if every prime ideal $\mathfrak{p}$ of height 1 is principal.*

Lemma 2.2.18. *Let R be a Noetherian domain and π a prime element in R. Then R is factorial if and only if R_π is factorial.*

Theorem 2.2.19 (Auslander–Buchsbaum–Nagata). *A regular local ring R is factorial.*

PROOF. We use induction on $\dim R$. If $\dim R = 0$, then R is a field, and there is nothing to prove. So suppose $\dim R > 0$, and choose $\pi \in \mathfrak{m} \setminus \mathfrak{m}^2$. Since $R/(\pi)$ is again a regular local ring, π is a prime element. According to the previous lemma, it is enough to show that $S = R_\pi$ is factorial.

Let $\mathfrak{p}$ be a prime ideal of S with $\operatorname{height} \mathfrak{p} = 1$. Every localization $S_\mathfrak{q}$ is a localization of R with respect to a prime ideal $\neq \mathfrak{m}$, and therefore a regular local ring by 2.2.9. By induction $S_\mathfrak{q}$ is factorial. If $\mathfrak{p} \not\subset \mathfrak{q}$, then $\mathfrak{p}S_\mathfrak{q} \cong S_\mathfrak{q}$ for trivial reasons, and if $\mathfrak{p} \subset \mathfrak{q}$, then also $\mathfrak{p}S_\mathfrak{q} \cong S_\mathfrak{q}$ as follows from 2.2.17 in conjunction with the factoriality of $S_\mathfrak{q}$. This implies that $\mathfrak{p}$ is a projective S-module of rank 1.

Of course $\mathfrak{p}$ is of the form $\mathfrak{P}S$ with a prime ideal $\mathfrak{P}$ of R. The R-module $\mathfrak{P}$ has a finite free resolution $F_\bullet$, whence $\mathfrak{p} = \mathfrak{P}_\pi$ has an augmented resolution

$$G_\bullet : 0 \longrightarrow G_s \xrightarrow{\varphi_s} G_{s-1} \longrightarrow \cdots \longrightarrow G_1 \xrightarrow{\varphi_1} G_0 \longrightarrow \mathfrak{p} \longrightarrow 0$$

by finite free S-modules. However, $\mathfrak{p}$ is a projective S-module, and its syzygy modules with respect to $G_.$ are likewise projective. In particular $\operatorname{Im} \varphi_{s-1} \oplus G_s \cong G_{s-1}$. If $s > 1$, we can modify the tail of $G_.$ to obtain the free resolution

$$G_. : 0 \longrightarrow \operatorname{Im} \varphi_{s-1} \oplus G_s \longrightarrow G_{s-2} \oplus G_s \longrightarrow \cdots \longrightarrow G_1 \xrightarrow{\varphi_1} G_0 \longrightarrow \mathfrak{p} \longrightarrow 0.$$

Therefore, by induction on the length of $G_.$, $\mathfrak{p}$ in fact has a free resolution

$$0 \longrightarrow S^n \xrightarrow{\varphi} S^{n+1} \longrightarrow \mathfrak{p} \longrightarrow 0.$$

The Hilbert–Burch theorem 1.4.17 yields that $\mathfrak{p} = aI_n(\varphi)$ with some $a \in S$, and furthermore that $\mathfrak{p} = aS$ since $\mathfrak{p}$ is projective. So $\mathfrak{p}$ is a principal ideal. □

A ring is *normal* if all its localizations are integrally closed domains; a Noetherian ring is normal if and only if it is the direct product of finitely many integrally closed domains (see [270] for a detailed discussion of normality).

Corollary 2.2.20. *A regular local ring is a normal domain. A regular ring is the direct product of regular domains.*

In fact, every factorial ring R is a normal domain. (One proves this just as for the special case $R = \mathbb{Z}$.) The 'classical' proof of the corollary uses 2.2.5 and the fact that a Noetherian local ring is a normal domain if $\operatorname{gr}_m(R)$ is a normal domain; see 4.5.9. There is even a third proof, as we shall see now.

Serre's normality criterion. A Noetherian ring R satisfies *Serre's condition* (R_n) if $R_\mathfrak{p}$ is a regular local ring for all prime ideals $\mathfrak{p}$ in R with $\dim R_\mathfrak{p} \le n$. (Note the similarity with (S_n); contrary to (S_n) however, (R_n) says nothing about localizations $R_\mathfrak{p}$ with $\dim R_\mathfrak{p} > n$.)

We leave it as an exercise for the reader to prove that the behaviour of (R_n) under flat local extensions is the same as that of (S_n):

Proposition 2.2.21. *Let $\varphi : R \to S$ be a flat homomorphism of Noetherian rings.*
(a) *Let $\mathfrak{q} \in \operatorname{Spec} S$ and $\mathfrak{p} = \mathfrak{q} \cap R$. If $S_\mathfrak{q}$ satisfies (R_n), then so does $R_\mathfrak{p}$.*
(b) *If R and all fibres $k(\mathfrak{p}) \otimes S$, $\mathfrak{p} \in \operatorname{Spec} R$, satisfy (R_n), then so does S.*

It is easy to see that a Noetherian ring R is reduced if and only if it satisfies (R_0) and (S_1). Serre characterized normality in a similar way:

Theorem 2.2.22 (Serre). *A Noetherian ring R is normal if and only if it satisfies (R_1) and (S_2).*

We refer the reader to Serre [334], IV.4, or [270], §23, for a proof of 2.2.22. The following corollary is an evident consequence of 2.1.16, 2.2.21, and 2.2.22:

Corollary 2.2.23. *Let* $\varphi: R \to S$ *be a flat homomorphism of Noetherian rings.*
(a) *Let* $q \in \operatorname{Spec} S$ *and* $p = q \cap R$. *If* S_q *is normal, then so is* R_p.
(b) *If* R *and all the fibres* $k(p) \otimes S$, $p \in \operatorname{Spec} R$, *are normal, then so is* S.

Suppose that $(R, \mathfrak{m})$ and $(S, \mathfrak{n})$ are local, and that φ is flat and local. Then, for S to be normal, it is not sufficient to have R and $S/\mathfrak{m}S$ normal: there are normal local domains whose completions are not even domains; see [284], p. 209, Example 7.

Exercises

2.2.24. Let R be a Noetherian graded ring. Show:
(a) For $p \in \operatorname{Spec} R$ the localization R_p is regular if and only if R_{p^*} is.
(b) The following are equivalent:
    (i) R is regular;
    (ii) R_p is regular for all graded prime ideals p;
    (iii) $R_{(p)}$ is regular for all graded prime ideals p.
(c) Suppose moreover that $(R, \mathfrak{m})$ is *local. Then R is regular if and only if $R_{\mathfrak{m}}$ is. Hint: Use 1.6.33.

2.2.25. Let R be a positively graded k-algebra over a field k. Prove the following are equivalent:
(a) R is regular;
(b) $R_{\mathfrak{m}}$ is regular where $\mathfrak{m}$ is the *maximal ideal;
(c) there exist homogeneous elements $x_1, \ldots, x_n$ of positive degree for which the assignment $X_i \mapsto x_i$ induces an isomorphism $k[X_1, \ldots, X_n] \cong R$.
Hint: For the non-trivial implication (b) $\Rightarrow$ (c) choose a minimal homogeneous system of generators $x_1, \ldots, x_n$ of $\mathfrak{m}$; then apply 1.5.15 and 1.5.4. The rest is a simple dimension argument.

2.2.26. In the situation of 2.2.11 characterize the Cohen–Macaulay R-modules by a property they have as S-modules.

2.2.27. Let R be a Noetherian ring over which every finite module has a finite free resolution. Show R is a factorial domain.

2.2.28. Let R be a regular local ring, and I an ideal of height 1. Prove that the following are equivalent:
(a) R/I is Cohen–Macaulay;
(b) height $p = 1$ for all prime ideals $p \in \operatorname{Ass} R/I$;
(c) I is a principal ideal.
Hint: For (b) $\Rightarrow$ (c) one uses primary decomposition and the factoriality of R.

2.2.29. Prove that a Noetherian ring R satisfies (R_i) and (S_{i+1}) if and only if R_p is regular for every prime ideal p such that depth $R_p \le i$.

2.2.30. (a) Show a Noetherian normal ring of dimension 2 is Cohen–Macaulay.
(b) A Cohen–Macaulay ring is normal if and only if it satisfies (R_1).

2.2.31. (a) Let R be a Noetherian complete local domain. Then R is a finite module over a regular local ring S contained in R; see A.22. Set $M = \operatorname{Hom}_S(R, S)$; then

M is an R-module in a natural way. Show that $\operatorname{depth}_R M = \operatorname{depth}_S M \geq \min(\dim R, 2)$.

(b) Prove that every Noetherian complete local ring of dimension 2 has a maximal Cohen–Macaulay module.

2.2.32. Let R be a Noetherian complete local domain. It is known that the integral closure of R in its field of fractions is a finite R-module ([284], (32.1) or [47], Ch. IX, §4). Use this to give a fresh proof of the fact that a Noetherian complete local ring of dimension 2 has a maximal Cohen–Macaulay module.

2.2.33. Let R be a Noetherian ring, and $x \in R$ an R-regular element.

(a) Assume that R_x fulfills (R_n) and (S_{n+1}), and that (R_{n-1}) and (S_n) hold for $R/(x)$. Show that R satisfies (R_n) and (S_{n+1}).

(b) Assume R_x is a normal domain, and $R/(x)$ is reduced. Show that R is a normal domain.

2.2.34. Let R be a Noetherian graded ring, and A its dehomogenization with respect to an element of degree 1 (see 1.5.18). Show that if R is a normal domain, then so is A.

2.2.35. We keep the notation of 2.2.34. Let $\mathfrak{p}$ be a graded prime ideal of R with $x \notin \mathfrak{p}$, and $\mathfrak{q} \in \operatorname{Spec} A$ its dehomogenization (see 1.5.26). Show that $R_\mathfrak{p}$ is a flat local extension of $A_\mathfrak{q}$, and determine its fibre. Compare $R_\mathfrak{p}$ and $S_\mathfrak{q}$ with respect to the following quantities and properties: dimension, depth, type; being reduced, an integral domain, Cohen–Macaulay, normal, regular.

2.3 Complete intersections

We observed that the homological relationship between a local ring S and a residue class ring $R = S/I$ is particularly strong if I is generated by an S-sequence. In this section we investigate such residue class rings of regular local rings. Slightly more generally we define:

Definition 2.3.1. A Noetherian local ring R is a *complete intersection* (*ring*) if its completion $\hat{R}$ is a residue class ring of a regular local ring S with respect to an ideal generated by an S-sequence.

Note that $\hat{R}$ is always a residue class ring of a regular local ring (see A.21). It follows immediately from 2.1.3, 2.1.8, and 2.2.6 that a complete intersection is Cohen–Macaulay.

The nomenclature 'complete intersection' comes from algebraic geometry. Suppose R is the coordinate ring of an affine variety over an algebraically closed field k. Then R has the form $R = S/I$ where S is a polynomial ring over k, and R is called a complete intersection if I is generated by the least possible number of elements, namely $\operatorname{codim} V = \operatorname{height} I$. Then V is the intersection of $\operatorname{codim} V$ hypersurfaces, and I is generated by an S-sequence.

Let $(S, \mathfrak{n})$ be a regular local ring, and $(R, \mathfrak{m})$ a residue class ring, $R = S/I$. Suppose that $I \not\subset \mathfrak{n}^2$. Then there exists $x \in I$, $x \notin \mathfrak{n}^2$, and we obtain a representation $R = S'/I'$ with $S' = S/(x)$, $I' = I/(x)$. The ring S' is regular again, and I is generated by a regular sequence if and only if I' is: the element x is part of a minimal system of generators of I, and I can be generated by an S-sequence if and only if every minimal system of generators is an S-sequence; see 1.6.19. Iterating this procedure we eventually obtain a *minimal presentation* $R = S''/I''$ in which S'' is regular and $I'' \subset (\mathfrak{n}'')^2$. It follows that $\mu(\mathfrak{m}) = \mu(\mathfrak{n}'') = \dim S''$. For an arbitrary local ring $(R, \mathfrak{m})$ the number $\mu(\mathfrak{m})$ is called the *embedding dimension* of R,

$$\text{emb}\dim R = \mu(\mathfrak{m}).$$

(This terminology is again to be illustrated by the geometric analogue.) The discussion above shows that we may freely assume that $I \subset \mathfrak{n}^2$ when it is only to be verified whether I is generated by an S-sequence or otherwise.

Nevertheless our definition has two flaws: first, it does not use *intrinsic* characteristics of R; second, it is not clear whether for an *arbitrary* presentation $\hat{R} = S/I$ with S regular local, the ideal I is generated by an S-sequence if R is a complete intersection. The intrinsic characteristics we are seeking are hidden in a Koszul complex. Let us fix some standard notation which we shall use frequently throughout this section: $(R, \mathfrak{m}, k)$ is a Noetherian local ring, and $x = x_1, \ldots, x_n$ is a minimal system of generators of $\mathfrak{m}$. If present, $(S, \mathfrak{n}, k)$ is a regular local ring such that $R = S/I$ with $I \subset \mathfrak{n}^2$; the ideal I is minimally generated by $a = a_1, \ldots, a_m$, and $y = y_1 \ldots, y_n$ is a regular system of parameters such that x_i is the residue class of y_i. Furthermore we write $a_i = \sum a_{ji} y_j$ with $a_{ji} \in \mathfrak{n}$ (necessarily).

Let $\varphi: S^m \to S^n$ be given by the matrix (a_{ji}), and $g: S^n \to S$ and $h: S^m \to S$ be the linear forms defined by y and a respectively. Then $h = g \circ \varphi$, and

$$\bigwedge \varphi: K_\bullet(a) \to K_\bullet(y)$$

is a complex homomorphism; see 1.6.8. By 1.6.7, $K_\bullet(y) \otimes R$ is just $K_\bullet(x)$, and $K_\bullet(a) \otimes R$ has zero differential; forgetting it, we write $\bigwedge R^m$ for $K_\bullet(a) \otimes R$. So we have a complex homomorphism

$$\bigwedge \varphi \otimes R: \bigwedge R^m \to K_\bullet(x).$$

Since $\bigwedge R^m$ has zero differential, this yields a map $\bigwedge R^m \to H_\bullet(x)$, and finally, as $\mathfrak{m} H_\bullet(x) = 0$, a map

$$\lambda: \bigwedge k^m \to H_\bullet(x).$$

As λ is induced by $\bigwedge \varphi$, it is a homomorphism of graded k-algebras.

Sometimes it will be necessary to use the canonical bases $f_1, \ldots, f_n$ of S^n, $e_1, \ldots, e_m$ of S^m, and the elements $u_i = \varphi(e_i) \in S^n$. By the choice of φ, $d_a(e_i) = a_i = d_y(u_i)$. Finally, $^-$ denotes residue classes mod I.

Theorem 2.3.2. *With the notation just introduced,*
(a) $\lambda_1 : k^m \to H_1(x)$ *is an isomorphism of k-vector spaces,*
(b) $\mu(I) = \dim_k H_1(x)$,
(c) $\beta_2(k) = \binom{\operatorname{emb\,dim} R}{2} + \dim_k H_1(x)$.

(Here $\beta_2(k)$ is the second Betti number of k as an R-module; see Section 1.3.)

PROOF. We constructed λ such that it maps the canonical basis of k^m to the homology classes of $u_1, \ldots, u_m$. A minimal free resolution of k starts as

$$R^n \xrightarrow{x} R \longrightarrow 0,$$

and therefore $\beta_2(k) = \mu(Z_1(x))$. So it is enough for (a), (b), and (c) to prove that

$$d_x(\bar{f}_p \wedge \bar{f}_q), \ 1 \le p < q \le n, \quad \text{and} \quad \bar{u}_i, \ i = 1, \ldots, m,$$

form a minimal system of generators of $Z_1(x)$.
Suppose that $\bar{b} \in Z_1(x)$, $b \in S^n$. Then $d_y(b) \in I$,

$$d_y(b) = c_1 a_1 + \cdots + c_m a_m = c_1 d_y(u_1) + \cdots + c_m d_y(u_m).$$

Since $K_{\boldsymbol{\cdot}}(y)$ is acyclic, $b - \sum c_i u_i$ is a linear combination of the elements $d_y(f_p \wedge f_q)$. That is, the elements considered generate $Z_1(x)$.

Now assume that $\sum \bar{\alpha}_{pq} d_x(\bar{f}_p \wedge \bar{f}_q) + \sum \bar{\beta}_i \bar{u}_i = 0$. We have to show that all the coefficients $\bar{\alpha}_{pq}$, $\bar{\beta}_i$ are in $\mathfrak{m}$. Lifting the equation to S^n gives

$$\sum \alpha_{pq} d_y(f_p \wedge f_q) + \sum \beta_i u_i \in I S^n,$$

and applying d_y yields $\sum \beta_i a_i \in \mathfrak{n} I$. So $\beta_i \in \mathfrak{n}$ by the choice of $\boldsymbol{a}$. As $I \subset \mathfrak{n}^2$ one obtains $\sum \alpha_{pq} d_y(f_p \wedge f_q) \in \mathfrak{n}^2 S^n$. Looking at the components of the elements $d_y(f_p \wedge f_q) \in S^n$ and since y is a minimal system of generators of $\mathfrak{n}$, one sees that $\alpha_{pq} \in \mathfrak{n}$ for all p, q. $\square$

The Koszul complexes with respect to different minimal systems of generators of $\mathfrak{m}$ are isomorphic R-algebras; see the discussion before 1.6.21. In particular $H_{\boldsymbol{\cdot}}(x)$ is essentially independent of x; this justifies the notation

$$H_{\boldsymbol{\cdot}}(R) = H_{\boldsymbol{\cdot}}(x),$$

and we call $H_{\boldsymbol{\cdot}}(R)$ the *Koszul algebra of R*. The number

$$\varepsilon_1(R) = \dim_k H_1(R)$$

is the *first deviation of R*. It follows immediately from 2.2.5 and 1.6.19 that R is regular if and only if $\varepsilon_1(R) = 0$. So $\varepsilon_1(R)$ may be considered a measure of how far R deviates from regularity.

The following theorem contains the desired intrinsic characterization of complete intersections.

Theorem 2.3.3. *Let* $(R, \mathfrak{m}, k)$ *be a Noetherian local ring.*
(a) *One has* $\varepsilon_1(R) = \varepsilon_1(\hat{R})$.
(b) *The following are equivalent:*
   (i) R *is a complete intersection;*
   (ii) $\varepsilon_1(R) = \operatorname{emb dim} R - \dim R$;
   (iii) $\beta_2(k) = \binom{\beta_1(k)}{2} + \beta_1(k) - \dim R$.
(c) *Suppose that* $R = S/I$ *with* S *regular and local. Then* R *is a complete intersection if and only if* I *is generated by an* S-*sequence.*

PROOF. (a) Choose a minimal system x of generators of $\mathfrak{m}$. We write $\hat{x}$ for x considered as a sequence in $\hat{R}$. Then $H_{\bullet}(\hat{R}) \cong H_{\bullet}(\hat{x}) \cong H_{\bullet}(x) \otimes \hat{R} \cong H_{\bullet}(R) \otimes \hat{R}$ by 1.6.7. Since $H_{\bullet}(R)$ has finite length, one has $H_{\bullet}(R) \otimes \hat{R} \cong H_{\bullet}(R)$.

(b) Because of (a) and the definition of complete intersection we may assume that R is complete and has a minimal presentation $R = S/I$. If R is a complete intersection, then there is such a presentation with I generated by an S-sequence, hence $\varepsilon_1(R) = \mu(I) = \dim S - \dim R$. Conversely, if $\varepsilon_1(R) = \operatorname{emb dim} R - \dim R$, then $\mu(I) = \dim S - \dim R$ in an arbitrary minimal presentation, and so I is generated by an S-sequence; see 1.6.19.

The equivalence of (ii) and (iii) follows immediately from 2.3.2.

(c) is proved along the same lines as (b). □

Permanence properties of complete intersections. As we did for the Cohen–Macaulay property and regularity we want to discuss how complete intersections behave under certain standard ring extensions.

Theorem 2.3.4. *Let* $(R, \mathfrak{m}, k)$ *be a Noetherian local ring.*
(a) *Suppose* x *is an* R-*sequence. Then*

$$\varepsilon_1(R/(x)) - (\operatorname{emb dim} R/(x) - \dim R/(x)) = \varepsilon_1(R) - (\operatorname{emb dim} R - \dim R);$$

in particular R *is a complete intersection if and only if* $R/(x)$ *is a complete intersection.*
(b) *Suppose* R *is a residue class ring of a regular local ring. Then if* R *is a complete intersection, so is* $R_\mathfrak{p}$ *for every* $\mathfrak{p} \in \operatorname{Spec} R$.

PROOF. Using induction we only need to prove (a) in the case in which $x = x \in R$. Suppose first that $x \notin \mathfrak{m}^2$. Then $\operatorname{emb dim} R/(x) = \operatorname{emb dim} R - 1$ and $\dim R/(x) = \dim R - 1$; furthermore $H_1(R) \cong H_1(R/(x))$ as k-vector spaces by 1.6.13(b). So $\varepsilon_1(R) = \varepsilon_1(R/(x))$. Now suppose that $x \in \mathfrak{m}^2$. Then $\operatorname{emb dim} R/(x) = \operatorname{emb dim} R$ and $\dim R/(x) = \dim R - 1$; moreover we have an exact sequence

$$0 \longrightarrow H_1(R) \longrightarrow H_1(R/(x)) \longrightarrow H_0(R) \cong k \longrightarrow 0$$

as in the proof of 1.6.16. Thus $\varepsilon_1(R/(x)) = \varepsilon_1(R) + 1$.

The proof of (b) is very easy; it uses 2.3.3 and basic properties of regular sequences. □

Remark 2.3.5. Having studied an 'abstract' characterization of complete intersections, the reader may expect an 'abstract' version of 2.3.4(b) without any restrictions. In fact, such an assertion holds for arbitrary complete intersections as was proved by Avramov [22]. Actually Avramov proved a stronger result, namely the analogue of 2.1.7: suppose $(R, \mathfrak{m}) \to (S, \mathfrak{n})$ is a flat homomorphism of Noetherian local rings; then S is a complete intersection if and only if R and $S/\mathfrak{m}S$ are complete intersections. It is not difficult to deduce the localization property from the theorem on flat extensions: there is (by faithful flatness) a prime ideal $\mathfrak{q} \subset \hat{R}$ such that $\mathfrak{p} = \mathfrak{q} \cap R$; the extension $R_{\mathfrak{p}} \to \hat{R}_{\mathfrak{q}}$ is local and flat, and $\hat{R}_{\mathfrak{q}}$ is a complete intersection by 2.3.4(b).

In [23], Avramov gave quantitatively precise results concerning flat extensions (and localizations): let $\delta(R) = \varepsilon_1(R) - (\mathrm{emb\,dim}\,R - \dim R)$ be the *complete intersection defect of* R; then, in the situation of a flat extension, $\delta(S) = \delta(R) + \delta(S/\mathfrak{m}S)$. (Limitation of space prevents us including a proof.) Using Avramov's theorem one can also remove the undesirable restrictions in 2.3.6 and 2.3.7 below.

In the following we say that a Noetherian ring is a *locally complete intersection* if all its localizations are complete intersections.

Theorem 2.3.6. *Let R be a Noetherian ring which is a residue class ring of a regular ring S. Then R is a locally complete intersection if and only if $R[X_1, \dots, X_n]$ is a locally complete intersection. The same holds for R and $R[\![X_1, \dots, X_n]\!]$.*

The proof follows the pattern of that of 2.1.9; one notes that $S[X_1, \dots, X_n]$ and $S[\![X_1, \dots, X_n]\!]$ are regular rings by 2.2.13, and replaces 2.1.7 by Exercise 2.3.20.

As with the Cohen–Macaulay property, the argument outlined really proves the stronger local version of 2.3.6: $R_{\mathfrak{p}}$ is a complete intersection if and only if $R[X_1, \dots, X_n]_{\mathfrak{q}}$ is a complete intersection for $\mathfrak{p} \in \mathrm{Spec}\,R$, $\mathfrak{q} \in \mathrm{Spec}\,R[X_1, \dots, X_n]$ with $R \cap \mathfrak{q} = \mathfrak{p}$. A similar remark applies to the following theorem and its proof.

Theorem 2.3.7. *Let k be a field, R a Noetherian k-algebra, and K an extension field of k. Suppose that R is a (ring of fractions of a) finitely generated k-algebra or K is finitely generated as an extension field. Moreover, suppose that R is a residue class ring of a regular ring. Then R is a locally complete intersection if and only if $R \otimes_k K$ is a locally complete intersection.*

PROOF. We saw in 2.1.10 that $R \otimes_k K$ is a Noetherian ring. Given a prime ideal $\mathfrak{q}$ in $R \otimes_k K$, we set $\mathfrak{p} = R \cap \mathfrak{q}$; conversely, by faithful flatness, for every $\mathfrak{p} \in \mathrm{Spec}\,R$ there exists $\mathfrak{q} \in \mathrm{Spec}\,R \otimes K$ such that $\mathfrak{p} = R \cap \mathfrak{q}$. Furthermore the extension $R \to (R \otimes K)_{\mathfrak{q}}$ factors through $R_{\mathfrak{p}} \otimes K$; so we may replace R by R_p. By hypothesis, $R = S/I$ with a regular local k-algebra S.

Let R be a complete intersection. Then I is generated by a regular sequence $g_1, \ldots, g_r$. Because of faithful flatness $g_1, \ldots, g_r$ is also a regular sequence in $S \otimes K$. So it suffices that $S \otimes K$ is a locally complete intersection, and this is immediate from 2.1.11 (in conjunction with 2.2.13).

As to the converse, we only do the more difficult case in which K is a finitely generated field extension. By 2.1.11 again, one has

$$K \cong (k[X_1, \ldots, X_n])_T / (h_1, \ldots, h_m)$$

where $h_1, \ldots, h_m$ is a regular sequence in $(k[X_1, \ldots, X_n])_T$ and T is a multiplicatively closed set. Therefore $h_1, \ldots, h_m$ is a regular sequence in the faithfully flat extension $R \otimes (k[X_1, \ldots, X_n])_T \cong (R[X_1, \ldots, X_n])_{T'}$; here T' is the image of the natural map $k[X_1, \ldots, X_n] \to R[X_1, \ldots, X_n]$. Moreover, $(R \otimes K)_q$ has the form $R[X_1, \ldots, X_n]_{\mathfrak{Q}} / (h_1, \ldots, h_m)_{\mathfrak{Q}}$ with $\mathfrak{Q} \in \operatorname{Spec} R[X_1, \ldots, X_n]$ such that $\mathfrak{Q} \cap R = p$ and $T' \cap \mathfrak{Q} = \emptyset$. By 2.3.4, $R[X_1, \ldots, X_n]_{\mathfrak{Q}}$ is a complete intersection. So we can apply the local version of 2.3.6. $\qquad\square$

The Koszul algebra of a complete intersection. Above we constructed an algebra homomorphism $\lambda : \bigwedge k^m \to H_\bullet(R)$, $m = \varepsilon_1(R)$, starting from a minimal presentation $R = S/I$, and we saw that $\lambda_1 : k^m \to H_1(R)$ is an isomorphism. Such a homomorphism is always present. In fact, $H_\bullet(R)$ is an alternating graded k-algebra; therefore, by the universal property of the exterior algebra, there exists a unique algebra homomorphism $\lambda' : \bigwedge H_1(R) \to H_\bullet(R)$ extending the identity on $H_1(R)$. Moreover, algebra homomorphisms $\lambda, \lambda' : \bigwedge H_1(R) \to H_\bullet(R)$ such that λ_1 and λ'_1 are isomorphisms, only differ by the automorphism $\bigwedge(\lambda'_1 \lambda_1^{-1})$ of $\bigwedge H_1(R)$. So we may replace the 'abstract' homomorphism λ' by the 'concrete' λ whenever we have a minimal presentation.

The situation under consideration can be generalized as follows: S is a ring, I and $\mathfrak{n}$ are ideals generated by sequences $\boldsymbol{a} = a_1, \ldots, a_m$ and $\boldsymbol{y} = y_1, \ldots, y_n$, and we have $I \subset \mathfrak{n}$. Then, as above 2.3.2, there is a homomorphism

$$\lambda : H_\bullet(\boldsymbol{a}, S/\mathfrak{n}) = \bigwedge(S/\mathfrak{n})^m \to H_\bullet(\boldsymbol{y}, S/I).$$

Choose S-free resolutions $F_\bullet$ of S/I and $G_\bullet$ of $S/\mathfrak{n}$. Then there exist complex homomorphisms $K_\bullet(\boldsymbol{a}) \to F_\bullet$ and $K_\bullet(\boldsymbol{y}) \to G_\bullet$. These in turn induce maps

$$\rho : H_\bullet(\boldsymbol{a}, S/\mathfrak{n}) \longrightarrow H_\bullet(F_\bullet \otimes S/\mathfrak{n}) \cong \operatorname{Tor}^S_\bullet(S/I, S/\mathfrak{n}),$$

$$\sigma : H_\bullet(\boldsymbol{y}, S/I) \longrightarrow H_\bullet(S/I \otimes G_\bullet) \cong \operatorname{Tor}^S_\bullet(S/I, S/\mathfrak{n}).$$

Hence there exist two maps from $H.(a, S/\mathfrak{n}) \cong \bigwedge (S/\mathfrak{n})^m$ to $\text{Tor}^S_{\bullet}(S/I, S/\mathfrak{n})$, namely ρ and $\sigma \circ \lambda$. It is crucial that these maps are essentially equal – of course, we must use the proper identification of $H.(F. \otimes S/\mathfrak{n})$ and $H.(S/I \otimes G.)$. To this end one forms the double complex $F. \otimes G.$ and considers S/I and $S/\mathfrak{n}$ as complexes concentrated in degree 0. Then one has complex homomorphisms $K.(a) \to F. \to S/I$ and $K.(y) \to G. \to S/\mathfrak{n}$. Taking tensor products yields a commutative diagram

$$K.(a) \otimes S/\mathfrak{n} \xleftarrow{\alpha} K.(a) \otimes K.(y) \xrightarrow{\beta} S/I \otimes K.(y)$$
$$\downarrow \qquad\qquad\qquad \downarrow \qquad\qquad\qquad \downarrow$$
$$F. \otimes S/\mathfrak{n} \longleftarrow F. \otimes G. \longrightarrow S/I \otimes G.$$

By a fundamental theorem of homological algebra ([318], Theorem 11.21) the bottom row induces an isomorphism

$$H.(F. \otimes S/\mathfrak{n}) \xleftarrow{\cong} H.(F. \otimes G.) \xrightarrow{\cong} H.(S/I \otimes G.).$$

It is this identification we need:

Lemma 2.3.8. *With the notation introduced,* $\rho_s = (-1)^s \sigma_s \circ \lambda_s$.

PROOF. Let $e_1, \ldots, e_m$ be a basis of S^m and choose elements $u_i \in S^n$ with $d_y(u_i) = d_a(e_i)$. One has $\lambda_s(\bar{e}_{i_1} \wedge \cdots \wedge \bar{e}_{i_s}) = \bar{u}_{i_1} \wedge \cdots \wedge \bar{u}_{i_s}$. Thus it is enough to show that $\rho_s(\bar{e}_{i_1} \wedge \cdots \wedge \bar{e}_{i_s}) = (-1)^s \sigma_s(\bar{u}_{i_1} \wedge \cdots \wedge \bar{u}_{i_s})$.

Let $z \in K.(a) \otimes K.(y)$ be a cycle. Then the commutativity of the diagram above implies that z, $\alpha(z)$, and $\beta(z)$ are all mapped to the same homology class. So it suffices to exhibit a cycle z with $\alpha(z) = \bar{e}_{i_1} \wedge \cdots \wedge \bar{e}_{i_s}$ and $\beta(z) = (-1)^s (\bar{u}_{i_1} \wedge \cdots \wedge \bar{u}_{i_s})$.

Simply take $z = (e_{i_1} \otimes 1 - 1 \otimes u_{i_1}) \cdots (e_{i_s} \otimes 1 - 1 \otimes u_{i_s})$. In order to see that it is a cycle, one uses the definition of the differential of $K.(a) \otimes K.(y)$ and the fact that a product of cycles is again a cycle. $\square$

Theorem 2.3.9. *Let S be a ring, and $a = a_1, \ldots, a_m$ and $y = y_1, \ldots, y_n$ be S-sequences such that $I = (a) \subset \mathfrak{n} = (y)$. Then $H.(y, S/I) \cong \text{Tor}^S_{\bullet}(S/I, S/\mathfrak{n})$ is (isomorphic with) the exterior algebra $\bigwedge (S/\mathfrak{n})^m$.*

PROOF. The isomorphism $H.(y, S/I) \cong \text{Tor}^S_{\bullet}(S/I, S/\mathfrak{n})$ results from the fact that $K.(y)$ is a free resolution of $S/\mathfrak{n}$; see 1.6.14. So, with the notation above, σ is an isomorphism. Similarly ρ is an isomorphism; hence λ, being an algebra homomorphism, is an isomorphism of graded algebras. (In order to remove the sign in 2.3.8 one would have to replace λ by $\bigwedge (-\lambda_1)$.) $\square$

Corollary 2.3.10. (a) *With the hypotheses of 2.3.9 suppose that $m = n$. Write $a_i = \sum a_{ji} y_j$, $i = 1, \ldots, n$. Then $I : \mathfrak{n} = I + S\Delta$, $\Delta = \det(a_{ji})$.*
(b) *In particular, suppose that y is a regular system of parameters in a regular local ring S. Then $\text{Soc}(S/I) = \Delta(S/I)$.*

PROOF. (a) That $(I : \mathfrak{n})/I \cong \mathrm{Hom}_S(S/\mathfrak{n}, S/I)$ and $H_n(y, S/I)$ can be identified was shown in the proof of 1.6.16. Now let $f_1, \ldots, f_n$ be a basis of S^n with $d_y(f_i) = y_i$, and set $u_i = \sum_{j=1}^n a_{ji} f_j$. Then if $e_1, \ldots, e_n$ is a basis of S^n with $d_a(e_i) = a_i$, one has $d_y(u_i) = a_i = d_a(e_i)$. The theorem implies that $H_n(y, S/I)$ is generated by $\bar{u}_1 \wedge \cdots \wedge \bar{u}_n = \Delta \bar{f}_1 \wedge \cdots \wedge \bar{f}_n$. So $\bar{u}_1 \wedge \cdots \wedge \bar{u}_n$ is mapped to the residue class of Δ in S/I by the homomorphism which sends $f_1 \wedge \cdots \wedge f_n$ to 1 (and thus gives the identification $(I : \mathfrak{n})/I = H_n(y, S/I)$).

(b) By definition, $\mathrm{Soc}(S/I) = (I : \mathfrak{n})/I$. $\square$

We have completed our preparations for the following beautiful characterization of complete intersections:

Theorem 2.3.11 (Tate, Assmus). *Let $(R, \mathfrak{m}, k)$ be a Noetherian local ring. Then the following are equivalent:*
(a) *R is a complete intersection;*
(b) *$H_\cdot(R)$ is (isomorphic to) the exterior algebra of $H_1(R)$;*
(c) *$H_\cdot(R)$ is generated by $H_1(R)$;*
(d) *$H_2(R) = H_1(R)^2$.*

Here $H_1(R)^2$ is the k-vector space generated by the products $w \wedge z$ with $w, z \in H_1(R)$.

PROOF. It was observed in the proof of 2.3.3 that $H_\cdot(R)$ is invariant under completion. So we may assume that R is complete and has a minimal presentation $R = S/I$.

The implication (a) $\Rightarrow$ (b) is a special case of 2.3.9, and (b) $\Rightarrow$ (c) $\Rightarrow$ (d) is trivial.

For (d) $\Rightarrow$ (a) we note first that the map σ above is an isomorphism. Next, (d) says that $\lambda_2 : K_2(a) \otimes S/\mathfrak{n} \to H_2(y, S/I) \cong \mathrm{Tor}_2^S(R, S/\mathfrak{n})$ is surjective. So ρ_2 is surjective. Choose $F_\cdot$ as a minimal free resolution of S/I. Then we have a commutative diagram

$$
\begin{array}{ccccccc}
\bigwedge^2 S^m & \longrightarrow & S^m & \longrightarrow & S & \longrightarrow & 0 \\
\downarrow{\scriptstyle \gamma} & & \downarrow{\scriptstyle \cong} & & \| & & \\
F_2 & \longrightarrow & F_1 & \longrightarrow & S & \longrightarrow & 0
\end{array}
$$

The map ρ_2 is just $\gamma \otimes k$, and $\gamma \otimes k$ being surjective, γ is surjective itself. It follows immediately that $H_1(a) = 0$, whence a is an S-sequence by 1.6.19. $\square$

Theorem 2.3.3 contains a characterization of complete intersections in terms of the numerical invariants $\dim R$, $\mathrm{emb\,dim}\, R = \beta_1(k)$, and $\beta_2(k)$. It is possible to remove the 'non-homological' Krull dimension, and to give a description of complete intersections using $\beta_1(k)$, $\beta_2(k)$, and $\beta_3(k)$.

In order to construct the first steps in a free resolution of k, we start with the Koszul complex

$$\bigwedge^2 R^n \longrightarrow R^n \longrightarrow R \longrightarrow 0;$$

unless R is regular, $H_1(R)$ is non-zero. So we add a free direct summand R^m with $m = \dim_k H_1(R) = \varepsilon_1(R)$, and send its generators $e_1, \ldots, e_m$ to cycles $u_1, \ldots, u_m$ whose homology classes generate $H_1(R)$:

$$R^m \oplus \bigwedge^2 R^n \xrightarrow{d_2} R^n \xrightarrow{d_1} R \longrightarrow 0.$$

The kernel of d_2 contains the Koszul cycles $d_x(f_i \wedge f_j \wedge f_l)$ as well as the elements $x_i e_p - f_i \wedge u_p$; again, $f_1, \ldots, f_n$ denotes a basis of R^n. In order to 'kill' at least these cycles we form the complex

$$T_. : (R^n \otimes R^m) \oplus \bigwedge^3 R^n \xrightarrow{d_3} R^m \oplus \bigwedge^2 R^n \xrightarrow{d_2} R^n \xrightarrow{d_1} R \longrightarrow 0$$

with

$$d_3(f_i \wedge f_j \wedge f_l) = d_x(f_i \wedge f_j \wedge f_l),$$
$$d_3(f_i \otimes e_p) = x_i e_p - f_i \wedge u_p = d_x(f_i)e_p - f_i \wedge d_2(e_p).$$

Part (a) of the following theorem shows that $\mu(H_2(T_.))$ is an invariant of R. One writes $\varepsilon_2(R) = \mu(H_2(T_.))$ and calls this number the *second deviation of R*.

Theorem 2.3.12. *Let $(R, \mathfrak{m}, k)$ be a Noetherian local ring. Then*

(a) $H_2(T_.) \cong H_2(R)/H_1(R)^2$,

(b) *R is a complete intersection if and only if $\varepsilon_2(R) = 0$,*

(c) $\varepsilon_2(R) = \beta_3(k) - \binom{\beta_1(k)}{3} - \beta_1(k)\left(\beta_2(k) - \binom{\beta_1(k)}{2}\right).$

PROOF. (a) Let $K_.$ be the complex

$$K_. : \bigwedge^3 R^n \longrightarrow \bigwedge^2 R^n \longrightarrow R^n \longrightarrow R \longrightarrow 0,$$

obtained by truncating the Koszul complex; $K_.$ is a subcomplex of $T_.$. The quotient $T_./K_.$ is isomorphic to

$$L_. : R^n \otimes R^m \xrightarrow{d_1 \otimes \mathrm{id}} R^m \longrightarrow 0$$

with $R^n \otimes R^m$ in degree 3. Consider the exact sequence of homology

$$H_3(L_.) \to H_2(K_.) \to H_2(T_.) \to H_2(L_.) \to H_1(K_.) \to H_1(T_.) = 0.$$

The map $H_2(L_{\bullet}) \to H_1(K_{\bullet})$ is an isomorphism since both vector spaces
have dimension m. Hence we have an exact sequence $H_3(L_{\bullet}) \to H_2(K_{\bullet}) \to$
$H_2(T_{\bullet}) \to 0$. The kernel of $R^n \otimes R^m \to R^m$ is obviously generated by the
elements $d_x(f_i \wedge f_j) \otimes e_q$ and $u_p \otimes e_q$. An analysis of the connecting
homomorphism shows that the class of $d_x(f_i \wedge f_j) \otimes e_q$ goes to that of
$d_x(f_i \wedge f_j) \wedge u_q$ which is a boundary in the Koszul complex (see the
formulas above 1.6.4); the class of $u_p \otimes e_q$ goes to that of $u_p \wedge u_q$ whence
the image of $H_3(L_{\bullet}) \to H_2(K_{\bullet})$ is $H_1(K_{\bullet})^2$.

(b) follows immediately from (a) and 2.3.11.

(c) The verification is similar to that of 2.3.2(c), and therefore left to
the reader.   □

Remark 2.3.13. The equation $d_3(f_i \otimes e_p) = d_x(f_i)e_p - f_i \wedge d_2(e_p)$ suggests
that d_3 is a component of an antiderivation of an alternating algebra. In
fact, the choice of d_3 is part of Tate's construction of resolutions with
algebra structures [369].

Suppose that $A_{\bullet}$ is an alternating graded R-algebra equipped with an
antiderivation ∂ of degree -1 such that $\partial^2 = 0$, and consider the homo-
logy $H_{\bullet}(A_{\bullet}) = \operatorname{Ker}\partial / \operatorname{Im}\partial$. Let $\bar{z} \in H_p(A_{\bullet})$ be a non-zero homology
element. Then one may adjoin a variable to 'kill' the cycle z representing
$\bar{z}$:

(i) If p is even, let $B_{\bullet}$ be the exterior algebra in a variable of degree $p + 1$,
i.e. $B_{\bullet} = R \oplus Re$ with R in degree 0 and $Re \cong R$ in degree $p + 1$, the
multiplication being defined by $e^2 = 0$.

(ii) For p odd let $B_{\bullet}$ be the 'divided power algebra' over R in a variable
of degree $p + 1$, i.e. $B_{\bullet} = \bigoplus_{j=0}^{\infty} Re_j$ with $Re_j \cong R$ in degree $j(p + 1)$, the
multiplication being defined by $e_j e_l = ((j + l)!/j!\,l!)e_{j+l}$.

In both cases $A_{\bullet} \otimes B_{\bullet}$ is again an alternating algebra, and there is a unique
antiderivation d on $A_{\bullet} \otimes B_{\bullet}$ such that $d|_{A_{\bullet} \otimes 1} = \partial$, $d(e) = z$ in case (i), and
$d(e_j) = ze_{j-1}$ for all j in case (ii); moreover, one has $d^2 = 0$. It follows
easily that $H_q(A_{\bullet} \otimes B_{\bullet}) \cong H_q(A_{\bullet})$ for $q < p$ and $H_p(A_{\bullet} \otimes B_{\bullet}) \cong H_p(A_{\bullet})/R\bar{z}$.

In order to resolve the residue class field k of a local ring $(R, \mathfrak{m}, k)$
one starts with the R-algebra $T_{\bullet}^{(0)} = R$. Let $\varepsilon_0(R) = \operatorname{emb\,dim}R = \mu(\mathfrak{m})$,
and successively adjoin $\varepsilon_0(R)$ variables of degree 1 'to kill the zero-cycles'.
The resulting algebra $T_{\bullet}^{(1)}$ is the Koszul complex of a minimal system
of generators of $\mathfrak{m}$. Next one adjoins $\varepsilon_1(R)$ variables of degree 2 to kill
$\varepsilon_1(R)$ cycles generating $H_1(T_{\bullet}^{(1)})$. The algebra $T_{\bullet}^{(2)}$ thus constructed has
$H_1(T_{\bullet}^{(2)}) = 0$. It is a theorem of Tate [369] that in the case of a complete
intersection the complex $T_{\bullet}^{(2)}$ is a minimal free resolution of k. However,
if R is not a complete intersection, then one has to adjoin $\varepsilon_2(R)$ variables
of degree 3 etc. A famous theorem of Gulliksen [146] and Schoeller [330]
says that the resolution of k obtained in this way is always minimal.

For a comprehensive study of resolutions with an algebra structure we refer the reader to Gulliksen and Levin [147].

As above, let $(R, \mathfrak{m}, k)$ be a Noetherian local ring. We saw in 2.3.11 that surjectivity of the natural homomorphism $\lambda \colon \bigwedge H_1(R) \to H_{\cdot}(R)$ is already sufficient for R to be a complete intersection. This is also true for injectivity, at least when R contains a field.

Theorem 2.3.14. *Let* $(R, \mathfrak{m}, k)$ *be a Noetherian local ring containing a field. Then*
(a) $H_1(R)^j = 0$ *for* $j > \operatorname{emb dim} R - \dim R$;
(b) *in particular,* R *is a complete intersection if* (*and only if*) *the natural map* $\lambda \colon \bigwedge H_1(R) \to H_{\cdot}(R)$ *is injective.*

PROOF. It is harmless to complete R so that we may assume that R has a minimal presentation $R = S/I$ as above. In order to prove (a) we must anticipate Corollary 9.5.3: it says that, with the notation of 2.3.8, $\rho_j = 0$ for $j > \operatorname{emb dim} R - \dim R$. Since, in the present circumstances, σ_j is an isomorphism, one has $\lambda_j = 0$ for $j > \operatorname{emb dim} R - \dim R$. As $H_1(R)^j = \lambda_j(\bigwedge^j H_1(R))$, one has $H_1(R)^j = 0$. This proves (a), and (b) is an obvious consequence of (a). $\square$

The restriction to local rings containing a field is forced upon us since there does not yet exist a proof of 9.5.3 without this restriction. However, one always has $H_1(R)^j = 0$ for $j > \operatorname{emb dim} R - \dim R + 1$ so that the gap in 2.3.14 is as small as it could be; see 9.5.7.

The reader may have noticed that 2.3.14 is trivial for Cohen–Macaulay rings R: if $j > \operatorname{emb dim} R - \operatorname{depth} R$, then even $H_j(R) = 0$ by 1.6.17. On the other hand, for Cohen–Macaulay R the non-vanishing of $H_1(R)^p$ for $p = \operatorname{emb dim} R - \dim R$ conveys the strongest possible information: R is a complete intersection. More generally, we have the following theorem.

Theorem 2.3.15. *Let* $(R, \mathfrak{m}, k)$ *be a Noetherian local ring. Then* R *is a complete intersection if* (*and only if*) $H_1(R)^p \neq 0$ *for* $p = \operatorname{emb dim} R - \operatorname{depth} R$.

PROOF. By virtue of 2.3.14 the hypothesis $H_1(R)^p \neq 0$ for $p = \operatorname{emb dim} R - \operatorname{depth} R$ forces R to be Cohen–Macaulay if it contains a field. Because of this restriction we give a proof not using 2.3.14.

First we reduce to the case $\operatorname{depth} R = 0$. So suppose that $\operatorname{depth} R > 0$. Then there exists an $x \in \mathfrak{m} \setminus \mathfrak{m}^2$ which is not a zero-divisor, and 1.6.13 furnishes us with an isomorphism $\alpha \colon H_{\cdot}(R) \cong H_{\cdot}(R')$, $R' = R/(x)$. It is not difficult to verify that α is a k-algebra isomorphism; after all, α is induced by $\bigwedge \pi$, π being the composition $R^n \to R^{n-1} \to (R')^{n-1}$ (see the proof of 1.6.12). Furthermore $\operatorname{emb dim} R' - \operatorname{depth} R' = \operatorname{emb dim} R - \operatorname{depth} R$, and R is a complete intersection if and only if this holds for R'.

It remains to show that R is a zero dimensional complete intersection if $H_1(R)^n \neq 0$ for $n = \operatorname{emb dim} R$. The complex $K_{\cdot}(R)$ has length n; so

$B_{n+1}(R) = 0$, and therefore

$$H_1(R)^n = (Z_1(R)^n + B_{n+1}(R))/B_{n+1}(R) = Z_1(R)^n.$$

Consider an exact sequence $R^m \oplus \bigwedge^2 R^n \xrightarrow{d_2} R^n \longrightarrow \mathfrak{m} \longrightarrow 0$ as above. Choose elements $v_1, \ldots, v_n \in \operatorname{Im} d_2 = Z_1(R)$, $v_i = \sum v_{ji} f_j$ where $f_1, \ldots, f_n$ is a basis of R^n. Then $v_1 \wedge \cdots \wedge v_n = \det(v_{ji}) f_1 \wedge \cdots \wedge f_n$, whence $H_1(R)^n \neq 0$ is equivalent to $I_n(d_2) \neq 0$. It remains to apply the next theorem. $\square$

Theorem 2.3.16 (Wiebe). *Let $(R, \mathfrak{m}, k)$ be a Noetherian local ring, and $R^r \xrightarrow{\varphi} R^n \to \mathfrak{m} \to 0$ a presentation of its maximal ideal. If $I_n(\varphi) \neq 0$, then R is a complete intersection of dimension zero (and conversely).*

PROOF. The ideal $I_n(\varphi)$ is the zeroth Fitting ideal of $\mathfrak{m}$, which is an invariant of $\mathfrak{m}$. Therefore it is enough to consider a special presentation. Moreover, we may assume that R is complete. Then $R = S/I$ where $(S, \mathfrak{n})$ is a regular local ring and $I \subset \mathfrak{n}^2$. Let $y = y_1, \ldots, y_n$ be a regular system of parameters of S, and $a = a_1, \ldots, a_m$ a minimal system of generators of I. Write $a_i = \sum a_{ji} y_j$.

The converse of the theorem is part of Corollary 2.3.10; it implies the following claim which is crucial in what follows. *Let $b = b_1, \ldots, b_n$ be a maximal S-sequence, and J' an ideal properly containing $J = (b)$; then $\det(b_{ji}) \in J'$ where the b_{ji} are chosen such that $b_i = \sum b_{ji} y_j$.* In fact, $\det(b_{ji}) S/J$ is the socle of S/J. Since it has dimension 1 over k, it is contained in every non-zero ideal of S/J.

Let $f_1, \ldots, f_n$ be a basis of S^n, and $e_1, \ldots, e_m$ a basis of S^m. Define $\psi : S^m \oplus \bigwedge^2 S^n \to S^n$ by the Koszul map $\bigwedge^2 S^n \to S^n$ with respect to y and $\psi(e_i) = \sum a_{ji} f_j$. We saw in 2.3.2 that $\operatorname{Coker} \psi \otimes S/I \cong \mathfrak{m}$. So the theorem claims that $I_n(\psi) \subset I$ unless S/I is a zero dimensional complete intersection.

Choose an $n \times n$ submatrix U of (a matrix of) ψ. If U involves a column corresponding to one of the elements $f_i \wedge f_j$, then $\det U \in I$, since, on the level of R, we are taking the exterior product of n cycles at least one of which is a boundary: $(\det U) \bar{f}_1 \wedge \cdots \wedge \bar{f}_n \in B_{n+1}(R) = 0$. Therefore it is enough to consider submatrices U of (a_{ji}). For simplicity of notation one may assume U consists of the first n columns.

If $a_1, \ldots, a_n$ is a regular sequence, then I contains $J = (a_1, \ldots, a_n)$ properly since S/I is not a complete intersection. So $\det U \subset I$ by the claim above.

If $a_1, \ldots, a_n$ is not a regular sequence, then $\dim S/J > 0$, and it is certainly enough to show that $\det U \in J$, that is, we may assume that $I = (a_1, \ldots, a_n)$.

We will show that $\det U \in I + \mathfrak{n}^p$ for all $p \in \mathbb{N}$. Then $\det U \in I$ follows from Krull's intersection theorem. Fix $p \in \mathbb{N}$. According to Exercise 2.3.17 one finds elements $a_i'' \in \mathfrak{n}^{p+1}$ such that $a_i' = a_i + a_i''$, $i = 1, \ldots, n$, is a regular sequence. Write $a'_i = \sum a'_{ji} y_j$; then $a_{ji} - a'_{ji} \in \mathfrak{n}^p$ follows from the quasi-regularity of the regular sequence y, in other words, from the fact that the associated graded ring $\mathrm{gr}_\mathfrak{n}(S)$ is a polynomial ring (see 1.1.8). Therefore

$$\det U - \det(a'_{ji}) \in \mathfrak{n}^p.$$

The ideal $I + \mathfrak{n}^p$ properly contains $(a'_1, \dots, a'_n) \subset I + \mathfrak{n}^{p+1}$. Once more the auxiliary claim above is applied, and it yields $\det(a'_{ji}) \in I + \mathfrak{n}^p$. □

Exercises

2.3.17. Let $(R, \mathfrak{m})$ be a Noetherian local ring of depth t, and $a_1, \dots, a_t \in \mathfrak{m}$. Then, given $p \in \mathbb{N}$, show there exist $a'_1, \dots, a'_t \in \mathfrak{m}^p$ such that $a_1 + a'_1, \dots, a_t + a'_t$ is an R-sequence.

2.3.18. Let S be a regular local ring of dimension 4, and $y_1, \dots, y_4$ a regular system of parameters. Let $I = (y_1 y_2, y_3 y_4, y_1 y_3 + y_2 y_4)$ and $R = S/I$.
(a) Construct a minimal free resolution of R.
(b) Prove depth $R = 0$ and dim $R = 2$.
(c) Show that the vector space $H_1(R)^2$ has dimension 3, the maximal value for an ideal generated by 3 elements, but R is not a complete intersection.

2.3.19. Prove all the claims in the second paragraph of 2.3.13.

2.3.20. Let $\varphi : (S_1, \mathfrak{n}_1) \to (S_2, \mathfrak{n}_2)$ be a flat local homomorphism of regular rings, and $I \subset S_1$ an ideal. Verify S_1/I is a complete intersection if and only if S_2/IS_2 is a complete intersection.

2.3.21. Let R be a Noetherian graded ring. Show:
(a) For $\mathfrak{p} \in \mathrm{Spec}\, R$ the localization $R_\mathfrak{p}$ is a complete intersection if and only if $R_{\mathfrak{p}^*}$ is.
(b) The following are equivalent:
    (i) R is locally a complete intersection;
    (ii) $R_\mathfrak{p}$ is a complete intersection for all graded prime ideals $\mathfrak{p}$;
    (iii) $R_{(\mathfrak{p})}$ is locally a complete intersection for all graded prime ideals $\mathfrak{p}$.
(c) Suppose in addition that $(R, \mathfrak{m})$ is *local. Then R is locally a complete intersection if and only if $R_\mathfrak{m}$ is a complete intersection.
Hint: 1.6.33.

2.3.22. Extend 2.3.2 and 2.3.12 to the following theorem which Serre [332] used to prove 2.2.7(c) $\Rightarrow$ (a): let $(R, \mathfrak{m}, k)$ be a Noetherian local ring, and x a minimal system of generators of $\mathfrak{m}$; then the natural map $K_\bullet(x, k) \to \mathrm{Tor}_\bullet^R(k, k)$ (see 1.6.9) is injective.

Notes

The origins of the theory of Cohen–Macaulay rings are the unmixedness theorems of Macaulay [263] and Cohen [71] and the notion of perfect ideals, which also goes back to Macaulay and was clarified by

Gröbner [141]. The present shape of the theory was formed by Auslander and Buchsbaum [18], Nagata [283], and Rees [303]. It seems that Cohen–Macaulay modules made their first appearance in Auslander and Buchsbaum [19].

The characterization 2.1.27 of graded Cohen–Macaulay rings is essentially due to Hochster and Ratliff [200] and Matijevic and Roberts [268].

By analogy to the desingularization, one can try to 'Macaulayfy' a Noetherian scheme. Special results in this direction were obtained by Brodmann [51] and Faltings [100]. Recently Kawasaki [234] has proved a general theorem on the existence of 'Macaulayfications'.

Of all the notions generalizing Cohen–Macaulay rings and modules, the concept of Buchsbaum ring or module is the most important; see Stückrad and Vogel [365] and Schenzel [329].

The 'classical' theory of regular local rings, to be found in Zariski and Samuel [397], Vol. II, was developed by Krull [242], Chevalley [68], Cohen [71], and Zariski [396]. It depends in an essential way on power series methods, and is therefore mainly restricted to local rings containing a field. The problems it could not solve were (i) the regularity of a localization of a regular local ring R (even if R contains a field), and (ii) the factoriality of such rings (because of the Cohen structure theorem this is easy if R contains a field).

The breakthrough was the theorem 2.2.7 of Auslander and Buchsbaum [17], [18] and Serre [332] which not only solved the localization problem: 'this resounding triumph of the new homological method marked a turning point of the subject of commutative Noetherian rings' (Kaplansky [230], p. 159). Theorem 2.2.8 was independently given by Ferrand [105] and Vasconcelos [379]; it generalizes Kaplansky's proof of 2.2.14(c) ⇒ (a) (see [270], §19).

The problem of factoriality was solved by Auslander and Buchsbaum [20], using results of Zariski and Nagata who reduced the theorem to the case of Krull dimension 3. See Nagata [284], p. 217 for a minute history. The proof we have reproduced is due to Kaplansky (except for the application of the Hilbert–Burch theorem). That regular local rings are factorial can be expressed by saying that every ideal I has a greatest common divisor: there is a regular element a and an ideal J of grade ≥ 2 such that $I = aJ$. MacRae [265] proved this fact for every ideal with a finite free resolution. Another (related) generalization is that every module with a finite free resolution over a normal domain has divisor class zero; see [47], Ch. VII, §4. The most concrete and computationally effective result is the factorization theorem of Buchsbaum and Eisenbud [64].

The notion of complete intersection is classical in algebraic geometry. An abstract definition in terms of local algebra was given by Scheja

[322], together with 2.3.3. Our definition is that of Grothendieck [142]. Avramov's contributions [22], [23] have been described in 2.3.5; they ultimately justified the abstract notion of complete intersection.

The program of Tate's seminal paper [369] has been outlined in Remark 2.3.13. Assmus [14] used Tate's method to give the description 2.3.11 of complete intersections in terms of their Koszul algebras. There are several papers devoted to the characterization of complete intersections by the vanishing of a deviation ε_i (which we defined only for $i = 1, 2$); the question was finally settled by Halperin [148] who showed that $\varepsilon_i > 0$ for all i if R is not a complete intersection. Wiebe's theorem 2.3.16 appeared in [395]; see Kunz [244], Hilfssatz 1, for a related result.

A driving force in this area of research was the problem (posed by Serre [334]) of whether the Poincaré series $\sum \beta_i(k)t^i$ of a Noetherian local ring $(R, \mathfrak{m}, k)$ is a rational function of t. After several special cases had been solved positively, the general question was answered negatively by Anick [9].

In several theorems we studied the behaviour of ring-theoretic properties under flat extensions $R \rightarrow S$. Avramov, Foxby, and Halperin [31] investigated the more general situation in which S is supposed of finite flat dimension over R. As we have seen, another 'homologically nice' type of extension is that of passing to a residue class modulo a regular sequence. Avramov and Foxby have essentially completed a program which aims at the unification of these types of extensions by introducing a suitable notion of fibre; see [27], [28], [29], [30].

3 The canonical module. Gorenstein rings

The concept of a canonical module is of fundamental importance in the study of Cohen–Macaulay local rings. The purpose of this chapter is to introduce the canonical module and derive its basic properties. By definition it is a maximal Cohen–Macaulay module of type 1 and of finite injective dimension.

In the first two sections we investigate the injective dimension of a module, and prove Matlis duality which plays a central role in Grothendieck's local duality theorem. Actually the canonical module has its origin in this theory. Here the canonical module is introduced independently of local cohomology which is an important notion in itself and will be treated later in this chapter.

A ring which is its own canonical module is called a Gorenstein ring. Next to regular rings and complete intersections, Gorenstein rings are in many ways the 'nicest' rings. Distinguished by the fact that they are of finite injective dimension, they have various symmetry properties, as reflected in their free resolution, their Koszul homology, and their Hilbert function. The last aspect will be discussed in the next chapter.

Gorenstein rings of embedding dimension at most two are complete intersections. The first non-trivial Gorenstein rings occur in embedding dimension three, and they are classified by the Buchsbaum–Eisenbud structure theorem.

In the final section the canonical module of a graded ring is introduced.

3.1 Finite modules of finite injective dimension

In this section we study injective resolutions of finite modules. We shall see that the injective dimension of a finite module M over a Noetherian local ring R either is infinite or equals the depth of R, and is bounded below by the dimension of M. Thus, quite contrary to the behaviour of projective dimension, the injective dimension, if it is finite, does not depend on the module. We introduce Gorenstein rings and show that Gorenstein rings are Cohen–Macaulay rings.

Definition 3.1.1. Let R be a ring. An R-module I is *injective* if the functor $\operatorname{Hom}_R(\_, I)$ is exact.

Notice that $\text{Hom}_R(\_, I)$ is always left exact. Thus the R-module I is injective if and only if $\text{Hom}_R(\_, I)$ is right exact as well.

We now list some useful characterizations of injective modules.

Proposition 3.1.2. *Let R be a ring and I an R-module. The following conditions are equivalent:*

(a) I *is injective;*

(b) *given a monomorphism* $\varphi : N \to M$ *of R-modules, and a homomorphism* $\alpha : N \to I$, *there exists a homomorphism* $\beta : M \to I$ *such that* $\alpha = \beta \circ \varphi$;

(c) *given R-modules $N \subset M$, and a homomorphism* $\alpha : N \to I$, *there exists a homomorphism* $\beta : M \to I$ *such that* $\beta|_N = \alpha$; *in other words,* $\alpha : N \to I$ *can be extended to a homomorphism* $\beta : M \to I$;

(d) *for all ideals $J \subset R$, every homomorphism $J \to I$ can be extended to R, that is,* $\text{Ext}_R^1(R/J, I) = 0$;

(e) *let M be an R-module with $I \subset M$; then I is a direct summand of M;*

(f) $\text{Ext}_R^1(M, I) = 0$ *for all R-modules M;*

(g) $\text{Ext}_R^i(M, I) = 0$ *for all R-modules M and all $i > 0$.*

PROOF. The $\text{Ext}_R^i(\_, I)$ are the right derived functors of $\text{Hom}_R(\_, I)$. The equivalence of (a), (f) and (g) follows therefore from the general properties of right derived functors. For details we refer to [318], Section 6.

(a) $\Rightarrow$ (b): The monomorphism $\varphi : N \to M$ induces the homomorphism

$$\text{Hom}(\varphi, I) : \text{Hom}_R(M, I) \longrightarrow \text{Hom}_R(N, I),$$

where $\text{Hom}(\varphi, I)(\beta) = \beta \circ \varphi$ for all $\beta \in \text{Hom}_R(M, I)$. By assumption, $\text{Hom}(\varphi, I)$ is an epimorphism, and so $\alpha \in \text{Hom}_R(N, I)$ is of the form $\beta \circ \varphi$ for some $\beta \in \text{Hom}_R(M, I)$. Similarly one proves (b) $\Rightarrow$ (a).

The implications (b) $\Longleftrightarrow$ (c) and (c) $\Rightarrow$ (d) are clear.

(d) $\Rightarrow$ (c): We consider the set of all pairs (U, φ), where U is a submodule of M with $N \subset U$, and where φ extends α. We order this set partially: $(U_1, \varphi_1) \leq (U_2, \varphi_2)$ if and only if $U_1 \subset U_2$ and $\varphi_1 = \varphi_2|_{U_1}$. By Zorn's lemma there exists a maximal element (U', φ') in this set. Suppose $U' \neq M$; then we may choose $x \in M \setminus U'$. Set $W = U' + Rx$; then $W/U' \cong R/J$ for some ideal J in R. Applying the functor $\text{Hom}_R(\_, I)$ to the exact sequence

$$0 \longrightarrow U' \longrightarrow W \longrightarrow R/J \longrightarrow 0,$$

we obtain the exact sequence

$$\text{Hom}_R(W, I) \longrightarrow \text{Hom}_R(U', I) \longrightarrow \text{Ext}_R^1(R/J, I).$$

Since by assumption $\text{Ext}_R^1(R/J, I) = 0$, it follows from the exact sequence that any homomorphism from U' to I can be extended to a homomorphism $W \to I$, contradicting the maximality of (U', φ'). Thus we have shown that $U' = M$.

(c) $\Rightarrow$ (e): There exists a homomorphism $\beta : M \to I$ with $\beta|_I = \mathrm{id}_I$. Therefore, $M = I \oplus \mathrm{Ker}\,\beta$.

(e) $\Rightarrow$ (b): Given a monomorphism $\varphi : N \to M$ and a homomorphism $\alpha : N \to I$, we want to find $\beta : M \to I$ such that $\alpha = \beta \circ \varphi$. In order to do this, we construct a commutative diagram

$$
\begin{array}{ccc}
N & \xrightarrow{\ \varphi\ } & M \\
\alpha \downarrow & & \downarrow \gamma \\
I & \xrightarrow{\ \psi\ } & W
\end{array}
$$

where ψ is injective. In fact, we may choose $W = (M \oplus I)/C$ with $C = \{(\varphi(x), -\alpha(x)) : x \in N\}$; γ and ψ are the natural homomorphisms arising from this situation. (This diagram is called the *pushout* of α and φ.)

Since ψ is injective by construction, it is split injective by our assumption (e). This means that there exists a homomorphism $\sigma : W \to I$ with $\sigma \circ \psi = \mathrm{id}_I$. The homomorphism $\beta : M \to I$, $\beta = \sigma \circ \gamma$, is the desired extension of α. $\qquad\square$

Corollary 3.1.3. *Let R be a Noetherian ring.*
(a) *If I is an injective R-module and S is a multiplicatively closed set of R, then I_S is an injective R_S-module.*
(b) *If $(I_\lambda)_{\lambda \in \Lambda}$ is a family of injective R-modules, then the direct sum*

$$
I = \bigoplus_{\lambda \in \Lambda} I_\lambda
$$

is an injective R-module.

PROOF. (a) Let J be an ideal of R. Since R is Noetherian one has

$$
\mathrm{Ext}^1_{R_S}(R_S/JR_S, I_S) \cong \mathrm{Ext}^1_R(R/J, I)_S = 0.
$$

Since every ideal of R_S is extended from R, 3.1.2 yields that I_S is an injective R_S-module.

(b) By 3.1.2 it is enough to show that for an ideal J of R, any homomorphism $\varphi : J \to \bigoplus_{\lambda \in \Lambda} I_\lambda$ extends to R. Since J is finitely generated there exists a finite subset $\{\lambda_1, \ldots, \lambda_n\}$ of Λ such that $\mathrm{Im}\,\varphi \subset \bigoplus_{i=1}^n I_{\lambda_i}$. We denote by φ_j the j-th component of φ. Since I_{λ_i} is injective we can extend $\varphi_{\lambda_i} : J \to I_{\lambda_i}$ to a homomorphism $\psi_i : R \to I_{\lambda_i}$. It is clear that $\psi : R \to \bigoplus_{\lambda \in \Lambda} I_\lambda$ with $\psi(a) = \sum_{i=1}^n \psi_i(a)$, $a \in R$, extends φ to all of R. $\qquad\square$

Remark 3.1.4. It is a simple exercise to see that for an arbitrary ring R any direct product of injective modules is injective. It is however essential to require that R is Noetherian (as we have done in 3.1.3) to obtain a similar result for direct sums. In fact, this property characterizes Noetherian rings; see [318], Theorem 4.10.

An R-module M is *divisible* if for every regular element $r \in R$, and every element $m \in M$, there exists an element $m' \in M$ such that $m = rm'$. Condition 3.1.2(d) has the following consequence.

Corollary 3.1.5. *Let R be a ring and I an R-module.*
(a) *If I is injective, then I is divisible.*
(b) *If R is a principal domain and I is divisible, then I is injective.*

PROOF. The property that I is divisible is equivalent to the property that every homomorphism $\alpha : (r) \to I$, r regular, can be extended to R. Therefore (a) and (b) follow from 3.1.2(d). □

For later applications we note the following result about change of rings.

Lemma 3.1.6. *Let $\varphi : R \to S$ be a ring homomorphism, and let I be an injective R-module. Then $\mathrm{Hom}_R(S, I)$ (equipped with the natural S-module structure) is an injective S-module.*

PROOF. Let M be an S-module. There is a natural isomorphism

$$\mathrm{Hom}_S(M, \mathrm{Hom}_R(S, I)) \cong \mathrm{Hom}_R(M, I)$$

of S-modules. Indeed, to $\psi \in \mathrm{Hom}_S(M, \mathrm{Hom}_R(S, I))$ one assigns $\psi' \in \mathrm{Hom}_R(M, I)$ where $\psi'(x) = \psi(x)(1)$ for all $x \in M$. Thus the exactness of the functor $\mathrm{Hom}_R(\_, I)$ on the category of S-modules (considered as R-modules via φ) implies the exactness of the functor $\mathrm{Hom}_S(\_, \mathrm{Hom}_R(S, I))$. This means that $\mathrm{Hom}_R(S, I)$ is an injective S-module. □

Definition 3.1.7. Let R be a ring and M an R-module. A complex

$$I^\bullet : 0 \longrightarrow I^0 \longrightarrow I^1 \longrightarrow I^2 \longrightarrow \cdots$$

with injective modules I^i is an *injective resolution* of M if $H^0(I^\bullet) \cong M$ and $H^i(I^\bullet) = 0$ for $i > 0$.

While it is obvious that every module has a projective resolution, it is less obvious that it has an injective resolution. It is however clear that an injective resolution can be constructed by resorting to the following result.

Theorem 3.1.8. *Let R be a ring. Every R-module can be embedded into an injective R-module.*

PROOF. The $\mathbb{Z}$-module $\mathbb{Q}$ is divisible, and hence injective. Therefore any free $\mathbb{Z}$-module F can be embedded into an injective $\mathbb{Z}$-module I. We just take sufficiently many copies of $\mathbb{Q}$. If G is an arbitrary $\mathbb{Z}$-module, then $G \cong F/U$, and we can embed G into I/U. It is immediate that I/U is again divisible, and hence injective. Thus the theorem is proved for $\mathbb{Z}$-modules.

Now let R be an arbitrary ring, and let M be an R-module. The map $\alpha: M \to \text{Hom}_{\mathbb{Z}}(R, M)$ with $\alpha(x)(a) = ax$ for all $x \in M$ and all $a \in R$ is an R-module monomorphism. By our considerations above, the R-module M can be embedded as a $\mathbb{Z}$-module into an injective $\mathbb{Z}$-module I. This inclusion induces a monomorphism

$$\beta: \text{Hom}_{\mathbb{Z}}(R, M) \longrightarrow \text{Hom}_{\mathbb{Z}}(R, I).$$

By 3.1.6, the R-module $J = \text{Hom}_{\mathbb{Z}}(R, I)$ is injective, and thus $\beta \circ \alpha: M \to J$ is the desired embedding. $\square$

Injective dimension. Let R be a ring and M an R-module. The *injective dimension* of M (denoted $\text{inj dim } M$ or $\text{inj dim}_R M$) is the smallest integer n for which there exists an injective resolution $I^\bullet$ of M with $I^m = 0$ for $m > n$. If there is no such n, the injective dimension of M is infinite.

The following observation is an immediate consequence of 3.1.3 and the exactness of localization.

Proposition 3.1.9. *Let R be a Noetherian ring, M an R-module and S a multiplicatively closed set. Then $\text{inj dim}_{R_S} M_S \leq \text{inj dim}_R M$.*

In the next proposition we characterize the injective dimension of a module homologically.

Proposition 3.1.10. *Let R be a ring and M an R-module. The following conditions are equivalent:*
(a) $\text{inj dim } M \leq n$;
(b) $\text{Ext}_R^{n+1}(N, M) = 0$ *for all R-modules N;*
(c) $\text{Ext}_R^{n+1}(R/J, M) = 0$ *for all ideals J of R.*

PROOF. (a) $\Rightarrow$ (b) follows from the fact that $\text{Ext}_R^{n+1}(N, M)$ can be computed from an injective resolution of M.

(b) $\Rightarrow$ (c) is trivial.

(c) $\Rightarrow$ (a): Let

$$0 \longrightarrow M \longrightarrow I^0 \longrightarrow I^1 \longrightarrow \cdots \longrightarrow I^{n-1} \longrightarrow C \longrightarrow 0$$

be an exact sequence, where the modules I^j are injective. From the fact that $\text{Ext}_R^i(R/J, I) = 0$ for $i > 0$ if I is an injective R-module, the above exact sequence yields the isomorphism

$$\text{Ext}_R^1(R/J, C) \cong \text{Ext}_R^{n+1}(R/J, M),$$

and so $\text{Ext}_R^1(R/J, C) = 0$ for all ideals J of R. This is condition (d) of 3.1.2, and so C is injective. $\square$

Proposition 3.1.10 can be sharpened if R is Noetherian. We first observe:

Lemma 3.1.11. *Let R be a Noetherian ring, M an R-module, N a finite R-module and $n > 0$ an integer. Suppose that $\operatorname{Ext}_R^n(R/\mathfrak{p}, M) = 0$ for all $\mathfrak{p} \in \operatorname{Supp} N$. Then $\operatorname{Ext}_R^n(N, M) = 0$.*

PROOF. N has a finite filtration whose factors are isomorphic to $R/\mathfrak{p}$ for certain $\mathfrak{p} \in \operatorname{Supp} N$. Hence the lemma follows from the 'additivity' of the vanishing of $\operatorname{Ext}_R^n(\_, M)$. □

Corollary 3.1.12. *Let R be Noetherian and M an R-module. The following conditions are equivalent:*
(a) $\operatorname{inj\,dim} M \leq n$;
(b) $\operatorname{Ext}_R^{n+1}(R/\mathfrak{p}, M) = 0$ for all $\mathfrak{p} \in \operatorname{Spec} R$.

Lemma 3.1.11 has another remarkable consequence.

Proposition 3.1.13. *Let $(R, \mathfrak{m}, k)$ be a Noetherian local ring, $\mathfrak{p}$ a prime ideal different from $\mathfrak{m}$, and M a finite R-module. If $\operatorname{Ext}_R^{n+1}(R/\mathfrak{q}, M) = 0$ for all prime ideals $\mathfrak{q} \in V(\mathfrak{p})$, $\mathfrak{q} \neq \mathfrak{p}$, then $\operatorname{Ext}_R^n(R/\mathfrak{p}, M) = 0$.*

PROOF. We choose an element $x \in \mathfrak{m} \setminus \mathfrak{p}$. The element is $R/\mathfrak{p}$-regular, and therefore we get the exact sequence

$$0 \longrightarrow R/\mathfrak{p} \overset{x}{\longrightarrow} R/\mathfrak{p} \longrightarrow R/(x, \mathfrak{p}) \longrightarrow 0$$

which induces the exact sequence

$$\operatorname{Ext}_R^n(R/\mathfrak{p}, M) \overset{x}{\longrightarrow} \operatorname{Ext}_R^n(R/\mathfrak{p}, M) \longrightarrow \operatorname{Ext}_R^{n+1}(R/(x, \mathfrak{p}), M).$$

Since $V(x, \mathfrak{p}) \subset \{\mathfrak{q} \in V(\mathfrak{p}): \mathfrak{q} \neq \mathfrak{p}\}$, Lemma 3.1.11 and our assumption imply

$$\operatorname{Ext}_R^{n+1}(R/(x, \mathfrak{p}), M) = 0,$$

so that multiplication by x on the finite R-module $\operatorname{Ext}_R^n(R/\mathfrak{p}, M)$ is a surjective homomorphism. The desired result follows from Nakayama's lemma. □

It is now easy to derive the following useful formula for the injective dimension of a finite module.

Proposition 3.1.14. *Let $(R, \mathfrak{m}, k)$ be a Noetherian local ring, and M a finite R-module. Then*

$$\operatorname{inj\,dim} M = \sup\{i: \operatorname{Ext}_R^i(k, M) \neq 0\}.$$

PROOF. We set $t = \sup\{i: \operatorname{Ext}_R^i(k, M) \neq 0\}$. It is clear that $\operatorname{inj\,dim} M \geq t$. To prove the converse inequality, note that the repeated application of 3.1.13 yields $\operatorname{Ext}_R^i(R/\mathfrak{p}, M) = 0$ for all $\mathfrak{p} \in \operatorname{Spec} R$ and all $i > t$. According to 3.1.12 this implies $\operatorname{inj\,dim} M \leq t$. □

Corollary 3.1.15. *Let $(R, \mathfrak{m}, k)$ be a Noetherian local ring and M a finite R-module. If $x \in \mathfrak{m}$ is an element which is R- and M-regular, then*

$$\text{inj dim}_{R/(x)} M/xM = \text{inj dim}_R M - 1.$$

The proof is an immediate consequence of 3.1.14 and the following result of Rees [302], Theorem 2.1.

Lemma 3.1.16. *Let R be a ring, and let M and N be R-modules. If x is an R- and M-regular element with $x \cdot N = 0$, then*

$$\text{Ext}_R^{i+1}(N, M) \cong \text{Ext}_{R/(x)}^i(N, M/xM)$$

for all $i \geq 0$.

PROOF. We set $\bar{R} = R/(x)$ and $\bar{M} = M/xM$, and show that the functors $\text{Ext}_R^{i+1}(\_, M)$, $i \geq 0$, from the category of $\bar{R}$-modules into itself are the right derived functors of $\text{Hom}_{\bar{R}}(\_, \bar{M})$. To see this, we have to verify

(1) the functors $\text{Ext}_R^{i+1}(\_, M)$, $i \geq 0$, are strongly connected,

(2) the functors $\text{Ext}_R^1(\_, M)$ and $\text{Hom}_{\bar{R}}(\_, \bar{M})$ are equivalent,

(3) $\text{Ext}_R^{i+1}(F, M) = 0$ for all $i > 0$ and every free $\bar{R}$-module F.

(An axiomatic description of the Ext groups as functors in the second variable is given in [318], Theorem 7.22. Similarly the Ext groups can be described axiomatically as functors in the first variable; see [318], Exercise 7.27.)

(1) is obvious. The exact sequence $0 \longrightarrow M \overset{x}{\longrightarrow} M \longrightarrow \bar{M} \longrightarrow 0$ yields the exact sequence

$$\text{Hom}_R(N, M) \longrightarrow \text{Hom}_R(N, \bar{M}) \longrightarrow \text{Ext}_R^1(N, M) \overset{x}{\longrightarrow} \cdots$$

Since $\text{Hom}_R(N, M) = 0$, and since x annihilates $\text{Ext}_R^1(N, M)$, we obtain the natural isomorphism

$$\text{Hom}_{\bar{R}}(N, \bar{M}) \cong \text{Ext}_R^1(N, M).$$

This proves (2). Finally, (3) is clear since $\text{proj dim}_R F \leq 1$ for every free $\bar{R}$-module F. $\square$

We now present the main result of this section.

Theorem 3.1.17. *Let $(R, \mathfrak{m}, k)$ be a Noetherian local ring, and let M be a finite R-module of finite injective dimension. Then*

$$\dim M \leq \text{inj dim} M = \text{depth} R.$$

PROOF. Let $\mathfrak{p}_0 \subset \mathfrak{p}_1 \subset \cdots \subset \mathfrak{p}_d = \mathfrak{m}$ be a maximal chain of prime ideals in Supp M. We show by induction on i that $\mathrm{Ext}^i_{R_{\mathfrak{p}_i}}(k(\mathfrak{p}_i), M_{\mathfrak{p}_i}) \neq 0$. In particular, it will follow that $\mathrm{Ext}^d_R(k, M) \neq 0$ for $d = \dim M$, so that $\dim M \leq \mathrm{inj\,dim}\, M$ by 3.1.14.

If $i = 0$, then $\mathfrak{p}_0 R_{\mathfrak{p}_0} \in \mathrm{Ass}\, M_{\mathfrak{p}_0}$, and therefore $\mathrm{Hom}_{R_{\mathfrak{p}_0}}(k(\mathfrak{p}_0), M_{\mathfrak{p}_0}) \neq 0$. Now suppose $i > 0$. We set $B = R_{\mathfrak{p}_i}$; then

$$\mathrm{Ext}^{i-1}_B(B/\mathfrak{p}_{i-1}B, M_{\mathfrak{p}_i})_{\mathfrak{p}_{i-1}} \cong \mathrm{Ext}^{i-1}_{R_{\mathfrak{p}_{i-1}}}(k(\mathfrak{p}_{i-1}), M_{\mathfrak{p}_{i-1}}) \neq 0,$$

by the induction hypothesis, and so $\mathrm{Ext}^{i-1}_B(B/\mathfrak{p}_{i-1}B, M_{\mathfrak{p}_i}) \neq 0$. It follows from 3.1.13 that

$$\mathrm{Ext}^i_B(k(\mathfrak{p}_i), M_{\mathfrak{p}_i}) \neq 0.$$

To prove the equality $\mathrm{inj\,dim}\, M = \mathrm{depth}\, R$, we set $r = \mathrm{inj\,dim}\, M$ and $t = \mathrm{depth}\, R$. Let $x = x_1, \ldots, x_t$ be a maximal R-sequence. Then the Koszul complex $K_{\bullet}(x)$ is a minimal free resolution of $R/(x)$ by 1.6.19 so that $\mathrm{proj\,dim}\, R/(x) = t$ and furthermore $\mathrm{Ext}^t_R(R/(x), M)$ is isomorphic to the t-th Koszul cohomology $H^t(x, M)$. It follows from 1.6.10 that $H^t(x, M) \cong H_0(x, M) = M/xM \neq 0$. This implies $r \geq t$.

On the other hand, since $\mathrm{depth}\, R/(x) = 0$, there is an embedding $k \to R/(x)$ which induces an epimorphism

$$\mathrm{Ext}^r_R(R/(x), M) \longrightarrow \mathrm{Ext}^r_R(k, M)$$

since $\mathrm{Ext}^{r+1}_R(N, M) = 0$ for all R-modules N. But $\mathrm{Ext}^r_R(k, M) \neq 0$ by 3.1.14, and so $\mathrm{Ext}^r_R(R/(x), M) \neq 0$. It follows that $t = \mathrm{proj\,dim}_R R/(x) \geq r$. $\square$

Gorenstein rings. We are now going to introduce an important class of local rings. As for regular rings, this class can be characterized in terms of homological algebra.

Definition 3.1.18. A Noetherian local ring R is a *Gorenstein ring* if $\mathrm{inj\,dim}_R R < \infty$. A Noetherian ring is a *Gorenstein ring* if its localization at every maximal ideal is a Gorenstein local ring.

The Gorenstein property is stable under standard ring operations. To begin with we show

Proposition 3.1.19. *Let R be a Noetherian ring.*

(a) *Suppose R is Gorenstein. Then for every multiplicatively closed set S in R the localized ring R_S is also Gorenstein. In particular, $R_{\mathfrak{p}}$ is Gorenstein for every $\mathfrak{p} \in \mathrm{Spec}\, R$.*

(b) *Suppose x is an R-regular sequence. If R is Gorenstein, then so is $R/(x)$. The converse holds when R is local.*

(c) *Suppose R is local. Then R is Gorenstein if and only if its completion $\hat{R}$ is Gorenstein.*

PROOF. (a) Let q be a maximal ideal of R_S. The ideal q is the extension of a prime ideal p in R, and so $(R_S)_q \cong R_p$. Let $\mathfrak{m}$ be a maximal ideal of R containing p. Then R_p is a localization of the Gorenstein local ring $R_\mathfrak{m}$. From 3.1.9 the conclusion follows.

(b) Without restriction we may assume that R is local. Thus (b) is an immediate consequence of 3.1.15.

(c) Let k be the residue field of R. Use that $\mathrm{Ext}_R^i(k, R)\widehat{} \cong \mathrm{Ext}_{\hat{R}}^i(k, \hat{R})$.
□

In concluding this section we clarify the position of the Gorenstein rings in the hierarchy of Noetherian local rings.

Proposition 3.1.20. *Let* $(R, \mathfrak{m}, k)$ *be a Noetherian local ring. Then we have the following implications:*

R is regular $\Rightarrow$ *R is a complete intersection* $\Rightarrow$ *R is Gorenstein*

$\Rightarrow$ *R is Cohen–Macaulay.*

PROOF. The first implication is trivial. If R is regular, then its global homological dimension is finite (see 2.2.7), and hence $\mathrm{Ext}_R^i(k, R) = 0$ for $i \gg 0$. It follows from 3.1.14 that R is Gorenstein. In view of 3.1.19(c) we may as well assume that R is complete. Now 3.1.19(b) implies that a complete intersection is Gorenstein. The last implication follows from 3.1.17.
□

All the implications of 3.1.20 are strict. This is clear for the first, and will be shown for the other implications in the next section (see 3.2.11), where we derive a different, more easily verifiable, characterization of Gorenstein rings.

Exercises

3.1.21. Let R be a principal ideal domain with field of fractions K. Prove that $0 \to K \to K/R \to 0$ is an injective resolution of R.

3.1.22. Let k be a field, and let R be a local k-algebra of finite k-dimension. Show that the R-module $\mathrm{Hom}_k(R, k)$ is an indecomposable (see the definition before 3.2.6) injective R-module.

3.1.23. Let R be a Noetherian local ring. If there exists a non-zero finite injective R-module, then deduce R is Artinian.

3.1.24. Let $(R, \mathfrak{m}, k)$ be a Noetherian local ring, $M \neq 0$ and $N \neq 0$ finite R-modules. If $\mathrm{inj\,dim}\, N < \infty$, then deduce the following result of Ischebeck [226]:

$$\mathrm{depth}\, R - \mathrm{depth}\, M = \sup\{i : \mathrm{Ext}_R^i(M, N) \neq 0\}.$$

In particular, if R admits a finite module of finite injective dimension, then show that the depth of any finite R-module does not exceed the depth of R. (Bass' conjecture claims more: in the above situation R is Cohen–Macaulay. In 9.6.2 a proof of this conjecture will be given, provided R contains a field.)

3.1.25. Let R be a Gorenstein local ring, and M a finite R-module. Show proj dim $M < \infty$ if and only if inj dim $M < \infty$. (Foxby [114] proved the following remarkable characterization of Gorenstein rings: if a Noetherian local ring possesses a finite module M for which inj dim $M < \infty$ and proj dim $M < \infty$, then it is Gorenstein.)

3.1.26. Let $(R, \mathfrak{m}, k)$ be a Noetherian local ring. If inj dim $k < \infty$, show R is regular.

3.2 Injective hulls. Matlis duality

We saw in Section 3.1 that any module M can be embedded into an injective module. Here we will show that such an embedding can be chosen minimal. In this case the corresponding injective module is unique up to isomorphism, and is called the injective hull of M.

We will see that for a Noetherian ring R an injective module can be uniquely written as a direct sum of indecomposable injective modules, and the indecomposable injective R-modules are just the injective hulls of the cyclic R-modules $R/\mathfrak{p}$, where $\mathfrak{p} \in \operatorname{Spec} R$. If $(R, \mathfrak{m}, k)$ is a complete Noetherian local ring, and E is the injective hull of k, then the functor $\operatorname{Hom}_R(\_, E)$ establishes an anti-equivalence between the category of Artinian R-modules and the category of finite R-modules. This result is known as the main theorem of Matlis duality.

Definition 3.2.1. Let R be a ring and let $N \subset M$ be R-modules. M is an *essential extension* of N if for any non-zero R-submodule U of M one has $U \cap N \neq 0$. An essential extension M of N is called *proper* if $N \neq M$.

The following proposition gives a new characterization of injective modules.

Proposition 3.2.2. *Let R be a ring. An R-module N is injective if and only if it has no proper essential extension.*

PROOF. Let $N \subset M$ be an extension. If N is injective, then N is a direct summand of M. Let W be a complement of N in M. Then $N \cap W = 0$, and so, if the extension is essential, $W = 0$. It follows that $N = M$.

Conversely, suppose that N has no proper essential extension. Given a monomorphism $\varphi: U \to V$ and a homomorphism $\alpha: U \to N$, we want to construct $\beta: V \to N$ such that $\alpha = \beta \circ \varphi$.

As in the proof of 3.1.2 we consider the pushout diagram

$$
\begin{array}{ccc}
U & \xrightarrow{\ \varphi\ } & V \\
{\scriptstyle\alpha}\downarrow & & \downarrow{\scriptstyle\gamma} \\
N & \xrightarrow{\ \psi\ } & W
\end{array}
$$

Here ψ is a monomorphism, since φ is a monomorphism. Thus we may consider N as a submodule of W. Employing Zorn's lemma one shows that there exists a maximal submodule $D \subset W$ such that $N \cap D = 0$, and so N may even be considered as a submodule of W/D; obviously, W/D is an essential extension of N. It follows that $N = W/D$, since N has no proper essential extension, and so $W = N \oplus D$. Let $\pi: W \to N$ be the natural projection of W onto the first summand. The composition $\pi \circ \gamma: V \to N$ is an extension of α. $\square$

Definition 3.2.3. Let R be a ring and M an R-module. An injective module E such that $M \subset E$ is an essential extension is called an *injective hull* of M. Our notation will be $E(M)$ or $E_R(M)$.

The next proposition justifies this name.

Proposition 3.2.4. *Let R be a ring and M an R-module.*
(a) *M admits an injective hull. Moreover, if $M \subset I$ and I is injective, then a maximal essential extension of M in I is an injective hull of M.*
(b) *Let E be an injective hull of M, let I be an injective R-module, and $\alpha: M \to I$ be a monomorphism. Then there exists a monomorphism $\varphi: E \to I$ such that the diagram*

is commutative, where $M \to E$ is the inclusion map. In other words, the injective hulls of M are the 'minimal' injective modules in which M can be embedded.
(c) *If E and E' are injective hulls of M, then there exists an isomorphism $\varphi: E \to E'$ such that the diagram*

$$M$$
$$\swarrow \quad \searrow$$
$$E \xrightarrow{\varphi} E'$$

commutes. Here $M \to E$ and $M \to E'$ are the inclusion maps.

PROOF. (a) We embed M into an injective R-module I. Consider the set $\mathcal{S}$ of all essential extensions $M \subset N$ with $N \subset I$. Zorn's lemma applied to $\mathcal{S}$ yields the existence of a maximal essential extension $M \subset E$ with $E \subset I$. We claim that E has no proper essential extension, and this together with 3.2.2 implies then that E is an injective hull of M. Indeed, assume that E has a proper essential extension E'. Since I is injective there exists $\psi: E' \to I$ extending the inclusion $E \subset I$. Suppose $\operatorname{Ker} \psi = 0$; then $\operatorname{Im} \psi \subset I$ is an essential extension of M (in I) properly containing E, a contradiction. On the other hand, since ψ extends the inclusion $E \subset I$ we have $E \cap \operatorname{Ker} \psi = 0$. But this contradicts the essentiality of the extension $E \subset E'$.

(b) Since I is injective, α can be extended to a homomorphism $\varphi : E \to I$. We have $\varphi|_M = \alpha$, and so $M \cap \operatorname{Ker} \varphi = \operatorname{Ker} \alpha = 0$. Thus, since the extension $M \subset E$ is essential, we even have $\operatorname{Ker} \varphi = 0$.

(c) By (b) there is a monomorphism $\varphi : E \to E'$ such that $\varphi|_M$ equals the inclusion $M \subset E'$. $\operatorname{Im} \varphi$ is injective and hence a direct summand of E'. However, since the extension $M \subset E'$ is essential, φ is surjective, and therefore an isomorphism. □

We may apply 3.2.4 to construct an injective resolution $E^\bullet(M)$ of a module M which for obvious reasons is called the *minimal injective resolution of M*: we let $E^0(M) = E(M)$, and denote by ∂^{-1} the embedding $M \to E^0(M)$. Suppose the injective resolution has already been constructed up to the i-th step:

$$0 \longrightarrow E^0(M) \xrightarrow{\partial^0} E^1(M) \longrightarrow \cdots \longrightarrow E^{i-1}(M) \xrightarrow{\partial^{i-1}} E^i(M).$$

We then define $E^{i+1}(M) = E(\operatorname{Coker} \partial^{i-1})$, and ∂^i is defined in the obvious way.

It is clear that any two minimal injective resolutions of M are isomorphic. Moreover, if $I^\bullet$ is an arbitrary injective resolution of M, then, as is readily seen, $E^\bullet(M)$ is isomorphic to a direct summand of $I^\bullet$.

We note a technical result about injective hulls which will be needed later in this section.

Lemma 3.2.5. *Let R be a Noetherian ring, $S \subset R$ a multiplicatively closed set and M an R-module. Then $E_R(M)_S \cong E_{R_S}(M_S)$.*

PROOF. We show that $E_R(M)_S$ is an injective hull of the R_S-module M_S. We know from 3.1.3 that $E_R(M)_S$ is an injective R_S-module. It remains to be shown that $E_R(M)_S$ is an essential extension of M_S. To simplify notation we set $N = E_R(M)$, and pick $x \in N_S$, $x \neq 0$. We want to prove that $R_S x \cap M_S \neq 0$.

There exists $y \in N$ such that $R_S y = R_S x$. Thus we may as well assume that $x \in N$. We consider the set of ideals $\mathscr{S} = \{\operatorname{Ann}(tx) : t \in S\}$. Since R is Noetherian this set has a maximal element, say $\operatorname{Ann}(sx)$, and since $R_S x = R_S(sx)$, we may replace x by sx, and thus may assume that $\operatorname{Ann}(x)$ is maximal in the set $\mathscr{S}$.

Since N is an essential extension of M, we have $Rx \cap M = Ix \neq 0$, where I is an ideal in R. Let $I = (a_1, \ldots, a_n)$, and assume that $a_i x = 0$ in N_S for $i = 1, \ldots, n$. Then there exists $t \in S$ such that $t(a_i x) = 0$ in N for $i = 1, \ldots, n$. But $\operatorname{Ann}(tx) = \operatorname{Ann}(x)$, by the choice of x, and so $Ix = 0$. This is a contradiction. Hence $a_i x \neq 0$ in N_S for some i, and it follows that $R_S x \cap M_S \neq 0$. □

In the next theorem we determine the indecomposable injective R-modules of a Noetherian ring R. Recall that an R-module M is *decomposable* if there exist non-zero submodules M_1, M_2 of M such that $M = M_1 \oplus M_2$; otherwise it is *indecomposable*.

Theorem 3.2.6. *Let R be a Noetherian ring.*
(a) *For all $\mathfrak{p} \in \operatorname{Spec} R$ the module $E(R/\mathfrak{p})$ is indecomposable.*
(b) *Let $I \neq 0$ be an injective R-module and let $\mathfrak{p} \in \operatorname{Ass} I$. Then $E(R/\mathfrak{p})$ is a direct summand of I. In particular, if I is indecomposable, then*

$$I \cong E(R/\mathfrak{p}).$$

(c) *Let $\mathfrak{p}, \mathfrak{q} \in \operatorname{Spec} R$. Then $E(R/\mathfrak{p}) \cong E(R/\mathfrak{q}) \Longleftrightarrow \mathfrak{p} = \mathfrak{q}$.*

PROOF. (a) Suppose $E(R/\mathfrak{p})$ is decomposable. Then there exist non-zero submodules N_1, N_2 of $E(R/\mathfrak{p})$ such that $N_1 \cap N_2 = 0$. It follows that $(N_1 \cap R/\mathfrak{p}) \cap (N_2 \cap R/\mathfrak{p}) = (N_1 \cap N_2) \cap R/\mathfrak{p} = 0$. On the other hand, since $R/\mathfrak{p} \subset E(R/\mathfrak{p})$ is an essential extension, we have $N_1 \cap R/\mathfrak{p} \neq 0 \neq N_2 \cap R/\mathfrak{p}$. This contradicts the fact that $R/\mathfrak{p}$ is a domain.

(b) $R/\mathfrak{p}$ may be considered as a submodule of I since $\mathfrak{p} \in \operatorname{Ass} I$. It follows from 3.2.4 that there exists an injective hull $E(R/\mathfrak{p})$ of $R/\mathfrak{p}$ such that $E(R/\mathfrak{p}) \subset I$. As $E(R/\mathfrak{p})$ is injective, it is a direct summand of I. Statement (c) follows from the next lemma.   □

Lemma 3.2.7. *Let R be a Noetherian ring, $\mathfrak{p} \in \operatorname{Spec} R$, and M a finite R-module. Then*
(a) $\operatorname{Ass} M = \operatorname{Ass} E(M)$; *in particular one has $\{\mathfrak{p}\} = \operatorname{Ass} E(R/\mathfrak{p})$;*
(b) $k(\mathfrak{p}) \cong \operatorname{Hom}_{R_\mathfrak{p}}(k(\mathfrak{p}), E(R/\mathfrak{p})_\mathfrak{p})$.

PROOF. (a) It is clear that $\operatorname{Ass} M \subset \operatorname{Ass} E(M)$. Conversely, suppose $\mathfrak{q} \in \operatorname{Ass} E(M)$. Then there exists a submodule $U \subset E(M)$ which is isomorphic to $R/\mathfrak{q}$. We have $U \cap M \neq 0$ since the extension $M \subset E(M)$ is essential, and so $\mathfrak{q} \in \operatorname{Ass}(U \cap M) \subset \operatorname{Ass} M$.

(b) Since $E(R/\mathfrak{p})_\mathfrak{p} \cong E_{R_\mathfrak{p}}(k(\mathfrak{p}))$, we assume that $(R, \mathfrak{m}, k)$ is local and $\mathfrak{p} = \mathfrak{m}$ is the maximal ideal. The k-vector space $\operatorname{Hom}_R(k, E(k))$ may be identified with $V = \{x \in E(k) : \mathfrak{m}x = 0\}$; it contains k. If $V \neq k$, then there exists a non-zero vector subspace W of V with $k \cap W = 0$. This, however, contradicts the essentiality of the extension $k \subset E(k)$.   □

The importance of the indecomposable injective R-modules results from the following:

Theorem 3.2.8. *Every injective module I over a Noetherian ring R is a direct sum of indecomposable injective R-modules, and this decomposition is unique in the following sense: for any $\mathfrak{p} \in \operatorname{Spec} R$ the number of indecomposable summands in the decomposition of I which are isomorphic to $E(R/\mathfrak{p})$ depends only on I and $\mathfrak{p}$ (and not on the particular decomposition). In fact, this number equals $\dim_{k(\mathfrak{p})} \operatorname{Hom}_{R_\mathfrak{p}}(k(\mathfrak{p}), I_\mathfrak{p})$.*

PROOF. Consider the set $\mathcal{S}$ of all subsets of the set of indecomposable injective submodules of I with the property: if $\mathcal{F} \in \mathcal{S}$, then the sum of all modules belonging to $\mathcal{F}$ is direct. The set $\mathcal{S}$ is partially ordered by inclusion. By Zorn's lemma it has a maximal element $\mathcal{F}'$. Let E be the sum of all the modules in $\mathcal{F}'$. The module E is a direct sum of injective modules, and hence by 3.1.3 is itself injective. Therefore E is a direct summand of I, and we can write $I = E \oplus H$, where H is injective since it is a direct summand of I. Suppose $H \neq 0$; then there exists $\mathfrak{p} \in \mathrm{Ass}\, H$, and so $E(R/\mathfrak{p})$ is a direct summand of H; see 3.2.6(b). Thus we may enlarge $\mathcal{F}'$ by $E(R/\mathfrak{p})$, contradicting the maximality of $\mathcal{F}'$. We conclude that $H = 0$ and $I = E$.

Suppose that $I = \bigoplus_{\lambda \in \Lambda} I_\lambda$ is the given decomposition. Then

$$\mathrm{Hom}_{R_\mathfrak{p}}(k(\mathfrak{p}), I_\mathfrak{p}) \cong \mathrm{Hom}_{R_\mathfrak{p}}\Big(k(\mathfrak{p}), \bigoplus_{\lambda \in \Lambda}(I_\lambda)_\mathfrak{p}\Big) \cong \bigoplus_{\lambda \in \Lambda} \mathrm{Hom}_{R_\mathfrak{p}}(k(\mathfrak{p}), (I_\lambda)_\mathfrak{p}).$$

By 3.2.7 we have

$$\bigoplus_{\lambda \in \Lambda} \mathrm{Hom}_{R_\mathfrak{p}}(k(\mathfrak{p}), (I_\lambda)_\mathfrak{p}) \cong \bigoplus_{\lambda \in \Lambda_0} \mathrm{Hom}_{R_\mathfrak{p}}(k(\mathfrak{p}), (I_\lambda)_\mathfrak{p}),$$

where $\Lambda_0 = \{\lambda \in \Lambda : I_\lambda \cong E(R/\mathfrak{p})\}$. If we again use 3.2.7, we finally get

$$\mathrm{Hom}_{R_\mathfrak{p}}(k(\mathfrak{p}), I_\mathfrak{p}) \cong \bigoplus_{\lambda \in \Lambda_0} \mathrm{Hom}_{R_\mathfrak{p}}(k(\mathfrak{p}), (I_\lambda)_\mathfrak{p}) \cong k(\mathfrak{p})^{(\Lambda_0)}. \qquad \square$$

Bass numbers. Let R be a Noetherian ring, M a finite R-module and $\mathfrak{p} \in \mathrm{Spec}\, R$. The (finite) number $\mu_i(\mathfrak{p}, M) = \dim_{k(\mathfrak{p})} \mathrm{Ext}^i_{R_\mathfrak{p}}(k(\mathfrak{p}), M_\mathfrak{p})$ is called the i-th *Bass number* of M with respect to $\mathfrak{p}$.

These numbers have an interpretation in terms of the minimal injective resolution of M.

Proposition 3.2.9. *Let R be a Noetherian ring, M a finite R-module, and $E^\bullet(M)$ the minimal injective resolution of M. Then*

$$E^i(M) \cong \bigoplus_{\mathfrak{p} \in \mathrm{Spec}\, R} E(R/\mathfrak{p})^{\mu_i(\mathfrak{p}, M)}.$$

PROOF. Let $0 \longrightarrow M \longrightarrow E^0(M) \overset{\partial^0}{\longrightarrow} E^1(M) \overset{\partial^1}{\longrightarrow} \cdots$ be the minimal injective resolution of M, and let $\mathfrak{p} \in \mathrm{Spec}\, R$. Since localization is exact, it follows from 3.2.5 that

$$0 \longrightarrow M_\mathfrak{p} \longrightarrow E^0(M)_\mathfrak{p} \overset{d^0}{\longrightarrow} E^1(M)_\mathfrak{p} \overset{d^1}{\longrightarrow} \cdots$$

is the minimal injective resolution of $M_\mathfrak{p}$; here d^i is the localization of ∂^i.

The complex $\operatorname{Hom}_{R_\mathfrak{p}}(k(\mathfrak{p}), E^{\bullet}(M)_\mathfrak{p})$ is isomorphic to the subcomplex $C^{\bullet}$ of $E^{\bullet}(M)_\mathfrak{p}$, where

$$C^i = \{x \in E^i(M)_\mathfrak{p} : \mathfrak{p}R_\mathfrak{p} \cdot x = 0\}.$$

Let x be a non-zero element of C^i. Since the extension $\operatorname{Im} d^{i-1} \subset E^i(M)_\mathfrak{p}$ is essential, there exists $a \in R_\mathfrak{p}$ with $ax \in \operatorname{Im} d^{i-1}$ and $ax \neq 0$. Since $\mathfrak{p}R_\mathfrak{p}$ annihilates x, we see that $a \notin \mathfrak{p}R_\mathfrak{p}$. Hence a is a unit in $R_\mathfrak{p}$, and $x \in \operatorname{Im} d^{i-1}$. It follows that $d^i(x) = 0$, and hence $d^i|_{C^i} = 0$ for all i. Consequently we get $\operatorname{Ext}^i_{R_\mathfrak{p}}(k(\mathfrak{p}), M_\mathfrak{p}) \cong \operatorname{Hom}_{R_\mathfrak{p}}(k(\mathfrak{p}), E^i(M)_\mathfrak{p})$, which by 3.2.8 implies the isomorphism asserted.   □

Among the Bass numbers the type of a module or a local ring is of particular importance. Let $(R, \mathfrak{m}, k)$ be a Noetherian local ring and M a finite module of depth t. In Chapter 1 we have already considered the Bass number $r(M) = \mu_t(\mathfrak{m}, M)$, and called it the type of M.

In the next theorem we give a new, extremely useful characterization of Gorenstein rings.

Theorem 3.2.10. *Let $(R, \mathfrak{m}, k)$ be a Noetherian local ring. The following conditions are equivalent:*
(a) *R is a Gorenstein ring;*
(b) *R is a Cohen–Macaulay ring of type 1.*

PROOF. Let $\boldsymbol{x}$ be a maximal R-sequence. By 3.1.19, R is Gorenstein if and only if $R/(\boldsymbol{x})$ is. Similarly the properties in (b) are stable under specialization modulo $\boldsymbol{x}$; see 1.2.19 and 2.1.3. Thus we may assume $\dim R = 0$.

(a) $\Rightarrow$ (b): By 3.1.17, R is an injective R-module. Since R is local, it is indecomposable as an R-module, and so, since $\operatorname{Ass} R = \{\mathfrak{m}\}$, we have that $R \cong E_R(k)$; see 3.2.6. It follows from 3.2.7 that R is of type 1.

(b) $\Rightarrow$ (a) follows from statement (e) in 3.2.12 below.   □

We use this new characterization of Gorenstein rings to give examples of Cohen–Macaulay rings which are not Gorenstein, and of Gorenstein rings which are not complete intersections.

Examples 3.2.11. (a) Let $(R, \mathfrak{m}, k)$ be an Artinian local ring for which $\mathfrak{m}^2 = 0$. For instance, $R = k[X_1, \ldots, X_n]/(X_1, \ldots, X_n)^2$ is such a ring. It is easily seen that $\mathfrak{m} = \operatorname{Soc} R$. Hence we have $r(R) = \operatorname{emb dim} R$, and conclude that R is Gorenstein if and only if $\operatorname{emb dim} R = 1$. When R is Gorenstein, it is even a complete intersection.

(b) In the following we present a method to produce a large class of Artinian Gorenstein rings: let k be a field, $S = k[X_1, \ldots, X_n]$ the polynomial ring in n variables over k, m an integer, S_m the m-th homogeneous part of S, and $\varphi : S_m \to k$ a non-trivial k-linear map.

For every $j \in \mathbb{N}$, we define $I_j = \{a \in S_j : \varphi(a \cdot S_{m-j}) = 0\}$. It is readily seen that $I = \bigoplus_{j \geq 0} I_j$ is a graded ideal with $I_j = S_j$ for $j > m$. Thus we conclude that $R = S/I$ is an Artinian (graded) local ring.

We claim that R is a Gorenstein ring. To see this, we determine the socle of R. For any element $a \in S$, we denote by $\bar{a}$ its residue class modulo I. Let $j \in \mathbb{N}$ with $0 < j < m$, and let $\bar{a} \in R_j, \bar{a} \neq 0$. Then by the definition of I, there exists $b \in S_{m-j}$ such that $\varphi(a \cdot b) \neq 0$, and so $\bar{a} \cdot \bar{b} \neq 0$. But since $\bar{b}$ belongs to the maximal ideal of R, it follows that $\bar{a} \notin \operatorname{Soc} R$. Therefore, $\operatorname{Soc} R = R_m$. As $\dim_k R_m = 1$, it follows that R is Gorenstein.

We give an explicit example for this construction: let $\varphi : S_2 \to k$ be the k-linear map with

$$\varphi(X_i X_j) = 0, \quad 1 \leq i < j \leq n, \qquad \varphi(X_i^2) = 1, \quad i = 1, \ldots, n.$$

For this linear form φ we get

$$I = (X_1^2 - X_2^2, \ldots, X_1^2 - X_n^2, X_1 X_2, X_1 X_3, \ldots, X_{n-1} X_n).$$

Therefore, $R = S/I$ is Gorenstein, and is a complete intersection if and only if $n \leq 2$.

Matlis duality. Let $(R, \mathfrak{m}, k)$ be a Noetherian local ring. We are going to study the functor which takes the dual M' of an R-module M with respect to the injective hull E of k. If M is a finite module, the dual M' need not be finite. Indeed, we know from Exercise 3.1.23 that $R' \cong E$ is finite only if R is Artinian. However, the E-dual of a module of finite length also has finite length, as we shall see now.

Proposition 3.2.12. *Let $(R, \mathfrak{m}, k)$ be a Noetherian local ring, E the injective hull of k, and N an R-module of finite length. For any R-module M we set $M' = \operatorname{Hom}_R(M, E)$. Then:*
(a) *one has*

$$\operatorname{Ext}^i_R(k, E) \cong \begin{cases} k & \text{for } i = 0, \\ 0 & \text{for } i > 0; \end{cases}$$

(b) $\ell(N) = \ell(N')$;
(c) *the canonical homomorphism $N \to N''$ is an isomorphism;*
(d) $\mu(N) = r(N')$ *and* $r(N) = \mu(N')$;
(e) *if R is Artinian, then E is a finite faithful R-module satisfying*
  (i) $\ell(E) = \ell(R)$,
  (ii) *the canonical homomorphism $R \to \operatorname{End}_R(E)$, $a \mapsto \varphi_a$, where $\varphi_a(x) = ax$ for all $x \in E$, is an isomorphism,*
  (iii) $r(E) = 1$ *and* $\mu(E) = r(R)$;
conversely, any finite faithful R-module of type 1 is isomorphic to E.

PROOF. (a) $\mathrm{Ext}^i_R(k, E) = 0$ for $i > 0$, as E is injective; furthermore $\mathrm{Hom}_R(k, E) \cong k$; see 3.2.7.

(b) We prove the equality asserted by induction on the length of N. If $\ell(N) = 1$, then $N \cong k$, and the equality follows from (a). Now suppose that $\ell(N) > 1$. Then there exists a proper submodule $U \subset N$, and we obtain an exact sequence

$$0 \longrightarrow U \longrightarrow N \longrightarrow W \longrightarrow 0$$

with $\ell(U) < \ell(N)$ and $\ell(W) < \ell(N)$.

Since E is injective this sequence yields the dual exact sequence

$$0 \longrightarrow W' \longrightarrow N' \longrightarrow U' \longrightarrow 0.$$

The induction hypothesis applies to U and W, and the additivity of length gives the result.

(c) Again we use induction on $\ell(N)$. If $\ell(N) = 1$, then $N \cong k$, and $N'' \cong k$ by (a). Therefore it suffices to show that the canonical homomorphism $\alpha : k \to \mathrm{Hom}_R(\mathrm{Hom}_R(k, E), E)$ is not the zero map. Let $x \in E$, $x \neq 0$, be a socle element of E. There exists $\varphi \in \mathrm{Hom}_R(k, E)$ with $\varphi(1) = x$. Then $\alpha(1)(\varphi) = x \neq 0$, and so $\alpha \neq 0$. If $\ell(N) > 1$, we choose as before an exact sequence

$$0 \longrightarrow U \longrightarrow N \longrightarrow W \longrightarrow 0$$

with $\ell(U) < \ell(N)$ and $\ell(W) < \ell(N)$.

The natural homomorphisms into the bidual modules induce a commutative diagram

$$
\begin{array}{ccccccccc}
0 & \longrightarrow & U & \longrightarrow & N & \longrightarrow & W & \longrightarrow & 0 \\
& & \downarrow & & \downarrow & & \downarrow & & \\
0 & \longrightarrow & U'' & \longrightarrow & N'' & \longrightarrow & W'' & \longrightarrow & 0
\end{array}
$$

where the outer vertical arrows are isomorphisms by our induction hypothesis. The snake lemma ([318], Theorem 6.5) applied to this diagram implies $N \to N''$ is an isomorphism.

(d) The module $(N/\mathfrak{m}N)'$ is the kernel of the linear map $N' \to (\mathfrak{m}N)'$ which assigns to every $\varphi \in N'$ its restriction to $\mathfrak{m}N$. Hence $\varphi \in (N/\mathfrak{m}N)'$ if and only if $\mathfrak{m}\varphi(N) = \varphi(\mathfrak{m}N) = 0$. In other words,

$$(N/\mathfrak{m}N)' = \{\varphi \in N' : \mathfrak{m} \cdot \varphi = 0\} = \mathrm{Soc}\, N'.$$

Thus we get $\mu(N) = \dim_k N/\mathfrak{m}N = \dim_k(N/\mathfrak{m}N)' = \dim_k \mathrm{Soc}\, N' = r(N')$. The second equality follows from the first by (c).

(e) By (b) we have $\ell(E) = \ell(R') = \ell(R) < \infty$. In particular, E is a finite R-module. Next it follows from (c) that the canonical homomorphism $\alpha : R \to \mathrm{Hom}_R(\mathrm{Hom}_R(R, E), E)$ is an isomorphism. If we identify

$\text{Hom}_R(R, E)$ with E, then α identifies with the canonical homomorphism $R \to \text{End}_R(E)$. A module whose endomorphism ring is R is necessarily faithful. Statement (e)(iii) follows from (d).

Finally, let N be a faithful R-module of type 1. Then N' is cyclic, and so $N \cong \text{Hom}_R(R/I, E)$ for some ideal I. Here we have used (c) and (d). But since N is faithful, $I = 0$ and so $N \cong E$. $\qquad\square$

Proposition 3.2.12 may be viewed as the Matlis duality theorem for finite Artinian modules. Now we prove its general form. It will be of crucial importance for the local duality theorem of Grothendieck, which we will discuss in Section 3.5.

Let $(R, \mathfrak{m}, k)$ be a complete local ring. We denote by $\mathcal{M}(R)$ the category of R-modules, by $\mathcal{A}(R)$ the full subcategory of Artinian R-modules and by $\mathcal{F}(R)$ the full subcategory of finite R-modules. Let E be an injective hull of k. We set $T(\_) = \text{Hom}_R(\_, E)$. The contravariant functor $T : \mathcal{M}(R) \to \mathcal{M}(R)$ is exact. Its restriction to $\mathcal{A}(R)$ or $\mathcal{F}(R)$ will again be denoted by T.

Theorem 3.2.13 (Matlis). *Let $(R, \mathfrak{m}, k)$ be a Noetherian complete local ring, $N \in \mathcal{A}(R)$ and $M \in \mathcal{F}(R)$. Then*
(a) $T(R) \cong E$ *and* $T(E) \cong R$,
(b) $T(M) \in \mathcal{A}(R)$ *and* $T(N) \in \mathcal{F}(R)$,
(c) *there are natural isomorphisms* $T(T(N)) \cong N$ *and* $T(T(M)) \cong M$,
(d) *the functor T establishes an anti-equivalence between the categories* $\mathcal{A}(R)$ *and* $\mathcal{F}(R)$.

PROOF. We proceed in several stages. (1) For all $n \in \mathbb{N}$ we set $E_n = \{x \in E : \mathfrak{m}^n x = 0\}$. Let $x \in E$, $x \neq 0$; then $\text{Ass}(Rx) \subset \text{Ass}\,E = \{\mathfrak{m}\}$, see 3.2.7. Hence there exists an integer n such that $\mathfrak{m}^n x = 0$. This proves that $E = \bigcup_{n \geq 0} E_n = \varinjlim E_n$.

(2) The natural homomorphism $R \to \text{End}_R(E) = T(E)$ is an isomorphism: by 3.2.12(e)(ii), the natural homomorphisms $\alpha_n : R/\mathfrak{m}^n \to T(E_n)$ are isomorphisms, and we obtain commutative diagrams such as

$$
\begin{array}{ccc}
R & \longrightarrow & T(E) \\
\downarrow & & \downarrow \\
R/\mathfrak{m}^n & \xrightarrow{\;\alpha_n\;} & T(E_n)
\end{array}
$$

in which the only homomorphisms are the natural ones. As R is complete, the map $R \to \varprojlim R/\mathfrak{m}^n$ is an isomorphism. Likewise $T(E) \to \varprojlim T(E_n)$ is an isomorphism since by [318], Theorem 2.27, and (1) we have $\varprojlim T(E_n) = T(\varinjlim E_n) = T(E)$. It follows that the natural homomorphism $R \to T(E)$ is an isomorphism as well. This proves (a).

(3) E is Artinian: let $E = U_0 \supset U_1 \supset U_2 \supset \cdots$ be a descending chain of submodules of E. This chain induces a sequence of epimorphisms

$$R = T(E) \longrightarrow T(U_1) \longrightarrow T(U_2) \longrightarrow \cdots$$

Thus we can write $T(U_i) = R/I_i$, where $(0) = I_0 \subset I_1 \subset I_2 \subset \cdots$ is an ascending chain of ideals. Since R is Noetherian this chain stabilizes, and so there exists an integer i_0 such that $T(U_i) = T(U_{i+1})$ for $i \geq i_0$. We will show that $U_i = U_{i+1}$ for $i \geq i_0$. Suppose that $U_i \neq U_{i+1}$, but $T(U_i) = T(U_{i+1})$. Let $V = U_i/U_{i+1}$; then $V \neq 0$, but $T(V) = 0$. However, V is a subquotient of E, and so $\operatorname{Ass} V = \{\mathfrak{m}\}$; see 3.2.7. In other words, there exists a monomorphism $k \to V$. Applying T, we obtain an epimorphism $0 = T(V) \to T(k) = k$, a contradiction.

(4) If N is Artinian, then there exists an embedding $N \to E^n$ for some integer n: $\operatorname{Soc} N$ is a finite dimensional k-vector space since N is Artinian. Moreover, the extension $\operatorname{Soc} N \subset N$ is essential. In fact, if $x \in N$, then Rx is a finite Artinian module, and therefore $Rx \cap \operatorname{Soc} N = \operatorname{Soc} Rx \neq 0$. Let $N \subset I$ be an embedding of N into an injective R-module. By 3.2.4, an injective hull $E(N)$ can be chosen as a maximal essential extension of N in I. Since the extension $\operatorname{Soc} N \subset N$ is essential, $E(N)$ is likewise an injective hull of $\operatorname{Soc} N$. Suppose $\operatorname{Soc} N \cong k^n$; then it follows that $N \subset E(\operatorname{Soc} N) \cong E^n$.

The remaining assertions of the theorem now follow easily.

(b) Let $N \in \mathscr{A}(R)$; then by (4) there exists an embedding $N \subset E^n$ which by (2) induces an epimorphism $R^n \to T(N)$; therefore $T(N) \in \mathscr{F}(R)$. Conversely, suppose $M \in \mathscr{F}(R)$. We choose an epimorphism $R^n \to M$. This epimorphism yields an embedding $T(M) \to E^n$. The module E is Artinian by (3), and so any submodule of E^n is Artinian. It follows that $T(M) \in \mathscr{A}(R)$.

(c) By (4) there exists an integer n and an exact sequence $0 \to N \to E^n \to W \to 0$ which we may complete to a commutative diagram

$$
\begin{array}{ccccccccc}
0 & \longrightarrow & N & \longrightarrow & E^n & \longrightarrow & W & \longrightarrow & 0 \\
& & \alpha \downarrow & & \beta \downarrow & & \gamma \downarrow & & \\
0 & \longrightarrow & T(T(N)) & \longrightarrow & T(T(E^n)) & \longrightarrow & T(T(W)) & \longrightarrow & 0
\end{array}
$$

whose vertical maps are just the canonical homomorphisms.

It follows from (2) that β is an isomorphism. Therefore, by the snake lemma, α is an isomorphism if and only if γ is a monomorphism. Let $x \in \operatorname{Ker} \gamma$; then $\varphi(x) = 0$ for all $\varphi \in \operatorname{Hom}_R(W, E)$. Suppose $x \neq 0$, and let $\psi : Rx \to E$ be the homomorphism which maps x to a non-zero socle element of E. Then $\psi(x) \neq 0$, and since E is injective, ψ can be extended to a homomorphism $\varphi : W \to E$. We then have $\varphi(x) \neq 0$, a contradiction.

Similarly one proves that the natural linear map $M \to T(T(M))$ is an isomorphism, starting with the exact sequence $0 \to U \to R^n \to M \to 0$ and using the fact that the natural homomorphism $R \to T(T(R))$ is an isomorphism (which is an immediate consequence of (2)). □

Exercises

3.2.14. Let $(R, \mathfrak{m}, k)$ be a Noetherian local ring, E an injective hull of k. Prove:
(a) The natural homomorphism $E \to E \otimes_R \hat{R}$ is an isomorphism. In particular, E is an $\hat{R}$-module.
(b) As an $\hat{R}$-module, $E \cong E_{\hat{R}}(k)$.
(c) For all finite R-modules N there exists a natural isomorphism

$$\operatorname{Hom}_R(N, E) \cong \operatorname{Hom}_{\hat{R}}(\hat{N}, E).$$

(If this problem seems to be too difficult, the reader may consult [270], Theorem 18.6.)

3.2.15. Let $(R, \mathfrak{m})$ be an Artinian local ring. Show the following conditions are equivalent:
(a) R is a Gorenstein ring;
(b) all finite R-modules are reflexive;
(c) $I = \operatorname{Ann}\operatorname{Ann} I$ for all ideals I of R;
(d) for all non-zero ideals I and J one has $I \cap J \neq 0$.

3.3 The canonical module

So far we have studied finite modules of finite injective dimension over Noetherian local rings, but we have ignored the question as to under what circumstances such modules actually exist. A Gorenstein ring R admits plenty of finite modules of finite injective dimension: any module of finite projective dimension has finite injective dimension as well, simply because R itself has finite injective dimension by definition. Also any Artinian local ring $(R, \mathfrak{m}, k)$, Gorenstein or not, admits a finite injective module – the injective hull of k. The question becomes more delicate for non-Gorenstein local rings of positive dimension. One of the main results of this section will be that any Cohen–Macaulay ring which is a homomorphic image of a Gorenstein ring has a finite module of finite injective dimension. Moreover, this module can be chosen to be a maximal Cohen–Macaulay module of type 1. It will be shown that such a module is unique up to isomorphism. It is called the *canonical module* of R. For a Gorenstein ring the canonical module is just the ring itself.

We shall study the behaviour of the canonical module under flat extensions, localizations, and specializations.

Definition 3.3.1. Let $(R, \mathfrak{m}, k)$ be a Cohen–Macaulay local ring. A maximal Cohen–Macaulay module C of type 1 and of finite injective dimension is called a *canonical module* of R.

It is immediate (see 1.2.8 and 3.1.14) that C is a canonical module of R if and only if

$$\dim_k \operatorname{Ext}_R^i(k, C) = \delta_{id}, \qquad d = \dim R.$$

Two questions arise: when does a canonical module exist, and is it uniquely determined up to isomorphism? This question has a simple answer in the case $\dim R = 0$: by 3.2.8, $E_R(k)$ is the uniquely determined canonical module. To prove uniqueness in general, we will need the following two results.

Lemma 3.3.2. Let $(R, \mathfrak{m}, k)$ be a Noetherian local ring, $\varphi : M \to N$ a homomorphism of finite R-modules, and x an N-sequence. If $\varphi \otimes R/(x)$ is an isomorphism, then φ is an isomorphism.

PROOF. The surjectivity of φ follows from Nakayama's lemma. In order to prove that φ is injective, we may assume without loss of generality that the sequence x consists of one element, say x. Let $K = \operatorname{Ker} \varphi$; since x is N-regular, the exact sequence

$$0 \longrightarrow K \longrightarrow M \longrightarrow N \longrightarrow 0$$

induces the exact sequence

$$0 \longrightarrow K/xK \longrightarrow M/xM \longrightarrow N/xN \longrightarrow 0.$$

By assumption, $K/xK = 0$, and hence $K = 0$, by Nakayama's lemma. $\square$

Proposition 3.3.3. Let $(R, \mathfrak{m}, k)$ be a Cohen–Macaulay local ring of dimension d, and C a maximal Cohen–Macaulay R-module.
(a) Suppose M is a maximal Cohen–Macaulay R-module with $\operatorname{Ext}_R^j(M, C) = 0$ for all $j > 0$. Then $\operatorname{Hom}_R(M, C)$ is a maximal Cohen–Macaulay module, and for any R-sequence x we have

$$\operatorname{Hom}_R(M, C) \otimes R/xR \cong \operatorname{Hom}_{R/xR}(M/xM, C/xC).$$

(b) Assume in addition that C has finite injective dimension, and M is a Cohen–Macaulay R-module of dimension t. Then
   (i) $\operatorname{Ext}_R^j(M, C) = 0$ for $j \neq d - t$,
   (ii) $\operatorname{Ext}_R^{d-t}(M, C)$ is a Cohen–Macaulay module of dimension t.

PROOF. (a) Let $x \in \mathfrak{m}$ be an R-regular element. Since C is a maximal Cohen–Macaulay module, the element x is C-regular as well, and one has the exact sequence

$$0 \longrightarrow C \xrightarrow{x} C \longrightarrow C/xC \longrightarrow 0$$

which by our assumtion induces the exact sequence

$$0 \longrightarrow \operatorname{Hom}_R(M, C) \xrightarrow{x} \operatorname{Hom}_R(M, C) \longrightarrow \operatorname{Hom}_R(M, C/xC) \longrightarrow 0.$$

Therefore,

$$\mathrm{Hom}_{R/xR}(M/xM, C/xC) \cong \mathrm{Hom}_R(M, C/xC)$$
$$\cong \mathrm{Hom}_R(M, C)/x\, \mathrm{Hom}_R(M, C)$$
$$\cong \mathrm{Hom}_R(M, C) \otimes R/xR.$$

For an arbitrary R-sequence x one proceeds by induction on the length of the sequence.

(b)(i) It follows from 1.2.10(e) that $\mathrm{Ext}_R^j(M, C) = 0$ for $j < d - t$. Next we show by induction on t that $\mathrm{Ext}_R^j(M, C) = 0$ for any t-dimensional Cohen–Macaulay module M and all $j > d - t$. If $t = 0$, then the claim follows from 3.1.17. Now suppose that $t > 0$, and let $x \in \mathfrak{m}$ be an M-regular element. The exact sequence

$$0 \longrightarrow M \overset{x}{\longrightarrow} M \longrightarrow M/xM \longrightarrow 0$$

induces the exact sequence

$$\mathrm{Ext}_R^j(M, C) \overset{x}{\longrightarrow} \mathrm{Ext}_R^j(M, C) \longrightarrow \mathrm{Ext}_R^{j+1}(M/xM, C).$$

M/xM is a $(t - 1)$-dimensional Cohen–Macaulay module. Hence by the induction hypothesis we have $\mathrm{Ext}_R^{j+1}(M/xM, C) = 0$ for $j > d - t$, and so Nakayama's lemma implies that $\mathrm{Ext}_R^j(M, C) = 0$ for $j > d - t$.

(ii) We proceed by induction on t. The assertion is trivial if $t = 0$. Assume now $\dim M = t > 0$, and let $x \in \mathfrak{m}$ be an M-regular element. By (i), the exact sequence

$$0 \longrightarrow M \overset{x}{\longrightarrow} M \longrightarrow M/xM \longrightarrow 0$$

yields the exact sequence

$$0 \longrightarrow \mathrm{Ext}_R^{d-t}(M, C) \overset{x}{\longrightarrow} \mathrm{Ext}_R^{d-t}(M, C) \longrightarrow \mathrm{Ext}_R^{d-(t-1)}(M/xM, C) \longrightarrow 0.$$

Thus x is regular on $\mathrm{Ext}_R^{d-t}(M, C)$, and so it follows from our induction hypothesis that $\mathrm{Ext}_R^{d-t}(M, C)$ is Cohen–Macaulay. $\square$

We are now ready to prove the uniqueness of the canonical module.

Theorem 3.3.4. *Let $(R, \mathfrak{m}, k)$ be a Cohen–Macaulay local ring, and let C and C' be canonical modules of R. Then*

(a) $C/xC \cong E_{R/(x)}(k)$ *for any maximal R-sequence x,*

(b) *the canonical modules C and C' are isomorphic,*

(c) $\mathrm{Hom}_R(C, C') \cong R$, *and any generator φ of $\mathrm{Hom}(C, C')$ is an isomorphism,*

(d) *the canonical homomorphism $R \to \mathrm{End}_R(C)$ is an isomorphism.*

PROOF. (a) By 3.1.15, C/xC is an injective $R/(x)$-module of type 1. Since $\mathrm{Spec}(R/(x)) = \{\mathfrak{m}/(x)\}$, 3.2.8 yields the assertion.

(b) and (c): It follows from (a) that

$$C/xC \cong E_{R/(x)}(k) \cong C'/xC'.$$

Now 3.3.3 and 3.2.12 imply that

$$\mathrm{Hom}_R(C, C') \otimes_R R/(x) \cong \mathrm{Hom}_{R/(x)}(C/xC, C'/xC') \cong R/(x),$$

and so $\mathrm{Hom}_R(C, C')$ is cyclic by Nakayama's lemma. Let φ be a generator of this module. Then the natural inclusion $R\varphi \to \mathrm{Hom}_R(C, C')$ induces the above isomorphism modulo x. By 3.3.3, $\mathrm{Hom}_R(C, C')$ is a maximal Cohen–Macaulay module. Thus 3.3.2 implies that $R\varphi \to \mathrm{Hom}_R(C, C')$ is an isomorphism. In particular it follows that $R\varphi$ is a maximal Cohen–Macaulay module. We may therefore apply 3.3.2 once again to conclude that $R \to R\varphi$ is an isomorphism, too.

Next we show that $\varphi: C \to C'$ is an isomorphism. Indeed, $\varphi \otimes R/(x)$ can be identified with a generator of $\mathrm{End}(E_{R/(x)}(k))$. It follows therefore from 3.2.12(e)(ii) that $\varphi \otimes R/(x)$ is an isomorphism. Since C' is a maximal Cohen–Macaulay module, 3.3.2 implies that φ is an isomorphism. It is clear that any other isomorphism $C \to C'$ is a generator of $\mathrm{Hom}_R(C, C')$, too.

(d) is proved similarly.   □

In view of this result we may talk of *the* canonical module of R provided it exists. From now on we will denote the canonical module of R by ω_R.

The next theorem lists some useful and often applied change of ring formulas for the canonical module.

Theorem 3.3.5. *Let $(R, \mathfrak{m}, k)$ be a Cohen–Macaulay local ring with canonical module ω_R. Then*

(a) $\omega_R/x\omega_R \cong \omega_{R/xR}$ *for all R-sequences x, that is, the canonical module specializes,*

(b) $(\omega_R)_\mathfrak{p} \cong \omega_{R_\mathfrak{p}}$ *for all $\mathfrak{p} \in \mathrm{Spec}\, R$, that is, the canonical module localizes,*

(c) $(\omega_R)\widehat{} \cong \omega_{\hat{R}}$.

PROOF. (a) First notice that x is an ω_R-sequence, too. The (R/xR)-module $\omega_R/x\omega_R$ has finite injective dimension; see 3.1.15. Since $r(\omega_R/x\omega_R) = r(\omega_R) = 1$, the module $\omega_R/x\omega_R$ is the canonical module of R/xR, by definition.

(b) The $R_\mathfrak{p}$-module $(\omega_R)_\mathfrak{p}$ has finite injective dimension (see 3.1.9), and is again a maximal Cohen–Macaulay module. It remains to be shown that $r((\omega_R)_\mathfrak{p}) = 1$. Let x be a sequence of elements of R whose image in $R_\mathfrak{p}$ is a maximal $R_\mathfrak{p}$-sequence. Then by 3.1.15,

$$M = (\omega_R)_\mathfrak{p}/x(\omega_R)_\mathfrak{p}$$

is an injective module over the Artinian local ring $A = R_\mathfrak{p}/xR_\mathfrak{p}$. It follows from 3.2.7 and 3.2.8 that

$$M \cong E_A(k(\mathfrak{p}))^r, \qquad r = r(M).$$

From 3.2.12 we get

(1) $$\operatorname{Hom}_A(M, M) \cong A^{r^2}.$$

On the other hand, from 3.3.5(a) we obtain

(2) $$\begin{aligned} \operatorname{Hom}_A(M, M) &\cong \operatorname{Hom}_{R/xR}(\omega_R/x\omega_R, \omega_R/x\omega_R)_\mathfrak{p} \\ &\cong \operatorname{Hom}_{R/xR}(\omega_{R/xR}, \omega_{R/xR})_\mathfrak{p} \\ &\cong (R/xR)_\mathfrak{p} = A. \end{aligned}$$

For the last isomorphism we used the fact that the endomorphism ring of the canonical module of $S = R/xR$ is isomorphic to S; see 3.3.4. A comparison of (1) and (2) yields $r = 1$, as desired.

(c) The fibre of $R \to \hat{R}$ is k, so that by flatness, $\operatorname{Ext}_R^i(k, \omega_R)\widehat{} \cong \operatorname{Ext}_{\hat{R}}^i(k, (\omega_R)\widehat{})$ for all i. This implies the assertion. $\qquad\square$

Existence of the canonical module. Our next goal is to clarify for which Cohen–Macaulay local rings the canonical module exists.

Theorem 3.3.6. *Let $(R, \mathfrak{m}, k)$ be a Cohen–Macaulay local ring. The following conditions are equivalent:*
(a) *R admits a canonical module;*
(b) *R is the homomorphic image of a Gorenstein local ring.*

One direction of the proof resorts to the principle of idealization due to Nagata: let R be a ring and M an R-module. We construct a ring extension $R \subset R * M$ of R, called the *trivial extension* of R by M. As an R-module, $R * M$ is just the direct sum of R and M. The multiplication is defined by

$$(a, x)(b, y) = (ab, ay + bx)$$

for all $a, b \in R$ and $x, y \in M$.

Some basic facts on trivial extensions are the subject of Exercise 3.3.22. Here we will only use that $R * M$ is a ring, and if M is finite and R is a Noetherian (or Artinian) local ring with maximal ideal $\mathfrak{m}$, then so is $R * M$ with maximal ideal $\mathfrak{m} * M = \{(a, x) \in R * M : a \in \mathfrak{m}\}$.

PROOF OF 3.3.6. (a) $\Rightarrow$ (b): The ring R is a homomorphic image of the trivial extension $R * \omega_R$. We will show that $R * \omega_R$ is a Gorenstein ring. Let x be an R-regular sequence of maximal length. It is easy to see that x is a maximal $(R * \omega_R)$-sequence as well, and that

$$(R * \omega_R)/x(R * \omega_R) \cong (R/xR) * (\omega_R/x\omega_R).$$

By 3.3.4, $\omega_R/x\omega_R \cong E_{R/xR}(k)$. Bearing in mind the characterization 3.2.10 of Gorenstein rings, we may assume that R is Artinian, and it remains to be shown that the type of the Artinian local ring $R' = R * E_R(k)$ is 1.

Let $(a, x) \in \operatorname{Soc} R'$; then $(b, 0)(a, x) = (ba, bx) = (0, 0)$ for all $b \in \mathfrak{m}$. This implies that $a \in \operatorname{Soc} R$ and $x \in \operatorname{Soc} E_R(k)$.

Assume that $a \neq 0$. The exact sequence

$$R \xrightarrow{a} R \longrightarrow R/(a) \longrightarrow 0$$

induces the exact sequence $0 \longrightarrow E_{R/(a)}(k) \longrightarrow E_R(k) \xrightarrow{a} E_R(k)$ (see 3.1.6).

As

$$\ell(E_{R/(a)}(k)) = \ell(R/(a)) < \ell(R) = \ell(E_R(k))$$

(see 3.2.12), multiplication by a on $E_R(k)$ cannot be the zero map. Therefore there exists $y \in E_R(k)$ with $ay \neq 0$, and so $(0, y)(a, x) = (0, ay) \neq (0, 0)$, a contradiction.

Our conclusion is that $\operatorname{Soc}(R * E_R(k)) \cong \operatorname{Soc} E_R(k)$, and therefore by 3.2.12, $r(R * E(k)) = 1$.

For the proof of 3.3.6(b) $\Rightarrow$ (a) we note the following more general result.

Theorem 3.3.7. *Let $(R, \mathfrak{m})$ be a Cohen–Macaulay local ring.*
(a) *The following conditions are equivalent:*
    (i) *R is Gorenstein;*
    (ii) *ω_R exists and is isomorphic to R.*
(b) *Let $\varphi : (R, \mathfrak{m}) \to (S, \mathfrak{n})$ be a local homomorphism of Cohen–Macaulay local rings such that S is a finite R-module. If ω_R exists, then ω_S exists and*

$$\omega_S \cong \operatorname{Ext}_R^t(S, \omega_R), \qquad t = \dim R - \dim S.$$

PROOF. (a)(i) $\Longleftrightarrow$ (ii) follows from 3.1.18 and 3.2.10.

(b) By virtue of 2.1.2(b), and since $\dim S = \dim(R/\operatorname{Ker} \varphi)$, there exists an R-sequence $x = x_1, \dots, x_t$ with $x_i \in \operatorname{Ker} \varphi$, $t = \dim R - \dim S$. Set $\bar{R} = R/(x)R$; as $\omega_R/(x)\omega_R \cong \omega_{\bar{R}}$ (see 3.3.5), we have $\operatorname{Ext}_R^t(S, \omega_R) \cong \operatorname{Hom}_{\bar{R}}(S, \omega_{\bar{R}})$, by 3.1.16. Thus we may assume from the beginning that $\dim R = \dim S$.

Let $d = \dim R$, and $y = y_1, \dots, y_d$ an R-sequence. Then y is ω_R-regular and $\operatorname{Hom}_R(S, \omega_R)$-regular as well, since both modules are Cohen–Macaulay modules of dimension d; see 3.3.3. It follows from 3.3.3(a) that

$$\operatorname{Hom}_R(S, \omega_R) \otimes_R R' \cong \operatorname{Hom}_{R'}(S', \omega_{R'}),$$

where $R' = R/(y)R$, and $S' = S/(y)S$. In view of Exercise 3.3.23 it suffices to show that $\operatorname{Hom}_{R'}(S', \omega_{R'})$ is the canonical module of S'. Since $\omega_{R'} \cong E_{R'}(k)$, 3.1.6 implies that $\operatorname{Hom}_{R'}(S', E_{R'}(k))$ is an injective S'-module, and so $\operatorname{Hom}_{R'}(S', E_{R'}(k)) \cong E_{S'}(k)^r$ for some $r > 0$. By 3.2.12(b) and (e)(i) we get $\ell(E_{S'}(k)) = \ell(S') = \ell(\operatorname{Hom}_{R'}(S', E_{R'}(k))) = r\,\ell(E_{S'}(k))$; therefore $r = 1$. □

A noteworthy case of 3.3.7 is the following: let k be a field, and R an Artinian local k-algebra. Then $\operatorname{Hom}_k(R, k)$ is the canonical module of R.

A Noetherian complete local ring is a homomorphic image of a regular local ring; see A.21. Regular local rings are Gorenstein (see 3.1.20), and so 3.3.6 implies

Corollary 3.3.8. *A complete Cohen–Macaulay local ring admits a canonical module.*

Corollary 3.3.9. *Let $(R, \mathfrak{m}, k)$ be a regular local ring and $I \subset \mathfrak{m}$ an ideal of height g such that $S = R/I$ is Cohen–Macaulay. Let*

$$F_\bullet : 0 \longrightarrow F_g \longrightarrow F_{g-1} \longrightarrow \cdots \longrightarrow F_0 \longrightarrow 0$$

be the minimal free R-resolution of S, and let $G_\bullet = \operatorname{Hom}_R(F_\bullet, R)$ be the dual complex

$$G_\bullet : 0 \longrightarrow G_g \longrightarrow G_{g-1} \longrightarrow \cdots \longrightarrow G_0 \longrightarrow 0,$$

where $G_i = F^*_{g-i}$ for $i = 0, \ldots, g$. Then $G_\bullet$ is a minimal free R-resolution of ω_S.

PROOF. Note that g is indeed the length of the minimal free resolution of S; see 2.2.10. One has $\operatorname{Ext}^i_R(S, R) \cong H^i(F^*_\bullet)$ for all $i \geq 0$. The corollary follows therefore from 3.3.3(b) and 3.3.7. □

Further properties of the canonical module. In the next theorem some useful characterizations of the canonical module will be given.

Theorem 3.3.10. *Let $(R, \mathfrak{m}, k)$ be a Cohen–Macaulay local ring of dimension d, and let C be a finite R-module. Then the following conditions are equivalent:*

(a) *C is the canonical module of R;*

(b) *$\mu_i(\mathfrak{p}, C) = \delta_{ih}$ for all $i \geq 0$ and all $\mathfrak{p} \in \operatorname{Spec} R$, where $h = \operatorname{height} \mathfrak{p}$;*

(c) *for all integers $t = 0, 1, \ldots, d$, and all Cohen–Macaulay R-modules M of dimension t one has*

    (i) *$\operatorname{Ext}^{d-t}_R(M, C)$ is a Cohen–Macaulay R-module of dimension t,*

    (ii) *$\operatorname{Ext}^i_R(M, C) = 0$ for all $i \neq d - t$,*

    (iii) *there exists an isomorphism $M \to \operatorname{Ext}^{d-t}_R(\operatorname{Ext}^{d-t}_R(M, C), C)$ which in the case $d = t$ is just the natural homomorphism from M into the bidual of M with respect to C;*

(d) *for all maximal Cohen–Macaulay R-modules M one has*

    (i) *$\operatorname{Hom}_R(M, C)$ is a maximal Cohen–Macaulay R-module,*

    (ii) *$\operatorname{Ext}^i_R(M, C) = 0$ for $i > 0$,*

    (iii) *the natural homomorphism $M \to \operatorname{Hom}_R(\operatorname{Hom}_R(M, C), C)$ is an isomorphism.*

PROOF. (a) $\Longleftrightarrow$ (b): The canonical module localizes; see 3.3.5; therefore (a) implies (b). Choosing $\mathfrak{p} = \mathfrak{m}$, we obtain (a) from (b).

(a) $\Rightarrow$ (c): (i) and (ii) have already been shown in 3.3.3. From the Rees lemma 3.1.16 and (i) one deduces that

$$\mathrm{Ext}_R^{d-t}(\mathrm{Ext}_R^{d-t}(M, C), C) \cong \mathrm{Hom}_{R/xR}(\mathrm{Hom}_{R/xR}(M, C/xC), C/xC)$$

for an R-sequence x of length $d - t$ which is contained in $\mathrm{Ann}_R M$. Replacing R by R/xR, we may as well assume that $t = d$. Since by 3.3.3, $\mathrm{Hom}_R(M, C) \otimes R/yR \cong \mathrm{Hom}_{R/yR}(M/yM, C/yC)$ for any R-sequence y, we may finally assume that $\dim R = 0$. In this case however $C \cong E_R(k)$, and the assertion follows from 3.2.12.

(d) is a special case of (c).

(d) $\Rightarrow$ (a): If we choose $M = R$, then it follows from (i) that C is a maximal Cohen–Macaulay module.

According to Exercise 2.1.26, for all $i \geq d$ the i-th syzygy module of the residue class field k of R is 0 or a maximal Cohen–Macaulay module. Therefore (ii) implies that $\mathrm{Ext}_R^i(k, C) = 0$ for $i > d$, and hence we have $\mathrm{inj\,dim}\, C < \infty$; see 3.1.14.

It remains to be shown that $r(C) = 1$. By 3.3.3(a), the conditions in (d) are stable under reduction modulo R-sequences. Thus, since the type of C is also stable under reduction modulo R-sequences, we may restrict ourselves to the case where R is Artinian. Then the module C is necessarily injective, and so it must be isomorphic to $E_R(k)^r$, $r = r(C)$. Now it follows from 3.2.12 that $R^{r^2} \cong \mathrm{Hom}_R(\mathrm{Hom}_R(R, C), C)$. Consequently condition (iii) implies $r(C) = 1$. $\square$

We complement the previous theorem with some extra information about the $\mathrm{Ext}^i(\_, \omega_R)$. Observe the analogy of the statements with 3.2.12. The canonical module takes the position of the injective hull when one deals with arbitrary Cohen–Macaulay local rings rather than Artinian local rings.

Proposition 3.3.11. *Let R be a Cohen–Macaulay local ring of dimension d with canonical module ω_R, and M a Cohen–Macaulay R-module of dimension t. Then*

(a) $\mu(M) = r(\mathrm{Ext}_R^{d-t}(M, \omega_R))$,

(b) $r(M) = \mu(\mathrm{Ext}_R^{d-t}(M, \omega_R))$,

(c) ω_R *is a faithful R-module, and*

   (i) $r(\omega_R) = 1$, $\mu(\omega_R) = r(R)$,

   (ii) $\mathrm{End}(\omega_R) = R$.

PROOF. There exists an ω_R-sequence x of length $d - t$ in $\mathrm{Ann}\, M$, and we get

$$\mathrm{Ext}_R^{d-t}(M, \omega_R) \cong \mathrm{Hom}_{R/xR}(M, \omega_{R/xR}).$$

So we may assume that $\dim R = \dim M$. By 3.3.3, we may further assume that $\dim R = 0$. Since the canonical module of an Artinian local ring is the injective hull of the residue class field, all assertions follow from 3.2.12. □

The previous proposition has an interesting application.

Corollary 3.3.12. *Let R be a Cohen–Macaulay local ring, M a Cohen–Macaulay R-module and $\mathfrak{p} \in \operatorname{Supp} M$. Then $r(M_\mathfrak{p}) \leq r(M)$.*

PROOF. Pick $\mathfrak{q} \in \operatorname{Ass}(\hat{R}/\mathfrak{p}\hat{R})$; then $\dim \hat{R}/\mathfrak{q} = \dim R/\mathfrak{p}$; see 2.1.15. Therefore we obtain a flat local homomorphism $R_\mathfrak{p} \to \hat{R}_\mathfrak{q}$ whose fibre is of dimension zero. From 1.2.16 it follows that $r(M_\mathfrak{p}) \leq r(\hat{M}_\mathfrak{q})$. Since $R \to \hat{R}$ is flat with fibre k, 1.2.16 once again applied gives $r(M) = r(\hat{M})$. We may therefore assume that R is complete. By A.11, R is the epimorphic image of complete regular local S, and by Exercise 1.2.26(c) we have $r_S(M) = r_R(M)$. Thus we may assume that R is regular. In particular R is Gorenstein. Hence, by 3.3.7, R has a canonical module and is isomorphic to R, and so 3.3.11 yields

$$r(M_\mathfrak{p}) = \mu(\operatorname{Ext}_{R_\mathfrak{p}}^{d-t}(M_\mathfrak{p}, R_\mathfrak{p})) = \mu(\operatorname{Ext}_R^{d-t}(M, R)_\mathfrak{p})$$
$$\leq \mu(\operatorname{Ext}_R^{d-t}(M, R)) = r(M),$$

where $d = \dim R$ and $t = \dim M$. Here we have used that, by 2.1.3,

$$d - t = \dim R - \dim M = \dim R - \dim R/\mathfrak{p} - (\dim M - \dim M/\mathfrak{p}M)$$
$$= \dim R_\mathfrak{p} - \dim M_\mathfrak{p}. \qquad \square$$

The canonical module and flat extensions. We will show that the canonical module behaves well under flat ring extensions. For the proof we need

Proposition 3.3.13. *Let $(R, \mathfrak{m})$ be a Cohen–Macaulay local ring, and C a finite R-module. The following conditions are equivalent:*
(a) C is the canonical module of R;
(b) C is a faithful maximal Cohen–Macaulay R-module of type 1.

PROOF. (a) $\Rightarrow$ (b): The canonical module is a maximal Cohen–Macaulay module of type 1, by definition, and faithful by 3.3.11.

(b) $\Rightarrow$ (a): Note first that C has one of the properties in (a) or (b) if and only if the completion $\hat{C}$ has this property. For instance, the property of being faithful means that the canonical homomorphism $\varphi: R \to \operatorname{Hom}_R(C, C)$ is injective. Since $R \to \hat{R}$ is faithfully flat, φ is injective if and only if its completion is injective. The reader should check the other properties.

We may now assume that R is complete. Then, by 3.3.8, R has a canonical module ω_R. By 3.3.11(b), $\operatorname{Hom}_R(C, \omega_R)$ is a cyclic module, say

R/I, so that by 3.3.10(d)(iii) we have $C \cong \mathrm{Hom}_R(R/I, \omega_R)$. It follows that I annihilates C. Since we assume that C is faithful, we get $I = 0$, and hence $C \cong \mathrm{Hom}_R(R, \omega_R) \cong \omega_R$. $\square$

Theorem 3.3.14. *Let $(R, \mathfrak{m})$ be a Cohen–Macaulay local ring, and $(R, \mathfrak{m}) \to (S, \mathfrak{n})$ a flat homomorphism of local rings.*
(a) *If ω_R exists and $S/\mathfrak{m}S$ is Gorenstein, then $\omega_S = \omega_R \otimes S$.*
(b) *If C is a finite R-module, and S a Cohen–Macaulay ring with canonical module $\omega_S = C \otimes S$, then $S/\mathfrak{m}S$ is Gorenstein and $C \cong \omega_R$.*

PROOF. A flat local homomorphism is faithfully flat. Thus we see as in the proof of the previous proposition that a finite R-module M is faithful if and only if $M \otimes_R S$ is a faithful S-module. By 1.2.16, we have

$$r(\omega_R \otimes S) = r(S/\mathfrak{m}S)r(\omega_R) = 1;$$

hence (a) follows from 3.3.13.

Part (b) is proved similarly: by 1.2.16, $1 = r(S/\mathfrak{m}S)r(C)$ and C is a maximal Cohen–Macaulay module. It follows that $r(S/\mathfrak{m}S) = 1$ and $r(C) = 1$. Therefore $S/\mathfrak{m}S$ is Gorenstein (see 3.2.10), and in view of 3.3.13, C is the canonical module of R. $\square$

Corollary 3.3.15. *Let $\varphi : (R, \mathfrak{m}) \to (S, \mathfrak{n})$ be a flat homomorphism of Noetherian local rings. Then S is Gorenstein if and only if R and $S/\mathfrak{m}S$ are Gorenstein.*

The canonical module for non-local rings. We saw in 3.3.5 that the canonical module localizes. This suggests the following

Definition 3.3.16. Let R be a Cohen–Macaulay ring. A finite R-module ω_R is a *canonical module* of R if $(\omega_R)_\mathfrak{m}$ is a canonical module of $R_\mathfrak{m}$ for all maximal ideals $\mathfrak{m}$ of R.

Remark 3.3.17. In contrast to the local case, a canonical module is in general not unique (up to isomorphism). Indeed, let R be a Cohen–Macaulay ring (not necessarily local), and let ω_R and ω'_R be canonical modules of R. We set $I = \mathrm{Hom}_R(\omega_R, \omega'_R)$. Localizing at a prime ideal and using 3.3.4 and 3.3.5 we see that $I_\mathfrak{p} \cong R_\mathfrak{p}$ for all $\mathfrak{p} \in \mathrm{Spec}\, R$.

We define an R-module homomorphism $\alpha : I \otimes \omega_R \to \omega'_R$ by $\alpha(\varphi \otimes x) = \varphi(x)$ for all $\varphi \in I$ and all $x \in \omega_R$. Then α is an isomorphism since it is locally an isomorphism.

Conversely, suppose ω_R is a canonical module of R and I is a locally free R-module of rank 1. Then $I \otimes \omega_R$ is locally isomorphic to ω_R, and so is a canonical module of R. Thus a canonical module of a Cohen–Macaulay ring is only unique up to a tensor product with a locally free module of rank 1.

Proposition 3.3.18. *Let R be a Cohen–Macaulay ring and ω_R a canonical module of R.*
(a) *The following conditions are equivalent:*
(i) ω_R *has a rank;*
(ii) $\operatorname{rank} \omega_R = 1$;
(iii) *R is generically Gorenstein, i.e. $R_\mathfrak{p}$ is Gorenstein for all minimal prime ideals $\mathfrak{p}$ of R.*
(b) *If the equivalent conditions of* (a) *hold, then ω_R can be identified with an ideal in R. For any such identification, ω_R is an ideal of height 1 or equals R. In the first case, the ring R/ω_R is Gorenstein.*

PROOF. (a)(i) $\Rightarrow$ (ii): Let $\mathfrak{p} \in \operatorname{Spec} R$ be a minimal prime ideal. Then $\omega_{R_\mathfrak{p}} \cong (\omega_R)_\mathfrak{p}$ is a free $R_\mathfrak{p}$-module, and $R_\mathfrak{p}$ is Artinian. The canonical module of an Artinian local ring is the injective hull E of the residue class field, and so the free $R_\mathfrak{p}$-module $(\omega_R)_\mathfrak{p}$ has rank 1 since E is indecomposable; see 3.2.6. The implications (ii) $\Rightarrow$ (iii) and (iii) $\Rightarrow$ (i) are clear in view of 3.3.7.

(b) The canonical module ω_R is torsion-free since all R-regular elements are ω_R-regular as well. According to Exercise 1.4.18, ω_R is isomorphic to a submodule of R. Therefore it may be identified with an ideal in R which we again denote by ω_R.

If $\dim R = 0$, then necessarily $\omega_R = R$. We may therefore assume that $\dim R > 0$, and that ω_R is a proper ideal of R. Then ω_R must contain an R-regular element since $\operatorname{rank} \omega_R = 1$. Let $\mathfrak{p}$ be a prime ideal containing ω_R. Using the fact that $\omega_R R_\mathfrak{p}$ is a maximal Cohen–Macaulay $R_\mathfrak{p}$-module we then get $\dim R_\mathfrak{p} - 1 \geq \dim(R_\mathfrak{p}/\omega_R R_\mathfrak{p}) \geq \operatorname{depth}(R_\mathfrak{p}/\omega_R R_\mathfrak{p}) \geq \operatorname{depth} R_\mathfrak{p} - 1 = \dim R_\mathfrak{p} - 1$. This shows that $\operatorname{height} \omega_R = 1$, and that R/ω_R is Cohen–Macaulay.

Finally we prove that R/ω_R is Gorenstein. To show this, we may assume that R is local. Applying the functor $\operatorname{Hom}_R(\_, \omega_R)$ to the exact sequence

$$0 \longrightarrow \omega_R \longrightarrow R \longrightarrow R/\omega_R \longrightarrow 0,$$

and using 3.3.4(d) we obtain the exact sequence

$$0 \longrightarrow \omega_R \longrightarrow R \longrightarrow \operatorname{Ext}_R^1(R/\omega_R, \omega_R) \longrightarrow \operatorname{Ext}_R^1(R, \omega_R) = 0.$$

This implies $R/\omega_R \cong \operatorname{Ext}_R^1(R/\omega_R, \omega_R)$. Thus the conclusion follows from 3.3.7. $\qquad\square$

Corollary 3.3.19. *Let R be a Cohen–Macaulay normal domain with canonical module ω_R. Then ω_R is isomorphic to a divisorial ideal. In particular, if R is factorial, then R is Gorenstein.*

PROOF. By 3.3.18, ω_R is an ideal. It satisfies the Serre condition (S_2), and moreover, $(\omega_R)_\mathfrak{p} \cong \omega_{R_\mathfrak{p}} \cong R_\mathfrak{p}$ for all prime ideals of height 1. This follows from 3.3.7 since, by normality, $R_\mathfrak{p}$ is regular for all prime ideals of height 1. Thus we have shown that ω_R is a reflexive ideal; see 1.4.1.

A reflexive ideal is divisorial. (We refer to Fossum [108] for the theory of divisorial ideals.) In a factorial ring all divisorial ideals are principal, and so ω_R is principal and R is Gorenstein; see 3.3.7.   □

In concluding these considerations we show that formula 3.3.14(a) for the canonical module under flat extensions has a non-local counterpart.

Proposition 3.3.20. *Let $\varphi \colon R \to S$ be a flat homomorphism of Noetherian rings whose fibres $S \otimes_R k(\mathfrak{p})$ are Gorenstein for all $\mathfrak{p} \in \operatorname{Spec} R$ for which there exists a maximal ideal $\mathfrak{q}$ in S with $\mathfrak{p} = \mathfrak{q} \cap R$. If ω_R is a canonical module of R, then $\omega_R \otimes_R S$ is a canonical module of S.*

PROOF. Let $\mathfrak{q}$ be a maximal ideal of S, $\mathfrak{p} = \mathfrak{q} \cap R$: then $R_\mathfrak{p} \to S_\mathfrak{q}$ is a flat local homomorphism whose fibre is a localization of $S \otimes_R k(\mathfrak{p})$, and thus is Gorenstein. It follows from 3.3.14(a) that $\omega_{R_\mathfrak{p}} \otimes_{R_\mathfrak{p}} S_\mathfrak{q}$ is a canonical module of $S_\mathfrak{q}$. Since $(\omega_R \otimes_R S)_\mathfrak{q} \cong \omega_{R_\mathfrak{p}} \otimes_{R_\mathfrak{p}} S_\mathfrak{q}$, the proposition is proved.   □

Corollary 3.3.21. *Let R be a Cohen–Macaulay ring with canonical module ω_R, and let S be either the polynomial ring $R[X_1, \ldots, X_n]$ or the formal power series ring $R[[X_1, \ldots, X_n]]$. Then $\omega_R \otimes_R S$ is a canonical module of S. In particular, if R is Gorenstein, then so is S.*

PROOF. We may assume that $n = 1$. The result then follows from 3.3.20 since in both cases the fibres considered there are regular rings; see the proof of A.12.   □

Exercises

3.3.22. Let R be a ring, M an R-module, and $R * M$ the trivial extension of R by M. (The definition of $R * M$ is given after Theorem 3.3.6.) Prove:
(a) $R * M$ is a ring.
(b) R can be identified with the subring $R * 0 = \{(a, x) \in R * M : x = 0\}$.
(c) $0 * M = \{(a, x) \in R * M : a = 0\}$ is an ideal in $R * M$ with $(0 * M)^2 = 0$. As R-modules, M and $0 * M$ are isomorphic.
(d) If $(R, \mathfrak{m})$ is local, then $R * M$ is local with maximal ideal $\mathfrak{m} * M = \{(a, x) \in R * M : a \in \mathfrak{m}\}$.
(e) The natural inclusion $R \to R * M$ composed with the natural epimorphism $R * M \to (R * M)/(0 * M)$ is an isomorphism.
(e) If R is Noetherian and M is a finite R-module, then $R * M$ is Noetherian and $\dim R = \dim R * M$.

3.3.23. Let R be a Cohen–Macaulay local ring, C a maximal Cohen–Macaulay R-module, and x an R-sequence. If C/xC is the canonical module of $R/(x)$, show that C is the canonical module of R.

3.3.24. Let $(R, \mathfrak{m})$ be a Gorenstein local ring and $I \subset R$ an ideal of grade g such that $S = R/I$ is a Cohen–Macaulay ring. Let $x = x_1, \ldots, x_n$ be a system of generators of I. Show that $\omega_S \cong H_{n-g}(x)$.

3.3.25. Let $(R, \mathfrak{m})$ be a Gorenstein local ring, $I \subset R$ a perfect ideal of grade g, and let

$$0 \longrightarrow F_g \xrightarrow{\partial_g} \cdots \xrightarrow{\partial_1} F_0 \longrightarrow 0$$

be a minimal free R-resolution of $S = R/I$. Prove:

(a) The dual complex $0 \longrightarrow F_0^* \xrightarrow{\partial_1^*} \cdots \xrightarrow{\partial_g^*} F_g^* \longrightarrow 0$ is acyclic, and Coker $\partial_g^* \cong \omega_S$.

(b) $\operatorname{Hom}_S(\omega_S, S) \cong \operatorname{Ker}(F_g \otimes S \to F_{g-1} \otimes S) \cong \operatorname{Tor}_g^R(S, S)$.

(c) S is Gorenstein if and only if $\operatorname{Tor}_g^R(S, S) \cong S$. (If you find this problem too difficult, consult 1.4.22 or [159], Section 7.5.)

(d) Suppose $g = 2$; then $\mu(I) = r(S) + 1$. In particular, if R is regular and S is Gorenstein, then S is a complete intersection

3.3.26. Let $(R, \mathfrak{m}, k)$ be a Gorenstein local ring of dimension d, and M a finite module of finite projective dimension. Show that

$$\operatorname{Tor}_i^R(k, M) \cong \operatorname{Ext}_R^{d-i}(k, M) \qquad \text{for all } i.$$

3.3.27. Let $(R, \mathfrak{m})$ be a Cohen–Macaulay local ring with canonical module ω_R. Suppose for all finite R-modules M there exist an integer n and an epimorphism $\omega_R^n \to M$. Prove R is a Gorenstein ring.

3.3.28. Let $(R, \mathfrak{m})$ be a Cohen–Macaulay local ring with canonical module ω_R.

(a) Suppose M is a maximal Cohen–Macaulay R-module of finite injective dimension. Show M is isomorphic to a direct sum of finitely many copies of ω_R.

(b) Let M be a finite R-module. Show inj dim $M < \infty$ if and only if M has a finite ω_R-resolution, that is, there exists an exact sequence

$$0 \longrightarrow \omega_R^{r_p} \xrightarrow{\varphi_p} \cdots \xrightarrow{\varphi_1} \omega_R^{r_0} \longrightarrow M \longrightarrow 0.$$

Hint: For all finite R-modules M there exists an exact sequence $0 \to Y \to X \to M \to 0$ where X is a maximal Cohen–Macaulay R-module, and Y a module of finite injective dimension; see [21]. Such an exact sequence is called a *Cohen–Macaulay approximation*.

(c) The ω_R-resolution is *minimal* if $\operatorname{Im} \varphi_i \subset \mathfrak{m} \omega_R^{r_{i-1}}$ for $i = 1, \ldots, p$. Show that a module M of finite injective dimension even has a minimal ω_R-resolution, and that $r_i = \mu_{d-i}(\mathfrak{m}, M)$ for all i when the resolution is minimal.

3.3.29. Let $(R, \mathfrak{m})$ be a Cohen–Macaulay local ring of dimension 1. A subset I of the total ring Q of fractions of R is called a *fractionary ideal* if there exist R-regular elements x, y such that $y \in xI \subset R$. The *inverse* of a fractionary ideal I is the set $I^{-1} = \{a \in Q : aI \subset R\}$. We denote by $\mathscr{F}$ the set of fractionary ideals of R. Show:

(a) If $I \in \mathscr{F}$, then $I^{-1} \in \mathscr{F}$ and $I \subset (I^{-1})^{-1}$.

(b) If $I \subset R$ is a fractionary ideal, then $\ell(R/I) < \infty$, $R \subset I^{-1}$ and $\ell(I^{-1}/R) < \infty$.

(c) The following conditions are equivalent:

    (i) R is a Gorenstein ring;

    (ii) $I = (I^{-1})^{-1}$ for all $I \in \mathscr{F}$,

    (iii) $\ell(R/I) = \ell(I^{-1}/R)$ for all $I \in \mathscr{F}, I \subset R$.

3.3.30. Let $R \to S$ be a faithfully flat homomorphism of Noetherian rings, and C a finite R-module. Show the following are equivalent:
(a) $C \otimes S$ is a canonical module of S;
(b) C is a canonical module of R, and for every prime ideal $q \in \operatorname{Spec} S$ the fibre $S_q/\mathfrak{p}S_q$, $\mathfrak{p} = q \cap R$, is Gorenstein.

3.3.31. Let k be a field, and R a k-algebra which is Cohen–Macaulay and admits a canonical module. Let K be a field, and suppose that either R is a finitely generated k-algebra or K is a finitely generated extension field of k. Show that $\omega_R \otimes_k K$ is a canonical module of $R \otimes_k K$.
Hint: apply 2.1.11.

3.4 Gorenstein ideals of grade 3. Poincaré duality

The Hilbert–Burch theorem 1.4.17 identifies perfect ideals of grade 2 as the ideals of maximal minors of certain matrices. For Gorenstein ideals of grade 3 there exists a similar 'structure theorem' due to Buchsbaum and Eisenbud [65].

Let R be a Noetherian local ring. An ideal $I \subset R$ is a *Gorenstein ideal* (*of grade g*) if I is perfect and $\operatorname{Ext}_R^g(R/I, R) \cong R/I$. Note that if R is Gorenstein and I is perfect, then I is Gorenstein if and only if R/I is Gorenstein. This follows from 3.3.7(b).

To describe the structure theorem we recall a few facts from linear algebra: let R be a commutative ring, and F a finite free R-module. An R-module homomorphism $\varphi: F \to F^*$ is said to be *alternating* if with respect to some (and therefore with respect to any) basis of F and the corresponding dual basis F^*, the matrix of φ is skew-symmetric and all its diagonal elements are 0.

Suppose now that φ is alternating, choose a basis of F and the basis dual to this, and identify φ with the corresponding matrix (a_{ij}). If rank F is odd, then $\det \varphi = 0$, and if rank F is even, there exists an element $\operatorname{pf}(\varphi) \in R$, called the *Pfaffian* of φ, which is a polynomial function of the entries of φ, such that $\det(\varphi) = \operatorname{pf}(\varphi)^2$. For more details about Pfaffians we refer the reader to [48], Ch. IX, §5, no. 2. We set $\operatorname{pf}(\varphi) = 0$ if rank F is odd. Just like determinants, Pfaffians can be developed along a row. Denote by φ_{ij} the matrix obtained from φ by deleting the i-th and j-th rows and columns of φ; then for all i,

$$\operatorname{pf}(\varphi) = \sum_{j=1}^n (-1)^{i+j-1} \sigma(i,j) a_{ij} \operatorname{pf}(\varphi_{ij})$$

($\sigma(i,j)$ is the sign of $j - i$). From now on we assume that the rank of F is odd, and consider the matrix ψ derived from φ by repeating the i-th row and column as indicated in the following picture

$$\psi = \begin{pmatrix} 0 & a_{i1} & \cdots & a_{in} \\ \hline -a_{i1} & & & \\ \vdots & & \varphi & \\ -a_{in} & & & \end{pmatrix}.$$

Expansion with respect to the first row of ψ yields the equations

$$0 = -\operatorname{pf}(\psi) = \sum_{j=1}^{n} (-1)^j a_{ij} \operatorname{pf}(\varphi_j)$$

for $i = 1, \ldots, n$, where φ_j is the matrix obtained from φ by deleting the j-th row and column. In other words, if we let $\gamma : R \to F$ be the linear map defined by $(p_1, \ldots, p_n)$ (with respect to the given basis) where $p_j = (-1)^j \operatorname{pf}(\varphi_j)$, $j = 1, \ldots, n$, are the submaximal Pfaffians of φ, then we obtain the complex

$$F_\bullet(\varphi) : 0 \longrightarrow R \xrightarrow{\gamma} F \xrightarrow{\varphi} F^* \xrightarrow{\gamma^*} R \longrightarrow 0.$$

Theorem 3.4.1 (Buchsbaum–Eisenbud). *Let $(R, \mathfrak{m})$ be a Noetherian local ring.*
(a) *Suppose $n \geq 3$ is an integer, F a free R-module of rank n, and $\varphi : F \to F^*$ an alternating map of rank $n-1$ whose image is contained in $\mathfrak{m}F^*$. Then n is odd. Moreover, if $\operatorname{Pf}(\varphi)$ denotes the ideal generated by the submaximal Pfaffians of φ, then $\operatorname{grade} \operatorname{Pf}(\varphi) \leq 3$. If $\operatorname{grade} \operatorname{Pf}(\varphi) = 3$, then $F_\bullet(\varphi)$ is acyclic and $\operatorname{Pf}(\varphi)$ is a Gorenstein ideal.*
(b) *Conversely, let I be a Gorenstein ideal of grade 3. Then there exist a free module F of odd rank and an alternating homomorphism $\varphi : F \to F^*$ such that $F_\bullet(\varphi)$ is a minimal free R-resolution of R/I. In particular, any Gorenstein ideal of grade 3 is minimally generated by an odd number of Pfaffians.*

Part (a) of the theorem is a consequence of the Buchsbaum–Eisenbud acyclicity criterion 1.4.13 and the following simple observation relating the ideal of $(n-1)$-minors of φ to the ideal of submaximal Pfaffians.

Lemma 3.4.2. *Let $(R, \mathfrak{m})$ be a Noetherian ring, F a free R-module of rank n, and $\varphi : F \to F^*$ an alternating map of rank $n-1$. Then $\operatorname{Rad} \operatorname{Pf}(\varphi) = \operatorname{Rad} I_{n-1}(\varphi)$, and n is odd.*

PROOF. For all $i, j = 1, \ldots, n$, we denote by α_{ij} the matrix which is obtained from φ by deleting the i-th row and j-th column. Then $I_{n-1}(\varphi)$ is generated by the elements $\det(\alpha_{ij})$, and since $\operatorname{pf}(\varphi_i)^2 = \det(\alpha_{ii})$ it follows right away that a power of $\operatorname{Pf}(\varphi)$ is contained in $I_{n-1}(\varphi)$. Conversely, we consider the matrix of $\bigwedge^{n-1} \varphi$ with entries $\det(\alpha_{ij})$,

$i, j = 1, \ldots, n$. It follows from Exercise 1.6.25 that $\operatorname{rank} \bigwedge^{n-1} \varphi = 1$ since, by assumption, $\operatorname{rank} \varphi = n - 1$. Now 1.4.11 implies that all 2-minors of $\bigwedge^{n-1} \varphi$ are zero. Therefore, since φ is skew-symmetric, we have $-\det(\alpha_{ij})^2 = \det(\alpha_{ij})\det(\alpha_{ji}) = \det(\alpha_{ii})\det(\alpha_{jj}) = \operatorname{pf}(\varphi_i)^2 \operatorname{pf}(\varphi_j)^2$ for all $i, j = 1, \ldots, n$. This implies that a power of $I_{n-1}(\varphi)$ is contained in $\operatorname{Pf}(\varphi)$.

Finally, since $I_{n-1}(\varphi) \neq 0$, we conclude that $\operatorname{Pf}(\varphi) \neq 0$. This is only possible if n is odd. $\qquad \square$

For the proof of part (b) of 3.4.1 a little excursion to resolutions with algebra structures is needed. Let R be a commutative ring, and let

$$P_\bullet : \cdots \longrightarrow P_2 \xrightarrow{\ d\ } P_1 \xrightarrow{\ d\ } P_0 = R \longrightarrow 0$$

be an acyclic complex of projective modules. We may consider $P_\bullet$ as a graded module equipped with an endomorphism $d : P_\bullet \to P_\bullet$ of degree -1 satisfying $d \circ d = 0$. The question is whether there can be defined an associative multiplication on $P_\bullet$ satisfying the following rules:

(a) $P_p P_q \subset P_{p+q}$ for all $p, q \geq 0$;

(b) $1 \in P_0$ acts as the unit element, i.e. $1a = a1 = a$ for all $a \in P_\bullet$;

(c) $ab = (-1)^{(\deg a)(\deg b)} ba$ for all homogeneous elements $a, b \in P_\bullet$;

(d) $aa = 0$ for all homogeneous elements $a \in P_\bullet$ of odd degree;

(e) $d(ab) = (da)b + (-1)^{\deg a} a(db)$ for all homogeneous elements $a, b \in P_\bullet$.

An example of a complex admitting such a multiplication is the Koszul complex. Unfortunately not all finite projective resolutions can be given an algebra structure with these properties. Avramov [25] found obstructions for this, and gave explicit examples of finite projective resolutions which fail to have such a structure. Nevertheless, if we do not insist on the associativity of the multiplication, we surprisingly have

Theorem 3.4.3 (Buchsbaum–Eisenbud). *Any projective resolution $P_\bullet$ with $P_0 = R$ admits a (possibly non-associative) multiplication satisfying the conditions (a)–(e).*

PROOF. We form the tensor product $P_\bullet \otimes P_\bullet$ of complexes, and define the second symmetric power $S_2(P_\bullet)$ of $P_\bullet$ to be

$$S_2(P_\bullet) = (P_\bullet \otimes P_\bullet)/U$$

where U is the graded submodule of $P_\bullet \otimes P_\bullet$ which is generated by the elements $a \otimes b - (-1)^{(\deg a)(\deg b)} b \otimes a$ with homogeneous $a, b \in P_\bullet$, and the elements $a \otimes a$ with homogeneous $a \in P_\bullet$ of odd degree. Let d again denote the differential of $P_\bullet \otimes P_\bullet$; then $d(U) \subset U$. This implies that d induces a differential on $S_2(P_\bullet)$, so that $S_2(P_\bullet)$ inherits a complex structure. We claim (and this is crucial for the proof) that the homogeneous components $S_2(P_\bullet)_k$ of this complex are all projective modules. Indeed, we have

$$S_2(P_\bullet)_k \cong (\bigoplus_{\substack{i+j=k \\ i<j}} P_i \otimes P_j) \oplus T_k$$

where

$$T_k \cong \begin{cases} 0 & \text{if } k \text{ is odd,} \\ \bigwedge^2 P_{k/2} & \text{if } k \text{ is of the form } 4n+2, \\ S_2(P_{k/2}) & \text{if } k \text{ is of the form } 4n. \end{cases}$$

Thus $S_2(P_\bullet)$ is a complex of projective R-modules which coincides with $P_\bullet$ in degrees 0 and 1. Therefore there exists a complex homomorphism $\Phi : S_2(P_\bullet) \to P_\bullet$ extending the identity in degrees 0 and 1, and we may assume that Φ is chosen such that its restriction to $R \otimes P_k$ is just the natural homomorphism to P_k.

For all homogeneous elements $a, b \in P_\bullet$ we denote by ab the image of $a \otimes b$ under the composition of the maps $P_\bullet \otimes P_\bullet \xrightarrow{\quad} S_2(P_\bullet) \xrightarrow{\Phi} P_\bullet$, and extend this multiplication by linearity to all other elements of $P_\bullet$. It is clear that it has all the desired properties. $\qquad \square$

Suppose now we are given a Noetherian local ring R and a Gorenstein ideal $I \subset R$ of grade g. Let

$$F_\bullet : 0 \longrightarrow F_g \longrightarrow \cdots \longrightarrow F_1 \longrightarrow F_0 \longrightarrow 0$$

be the minimal free resolution of R/I. The dual complex $F_\bullet^*$ is a minimal free resolution of $\operatorname{Ext}_R^g(R/I, R) \cong R/I$, and hence must be isomorphic to $F_\bullet$. Such an isomorphism is unique up to homotopy, and we are now choosing one which is derived from the multiplicative structure on $F_\bullet$ as given by 3.4.3. Observe that the multiplication defines maps $F_i \otimes F_{g-i} \to F_g \cong R$ which in turn induce R-module homomorphisms $s_i : F_i \to F_{g-i}^*$.

For $i = 0, \ldots, g$ we let

$$t_i = \begin{cases} s_i & \text{if } i \equiv 0, 1 \bmod 4, \\ -s_i & \text{if } i \equiv 2, 3 \bmod 4. \end{cases}$$

Proposition 3.4.4. $t_\bullet : F_\bullet \to F_\bullet^*$ *is an isomorphism of complexes. In particular, $s_i : F_i \to F_{g-i}^*$ is an isomorphism for $i = 0, \ldots, g$.*

PROOF. We denote by d the differential of $F_\bullet$. Let $a \in F_i$ and $b \in F_{g+1-i}$. Then $ab = 0$, and therefore $0 = d(ab) = d(a)b + (-1)^i ad(b)$, or $d(a)b = (-1)^{i+1} ad(b)$. It follows that

$$s_{i-1}(d(a))(b) = d(a)b = (-1)^{i+1} ad(b) = (-1)^{i+1} s_i(a)(d(b))$$
$$= (-1)^{i+1} d^*(s_i(a))(b).$$

Thus $s_{i-1} \circ d = (-1)^{i+1} d^* \circ s_i$ which implies that $t.$ is a homomorphism of complexes. The induced homomorphism $H_0(t.): R/I \to R/I$ must be an isomorphism since $t_0 = s_0$ is an isomorphism. Since $t.$ extends $H_0(t.)$ it must be an isomorphism as well. $\square$

We are now ready to prove 3.4.1(b): let

$$F. : 0 \longrightarrow F_3 \xrightarrow{\psi} F_2 \xrightarrow{\varphi} F_1 \xrightarrow{\rho} R \longrightarrow 0$$

be the minimal free resolution of R/I equipped with a multiplication as in 3.4.3. Let $e_1, \dots, e_n$ be a basis of F_2. Then, as we have just seen, $s_2(e_1), \dots, s_2(e_n)$ is a basis of F_1^*, and we may choose basis elements $f_1, \dots, f_n$ of F_1 such that $f_i^* = s_2(e_i)$ for $i = 1, \dots, n$. Then $e_i f_j = \delta_{ij} g$ for all $i, j = 1, \dots, n$ where g is a basis element of F_3 and δ_{ij} denotes the Kronecker symbol. Let $\varphi(e_i) = \sum_{j=1}^n a_{ij} f_j$; we claim that (a_{ij}) is skew-symmetric, and all its diagonal elements are 0. To see this, notice that $e_j \varphi(e_i) = a_{ij} g$. Therefore,

$$a_{ij} \psi(g) = \psi(e_j \varphi(e_i)) = \varphi(e_j) \varphi(e_i).$$

The claim follows since ψ is injective, and since $\varphi(e_i) \varphi(e_j) = -\varphi(e_j) \varphi(e_i)$ and $\varphi(e_i) \varphi(e_i) = 0$ according to the multiplication rules.

Now let $\psi(g) = \sum_{i=1}^n a_i e_i$; since $F. \cong F.^*$, ψ is isomorphic to the transpose of ρ, and we conclude that $I = (a_1, \dots, a_n)$. On the other hand, rank $\varphi = n - 1$ and grade $\mathrm{Pf}(\varphi) = \mathrm{grade}\, I_{n-1}(\varphi)$, by 3.4.2. Now grade $I_{n-1}(\varphi) \geq 3$ since $F.$ becomes split exact after localizations at prime ideals of height ≤ 2. Thus part (a) of Theorem 3.4.1 implies that $F.(\varphi)$ is acyclic. In particular it follows that $\mathrm{Ker}\,\varphi$ is generated by $\sum_{i=1}^n p_i e_i$ where, up to signs, the p_i are the submaximal Pfaffians of φ. Thus we have $\mathrm{Pf}(\varphi) = I$, as desired.

Poincaré duality. Buchsbaum and Eisenbud [65] remark that the multiplication defined on $F.$ induces a multiplication on $\mathrm{Tor}.(k, R/I)$ giving it the structure of an associative graded alternating algebra. They further point out that in view of 3.4.4, $\mathrm{Tor}.(k, R/I)$ is a Poincaré algebra if I is a Gorenstein ideal. Recall that an associative graded alternating algebra $A = \bigoplus_{i=0}^g A_i$ is a *Poincaré algebra* if for all $i = 0, \dots, g$ the A_0-homomorphisms $A_i \to \mathrm{Hom}_{A_0}(A_{g-i}, A_g)$, $a \mapsto \varphi_a$ with $\varphi_a(b) = ab$, are isomorphisms.

Notice that if R is regular, then there is a natural isomorphism between $\mathrm{Tor}.(k, R/I)$ and the Koszul homology $H.(R/I) = H.(x, R/I)$ where x is a minimal set of generators of the maximal ideal of R; see 1.6.9. It can be shown that this is an isomorphism of algebras. In particular, the Koszul homology $H.(R)$ of a Gorenstein ring is a Poincaré algebra. This is one direction of the theorem of Avramov and Golod [32] which asserts that a Gorenstein ring is characterized by its Koszul homology. Their theorem complements the result 2.3.11 of Tate and Assmus according to which the Koszul algebra of a complete intersection is an exterior algebra. We will present their proof which is independent of the above considerations.

Theorem 3.4.5 (Avramov–Golod). *Let $(R, \mathfrak{m}, k)$ be a Noetherian local ring, and let $n = \operatorname{emb\,dim} R - \operatorname{depth} R$. The following conditions are equivalent:*
(a) *R is a Gorenstein ring;*
(b) *$H_{\textbf{.}}(R)$ is a Poincaré algebra;*
(c) *the k-linear map $H_{n-1}(R) \to \operatorname{Hom}_k(H_1(R), H_n(R))$ induced by the multiplication on $H_{\textbf{.}}(R)$ is a monomorphism.*

We begin with a few preliminary remarks. Suppose $t = \operatorname{depth} R > 0$. By 1.2.2 (choose $M = R$ and $N = \mathfrak{m}$), there exists an R-regular element $y_1 \in \mathfrak{m} \setminus \mathfrak{m}^2$. Hence by induction on t we may construct an R-sequence $y = y_1, \ldots, y_t$ such that y is part of a minimal system of generators of $\mathfrak{m}$. By 1.6.13, one has $H_{\textbf{.}}(R) \cong H_{\textbf{.}}(R/yR)$ as graded k-vector spaces. Inspecting this isomorphism we see that it is actually a k-algebra isomorphism. On the other hand, R is Gorenstein if and only if R/yR is too. Thus we may assume that $\operatorname{depth} R = 0$.

Let $x = x_1, \ldots, x_n$ be a minimal system of generators of $\mathfrak{m}$, and $K_{\textbf{.}} = K_{\textbf{.}}(x)$ the Koszul complex of this sequence; then, by definition, $H_{\textbf{.}}(R) = H_{\textbf{.}}(K_{\textbf{.}})$. Note that $K_{\textbf{.}}$ is a Poincaré algebra: let $e_1, \ldots, e_n$ be an R-basis of K_1; then, in the terminology of Section 1.6, the e_I, $I \subset \{1, \ldots, n\}$, $|I| = i$, form an R-basis of K_i, and we have $e_I \wedge e_J = \sigma(I, J) e_1 \wedge \cdots \wedge e_n$ for $J \subset \{1, \ldots, n\}$, $|J| = n - i$. Here $\sigma(I, J) = \pm 1$ if $I \cap J = \emptyset$, and 0 otherwise. This clearly proves that the maps $\omega_i : K_i \to \operatorname{Hom}_R(K_{n-i}, K_n)$, $\omega_i(a) = \varphi_a$ with $\hat{\varphi}_a(b) = a \wedge b$, are isomorphisms as asserted.

We denote by $d_{\textbf{.}}$ the differential of $K_{\textbf{.}}$. Then $d_{\textbf{.}}$ and its dual anticommute with $\omega_{\textbf{.}}$. In other words, we have

$$\omega_{i-1} \circ d_i = (-1)^{i-1} \operatorname{Hom}(d_{n-i+1}, K_n) \circ \omega_i$$

for all $i = 0, \ldots, n$. This equation is stated in 1.6.10, the only difference being that there K_n is identified with R. It follows that the isomorphisms ω_i induce isomorphisms $\tilde{\omega}_i : H_i(R) \to H^{n-i}(R)$ where we identify $H^{n-i}(R)$ with $H^{n-i}((K_{\textbf{.}})^{\textbf{*}})$, and where $(K_{\textbf{.}})^{\textbf{*}} = \operatorname{Hom}_R(K_{\textbf{.}}, K_n)$.

Consider the diagram

$$
\begin{array}{ccc}
H_i(R) & \xrightarrow{\;\;\Delta_i\;\;} & \operatorname{Hom}_k(H_{n-i}(R), H_n(R)) \\
\tilde{\omega}_i \downarrow & & \downarrow \gamma_i \\
H^{n-i}(R) & \xrightarrow{\;\;\beta_i\;\;} & \operatorname{Hom}_R(H_{n-i}(R), K_n)
\end{array}
$$

Here the upper map Δ_i is induced by the multiplication on $H_{\textbf{.}}(R)$. We have just seen that $\tilde{\omega}_i$ is an isomorphism. The lower map β_i is the natural homomorphism which assigns to a homology class $\bar{\psi}$, ψ an $(n-i)$-cycle in

$(K_.)^*$, the (well defined) homomorphism $\beta_i(\bar{\psi}) \in \operatorname{Hom}_R(H_{n-i}(R), K_n)$ with $\beta_i(\bar{\psi})(\bar{a}) = \psi(a)$, $a \in K_{n-i}$ a cycle. Next note that $H_n(R) = \operatorname{Soc} K_n \subset K_n$. We define γ_i to be $\operatorname{Hom}(H_{n-i}(R), \iota)$ where $\iota: \operatorname{Soc} K_n \to K_n$ is the natural inclusion. It is clear that γ_i is an isomorphism. In fact, since $H_{n-i}(R)$ is annihilated by $\mathfrak{m}$, any homomorphism $H_{n-i}(R) \to K_n$ necessarily maps $H_{n-i}(R)$ into $\operatorname{Soc} K_n$.

We leave it to the reader to check the commutativity of the diagram. In conclusion we have that Δ_i is a mono-, epi-, or isomorphism if and only if β_i is too. We will determine the kernel and cokernel of β_i.

Lemma 3.4.6. *Let $B_.$ denote the boundaries of $K_.$. Then for any i we have a long exact sequence*

$$0 \longrightarrow \operatorname{Ext}^1_R(K_{i-1}/B_{i-1}, K_n) \longrightarrow H^i(R) \xrightarrow{\beta_{n-i}} \operatorname{Hom}_R(H_i(R), K_n)$$
$$\longrightarrow \operatorname{Ext}^1_R(B_{i-1}, K_n) \longrightarrow \cdots$$

PROOF. The short exact sequence $0 \to H_i(R) \to K_i/B_i \to B_{i-1} \to 0$ gives rise to the long exact sequence

$$0 \longrightarrow \operatorname{Hom}_R(B_{i-1}, K_n) \longrightarrow \operatorname{Hom}_R(K_i/B_i, K_n) \xrightarrow{\sigma_i} \operatorname{Hom}_R(H_i(R), K_n)$$
$$\longrightarrow \operatorname{Ext}^1_R(B_{i-1}, K_n) \longrightarrow \cdots$$

It is immediate to see that $\operatorname{Im} \beta_{n-i} = \operatorname{Im} \sigma_i$, so that the sequence

$$H^i(R) \xrightarrow{\beta_{n-i}} \operatorname{Hom}_R(H_i(R), K_n) \longrightarrow \operatorname{Ext}^1_R(B_{i-1}, K_n) \longrightarrow \cdots$$

is exact.

The module U of i-cycles of $(K_.)^*$ whose homology classes belong to $\operatorname{Ker} \beta_{n-i}$ is the module of homomorphisms $\psi \in (K_i)^*$ for which $\psi|_{Z_i} = 0$ (Z_i cycles in K_i). Therefore U is isomorphic to $\operatorname{Hom}_R(B_{i-1}, K_n)$. Under this identification $\operatorname{Ker} \beta_{n-i}$ equals U/V where V is the module of homomorphisms $\psi: B_{i-1} \to K_n$ which can be extended to K_{i-1}. This means that $\operatorname{Ker} \beta_{n-i} \cong \operatorname{Ext}^1_R(K_{i-1}/B_{i-1}, K_n)$. $\square$

In order to complete the proof of 3.4.5 we need the following

Lemma 3.4.7. *Let $(R, \mathfrak{m}, k)$ be a Noetherian local ring of depth 0. If $\operatorname{Ext}^1_R(k, R) = 0$, then R is Gorenstein.*

PROOF. The hypothesis implies that the functor $\operatorname{Hom}_R(\_, R)$ is exact on the category of R-modules of finite length. This yields $\ell(\operatorname{Hom}_R(M, R)) = \ell(M)\,\ell(\operatorname{Hom}_R(k, R))$ for any R-module of finite length M.

Now assume $\dim R > 0$. Then $\ell(R/\mathfrak{m}^n)$, and so $\ell(\operatorname{Hom}_R(R/\mathfrak{m}^n, R))$, tends to infinity with n. On the other hand, $\operatorname{Hom}_R(R/\mathfrak{m}^n, R) \cong 0 : \mathfrak{m}^n$. Since $0 : \mathfrak{m} \subset 0 : \mathfrak{m}^2 \subset \cdots$ is an ascending chain of ideals, and since R

is Noetherian, this chain stabilizes. Consequently, $\ell(\mathrm{Hom}_R(R/\mathfrak{m}^n, R))$ is bounded, a contradiction.

Thus R is a zero dimensional ring for which $\mathrm{Hom}_R(_-, R)$ is an exact functor. Hence R is an injective R-module, and so R is Gorenstein by definition. □

End of the proof of 3.4.5: We have already accomplished the reduction to the case depth $R = 0$.

(a) $\Rightarrow$ (b): Since R is Gorenstein and $K_n \cong R$, all Ext groups in the exact sequence 3.4.6 vanish, and so β_i is an isomorphism for all i.

(b) $\Rightarrow$ (c) is trivial.

(c) $\Rightarrow$ (a): By assumption Δ_{n-1} is injective, and this implies that β_{n-1} is injective. Thus it follows from 3.4.6 that $\mathrm{Ext}_R^1(K_0/B_0, R) = 0$. Now 3.4.7 completes the proof since $K_0/B_0 = k$.

Corollary 3.4.8. *Let R be a Gorenstein local ring which is not a complete intersection. Then $H_1(R)^{n-1} = 0$ for $n = \mathrm{emb}\dim R - \dim R$.*

PROOF. Suppose the vector subspace $H_1(R)^{n-1}$ of $H_{n-1}(R)$ is not zero. Then $H_1(R)^n \neq 0$ since $\Delta_1 : H_1(R) \to \mathrm{Hom}_k(H_{n-1}(R), H_n(R))$ is an isomorphism. This contradicts 2.3.15. □

Exercises

3.4.9. Let k be a field and $I \subset k[\![X_1, X_2, X_3]\!]$ the ideal generated by the polynomials $X_1^2 - X_2^2$, $X_1^2 - X_3^2$, $X_1 X_2$, $X_1 X_3$, $X_2 X_3$. By 3.2.11 it is a Gorenstein ideal of grade 3. Compute its free resolution (as a $k[\![X_1, X_2, X_3]\!]$-module).

3.4.10. Let $(R, \mathfrak{m}, k)$ be a Cohen–Macaulay local ring with canonical module ω_R, and x a minimal set of generators of $\mathfrak{m}$. We denote by $H_\bullet(M)$ the Koszul homology of an R-module M with respect to x. Recall that $H_\bullet(M)$ is an $H_\bullet(R)$-module. Let $n = \mathrm{emb}\dim R - \dim R$. Show that for all i, $0 \leq i \leq n$, the k-linear map $H_i(R) \to \mathrm{Hom}_k(H_{n-i}(\omega_R), H_n(\omega_R))$ which is induced by the scalar multiplication of $H_\bullet(R)$ on $H_\bullet(\omega_R)$ is an isomorphism.

3.5 Local cohomology. The local duality theorem

The canonical module was introduced by Grothendieck in connection with the local duality theorem which relates local cohomology with certain Ext functors. We will describe this approach to the canonical module in this section. First local cohomology functors will be introduced, and it will be shown that the depth and the dimension of a module can be expressed in terms of their vanishing and non-vanishing. We end with the local duality theorem.

Let $(R, \mathfrak{m}, k)$ be a Noetherian local ring and M an R-module. Denote by $\Gamma_\mathfrak{m}(M)$ the submodule of M consisting of all elements of M with support in $\{\mathfrak{m}\}$. That is,

$$\Gamma_{\mathfrak{m}}(M) = \{x \in M : \mathfrak{m}^k x = 0 \text{ for some } k \geq 0\}.$$

Let $\mathscr{F} = (I_k)_{k \geq 0}$ be a family of ideals of R such that $I_j \subset I_k$ for all $j > k$. Then $\mathscr{F}$ defines a topology on R; see [270], Section 8. $\mathscr{F}$ gives the $\mathfrak{m}$-adic topology on R if and only if for each I_k there is a $j \in \mathbb{N}$ such that $\mathfrak{m}^j \subset I_k$, and for each $\mathfrak{m}^i$ there is an $l \in \mathbb{N}$ such that $I_l \subset \mathfrak{m}^i$.

It is clear that for any such family one has

$$\Gamma_{\mathfrak{m}}(M) = \{x \in M : I_k x = 0 \text{ for some } k \geq 0\}.$$

Let $x = x_1, \ldots, x_n$ be a sequence of elements in R generating an $\mathfrak{m}$-primary ideal. We set

$$x^k = x_1^k, \ldots, x_n^k \qquad \text{for all} \quad k \geq 0.$$

The family (x^k) gives the $\mathfrak{m}$-adic topology on R, and so

$$\Gamma_{\mathfrak{m}}(M) = \{y \in M : (x^k)y = 0 \text{ for some } k \geq 0\}.$$

Noting that $\operatorname{Hom}_R(R/I, M) = \{x \in M : Ix = 0\}$ for any ideal I of R, we obtain natural isomorphisms

$$\Gamma_{\mathfrak{m}}(M) \cong \varinjlim \operatorname{Hom}_R(R/\mathfrak{m}^k, M) \cong \varinjlim \operatorname{Hom}_R(R/(x^k), M).$$

Proposition 3.5.1. $\Gamma_{\mathfrak{m}}(\_)$ *is a left exact additive functor.*

PROOF. The additivity of $\Gamma_{\mathfrak{m}}(\_)$ is trivial. We show that $\Gamma_{\mathfrak{m}}(\_)$ is left exact. If

$$0 \longrightarrow M_1 \stackrel{\alpha}{\longrightarrow} M_2 \stackrel{\beta}{\longrightarrow} M_3$$

is exact, then we have a sequence $0 \longrightarrow \Gamma_{\mathfrak{m}}(M_1) \stackrel{\alpha'}{\longrightarrow} \Gamma_{\mathfrak{m}}(M_2) \stackrel{\beta'}{\longrightarrow} \Gamma_{\mathfrak{m}}(M_3)$, where $\alpha' = \Gamma_{\mathfrak{m}}(\alpha) = \alpha|_{\Gamma_{\mathfrak{m}}(M_1)}$ and $\beta' = \Gamma_{\mathfrak{m}}(\beta) = \beta|_{\Gamma_{\mathfrak{m}}(M_2)}$.

It is obvious that α' is injective. Let $x \in \operatorname{Ker} \beta'$; then $\beta(x) = 0$, and so there exists $y \in M_1$ such that $x = \alpha(y)$. Since $x \in \Gamma_{\mathfrak{m}}(M_2)$, there exists an integer $k \geq 0$ such that $\mathfrak{m}^k x = 0$. It follows that $\mathfrak{m}^k \alpha(y) = \alpha(\mathfrak{m}^k y) = 0$. But α is injective, and so $y \in \Gamma_{\mathfrak{m}}(M_1)$ and $\alpha'(y) = x$. $\qquad \square$

Definition 3.5.2. The *local cohomology functors*, denoted by $H_{\mathfrak{m}}^i(\_)$, are the right derived functors of $\Gamma_{\mathfrak{m}}(\_)$. In other words, if $I^\bullet$ is an injective resolution of the R-module M, then $H_{\mathfrak{m}}^i(M) \cong H^i(\Gamma_{\mathfrak{m}}(I^\bullet))$ for all $i \geq 0$.

Remarks 3.5.3. (a) Let M be an R-module; then $H_{\mathfrak{m}}^0(M) \cong \Gamma_{\mathfrak{m}}(M)$ and $H_{\mathfrak{m}}^i(M) = 0$ for $i < 0$.
(b) If I is an injective R-module, then $H_{\mathfrak{m}}^i(I) = 0$ for all $i > 0$.
(c) For any R-module M and all $i \geq 0$ one has

$$H_{\mathfrak{m}}^i(M) \cong \varinjlim \operatorname{Ext}_R^i(R/\mathfrak{m}^k, M) \cong \varinjlim \operatorname{Ext}_R^i(R/(x^k), M),$$

where x is a sequence in R generating an $\mathfrak{m}$-primary ideal.
(d) A short exact sequence of R-modules

$$0 \longrightarrow M_1 \longrightarrow M_2 \longrightarrow M_3 \longrightarrow 0$$

gives rise to a long exact sequence

$$0 \longrightarrow \Gamma_\mathfrak{m}(M_1) \longrightarrow \Gamma_\mathfrak{m}(M_2) \longrightarrow \Gamma_\mathfrak{m}(M_3) \longrightarrow H^1_\mathfrak{m}(M_1) \longrightarrow \cdots$$
$$\longrightarrow H^{i-1}_\mathfrak{m}(M_3) \longrightarrow H^i_\mathfrak{m}(M_1) \longrightarrow H^i_\mathfrak{m}(M_2) \longrightarrow \cdots$$

Only (c) needs some explanation: $\varinjlim$ is an exact functor; see [318], Theorem 2.18. Therefore if $I^\bullet$ is an injective resolution of M, then

$$H^i_\mathfrak{m}(M) \cong H^i(\varinjlim \mathrm{Hom}_R(R/\mathfrak{m}^k, I^\bullet)) \cong \varinjlim H^i(\mathrm{Hom}_R(R/\mathfrak{m}^k, I^\bullet))$$
$$\cong \varinjlim \mathrm{Ext}^i_R(R/\mathfrak{m}^k, M).$$

Note that

$$\Gamma_\mathfrak{m}(E(R/\mathfrak{p})) = \begin{cases} E(k) & \text{if } \mathfrak{p} = \mathfrak{m}, \\ 0 & \text{otherwise}; \end{cases}$$

see 3.2.7 and part (1) of the proof of 3.2.13. Using the structure of the minimal injective resolution $E^\bullet(M)$ of M given in 3.2.9, we conclude that $\Gamma_\mathfrak{m}(E^\bullet(M))$ is a complex of the form

$$0 \longrightarrow E(k)^{\mu_0(\mathfrak{m},M)} \longrightarrow E(k)^{\mu_1(\mathfrak{m},M)} \longrightarrow \cdots \longrightarrow E(k)^{\mu_i(\mathfrak{m},M)} \longrightarrow \cdots$$

This entails

Proposition 3.5.4. *Let $(R, \mathfrak{m}, k)$ be a Noetherian local ring and M a finite R-module.*
(a) *The modules $H^i_\mathfrak{m}(M)$ are Artinian.*
(b) *One has $H^i_\mathfrak{m}(M) = 0$ if and only if $i < \mathrm{depth}\, M$.*
(c) *If R is Gorenstein, then*

$$H^i_\mathfrak{m}(R) \cong \begin{cases} E(k) & \text{for } i = \dim R, \\ 0 & \text{otherwise}. \end{cases}$$

(d) *Let $\hat{N}$ denote the $\mathfrak{m}$-adic completion of an R-module N. Then*

$$H^i_\mathfrak{m}(M) \cong H^i_\mathfrak{m}(M) \otimes_R \hat{R} \cong H^i_{\hat{\mathfrak{m}}}(\hat{M}) \qquad \text{for all} \quad i \geq 0.$$

PROOF. (a), (b) and (c) follow from the structure of $\Gamma_\mathfrak{m}(E^\bullet(M))$ and the fact that $\mathrm{depth}\, M = \inf\{i : \mu_i(\mathfrak{m}, M) \neq 0\}$.

(d) As $H^i_\mathfrak{m}(M)$ is Artinian, it is the direct limit of submodules U_j of finite length. For each U_j one has $U_j \otimes_R \hat{R} \cong U_j$, and so

$$H^i_\mathfrak{m}(M) \cong \varinjlim(U_j \otimes_R \hat{R}) \cong (\varinjlim U_j) \otimes_R \hat{R} \cong H^i_\mathfrak{m}(M) \otimes_R \hat{R}.$$

Using the R-flatness of $\hat{R}$, we get

$$H^i_\mathfrak{m}(M) \otimes_R \hat{R} \cong \varinjlim \mathrm{Ext}^i_R(R/\mathfrak{m}^j, M) \otimes_R \hat{R} \cong \varinjlim \mathrm{Ext}^i_{\hat{R}}(\hat{R}/\hat{\mathfrak{m}}^j, \hat{M})$$
$$\cong H^i_{\hat{\mathfrak{m}}}(\hat{M}). \qquad \square$$

Local cohomology and the Koszul complex. Our next goal is to construct a more explicit complex whose cohomology gives us $H_{\mathfrak{m}}^{\bullet}(M)$. Let $x = x_1, \ldots, x_n$ be a system of parameters of R. For all $l \geq 0$ we get a commutative diagram

$$
\begin{array}{ccc}
K_1(x^{l+1}) & \xrightarrow{\;\varphi_1^{(l)}\;} & K_1(x^l) \\
\downarrow & & \downarrow \\
K_0(x^{l+1}) & =\!=\!=\!= & K_0(x^l)
\end{array}
$$

with $\varphi_1^{(l)}(e_i) = x_i e_i$ for $i = 1, \ldots, n$. (In both Koszul complexes we denote the natural basis of $K_1 \cong R^n$ by $e_1, \ldots, e_n$).

Let $\varphi_i^{(l)} = \bigwedge^i \varphi_1^{(l)}$; then $\varphi_{\bullet}^{(l)} : K_{\bullet}(x^{l+1}) \longrightarrow K_{\bullet}(x^l)$ is a complex homomorphism; see 1.6.8. We denote by

$$
\varphi_l^{\bullet} : K^{\bullet}(x^l) \longrightarrow K^{\bullet}(x^{l+1})
$$

the dual complex homomorphism. This can be done for each l, and so we obtain a direct system of complexes. Thus we may form the complex

$$
\varinjlim K^{\bullet}(x^l).
$$

On the other hand, one defines a complex

$$
C^{\bullet} : 0 \longrightarrow C^0 \longrightarrow C^1 \longrightarrow \cdots \longrightarrow C^n \longrightarrow 0,
$$

$$
C^t = \bigoplus_{1 \leq i_1 < i_2 < \cdots < i_t \leq n} R_{x_{i_1} x_{i_2} \cdots x_{i_t}},
$$

where the differentiation $d^t : C^t \to C^{t+1}$ is given on the component

$$
R_{x_{i_1} \cdots x_{i_t}} \longrightarrow R_{x_{j_1} \cdots x_{j_{t+1}}}
$$

to be the homomorphism $(-1)^{s-1} \cdot \mathrm{nat} : R_{x_{i_1} \cdots x_{i_t}} \to (R_{x_{i_1} \cdots x_{i_t}})_{x_{j_s}}$ if $\{i_1, \ldots, i_t\} = \{j_1, \ldots, \widehat{j_s}, \ldots, j_{t+1}\}$ and 0 otherwise.

The complex $C^{\bullet}$ is called the *modified Čech complex*. In the usual Čech complex, C^0 is replaced by 0 and the homological degree is shifted by 1.

Proposition 3.5.5. $\varinjlim K^{\bullet}(x^l) \cong C^{\bullet}$.

PROOF. For all $l \geq 0$ we define a complex homomorphism

$$
\psi_l^{\bullet} : K^{\bullet}(x^l) \longrightarrow C^{\bullet} \qquad \text{by} \qquad \psi_l^t((e_{j_1} \wedge \cdots \wedge e_{j_t})^{\bullet}) = \frac{1}{(x_{j_1} x_{j_2} \cdots x_{j_t})^l};
$$

here $(e_{j_1} \wedge \cdots \wedge e_{j_t})^{\bullet}$ is an element of the basis of $(\bigwedge^t R^n)^{\bullet}$ which is dual to the standard basis of $\bigwedge^t R^n$. A straightforward calculation shows

(i) $\psi_l^\bullet$ is indeed a complex homomorphism,

(ii) $\psi_l^\bullet = \psi_{l+1}^\bullet \circ \varphi_l^\bullet$ for all $l \geq 0$, and therefore the family $(\psi_l^\bullet)$ induces a complex homomorphism $\psi^\bullet \colon \varinjlim K^\bullet(x^l) \longrightarrow C^\bullet$,

(iii) $\psi^\bullet$ is an isomorphism.

Note that for (iii) one essentially has to verify that $R_{x_{i_1} \cdots x_{i_t}}$ is the limit of the direct system $(F_i)_{i \geq 0}$ in which $F_i = R$ for all i and the map $F_i \to F_{i+1}$ is just multiplication by $x_{i_1} \cdots x_{i_t}$. $\qquad\qquad\square$

The importance of these complexes results from

Theorem 3.5.6. *Let M be an R-module. Then*

$$H_{\mathfrak{m}}^i(M) \cong H^i(M \otimes_R C^\bullet) \cong \varinjlim H^i(x^l, M) \qquad \text{for all} \quad i \geq 0.$$

PROOF. The second isomorphism follows from the fact that $\varinjlim$ is an exact functor, and hence commutes with cohomology. In order to prove the first isomorphism, we show that the functors $H^i(\_ \otimes C^\bullet)$ are the right derived functors of $\Gamma_{\mathfrak{m}}(\_)$.

Identifying $M \otimes R_{x_i}$ with M_{x_i} we have

$$H^0(M \otimes_R C^\bullet) = \operatorname{Ker}(M \longrightarrow \bigoplus_{j=1}^{n} M_{x_j}).$$

The kernel consists of all $m \in M$ for which there exist integers l_j, $j = 1, \ldots, n$, such that $x_j^{l_j} m = 0$, and this set is obviously equal to $\Gamma_{\mathfrak{m}}(M)$.

Since $C^\bullet$ is a complex of flat R-modules, the exact sequence of R-modules

$$0 \longrightarrow M_1 \longrightarrow M_2 \longrightarrow M_3 \longrightarrow 0$$

yields the exact sequence

$$0 \longrightarrow M_1 \otimes_R C^\bullet \longrightarrow M_2 \otimes_R C^\bullet \longrightarrow M_3 \otimes_R C^\bullet \longrightarrow 0,$$

from which we obtain the long exact sequence

$$0 \longrightarrow H^0(M_1 \otimes_R C^\bullet) \longrightarrow H^0(M_2 \otimes_R C^\bullet) \longrightarrow H^0(M_3 \otimes_R C^\bullet)$$
$$\longrightarrow H^1(M_1 \otimes_R C^\bullet) \longrightarrow H^1(M_2 \otimes_R C^\bullet) \longrightarrow \cdots$$

It remains to show that $H^i(I \otimes_R C^\bullet) = 0$ for $i > 0$ and any injective R-module I. Of course we may assume that I is indecomposable.

Let $I = E(k)$ and $a \in E(k)$. Then for $j = 1, \ldots, n$ there exist integers $l_j > 0$ such that $x_j^{l_j} a = 0$, and so $E(k) \otimes C^i = 0$ for $i > 0$.

Next assume $I = E(R/\mathfrak{p})$, $\mathfrak{p} \neq \mathfrak{m}$; then there exists $j \in \{1, \ldots, n\}$ such that $x_j \notin \mathfrak{p}$. We claim that multiplication by x_j on $E(R/\mathfrak{p})$ is an isomorphism. It is certainly a monomorphism since $\operatorname{Ass} E(R/\mathfrak{p}) = \{\mathfrak{p}\}$ (see 3.2.7) and since $x_j \notin \mathfrak{p}$. The submodule $x_j E(R/\mathfrak{p})$ of $E(R/\mathfrak{p})$ is an injective module since $x_j E(R/\mathfrak{p}) \cong E(R/\mathfrak{p})$, and hence is a direct summand of $E(R/\mathfrak{p})$. But $E(R/\mathfrak{p})$ is indecomposable, and so $x_j E(R/\mathfrak{p}) = E(R/\mathfrak{p})$. We can then define a homotopy $\sigma^\bullet$ of the complex $E(R/\mathfrak{p}) \otimes_R C^\bullet$:

$$\sigma^l : E(R/\mathfrak{p}) \otimes_R C^l \longrightarrow E(R/\mathfrak{p}) \otimes_R C^{l-1}$$

is defined on the component $E(R/\mathfrak{p})_{x_{i_1} \cdots x_{i_l}} \longrightarrow E(R/\mathfrak{p})_{x_{j_1} \cdots x_{j_{l-1}}}$ to be $(-1)^{s-1} \cdot$ nat, if $\{j_1, \ldots, j_{l-1}\} = \{i_1, \ldots, \widehat{i_s}, \ldots, i_l\}$ and $i_s = j$, and 0 otherwise. It is easily verified that $\sigma^\bullet$ is a contracting homotopy, that is, the identity and the zero-map of the complex are homotopic via $\sigma^\bullet$. This implies that $E(R/\mathfrak{p}) \otimes_R C^\bullet$ is exact. $\qquad\square$

Grothendieck's theorems. We are now in the position to prove the following important vanishing theorem.

Theorem 3.5.7 (Grothendieck). *Let $(R, \mathfrak{m}, k)$ be a Noetherian local ring and M a finite R-module of depth t and dimension d. Then*
(a) $H^i_\mathfrak{m}(M) = 0$ *for $i < t$ and $i > d$,*
(b) $H^t_\mathfrak{m}(M) \neq 0$ *and $H^d_\mathfrak{m}(M) \neq 0$.*

PROOF. We first note the following rule which will be used several times in the proof of (a) and (b):

Let $\varphi : (R, \mathfrak{m}, k) \to (R', \mathfrak{m}', k')$ be a local ring homomorphism such that $\mathfrak{m} R'$ is an $\mathfrak{m}'$-primary ideal. Then for any R'-module M one has

$$(3) \qquad\qquad H^i_\mathfrak{m}(M) \cong H^i_{\mathfrak{m}'}(M) \qquad \text{for all} \quad i \geq 0.$$

Of course, on the left hand side of this formula M is considered as an R-module.

In fact, if $\mathfrak{m} = (x)$ with $x = x_1, \ldots, x_n$ and if $x' = \varphi(x_1), \ldots, \varphi(x_n)$, then $C^\bullet \otimes_R M \cong C'^\bullet \otimes_{R'} M$, where $C^\bullet$ and $C'^\bullet$ are the complexes of 3.5.6 defined with respect to x and x'. The isomorphism (3) follows from 3.5.6.

(a) We only need to prove that $H^i_\mathfrak{m}(M) = 0$ for $i > d$. The other part of statement (a) has already been shown in 3.5.4.

Let $R \to R' = R/\operatorname{Ann} M$ be the canonical epimorphism. Then M is an R'-module with $\dim M = \dim R'$. Using (3) we may therefore assume that $\dim R = \dim M = d$. Let $x = x_1, \ldots, x_d$ be a system of parameters of R, and let $C^\bullet$ be the complex 3.5.6 defined with respect to x. Then $C^i = 0$ for $i > d$, and so $H^i_\mathfrak{m}(M) \cong H^i(M \otimes_R C^\bullet) = 0$ for $i > d$; see 3.5.6.

(b) We proceed by induction on t in order to show that $H^t_\mathfrak{m}(M) \neq 0$. If $t = 0$, then $0 \neq \operatorname{Soc} M \subset H^0_\mathfrak{m}(M)$. Now suppose $t > 0$; then there exists an M-regular element $x \in \mathfrak{m}$. The exact sequence

$$0 \longrightarrow M \xrightarrow{\ x\ } M \longrightarrow M/xM \longrightarrow 0$$

yields the exact sequence $0 = H^{t-1}_\mathfrak{m}(M) \longrightarrow H^{t-1}_\mathfrak{m}(M/xM) \longrightarrow H^t_\mathfrak{m}(M)$. By our induction hypothesis we have $H^{t-1}_\mathfrak{m}(M/xM) \neq 0$; this implies $H^t_\mathfrak{m}(M) \neq 0$.

Finally we show that $H_{\mathfrak{m}}^d(M) \neq 0$. Using 3.5.4 and the fact that $\dim \hat{M} = \dim M$ for the $\mathfrak{m}$-adic completion $\hat{M}$ of M, we may assume that R is complete.

Let $\mathfrak{p} \in \operatorname{Supp} M$ with $\dim M = \dim R/\mathfrak{p}$. Then $\dim M/\mathfrak{p}M = \dim M = d$, and we get an exact sequence of R-modules

$$0 \longrightarrow U \longrightarrow M \longrightarrow M/\mathfrak{p}M \longrightarrow 0,$$

inducing the exact sequence $H_{\mathfrak{m}}^d(M) \longrightarrow H_{\mathfrak{m}}^d(M/\mathfrak{p}M) \longrightarrow H_{\mathfrak{m}}^{d+1}(U)$. According to (a) we have $H_{\mathfrak{m}}^{d+1}(U) = 0$, and so, if $H_{\mathfrak{m}}^d(M/\mathfrak{p}M) \neq 0$, then $H_{\mathfrak{m}}^d(M) \neq 0$. As $M/\mathfrak{p}M$ is an $R/\mathfrak{p}$-module we may as well assume, by (3), that R is a domain and $\dim R = \dim M$.

Any complete Noetherian domain has a Noether normalization: there exists a regular local subring $(S, \mathfrak{n})$ such that R is a finite S-module; see A.22. In particular, the extension ideal $\mathfrak{n}R$ is $\mathfrak{m}$-primary. Again using (3) we may replace R by S, and so may assume that R itself is regular. Let K be the fraction field of R, and let $\alpha : M \to K \otimes_R M$ be the canonical homomorphism. We set $U = \operatorname{Ker} \alpha$ and $N = \operatorname{Im} \alpha$. Then we obtain the exact sequence

$$(4) \qquad\qquad 0 \longrightarrow U \longrightarrow M \longrightarrow N \longrightarrow 0,$$

and, as a consequence of Exercise 1.4.18, an exact sequence

$$(5) \qquad\qquad 0 \longrightarrow N \longrightarrow R^s \longrightarrow W \longrightarrow 0,$$

where $s = \operatorname{rank} M = \operatorname{rank} N$ and consequently $\dim W < \dim R = d$. As $\dim W < d$, (5) yields the exact sequence

$$H_{\mathfrak{m}}^d(N) \longrightarrow H_{\mathfrak{m}}^d(R^s) \longrightarrow H_{\mathfrak{m}}^d(W) = 0.$$

We have $H_{\mathfrak{m}}^d(R^s) \cong H_{\mathfrak{m}}^d(R)^s \cong E(k)^s$ (see 3.5.4) and so $H_{\mathfrak{m}}^d(N) \neq 0$. Finally, from the exact sequence (4) it follows that $H_{\mathfrak{m}}^d(M) \neq 0$. $\qquad \square$

The next theorem is known as the local duality theorem.

Theorem 3.5.8 (Grothendieck). *Let $(R, \mathfrak{m}, k)$ be a Cohen–Macaulay complete local ring of dimension d. Then for all finite R-modules M and all integers i there exist natural isomorphisms*

$$H_{\mathfrak{m}}^i(M) \cong \operatorname{Hom}_R(\operatorname{Ext}_R^{d-i}(M, \omega_R), E(k)), \quad and$$

$$\operatorname{Ext}_R^i(M, \omega_R) \cong \operatorname{Hom}_R(H_{\mathfrak{m}}^{d-i}(M), E(k)).$$

PROOF. The first isomorphisms result from the second by Matlis duality 3.2.13. For the proof of the second isomorphisms note that both sides vanish for $i < 0$; see 3.5.7. For $i \geq 0$ we set $T^i(\_) = \operatorname{Hom}_R(H_{\mathfrak{m}}^{d-i}(\_), E(k))$. It is clear that $T^0(\_)$ is a contravariant left exact functor which maps direct sums to direct products. Hence there exists an R-module C such that

$$T^0(\_) \cong \text{Hom}_R(\_, C);$$

see [318], Theorem 3.36. It follows that $C \cong T^0(R)$. As $H_{\mathfrak{m}}^d(R)$ is an Artinian module, Matlis duality 3.2.13 implies that C is a finite R-module.

In order to conclude the proof we will show that the functors $T^i(\_)$ are the right derived functors of $T^0(\_)$, and that $C \cong \omega_R$.

Remark 3.5.3 implies immediately that the functors $T^i(\_)$ are strongly connected (see [318], p. 212). Thus the $T^i(\_)$ are the right derived functors of $T^0(\_)$, once we have shown that $T^i(F) = 0$ for every free R-module F and all $i \geq 1$.

The functors $T^i(\_)$ map direct sums to direct products, and so it suffices to show that $T^i(R) = 0$ for $i \geq 1$, or equivalently that $H_{\mathfrak{m}}^i(R) = 0$ for $i < d$. This however follows from 3.5.7 since R is Cohen–Macaulay.

Summing up we have

$$(6) \qquad \text{Hom}_R(H_{\mathfrak{m}}^i(M), E(k)) \cong \text{Ext}_R^{d-i}(M, C)$$

for all i and all R-modules M. Now 3.5.7 yields

$$H_{\mathfrak{m}}^i(k) \cong \begin{cases} k & \text{for } i = 0, \\ 0 & \text{for } i > 0, \end{cases} \quad \text{and therefore} \quad \text{Ext}_R^i(k, C) \cong \begin{cases} k & \text{for } i = d, \\ 0 & \text{for } i \neq d, \end{cases}$$

by (6). Thus it follows from the remark after 3.3.1 that $C \cong \omega_R$. $\qquad \square$

Grothendieck's duality theorem has the following often applied variant:

Corollary 3.5.9. *Let $(R, \mathfrak{m}, k)$ be a Cohen–Macaulay local ring of dimension d which is the homomorphic image of a Gorenstein local ring. Then R has a canonical module, and for all finite R-modules M and all integers i there exist natural isomorphisms*

$$H_{\mathfrak{m}}^i(M) \cong \text{Hom}_R(\text{Ext}_R^{d-i}(M, \omega_R), E(k)).$$

PROOF. For the proof we apply 3.5.4, 3.3.5, and Exercise 3.2.14. Then

$$H_{\mathfrak{m}}^i(M) \cong H_{\hat{\mathfrak{m}}}^i(\hat{M}) \cong \text{Hom}_{\hat{R}}(\text{Ext}_{\hat{R}}^{d-i}(\hat{M}, \omega_{\hat{R}}), E(k))$$

$$\cong \text{Hom}_R(\text{Ext}_R^{d-i}(M, \omega_R), E(k)). \qquad \square$$

Remark 3.5.10. Let $(R, \mathfrak{m}, k)$ be a complete local ring. The proof of 3.5.8 shows that the functor $\text{Hom}_R(H_{\mathfrak{m}}^d(\_), E(k))$ is representable, even if R is not a Cohen–Macaulay ring. In other words, there exist a unique R-module K_R (in the proof of 3.5.8 this module was denoted by C) and a canonical isomorphism

$$\text{Hom}_R(H_{\mathfrak{m}}^d(M), E(k)) \cong \text{Hom}_R(M, K_R)$$

for all R-modules M.

Of course, $K_R \cong \omega_R$ if R is Cohen–Macaulay. Even in the more general situation when the ring is not Cohen–Macaulay, the module K_R is often called the canonical module of R. Its properties have been investigated by Aoyama [10]. Schenzel [329] has introduced the canonical module K_M of an R-module M.

The local duality theorem combined with 3.5.7 allows us to generalize 3.3.10(d).

Corollary 3.5.11. *Let $(R, \mathfrak{m}, k)$ be a Cohen–Macaulay local ring of dimension n with canonical module ω_R, and M a finite R-module of depth t and dimension d. Then*
(a) $\operatorname{Ext}_R^i(M, \omega_R) = 0$ *for* $i < n - d$ *and* $i > n - t$,
(b) $\operatorname{Ext}_R^i(M, \omega_R) \neq 0$ *for* $i = n - d$ *and* $i = n - t$,
(c) $\dim \operatorname{Ext}_R^i(M, \omega_R) \leq n - i$ *for all* $i \geq 0$.

PROOF. We have $\operatorname{Ext}_R^i(M, \omega_R)\widehat{} \cong \operatorname{Ext}_{\hat{R}}^i(\hat{M}, \omega_{\hat{R}})$ for all $i \geq 0$, since $(\omega_R)\widehat{} \cong \omega_{\hat{R}}$ (see 3.3.5). Under completion depth and dimension of a module are preserved. We may therefore assume that R is complete, and so (a) and (b) follow from 3.5.7 and 3.5.8.

To prove (c), we choose $\mathfrak{p} \in \operatorname{Supp} \operatorname{Ext}_R^i(M, \omega_R)$ such that
$$\dim \operatorname{Ext}_R^i(M, \omega_R) = \dim R/\mathfrak{p} = \dim R - \dim R_\mathfrak{p}.$$

(The last equality holds since R is Cohen–Macaulay; see 2.1.3.) By the choice of $\mathfrak{p}$ we have
$$0 \neq \operatorname{Ext}_R^i(M, \omega_R)_\mathfrak{p} \cong \operatorname{Ext}_{R_\mathfrak{p}}^i(M_\mathfrak{p}, \omega_{R_\mathfrak{p}}),$$

and so (a) yields $i \leq \dim R_\mathfrak{p} = n - \dim \operatorname{Ext}_R^i(M, \omega_R)$. $\square$

Exercises

3.5.12. Let $(R, \mathfrak{m})$ be a Noetherian local ring, and M a finite R-module. Prove:
(a) If $\mu_{i-1}(\mathfrak{m}, M) = 0$ and $\mu_i(\mathfrak{m}, M) \neq 0$, then $H_\mathfrak{m}^i(M) \neq 0$.
(b) Suppose inj dim $M = \infty$; then $\mu_i(\mathfrak{m}, M) \neq 0$ for all $i > \dim M$.

3.5.13. Find a Noetherian local ring $(R, \mathfrak{m})$ of dimension d and depth t with
(a) $H_\mathfrak{m}^i(R) \neq 0$ for $i = t, \ldots, d$,
(b) $H_\mathfrak{m}^i(R) = 0$ for $i \neq t$ and $i \neq d$.

3.5.14. Let $(S, \mathfrak{n}, k)$ be a complete Cohen–Macaulay local ring, $(R, \mathfrak{m}, k)$ a residue class ring of S, and M a finite R-module. Show that
$$\operatorname{Hom}_R(H_\mathfrak{m}^i(M), E_R(k)) \cong \operatorname{Hom}_S(H_\mathfrak{n}^i(M), E_S(k)),$$

and derive the following version of the local duality theorem: for all integers i there exist natural isomorphisms
$$\operatorname{Hom}_R(H_\mathfrak{m}^i(M), E_R(k)) \cong \operatorname{Ext}_S^{d-i}(M, \omega_S), \qquad d = \dim S.$$

Hint: See the first step in the proof of 3.5.7, and use 3.1.6.

3.5.15. Let $(R, \mathfrak{m}, k)$ be a regular local ring of dimension $d \geq 2$. Let E be a finite R-module which is locally free on the punctured spectrum of R. That is, $E_\mathfrak{p}$ is free for all $\mathfrak{p} \in \operatorname{Spec} R$, $\mathfrak{p} \neq \mathfrak{m}$. Show

(a) $\ell(H_\mathfrak{m}^i(E)) < \infty$ for all $i < d$,

(b) the R-dual E^* of E is again locally free on the punctured spectrum of R, and $H_\mathfrak{m}^i(E^*) = 0$ for $i = 0, 1$,

(c) $H_\mathfrak{m}^{i+1}(E^*) \cong \operatorname{Hom}_R(H_\mathfrak{m}^{d-i}(E), E(k))$ for $i = 1, \ldots, d - 2$.

3.6 The canonical module of a graded ring

For a graded ring R we define the canonical module in the category of graded R-modules and establish the graded version of the local duality theorem. Under certain restrictive assumptions on R the degrees of the generators in a minimal set of generators of the canonical module are uniquely determined, and one defines the a-invariant of R to be the smallest of these degrees, multiplied by -1.

We adopt the assumptions and notation of Section 1.5. Thus R will be a Noetherian graded ring, and $\mathscr{M}_0(R)$ the category of graded R-modules. $\mathscr{M}_0(R)$ is an Abelian category which has direct sums and direct products; see 1.5.19. Likewise limits and colimits exist in $\mathscr{M}_0(R)$. As we have already mentioned in Section 1.5, $\mathscr{M}_0(R)$ has enough projectives. Our next concern will be to show that $\mathscr{M}_0(R)$ has enough injectives as well.

Injective modules. A graded R-module M is called *injective* if it is an injective object in $\mathscr{M}_0(R)$. One sees easily that this is the case if and only if the functor

$$*\operatorname{Hom}_R(\_, M) \colon \mathscr{M}_0(R) \longrightarrow \mathscr{M}_0(R)$$

is exact.

A *injective module M need not be injective (in the category $\mathscr{M}(R)$; see 3.6.5). Just as in the category of all R-modules, one calls an extension $N \subset M$ of graded R-modules *essential* if for any graded submodule $0 \neq U \subset M$ one has $U \cap N \neq 0$. If, in addition, $N \neq M$, the extension is called a *proper *essential extension*. Similarly as in the non-graded case (see 3.2.2) one shows

Proposition 3.6.1. *A graded module is *injective if and only if it has no proper *essential extension.*

We now prove that any graded R-module has a *injective hull. In analogy to the definition in the non-graded case, E is called a *injective hull* of M if it is *injective and a *essential extension of M.

Theorem 3.6.2. *Any graded R-module M admits a *injective hull, and any two *injective hulls of M are isomorphic.*

PROOF. We embed M into a (not necessarily graded) injective R-module I. According to 3.1.8 this is possible. Similarly as in the proof of 3.2.4 we consider the set $\mathcal{S} = \{N : M \subset N \subset I,\ M \subset N \text{ is } ^*\text{essential}\}$. We define a partial order $\leq$ on $\mathcal{S}$ by setting $N_1 \leq N_2$ if N_1 is a graded submodule of N_2. Zorn's lemma applied to this set yields a maximal *essential extension $M \subset E$ with $E \subset I$. Suppose E is not *injective; then E has a proper *essential extension $E \subset E'$ by 3.6.1. As I is injective, there exists an R-module homomorphism $\varphi : E' \to I$ (not necessarily homogeneous), extending the inclusion $E \subset I$. We claim that φ is injective. In fact, assume that there is a non-zero element $x \in \operatorname{Ker} \varphi$, say $x = x_r + \cdots + x_s$, with x_i homogeneous of degree i, $r \leq s$, and $x_r \neq 0$. We show by induction on $s - r$ that there exists a homogeneous element $a \in R$ such that $ax \in E$ and $ax \neq 0$. Since $\varphi|_E$ is injective, this gives a contradiction.

If $s - r = 0$, x is homogeneous, and the assertion follows since the extension $E \subset E'$ is *essential. Now suppose that $s - r > 0$. We choose a homogeneous element $a \in R$ such that $ax_r \in E \setminus \{0\}$. Let

$$x' = x - x_r = x_{r+1} + \cdots + x_s.$$

If $ax' = 0$, then $ax = ax_r \in E \setminus \{0\}$ and we are done. Otherwise $ax' \neq 0$, and by our induction hypothesis we may choose a homogeneous element $b \in R$ such that $bax' \in E$ and $bax' \neq 0$. Then $bax = bax' + bax_r \in E$, and $bax \neq 0$.

Next let $\widetilde{E} = \operatorname{Im} \varphi$. As φ is injective we may give $\widetilde{E}$ a natural graded structure ($\widetilde{E}_i = \varphi(E_i)$ for all $i \in \mathbb{Z}$). Then $E \subset \widetilde{E}$ is a proper *essential extension with $\widetilde{E} \subset I$, contradicting the maximality of E.

The uniqueness of the *injective hull is proved as in the non-graded case. $\qquad\Box$

We denote the *injective hull of a graded R-module M by $^*E(M)$ or $^*E_R(M)$.

The preceding theorem implies in particular that any graded R-module N has a *injective resolution. That is, there exists a complex

$$I^{\bullet} : 0 \longrightarrow I^0 \longrightarrow I^1 \longrightarrow I^2 \longrightarrow \cdots$$

with *injective modules I^i such that $H^0(I^{\bullet}) \cong N$ and $H^i(I^{\bullet}) = 0$ for $i > 0$. Given such a *injective resolution $I^{\bullet}$ of N we have

$$^*\operatorname{Ext}_R^i(M, N) \cong H^i(^*\operatorname{Hom}_R(M, I^{\bullet}))$$

for all $i \geq 0$ and all graded R-modules M.

We omit the proofs of the following two results which have their analogues in 3.2.6, 3.2.7, 3.2.8, and 3.2.9, and which can be proved along the same lines as in the corresponding local case.

Theorem 3.6.3. *Let R be a Noetherian graded ring. Then*
(a) $\mathrm{Ass}\,{}^*E(M) = \mathrm{Ass}\,M$ *for all* $M \in \mathcal{M}_0(R)$,
(b) $E \in \mathcal{M}_0(R)$ *is a* *indecomposable* *injective module if and only if*

$$E \cong {}^*E(R/\mathfrak{p})(n)$$

for some graded prime ideal $\mathfrak{p} \in R$ *and some integer* $n \in \mathbb{Z}$,
(c) *every* *injective module can be decomposed into a direct sum of* *indecomposable* *injective modules, and this decomposition is unique up to homogeneous isomorphism.*

Proposition 3.6.4. *Let R be a Noetherian graded ring, and M a graded R-module. Consider the minimal* *injective resolution*

$$0 \longrightarrow M \longrightarrow {}^*E^0(M) \xrightarrow{d^0} {}^*E^1(M) \xrightarrow{d^1} \cdots$$

of M (which is obtained recursively by setting ${}^*E^i(M) = {}^*E(\mathrm{Im}\,d^{i-1})$). *Then, for every graded prime ideal $\mathfrak{p}$ of R and for every integer $i \geq 0$, the Bass number $\mu_i(\mathfrak{p}, M)$ equals the number of graded R-modules of the form* ${}^*E(R/\mathfrak{p})(n)$, $n \in \mathbb{Z}$, *that appear in* ${}^*E^i(M)$ *as direct summands.*

For any $M \in \mathcal{M}_0(R)$ we denote by ${}^*\mathrm{inj\,dim}\,M$ the *injective dimension.

Theorem 3.6.5. *Let R be a Noetherian graded ring and $M \in \mathcal{M}_0(R)$. Then*
(a) $\mathrm{inj\,dim}\,M \leq {}^*\mathrm{inj\,dim}\,M + 1$,
(b) *if M is* *injective, then* $\mathrm{inj\,dim}\,M = 1$ *if and only if* $\mathfrak{p}^* \in \mathrm{Ass}\,M$ *for some non-graded prime ideal $\mathfrak{p}$ of R.*

For the proof of 3.6.5 we will use

Proposition 3.6.6. *Let $M \in \mathcal{M}_0(R)$ and $\mathfrak{p}$ a non-graded prime ideal in R. Then $\mu_0(\mathfrak{p}, M) = 0$, and $\mu_{i+1}(\mathfrak{p}, M) = \mu_i(\mathfrak{p}^*, M)$ for every integer $i \geq 0$.*

The proof of this proposition is already given in 1.5.9, where we actually prove more than is stated in that theorem itself.

PROOF OF 3.6.5. (a) We may assume that ${}^*\mathrm{inj\,dim}\,M = t < \infty$. Let $\mathfrak{p} \in \mathrm{Spec}\,R$; we want to prove that $\mu_i(\mathfrak{p}, M) = 0$ for $i \geq t + 2$. This is certainly true when $\mathfrak{p}$ is a graded prime ideal; see 3.6.4. Now suppose that $\mathfrak{p}$ is not a graded prime ideal. Then $\mu_{i+1}(\mathfrak{p}, M) = \mu_i(\mathfrak{p}^*, M)$ by 3.6.6 and $\mu_i(\mathfrak{p}^*, M) = 0$ for $i \geq t + 1$, hence the assertion follows.
  (b) $\mathrm{inj\,dim}\,M = 1$ happens if and only if $\mu_1(\mathfrak{p}, M) \neq 0$ for some non-graded prime ideal $\mathfrak{p}$ of R. But $\mu_1(\mathfrak{p}, M) = \mu_0(\mathfrak{p}^*, M)$, and so $\mu_1(\mathfrak{p}, M) \neq 0$ if and only if $\mathfrak{p}^* \in \mathrm{Ass}\,M$. $\qquad\square$

Corollary 3.6.7. *Let R be a Noetherian graded ring and $\mathfrak{m}$ a graded maximal ideal of R. Then*

$${}^*E(R/\mathfrak{m}) \cong E(R/\mathfrak{m}).$$

PROOF. By 3.6.3 we have $\mathrm{Ass}\,{}^*E(R/\mathfrak{m}) = \{\mathfrak{m}\}$, and so 3.6.5 implies that ${}^*E(R/\mathfrak{m})$ is injective as an object in $\mathcal{M}(R)$. Since $\mu_0(\mathfrak{m}, {}^*E(R/\mathfrak{m})) = 1$ (see

3.6.4), we conclude that $^*E(R/\mathfrak{m})$ is indecomposable in $\mathcal{M}(R)$, and hence by 3.2.6 it must be isomorphic to $E(R/\mathfrak{m})$. $\square$

*The *canonical module of a graded ring.* Recall from 1.5 that a graded ring is a *local ring if it has a unique *maximal ideal, that is, a graded ideal $\mathfrak{m}$ which is not properly contained in any graded ideal $\neq R$.

Definition 3.6.8. Let $(R, \mathfrak{m})$ be a Cohen–Macaulay *local ring of *dimension d. A finite graded R-module C is a *canonical module* of R if there exist homogeneous isomorphisms

$$^*\mathrm{Ext}^i_R(R/\mathfrak{m}, C) \cong \begin{cases} 0 & \text{for } i \neq d, \\ R/\mathfrak{m} & \text{for } i = d. \end{cases}$$

For a finite graded R-module M it may happen that there exists a homogeneous isomorphism $M \cong M(i)$ with $i \neq 0$. To avoid this phenomenon, one has to require that R has no homogeneous units of positive degree. For a *local ring $(R, \mathfrak{m})$ this is the case if and only if $\mathfrak{m}$ is maximal (in the usual sense); see 1.5.16.

Proposition 3.6.9. *Let $(R, \mathfrak{m})$ be a Cohen–Macaulay *local ring, and C be a *canonical module of R. Then*
(a) *C is a canonical module of R,*
(b) *C is uniquely determined up to homogeneous isomorphism, provided $\mathfrak{m}$ is maximal.*

PROOF. (a) We need to show that $C_\mathfrak{p}$ is a canonical module of $R_\mathfrak{p}$ for all $\mathfrak{p} \in \mathrm{Spec}\, R$. First, let $\mathfrak{p} \in \mathrm{Spec}\, R$ be a graded prime ideal; then $\mathfrak{p} \subset \mathfrak{m}$. The definition of the *canonical module implies that $C_\mathfrak{m}$ is a canonical module of $R_\mathfrak{m}$, and so $C_\mathfrak{p}$ is a canonical module of $R_\mathfrak{p}$; see 3.3.5. Now let $\mathfrak{p} \in \mathrm{Spec}\, R$ be a non-graded prime ideal. Then $\mu_{i+1}(\mathfrak{p}, C) = \mu_i(\mathfrak{p}^*, C)$, by 3.6.6, and the assertion follows again.

(b) Let C' be another *canonical module. Remark 3.3.17 implies that $^*\mathrm{Hom}_R(C, C')$ is a projective module of rank 1, and hence, as a graded module, is free (see 1.5.15). Therefore, $^*\mathrm{Hom}_R(C, C') \cong R(i)$ for some $i \in \mathbb{Z}$. This implies $^*\mathrm{Hom}_R(C, C'(-i)) \cong R$. Let $\varphi \in {}^*\mathrm{Hom}_R(C, C'(-i))$ be an element corresponding to 1 under this identification. Then, since $^*\mathrm{Hom}_R(C, C'(-i)) = \mathrm{Hom}_R(C, C')$ by Exercise 1.5.19(f), it follows from (a) and 3.3.4(c) that φ is locally an isomorphism. But then φ is a homogeneous isomorphism, and we have

$$R/\mathfrak{m} \cong {}^*\mathrm{Ext}^d_R(R/\mathfrak{m}, C) \cong {}^*\mathrm{Ext}^d_R(R/\mathfrak{m}, C'(-i))$$
$$\cong {}^*\mathrm{Ext}^d_R(R/\mathfrak{m}, C')(-i) \cong (R/\mathfrak{m})(-i).$$

Therefore $i = 0$, and $C \cong C'$. $\square$

Example 3.6.10. Let $R = k[X_1, \ldots, X_n]$ be a polynomial ring over a field, and assign to the indeterminates the degree $\deg X_i = a_i > 0$ for $i = 1, \ldots, n$. The *maximal ideal of R is $\mathfrak{m} = (X_1, \ldots, X_n)$, and the Koszul complex of $X_1, \ldots, X_n$ yields a homogeneous free resolution of $R/\mathfrak{m}$ whose last term is $R(-\sum_{i=1}^n a_i)$. From this one concludes that $^*\mathrm{Ext}^i(R/\mathfrak{m}, R) = 0$ for $i \neq n$, and $^*\mathrm{Ext}^n(R/\mathfrak{m}, R) = (R/\mathfrak{m})(\sum_{i=1}^n a_i)$. In other words, the *canonical module of R is $R(-\sum_{i=1}^n a_i)$.

Proposition 3.6.11. *Let $(R, \mathfrak{m})$ be a Cohen–Macaulay *local ring with *canonical module ω_R. The following conditions are equivalent:*
(a) *R is a Gorenstein ring;*
(b) *$\omega_R \cong R(a)$ for some integer $a \in \mathbb{Z}$.*

PROOF. R is Gorenstein if and only if ω_R is locally free. By 1.5.15(d) this is the case if and only if $\omega_R \cong R(a)$ for some $a \in \mathbb{Z}$. $\square$

The number a occurring in 3.6.11 is a numerical invariant of the Gorenstein *local ring $(R, \mathfrak{m})$, provided $\mathfrak{m}$ is maximal. In the case of a positively graded algebra over a field it will be given a special name; see 3.6.13.

Let $(R, \mathfrak{m})$ be a Cohen–Macaulay *local ring with *canonical module ω_R. The *canonical module is a graded module, and by 1.5.15, every minimal system of homogeneous generators of ω_R has exactly $\mu((\omega_R)_\mathfrak{m})$ elements. In analogy to the local case we define this number to be the *type of R*, and denote it by $r(R)$.

In view of 3.6.11 it is clear that R is Gorenstein if and only if $r(R) = 1$.

For the sake of completeness we list a few change of rings properties of the *canonical module. While part (a) of the next proposition follows easily from the results proved so far, it is best to use the change of rings spectral sequence

$$^*\mathrm{Ext}_S^p(k, {}^*\mathrm{Ext}_R^q(S, \omega_R)) \underset{p}{\Longrightarrow} {}^*\mathrm{Ext}_R^n(k, \omega_R)$$

for (b) (see [318], 11.66); the reduction we used for the corresponding local result 3.3.7 is only possible if homogeneous systems of parameters are available.

Proposition 3.6.12. *Let $(R, \mathfrak{m})$ be a Cohen–Macaulay *local ring with *canonical module ω_R.*
(a) *If $\mathfrak{p}$ is a graded prime ideal of R, then $\omega_{R_{(\mathfrak{p})}} \cong (\omega_R)_{(\mathfrak{p})}$ up to a shift.*
(b) *Let $\varphi \colon (R, \mathfrak{m}) \to (S, \mathfrak{n})$ be a ring homomorphism of Cohen–Macaulay *local rings satisfying*
   (i) *$\varphi(R_i) \subset S_i$ for all $i \in \mathbb{Z}$,*
   (ii) *$\varphi(\mathfrak{m}) \subset \mathfrak{n}$,*
   (iii) *S is a finite graded R-module.*
Then $\omega_S \cong {}^*\mathrm{Ext}_R^t(S, \omega_R)$, where $t = {}^*\dim R - {}^*\dim S$.

Example 3.6.10 and 3.6.12 imply that any Cohen–Macaulay positively graded algebra over a field admits a *canonical module. Following Goto and Watanabe [134] we define:

Definition 3.6.13. Let k be a field, and R a Cohen–Macaulay positively graded k-algebra. Then

$$a(R) = -\min\{i : (\omega_R)_i \neq 0\}$$

is called the *a-invariant* of R.

As a consequence of 3.6.12 we have

Corollary 3.6.14. *Let R be a Cohen–Macaulay *local ring with *canonical module ω_R, and let $x = x_1, \ldots, x_n$ be an R-sequence of homogeneous elements with $\deg x_i = a_i$ for $i = 1, \ldots, n$. Then*

$$\omega_{R/xR} \cong (\omega_R/x\omega_R)(\sum_{i=1}^{n} a_i).$$

In particular, if k be a field, and R a Cohen–Macaulay positively graded k-algebra, then $a(R/xR) = a(R) + \sum_{i=1}^{n} a_i$.

PROOF. The Koszul complex $K_.(x; R)$ is a graded free R-resolution of R/xR, and $K_n(x; R) \cong R(-\sum_{i=1}^{n} a_i)$. From 3.6.12 we obtain

$$\omega_{R/xR} \cong {}^*\mathrm{Ext}_R^n(R/xR, \omega_R) \cong H^n(x, \omega_R) \cong (\omega_R/x\omega_R)(\sum_{i=1}^{n} a_i). \quad \square$$

Examples 3.6.15. (a) A graded polynomial ring $R = k[X_1, \ldots, X_n]$ over a field k with $\deg X_i = a_i > 0$ has the *a*-invariant $a(R) = -\sum_{i=1}^{n} \deg a_i$.

(b) Let k be a field, and $R \to S$ a homomorphism of Cohen–Macaulay positively graded rings with *maximal ideals $\mathfrak{m}$ and $\mathfrak{n}$, respectively. Suppose the homomorphism satisfies the conditions of 3.6.12(b), and suppose further that S has a finite free homogeneous R-resolution

$$0 \longrightarrow F_t \longrightarrow F_{t-1} \longrightarrow \cdots \longrightarrow F_0 \longrightarrow S \longrightarrow 0,$$

where $t = \dim R - \dim S$. Write $F_t = \bigoplus_{i \in \mathbb{Z}} R(-a_i)$; then $a(S) = a(R) + \max\{a_i : i \in \mathbb{Z}\}$. This is proved exactly as in the special case 3.6.14.

Local Duality. Our final objective is to derive the graded version of the local duality theorem. We begin with Matlis duality.

Let $(R, \mathfrak{m})$ be a Noetherian *local ring; then R_0 is local with maximal ideal $\mathfrak{m}_0$. We consider R_0 as a graded ring by defining $(R_0)_i = 0$ for $i \neq 0$. Similarly any R_0-module may be considered a graded R_0-module concentrated in degree 0. Moreover, if M is a graded R-module, it may be viewed as a graded R_0-module as well. Thus we can define

$$M^\vee = {}^*\mathrm{Hom}_{R_0}(M, E_{R_0}(R_0/\mathfrak{m}_0)).$$

A priori, $M^\vee$ is a graded R_0-module whose grading is given by

$$(M^\vee)_i = \operatorname{Hom}_{R_0}(M_{-i}, E_{R_0}(R_0/\mathfrak{m}_0))$$

for all $i \in \mathbb{Z}$. But it is obvious that $M^\vee$ has a natural structure as a graded R-module.

The Noetherian *local ring $(R, \mathfrak{m})$ is said to be *complete* if $(R_0, \mathfrak{m}_0)$ is complete. If $(R, \mathfrak{m})$ is *complete and M is a finite graded R-module, then all homogeneous components M_i of M are complete R_0-modules (since they are finite R_0-modules).

Proposition 3.6.16. *Let $(R, \mathfrak{m})$ be a Noetherian *complete *local ring. Then*
(a) *the additive contravariant functor* $(\_)^\vee : \mathcal{M}_0(R) \longrightarrow \mathcal{M}_0(R)$ *is exact;*
(b) $M^\vee \cong {}^*\operatorname{Hom}_R(M, R^\vee)$ *for all graded R-modules M;*
(c) *one has* $R^\vee \cong {}^*E_R(R/\mathfrak{m})$.

PROOF. (a) is obvious.

(b) We define $\varphi : {}^*\operatorname{Hom}_R(M, {}^*\operatorname{Hom}_{R_0}(R, E)) \to {}^*\operatorname{Hom}_{R_0}(M, E)$ by setting $\varphi(\alpha)(x) = \alpha(x)(1)$ for all $\alpha \in {}^*\operatorname{Hom}(M, {}^*\operatorname{Hom}_{R_0}(R, E))$ and all $x \in M$. It is readily seen that φ is an isomorphism.

(c) It follows from (a) and (b) that ${}^*\operatorname{Hom}_R(\_, R^\vee)$ is an exact functor, and so $R^\vee$ is *injective. $R^\vee$ is *indecomposable since $R^{\vee\vee} \cong R$. Note further that $(R/\mathfrak{m})^\vee \cong R/\mathfrak{m}$. This is clear in the case where $R/\mathfrak{m} \cong k$ is a field, and it is easy to see in the case $R/\mathfrak{m} \cong k[t, t^{-1}]$, since then all homogeneous components of $k[t, t^{-1}]$ are isomorphic to k. Therefore the canonical epimorphism $R \to R/\mathfrak{m}$ yields a monomorphism $R/\mathfrak{m} \cong (R/\mathfrak{m})^\vee \longrightarrow R^\vee$, and the assertion follows from 3.6.3. $\square$

$M \in \mathcal{M}_0(R)$ is called *Artinian* if every descending chain of graded submodules terminates. The *homogeneous socle* of a graded R-module M is defined to be ${}^*\operatorname{Soc} M = {}^*\operatorname{Hom}_R(R/\mathfrak{m}, M)$. It is an $R/\mathfrak{m}$-module and can be viewed as a graded submodule of M. As an $R/\mathfrak{m}$-module it is free (see Exercise 1.5.20), and so ${}^*\operatorname{Soc} M \cong \bigoplus_{i \in I}(R/\mathfrak{m})(a_i)$. If M is Artinian, then ${}^*\operatorname{Soc} M$ can have only finitely many summands $(R/\mathfrak{m})(a_i)$. Hence we may write

$$^*\operatorname{Soc} M \cong \bigoplus_{i=1}^{n}(R/\mathfrak{m})(a_i).$$

As in the proof of 3.2.13 we conclude that M is *Artinian if and only if there exist an integer $n \geq 0$ and integers $a_1, \ldots, a_n$ such that

$$(7) \qquad\qquad M \subset \bigoplus_{i=1}^{n} R^\vee(a_i).$$

Let $\mathcal{A}_0(R)$ denote the full subcategory of $\mathcal{M}_0(R)$ consisting of all *Artinian R-modules and $\mathcal{F}_0(R)$ the full subcategory of all finite graded R-modules.

Theorem 3.6.17 (Matlis duality for graded modules). *Let* $(R, \mathfrak{m})$ *be a Noetherian* **complete* **local ring, and let* $M \in \mathscr{F}_0(R)$ *and* $N \in \mathscr{A}_0(R)$. *Then*
(a) $M^\vee \in \mathscr{A}_0(R)$ *and* $N^\vee \in \mathscr{F}_0(R)$,
(b) $M^{\vee\vee} \cong M$ *and* $N^{\vee\vee} \cong N$,
(c) *the functor* $(\_)^\vee : \mathscr{F}_0(R) \longrightarrow \mathscr{A}_0(R)$ *establishes an anti-equivalence of categories.*

PROOF. Using (7) one proves the theorem in the same way as 3.2.13. For example, in order to show (b) we set $E = E_{R_0}(R_0/\mathfrak{m}_0)$. Then we have

$$(M^{\vee\vee})_i = \mathrm{Hom}_{R_0}(\mathrm{Hom}_{R_0}(M_i, E), E) \cong M_i$$

by Matlis duality; see 3.2.13. □

Now let $(R, \mathfrak{m})$ be a Noetherian *local ring. For $M \in \mathscr{M}_0(R)$ we define

$$^*H^i_{\mathfrak{m}}(M) = {}^*\varinjlim {}^*\mathrm{Ext}^i_R(R/\mathfrak{m}^k, M);$$

it is called the *i-th* **local cohomology functor*. $^*H^0_{\mathfrak{m}}(\_)$ is left exact and the functors $^*H^i_{\mathfrak{m}}(\_)$, $i \geq 0$, are the right derived functors of $^*H^0_{\mathfrak{m}}(\_)$.

Remark 3.6.18. Assume in addition that the *maximal ideal $\mathfrak{m}$ of R is maximal. Since for all i and j, and all $M \in \mathscr{M}_0(R)$, we have $^*\mathrm{Ext}^i_R(R/\mathfrak{m}^j, M) \cong \mathrm{Ext}^i_R(R/\mathfrak{m}^j, M) \cong \mathrm{Ext}^i_{R_{\mathfrak{m}}}(R_{\mathfrak{m}}/\mathfrak{m}^j R_{\mathfrak{m}}, M_{\mathfrak{m}})$, we see that in this case $^*H^i_{\mathfrak{m}}(M) \cong H^i_{R_{\mathfrak{m}}}(M_{\mathfrak{m}})$.

Theorem 3.6.19 (The local duality theorem for graded modules). *Let* $(R, \mathfrak{m})$ *be a Cohen–Macaulay* **complete* **local ring of* **dimension d. Then*
(a) R *has a* **canonical module* ω_R, *and* $\omega_R \cong ({}^*H^d_{\mathfrak{m}}(R))^\vee$,
(b) *for all finite graded R-modules M and all integers i there exist natural homogeneous isomorphisms*

$$({}^*H^i_{\mathfrak{m}}(M))^\vee \cong {}^*\mathrm{Ext}^{d-i}_R(M, \omega_R).$$

The proof follows as in the non-graded case; see 3.5.8.

Let R be a positively graded k-algebra. Then it follows from 3.6.19(a) that

$$a(R) = \max\{i : {}^*H^d_{\mathfrak{m}}(R)_i \neq 0\}.$$

If in addition $\dim R = 0$, then $a(R) = \max\{i : R_i \neq 0\}$.

Exercises

3.6.20. Let R be a Noetherian graded ring.
(a) For $\mathfrak{p} \in \mathrm{Spec}\, R$ show that $R_{\mathfrak{p}}$ is Gorenstein if and only if $R_{\mathfrak{p}^*}$ is Gorenstein.
(b) Show the following conditions are equivalent:
   (i) R is a Gorenstein ring;

(ii) $R_{\mathfrak{p}}$ is a Gorenstein ring for all graded prime ideals $\mathfrak{p} \in \operatorname{Spec} R$;
(iii) $R_{(\mathfrak{p})}$ is a Gorenstein ring for all graded prime ideals $\mathfrak{p} \in \operatorname{Spec} R$.
(c) Let $(R, \mathfrak{m})$ be *local ring. Deduce that R is Gorenstein if and only if $R_{\mathfrak{m}}$ is Gorenstein.

3.6.21. The purpose of this exercise is to re-prove a few results of Goto and Watanabe.

Let R be a graded ring, d a positive integer. The ring $R^{(d)} = \bigoplus_{i \in \mathbb{Z}} R_{id}$ is called the *d-th Veronese subring* of R. It is a graded subring of R with grading $(R^{(d)})_i = R_{id}$ for all $i \in \mathbb{Z}$. For $j = 0, \ldots, d-1$ we consider the graded $R^{(d)}$-modules $M_j = \bigoplus_{i \in \mathbb{Z}} R_{id+j}$ with grading $(M_j)_i = R_{id+j}$ for all $i \in \mathbb{Z}$. We assume that R is Noetherian, and for (d) and (e) that it is a positively graded algebra over a field. Show:
(a) $R = \bigoplus_{j=0}^{d-1} M_j$ (as $R^{(d)}$-module). In particular, $R^{(d)}$ is a direct summand of R; $R^{(d)}$ is Noetherian, and the M_j are finite $R^{(d)}$-modules. (Hint: Compare the proof of 1.5.5.)
(b) R is Cohen–Macaulay if and only if all M_j are maximal Cohen–Macaulay $R^{(d)}$-modules.
(c) If R is Cohen–Macaulay, then $\omega_{R^{(d)}} \cong \bigoplus_{i \in \mathbb{Z}} (\omega_R)_{id}$.
(d) If R is Gorenstein and $a(R) = bd + j$, $0 \le j \le d - 1$, then $\omega_{R^{(d)}} \cong M_j(b)$.
(e) If R is Gorenstein and $a(R) \equiv 0 \bmod d$, then $R^{(d)}$ is Gorenstein. Is $a(R) \equiv 0 \bmod d$ if R and $R^{(d)}$ are Gorenstein?

3.6.22. Let k be a field. We consider $R = k[X_1, \ldots, X_n]$ as a graded k-algebra with $\deg X_i = a_i > 0$ for $i = 1, \ldots, n$. Determine all Veronese subrings of R which are Gorenstein.

3.6.23. Let R be a homogeneous k-algebra, k a field. Express $a(R^{(d)})$ in terms of $a(R)$ and d.

Notes

Grothendieck introduced the canonical module (often called dualizing module) and proved the local duality theorem. A comprehensive presentation of this theory including local cohomology is given in [143]. Equally fundamental is the famous paper of Bass [37]. The interested reader can find some more historical background there.

We were guided by the books of Kaplansky [231] and Matsumura [270] in Sections 3.1 and 3.2. In Sections 3.3 and 3.5 we follow partly the lecture notes of Herzog and Kunz [159]. Part of Section 3.5 has been influenced by the notes of P. Roberts [311]. In particular the description of the modified Čech complex has been taken from this source. In Section 3.6 we follow to a large extent the papers [134] by Goto and Watanabe and [109] by Fossum and Foxby.

The main result 3.1.17 of Section 3.1 is due to Bass [37]. The characterization 3.2.10 of Gorenstein rings in terms of the type was first proved by Bass [36]. Bass [37] gives a list of other equivalent conditions for the Gorenstein property. In [115] Foxby proves the following conjecture of

Vasconcelos, which is a remarkable characterization of Gorenstein rings. Suppose $(R, \mathfrak{m}, k)$ is a Noetherian local ring of dimension d containing a field; then R is a Gorenstein ring if $\mu_d(\mathfrak{m}, R) = 1$. (We will give a proof of this result in 9.6.3.) The main theorem of Matlis duality and the structure theorem for injective modules are proved by Matlis in [269], and can also be found in the more general framework of Abelian categories in Gabriel [122].

Theorem 3.3.10 which characterizes the canonical module is taken from [159]. However in the proof given here we do not use local cohomology. Theorem 3.3.6 on the existence of the canonical module is independently due to Foxby [113] and Reiten [307], and 3.3.19 is a theorem of Murthy [281]. It says that every factorial Cohen–Macaulay ring with a canonical module is Gorenstein. In [372] Ulrich proves a certain converse of Murthy's theorem: any Gorenstein ring which is a factor ring of a regular local ring and which is locally a complete intersection in codimension one can be realized as a specialization of a Cohen–Macaulay factorial domain.

For a while it had been open whether or not there exist non–Cohen–Macaulay factorial local rings. Such examples were found by Bertin [41] (see 6.4.7) and also by Fossum and Griffith [111]) in characteristic p and by Freitag and Kiehl [118] in characteristic 0.

There are two remarkable extensions of the theory of the canonical module in its basic form as presented here: Sharp introduced Gorenstein modules in [337] as those finite modules G whose Cousin complex provides a (minimal) injective resolution for G. A Gorenstein module shares many properties with the canonical module. It is a Cohen–Macaulay module of finite injective dimension whose type and rank, however, may be bigger than one. We refer the reader to the papers on Cousin complexes and Gorenstein modules [336], [337], [338], and [341] by Sharp, and the article [113] by Foxby. It is shown in [339] that for a Noetherian local ring admitting a canonical module ω_R, any Gorenstein module is a direct sum of copies of ω_R. However there exist Cohen–Macaulay local rings not admitting a canonical module. A first example of a one dimensional ring with this property was given by Ferrand and Raynaud [106], and an example of a factorial Cohen–Macaulay ring without a canonical module is due to Ogoma [294]. In [392] Weston gives an example of a ring with a Gorenstein module of rank 2, admitting no canonical module.

The second extension of the basic concept gives a duality theory even for non–Cohen–Macaulay rings. In this theory the canonical module has to be replaced by the so-called dualizing complex, and duality is obtained in the derived category. We refer the reader to the book of Hartshorne [151]. A more elementary account of the theory can be found in Sharp [340].

As a consequence of the structure theorem 3.4.1 for Gorenstein ideals of grade three, these ideals have an odd number of generators. This had been observed before by J. Watanabe [386]. He uses linkage arguments in his proof. Linkage had already been considered in 1945 by Apéry [11] and by Gaeta [123] in 1952. It has become popular as a result of the paper of Peskine and Szpiro [298]. Linkage provides a technique to construct large and interesting classes of perfect ideals or of Gorenstein ideals whose structure is well understood. Of particular interest are the ideals in the linkage class of a complete intersection, called *licci ideals*. The simplest examples are the so-called Northcott ideals [288] and the Gorenstein ideals defined in [156]. More important is the fact that perfect ideals of grade two [298] and Gorenstein ideals of grade three [386] are in the linkage class of a complete intersection. They are in a sense the archetypes of licci ideals as shown by Huneke and Ulrich [220]. For further study of linkage theory we refer the reader to the papers of Huneke [210], [213], Huneke and Ulrich [219], [221], [222], Kustin and Miller [251], [252], [254], and Ulrich [374], [373].

The height 3 monomial Gorenstein ideals have been completely classified by Bruns and Herzog [58].

There have been attempts to obtain structure theorems for non-Gorenstein ideals of grade 3 or even for ideals of grade higher than 3. The next case of interest is Gorenstein ideals of grade 4. As a first approach to the problem one may try to classify the Tor-algebras of these ideals I, i.e. $\mathrm{Tor}_R(k, R/I)$ when I is an ideal in the local ring R with residue class field k. For ideals I such that $\mathrm{proj\,dim}\, R/I = 3$ this has been done by Weyman [393] in characteristic 0 and by Avramov, Kustin, and Miller [33] in all characteristics. The next case of interest is Gorenstein ideals of grade 4. At the moment a general structure theorem for these ideals seems to be out of range. Kustin and Miller [253] succeeded in classifying their Tor-algebras.

A remarkable result, valid for ideals of arbitrary grade, is due to Kunz [246]: if I is a Gorenstein ideal, then $\mu(I) \neq \mathrm{grade}\, I + 1$.

Duality theory is a classical and fundamental topic in algebraic geometry, and has also several algebraic aspects we have not even touched upon. We must content ourselves with a list of keywords and references: Riemann–Roch theorem, Serre duality, modules of regular differentials, residue symbols, trace maps; see Hartshorne [151], [152], Kunz [247], Kunz and Waldi [250], Lipman [259], Scheja and Storch [324], [325], and Serre [333].

4 Hilbert functions and multiplicities

The Hilbert function $H(M, n)$ measures the dimension of the n-th homogeneous piece of a graded module M. In the first section of this chapter we study the Hilbert function of modules over homogeneous rings, prove that it is a polynomial for large values of n, and introduce the Hilbert series and multiplicity of a graded module. The next section is devoted to the proof of Macaulay's theorem which describes the possible Hilbert functions. The third section complements these results by Gotzmann's regularity and persistence theorem.

The Hilbert function behaves quite regular, even for graded, non-homogeneous rings. Such rings will be considered in the fourth section, where we will also investigate the Hilbert function of the canonical module.

The passage to the associated graded ring with respect to a filtration allows us to extend some concepts for graded rings like 'Hilbert function' or 'multiplicity' to non-graded rings, and leads to the Hilbert–Samuel function and the multiplicity of a finite module with respect to an ideal of definition. We shall study basic properties of filtrations and their associated Rees rings and modules, and sketch the theory of reduction ideals. Finally we prove Serre's theorem which interprets multiplicity as the Euler characteristic of a certain Koszul homology.

4.1 Hilbert functions over homogeneous rings

We begin by studying numerical properties of finite graded modules over a graded ring R. Our standard assumption in this section will be that R_0 is an Artinian local ring, and that R is finitely generated over R_0. Notice that for each finite graded R-module M, the homogeneous components M_n of M are finite R_0-modules, and hence have finite length.

Definition 4.1.1. Let M be a graded R-module whose graded components M_n have finite length for all n. The numerical function $H(M, \_): \mathbb{Z} \to \mathbb{Z}$ with $H(M, n) = \ell(M_n)$ for all $n \in \mathbb{Z}$ is the *Hilbert function*, and $H_M(t) = \sum_{n \in \mathbb{Z}} H(M, n) t^n$ is the *Hilbert series* of M.

For the rest of this section we will assume that R is generated over R_0 by elements of degree 1, that is, $R = R_0[R_1]$. Recall that such a ring is said to be homogeneous.

147

We say that a numerical function $F : \mathbb{Z} \to \mathbb{Z}$ is of *polynomial type* (*of degree* d) if there exists a polynomial $P(X) \in \mathbb{Q}[X]$ (of degree d) such that $F(n) = P(n)$ for all $n \gg 0$. By convention the zero polynomial has degree -1.

We define the *difference operator* Δ on the set of numerical functions by setting $(\Delta F)(n) = F(n+1) - F(n)$ for all $n \in \mathbb{Z}$. Notice that Δ maps polynomial functions to polynomial functions, lowering the degree of non-zero polynomials by 1. The d times iterated Δ-operator will be denoted by Δ^d. We further set $\Delta^0 F = F$.

Lemma 4.1.2. *Let* $F : \mathbb{Z} \to \mathbb{Z}$ *be a numerical function, and* $d \geq 0$ *an integer. The following conditions are equivalent:*
(a) $\Delta^d F(n) = c$, $c \neq 0$, *for all* $n \gg 0$;
(b) F *is of polynomial type of degree* d.

PROOF. (b) $\Rightarrow$ (a) is easy. We prove the other implication by induction on d. The assertion is trivial for $d = 0$. Now assume that $d > 0$, and $\Delta^d F(n) = \Delta^{d-1}(F(n+1) - F(n)) = c$, $c \neq 0$, for all $n \gg 0$. By the induction hypothesis it then follows that there exist an integer n_0 and a polynomial $P(X) \in \mathbb{Q}[X]$ of degree $d - 1$ such that $F(n+1) - F(n) = P(n)$ for all $n \geq n_0$. Then $F(n+1) = F(n_0) + \sum_{k=n_0}^{n} P(k)$, and this last sum is a polynomial function in n of degree d. $\qquad\qquad\qquad\qquad\square$

After these preparations we can state the main result of this section as follows:

Theorem 4.1.3 (Hilbert). *Let* M *be a finite graded R-module of dimension* d. *Then* $H(M,n)$ *is of polynomial type of degree* $d - 1$.

PROOF. We prove the theorem by induction on the dimension d of M. First note that there is a chain $0 = N_0 \subset N_1 \subset \cdots \subset N_n = M$ of graded submodules of M such that for each i we have $N_{i+1}/N_i \cong (R/\mathfrak{p}_i)(a_i)$ where $\mathfrak{p}_i$ is a graded prime ideal. Indeed, we may assume that $M \neq 0$. Choose $\mathfrak{p}_1 \in \mathrm{Ass}(M)$. The prime ideal $\mathfrak{p}_1$ is graded; see 1.5.6. There exists a graded submodule $N_1 \subset M$ with $N_1 \cong (R/\mathfrak{p}_1)(a_1)$. If $N_1 \neq M$ we choose $\mathfrak{p}_2 \in \mathrm{Ass}(M/N_1)$. Then there exists a graded submodule $N_2 \subset M$ with $N_2/N_1 \cong (R/\mathfrak{p}_2)(a_2)$. If $N_2 \neq M$, we may proceed in the same way. But M is Noetherian, and so this process terminates eventually.

Now, since the Hilbert function is additive on short exact sequences, it follows that $H(M,n) = \sum_i H((R/\mathfrak{p}_i)(a_i), n)$. Notice that d is the supremum of the numbers $\dim R/\mathfrak{p}_i$. Hence the theorem will follow once we have shown it for $M = R/\mathfrak{p}$, $\mathfrak{p}$ a graded prime ideal. (Here, of course, one has to observe that the polynomials describing Hilbert functions are zero or have positive leading coefficients since their values are non-negative for $n \gg 0$. As a consequence, the degree of the sum of such polynomials is the maximum of their degrees.)

If $\dim R/\mathfrak{p} = 0$, then $\mathfrak{p}$ is the unique graded maximal ideal $\mathfrak{m}_0 \oplus \bigoplus_{n>0} R_n$ of R, where $\mathfrak{m}_0$ is the maximal ideal of R_0. It follows that $H(R/\mathfrak{p}, n) = 0$ for $n > 0$.

If $\dim R/\mathfrak{p} > 0$, we may choose a homogeneous element $x \in R/\mathfrak{p}$, $x \neq 0$, of degree 1. Here we use the fact that R is homogeneous. The exact sequence

$$0 \longrightarrow (R/\mathfrak{p})(-1) \overset{x}{\longrightarrow} R/\mathfrak{p} \longrightarrow R/(x,\mathfrak{p}) \longrightarrow 0$$

gives the equation

$$\Delta H(R/\mathfrak{p}, n) = H(R/\mathfrak{p}, n+1) - H(R/\mathfrak{p}, n) = H(R/(x,\mathfrak{p}), n+1).$$

As $\dim R/(x,\mathfrak{p}) = d - 1$, our induction hypothesis implies that $\Delta H(R/\mathfrak{p}, n)$ is of polynomial type of degree $d - 2$. Hence if $d > 1$, then 4.1.2 implies that $\Delta^{d-1} H(R/\mathfrak{p}, n) = \Delta^{d-2}(\Delta H(R/\mathfrak{p}, n))$ is a non-zero constant function for large n, and if $d = 1$, then $\Delta^{d-1} H(R/\mathfrak{p}, n) = H(R/\mathfrak{p}, n) = H(R/\mathfrak{p}, 0) + \sum_{i=1}^{n} H(R/(\mathfrak{p}, x), i)$ is constant for large n since $H(R/(\mathfrak{p}, x), i) = 0$ for $i \gg 0$. Again this constant is not zero since $H(R/\mathfrak{p}, 0) \neq 0$. Now 4.1.2, (a) $\Rightarrow$ (b) yields the assertion. $\square$

Hilbert's original proof (see [171]) makes use of his syzygy theorem 2.2.14. This approach will be described in 4.1.13.

The next lemma clarifies which polynomials in $\mathbb{Q}[X]$ have integer values.

Lemma 4.1.4. *Let $P(X) \in \mathbb{Q}[X]$ be a polynomial of degree $d-1$. Then the following conditions are equivalent:*
(a) *$P(n) \in \mathbb{Z}$ for all $n \in \mathbb{Z}$;*
(b) *there exist integers $a_0, \ldots, a_{d-1}$ such that*

$$P(X) = \sum_{i=0}^{d-1} a_i \binom{X+i}{i}.$$

PROOF. (b) $\Rightarrow$ (a) is trivial. For the converse observe that the polynomials $\binom{X+i}{i}$, $i \in \mathbb{N}$, form a $\mathbb{Q}$-basis of $\mathbb{Q}[X]$. Therefore $P(X) = \sum_{i=0}^{d-1} a_i \binom{X+i}{i}$ with $a_i \in \mathbb{Q}$. The identity $\binom{X+i+1}{i} - \binom{X+i}{i} = \binom{X+i}{i-1}$ immediately implies that $a_i = \Delta^i P(-i-1) \in \mathbb{Z}$ for $i = 0, \ldots, d-1$. $\square$

Definition 4.1.5. Let M be a finite graded R-module of dimension d. The unique polynomial $P_M(X) \in \mathbb{Q}[X]$ for which $H(M, n) = P_M(n)$ for all $n \gg 0$ is called the *Hilbert polynomial* of M. We write

$$P_M(X) = \sum_{i=0}^{d-1} (-1)^{d-1-i} e_{d-1-i} \binom{X+i}{i}.$$

Then the *multiplicity* of M is defined to be

$$e(M) = \begin{cases} e_0 & \text{if } d > 0, \\ \ell(M) & \text{if } d = 0. \end{cases}$$

Remark 4.1.6. The *higher iterated Hilbert functions* $H_i(M, n)$, $i \in \mathbb{N}$, of a finite graded R-module M are defined recursively as follows:

$$H_0(M, n) = H(M, n), \quad \text{and} \quad H_i(M, n) = \sum_{j \leq n} H_{i-1}(M, j)$$

for $i > 0$. Occasionally the functions $H_i(M, \_)$ are called the *sum transforms of $H(M, \_)$.*

It follows from 4.1.2 and 4.1.3 that $H_i(M, n)$ is of polynomial type of degree $d + i - 1$, where $d = \dim M$. In particular, for all $n \gg 0$ there is a representation $H_1(M, n) = \sum_{i=0}^{d} a_i \binom{n+i}{i}$ with $a_i \in \mathbb{Z}$, and it is easy to see that $a_d = e(M)$. Another formula for the multiplicity will be given in 4.1.9.

Theorem 4.1.3 together with the next lemma yields a structural result about Hilbert series.

Lemma 4.1.7. *Let $H(t) = \sum a_n t^n$ be a formal Laurent series with integer coefficients, and $a_i = 0$ for $i \ll 0$. Further, let $d > 0$ be an integer. Then the following conditions are equivalent:*
(a) *there exists a polynomial $P(X) \in \mathbb{Q}[X]$ of degree $d - 1$ such that $P(n) = a_n$ for large n;*
(b) *$H(t) = Q(t)/(1 - t)^d$ where $Q(t) \in \mathbb{Z}[t, t^{-1}]$ and $Q(1) \neq 0$.*

PROOF. Assume (a), and set $F(n) = a_n$ for all $n \in \mathbb{Z}$. Then

$$(1 - t)^d H(t) = \sum_n \Delta^d F(n - d) t^n,$$

and it follows from 4.1.2 that $(1 - t)^d H(t) \in \mathbb{Z}[t, t^{-1}]$. We set $Q(t) = \sum_n \Delta^d F(n - d) t^n$. Suppose $Q(1) = 0$; then

$$0 = \sum_n \Delta^d F(n - d) = \sum_n (\Delta^{d-1} F(n + 1 - d) - \Delta^{d-1} F(n - d)) = \Delta^{d-1} F(m)$$

for large m. This contradicts 4.1.2, and thus proves the implication (a) $\Rightarrow$ (b). The converse is proved similarly. $\qquad\square$

Corollary 4.1.8. *Let $M \neq 0$ be a finite graded R-module of dimension d. Then there exists a unique $Q_M(t) \in \mathbb{Z}[t, t^{-1}]$ with $Q_M(1) \neq 0$ such that*

$$H_M(t) = \frac{Q_M(t)}{(1 - t)^d}.$$

Moreover, if $Q_M(t) = \sum_i h_i t^i$, then $\min\{i : h_i \neq 0\}$ is the least number such that $M_i \neq 0$.

PROOF. The first part of the assertion is clear for $d = 0$, and for $d > 0$ it follows from 4.1.3 and 4.1.7. In order to prove the second part multiply both sides of $H_M(t) = Q_M(t)/(1-t)^d$ by $(1-t)^d$ and compare coefficients. $\square$

In the next proposition we show how one can recover the coefficients e_i of the Hilbert polynomial of a module M from Q_M. We will denote by $P^{(i)}$ the i-th formal derivative of an element $P \in \mathbb{Z}[t, t^{-1}]$.

Proposition 4.1.9. *Under the assumptions of* 4.1.8 *the following formulas hold:*

$$e_i = \frac{Q_M^{(i)}(1)}{i!}$$

for $i = 0, \ldots, d-1$. *Moreover,* $e(M) = Q_M(1)$.

PROOF. We write

$$H_M(t) - \sum_{i=0}^{d-1} \frac{(-1)^i}{i!} \frac{Q_M^{(i)}(1)}{(1-t)^{d-i}} = \frac{D(t)}{(1-t)^d}$$

where $D(t) = Q_M(t) - \sum_{i=0}^{d-1} \frac{(-1)^i}{i!} Q_M^{(i)}(1)(1-t)^i$ is the remainder of the Taylor expansion of $Q_M(t)$ up to degree $d-1$. The element $D(t) \in \mathbb{Z}[t, t^{-1}]$ is divisible by $(1-t)^d$ since $D^{(j)}(1) = 0$ for $j = 0, \ldots, d-1$. It follows that the coefficients of $H_M(t)$ and $\sum_{i=0}^{d-1}(\frac{(-1)^i}{i!} Q_M^{(i)}(1)/(1-t)^{d-i})$ coincide for large n. Hence

$$\sum_{i=0}^{d-1} \frac{(-1)^i}{i!} \frac{Q_M^{(i)}(1)}{(1-t)^{d-i}} = \sum_{n \geq 0} P_M(n)t^n,$$

since the coefficients of both series are polynomial functions in n which are equal for large n (and hence must be equal for all n). Expanding the left hand side of the equation as a power series, and comparing coefficients we get $e_i = Q_M^{(i)}(1)/i!$.

Finally, by what we have just proved, we have $e(M) = e_0 = Q_M(1)$ if $d > 0$, and, if $d = 0$, $e(M) = \ell(M) = \sum_n H(M, n) = H_M(1) = Q_M(1)$, since in this case $H_M(t) = Q_M(t)$. $\square$

Corollary 4.1.10. *Assume that in addition to the assumptions of* 4.1.8 *the module M is Cohen–Macaulay. Let* $Q_M(t) = \sum h_i t^i$. *Then* $h_i \geq 0$ *for all i. Moreover,* $e_i \geq 0$ *for all i if* $M_j = 0$ *for all* $j < 0$.

PROOF. Without loss of generality we may assume that the residue class field of R_0 is infinite. Otherwise we resort to a standard trick: we replace R by $R' = R \otimes_{R_0} R_0(Y)$ and M by $M' = M \otimes_{R_0} R_0(Y)$ where Y is an indeterminate over R_0, and where $R_0(Y)$ is the local ring $R_0[Y]_S$, S being the multiplicatively closed set of polynomials $P(Y) \in R_0[Y]$ which have at least one unit among their coefficients. The natural ring

homomorphism $R_0 \to R_0(Y)$ is local and flat, and its fibre is the residue
class field of $R_0(Y)$, namely the field $k(Y)$ of rational functions over
$k = R_0/\mathfrak{m}_0$. If we assign the degree 0 to the elements of $R_0(Y)$, then
both R' and M' are naturally graded, and because of flatness, M' is a
Cohen–Macaulay R'-module of dimension d with $H_{M'}(t) = H_M(t)$; see
2.1.7 and 1.2.25. Moreover, $R'_0 = R_0(Y)$, and hence has an infinite residue
class field.

In view of 4.1.9 it suffices to show that $h_i \geq 0$ for all i. We prove the
assertion by induction on d. If $d = 0$, then $Q_M(t) = H_M(t)$ (see 4.1.8), and
so all coefficients of $Q_M(t)$ are non-negative.

Suppose now that $d > 0$. The unique homogeneous maximal ideal
$\mathfrak{M} = \mathfrak{m}_0 \oplus \bigoplus_{n>0} R_n$ of R does not belong to $\operatorname{Ass} M$, and the ideal
$I = \bigoplus_{n>0} R_n$, generated by the elements of R_1, is $\mathfrak{M}$-primary. Thus, since
$R/\mathfrak{M}$ is infinite there exists an element $a \in R_1$ which is M-regular; see
1.5.12. Let $N = M/aM$; then N is a Cohen–Macaulay graded R-module
of dimension $d - 1$, and the exact sequence

$$0 \longrightarrow M(-1) \overset{a}{\longrightarrow} M \longrightarrow N \longrightarrow 0$$

gives the equation $(1 - t)H_M(t) = H_N(t)$. It follows that $Q_M(t) = Q_N(t)$,
which by our induction hypothesis yields the conclusion. $\square$

Remark 4.1.11. The arguments in the previous proof show the following
notable result: suppose M is a finite graded R-module, and x is an
M-sequence of elements of degree 1; then $Q_M(t) = Q_{M/(x)M}(t)$.

Hilbert's theorem tells us that the Hilbert function of a finite graded
module is a polynomial function for large n. We will determine from
which integer n onwards this happens.

Proposition 4.1.12. *Let $M \neq 0$ be a finite graded R-module of dimension
d, and $Q_M(t) = \sum_{i=a}^{b} h_i t^i$ with $h_b \neq 0$. Then $H(M, b - d) \neq P_M(b - d)$ and
$H(M, i) = P_M(i)$ for all $i \geq b - d + 1$.*

PROOF. For $i = a, \ldots, b$ we set $H_i(t) = h_i t^i/(1 - t)^d$ and $P_i(n) = h_i \binom{n-i+d-1}{d-1}$.
Then $H_i(t) = \sum_{n=i}^{\infty} P_i(n)t^n$, but since $P_i(n) = 0$ for $n = i - (d-1), \ldots, i - 1$
we even have $H_i(t) = \sum_{n=i-(d-1)}^{\infty} P_i(n)t^n$. Furthermore $P_M(n) = \sum_{i=a}^{b} P_i(n)$
for all $n \in \mathbb{Z}$. For $n \geq b - d + 1$ one has $H(M, n) = \sum_{i=a}^{b} P_i(n)$, whereas
$H(M, b - d) = \sum_{i=a}^{b-1} P_i(b - d)$ and $P_b(b - d) \neq 0$. Thus $H(M, b - d) \neq P_M(b - d)$. $\square$

In Section 4.4 we will give a homological interpretation of the difference between the Hilbert function and the Hilbert polynomial.

Hilbert series and free resolutions. The Hilbert series of a graded module
can be expressed in terms of its graded resolution.

Lemma 4.1.13. *Let M be a finite graded R-module of finite projective dimension, and let*

$$0 \longrightarrow \bigoplus_j R(-j)^{\beta_{pj}} \longrightarrow \cdots \longrightarrow \bigoplus_j R(-j)^{\beta_{0j}} \longrightarrow M \longrightarrow 0$$

be a graded free resolution of M. Then

$$H_M(t) = S_M(t)H_R(t)$$

where $S_M(t) = \sum_{i,j}(-1)^i \beta_{ij} t^j$. In particular, if $R = k[X_1, \ldots, X_n]$ is the polynomial ring over the field k, then

$$H_M(t) = \frac{S_M(t)}{(1-t)^n}.$$

PROOF. For the proof we simply note that the Hilbert function is additive on short exact sequences, so that $H_M(t) = \sum_i (-1)^i \beta_{ij} H_{R(-j)}(t)$. Taking into account that $H_{R(-j)}(t) = t^j H_R(t)$, we obtain the required formula.

If $R = k[X_1, \ldots, X_n]$ is the polynomial ring, then $H(R, i)$ equals the number of monomials in degree i. One easily proves by induction on n that this number is $\binom{n+i-1}{n-1}$, whence $H_R(t) = \sum_i \binom{n+i-1}{n-1} t^i = 1/(1-t)^n$. $\qquad\square$

Corollary 4.1.14. *Let $R = k[X_1, \ldots, X_n]$ be a polynomial ring over a field k, and let M be a finite graded R-module of dimension d. Then*

(a) $S_M(t) = (1-t)^{n-d} Q_M(t)$,

(b) $n - d = \inf\{i \colon S_M^{(i)}(1) \neq 0\}$,

(c) $S_M^{(n-d+i)}(1) = (-1)^{n-d} \binom{n-d+i}{i} e_i$.

PROOF. (a) and (b) are immediate while (c) follows from (a) and 4.1.9. $\qquad\square$

We conclude this section with an application to a special class of graded rings. Let $R = k[X_1, \ldots, X_n]$ be a polynomial ring over a field k, $I \subset R$ a graded ideal. We say that R/I has a *pure resolution* of type $(d_1, \ldots, d_p)$ if its minimal resolution has the form

$$0 \longrightarrow R(-d_p)^{\beta_p} \longrightarrow \cdots \longrightarrow R(-d_1)^{\beta_1} \longrightarrow R \longrightarrow R/I \longrightarrow 0.$$

Note that $d_1 < d_2 < \cdots < d_p$.

Theorem 4.1.15. *Suppose R/I is Cohen–Macaulay and has a pure resolution of type $(d_1, \ldots, d_p)$. Then*

$$\text{(a)} \quad \beta_i = (-1)^{i+1} \prod_{j \neq i} \frac{d_j}{(d_j - d_i)}, \qquad \text{(b)} \quad e(R/I) = \frac{1}{p!} \prod_{i=1}^{p} d_i.$$

PROOF. As R/I is Cohen–Macaulay, the Auslander–Buchsbaum formula 1.3.3 (in conjunction with 1.5.15) implies that $p = \dim R - \dim R/I$; therefore, with $\beta_0 = 1$, $d_0 = 0$, we have

$$S_{R/I}(t) = \sum_{i=0}^{p}(-1)^i \beta_i t^{d_i} \quad \text{and} \quad S_{R/I}^{(j)}(1) = 0$$

for $j = 0,\dots,p-1$; see 4.1.14. We obtain the following system of linear equations:

$$\sum_{i=1}^{p}(-1)^i \beta_i = -1,$$

$$\sum_{i=1}^{p}(-1)^i \beta_i d_i(d_i - 1)\cdots(d_i - j + 1) = 0 \quad \text{for} \quad j = 1,\dots,p-1.$$

Upon applying elementary row operations, which do not affect the solution of this system of linear equations with coefficient matrix

$$(d_i!/(d_i - j)!)_{\substack{j=0,\dots,p-1 \\ i=1,\dots,p}},$$

we are led to the Vandermonde matrix whose determinant is $\prod_{i>j}(d_i - d_j)$. Now Cramer's rule gives the stated solutions for the β_i.

(b) According to 4.1.14 we have

$$e(R/I) = (-1)^p \frac{S_{R/I}^{(p)}(1)}{p!} = \sum_{i=0}^{p}(-1)^{p+i}\beta_i \binom{d_i}{p}.$$

Thus (a) implies that

$$e(R/I) = \frac{1}{p!}\prod_{i=1}^{p}d_i \sum_{i=1}^{p}\frac{\prod_{j=1}^{p-1}(d_i - j)}{\prod_{j\neq i}(d_i - d_j)}.$$

It remains to show that the sum in this expression equals 1. We introduce the rational complex function

$$f(z) = \frac{\prod_{j=1}^{p-1}(z - j)}{\prod_{j=1}^{p}(z - d_j)}.$$

This function has simple poles at worst in the points $d_1,\dots,d_p$, and the residues in these points are

$$\operatorname{Res}_{d_i} f(z) = \frac{\prod_{j=1}^{p-1}(d_i - j)}{\prod_{j\neq i}(d_i - d_j)}.$$

The sum of all residues of a rational function at all points including ∞ is zero, and $\mathrm{Res}_\infty f(z) = -\mathrm{Res}_0 f(1/z)/z^2$. Therefore

$$\sum_{i=1}^{p} \left(\prod_{j=1}^{p-1}(d_i - j) \prod_{j \neq i}(d_i - d_j)^{-1} \right) = \sum_{i=1}^{p} \mathrm{Res}_{d_i} f(z)$$

$$= \mathrm{Res}_0 \frac{f(1/z)}{z^2} = \mathrm{Res}_0 \left(\frac{1}{z} \prod_{j=1}^{p-1}(1 - jz) \prod_{j=1}^{p}(1 - d_j z)^{-1} \right) = 1. \quad \square$$

Exercises

4.1.16. Let k be a field, and M a finite graded module over the polynomial ring $R = k[X_1, \ldots, X_n]$ with minimal graded resolution

$$0 \longrightarrow \bigoplus_j R(-j)^{\beta_{pj}} \longrightarrow \cdots \longrightarrow \bigoplus_j R(-j)^{\beta_{0j}} \longrightarrow M \longrightarrow 0.$$

We say that two modules have *numerically the same resolution* if their graded Betti numbers β_{ij} are the same. Show:
(a) The homogeneous rings $k[X, Y]/(X^2, Y^2)$ and $k[X, Y]/(X^2, XY, Y^3)$ have the same Hilbert series, but their minimal graded free $k[X, Y]$-resolutions are numerically different.
(b) The homogeneous rings $k[X, Y]/(X^2, Y^2)$ and $k[X, Y]/(XY, X^2 - Y^2)$ have numerically the same graded $k[X, Y]$-resolution, but are not isomorphic when $k = \mathbb{R}$.

4.1.17. Let k be a field, and let $R = k[X_1, \ldots, X_n]/I$ be a homogeneous Cohen–Macaulay ring. The ring R has an *m-linear resolution* if it has a pure resolution of type $(m, m + 1, \ldots, m + p - 1)$, $p = n - \dim R$.
(a) Suppose R has an m-linear resolution. What are the ranks of the free modules in the free resolution of R, and what is the multiplicity of R?
(b) Suppose $\dim R = 0$; prove R has an m-linear resolution if and only if $I = (X_1, \ldots, X_n)^m$.
Hint: relate the last shifts in the resolution of R with the degrees of the socle elements of R.
(c) Prove the homogeneous Cohen–Macaulay ring $R = k[X_1, \ldots, X_n]/I$ has an m-linear resolution if and only if $I_j = 0$ for $j < m$, and $\dim_k I_m = \binom{m+g-1}{m}$, where $g = \mathrm{height}\, I$.
Hint: reduce to dimension zero.

4.1.18. Let k be a field, and let $R = k[X_1, \ldots, X_n]/I$ be a homogeneous Gorenstein ring of dimension 0. Assume that all generators of I have the same degree c.
(a) Show $a(R) \geq 2c - 2$.
(b) Show $a(R) = 2c - 2$ if and only if R has a pure resolution of type $(c, c + 1, \ldots, c + n - 2, 2c + n - 2)$. In this case R is called an *extremal Gorenstein ring*. This class of Gorenstein rings was first considered by Schenzel [327].
(c) Compute the Betti numbers $\beta_i(R)$ of an extremal Gorenstein ring R in a minimal graded free $k[X_1, \ldots, X_n]$-resolution of R.

4.2 Macaulay's theorem on Hilbert functions

This section is devoted to a theorem of Macaulay describing exactly those numerical functions which occur as the Hilbert function $H(R, n)$ of a homogeneous k-algebra R, k a field. Macaulay's theorem says that for each n there is an upper bound for $H(R, n + 1)$ in terms of $H(R, n)$, and this bound is sharp in the sense that any numerical function satisfying it can indeed be realized as the Hilbert function of a suitable homogeneous k-algebra. One part of the proof of Macaulay's theorem will be based on a theorem of Green which relates the Hilbert function of a homogeneous ring R with the Hilbert function of the factor ring R/hR by a general linear form h.

Let $R = \bigoplus_{n \geq 0} R_n$ be a homogeneous k-algebra, where $R_0 = k$ is a field. We will show that R has a k-basis consisting of monomials in a basis $x_1, \ldots, x_m$ of R_1. We are going to define this basis of monomials on the level of the polynomial ring. So let

$$\pi : k[X_1, \ldots, X_m] \longrightarrow R$$

be the surjective k-algebra homomorphism with $\pi(X_i) = x_i$.

Definition 4.2.1. A non-empty set $\mathfrak{M}$ of monomials in the indeterminates $X_1, \ldots, X_m$ is called an *order ideal of monomials* if the following holds: whenever $m \in \mathfrak{M}$ and a monomial m' divides m, then $m' \in \mathfrak{M}$. Equivalently, if $X_1^{a_1} \cdots X_m^{a_m} \in \mathfrak{M}$ and $0 \leq b_i \leq a_i$ for $i = 1, \ldots, m$, then $X_1^{b_1} \cdots X_m^{b_m} \in \mathfrak{M}$.

Remarks 4.2.2. (a) In Chapter 9 we introduce the order ideal of an element in a module. This notion has nothing to do with the order ideal of monomials, and they should not be confused.

(b) Of course an order ideal of monomials $\mathfrak{M}$ is not a k-basis of an ideal, let alone an ideal. Quite the contrary, if we let $C\mathfrak{M}$ be the complement of $\mathfrak{M}$ in the set of all monomials, then $C\mathfrak{M}$ is a k-basis of the ideal generated by the monomials $m \in C\mathfrak{M}$.

Theorem 4.2.3 (Macaulay). *Let R be a homogeneous k-algebra, k a field. Further let $x_1, \ldots, x_m$ be a k-basis of R_1, and $\pi : k[X_1, \ldots, X_m] \to R$ the k-algebra homomorphism with $\pi(X_i) = x_i$ for $i = 1, \ldots, m$. Then there exists an order ideal $\mathfrak{M}$ of monomials such that $\pi(\mathfrak{M})$ is a k-basis of R.*

PROOF. Let $\mathscr{S}$ denote the set of all monomials in the indeterminates $X_1, \ldots, X_m$. We define a total order, the so-called *reverse degree-lexico-graphical order*, on $\mathscr{S}$: if $u = X_1^{a_1} \cdots X_m^{a_m}$ and $v = X_1^{b_1} \cdots X_m^{b_m}$, then $u < v$ if the last non-zero component of $(b_1 - a_1, \ldots, b_m - a_m, \sum b_i - \sum a_i)$ is positive. (The usage of the term 'reverse degree-lexicographical' is not coherent in the literature.)

It is clear that $(\mathscr{S}, <)$ is an ordered semigroup, i.e. $u < v \Rightarrow mu < mv$ for all $u, v, m \in \mathscr{S}$. Moreover, any descending chain $v_1 > v_2 > \cdots$ of elements of $\mathscr{S}$ must stop after a finite number of steps. Equivalently, every non-empty set of elements in $\mathscr{S}$ has a minimal element – a fact that will be used later.

Now we define recursively a sequence of monomials $u_1, u_2, \ldots$ according to the following rule. We set $u_1 = 1$; assume $u_1, \ldots, u_i$ have been defined; then we let u_{i+1} be the least element in the reverse degree-lexicographical order such that $\pi(u_1), \ldots, \pi(u_i)$, $\pi(u_{i+1})$ are linearly independent over k. If such u_{i+1} does not exist, the sequence terminates with u_i.

We claim that $\mathfrak{M} = \{u_1, u_2, \ldots\}$ is the required order ideal of monomials. By construction $\pi(\mathfrak{M})$ is a k-basis of R. Assume $\mathfrak{M}$ is not an order ideal of monomials. Then there exist $u_{i_0} \in \mathfrak{M}$ and $u \in \mathscr{S} \setminus \mathfrak{M}$ such that $u_{i_0} = u \cdot X_j$ for some X_j. As $u \notin \mathfrak{M}$, we can write $\pi(u) = \sum \lambda_i \pi(u_i)$ with $u_i \in \mathfrak{M}$, $u_i < u$, and $\lambda_i \in k$. Then $\pi(u_{i_0}) = \sum \lambda_i \pi(u_i X_j)$, and $u_i X_j < u_{i_0}$, for all i in the sum, a contradiction. □

We saw in the proof that our k-basis $\pi(u_1), \pi(u_2), \ldots$ of R has a remarkable property: let $u \in \mathscr{S}$, and write $\pi(u) = \sum_{\lambda_i \neq 0} \lambda_i \pi(u_i)$. Then $u_i \leq u$ for all i, and if $u \notin \mathfrak{M}$, these inequalities are strict.

The previous theorem and 4.2.2(b) immediately imply

Corollary 4.2.4. *Let J be the ideal which is generated by the monomials in $C\mathfrak{M}$. Then the homogeneous k-algebra R and $k[X_1, \ldots, X_m]/J$ have the same Hilbert function. In particular, all Hilbert functions of homogeneous rings arise as Hilbert functions of homogeneous rings whose defining ideal is generated by monomials.*

The set of monomials $C\mathfrak{M}$ associated with R can be described differently. Let $I = \operatorname{Ker} \pi$, and set

$$L(I) = \{L(f) : f \in I\}, \quad \text{and} \quad I^* = L(I)R$$

where $L(f)$ denotes the *leading monomial* of f, that is, the monomial occurring in f which is maximal in the reverse degree-lexicographical order. Then $L(I) = C\mathfrak{M}$. Indeed, let $v \in L(I)$, and choose $f \in I$, $f = \sum_{i=1}^{n} \lambda_i v_i$ with monomials v_i such that $v = L(f) = v_n$. Assume $v_n \notin C\mathfrak{M}$; then $v_n \in \mathfrak{M}$, and so

$$0 \neq \pi(v_n) = -\sum_{i=1}^{n-1} \lambda_n^{-1} \lambda_i \pi(v_i).$$

Each $\pi(v_i)$ is a linear combination $\sum \alpha_{ij} \pi(u_j)$, $\alpha_{ij} \in k$, $u_j \in \mathfrak{M}$, $u_j \leq v_i < v_n$. Replacing the $\pi(v_i)$ in the above equation by their linear combinations gives a representation as a non-trivial linear combination of elements in $\pi(\mathfrak{M})$. This contradicts 4.2.3.

Conversely, suppose $v \in C\mathfrak{M}$. Then $\pi(v) = \sum \lambda_i \pi(u_i)$ with $u_i \in \mathfrak{M}$, $u_i < v$. Hence, if we set $f = v - \sum \lambda_i u_i$, then $\pi(f) = 0$ and $L(f) = v$.

The ideal I^* is finitely generated. Therefore there exist polynomials $f_1, \ldots, f_n \in I$ such that $I^* = (L(f_1), \ldots, L(f_n))$. Any such subset of I is called a *Gröbner* or *standard basis* of I.

Note that any Gröbner basis of I generates I: let $f \in I$; then $L(f) = \sum g_i L(f_i)$ for some $g_i \in k[X_1, \ldots, X_m]$, and it follows that either $f = \lambda \sum g_i f_i$, or $L(f - \lambda \sum g_i f_i) < L(f)$ for a suitable $\lambda \in k$. In the first case, f is an element of $(f_1, \ldots, f_n)$. In the second case we apply the same procedure to $f' = f - \lambda \sum g_i f_i$ to obtain an element f'' which is either zero, in which case $f \in (f_1, \ldots, f_n)$, or which has $L(f'') < L(f')$. Since any descending sequence of elements in $\mathscr{S}$ terminates, we eventually arrive at the required conclusion.

We should warn the reader that the converse is not true. Consider for example the ideal $I = (f_1, f_2)$ with $f_1 = X_1 X_2 + X_3^2$, $f_2 = X_2 X_3$. Then $X_1 X_2^2 = X_2 f_1 - X_3 f_2$ is an element of I^*, but not of $(L(f_1), L(f_2)) = (X_3^2, X_2 X_3)$.

Even though a Gröbner basis of an ideal I is not simply given by the leading forms of a system of generators of I, there does exist an algorithm to compute a Gröbner basis – the so-called *Buchberger algorithm*. This, and the fact that most explicit calculations in commutative algebra are performed using Gröbner bases, explain their importance. Buchberger's algorithm has been implemented in various computer algebra programs.

Macaulay representations and lexsegment ideals. Let $S = k[X_1, \ldots, X_m]$ denote the polynomial ring over a field k. The problem of determining the Hilbert function of a homogeneous factor ring R of S boils down to the following question: given a subspace $V \subset S_d$, what can be said about the k-dimension of the subspace $S_1 V$ in S_{d+1}?

We will give the answer in a special but important case. Let $u \in S_d$ be a monomial. We define the sets

$$\mathscr{L}_u = \{v \in S_d : v < u\} \quad \text{and} \quad \mathscr{R}_u = \{v \in S_d : v \geq u\}$$

of the monomials of degree d which are 'left' and 'right' of u. Note that $\mathscr{R}_{X_1} = \{X_1, \ldots, X_n\}$. Monomial sets of the form $\mathscr{R}_u \subset S_d$ are called *lexsegments* (*of degree* d). The next lemma says that a lexsegment of degree d spans a lexsegment of degree $d + 1$.

Lemma 4.2.5. $\mathscr{R}_{X_1} \mathscr{R}_u = \mathscr{R}_{X_1 u}$.

PROOF. Let $v \in \mathscr{R}_u$; then $X_i v \geq X_1 v \geq X_1 u$. Conversely, let $v \in \mathscr{R}_{X_1 u}$. We may assume that X_1 does not divide v; then $v > X_1 u$. Let $u = X_1^{a_1} \cdots X_m^{a_m}$, $v = X_2^{b_2} \cdots X_m^{b_m}$, and i be the largest integer such that $b_i > a_i$. If there exists $j < i$ with $b_j > 0$, then $X_j^{-1} v \in \mathscr{R}_u$; otherwise, $X_i^{-1} v \in \mathscr{R}_u$. In both cases it follows that $v \in \mathscr{R}_{X_1} \mathscr{R}_u$. $\qquad\square$

The sets $\mathscr{L}_u$ admit a natural decomposition: let i be the largest integer such that X_i divides u. Then we can write

$$\mathscr{L}_u = \mathscr{L}'_u \cup \mathscr{L}''_u X_i,$$

where X_i does not divide any element in $\mathscr{L}'_u$. It is clear that this union is disjoint, that $\mathscr{L}'_u$ consists of all monomials of degree d in the variables $X_1, \ldots, X_{i-1}$, and that $\mathscr{L}''_u = \mathscr{L}_{X_i^{-1} u}$.

An example illustrates this decomposition. Let $S = k[X_1, \ldots, X_4]$ and $u = X_2^2 X_3$. Then

$$\mathscr{L}_u = \{X_1^3, \, X_1^2 X_2, \, X_1 X_2^2, \, X_2^3, \, X_1^2 X_3, \, X_1 X_2 X_3\},$$

$$\mathscr{L}'_u = \{X_1^3, \, X_1^2 X_2, \, X_1 X_2^2, \, X_2^3\},$$

$$\mathscr{L}''_u = \{X_1^2, \, X_1 X_2\} = \mathscr{L}_{X_2^2}.$$

It is convenient to denote the set of all monomials of degree d in the variables $X_1, \ldots, X_i$ by $[X_1, \ldots, X_i]_d$. We may again decompose $\mathscr{L}''_u$, etc. Thus if we write

$$u = X_{j(1)} X_{j(2)} \cdots X_{j(d)}$$

with $1 \leq j(1) \leq j(2) \leq \cdots \leq j(d)$, then

$$\mathscr{L}_u = [X_1, \ldots, X_{j(d)-1}]_d \cup \mathscr{L}_{X_{j(d)}^{-1} u} X_{j(d)}$$

$$= [X_1, \ldots, x_{j(d)-1}]_d \cup [X_1, \ldots, X_{j(d-1)-1}]_{d-1} X_{j(d)}$$

$$\cup \mathscr{L}_{X_{j(d-1)}^{-1} X_{j(d)}^{-1} u} X_{j(d-1)} X_{j(d)}$$

$$= \cdots,$$

and we end up with the disjoint union

$$\mathscr{L}_u = \bigcup_{i=1}^{d} [X_1, \ldots, X_{j(i)-1}]_i X_{j(i+1)} \cdots X_{j(d)},$$

called the *natural decomposition* of $\mathscr{L}_u$.

It follows that

$$|\mathscr{L}_u| = \sum_{i=1}^{d} \binom{k(i)}{i}$$

with $k(i) = j(i) + i - 2$. Note that $k(d) > k(d-1) > \cdots > k(1) \geq 0$. Here and in the sequel we use that $\binom{k}{l} = 0$ for $0 \leq k < l$.

The above considerations show that any non-negative integer has such a binomial sum expansion. We prove this directly.

Lemma 4.2.6. *Let d be a positive integer. Any $a \in \mathbb{N}$ can be written uniquely in the form*

$$a = \binom{k(d)}{d} + \binom{k(d-1)}{d-1} + \cdots + \binom{k(1)}{1},$$

where $k(d) > k(d-1) > \cdots > k(1) \geq 0$.

PROOF. In order to prove the existence, we choose $k(d)$ maximal such that $\binom{k(d)}{d} \leq a$. If $a = \binom{k(d)}{d}$, then $a = \sum_{i=1}^{d} \binom{k(i)}{i}$ with $k(i) = i - 1$ for $i = 1, \ldots, d - 1$. Now assume that $a' = a - \binom{k(d)}{d} > 0$. By the induction hypothesis we may assume that $a' = \sum_{i=1}^{d-1} \binom{k(i)}{i}$ with $k(d-1) > k(d-2) > \cdots > k(1) \geq 0$. It remains to show that $k(d) > k(d-1)$: since $\binom{k(d)+1}{d} > a$, it follows that

$$\binom{k(d)}{d-1} = \binom{k(d)+1}{d} - \binom{k(d)}{d} > a' \geq \binom{k(d-1)}{d-1}.$$

Hence $k(d) > k(d-1)$.

The uniqueness follows by induction on a, once we have shown the following: if $a = \sum_{i=1}^{d} \binom{k(i)}{i}$ with $k(d) > k(d-1) > \cdots > k(1) \geq 0$, then $k(d)$ is the largest integer with $\binom{k(d)}{d} \leq a$. Again we prove this statement by induction on a. For $a = 1$ the assertion is trivial. Now assume that $a > 1$, and $\binom{k(d)+1}{d} \leq a$. Then

$$\sum_{i=1}^{d-1} \binom{k(i)}{i} \geq \binom{k(d)+1}{d} - \binom{k(d)}{d} = \binom{k(d)}{d-1} \geq \binom{k(d-1)+1}{d-1},$$

and this contradicts the induction hypothesis.   □

Following Green [138] we refer to the sum 4.2.6 as the d-th Macaulay representation of a, and call $k(d), \ldots, k(1)$ the d-th Macaulay coefficients of a.

Note that for all $i \leq d$ the coefficient $k(i)$ is determined by the property of being the maximal integer j such that

$$\binom{j}{i} \leq a - \binom{k(d)}{d} - \cdots - \binom{k(i+1)}{i+1}.$$

The d-th Macaulay coefficients have the following nice property.

Lemma 4.2.7. Let $k(d), \ldots, k(1)$, respectively $k'(d), \ldots, k'(1)$, be the d-th Macaulay coefficients of a, respectively a'. Then $a > a'$ if and only if

$$(k(d), \ldots, k(1)) > (k'(d), \ldots, k'(1))$$

in the lexicographical order.

PROOF. We prove both implications by induction on d. For $d = 1$ the assertion is trivial. We now assume that $d > 1$. If $k(d) = k'(d)$, then $k(d-1), \ldots, k(1)$ (respectively $k'(d-1), \ldots, k'(1)$) are the $(d-1)$-th Macaulay coefficients of $a - \binom{k(d)}{d}$ (respectively $a' - \binom{k'(d)}{d}$), and we may apply the induction hypothesis. If $k(d) \neq k'(d)$, then $k(d) > k'(d)$ if and only if $a > a'$. This follows from the characterization of the d-th Macaulay coefficients preceding this lemma. □

Skipping the summands which are zero in the d-th Macaulay representation of a we get the following unique sum expansion:

$$a = \binom{k(d)}{d} + \binom{k(d-1)}{d-1} + \cdots + \binom{k(j)}{j}$$

where $k(d) > k(d-1) > \cdots > k(j) \geq j \geq 1$. We define

$$a^{\langle d \rangle} = \binom{k(d)+1}{d+1} + \binom{k(d-1)+1}{d} + \cdots + \binom{k(1)+1}{2}$$

$$= \binom{k(d)+1}{d+1} + \binom{k(d-1)+1}{d} + \cdots + \binom{k(j)+1}{j+1},$$

and set $0^{\langle d \rangle} = 0$.

Proposition 4.2.8. *Let u be a monomial of degree d in the polynomial ring S. Then $|\mathscr{L}_{X_1 u}| = |\mathscr{L}_u|^{\langle d \rangle}$.*

PROOF. Let $\mathscr{L}_u = \bigcup_{i=1}^{d} [X_1, \ldots, X_{j(i)-1}]_i X_{j(i+1)} \cdots X_{j(d)}$ be the canonical decomposition of $\mathscr{L}_u$. We claim that

$$\bigcup_{i=1}^{d} [X_1, \ldots, X_{j(i)-1}]_{i+1} X_{j(i+1)} \cdots X_{j(d)}$$

is the canonical decomposition of $\mathscr{L}_{X_1 u}$. Indeed, the canonical decomposition of $\mathscr{L}_u$ is completely determined by the sequence $j(1), \ldots, j(d)$ attached to u. Let $l(1), \ldots, l(d+1)$ be the corresponding sequence for $X_1 u$. Then $l(1) = 1$, and $l(i) = j(i-1)$ for $i = 2, \ldots, d+1$. This proves the claim and the proposition. □

Let $S = k[X_1, \ldots, X_m]$ be a polynomial ring over a field, and $u \in S$ a monomial. An ideal which in each degree is spanned by a lexsegment will be called a *lexsegment ideal*. In view of 4.2.5 and 4.2.8 we obtain

Corollary 4.2.9. *Let $I \subset S$ be a lexsegment ideal, and set $R = S/I$. Then*

$$H(R, n+1) \leq H(R, n)^{\langle n \rangle} \qquad \text{for all} \qquad n.$$

Equality holds for a given n if and only if $I_{n+1} = (X_1, \ldots, X_m) I_n$.

Macaulay's theorem. We now come to the main result of this section. It will follow that the growth of the Hilbert function of a homogeneous ring defined by a lexsegment ideal is, in a sense, the maximum possible.

Theorem 4.2.10 (Macaulay).   *Let k be a field, and let $h: \mathbb{N} \to \mathbb{N}$ be a numerical function. The following conditions are equivalent:*

(a) *there exists a homogeneous k-algebra R with Hilbert function $H(R, n) = h(n)$ for all $n \geq 0$;*

(b) *there exists a homogeneous k-algebra R with monomial relations and with Hilbert function $H(R, n) = h(n)$ for all $n \geq 0$;*

(c) *one has $h(0) = 1$, and $h(n + 1) \leq h(n)^{\langle n \rangle}$ for all $n \geq 1$;*

(d) *let $m = h(1)$, and for each $n \geq 0$ let $\mathcal{M}_n$ be the first $h(n)$ monomials in the variables $X_1, \ldots, X_m$ of degree n in the reverse degree-lexicographical order; set $\mathcal{M} = \bigcup_{n \geq 0} \mathcal{M}_n$; then $\mathcal{M}$ is an order ideal of monomials.*

The following example demonstrates the effectiveness of Macaulay's theorem: let us check that $1 + 3t + 5t^2 + 8t^3$ is not the Hilbert series of a homogeneous ring. In fact, condition (c) is violated since $5 = \binom{3}{2} + \binom{2}{1}$, and $5^{\langle 2 \rangle} = \binom{4}{3} + \binom{3}{2} = 7 < 8$. Instead we also could apply (d), and get $\mathcal{M}_1 = \{X_1, X_2, X_3\}$, $\mathcal{M}_2 = \{X_1^2, X_1X_2, X_2^2, X_1X_3, X_2X_3\}$, $\mathcal{M}_3 = \{X_1^3, X_1^2X_2, X_1X_2^2, X_2^3, X_1^2X_3, X_1X_2X_3, X_2^2X_3, X_1X_3^2\}$. Thus we see that $X_1X_3^2 \in \mathcal{M}_3$, but $X_1^{-1}(X_1X_3^2) \notin \mathcal{M}_2$. Therefore $\mathcal{M}$ is not an order ideal of monomials.

Most parts of the theorem have already been shown: the equivalence of (a) and (b) is the content of 4.2.4, and the implication (d) $\Rightarrow$ (b) is trivial. For the proof of (c) $\Rightarrow$ (d) we assume that $h(1) = m$. Then condition (c) implies that $h(n) \leq \binom{n+m-1}{n}$. Suppose that $h(n + 1) = \binom{n+m}{n+1}$, then $h(n) = \binom{n+m-1}{n}$, and so $\mathcal{M}_i = [X_1, \ldots, X_m]_i$ for $i = n$ and $i = n + 1$. Therefore, if $u \in \mathcal{M}_{n+1}$, and X_i divides u, then trivially $X_i^{-1}u \in \mathcal{M}_n$. Now we suppose that $h(n + 1) < \binom{n+m}{n+1}$, then there exist a monomial u_{n+1} such that $\mathcal{M}_{n+1} = \mathcal{L}_{u_{n+1}}$. If, as before, $\mathcal{M}_n = [X_1, \ldots, X_m]_n$, there is nothing to show. Otherwise, there exists a monomial u_n such that $\mathcal{M}_n = \mathcal{L}_{u_n}$. Condition (c) and 4.2.8 imply that $\mathcal{R}_{X_1}\mathcal{R}_{u_n} \subset \mathcal{R}_{u_{n+1}}$. Therefore, if $u \in \mathcal{M}_{n+1}$, and X_i divides u, then $X_i^{-1}u \in \mathcal{M}_n$. In other words, $\mathcal{M} = \bigcup_{n \geq 0} \mathcal{M}_n$ is an order ideal of monomials.

For the most difficult implication (a) $\Rightarrow$ (c) we present the elegant proof of Green [138]. This needs some preparations.

If a positive integer a has d-th Macaulay coefficients $k(d), \ldots, k(1)$, then let

$$a_{\langle d \rangle} = \binom{k(d) - 1}{d} + \binom{k(d - 1) - 1}{d - 1} + \cdots + \binom{k(1) - 1}{1}$$

$$= \binom{k(d) - 1}{d} + \cdots + \binom{k(j) - 1}{j},$$

where $j = \min\{i : k(i) \geq i\}$.

Note that $a_{\langle d \rangle}$ has d-th Macaulay coefficients $k(d) - 1, \ldots, k(j) - 1, j - 2, j - 3, \ldots, 0$.

Lemma 4.2.11. (a) *If $a \leq a'$, then $a_{\langle d \rangle} \leq a'_{\langle d \rangle}$.*
(b) *If $k(j) \neq j$ for $j = \min\{i : k(i) \geq i\}$, then $(a-1)_{\langle d \rangle} < a_{\langle d \rangle}$.*

PROOF. (a) follows from the observation preceding this lemma and 4.2.7.
For (b) let $k(d), k(d-1), \ldots, k(1)$ be the d-th Macaulay coefficients of a,
and $k'(d), k'(d-1), \ldots, k'(1)$ the d-th Macaulay coefficients of $a-1$; then
$k'(d) \leq k(d)$ by 4.2.7. If $k'(d) = k(d)$, we set $a' = a - \binom{k(d)}{d}$. Convince
yourself that $a' - 1 > 0$. Then it follows that a' (respectively $a' - 1$) has
$(d-1)$-th Macaulay coefficients $k(d-1), k(d-2), \ldots, k(1)$ (respectively
$k'(d-1), k'(d-2), \ldots, k'(1)$). Moreover, a' satisfies the hypothesis of
(b). Therefore, if we argue by induction on d, we may assume that
$(a'-1)_{\langle d-1 \rangle} < a'_{\langle d-1 \rangle}$. Hence the required inequality in the case in which
$k'(d) = k(d)$ follows from the equalities $a_{\langle d \rangle} = a'_{\langle d-1 \rangle} + \binom{k(d)-1}{d}$ and $(a-1)_{\langle d \rangle} = (a' - 1)_{\langle d-1 \rangle} + \binom{k(d)-1}{d}$.

Now suppose that $k'(d) < k(d)$. Our assumption implies that the
d-th Macaulay coefficient of $a_{\langle d \rangle}$ is $k(d) - 1$, and that the d-th Macaulay
coefficient of $(a-1)_{\langle d \rangle}$ is less than or equal to $k'(d) - 1$. Therefore the
conclusion follows from 4.2.7. $\qquad \square$

Theorem 4.2.12, interesting in its own right, is the key to the still
unproved implication (a) $\Rightarrow$ (c) of 4.2.10.

Let R be a homogeneous k-algebra, k an infinite field. The affine
k-space R_1 is irreducible, and so any non-empty (Zariski-) open subset is
dense in R_1. This suggests the following terminology: a property $\mathscr{P}$ holds
for a *general linear form* of R_1 if there exists a non-empty open subset U
of R_1 such that $\mathscr{P}$ holds for all $h \in U$.

Theorem 4.2.12 (Green). *Let R be a homogeneous k-algebra, k an infinite
field, and let $n \geq 1$ be an integer. Then*
$$H(R/hR, n) \leq H(R, n)_{\langle n \rangle}$$
for a general linear form h.

PROOF. Let $s = \sup\{\dim_k hR_{n-1} : h \in R_1\}$; then $\dim_k hR_{n-1} = s$ for a
general linear form. Indeed, let $U \subset R_1$ be the subset of elements $h \in R_1$
such that $\dim_k hR_{n-1} = s$. It is obvious that $U \neq \emptyset$. In order to see
that U is open, we choose a basis $a_1, \ldots, a_m$ of R_1 and bases of R_{n-1} and
R_n. Then the multiplication map $h : R_{n-1} \to R_n$, $h = \sum_{i=1}^m x_i a_i$, can be
described by a matrix of linear forms in $x_1, \ldots, x_m$. Replacing the x_i by
indeterminates yields a matrix A of linear polynomials with coefficients
in k, and it is clear that U is the complement of $V(I_s(A))$ in R_1.

Let $h \in U$, and set $S = R/hR$. We claim that $H(S, n) \leq H(R, n)_{\langle n \rangle}$, and
prove it by induction on $\min\{n, \dim_k R_1\}$. If either $n = 1$ or $\dim_k R_1 = 1$,
then the assertion is trivial. Now assume that $n > 1$ and $\dim_k R_1 > 1$. Let
$V \subset S_1$ be the subset of linear forms g for which $\dim_k g S_{n-1}$ is maximal,

and denote by φ the canonical epimorphism $R \to S$. We consider the open subset

$$W = (U \setminus kh) \cap \varphi^{-1}(V)$$

of R_1. The set W is non-empty since R_1 is irreducible and both $\varphi^{-1}(V)$ and $U \setminus kh$ are non-empty. In fact, assume that $U \subset kh$. Then, since U is a dense and kh is a closed subset in R_1 it follows that $R_1 = kh$, contradicting the assumption $\dim_k R_1 > 1$. Now we choose $h^* \in W$, and get

$$H(S, n) = \dim_k(S_n/h^* S_{n-1}) + \dim_k h^* S_{n-1}.$$

By our choice of h^*, the induction hypothesis yields the inequality

$$\dim_k(S_n/h^* S_{n-1}) \leq H(S, n)_{\langle n \rangle}.$$

To obtain an upper bound for the second summand note first that

$$\dim_k h^* S_{n-1} \leq \dim_k(h^* R_{n-1}/h(h^* R_{n-2})) = \dim_k(h R_{n-1}/h^*(h R_{n-2})).$$

The last equality holds true since the difference of both sides equals $\dim_k(R_n/h R_{n-1}) - \dim_k(R_n/h^* R_{n-1})$, and this difference is zero since both h and h^* belong to U.

Let $W^* \subset R_1$ be the (non-empty) open set of linear forms l for which $l(h R_{n-2})$ has maximal dimension. Then, if we actually choose $h^* \in W \cap W^*$, noting that $h R_{n-1}$ may be viewed as the $(n-1)$-th homogeneous component of $P = R/(\text{Ann } h)$, we may apply our induction hypothesis to conclude that $\dim_k h^* S_{n-1} \leq (H(R, n) - H(S, n))_{\langle n-1 \rangle}$. The rest of the proof is a purely numerical argument. What we need is this: given integers $0 < b < a$ such that

$$b \leq b_{\langle n \rangle} + (a - b)_{\langle n-1 \rangle},$$

then $b \leq a_{\langle n \rangle}$.

Assume this fails, and write $b = \binom{k(n)}{n} + \cdots + \binom{k(j)}{j}$ with $k(n) > \cdots > k(j) \geq j > 0$. Then $a < \binom{k(n)+1}{n} + \cdots + \binom{k(j)+1}{j}$, and so $a - b < \binom{k(n)}{n-1} + \cdots + \binom{k(j)}{j-1}$.

We distinguish two cases. If $j = 1$, then $a - b \leq \binom{k(n)}{n-1} + \cdots + \binom{k(2)}{1}$, and hence $(a - b)_{\langle n-1 \rangle} \leq \binom{k(n)-1}{n-1} + \cdots + \binom{k(2)-1}{1}$, and $b_{\langle n \rangle} = \binom{k(n)-1}{n} + \cdots + \binom{k(1)-1}{1}$. Thus our hypothesis implies $b \leq \binom{k(n)}{n} + \cdots + \binom{k(2)}{2} + \binom{k(1)-1}{1} < b$, a contradiction.

If $j > 1$, then $(a - b)_{\langle n-1 \rangle} < \binom{k(n)-1}{n-1} + \cdots + \binom{k(j)-1}{j-1}$, and this together with our assumption again yields a contradiction.   □

In order to complete the proof of Macaulay's theorem another numerical result is needed.

Lemma 4.2.13. *Let a, a', and d be positive integers.*
(a) *If $a \leq a'$, then $a^{\langle d \rangle} \leq a'^{\langle d \rangle}$.*
(b) *Let $k(d), \ldots, k(1)$ be the d-th Macaulay coefficients of a, and $j = \min\{i : k(i) \geq i\}$. Then*

$$(a+1)^{\langle d \rangle} = \begin{cases} a^{\langle d \rangle} + k(1) + 1 & \text{if } j = 1, \\ a^{\langle d \rangle} + 1 & \text{if } j > 1. \end{cases}$$

PROOF. Claim (a) follows from 4.2.7, and (b) is immediate for $j > 1$. Now assume that $j = 1$, and let i be the maximal integer such that $k(i) = k(1) + i - 1$. Then

$$a = \binom{k(d)}{d} + \cdots + \binom{k(i+1)}{i+1} + \sum_{r=1}^{i} \binom{k(1)+r-1}{r}$$

$$= \binom{k(d)}{d} + \cdots + \binom{k(i+1)}{i+1} + \binom{k(1)+i}{i} - 1,$$

and hence

$$a+1 = \binom{k(d)}{d} + \cdots + \binom{k(i+1)}{i+1} + \binom{k(1)+i}{i}$$

is the d-th Macaulay expansion of $a+1$ since $k(i+1) > k(1)+i$.
   Now we get

$$a^{\langle d \rangle} = \binom{k(d)+1}{d+1} + \cdots + \binom{k(i+1)+1}{i+2} + \sum_{r=1}^{i} \binom{k(1)+r}{r+1}$$

$$= \binom{k(d)+1}{d+1} + \cdots + \binom{k(i+1)+1}{i+2} + \sum_{r=2}^{i+1} \binom{k(1)+r-1}{r}$$

$$= \binom{k(d)+1}{d+1} + \cdots + \binom{k(i+1)+1}{i+2} + \binom{k(1)+i+1}{i+1} - k(1) - 1,$$

and so

$$(a+1)^{\langle d \rangle} = \binom{k(d)+1}{d+1} + \cdots + \binom{k(i+1)+1}{i+2} + \binom{k(1)+i+1}{i+1}$$

$$= a^{\langle d \rangle} + k(1) + 1,$$

as asserted. □

PROOF OF 4.2.10, (a) $\Rightarrow$ (c). We may assume that k is infinite: if necessary replace R by $l \otimes_k R$ where l is an infinite extension field of k.
   Let g be a linear form, and set $S = R/gR$. The exact sequence

$$0 \longrightarrow gR_n \longrightarrow R_{n+1} \longrightarrow S_{n+1} \longrightarrow 0$$

yields the inequality $H(R, n+1) \le H(R, n) + H(S, n+1)$. Set $a = H(R, n)$ and $b = H(R, n+1)$. For a general linear form g the inequality and 4.2.12 give $b \le a + b_{\langle n+1 \rangle}$. Let $k(n+1), \ldots, k(1)$ be the $(n+1)$-th Macaulay coefficients of b. Then

$$b_{\langle n+1 \rangle} = \binom{k(n+1) - 1}{n+1} + \cdots + \binom{k(1) - 1}{1},$$

and so

$$a \ge \binom{k(n+1) - 1}{n} + \cdots + \binom{k(2) - 1}{1} + \binom{k(1) - 1}{0}.$$

Let, as before, $j = \min\{i : k(i) \ge i\}$. If $j > 1$, then $k(1) = 0$, and

$$a^{\langle n \rangle} \ge \binom{k(n+1)}{n+1} + \cdots + \binom{k(2)}{2} = b.$$

If $j = 1$, then

$$a^{\langle n \rangle} \ge \binom{k(n+1)}{n+1} + \cdots + \binom{k(3)}{3} + \binom{k(2)}{2} + k(2),$$

by 4.2.13. But $k(2) > k(1)$, and hence $a^{\langle n \rangle} > b$.   □

Corollary 4.2.14. Let R be a homogeneous k-algebra, k a field. Then $H(R, n+1) = H(R, n)^{\langle n \rangle}$ for $n \gg 0$.

PROOF. Write $R = S/I$, $S = k[X_1, \ldots, X_m]$. According to 4.2.10(d) there exists an order ideal of monomials $\mathcal{M} = \bigcup_{n \ge 0} \mathcal{M}_n$ in S such that $H(R, n) = H(S/J, n)$ for all n, where J is the ideal generated by all the monomials not in $\mathcal{M}$. Moreover, the choice of $\mathcal{M}$ was such that $\mathcal{M}_\backslash$ consists of all monomials of degree n if $I_n = 0$ and $\mathcal{M}_n = \mathcal{L}_{u_n}$ for a suitable monomial u_n otherwise. Since J is finitely generated, there exists an integer r such that $\mathcal{R}_{u_{n+1}} = \mathcal{R}_{X_1}\mathcal{R}_{u_n}$ for all $n \ge r$. Thus the assertion follows from 4.2.8.   □

If we combine 1.5.12 with Macaulay's theorem, we obtain the following characterization of the Hilbert series of Cohen–Macaulay homogeneous algebras.

Proposition 4.2.15. Let k be a field, and $h_0, \ldots, h_s$ a finite sequence of positive integers. The following conditions are equivalent:
(a) there exist an integer d, and a Cohen–Macaulay (reduced) homogeneous k-algebra R of dimension d (whose defining ideal is generated by squarefree monomials) such that

$$H_R(t) = \frac{\sum_{i=0}^{s} h_i t^i}{(1-t)^d};$$

(b) $h_0 = 1$, and $0 \le h_{i+1} \le h_i^{\langle i \rangle}$ for all $i = 1, \ldots, s-1$.

PROOF. (a) $\Rightarrow$ (b): By 1.5.11 there exists an R-sequence $x = x_1, \ldots, x_d$ of degree 1 elements. According to 4.1.8 we have

$$H_R(t) = \frac{Q_R(t)}{(1-t)^d}, \qquad Q_R(t) = \sum_{i=0}^{s} h_i t^i.$$

Let $\bar{R} = R/xR$; then $H_{\bar{R}}(t) = (1-t)^d H_R(t) = Q_R(t)$. It follows that $H(\bar{R}, n) = h_n$ for all $n \geq 0$. Therefore 4.2.10 yields the assertion.

(b) $\Rightarrow$ (a): By 4.2.10 there exists a homogeneous k-algebra $R = k[X_1, \ldots, X_m]/I$, where I is generated by monomials, such that $H_R(t) = \sum_{i=0}^{s} h_i t^i$. The k-algebra R is Cohen–Macaulay, simply because R is of dimension zero. In order to get a reduced such k-algebra with the required Hilbert series we consider a certain 'deformation' of R as described in the next lemma.   □

Lemma 4.2.16. *Let* $R = k[X_1, \ldots, X_m]/I$ *be a homogeneous k-algebra, where k is a field and I is generated by monomials. Then there exist a reduced homogeneous k-algebra S whose defining ideal is generated by squarefree monomials, and an S-sequence y of elements of degree 1 such that* $R \cong S/yS$.

PROOF. Assume $I = (u_1, \ldots, u_n)$, $u_i = X_1^{a_{i1}} \cdots X_m^{a_{im}}$ for $i = 1, \ldots, n$. If all exponents a_{ij} are at most 1, then I is a radical ideal; see Exercise 4.4.17. Suppose now that at least one $a_{ij} > 1$, say $a_{i1} > 1$ for some i. We introduce a new indeterminate Y, and set

$$v_k = Y^{a_{k1}-1} X_1 X_2^{a_{k2}} \cdots X_m^{a_{km}}$$

if $a_{k1} > 1$, and $v_k = u_k$ otherwise. The v_i satisfy the following conditions:
(i) if Y divides v_i, then X_1 divides v_i;
(ii) the indeterminate X_1 occurs in each v_i with multiplicity at most 1.
We claim that $Y - X_1$ is regular modulo the ideal $J = (v_1, \ldots, v_n)$. Indeed, assume the contrary is true. Then there exists an associated prime ideal $\mathfrak{p}$ of $k[X_1, \ldots, X_m, Y]/J$ with $Y - X_1 \in \mathfrak{p}$. By Exercise 4.4.15, $\mathfrak{p}$ is generated by a set of variables, and so $X_1, Y \in \mathfrak{p}$. It follows that there exists $w \in k[X_1, \ldots, X_m, Y]$, $w \notin J$, with $X_1 w \in J$ and $Y w \in J$. As J is generated by monomials, we may assume that w is a monomial. Then there exist integers i, j and monomials u_1, u_2 such that $X_1 w = v_i u_1$ and $Y w = v_j u_2$. As Y divides v_j, it follows from (i) that X_1 does also, and so X_1 divides w. But then the multiplicity of X_1 in v_i is at least 2, a contradiction.

If all variables in the v_i occur with multiplicity one, then J is a radical ideal. Otherwise we repeat this construction, and eventually reach the goal, since at each step we lower the multiplicities of the variables in the generators.   □

Exercises

4.2.17. Let R be a homogeneous k-algebra, k a field.
(a) Establish from 4.2.14 that there exist integers $a_1 \geq a_2 \geq \cdots \geq a_j \geq 0$ such that

$$P_R(n) = \binom{n + a_1}{a_1} + \binom{n + a_2 - 1}{a_2} + \cdots + \binom{n + a_j - (j - 1)}{a_j}.$$

(b) Determine the dimension and the multiplicity of R in terms of the integers $a_1, \ldots, a_j$.

4.2.18. Let k be a field, $S = k[X_1, \ldots, X_m]$, and

$$u = X_1^{a_1} X_2^{a_2} \cdots X_m^{a_m} = X_{j(1)} X_{j(2)} \cdots X_{j(d)}, \qquad j(1) \leq j(2) \leq \cdots \leq j(d),$$

a monomial of degree d. Set $R = S/I$, where I is the lexsegment ideal generated by $\mathscr{R}_u$. Then deduce
(a) $\dim R = j(d) - 1$, and
(b) $e(R) = a_i$ where $i = \max\{j : a_j \neq 0\}$, provided $\dim R > 0$.

4.3 Gotzmann's regularity and persistence theorem

Gotzmann's [136] regularity and persistence theorems give some deeper insight into the nature of the Hilbert polynomial and the Hilbert function.

As before let $S = k[X_1, \ldots, X_m]$ be the polynomial ring in m variables defined over a field k, $I \subset S$ a graded ideal, and $R = S/I$. The regularity theorem is a statement about the regularity of the ideal sheaf $\mathscr{I}$ associated with I in projective space. Note that different ideals may yield the same ideal sheaf. The ideal $\bar{I} = \mathrm{Ker}(S \to R/^*H_{\mathfrak{m}}^0(R))$ is called the *saturation of* I; the sheafs associated with ideals I and J coincide if and only if $\bar{I} = \bar{J}$.

We will formulate Gotzmann's theorems in the language of commutative algebra. So we define the (*Castelnuovo–Mumford*) *regularity* of a finite graded S-module M, rather than that of a sheaf. It is the number

$$\mathrm{reg}\, M = \max\{i + j : {}^*H_{\mathfrak{m}}^i(M)_j \neq 0\}.$$

Let q be an integer. Then M is called q-*regular* if $q \geq \mathrm{reg}(M)$, equivalently, if ${}^*H_{\mathfrak{m}}^i(M)_{j-i} = 0$ for all i and all $j > q$.

Before we set out for Gotzmann's theorems, we include an interesting description of regularity in terms of graded Betti numbers. It shows that $\mathrm{reg}(M)$ measures the 'complexity' of the minimal graded free resolution of M. Therefore regularity plays an important rôle in algorithmic commutative algebra. Denoting by $M_{\geq q}$ the truncated graded R-module $\bigoplus_{j \geq q} M_j$, one has

Theorem 4.3.1 (Eisenbud–Goto). *The following conditions are equivalent:*
(a) *M is q-regular;*
(b) *${}^*\mathrm{Tor}_i^S(M, k)_{j+i} = 0$ for all i and all $j > q$;*

(c) $M_{\geq q}$ admits a linear S-resolution, i.e., a graded resolution of the form

$$0 \longrightarrow S(-q-l)^{c_l} \longrightarrow \cdots \longrightarrow S(-q-1)^{c_1} \longrightarrow S(-q)^{c_0} \longrightarrow M_{\geq q} \longrightarrow 0.$$

PROOF. (b) $\Longleftrightarrow$ (c): By definition, the module $M_{\geq q}$ has a linear resolution if and only if

$$^*\mathrm{Tor}_i^S(M_{\geq q}, k)_r = H_i(x; M_{\geq q})_r = 0 \qquad \text{for all } i, r, \ r \neq i + q.$$

Here $H_*(x; M)$ is the Koszul homology of M with respect to the sequence $x = X_1, \ldots, X_m$.

Since $(M_{\geq q})_j = 0$ for $j < q$, we always have $H_i(x; M_{\geq q})_r = 0$ for $r < i + q$, while for $r > i + q$

$$H_i(x; M_{\geq q})_r = H_i(x; M)_r = {}^*\mathrm{Tor}_i^S(M, k)_r.$$

Thus the desired result follows.

(a) $\Rightarrow$ (c): We may assume $q = 0$ and $M = M_{\geq 0}$. Then it is immediate that $^*H_{\mathfrak{m}}^0(M)$ is concentrated in degree 0. This implies $M = {}^*H_{\mathfrak{m}}^0(M) \oplus M/{}^*H_{\mathfrak{m}}^0(M)$. The first summand is a direct summand of copies of k. Hence M is 0-regular if and only if $M/{}^*H_{\mathfrak{m}}^0(M)$ is 0-regular. In other words, we may assume that depth $M > 0$. We may further assume that k is infinite. Then there exists an element $y \in S$ of degree 1 which is M-regular. From the cohomology exact sequence associated with

$$0 \longrightarrow M(-1) \overset{y}{\longrightarrow} M \longrightarrow M/yM \longrightarrow 0$$

it follows that M/yM is 0-regular. By induction on the dimension on M, we may suppose that M/yM has a linear S/yS-resolution. But if F_* is a minimal graded free S-resolution, then F_*/yF_* is a minimal graded S/yS-resolution of M/yM. This implies that F_* is a linear S-resolution of M.

(c) $\Rightarrow$ (a): Again one may assume $q = 0$ and $M = M_{\geq 0}$. Then M has a linear resolution

$$\cdots \longrightarrow S(-2)^{c_2} \longrightarrow S(-1)^{c_1} \longrightarrow S^{c_0} \longrightarrow M \longrightarrow 0.$$

Computing $^*\mathrm{Ext}_S^i(M, S)$ from this resolution we see that $^*\mathrm{Ext}_S^i(M, S)_j = 0$ for $j < -i$. By duality (see 3.6.19) there exists an isomorphism of graded R-modules

$$^*H_{\mathfrak{m}}^i(M) \cong \mathrm{Hom}_k\big({}^*\mathrm{Ext}_S^{m-i}(M, S(-m)), k\big).$$

Therefore $^*H_{\mathfrak{m}}^i(M)_{j-i} = 0$ for all $j > 0$, as desired. $\qquad \Box$

The regularity theorem says that the regularity of the saturation of an ideal I can be read off the Hilbert polynomial of S/I.

Theorem 4.3.2 (Gotzmann). *Write the Hilbert polynomial $P_R(n)$ of $R = S/I$ in the unique form*

$$P_R(n) = \binom{n + a_1}{a_1} + \binom{n + a_2 - 1}{a_2} + \cdots + \binom{n + a_s - (s-1)}{a_s}$$

with $a_1 \geq a_2 \geq \cdots \geq a_s \geq 0$, as described in 4.2.17. Then the saturation $\bar{I}$ of I is s-regular.

PROOF. We prove the theorem by induction on the dimension of S. For $m = 1$ the assertion is trivial. Now let $m > 1$, and choose be a general linear form h. Since $P_R(n) = P_{S/\bar{I}}(n)$ we may assume that $I = \bar{I}$. We may further assume that $\bar{I} \neq S$. Then depth $R > 0$, and h is R-regular. Hence we get an exact sequence

$$0 \longrightarrow R(-1) \overset{h}{\longrightarrow} R \longrightarrow R/hR \longrightarrow 0$$

yielding the equation

$$(1) \qquad\qquad P_{R/hR}(n) = P_R(n) - P_R(n-1).$$

Set $\widetilde{R} = R/hR$, $\widetilde{S} = S/hS$; then $\widetilde{R} = \widetilde{S}/J$ for some ideal $J \subset \widetilde{S}$. Furthermore $P_{\widetilde{R}}(n) = P_{\widetilde{S}/J}(n)$. Suppose that

$$(2)\quad P_{\widetilde{S}/J}(n) = \binom{n + b_1}{b_1} + \binom{n + b_2 - 1}{b_2} + \cdots + \binom{n + b_r - (r-1)}{b_r};$$

then $\bar{J}$ is r-regular by the induction hypothesis.

(1) and (2) imply

$$P_R(n) = \binom{n + a_1}{a_1} + \binom{n + a_2 - 1}{a_2} + \cdots + \binom{n + a_r - (r-1)}{a_r} + c,$$

where c is a constant and $a_i = b_i + 1$ for all i.

We claim that $c \geq 0$, and that $\bar{I}$ is s-regular for $s = r + c$. These two claims complete the proof. Indeed, we may set $a_{r+1} = \cdots = a_{r+c} = 0$.

In order to derive the first claim, assume that $c < 0$. For $n \gg 0$ we then have

$$(3)\quad H(R,n) < \binom{n + a_1}{a_1} + \binom{n + a_2 - 1}{a_2} + \cdots + \binom{n + a_r - (r-1)}{a_r}.$$

The right hand side b of this inequality satisfies the equation

$$b = \binom{n + a_1}{n} + \binom{n + a_2 - 1}{n-1} + \cdots + \binom{n + a_r - (r-1)}{n - (r-1)},$$

so that

$$b_{\langle n \rangle} = \binom{n+b_1}{n} + \binom{n+b_2-1}{n-1} + \cdots + \binom{n+b_r-(r-1)}{n-(r-1)}$$

$$= \binom{n+b_1}{b_1} + \binom{n+b_2-1}{b_2} + \cdots + \binom{n+b_r-(r-1)}{b_r}.$$

Observing that $n + a_r - (r-1) > n - (r-1)$, one deduces from (3) (see 4.2.11) that $H(R,n)_{\langle n \rangle} < b_{\langle n \rangle}$. Therefore by Green's theorem,

$$H(\widetilde{R}, n) < \binom{n+b_1}{b_1} + \binom{n+b_2-1}{b_2} + \cdots + \binom{n+b_r-(r-1)}{b_r}.$$

This contradicts (2).

For the proof of the second claim note first that ${}^{\bullet}H_{\mathfrak{m}}^i(\overline{J}) = {}^{\bullet}H_{\mathfrak{m}}^i(J)$ for $i > 1$. Therefore and since $\overline{J}$ is r-regular we deduce from the local cohomology sequence associated with

$$0 \longrightarrow I(-1) \xrightarrow{\ h\ } I \longrightarrow J \longrightarrow 0$$

that $H_{\mathfrak{m}}^i(I)_{j-i} = 0$ for all $i > 2$ and $j > r$ (and thus for $j > s$).

It remains to be shown that ${}^{\bullet}H_{\mathfrak{m}}^2(I)_{j-2} = 0$ for $j > s$. Suppose this is not the case, and let j be the largest number with ${}^{\bullet}H_{\mathfrak{m}}^2(I)_{j-2} \neq 0$. Then by 4.4.3(b) below,

$$H(R, j-2) - P_R(j-2) = -{}^{\bullet}H_{\mathfrak{m}}^1(R)_{j-2} < 0$$

since ${}^{\bullet}H_{\mathfrak{m}}^0(R) = 0$ and ${}^{\bullet}H_{\mathfrak{m}}^{i-1}(R)_{j-2} = {}^{\bullet}H_{\mathfrak{m}}^i(I)_{j-2} = 0$ for $i > 2$, as we have already seen. By our choice of j we have ${}^{\bullet}H_{\mathfrak{m}}^1(R)_{j-1} = 0$, so that

$$H(R, j-2) < P_R(j-2), \quad \text{but} \quad H(R, j-1) = P_R(j-1).$$

If $j = s+1$, then $j-2 = s-1$, and

$$P_R(s-1) = \binom{s-1+a_1}{s-1} + \cdots + \binom{1+a_{s-1}}{1} + 1,$$

whence

$$H(R, j-2) = H(R, s-1) \leq \binom{s-1+a_1}{s-1} + \cdots + \binom{1+a_{s-1}}{1}.$$

Thus Macaulay's theorem implies

$$H(R, j-1) \leq H(R, j-2)^{\langle j-2 \rangle} \leq \binom{s+a_1}{s} + \cdots + \binom{2+a_{s-1}}{2}$$

$$= P_R(s) - (a_s+1) < P_R(s) = P_R(j-1),$$

which is a contradiction.

If $j > s + 1$, then $P_R(j - 2)^{\langle j-2 \rangle} = P_R(j - 1)$ (see 4.2.14). We apply Macaulay's theorem again, and get

$$H(R, j - 1) \leq H(R, j - 2)^{\langle j-2 \rangle} < P_R(j - 1),$$

leading to the same contradiction. □

By 4.2.14 we have $H(R, n + 1) = H(R, n)^{\langle n \rangle}$ for all large n. But could it happen that $H(R, n + 1) = H(R, n)^{\langle n \rangle}$, and $H(R, r + 1) < H(R, r)^{\langle r \rangle}$ for some r and n with $r > n$? The following persistence theorem answers this question.

Theorem 4.3.3 (Gotzmann). *Suppose that $H(R, n+1) = H(R, n)^{\langle n \rangle}$ for some n and that I is generated by elements of degree $\leq n$. Then $H(R, r + 1) = H(R, r)^{\langle r \rangle}$ for all $r \geq n$.*

PROOF. We prove the theorem by induction on $m = \dim S$. If $m = 1$, I is principal, and the assertion is trivial. Now let us assume that $m > 1$. Let

$$H(R, n) = \binom{k(n)}{n} + \cdots + \binom{k(1)}{1}$$

be the n-th Macaulay expansion of $H(R, n)$. Macaulay's theorem implies

$$(4) \qquad H(R, r) \leq \binom{r - n + k(n)}{r} + \cdots + \binom{r - n + k(1)}{r - n + 1}$$

for all $r \geq n$, and it remains to be shown that equality holds.

Let h be a general linear form. Then

$$(5) \qquad (H(R, n)_{\langle n \rangle})^{\langle n \rangle} \geq H(R/hR, n)^{\langle n \rangle} \geq H(R/hR, n + 1)$$
$$\geq H(R, n + 1) - H(R, n) = (H(R, n)_{\langle n \rangle})^{\langle n \rangle}.$$

The first inequality is Green's theorem, the second is Macaulay's, the third follows from the exact sequence

$$R(-1) \xrightarrow{\ h\ } R \longrightarrow R/hR \longrightarrow 0,$$

and the last equality results from the hypothesis that $H(R, n + 1) = H(R, n)^{\langle n \rangle}$.

Since the first and last term in this chain of inequalities coincide, we must have equality everywhere. In particular, $H(R/hR, n + 1) = H(R/hR, n)^{\langle n \rangle}$. Since the defining ideal of R/hR is again generated by elements of degree $\leq n$, the induction hypothesis applies and yields $H(R/hR, r + 1) = H(R/hR, r)^{\langle r \rangle}$ for all $r \geq n$.

One also deduces from (5) that

$$H(R/hR, n + 1) = (H(R, n)_{\langle n \rangle})^{\langle n \rangle} = \binom{k(n)}{n + 1} + \cdots + \binom{k(1)}{2}.$$

Therefore

$$P_{R/hR}(r) = \binom{r + (k(n) - n - 1)}{k(n) - n - 1} + \cdots + \binom{r + (k(1) - 2) - (n - 1)}{k(1) - 2}$$

for all r.

Hence the saturation J of the defining ideal of R/hR is n-regular by the regularity theorem. Let $\overline{I}$ again denote the saturation of I and set $\overline{R} = S/\overline{I}$. It follows just as in the proof of the regularity theorem that

$$(6) \quad P_{\overline{R}}(r) = \binom{r + (k(n) - n)}{k(n) - n} + \cdots + \binom{r + (k(1) - 1) - (n - 1)}{k(1) - 1} + c$$

with $c \geq 0$.

Suppose $c > 0$; since $P_R(n) = P_{\overline{R}}(n)$, the inequality (4) then implies

$$P_R(r) = \binom{r - n + k(n)}{r} + \cdots + \binom{r - n + k(1)}{r - n + 1} + c$$
$$\geq H(R, r) + c > H(R, r)$$

for all $r \geq n$. This is a contradiction.

Now (6) and Gotzmann's regularity theorem entail that $\overline{I}$ is n-regular, whence $H(\overline{R}, r) = P_{\overline{R}}(r)$ for all $r \geq n$.

Thus for all $r \geq n$ we obtain the following string of inequalities:

$$H(\overline{R}, r) \leq H(R, r) \leq P_R(r) = P_{\overline{R}}(r) = H(\overline{R}, r).$$

Hence equality holds everywhere, and this proves the theorem. $\square$

Exercise

4.3.4. Let $S = k[X_1, \ldots, X_m]$ be a polynomial ring over a field k, and let $n \geq 1$ be an integer. A subspace V of the k-vector space S_n is called a *Gotzmann space* if the ideal I generated by V satisfies $H(S/I, n + 1) = H(S/I, n)^{\langle n \rangle}$.

(a) According to 4.2.9, lexsegments span Gotzmann spaces. Give an example of a set of monomials which is not a lexsegment (even after a permutation of the variables), but spans a Gotzmann space.

(b) Let I be the ideal generated by a Gotzmann space $V \subset S_n$. It can be shown that the ideal I has a linear resolution. Compute the Betti numbers of I.

(c) Suppose that $\dim R/I = 0$. Show I is generated by a Gotzmann space if and only if $I = \mathfrak{m}^n$ for some $n > 0$ where $\mathfrak{m} = (X_1, \ldots, X_m)$.

4.4 Hilbert functions over graded rings

In this section we consider positively graded k-algebras, that is, graded k-algebras of the form $R = \bigoplus_{i \geq 0} R_i$ where $R_0 = k$ and R is finitely generated over k. For simplicity we will assume that k is a field. In contrast to a homogeneous k-algebra the generators of a positively graded k-algebra may be of arbitrarily high degree.

In analogy with 4.1.8 we have

Proposition 4.4.1. *Let R be a positively graded k-algebra, k a field, and $M \neq 0$ a finite graded R-module of dimension d. Then there exist positive integers $a_1, \ldots, a_d$, and $Q(t) \in \mathbb{Z}[t, t^{-1}]$ such that*

$$H_M(t) = \frac{Q(t)}{\prod_{i=1}^{d}(1 - t^{a_i})} \qquad with \quad Q(1) > 0.$$

PROOF. We prove the assertion by induction on the dimension d of M. If $d = 0$, then $\dim_k M < \infty$, and so $M_n = 0$ for $n \gg 0$. Therefore $H_M(t) \in \mathbb{Z}[t, t^{-1}]$, and we set $Q(t) = H_M(t)$. It is clear that $Q(1) = \dim_k M > 0$.

Now assume that $d > 0$, and let $U = {}^*H_{\mathfrak{m}}^0(M)$, where $\mathfrak{m} = \bigoplus_{i>0} R_i$ is the unique graded maximal ideal of R. Note that U is a graded submodule of M with $\dim_k U < \infty$, and that $\mathfrak{m} \notin \mathrm{Ass}(M/U)$. We may assume that k is infinite (see the proof of 4.1.10). Then, according to 1.5.11, there exists a homogeneous (M/U)-regular element $x \in \mathfrak{m}$, say of degree a_1. The exact sequence

$$0 \longrightarrow (0 : x)_M(-a_1) \longrightarrow M(-a_1) \overset{x}{\longrightarrow} M \longrightarrow M/xM \longrightarrow 0,$$

where $(0 : x)_M = \{u \in M : xu = 0\}$, gives rise to the equation

$$H_M(t)(1 - t^{a_1}) = H_{M/xM}(t) - P(t),$$

where $P(t)$ is the Hilbert series of $(0 : x)_M(-a_1)$. The series $P(t)$ actually belongs to $\mathbb{Z}[t, t^{-1}]$ since $(0 : x)_M \subset U$, and U is of finite length. By the induction hypothesis there exist $\bar{Q}(t) \in \mathbb{Z}[t, t^{-1}]$, and positive integers $a_2, \ldots, a_d$ such that

$$H_{M/xM}(t) = \frac{\bar{Q}(t)}{\prod_{i=2}^{d}(1 - t^{a_i})}, \qquad \bar{Q}(1) > 0.$$

Set $Q(t) = \bar{Q}(t) - P(t) \prod_{i=2}^{d}(1 - t^{a_i})$; then, as required, we have $H_M(t) = Q(t)/\prod_{i=1}^{d}(1 - t^{a_i})$ with $Q(1) > 0$. $\qquad\qquad\square$

Remarks 4.4.2. (a) Proposition 4.4.1 is analogously valid in the case where R_0 is an Artinian local ring.
(b) It can easily be verified that the integers $a_1, \ldots, a_d$ found in the proof of 4.4.1 are the degrees of elements generating a Noether normalization of $R/\mathrm{Ann}\,M$. (Also see Exercise 4.4.12.)

A function $P : \mathbb{Z} \to \mathbb{C}$ is called a *quasi-polynomial* (*of period g*) if there exist a positive integer g and polynomials P_i, $i = 0, \ldots, g - 1$, such that for all $n \in \mathbb{Z}$ one has $P(n) = P_i(n)$ where $n = mg + i$ with $0 \leq i \leq g - 1$.

In the following theorem we consider the graded components of the modules ${}^*H_{\mathfrak{m}}^i(M)$. Note that they are finite dimensional k-vector spaces (why?).

Theorem 4.4.3 (Serre). *Let R be a positively graded k-algebra, k a field, and $M \neq 0$ a finite graded R-module of dimension d, and denote the *maximal ideal of R by $\mathfrak{m}$. Then*
(a) *there exists a uniquely determined quasi-polynomial P_M with $H(M, n) = P_M(n)$ for all $n \gg 0$,*
(b) $H(M, n) - P_M(n) = \sum_{i=0}^{d} (-1)^i \dim_k {}^*H_{\mathfrak{m}}^i(M)_n$ *for all $n \in \mathbb{Z}$,*
(c) *one has*

$$\deg H_M(t) = \max\{n : H(M, n) \neq P_M(n)\}$$

$$= \max\{n : \sum_{i=0}^{d} (-1)^i \dim_k {}^*H_{\mathfrak{m}}^i(M)_n \neq 0\}.$$

(*Here* $\deg H_M(t)$ *denotes the degree of the rational function* $H_M(t)$.)

PROOF. (a) follows from Exercise 4.4.10 or Exercise 4.4.11.

(b) holds when $d = 0$, since then $P_M = 0$ and $M = {}^*H_{\mathfrak{m}}^0(M)$ whereas ${}^*H_{\mathfrak{m}}^i(M) = 0$ for $i > 0$. Next one notes that both sides of the equation change by the same amount, namely $\dim_k {}^*H_{\mathfrak{m}}^0(M)_n$, if one replaces M by $M/{}^*H_{\mathfrak{m}}^0(M)$. As in the proof of 4.4.1 we may thus assume that ${}^*H_{\mathfrak{m}}^0(M) = 0$ and that $\mathfrak{m}$ contains a homogeneous M-regular element x of degree e. Then we have an exact sequence

$$0 \longrightarrow M(-e) \overset{x}{\longrightarrow} M \longrightarrow M/xM \longrightarrow 0.$$

Set $H'_M(t) = \sum_{n \in \mathbb{Z}} (H(M, n) - P_M(n))t^n$ and

$$H''_M(t) = \sum_{n \in \mathbb{Z}} (\sum_{i=0}^{d} (-1)^i \dim_k {}^*H_{\mathfrak{m}}^i(M)_n)t^n.$$

As $H_{M/xM}(t) = (1 - t^e)H_M(t)$, it follows that $P_{M/xM}(n) = P_M(n) - P_M(n-e)$ for all $n \gg 0$, and, hence, $P_{M/xM}(n) = P_M(n) - P_M(n - e)$ for all $n \in \mathbb{Z}$. Therefore $H'_{M/xM}(t) = (1 - t^e)H'_M(t)$. The long exact sequence of graded local cohomology derived from the exact sequence above easily yields that likewise $H''_{M/xM}(t) = (1 - t^e)H''_M(t)$. By induction, $H'_{M/xM}(t) = H''_{M/xM}(t)$, so $H'_M(t) = H''_M(t)$ as well.

(c) follows immediately from (b) and Exercise 4.4.10. $\square$

The previous theorem generalizes Hilbert's theorem 4.1.3, and consequently P_M is termed the *Hilbert quasi-polynomial of M*.

Suppose that $M = R$ in 4.4.3 and that R is Cohen–Macaulay. Then $\deg H_R(t)$ equals $\max\{n : {}^*H_{\mathfrak{m}}^d(R)_n \neq 0\}$, and thus is the a-invariant of R introduced in Section 3.6; see the remark following 3.6.19. This fact motivates the following extension of the notion of a-invariant:

Definition 4.4.4.   Let R be a positively graded k-algebra where k is a field. Then the degree of the Hilbert function of R is denoted by $a(R)$ and called the *a-invariant* of R.

Observe that $a(R) < 0$ if and only if $H(R, n) = P_R(n)$ for all $n \geq 0$. That this condition has structural implications, is exhibited by a theorem of Flenner [107] and Watanabe [389]: if k is an algebraically closed field of characteristic 0, and R is a normal Cohen–Macaulay positively graded k-algebra with negative a-invariant, then R has rational singularities, provided $R_\mathfrak{p}$ has rational singularities for all prime ideals $\mathfrak{p}$ different from the *maximal ideal of R. In Chapter 10 we will again encounter the condition $a(R) < 0$.

The Hilbert function of the canonical module. Stanley's theorem 4.4.6 analyzes how the Gorenstein property of a positively graded k-algebra is reflected by its Hilbert series. It will be deduced from the next result which asserts that the *canonical module of a Cohen–Macaulay positively graded k-algebra is determined by its Hilbert series, provided R is a domain. Occasionally one can use this fact to identify the *canonical module; see for example 6.4.9.

The automorphism $\varphi \colon \mathbb{Z}[t, t^{-1}] \to \mathbb{Z}[t, t^{-1}]$, $\varphi(t) = t^{-1}$, can be extended to all rational functions $F(t)$, and we set $F(t^{-1}) = \varphi(F(t))$.

Theorem 4.4.5.   *Let k be a field, R a d-dimensional Cohen–Macaulay positively graded k-algebra, M a Cohen–Macaulay graded R-module of dimension n, and $M' = {}^*\mathrm{Ext}_R^{d-n}(M, \omega_R)$. Then*
(a) $H_{M'}(t) = (-1)^n H_M(t^{-1})$,
(b) *if R is a domain, $\dim M = d$, and $H_M(t) = t^q H_{\omega_R}(t)$ for some q, then $M(q) \cong \omega_R$.*

PROOF. (a) We set

$$V_M(t) = \sum_{i \in \mathbb{Z}} \dim_k({}^*H_\mathfrak{m}^n(M)_{-i}) t^i.$$

By the graded local duality theorem 3.6.19 one has $V_M(t) = H_{M'}(t)$. Furthermore $H_M(t) = V_M(t^{-1})$ if $\dim M = 0$.

Let $a \in R$ be an M-regular homogeneous element of degree g. Then the exact sequence

$$0 \longrightarrow M(-g) \overset{a}{\longrightarrow} M \longrightarrow M/aM \longrightarrow 0$$

induces an exact sequence

$$0 \longrightarrow {}^*H_\mathfrak{m}^{n-1}(M/aM) \longrightarrow {}^*H_\mathfrak{m}^n(M(-g)) \longrightarrow {}^*H_\mathfrak{m}^n(M) \longrightarrow 0.$$

Since ${}^*H_\mathfrak{m}^n(M(-g)) \cong {}^*H_\mathfrak{m}^n(M)(-g)$, one obtains $V_{M/aM}(t) = (t^{-g} - 1) V_M(t)$.

By 1.5.11 there exists a maximal M-sequence x of homogeneous elements. Set $b_i = \deg x_i$. An iterated application of the previous argument then yields

$$H_M(t) = \frac{H_{M/xM}(t)}{\prod_{i=1}^{n}(1 - t^{b_i})} = \frac{V_{M/xM}(t^{-1})}{\prod_{i=1}^{n}(1 - t^{b_i})} = (-1)^n V_M(t^{-1}) = (-1)^n H_{M'}(t^{-1}).$$

(b) We may assume $q = 0$. Then $H_{M'}(t) = H_{\omega'_R}(t) = H_R(t)$. It follows that there exists an element $x \in M'$ of degree 0, $x \neq 0$. Let $\varphi : R \to M'$ be the homogeneous R-module homomorphism mapping 1 to x. Since R is a domain and M' a Cohen–Macaulay R-module of maximal dimension, the homomorphism φ is injective. But since R and M' have the same Hilbert series, φ must actually be an isomorphism, and it follows that $M \cong M'' \cong R' \cong \omega_R$. $\qquad\square$

Corollary 4.4.6 (Stanley). *With the notation and hypothesis of* 4.4.5 *suppose that R has the Hilbert series $H_R(t) = \sum_{i=0}^{s} h_i t^i / \prod_{j=1}^{d}(1 - t^{a_j})$.*
(a) *Then $H_{\omega_R}(t) = (-1)^d H_R(t^{-1})$, equivalently,*

$$H_{\omega_R}(t) = \frac{t^{\sum a_j - s} \sum_{i=0}^{s} h_{s-i} t^i}{\prod_{j=1}^{d}(1 - t^{a_j})}.$$

(b) *If R is Gorenstein, then $H_R(t) = (-1)^d t^{a(R)} H_R(t^{-1})$.*
(c) *Suppose R is a domain, and $H_R(t) = (-1)^d t^q H_R(t^{-1})$ for some integer q. Then R is Gorenstein.*

PROOF. (a) follows immediately from 4.4.5, and according to 3.6.11 we have $\omega_R = R(a(R))$. This implies (b). With $R = M$, (c) results from 4.4.5(b). $\qquad\square$

Remarks 4.4.7. (a) Assume the positively graded k-algebra R is Gorenstein, and write $H_R(t) = Q_R(t) / \prod_{i=1}^{d}(1 - t^{a_i})$. Then the functional equation 4.4.6(b) for $H_R(t)$ is equivalent to the equation $Q_R(t) = t^{\deg Q_R} Q_R(t^{-1})$, that is, to the symmetry of the polynomial $Q_R(t)$.
(b) Consider the homogeneous k-algebra $R = k[X, Y]/(X^3, XY, Y^2)$. Then $H_R(t) = 1 + 2t + t^2$, but R is not Gorenstein. Applying 4.2.16, we derive from R a reduced non-Gorenstein Cohen–Macaulay ring S satisfying $H_S(t) = (-1)^d t^q H_S(t^{-1})$. Thus for 4.4.6(c) it is essential to require that R be a domain.

On the other hand, suppose the Hilbert series of the positively graded k-algebra R satisfies 4.4.6(b), but R is not necessarily a domain. Instead suppose there exist a positively graded algebra S which is a Cohen–Macaulay domain, and a homogeneous S-sequence x such that $S/xS \cong R$. Since $Q_S(t)$ is symmetric if and only if $Q_R(t)$ is, we conclude that R is Gorenstein.

In particular it follows that the above Artinian algebra cannot be the residue class ring of a homogeneous domain S by a homogeneous S-sequence.

Stanley observed that the following result on numerical semigroup rings, due to Herzog and Kunz [160], can be derived easily from the previous corollary. A *numerical semigroup* is a subsemigroup S of the additive semigroup $\mathbb{N}$ such that $0 \in S$ and $\mathbb{N} \setminus S$ is finite. The last condition is equivalent to the requirement that the greatest common divisor of all the elements of S is 1. If S is a numerical semigroup, then there exist integers $0 < a_1 < \cdots < a_n$ such that S is the set of linear combinations

$$z_1 a_2 + z_2 a_2 + \cdots + z_n a_n \quad \text{with} \quad z_i \in \mathbb{N}.$$

Any such set of integers is called a *set of generators* of S, and we write $S = \langle a_1, \ldots, a_n \rangle$. It is clear that a minimal set of generators of S is uniquely determined.

The *conductor* $c = c(S)$ of S is defined by $c = \max\{a \in \mathbb{N} : a - 1 \notin S\}$. For example, $S = \langle 4, 7 \rangle$ has the conductor $c(S) = 18$.

If k is a field, $k[S]$ denotes the k-subalgebra of the polynomial ring $k[X]$ generated by all monomials X^a, $a \in S$. Note that $k[S] = k[X^{a_1}, \ldots, X^{a_n}]$ if $S = \langle a_1, \ldots, a_n \rangle$. Thus, if we set $\deg X = 1$, then $k[S]$ is a positively graded k-algebra with k-basis X^a, $a \in S$. Moreover, $k[S]$ is Cohen–Macaulay since it is a one dimensional domain.

Theorem 4.4.8. *Let S be a numerical semigroup with conductor c. The following conditions are equivalent:*
(a) $k[S]$ is Gorenstein;
(b) the semigroup S is symmetric, that is, for all i with $0 \le i \le c - 1$ one has $i \in S$ if and only if $c - i - 1 \notin S$.

PROOF. Write $R = k[S]$. Then

$$H_R(t) = \sum_{j \in S} t^j = 1/(1-t) - \sum_{i \in \mathbb{N} \setminus S} t^i,$$

and so

$$-H_R(t^{-1}) = t/(1-t) + \sum_{i \in \mathbb{N} \setminus S} t^{-i}.$$

Suppose $H_R(t) = -t^r H_R(t^{-1})$; then necessarily $r = c - 1$, and

$$1/(1-t) - \sum_{i \in \mathbb{N} \setminus S} t^i = t^c/(1-t) + \sum_{i \in \mathbb{N} \setminus S} t^{c-1-i}.$$

Hence $H_R(t) = -t^{c-1} H_R(t^{-1})$ if and only if S is symmetric, and the assertion follows from 4.4.6. $\square$

A homogeneous Cohen–Macaulay k-algebra R is called a *level ring* if all elements in a minimal set of generators of ω_R have the same degree. When R is Artinian, then $\mu(\omega_R) = \dim_k {}^*\mathrm{Soc}\,R$, and therefore R is a level ring if and only if the homogeneous socle of R equals R_s where $s = \max\{i : R_i \neq 0\}$.

Recall that a Cohen–Macaulay ring R is *generically Gorenstein* if $R_\mathfrak{p}$ is Gorenstein for all minimal prime ideals $\mathfrak{p}$ of R.

Theorem 4.4.9 (Stanley). *Let R be a homogeneous Cohen–Macaulay k-algebra. Suppose that R is a domain, or generically Gorenstein and a level ring. Let $H_R(t) = \sum_{i=0}^s h_i t^i / (1-t)^d$; then*

$$\sum_{i=0}^j h_i \leq \sum_{i=0}^j h_{s-i} \quad \text{for all} \quad j = 0, \ldots, s.$$

PROOF. Note that the least degree of a homogeneous non-zero element of ω_R is $b = -a(R)$. Our assumptions guarantee the existence of a homogeneous element $x \in \omega_R$ of degree b such that $Rx \cong R(-b)$. This is clear if R is a domain. Next assume that R is generically Gorenstein and a level ring. Then the natural homomorphism $\varphi : \omega_R \to (\omega_R)^{**}$ (which is homogeneous) is a monomorphism; see 1.4.1. Let $G \to (\omega_R)^*$ be a homogeneous epimorphism, where G is free. Then the dual homomorphism $(\omega_R)^{**} \to F = G^*$ is a monomorphism which, composed with φ, yields a homogeneous monomorphism $\psi : \omega_R \to F$. We identify ω_R with its image in F. Suppose that the k-vector space $(\omega_R)_b$ is contained in $\bigcup_{\mathfrak{p} \in \mathrm{Ass}\,R} \mathfrak{p} F$. Then, since $(\omega_R)_b \cap \mathfrak{p} F$ is a subspace of $(\omega_R)_b$, and as we may assume that k is infinite (the reader should check this), it follows that $(\omega_R)_b \subset \mathfrak{p} F$ for some $\mathfrak{p} \in \mathrm{Ass}\,R$. However the elements of $(\omega_R)_b$ generate the canonical module, and so $\omega_R \subset \mathfrak{p} F$. This is impossible, because it would imply that $\mathfrak{p} F_\mathfrak{p}$ contains a free $R_\mathfrak{p}$-module of rank 1. Now we choose $x \in (\omega_R)_b \setminus \bigcup_{\mathfrak{p} \in \mathrm{Ass}\,R} \mathfrak{p} F$; then $Rx \cong R(-b)$.

Thus, in any case, there exists an exact sequence of graded R-modules

$$0 \longrightarrow R \overset{\varphi}{\longrightarrow} \omega_R(b) \longrightarrow N \longrightarrow 0,$$

where $\varphi(1) = x$ is a non-zero homogeneous element of degree b in ω_R.

Let $H_R(t) = \sum_{i=0}^s h_i t^i / (1-t)^d$, $d = \dim R$, be the Hilbert series of R. By 4.4.6, the exact sequence implies that

$$H_N(t) = (1-t)^{-d} \sum_{i=0}^s (h_{s-i} - h_i) t^i.$$

The module N has rank 0 since $\mathrm{rank}\,\omega_R = 1$, and so $\dim N < d$. On the other hand, the exact sequence shows that $\mathrm{depth}\,N \geq d-1$, and thus we conclude that N is a graded Cohen–Macaulay module of dimension

$d-1$. Therefore $H_N(t) = \sum_{i=0}^{r} a_i t^i/(1-t)^{d-1}$ with $a_i \geq 0$ for $i = 1,\ldots,r$, see 4.1.10. It follows that

$$\sum_{i=0}^{s} (h_{s-i} - h_i)t^i = (1-t)\sum_{i=0}^{r} a_i t^i.$$

Thus we obtain the following set of equations:

$$h_s - h_0 = a_0, \quad h_{s-1} - h_1 = a_1 - a_0, \quad \ldots, \quad h_0 - h_s = a_s - a_{s-1}.$$

Here we have set $a_i = 0$ for $i > r$. Adding up the first $j + 1$ equations gives

$$\sum_{i=0}^{j} h_{s-i} - \sum_{i=0}^{j} h_i = a_j \geq 0,$$

as asserted.   □

Consider the sequence $(1, 2, 1, 1)$. Proposition 4.2.15 implies that this is the h-sequence of a Cohen–Macaulay reduced homogeneous k-algebra R. But 4.4.9 implies that such an R is not a domain.

Exercises

4.4.10. Let $F(t) = Q(t)/\prod_{i=1}^{d}(1-t^{a_i}) = \sum_{i=a}^{\infty} f_i t^i$ with $Q(t) \in \mathbb{Z}[t, t^{-1}]$ and positive integers $a_1,\ldots,a_d$. Let $\sum_{n=a}^{\infty} f_n t^n$ be the Laurent expansion of F at 0. Show
(a) there exists a unique quasi-polynomial P with $P(n) = f_n$ for $n \gg 0$,
(b) $\max\{n : f_n \neq P(n)\} = \deg F$.
Hint: For (b) one argues similarly as in the proof of 4.1.12.

4.4.11. Let R be a Noetherian positively graded k-algebra over a field k, generated by homogeneous elements $x_1,\ldots,x_m$ of degrees $e_1,\ldots,e_m$. Let e be the least common multiple of $e_1,\ldots,e_m$ and define S to be the k-subalgebra generated by the degree e homogeneous elements of R. A finite graded R-module obviously decomposes into the direct sum of its S-submodules $M_i = \bigoplus_{j\in\mathbb{Z}} M_{je+i}$, $i = 0,\ldots,e-1$.
(a) Show that the M_i are finite S-modules.
(b) By considering S as a homogeneous k-algebra in the appropriate way, deduce that the Hilbert function $H(M, n)$ is a quasi-polynomial of period e for $n \gg 0$.

4.4.12. Let R be a Noetherian positively graded k-algebra over a field k and $M \neq 0$ a finite R-module. Furthermore let S be a graded Noether normalization of $R/\operatorname{Ann} M$ generated by elements of degrees $a_1,\ldots,a_d$, $d = \dim M$.
(a) Derive 4.4.1 from Hilbert's syzygy theorem 2.2.14 by computing the polynomial $Q(t) \in \mathbb{Z}[t, t^{-1}]$ with $H_M(t) = Q(t)/\prod_{i=1}^{d}(1-t^{a_i})$; moreover, show $Q(1) = \operatorname{rank}_S M > 0$.
(b) Prove that the coefficients of $Q(t)$ are non-negative if M is a Cohen–Macaulay module.

4.4.13. (a) Let R be a Noetherian positively graded k-algebra of dimension 1, where k is an algebraically closed field. If $H(R,n) > 1$ for some n, show R is not a domain.

(b) Find an example of a 1-dimensional homogeneous $\mathbb{R}$-algebra R which is a domain, and for which $H(R,n) = 2$ for all $n > 0$.

4.4.14. Let k be a field, and $P(t)$ a formal power series with integer coefficients. Demonstrate the following conditions are equivalent:

(a) there exists a d-dimensional homogeneous k-algebra which is a complete intersection, and which has the Hilbert series $P(t)$;

(b) there exist an integer n, $n \le d$, and integers $a_i > 0$, $i = 1,\dots,n$, such that $P(t) = (1-t)^{-d} \prod_{i=1}^{n} (1 + t + t^2 + \cdots + t^{a_i})$.

4.4.15. Let k be a field. In this exercise we want to specify the associated prime ideals of an ideal $I \subset R = k[X_1,\dots,X_n]$ which is generated by monomials in the variables $X_1,\dots,X_n$. We order the monomials in the reverse degree-lexicographical order, and denote by $L(a)$ the leading monomial of a; see the proof of 4.2.3 and the discussion following 4.2.4.

(a) Let $a \in R$, and write $a = \sum_i \lambda_i v_i$ with $\lambda_i \in k$ and v_i monomials for all i. Then $L(a) \in I$ if $a \in I$. Conclude from this that $a \in I$ if and only if $v_i \in I$ for all i with $\lambda_i \neq 0$.

(b) Let $a \in R$ be a monomial. Show the ideal $J = \{b \in R : ba \in I\}$ is generated by monomials.

(c) Prove that an ideal generated by monomials is a prime ideal if and only if it is generated by a subset of $\{X_1,\dots,X_n\}$.

(d) Prove that the associated prime ideals of R/I are all generated by subsets of $\{X_1,\dots,X_n\}$

(e) Show an ideal I generated by monomials is a primary ideal if and only if it satisfies the following condition: for every variable X_i which divides a monomial $m \in I$ such that $m/X_i \notin I$, some power of X_i belongs to I.

4.4.16. (a) Let k be a field, and $I_1, I_2, I_3 \subset k[X_1,\dots,X_n]$ ideals generated by monomials. Show $I_1 \cap (I_2 + I_3) = (I_1 \cap I_2) + (I_1 \cap I_3)$.

(b) Let $v_1,\dots,v_m \subset k[X_1,\dots,X_n]$ monomials in $X_1,\dots,X_n$. Suppose $v_1 = ab$ is the product of monomials a and b with greatest common divisor 1, then show $(ab, v_2,\dots,v_m) = (a, v_2,\dots,v_m) \cap (b, v_2,\dots,v_m)$.

(c) Describe an algorithm to determine the primary components of an ideal generated by monomials.

4.4.17. Let k be a field, and $I \subset k[X_1,\dots,X_n]$ an ideal generated by squarefree monomials. Demonstrate that $k[X_1,\dots,X_n]/I$ is reduced.

4.4.18. Let k be a field and $I \subset R = k[X_1,\dots,X_n]$ the ideal generated by the monomials $X_i X_j$, $1 \le i < j \le n$. Determine $\operatorname{Ass} R/I$.

4.4.19. Let k be a field, $I \subset k[X_1,\dots,X_n]$ an ideal generated by monomials, and $R = k[X_1,\dots,X_n]/I$. Show

(a) Show that a 0-dimensional Gorenstein ring is a complete intersection. (This is also true in dimension 1; see Bruns and Herzog [58].)

(b) Give a 2-dimensional Gorenstein example that is not a complete intersection.

4.4.20. Prove the graded version of 3.1.16: let R be a graded ring, and M and N graded R-modules; if $x \in R$ is a homogeneous element of degree a which is R- and M-regular and annihilates N, then ${}^*\mathrm{Ext}_R^{i+1}(N,M)(-a) \cong {}^*\mathrm{Ext}_{R/(x)}^i(N,M/xM)$.

4.4.21. Let k be a field, and let R be a homogeneous Gorenstein k-algebra of dimension d. Prove $2e_1 = (a(R) + d)e_0$.

4.5 Filtered rings

In this section we introduce the extended Rees ring and associated graded ring of a filtered ring. We will compute their dimensions, and show that a filtered ring inherits many good properties from its associated graded ring. The results will be used in the next section where we consider the Hilbert–Samuel function, and in Chapter 7 for the study of graded Hodge algebras.

Definition 4.5.1. Let R be a ring. A *filtration F on R* is a descending chain $R = I_0 \supset I_1 \supset I_2 \supset \cdots$ of ideals such that $I_i I_j \subset I_{i+j}$ for all i and j. A *filtered ring* is a pair (R, F) where R is ring and F is a filtration on R.

The most common filtration is the one given by the powers of an ideal I, called the *I-adic filtration*.

Let R be a filtered ring with filtration $F = (I_i)_{i \geq 0}$. We define the *extended Rees ring of R with respect to F* by

$$\mathscr{R}(F) = \bigoplus_{i \in \mathbf{Z}} I_i t^i.$$

Here $I_i = R$ for $i \leq 0$, and $\mathscr{R}(F)$ is viewed as a graded subring of $R[t, t^{-1}]$. Moreover, we define *the associated graded ring of R with respect to F* by

$$\mathrm{gr}_F(R) = \bigoplus_{i=0}^{\infty} I_i / I_{i+1}.$$

It is a graded ring with multiplication induced by the multiplication map $I_i \times I_j \to I_{i+j}$.

Given an R-module M, $\mathscr{R}(F, M) = \bigoplus_{i \in \mathbf{Z}} I_i M t^i$ (respectively $\mathrm{gr}_F(M) = \bigoplus_{i=0}^{\infty} I_i M / I_{i+1} M$) is in a natural way a graded module over $\mathscr{R}(F)$ (respectively $\mathrm{gr}_F(R)$). In the case where F is the I-adic filtration we denote by $\mathscr{R}(I)$ the extended Rees ring, and in accordance with Section 1.1 by $\mathrm{gr}_I(R)$ the associated graded ring. Further we write $\mathscr{R}(I, M)$ for $\mathscr{R}(F, M)$ and $\mathrm{gr}_I(M)$ for $\mathrm{gr}_F(M)$.

We will also encounter the *Rees ring* $\mathscr{R}_+(F) = \bigoplus_{i=0}^{\infty} I_i t^i$ and the graded $\mathscr{R}_+(F)$-modules $\mathscr{R}_+(F, M) = \bigoplus_{i=0}^{\infty} I_i M t^i$. The notations in the case of I-adic filtrations are to be modified accordingly.

The following observation, whose proof is left to the reader, is of crucial importance in the study of the extended Rees ring.

Lemma 4.5.2. *Let R be a filtered ring with filtration F. Then*
(a) *the element $t^{-1} \in R[t, t^{-1}]$ belongs to $\mathcal{R}(F)$ and is $\mathcal{R}(F)$-regular,*
(b) $\mathcal{R}(F)/t^{-1}\mathcal{R}(F) \cong \mathrm{gr}_F(R)$,
(c) $\mathcal{R}(F)_{t^{-1}} \cong R[t, t^{-1}]$.

We call F *Noetherian* if $\mathcal{R}(F)$ is Noetherian. For example, I-adic filtrations on a Noetherian ring are Noetherian. It is clear that if R is Noetherian and $\mathcal{R}(F)$ is finitely generated over R, then F is Noetherian. But the converse is true as well:

Proposition 4.5.3. *Let R be a filtered ring with filtration $F = (I_i)_{i \geq 0}$. The following conditions are equivalent:*
(a) *F is Noetherian;*
(b) *R is Noetherian, and $\mathcal{R}(F)$ is finitely generated over R;*
(c) *R is Noetherian, and $\mathcal{R}_+(F)$ is finitely generated over R;*
(d) *R is Noetherian, and there exist positive integers $j(1), \ldots, j(n)$ and $x_i \in I_{j(i)}$, $i = 1, \ldots, n$, such that $I_k = \sum_{i=1}^n x_i I_{k-j(i)}$ for all $k > 0$.*

PROOF. The equivalence of the statements (a), (b) and (c) follows from 1.5.5.

(c) $\Rightarrow$ (d): Let $\mathcal{R}_+(F) = R[a_1, \ldots, a_n]$. We may assume that $a_1, \ldots, a_n$ are homogeneous elements of positive degree. Then $a_i = x_i t^{j(i)}$ for some $j(i) > 0$ and $x_i \in I_{j(i)}$, $i = 1, \ldots, n$. These x_i satisfy the conditions in (d).

(d) $\Rightarrow$ (a) is proved similarly. $\qquad\square$

Note that if R/I_1 is Noetherian and $\mathrm{gr}_F(R)$ is finitely generated over R/I_1, it does not follow in general that F is Noetherian. For example, let $(R, \mathfrak{m})$ be a local ring and let $F = (I_i)_{i \geq 0}$ with $I_i = \mathfrak{m}$ for all $i > 0$. Then $\mathrm{gr}_F(R) = R/\mathfrak{m}$, but $\mathcal{R}(F)$ is not Noetherian. Thus we have to pose an extra condition on F: the filtration $F = (I_i)_{i \geq 0}$ is *separated* if $\bigcap_{i \geq 0} I_i = 0$, and F is *strongly separated* if $\bigcap_{i \geq 0}(I + I_i) = I$ for all ideals $I \subset R$. By Krull's intersection theorem, I-adic filtrations on local rings are strongly separated, provided $I \neq R$.

Recall that the filtration $F = (I_i)_{i \geq 0}$ defines a topology on R whose base is given by the sets $a + I_i$, $a \in R$ and $i \geq 0$; see [289]. With this topology R is a Hausdorff space if and only if F is separated. The closure of an ideal I is given by $\bigcap_{i \geq 0}(I + I_i)$; hence F is strongly separated if and only if all ideals of R are closed subsets.

Let us denote by $\hat{M}$ the completion of an R-module M with respect to F (see [289], 9.5). Then $\hat{R}$ is complete with respect to the filtration $\hat{F} = (\hat{I}_i)_{i \geq 0}$, and $\hat{I}_i$ is the closure of $I_i \hat{R}$ in $\hat{R}$. If the filtration is separated, then the canonical homomorphism $R \to \hat{R}$ is injective, and $\hat{I}_i \cap R = I_i$; see [289], Theorem 3, p. 390. Further, if $\mathrm{gr}_F(R)$ is Noetherian, then $\hat{R}$ is Noetherian and for all ideals $I \subset R$, $I\hat{R}$ is the closure of I in $\hat{R}$ ([289], Theorem 15 and Corollary 1, p. 413).

Proposition 4.5.4. *If F is Noetherian, then R is Noetherian and* $\mathrm{gr}_F(R)$ *is finitely generated over* R/I_1. *Conversely, if F is strongly separated,* R/I_1 *is Noetherian, and* $\mathrm{gr}_F(R)$ *is finitely generated over* R/I_1, *then F is Noetherian.*

PROOF. The first part of the assertion is obvious in view of 4.5.3. For the converse we show that R is Noetherian, and that $\mathcal{R}(F)$ is finitely generated over R. Then, by 4.5.3, F is Noetherian.

Let $I \subset R$ be an ideal of R, and $a \in I\hat{R} \cap R$. Since $I\hat{R}$ is the closure of I in $\hat{R}$, there exist, for all $i \geq 0$, elements $a_i \in I$ such that $a_i - a \in \hat{I}_i \cap R = I_i\hat{R} \cap R = I_i$. Therefore $a \in \bigcap_{i \geq 0}(I + I_i) = I$. Thus we have $I\hat{R} \cap R = I$, and this proves that R is Noetherian since $\hat{R}$ is Noetherian.

In order to prove that $\mathcal{R}(F)$ is finitely generated we may assume that $\mathrm{gr}_F(R) = R/I_1[\bar{x}_1, \dots, \bar{x}_n]$ where the $\bar{x}_i$ are homogeneous of positive degree, say $\bar{x}_i = x_i + I_{j(i)+1}$, $x_i \in I_{j(i)}$ for $i = 1, \dots, n$. Let $A = R[t^{-1}, x_1 t^{j(1)}, \dots, x_n t^{j(n)}]$; then A is a graded subalgebra of $\mathcal{R}(F)$, and we claim that indeed $A = \mathcal{R}(F)$: let $k > 0$ and $x \in I_k t^k$; then $x = a_0 + b_0 t^k$ with $a_0 \in A_k$ and $b_0 \in I_{k+1}$, by the definition of A. For the same reason we have $b_0 t^{k+1} = a_1 + b_1 t^{k+1}$ with $a_1 \in A_{k+1}$ and $b_1 \in I_{k+2}$. It follows that $x = a_0 + a_1 t^{-1} + b_1 t^k$, and hence $x \in A_k + I_{k+2} t^k$. By induction on j one shows that $x \in A_k + I_{k+j} t^k$ for all $j \geq 1$. Thus $x \in \bigcap_{j \geq 1}(A_k + I_{k+j} t^k) = A_k$ since F is strongly separated. $\square$

In the next theorem we compare the dimension of a module M with the dimension of $\mathcal{R}(F, M)$ and $\mathrm{gr}_F(M)$ where F is a filtration on R. For the proof we will have to identify the minimal prime ideals of $\mathcal{R}(F)$.

Let $\mathfrak{p} \in \mathrm{Spec}\, R$; then $\mathfrak{p}' = \mathfrak{p}R[t, t^{-1}] \cap \mathcal{R}(F)$ is a prime ideal of $\mathcal{R}(F)$ and $\mathfrak{p}' \cap R = \mathfrak{p}$. It is clear that $\mathfrak{p}'$ belongs to the set $^*D(t^{-1})$ of graded prime ideals of $\mathcal{R}(F)$ which do not contain t^{-1}.

Lemma 4.5.5. *Let F be a Noetherian filtration.*
(a) *The map* $\alpha: \mathrm{Spec}\, R \to {}^*D(t^{-1})$, $\mathfrak{p} \mapsto \mathfrak{p}'$, *is an inclusion preserving bijection;*
(b) *height* $\mathfrak{p} =$ *height* $\mathfrak{p}'$ *for all* $\mathfrak{p} \in \mathrm{Spec}\, R$;
(c) α *induces a bijection between the minimal prime ideals of R and* $\mathcal{R}(F)$.

PROOF. (a) It is clear that α is injective and inclusion preserving. Let $\mathfrak{P} \in {}^*D(t^{-1})$; then $\mathfrak{P}\mathcal{R}(F)_{t^{-1}} = \mathfrak{P}R[t, t^{-1}]$ is a graded prime ideal of $R[t, t^{-1}]$, and hence of the form $\mathfrak{p}R[t, t^{-1}]$ for some $\mathfrak{p} \in \mathrm{Spec}\, R$. It follows that $\mathfrak{P} = \mathfrak{P}\mathcal{R}(F)_{t^{-1}} \cap \mathcal{R}(F) = \mathfrak{p}R[t, t^{-1}] \cap \mathcal{R}(F) = \mathfrak{p}'$.

(b) Obviously we have height $\mathfrak{p}' \geq$ height $\mathfrak{p}$. Suppose height $\mathfrak{p}' = h$. By 1.5.8, there exists a strictly descending chain of graded prime ideals $\mathfrak{p}' = \mathfrak{P}_0 \supset \mathfrak{P}_1 \supset \cdots \supset \mathfrak{P}_h$. Since all $\mathfrak{P}_i \in {}^*D(t^{-1})$, there exist $\mathfrak{p}_i \in \mathrm{Spec}\, R$ with $\mathfrak{p}'_i = \mathfrak{P}_i$. Then $\mathfrak{p} = \mathfrak{p}_0 \supset \mathfrak{p}_1 \supset \cdots \supset \mathfrak{p}_h$ is a strictly descending chain of prime ideals in R; thus height $\mathfrak{p} \geq h$.

(c) Let $\mathfrak{P}$ be a minimal prime ideal of $\mathscr{R}(F)$. Then $t^{-1} \notin \mathfrak{P}$ since t^{-1} is $\mathscr{R}(F)$-regular. According to 1.5.6, $\mathfrak{P}$ is graded, and so belongs to $^*D(t^{-1})$. The rest follows from (a) and (b). $\qquad\square$

Theorem 4.5.6. *Let R be a filtered ring with Noetherian filtration $F = (I_i)_{i \geq 0}$, and M a finite R-module. Then $\mathscr{R}(F, M)$ is a finite $\mathscr{R}(F)$-module, and*

(a) $\dim \mathscr{R}(F, M) = \dim M + 1$,

(b) $\dim \mathrm{gr}_F(M) = \sup\{\dim M_\mathfrak{m} : \mathfrak{m} \in \mathrm{Supp}(M/I_1 M), \mathfrak{m}\ maximal\}$. In particular, $\dim \mathrm{gr}_F(M) \leq \dim M$, and $\dim \mathrm{gr}_F(M) = \dim M$ if I_1 is contained in all maximal ideals of R.

PROOF. (a) It is clear that $\mathscr{R}(F, M)$ is a finite $\mathscr{R}(F)$-module. Let $J = \mathrm{Ann}\, M$, and set $R' = R/J$ and $F' = (I_i R')_{i \geq 0}$. Then M is an R'-module and $\mathscr{R}(F, M) \cong \mathscr{R}(F', M)$. Thus we may as well assume that M is a faithful R-module. But then $\mathscr{R}(F, M)$ is a faithful $\mathscr{R}(F)$-module, too, so that $\dim \mathscr{R}(F, M) = \dim \mathscr{R}(F)$. Therefore it suffices to prove the assertion for $M = R$.

Let $\mathfrak{P} \in \mathrm{Spec}\, \mathscr{R}(F)$, and set $\mathfrak{p} = \mathfrak{P} \cap R$. We choose a minimal prime ideal $\mathfrak{Q} \subset \mathfrak{P}$ such that $\mathrm{height}(\mathfrak{P}/\mathfrak{Q}) = \mathrm{height}\, \mathfrak{P}$. By 4.5.5, there exists $\mathfrak{q} \in \mathrm{Spec}\, R$ such that $\mathfrak{Q} = \mathfrak{q}'$. Thus we obtain the finitely generated extensions $R/\mathfrak{q} \subset \mathscr{R}(F)/\mathfrak{q}' \subset R[t, t^{-1}]/\mathfrak{q}R[t, t^{-1}]$ of integral domains, and it follows that the transcendence degree of the fraction field $Q(\mathscr{R}(F)/\mathfrak{q}')$ over $Q(R/\mathfrak{q})$ is one. Thus A.19 yields $\mathrm{height}\, \mathfrak{P} = \mathrm{height}(\mathfrak{P}/\mathfrak{q}') \leq \mathrm{height}(\mathfrak{p}/\mathfrak{q}) + 1 \leq \mathrm{height}\, \mathfrak{p} + 1$. In particular we conclude that $\dim \mathscr{R}(F) \leq \dim R + 1$.

Conversely, $\dim \mathscr{R}(F) \geq \dim \mathscr{R}(F)_{t^{-1}} = \dim R[t, t^{-1}] = \dim R + 1$. The reader may check the last equality.

(b) As for (a) we may reduce the assertion to the case in which M is faithful. Then $\mathscr{R}(F, M)$ is a faithful $\mathscr{R}(F)$-module, and therefore $\mathrm{gr}_F(M) \cong \mathscr{R}(F, M)/t^{-1}\mathscr{R}(F, M)$ is a faithful $\mathrm{gr}_F(R)$-module. Thus we may assume $M = R$.

According to Exercise 1.5.25, the dimension of $\mathrm{gr}_F(R)$ is the supremum of all numbers $\dim \mathrm{gr}_F(R)_\mathfrak{N}$ where the supremum is taken over all graded maximal ideals $\mathfrak{N} \in \mathrm{Spec}\, \mathrm{gr}_F(R)$. Let $\mathfrak{N}$ be such an ideal and $\mathfrak{M}$ its preimage in $\mathscr{R}(F)$. Then $\mathfrak{M}$ is a graded maximal ideal, and (hence) contains t^{-1}. Let $\mathfrak{m} = \mathfrak{M} \cap R$. As $\mathscr{R}(F)/\mathfrak{M}$ is a graded ring and a field, it is isomorphic with its degree zero homogeneous component $R/\mathfrak{m}$. Thus

$$\mathfrak{M} = \bigoplus_{i<0} Rt^i \oplus \mathfrak{m} \oplus \bigoplus_{i>0} I_i t^i.$$

In particular $\mathfrak{m}$ is a maximal ideal, and $I_i = (I_i t^i)t^{-i} \subset \mathfrak{m}$ for $i > 0$. Now the decomposition of $\mathfrak{M}$ shows that $\mathfrak{M} = (\mathfrak{m}', t^{-1})$. Since $\mathfrak{m} = \mathfrak{M} \cap R$, we have, as in (a), that $\mathrm{height}\, \mathfrak{m} = \mathrm{height}\, \mathfrak{m}' < \mathrm{height}\, \mathfrak{M} \leq \mathrm{height}\, \mathfrak{m} + 1$, so that $\mathrm{height}\, \mathfrak{N} = \mathrm{height}\, \mathfrak{M} - 1 = \mathrm{height}\, \mathfrak{m}$.

Conversely, let $\mathfrak{m} \supset I_1$ be a maximal ideal of R. Then $\mathscr{R}(F)/(\mathfrak{m}', t^{-1}) \cong R/\mathfrak{m}$, whence $\mathfrak{M} = (\mathfrak{m}', t^{-1})$ is a maximal ideal of $\mathscr{R}(F)$ with $\mathfrak{m} = \mathfrak{M} \cap R$. As above, it follows that height $\mathfrak{N} =$ height $\mathfrak{m}$ for the graded maximal ideal $\mathfrak{N} = \mathfrak{M}/(t^{-1})$ of $\mathrm{gr}_F(R)$. $\square$

The next series of results demonstrate that 'good' properties of $\mathrm{gr}_F(R)$ descend to R.

Theorem 4.5.7. *Let R be a filtered ring with Noetherian filtration $F = (I_i)_{i \geq 0}$.*
(a) *If $\mathrm{gr}_F(R)$ is Cohen–Macaulay, then so is $R_\mathfrak{p}$ for all $\mathfrak{p} \in V(I_1)$.*
(b) *If $\mathrm{gr}_F(R)$ is Gorenstein, then so is $R_\mathfrak{p}$ for all $\mathfrak{p} \in V(I_1)$.*

PROOF. (a) Let $\mathfrak{p} \in V(I_1)$. The filtration $F = (I_i)_{i \geq 0}$ on R induces the filtration $F' = (I_i R_\mathfrak{p})_{i \geq 0}$ on $R_\mathfrak{p}$, and we have $R_\mathfrak{p} \otimes_R \mathrm{gr}_F(R) \cong \mathrm{gr}_{F'}(R_\mathfrak{p})$. Thus we may as well assume that $(R, \mathfrak{m})$ is local, $I_1 \subset \mathfrak{m}$, and $\mathfrak{p} = \mathfrak{m}$. Then $\mathscr{R}(F)$ is *local, and t^{-1} belongs to the unique graded maximal ideal of $\mathscr{R}(F)$. Therefore $\mathscr{R}(F)$ is Cohen–Macaulay by 4.5.2(a), (b) and Exercise 2.1.28. Applying 4.5.2(c) we see that $R[t, t^{-1}]$ is Cohen–Macaulay. Since the extension $R \to R[t, t^{-1}]$ is faithfully flat, R is Cohen–Macaulay; see 2.1.23.

(b) is proved in a similar manner. $\square$

Theorem 4.5.8. *Let R be a filtered ring with separated filtration F. If $\mathrm{gr}_F(R)$ is reduced or a domain, then so is R.*

The proof, whose details we leave to the reader, follows easily from 4.5.10.

We close this section by showing that, under mild hypotheses, normality of the associated graded ring implies normality of the ring itself. Let R be a Noetherian domain with fraction field K. Recall that $x \in K$ is *completely integral over* R if there exists an element $a \neq 0$ in R such that $ax^n \in R$ for all $n \geq 0$. Note that R is normal (integrally closed in K) if and only if R is *completely integrally closed*, i.e. every element $x \in K$ which is completely integral over R is an element of R. Indeed, suppose $x \in K$, $x = c/d$, is integral over R. Then there exists an equation $x^m + a_1 x^{m-1} + \cdots + a_{m-1} x + a_m = 0$ with $a_i \in R$, and it is clear that $d^m x^n \in R$ for all $n \geq 0$. Conversely, if $x \in K$ such that $ax^n \in R$ for some $a \in R$, $a \neq 0$, and all $n \geq 0$, then $R[x] \subset a^{-1}R$. Since $a^{-1}R$ is a finite R-module this implies that x is integral over R; see [270], Theorem 9.1.

We introduce a notation which is useful in the proof of the following theorem. Let R be a filtered ring with separated filtration $F = (I_i)_{i \geq 0}$. For each non-zero $g \in R$ there exists a unique integer $i \geq 0$ such that $g \in I_i \setminus I_{i+1}$. We set $g^* = g + I_{i+1}$ and call it the *initial form* of g in $\mathrm{gr}_F(R)$; of course, $0^* = 0$.

Theorem 4.5.9. *Let R be a filtered ring with Noetherian filtration $F = (I_i)_{i \geq 0}$ satisfying $\bigcap_{i \geq 0}(aR + I_i) = aR$ for all $a \in R$. If $\mathrm{gr}_F(R)$ is a normal domain, then so is R.*

PROOF. The assumptions imply that F is separated. Hence by 4.5.8, R is a domain. Let K be the field of fractions of R, and $x = c/d$ an element in K which is completely integral over R. We want to show that $c \in Rd$. It suffices to prove that $c \in Rd + I_i$ for all $i \geq 0$, since $Rd = \bigcap_{i \geq 0}(Rd + I_i)$ by assumption. We prove this by induction on i, the case $i = 0$ being trivial. Suppose $c \in Rd + I_i$; then $c = ud + w$, $u \in R$, $w \in I_i$. As x is completely integral over R there exists $a \in R$, $a \neq 0$, such that $a(x - u)^n \in R$ for all $n \geq 0$, and this implies $a(w/d)^n \in R$ for all $n \geq 0$. In other words, there exist elements $w_n \in R$ such that $aw^n = w_n d^n$ for all $n \geq 0$.

We have $(gh)^\star = g^\star h^\star$ for all $g, h \in R$ since $\mathrm{gr}_F(R)$ is a domain; see Exercise 4.5.10. Applied to the above equation we obtain $a^\star(w^\star)^n = w_n^\star(d^\star)^n$. This means that $w^\star/d^\star$ is completely integral over $\mathrm{gr}_F(R)$. By assumption, $\mathrm{gr}_F(R)$ is an integrally closed domain. Therefore, $w^\star/d^\star \in \mathrm{gr}_F(R)$, or equivalently, $w^\star = v^\star d^\star$ for some $v \in R$. Since $w \in I_i$, the last equation yields $w - vd \in I_{i+1}$. But then $c = (u + v)d + (w - vd)$, as desired. □

Exercises

4.5.10. Let R be a filtered ring with separated filtration. Show:
(a) $a^\star b^\star = (ab)^\star$ or $a^\star b^\star = 0$;
(b) $a^\star b^\star = (ab)^\star$ if $\mathrm{gr}_F(R)$ is a domain.

4.5.11. Let R be a filtered ring with Noetherian filtration $F = (I_i)_{i \geq 0}$. Prove
(a) $\dim R \leq \dim \mathscr{R}_+(F) \leq \dim R + 1$,
(b) $\dim \mathscr{R}_+(F) = \dim R + 1 \Longleftrightarrow I_1 \not\subset \bigcap\{\mathfrak{p} \in \mathrm{Ass}\, R : \dim R/\mathfrak{p} = \dim R\}$.

4.5.12. Let R be a filtered ring with filtration $F = (I_i)_{i \geq 0}$. The s-th Veronese subring $\mathscr{R}_+^{(s)}(F)$ of $\mathscr{R}_+(F)$ is again a Rees ring which is defined by the filtration $F^{(s)} = (I_{si})_{i \geq 0}$. Show the following conditions are equivalent if R is Noetherian:
(a) $\mathscr{R}_+(F)$ is a finitely generated R-algebra;
(b) $\mathscr{R}_+^{(s)}(F)$ is a finitely generated R-algebra for all $s \geq 1$;
(c) there exists an integer $s \geq 1$ such that $\mathscr{R}_+^{(s)}(F)$ is a finitely generated R-algebra;
(d) there exists an integer $s \geq 1$ such that $I_{i+s} = I_i I_s$ for all $i \geq s$;
(e) there exists an integer $s \geq 1$ such that $I_{is} = (I_i)^s$.
Hints: for the proof of (c) $\Rightarrow$ (a) consider the ideals $M_j = \bigoplus_{i \geq 0} I_{is+j} t^{is}$ of $\mathscr{R}_+^{(s)}(F)$, $j = 0, \ldots, s-1$, and for (a) $\Rightarrow$ (d) use that $\mathscr{R}_+(F) = R[I_1 t, \ldots, I_r t^r]$ for some $r \geq 1$; then choose $s = (r - 1)r!$. (The implication (a) $\Rightarrow$ (d) is due to Rees [304].)

4.5.13. Let k be a field, $R = k[X^6, X^7, X^{15}]$ and $\mathfrak{m}$ the ideal in R generated by X^6, X^7 and X^{15}. Show R is Gorenstein, but $\mathrm{gr}_\mathfrak{m}(R)$ is not even Cohen–Macaulay.

4.6   The Hilbert–Samuel function and reduction ideals

Let $(R, \mathfrak{m})$ be a Noetherian local ring, and $M \neq 0$ a finite R-module. In order to define the *multiplicity* of M one passes to the associated graded module $\mathrm{gr}_{\mathfrak{m}}(M)$, and defines

$$e(M) = e(\mathrm{gr}_{\mathfrak{m}}(M)).$$

To be more flexible we may as well consider an ideal $I \subset \mathfrak{m}$ such that $\mathfrak{m}^n M \subset IM$ for some n. Any such ideal is called an *ideal of definition* of M.

The associated graded ring $\mathrm{gr}_I(R)$ is a homogeneous algebra, and $\mathrm{gr}_I(M)$ is a graded $\mathrm{gr}_I(R)$-module.

Definition 4.6.1. The first iterated Hilbert function

$$\chi_M^I(n) = H_1(\mathrm{gr}_I(M), n) = \sum_{i=0}^{n} H(\mathrm{gr}_I(M), i)$$

$$= \sum_{i=0}^{n} \ell(I^i M / I^{i+1} M) = \ell(M / I^{n+1} M)$$

is called the *Hilbert–Samuel function* of M, and $e(I, M) = e(\mathrm{gr}_I(M))$ the *multiplicity* of M with respect to I.

As an immediate consequence of 4.1.6 we obtain

Proposition 4.6.2. *Let $(R, \mathfrak{m})$ be a Noetherian local ring, $M \neq 0$ a finite R-module of dimension d, and I an ideal of definition of M. Then*
(a) *the Hilbert–Samuel function $\chi_M^I(n)$ is of polynomial type of degree d,*
(b) $e(I, M) = \lim_{n \to \infty} (d!/n^d) \ell(M/I^{n+1}M)$.

PROOF. By 4.5.6, we have $\dim M = \dim \mathrm{gr}_I(M)$. Thus (a) follows from 4.1.6.

(b) For large n we have $\chi_M^I(n) = (e(I, M)/d!)n^d +$terms in lower powers of n. This yields the desired result.   □

The polynomial $\Sigma_M^I(X) \in \mathbb{Q}[X]$ with $\Sigma_M^I(n) = \chi_M^I(n)$ for $n \gg 0$ is called the *Hilbert–Samuel polynomial* of M with respect to I. When $I = \mathfrak{m}$, we simply write $\Sigma_M(X)$ instead of $\Sigma_M^I(X)$. (Note that $\Sigma_M(X)$ is *not* the Hilbert polynomial of $\mathrm{gr}_{\mathfrak{m}}(M)$.)

Examples 4.6.3. (a) Let $(R, \mathfrak{m}, k)$ be a regular local ring of dimension d. Then the homogeneous k-algebra $\mathrm{gr}_{\mathfrak{m}}(R)$ is isomorphic to the polynomial ring $k[X_1, \ldots, X_d]$; see 2.2.5. Thus $\Sigma_R(X) = \binom{X+d}{d}$ and $e(R) = 1$.

(b) Let $(R, \mathfrak{m}, k)$ be a regular local ring, $I \subset R$ a proper ideal, and $S = R/I$. We denote by $\mathfrak{n}$ the maximal ideal of S. The canonical epimorphism $\varepsilon \colon R \to S$ induces a surjective homomorphism of graded k-algebras $\mathrm{gr}(\varepsilon) \colon \mathrm{gr}_{\mathfrak{m}}(R) \to \mathrm{gr}_{\mathfrak{n}}(S)$. Indeed, let $a \in R$, $a \neq 0$, and

$a^* = a + \mathfrak{m}^{j+1}$ in $\mathrm{gr}_\mathfrak{m}(R)$ its initial form; see the definition above 4.5.9. It is clear that the homogeneous elements of $\mathrm{gr}_\mathfrak{m}(R)$ are just the initial forms of elements of R. For $a \in \mathfrak{m}^j \setminus \mathfrak{m}^{j+1}$ we define $\mathrm{gr}_\mathfrak{m}(\varepsilon)(a^*) = \varepsilon(a) + \mathfrak{n}^{j+1}$.

Let I^* be the ideal generated by the elements a^*, $a \in I$. Then $I^* = \mathrm{Ker}(\mathrm{gr}_\mathfrak{m}(\varepsilon))$ because if $a^* = a + \mathfrak{m}^{j+1} \in I^*$, then $\varepsilon(a) \in \mathfrak{n}^{j+1}$. Hence there exists $b \in \mathfrak{m}^{j+1}$ such that $\varepsilon(b) = \varepsilon(a)$. It follows that $c = a - b \in I$, and $c^* = a^*$. The converse inclusion is obvious.

We conclude that $\mathrm{gr}_\mathfrak{n}(S) = k[X_1, \dots, X_d]/I^*$. Thus $\Sigma_S(X)$ and $e(S)$ may be computed once I^* and its graded resolution are known; see 4.1.13.

Assume $I = (a_1, \dots, a_m)$; then $(a_1^*, \dots, a_m^*) \subset I^*$ with equality if $m = 1$. In general, however, we have $(a_1^*, \dots, a_m^*) \neq I^*$ (Exercise 4.6.12).

Computing $e(I, M)$ may be a painful and often impossible task. We will show that an arbitrary ideal of definition of M may be replaced by an ideal J which is generated by a system of parameters of M such that $e(J, M) = e(I, M)$, provided the residue class field k of R is infinite.

Definition 4.6.4. Let R be a Noetherian ring, I a proper ideal, and M a finite R-module. An ideal $J \subset I$ is called a *reduction ideal of I with respect to M* if $JI^n M = I^{n+1} M$ for some (or equivalently all) $n \gg 0$.

The definition of a reduction ideal almost immediately yields

Lemma 4.6.5. *Let $(R, \mathfrak{m})$ be a Noetherian local ring, M a finite R-module, I an ideal of definition of M, and J a reduction ideal of I with respect to M. Then J is an ideal of definition of M, and $e(J, M) = e(I, M)$.*

PROOF. For large n we have $I^{n+1} M = JI^n M \subset JM$, and this shows that J is an ideal of definition of M. Moreover, we get the inequalities

$$\ell(M/I^{m+n+1} M) \geq \ell(M/J^m M) \geq \ell(M/I^m M)$$

for all $m \geq 1$. Thus 4.6.2 implies the assertion. □

In the framework of Rees rings and Rees modules, reduction ideals can be characterized as follows.

Proposition 4.6.6. *Let R be a Noetherian ring, $J \subset I$ proper ideals of R, and M a finite R-module. The following conditions are equivalent:*
(a) *J is a reduction ideal of I with respect to M;*
(b) *$\mathscr{R}_+(I, M)$ is a finite $\mathscr{R}_+(J)$-module.*

PROOF. (a) $\Rightarrow$ (b): Suppose $I^{n+1} M = JI^n M$; then $\mathscr{R}_+(I, M)$ is generated over $\mathscr{R}_+(J)$ by the elements of degree $\leq n$, and hence is finitely generated.

(b) $\Rightarrow$ (a): We may choose a homogeneous set of generators $x_1, \dots, x_r$ of $\mathscr{R}_+(I, M)$. Let n be the maximal degree of the elements x_i, and let $x \in I^{n+1} M$. There exist elements $a_i \in J^{b_i}$, $b_i = n + 1 - \deg x_i$, such that $x = \sum_{i=1}^r a_i x_i$. Since $a_i x_i \in J^{b_i} I^{n+1-b_i} M \subset JI^n M$, we have $x \in JI^n M$. Thus $I^{n+1} M \subset JI^n M$. The converse inclusion is trivial. □

In terms of Rees rings we now introduce an invariant which gives a lower bound for the number of generators of a reduction.

Definition 4.6.7. Let $(R, \mathfrak{m})$ be a Noetherian local ring, I a proper ideal of R, and M a finite R-module. The number

$$\lambda(I, M) = \dim\big(\mathscr{R}_+(I, M)/\mathfrak{m}\mathscr{R}_+(I, M)\big) = \dim\big(\mathrm{gr}_I(M)/\mathfrak{m}\,\mathrm{gr}_I(M)\big)$$

is the *analytic spread of I with respect to M*. We set $\lambda(I) = \lambda(I, R)$ and call it the *analytic spread of I*.

Proposition 4.6.8. *Under the hypothesis of 4.6.7 we have $\mu(J) \geq \lambda(I, M)$ for any reduction ideal J of I with respect to M. Suppose in addition that $R/\mathfrak{m}$ is infinite. Then there exists a reduction ideal J of I with respect to M such that $\mu(J) = \lambda(I, M)$.*

PROOF. The module $\mathscr{R}_+(I, M)/\mathfrak{m}\mathscr{R}_+(I, M)$ is finite over $\mathscr{R}_+(J)/\mathfrak{m}\mathscr{R}_+(J) = \bigoplus_{i \geq 0} J^i/\mathfrak{m}J^i$ which in turn is a factor ring of $k[X_1, \ldots, X_m]$, where $m = \dim_k J/\mathfrak{m}J = \mu(J)$. Therefore $\dim\big(\mathscr{R}_+(I, M)/\mathfrak{m}\mathscr{R}_+(I, M)\big) \leq m$. This proves the first part of the proposition.

Now let $A = \mathscr{R}_+(I)/\mathfrak{a}$, where $\mathfrak{a}$ is the annihilator of the $\mathscr{R}_+(I)$-module $\mathscr{R}_+(I, M)/\mathfrak{m}\mathscr{R}_+(I, M)$. The ideal $\mathfrak{a}$ is graded and contains $\mathfrak{m}\mathscr{R}_+(I)$. Consequently A is a homogeneous $R/\mathfrak{m}$-algebra, and $\dim A = \lambda(I, M)$. Since $R/\mathfrak{m}$ is infinite, the Noether normalization theorem says that there exist elements $y_1, \ldots, y_d \in A$ of degree 1, $d = \lambda(I, M)$, such that A is a finite B-module, where $B = k[y_1, \ldots, y_d]$; see 1.5.17. It follows that $\mathscr{R}_+(I, M)/\mathfrak{m}\mathscr{R}_+(I, M)$ is a finite graded B-module.

For each y_i we choose $z_i \in I$ such that z_i is mapped to y_i under the canonical map $I \to \mathscr{R}_+(I)/\mathfrak{a}$. Let $J = (z_1, \ldots, z_d)$; then $\mu(J) = \lambda(I, M)$, and $\mathscr{R}_+(I, M)/\mathfrak{m}\mathscr{R}_+(I, M)$ is a finite $\big(\mathscr{R}_+(J)/\mathfrak{m}\mathscr{R}_+(J)\big)$-module. Now the graded version 1.5.24 of Nakayama's lemma implies that $\mathscr{R}_+(I, M)$ is a finite $(\mathscr{R}_+(J))$-module, and this completes the proof; see 4.6.6.   □

Remark 4.6.9. Let $(R, \mathfrak{m}, k)$ be a Noetherian local ring, and I a proper ideal of R. Northcott and Rees [291] call an ideal J a *minimal reduction of I* if J is a reduction ideal of I, and J itself does not have any proper reductions, and they prove that minimal reductions exist – a fact which we will not use explicitly here. In the case where k is infinite one has the following result: let J be a reduction of I, and suppose that J is minimally generated by $x_1, \ldots, x_n$. Then J is a minimal reduction of I if and only if the elements $x_1, \ldots, x_n$ are analytically independent in I and $n = \lambda(I)$.

Recall that $x_1, \ldots, x_n$ are *analytically independent in I* if whenever $f(X_1, \ldots, X_n)$ is a homogeneous polynomial of degree m in $R[X_1, \ldots, X_n]$ (m arbitrary) such that $f(x_1, \ldots, x_n) \in I^m\mathfrak{m}$, then all the coefficients of f are in $\mathfrak{m}$.

It is clear from this description and the proof of 4.6.8 that the ideal J constructed there is a minimal reduction of I when $M = R$.

Corollary 4.6.10. *Let $(R, \mathfrak{m})$ be a Noetherian local ring with infinite residue class field, M a finite R-module, and I an ideal of definition of M. Then there exists a system of parameters x of M such that (x) is a reduction ideal of I with respect to M. In particular $e(I, M) = e((x), M)$.*

PROOF. We show that $\dim\big(\mathscr{R}_+(I, M)/\mathfrak{m}\mathscr{R}_+(I, M)\big) = \dim M$. This, in view of 4.6.5 and 4.6.8, implies the assertion. Note that $\ell(I^n M/\mathfrak{m}I^n M) = \mu(I^n M)$, and $\ell(I^n M/I^{n+1}M) \leq \mu(I^n M)\,\ell(M/IM)$. Indeed, let $y_1, \dots, y_m$, $m = \mu(I^n M)$, be a system of generators of $I^n M$. We may write $y_j = a_j x_j$ with $a_j \in I^n$, $x_j \in M$. Thus there exists an epimorphism $\bigoplus Rx_j \to I^n M$ which yields an epimorphism $\bigoplus R\bar{x}_j \to I^n M/I^{n+1}M$ where $\bar{x}_j$ denotes the residue class of x_j modulo IM. Since $R\bar{x}_j \subset M/IM$, the desired inequality follows. We therefore obtain the inequalities

$$H\big(\mathscr{R}_+(I, M)/\mathfrak{m}\mathscr{R}_+(I, M), n\big) \leq H(\mathrm{gr}_I(M), n)$$
$$\leq \ell(M/IM)\, H\big(\mathscr{R}_+(I, M)/\mathfrak{m}\mathscr{R}_+(I, M), n\big).$$

By 4.1.3 and 4.5.6, the Hilbert function $H(\mathrm{gr}_I(M), n)$ is a polynomial of degree $\dim M - 1$ for large n, and so is $H\big(\mathscr{R}_+(I, M)/\mathfrak{m}\mathscr{R}_+(I, M), n\big)$ by the above inequalities. Hence, if we again apply 4.1.3, the conclusion follows. $\square$

Exercises

4.6.11. Let R be a Noetherian ring, I a proper ideal of R, M a finite R-module, M' a submodule, and M'' a factor module of M. If J is a reduction ideal of I with respect to M, show it is a reduction ideal of I with respect to M' and M'', too. (Hint: Use 4.6.6).

4.6.12. (a) Let $(R, \mathfrak{m}, k)$ be a regular local ring, and f an element in $\mathfrak{m}$ whose initial form f^* has degree a; see 4.6.3. Set $S = R/(f)$ and prove that $\mathrm{gr}_{\mathfrak{m}}(S) \cong k[X_1, \dots, X_d]/(f^*)$, and $e(S) = a$.
(b) Let $I = (X^2, XY + Z^3) \subset k[\![X, Y, Z]\!]$. Show the ideal I^* of initial forms of I is not generated by the initial forms X^2 and XY of the generators of I.

4.6.13. Let $(R, \mathfrak{m})$ be a Noetherian local ring and I a proper ideal of R. Show that the analytic spread of an ideal has the following properties:
(a) $\lambda(I R_\mathfrak{p}) \leq \lambda(I)$ for all $\mathfrak{p} \in \mathrm{Spec}\, R$;
(b) if I is $\mathfrak{m}$-primary, then $\lambda(I) = \dim R$;
(c) $\mathrm{height}\, I \leq \lambda(I) \leq \dim R$.

4.6.14. Let $(R, \mathfrak{m}, k)$ be a d-dimensional Cohen–Macaulay local ring. Prove:
(a) If k is infinite, then there exists an R-sequence $x = x_1, \dots, x_d$ such that $e(R) = \ell(R/(x))$.
Hint: Proceed by induction on d; choose x_1 such that its initial form in $\mathrm{gr}_{\mathfrak{m}}(R)$ is an element of degree 1 whose annihilator has finite length.

(b) $e(R) \geq \operatorname{emb} \dim R - \dim R + 1$; if equality holds, then R is said to have *minimal multiplicity*.

(c) If k is infinite, then R has minimal multiplicity if and only if there exists an R-sequence x such that $\mathfrak{m}^2 = (x)\mathfrak{m}$.

4.6.15. Let $(R, \mathfrak{m}, k)$ be a one dimensional local ring. Prove:

(a) If k is infinite, there exists an element x such that $\mathfrak{m}^{n+1} = x\mathfrak{m}^n$ for all $n \gg 0$; any such element is called *superficial*.

(b) If x is a superficial element, then $x \notin \mathfrak{m}^2$.

(c) $e(R) = \mu(\mathfrak{m}^n)$ for large n.

4.6.16. Let $(R, \mathfrak{m}, k)$ be a one dimensional Cohen–Macaulay local ring, and x a superficial element of R.

(a) Suppose I is an ideal of height 1 in R. Show that $\ell(I/xI) = e(R)$.

(b) Prove that $\mu(I) \leq e(R)$ for all ideals of height 1 of R.

4.6.17. Let $(R, \mathfrak{m}, k)$ be a Noetherian local ring. Suppose there exists an integer n such that $\mu(I) \leq n$ for all ideals I of R. Show that $\dim R = 1$.

4.6.18. Let $R = k[\![t]\!]$, k a field. Let $f_1(t), \ldots, f_n(t) \in R$. We denote by $k[\![f_1(t), \ldots, f_n(t)]\!]$ the subring

$$A = \{F(f_1(t), \ldots, f_n(t)) : F \in k[\![X_1, \ldots, X_n]\!]\}$$

of R. Suppose the integral closure of A is R; prove $e(A)$ is the minimum of the initial degrees of the $f_i(t)$. Hint: use the fact that R is a finite A-module; see [270], §33.

4.7 The multiplicity symbol

In the previous section we saw that the computation of the multiplicity $e(I, M)$ of a finite module M with respect to an ideal of definition I can be reduced to the case when I is generated by a system of parameters of M. The advantage of this reduction will become apparent when we show that the multiplicity of a module M with respect to an ideal generated by a system of parameters x can be expressed in terms of the Koszul homology $H_\bullet(x, M)$. We approach this goal by introducing the *multiplicity symbol* $e(x, M)$, due to Northcott.

Let $(R, \mathfrak{m})$ be a Noetherian local ring, and M a finite R-module. A sequence of elements $x = x_1, \ldots, x_n$ in $\mathfrak{m}$ is a *multiplicity system* of M if $\ell(M/(x)M)$ is finite, equivalently, if (x) is an ideal of definition of M.

Lemma 4.7.1. *Let $(R, \mathfrak{m})$ be a Noetherian local ring, x a sequence of elements in R, and $0 \to M' \to M \to M'' \to 0$ an exact sequence of finite R-modules. The sequence x is a multiplicity system of M if and only if it is a multiplicity system of M' and M''.*

PROOF. The exactness of $M'/(x)M' \to M/(x)M \to M''/(x)M'' \to 0$ implies that

$$\ell(M''/(x)M'') \leq \ell(M/(x)M) \leq \ell(M''/(x)M'') + \ell(M'/(x)M').$$

It therefore remains to show that $\ell(M'/(x)M') < \infty$ if $\ell(M/(x)M)$ is. According to the Artin–Rees lemma ([270], Theorem 8.5) there exists an integer m such that $(x)^m M \cap M' \subset (x)M'$, and this implies that $\ell(M'/(x)M') \leq \ell(M'/(x)^m M \cap M') \leq \ell(M/(x)^m M)$. □

Corollary 4.7.2. *Let* $(R, \mathfrak{m})$ *be a Noetherian local ring,* M *a finite R-module, and* $x = x_1, \ldots, x_n$ *a multiplicity system of* M. *Then* $x' = x_2, \ldots, x_n$ *is a multiplicity system of* $M/x_1 M$ *and* $(0 : x_1)_M$.

PROOF. Note that $x(M/x_1 M) = x'(M/x_1 M)$, and $x(0 : x_1)_M = x'(0 : x_1)_M$. □

This corollary allows an inductive definition of the multiplicity symbol.

Definition 4.7.3. Let $(R, \mathfrak{m})$ be a Noetherian local ring, M a finite R-module, and $x = x_1, \ldots, x_n$ a multiplicity system of M. If $n = 0$, then $\ell(M) < \infty$, and we set $e(x, M) = \ell(M)$; if $n > 0$, we set $e(x, M) = e(x', M/x_1 M) - e(x', (0 : x_1)_M)$, $x' = x_2, \ldots, x_n$. We call $e(x, M)$ the *multiplicity symbol*.

At first glance it seems as if the multiplicity symbol depends on the order of the elements of x. That this is not the case will follow from the next theorem.

Note that the homology $H_\bullet(x, M)$ of the Koszul complex of a multiplicity system x of M has finite length as follows from 1.6.5. Hence we may consider the *Euler characteristic*

$$\chi(x, M) = \sum_i (-1)^i \ell(H_i(x, M))$$

of the Koszul homology.

Theorem 4.7.4 (Auslander–Buchsbaum). *Let* $(R, \mathfrak{m})$ *be a Noetherian local ring,* M *a finite R-module, and* x *a multiplicity system of* M. *Then*

$$e(x, M) = \chi(x, M).$$

The proof of the theorem is based on

Lemma 4.7.5. *Let* $(R, \mathfrak{m})$ *be a Noetherian local ring, and* $x = x_1, \ldots, x_n$ *a sequence of elements in* $\mathfrak{m}$. *Whenever the Euler characteristic is defined it has the following properties:*
(a) $\chi(x, \_)$ *is additive on short exact sequences, that is, for any short exact sequence* $0 \to M' \to M \to M'' \to 0$ *for which* x *is a multiplicity system of* M, *one has*

$$\chi(x, M) = \chi(x, M') + \chi(x, M'');$$

(b) *if* $x_1 M = 0$, *then* $\chi(x, M) = 0$;
(c) *if* x_1 *is M-regular, then* $\chi(x, M) = \chi(x_2, \ldots, x_n, M/x_1 M)$.

PROOF. (a) By the additivity of length, the alternating sum of the lengths of the homology modules in the long exact sequence

$$\cdots \longrightarrow H_i(x, M') \longrightarrow H_i(x, M) \longrightarrow H_i(x, M'') \longrightarrow \cdots$$

is zero. This yields the desired result.

(b) Let $x' = x_2, \ldots, x_n$. If $x_1 M = 0$, then

$$H_i(x, M) = H_i(0, x', M) \cong H_i(x', M) \oplus H_{i-1}(x', M),$$

for all i; see 1.6.21. Thus

$$\chi(x, M) = \sum_i (-1)^i \big(\ell(H_i(x', M)) + \ell(H_{i-1}(x', M)) \big) = 0.$$

(c) If x_1 is an M-regular element, then $H_i(x, M) \cong H_i(x', M/x_1 M)$ by 1.6.13. This implies the assertion.   □

PROOF OF 4.7.4. Let $x = x_1, \ldots, x_n$ and $x' = x_2, \ldots, x_n$. We show that

(7) $\chi(x, M) = \chi(x', M/x_1 M) - \chi(x', (0 : x_1)_M).$

The ascending chain $0 \subset (0 : x_1)_M \subset (0 : x_1^2)_M \subset \cdots$ of submodules of M stabilizes since M is Noetherian. Let a be an integer such that $(0 : x_1^a)_M = (0 : x_1^{a+1})_M$. We leave it to the reader to verify that x_1 is regular on $N = M/(0 : x_1^a)_M$, and that x' is a multiplicity system of $(0 : x_1^a)_M$.

Consider the following commutative diagram with exact rows and columns:

From 4.7.5(a) it follows that $\chi(x', N/x_1N) = \chi(x', M/x_1M) - \chi(x', C)$, and $\chi(x', C) = \chi(x', (0 : x_1)_M)$, and thus

(8) $\qquad \chi(x', N/x_1N) = \chi(x', M/x_1M) - \chi(x', (0 : x_1)_M)$.

Now we apply 4.7.5(a) and (c) to see that

(9) $\qquad \chi(x', N/x_1N) = \chi(x, N) = \chi(x, M) - \chi(x, (0 : x_1^a)_M)$.

Finally, by induction on i, it follows from 4.7.5(a) and (b), and the exact sequences

$$0 \longrightarrow (0 : x_1^{i-1})_M \longrightarrow (0 : x_1^i)_M \longrightarrow (0 : x_1^i)_M/(0 : x_1^{i-1})_M \longrightarrow 0$$

that $\chi(x, (0 : x_1^i)_M) = 0$ for all i. This, together with (8) and (9), completes the proof. $\qquad \square$

If the sequence x generates the ideal (x) minimally, then the Koszul homology $H_\bullet(x, M)$ depends only on the ideal (x). By 4.7.4, the same holds for the multiplicity symbol. Much more is true:

Theorem 4.7.6 (Serre). *Let $(R, \mathfrak{m})$ be a Noetherian local ring, M a finite R-module, $x = x_1, \ldots, x_n$ a multiplicity system of M, and I the ideal generated by x. Then*

$$\chi(x, M) = \begin{cases} e(I, M) & \text{if } x \text{ is a system of parameters of } M, \\ 0 & \text{otherwise.} \end{cases}$$

Taking into account 4.7.4 we see that for any system of parameters x of M the numbers $e(x, M)$, $e((x), M)$ and $\chi(x, M)$ are all the same.

PROOF OF 4.7.6. Let $K_\bullet = K_\bullet(x, M)$ be the Koszul complex, and for each integer m let $K_\bullet^{(m)}$ be the subcomplex

$$0 \longrightarrow I^m K_n \longrightarrow I^{m+1} K_{n-1} \longrightarrow \cdots \longrightarrow I^{m+n} K_0 \longrightarrow 0$$

of $K_\bullet$. We first claim that $K_\bullet^{(m)}$ is exact for all $m \gg 0$: for a fixed integer i its i-cycles are $Z_i(K_\bullet^{(m)}) = Z_i(K_\bullet) \cap I^{m+n-i} K_i$. By the Artin–Rees lemma ([270], Theorem 8.5) we have

$$Z_i(K_\bullet) \cap I^{m+n-i} K_i = I \cdot \left(Z_i(K_\bullet) \cap I^{m+n-i-1} K_i \right)$$

for all $m \gg 0$. We may pick m_0 large enough for this equality to hold simultaneously for all i and all $m \geq m_0$.

Now let $m \geq m_0$, and $z \in Z_i(K_\bullet^{(m)})$; then $z = \sum_{i=1}^n x_i z_i$ with $z_i \in Z_i(K_\bullet) \cap I^{m+n-i-1} K_i$. Let $e_1, \ldots, e_n$ be a basis of $K_1(x, R)$ with $d_x(e_i) = x_i$ for $i = 1, \ldots, n$, where d_x denotes the differential of $K_\bullet(x, R)$. Then $w = \sum_{i=1}^n e_i.z_i \in I^{m+n-i-1} K_{i+1}$ and $d_{x,M}(w) = z$. Thus $K_\bullet^{(m)}$ is indeed exact.

It follows from this, the exact sequence of complexes

$$0 \longrightarrow K_\bullet^{(m)} \longrightarrow K_\bullet \longrightarrow K_\bullet/K_\bullet^{(m)} \longrightarrow 0,$$

and the exactness of $K_\bullet^{(m)}$, that $H_\bullet(K_\bullet) \cong H_\bullet(K_\bullet/K_\bullet^{(m)})$; hence $\chi(x, M) = \sum_{i=0}^{n}(-1)^i \ell(H_i(K_\bullet/K_\bullet^{(m)}))$. However, since $K_i/K_i^{(m)}$ is of finite length for all i – its length is actually $\binom{n}{i} \ell(M/I^{m+n-i}M)$ – we have

$$\sum_{i=0}^{n}(-1)^i \ell(H_i(K_\bullet/K_\bullet^{(m)})) = \sum_{i=0}^{n}(-1)^i \ell(K_i/K_i^{(m)}),$$

and thus for $m \gg 0$,

$$\chi(x, M) = \sum_{i=0}^{n}(-1)^i \binom{n}{i} \chi_M^I(m + n - i - 1) = \Delta^n \chi_M^I(m - 1)$$

$$= \begin{cases} e(I, M) & \text{if } \dim M = n, \\ 0 & \text{if } \dim M < n; \end{cases}$$

see 4.6.2, and use that the application of Δ decreases the degree of a polynomial function by 1. □

Let $(R, \mathfrak{m})$ be a Noetherian local ring, and I an ideal of definition of R. We fix an integer q, and denote by $\mathscr{K}_q(R)$ the full subcategory of the category $\mathscr{M}(R)$ of finite R-modules whose dimension is at most q. We define

$$e_q(I, M) = \begin{cases} e(I, M) & \text{if } \dim M = q, \\ 0 & \text{if } \dim M < q. \end{cases}$$

Corollary 4.7.7. *The (modified) multiplicity $e_q(I, M)$ is an additive function on the category $\mathscr{K}_q(R)$, that is, $e_q(I, M) = e_q(I, M') + e_q(I, M'')$ for all exact sequences $0 \longrightarrow M' \longrightarrow M \longrightarrow M'' \longrightarrow 0$ in $\mathscr{K}_q(R)$.*

PROOF. Without loss of generality we may assume that $R/\mathfrak{m}$ is infinite. For otherwise we may extend the residue class field of R; see the proof of 4.1.10.

We may further assume that the module M in the above exact sequence has dimension q. By 4.6.10 there exists a system of parameters $x = x_1, \ldots, x_q$ of M such that (x) is a reduction ideal of I with respect to M. Now 4.6.10 and 4.7.6 imply that $e_q(I, M) = \chi(x, M)$. According to Exercise 4.6.11 the ideal (x) is a reduction ideal of I with respect to M' and M'' as well. Hence we also have $e_q(I, M') = \chi(x, M')$ and $e_q(I, M'') = \chi(x, M'')$. Thus the result follows from 4.7.5. □

Corollary 4.7.8. *Let $(R, \mathfrak{m})$ be a Noetherian local ring, I an ideal of definition of R, and M a finite R-module of dimension $\leq q$. Then*

$$e_q(I, M) = \sum_{\mathfrak{p}} \ell(M_\mathfrak{p}) e_q(I, R/\mathfrak{p}),$$

where the sum is taken over all prime ideals $\mathfrak{p}$ with $\dim R/\mathfrak{p} = q$.

PROOF. The module M has a filtration $0 = M_0 \subset M_1 \subset \cdots \subset M_{r-1} \subset M_r = M$ such that $M_i/M_{i-1} \cong R/\mathfrak{p}_i$ for $i = 1, \ldots, r$. Of course, $\dim R/\mathfrak{p}_i \leq q$ for all i. Thus by the previous corollary we have $e_q(I, M) = \sum_{i=1}^{r} e_q(I, R/\mathfrak{p}_i)$. Only those summands contribute to the sum for which $\dim R/\mathfrak{p}_i = q$. Fix a prime ideal $\mathfrak{p}$ with $\dim R/\mathfrak{p} = q$. Then the number of integers i for which $\mathfrak{p} = \mathfrak{p}_i$ equals the length of $M_\mathfrak{p}$, as can be easily seen by localization at $\mathfrak{p}$. This proves the formula asserted. □

As an important special case of the previous result we have

Corollary 4.7.9. *Let $(R, \mathfrak{m})$ be a Noetherian local ring, M a finite module of positive rank, and I an $\mathfrak{m}$-primary ideal of R. Then*

$$e(I, M) = e(I, R) \operatorname{rank} M.$$

In particular, $e(M) = e(R) \operatorname{rank} M$.

PROOF. Let $r = \operatorname{rank} M$. By virtue of 1.4.3 we have $M_\mathfrak{p} \cong R_\mathfrak{p}^r$ for all prime ideals $\mathfrak{p}$ of R with $\dim R/\mathfrak{p} = d$. In particular M has maximal dimension, and so $e(I, M) = e_d(I, M)$, $d = \dim R$. Therefore 4.7.8 yields

$$e(I, M) = \sum_\mathfrak{p} \ell(M_\mathfrak{p}) e(I, R/\mathfrak{p}) = \sum_\mathfrak{p} r \, \ell(R_\mathfrak{p}) e(I, R/\mathfrak{p}) = e(I, R) \operatorname{rank} M.$$

Here the sums are taken over the prime ideals $\mathfrak{p}$ with $\dim R/\mathfrak{p} = d$. □

Partial Euler characteristics. One remarkable consequence of 4.7.4 is the following: let $(R, \mathfrak{m})$ be a Noetherian local ring, M a finite R-module, and x a multiplicity system of M. Then $\chi(x, M) = \sum_i (-1)^i \ell(H_i(x, M)) \geq 0$.

One defines for all $j \geq 0$ the *partial Euler characteristics*

$$\chi_j(x, M) = \sum_{i \geq j} (-1)^{i-j} \ell(H_i(x, M))$$

of M with respect to x. Surprisingly, all the partial Euler characteristics are non-negative, as shown by Serre ([334], Appendice II). We only prove the result for χ_1 (see however Remark 4.7.12).

Theorem 4.7.10 (Serre). *Let $(R, \mathfrak{m})$ be a Noetherian local ring, M a finite R-module, and x a multiplicity system of M.*
(a) *$\chi_1(x, M) \geq 0$, or equivalently, $\ell(M/xM) \geq \chi(x, M)$.*
(b) *Assume in addition that x is a system of parameters of M. Then the following conditions are equivalent:*
  (i) *$\chi_1(x, M) = 0$;*
  (ii) *$H_1(x, M) = 0$;*
  (iii) *$H_i(x, M) = 0$ for $i \geq 1$;*
  (iv) *x is an M-sequence;*
  (v) *M is Cohen–Macaulay.*

PROOF. Let $x = x_1, \ldots, x_n$. We prove (a) by induction on n: if $n = 1$, then $\chi_1(x_1, M) = \ell(H_1(x_1; M))$, and the assertion is trivial. Now let $n > 1$, and set $x' = x_2, \ldots, x_n$. Notice that $\chi(x, M) = \ell(M/xM) - \chi_1(x, M)$, whence

(10) $\qquad \chi_1(x, M) = \chi_1(x', M/x_1 M) + \chi(x', (0 : x_1)_M).$

in view of equation (7) above. By induction $\chi_1(x', M/x_1 M) \geq 0$, and since $\chi(x', (0 : x_1)_M) \geq 0$, the assertion follows.

(b) The equivalence of the statements (ii)–(v) was shown in 1.6.19 and 2.1.2, and (iii) $\Rightarrow$ (i) is obvious. We now prove the implication (i) $\Rightarrow$ (v). Suppose that $\chi_1(x, M) = 0$; then (10) implies

$$\chi_1(x', M/x_1 M) = 0 \quad \text{and} \quad \chi(x', (0 : x_1)_M) = 0.$$

By induction we may assume that $M/x_1 M$ is a Cohen–Macaulay module of dimension $n - 1$. It remains to show that $(0 : x_1)_M = 0$. Set $M_1 = M/(0 : x_1)_M$; then the snake lemma applied to the commutative diagram

$$
\begin{array}{ccccccccc}
0 & \longrightarrow & (0 : x_1)_M & \longrightarrow & M & \longrightarrow & M_1 & \longrightarrow & 0 \\
& & {\scriptstyle x_1}\downarrow & & {\scriptstyle x_1}\downarrow & & {\scriptstyle x_1}\downarrow & & \\
0 & \longrightarrow & (0 : x_1)_M & \longrightarrow & M & \longrightarrow & M_1 & \longrightarrow & 0
\end{array}
$$

yields the exact sequence

$$0 \longrightarrow (0 : x_1)_M \overset{\varphi}{\longrightarrow} (0 : x_1)_M \longrightarrow (0 : x_1)_{M_1}$$
$$\longrightarrow (0 : x_1)_M \overset{\psi}{\longrightarrow} M/x_1 M \longrightarrow M_1/x_1 M_1 \longrightarrow 0.$$

It is clear that φ is an isomorphism. We claim that $\psi = 0$. Indeed, it follows from 4.7.4 and 4.7.6 that $\dim(0 : x_1)_M \leq n - 2$ since $\chi(x', (0 : x_1)_M) = 0$. On the other hand, $\dim R/\mathfrak{p} = n - 1$ for all $\mathfrak{p} \in \mathrm{Ass}(M/x_1 M)$ since $M/x_1 M$ is Cohen–Macaulay; see 2.1.2. Therefore $\mathrm{Hom}((0 : x_1)_M, M/x_1 M) = 0$ by 1.2.3.

We obtain the isomorphisms

$$M/x_1 M \cong M_1/x_1 M_1 \quad \text{and} \quad (0 : x_1)_M \cong (0 : x_1)_{M_1}.$$

It follows from (7) that

$$\chi_1(x, M) = \ell(M/xM) - \chi(x', M/x_1 M) + \chi(x', (0 : x_1)_M),$$

and hence the analogous equation for M_1 and the isomorphisms give us $\chi_1(x, M_1) = \chi_1(x, M) = 0$.

Repeating these arguments we obtain a sequence of modules M_n, defined recursively by $M_n = M_{n-1}/(0 : x_1)_{M_{n-1}}$ with

$$M_n/x_1 M_n \cong M_{n-1}/x_1 M_{n-1} \quad \text{and} \quad (0 : x_1)_{M_n} \cong (0 : x_1)_{M_{n-1}}.$$

Consider the composition $M \to M_1 \to \cdots \to M_{n-1} \to M_n$ of the canonical epimorphisms. A simple inductive argument shows its kernel is $(0 : x_1^n)_M$. Since M is Noetherian there exists an integer m such that $(0 : x_1^m)_M = (0 : x_1^{m+1})_M$, and so the canonical epimorphism $M_m \to M_{m+1}$ must be an isomorphism; therefore $(0 : x_1)_{M_m} = 0$. But then, as required,

$$(0 : x_1)_M \cong (0 : x_1)_{M_1} \cong \cdots \cong (0 : x_1)_{M_m} = 0. \qquad \square$$

Combining 4.7.9 and 4.7.10 we obtain the following Cohen–Macaulay criterion for modules:

Corollary 4.7.11. *Let $(R, \mathfrak{m})$ be a Noetherian local ring, M a finite R-module of positive rank, and I an ideal generated by a system of parameters of R.*
(a) $\ell(M/IM) \geq e(I, R) \operatorname{rank} M$.
(b) *M is Cohen–Macaulay if and only if $\ell(M/IM) = e(I, R) \operatorname{rank} M$.*
(c) *Suppose R is Cohen–Macaulay; then M is Cohen–Macaulay if and only if $\ell(M/IM) = \ell(R/I) \operatorname{rank} M$.*

Remark 4.7.12. The positivity of the partial Euler characteristics can be easily proved in an important special case. Let $(R, \mathfrak{m}, k)$ be a Noetherian local ring, M a finite R-module, and x a multiplicity system of M. Then $\chi_j(x, M) \geq 0$ for all $j \geq 0$, and if $\chi_j(x, M) = 0$ for some j, then $H_i(x, M) = 0$ for all $i \geq j$.

For the proof we may assume that R is complete since homology commutes with completion, so that $H_\bullet(x, \hat{M}) \cong H_\bullet(x, M)\hat{} \cong H_\bullet(x, M)$. The last isomorphism is valid since $H_\bullet(x, M)$ has finite length.

Now, as we assume that R is complete and contains a field, the ring R even contains its residue class field; see A.20. Let $A = k[[X_1, \ldots, X_n]]$, and define a ring homomorphism $\varphi : A \to R$ by $\varphi(X_i) = x_i$ for $i = 1, \ldots, n$. We may view M as an A-module via φ. It is then clear that M is a finite A-module, and that $H_\bullet(X_1, \ldots, X_n; M) \cong H_\bullet(x, M)$. In other words, we may assume that x is an R-sequence (replace R by A).

We prove the assertions by induction on j. For $j = 0$ and $j = 1$ we know the result from 4.7.6 and 4.7.10. Now we let $j > 1$, and consider an exact sequence

$$0 \longrightarrow U \longrightarrow F \longrightarrow M \longrightarrow 0$$

where F is a finite free R-module. Then, owing to 1.6.11 and the assumption that x is R-regular, it follows that $H_i(x, M) \cong H_{i-1}(x, U)$ for all $i > 1$. Therefore $\chi_j(x, M) = \chi_{j-1}(x, U)$, and the proof is complete by our induction hypothesis.

Exercises

4.7.13. (a) Let $(R, \mathfrak{m})$ be a Noetherian reduced local ring. Verify $e(R) = \sum_{\mathfrak{p}} e(R/\mathfrak{p})$ where the sum is taken over all prime ideals $\mathfrak{p}$ with $\dim R/\mathfrak{p} = \dim R$.

(b) Let k be a field. Compute the multiplicity of

$$k[[X_1,\ldots,X_n]]/(X_1X_2,X_2X_3,\ldots,X_{n-1}X_n,X_nX_1).$$

Hint: apply 4.4.17.

4.7.14. Let $(R,\mathfrak{m})$ be a Noetherian local ring of dimension d, M a maximal Cohen–Macaulay R-module, x a system of parameters of R, and n an integer such that $\mathfrak{m}^n \subset (x)$. Show that $\ell(M/xM) \leq n^d e(M)$.

4.7.15. (a) Let k be a field, and assume that $R = k[[X_1,\ldots,X_n]]/I$ is a domain of dimension d with quotient field L. Let $y = y_1,\ldots,y_d$ be a system of parameters of R. The subring $A = k[[y_1,\ldots,y_d]]$ of R is regular and R is a finite A-module; see A.22. We denote by K the quotient field of A.
(a) Show that $[L : K] \leq \ell(R/(y))$. Equality holds if and only if R is Cohen–Macaulay.
(b) Formulate and prove a similar statement for graded k-algebras.

4.7.16. In this exercise we want to use the criterion 4.7.15 in a concrete situation.
(a) Let k be a field, and $k(X_1,\ldots,X_n)$ the rational function field in n variables over k. For a vector $v = (a_1,\ldots,a_n)$ in $\mathbb{Z}^n$ we set $X^v = X_1^{a_1}\cdots X_n^{a_n}$. If $v_1,\ldots,v_n$ are vectors in $\mathbb{Z}^n$, show that

$$[k(X_1,\ldots,X_n) : k(X^{v_1},\ldots,X^{v_n})] = |\det(v_1,\ldots,v_n)|.$$

Hint: use the theory of elementary divisors.
(b) Let $m_1,\ldots,m_r$ be monomials in $X_1,\ldots,X_n$, and consider the subring $R = k[m_1,\ldots,m_r]$ of the polynomial ring $k[X_1,\ldots,X_n]$. (Such a ring is called an affine semigroup ring and will be studied more systematically in Chapter 6.) Assume that
   (i) $Q(R) = k(X_1,\ldots,X_n)$,
   (ii) there are monomials $w_1,\ldots,w_n \in R$ such that $\ell(R/(w_1,\ldots,w_n)R) < \infty$.
Let $w_i = X^{v_i}$ for $i = 1,\ldots,n$; prove that R is a Cohen–Macaulay ring if and only if $|\det(v_1,\ldots,v_n)|$ equals the number of all monomials in R not belonging to the ideal of monomials $(w_1,\ldots,w_n)$.
(c) Apply this criterion to show that the ring $k[X^3Y,X^2Y,XY,XY^2,XY^3]$ is Cohen–Macaulay, but that $k[X^3,X^2Y,XY,XY^2,Y^3]$ is not.

4.7.17. Let $(A,\mathfrak{m})$ be a regular local ring, $I \subset \mathfrak{m}$ an ideal and $R = A/I$. The R-module I/I^2 is called the *cotangent module* of R.
(a) Suppose R is Cohen–Macaulay and generically a complete intersection, that is, $R_\mathfrak{p}$ is a complete intersection for all minimal prime ideals $\mathfrak{p} \in \operatorname{Spec} R$. Prove that $\operatorname{rank} I/I^2 = \operatorname{height} I$.
(b) Let B be a local ring, and $J \subset B$ a proper ideal. The pair (J,B) is called an *embedded deformation of* (I,A) if there exists a B-sequence x such that $A \cong B/xB$, $I = JA$, and x is a B/J-sequence too.
   Suppose $\dim A/I = 0$ and $\operatorname{emb\,dim} A/I = n$. If (I,A) has an embedded deformation (J,B) such that B/J is generically a complete intersection, show $\ell(I/I^2) \geq n\,\ell(A/I)$.
(c) Let k be a field, and $I \subset k[X_1,\ldots,X_n]$ an ideal generated by monomials containing a power of each indeterminate. Show the number of monomials not contained in I^2 is greater than or equal to $n + 1$ times the number of monomials not contained in I. Equality holds if I is generated by powers of the X_i.

Notes

In his famous paper 'Über die Theorie der algebraischen Formen' [171] published a century ago, Hilbert proved that a graded module over a polynomial ring has a finite graded free resolution, and concluded from this fact that the function (which we now call the Hilbert function) is of polynomial type. The influence of this paper on commutative algebra has been tremendous. Till today both free resolutions and Hilbert functions have fascinated mathematicians, and many problems still remain open.

For many applications it is more convenient to consider the so-called Hilbert series of a graded module. This point of view is stressed in Section 4.1. Stanley calls the (finite) coefficient vector of the numerator of this rational function the h-vector of the module. Its significance became apparent in Stanley's work on combinatorics. An introduction to this aspect of commutative algebra is given in Stanley's monograph [363] which is well-known as the 'green book'. Certainly Stanley's work initiated a new interest in Hilbert functions; other important motivations come from algebraic geometry.

Graded free resolutions determine the Hilbert function, but the converse is not true, except when the module has a pure resolution. This is the content of 4.1.15 which is taken from Herzog and Kühl [158] and Huneke and Miller [217].

Section 4.2 is based on the paper [357] in which Stanley states Macaulay's theorem on Hilbert functions in the form presented in this book. We also took 4.2.15 and 4.4.6 from this article. The latter result is Stanley's beautiful theorem characterizing graded Gorenstein domains by their Hilbert function. (A generalization of 4.4.5 has recently been proved by Avramov, Buchweitz, and Sally [26]. Theorem 4.4.9 appears in an article of Stanley [362] with the hypothesis that R be a domain. Our slightly more general version was given by Hibi [170].

Macaulay's article 'Some properties of enumeration in the theory of modular systems' [263] appeared in 1927, and has become a source of inspiration in commutative algebra and combinatorics; see for instance Sperner [354], Whipple [394], Clements and Lindström [70], Elias and Iarrobino [96], Stanley [357], Hibi [167], and Green [138].

In the first part of his paper Macaulay shows that the Hilbert function of a homogeneous ring arises as the Hilbert function of a polynomial ring modulo an ideal which is defined by monomials. For his proof Macaulay ordered the monomials, and thereby introduced implicitly (and possibly for the first time) what nowadays is called a Gröbner basis. Buchberger [62] was the first to describe an algorithm computing the Gröbner basis of an ideal. Robbiano [308] is an 'early' survey of this topic. Meanwhile effective computation has become an important area of research in commutative algebra, and we recommend especially the books of Eisenbud [91] and Vasconcelos [384] for a detailed account. The importance of the Castelnuovo–Mumford regularity has been briefly indicated by Theorem 4.3.1, which is due to Eisenbud and Goto [93]. More information is provided by the survey of Bayer and Mumford [38] and by Eisenbud's book [91].

Macaulay's main result in [263] however is the inequality $H(R, n+1) \leq H(R, n)^{\langle n \rangle}$ which characterizes the Hilbert functions of homogeneous k-algebras. As a note preceding it Macaulay writes: 'This proof of the theorem which has been assumed earlier is given only to place it on record. It is too long and complicated to provide any but the most tedious reading.' We present Green's proof [138] of Macaulay's theorem which is less computational than the original. The proofs of Gotzmann's theorems [136] have also been drawn from Green [138].

Theorems of Gotzmann type for exterior algebras have recently been proved by Aramova, Herzog, and Hibi [12].

The lexsegment ideals that appear in the proof of Macaulay's theorem have a remarkable 'extremal' property: if J is the lexsegment ideal with the same Hilbert function as a given ideal I, then each graded Betti number $\beta_{ij}(I)$ is bounded above by $\beta_{ij}(J)$. This was shown independently by Bigatti [42] and Hulett [207] in characteristic 0 and by Pardue in positive characteristic [296]. (Macaulay's theorem states this inequality for $i = 0$!) See Valla [377] for a related result.

In his article [320], Samuel laid the foundation of modern multiplicity theory. He was the first to apply Hilbert's theory to the associated graded ring of an $\mathfrak{m}$-primary ideal I in a Noetherian local ring $(R, \mathfrak{m})$. This led to the so-called Hilbert–Samuel function, and provided the definition of the multiplicity of R with respect to I. In this context the notion of reduction ideals, invented and investigated by Northcott and Rees in [291], plays an important role. Our Proposition 4.6.8, though formulated for modules, is taken from this paper. In a special case 4.6.8 says that the multiplicity of a module with respect to an ideal equals the multiplicity of the module with respect to a suitable system of parameters. This had already been observed by Samuel in [320]. More information on reduction ideals and related questions can be found in Sally's book [319].

As a measure of the complexity of an ideal I may serve the *analytic deviation* $\lambda(I) -$ height I. Ideals of analytic deviation zero are called *equimultiple*. The interested reader may consult the monograph by Herrmann, Ikeda, and Orbanz [155]. Ideals with small analytic deviation have been studied by Huckaba and Huneke [205], [206], and Vasconcelos [383].

The question of when the Rees ring or the associated graded ring of an ideal is Cohen–Macaulay has been of central interest in commutative algebra. The problem is well understood for ideals generated by *d-sequences*. This notion, introduced by Huneke, generalizes the notion of a regular sequence considerably, but still guarantees that the Rees ring of an ideal I generated by a d-sequence is isomorphic to the symmetric algebra of I; see Huneke [209] and Valla [376]. The reader who wants more information on d-sequences is referred to the articles by Huneke [212], and Herzog, Simis, and Vasconcelos [161], [162].

Other approaches to the Rees ring and associated graded ring of an ideal can be found in the papers by Bruns, Simis, and Trung [60], Eisenbud and Huneke [94], Goto and Shimoda [132], Huneke [211], Ikeda [223], [224], Trung and Ikeda [225], Valla [375] and Vasconcelos [381]. A comprehensive account of the recent developments in this area is given in Vasconcelos' monograph [382].

Most important is Serre's theorem 4.7.6 which relates the multiplicity to the Euler characteristic of the Koszul complex. Serre proved this result in the mid-fifties. The notes [334] by Gabriel of Serre's course at the Collège de France were published 1965. Auslander and Buchsbaum, in their classic paper [19], proved a version of Serre's theorem for arbitrary Noetherian rings, and gave an axiomatic description of the multiplicity. In Section 4.7 we follow this axiomatic approach, and introduce the multiplicity symbol. This terminology stems from Northcott who, in his book [289], systematically developed multiplicity theory from the formal properties of this symbol.

Corollary 4.7.11 is taken from [19]. In our presentation it is a consequence of the fact that the first truncated Euler characteristic χ_1 of the Koszul complex is non-negative. Serre [334] proves this not just for the first but also for the higher truncated Euler characteristics. We only show the non-negativity of χ_1 (see 4.7.10), following Lichtenbaum [258]. In writing this part of the section we consulted the article [346] of Simis and Vasconcelos.

The Koszul homology can be interpreted as a Tor of modules, and this leads to a far reaching generalization: the intersection multiplicity of modules introduced by Serre [334]; see Remark 9.4.8.

Part II

Classes of Cohen–Macaulay rings

5 Stanley–Reisner rings

This chapter is an introduction to 'combinatorial commutative algebra', a fascinating new branch of commutative algebra created by Hochster and Stanley in the mid-seventies. The combinatorial objects considered are simplicial complexes to which one assigns algebraic objects, the Stanley–Reisner rings. We study how the face numbers of a simplicial complex are related to the Hilbert series of the corresponding Stanley–Reisner ring. This is the basis of all further investigations which culminate in Stanley's proof of the upper bound theorem for simplicial spheres. It turns out that most of the important algebraic notions introduced in the earlier chapters, such as 'Cohen–Macaulay', 'Gorenstein', 'local cohomology', and 'Hilbert series', are the proper concepts in solving purely combinatorial problems. Other applications of commutative algebra to combinatorics will be given in the next chapter.

5.1 Simplicial complexes

The present section is devoted to introducing the Stanley–Reisner ring associated with a simplicial complex, and studying its Hilbert series. The most important invariant of a simplicial complex, its f-vector, can be easily transformed into the h-vector, an invariant encoded by the Hilbert function of the associated Stanley–Reisner ring. It is of interest to know when a Stanley–Reisner ring is Cohen–Macaulay, because then the results about Hilbert functions of Chapter 4 may be employed to get information about the f-vector. In concluding this section we show that the Stanley–Reisner ring of a shellable simplicial complex is Cohen–Macaulay, and study systems of parameters of such a ring.

Definition 5.1.1. Let $V = \{v_1, \ldots, v_n\}$ be a finite set. A (*finite*) *simplicial complex* Δ on V is a collection of subsets of V such that $F \in \Delta$ whenever $F \subset G$ for some $G \in \Delta$, and such that $\{v_i\} \in \Delta$ for $i = 1, \ldots, n$.

The elements of Δ are called *faces*, and the *dimension*, $\dim F$, of a face F is the number $|F| - 1$. The *dimension* of the simplicial complex Δ is $\dim \Delta = \max\{\dim F : F \in \Delta\}$.

Note that the empty set $\emptyset$ is a face (of dimension -1) of any non-empty simplicial complex. Faces of dimension 0 and 1 are called *vertices* and *edges*, respectively. The maximal faces under inclusion are called the *facets* of the simplicial complex.

Given an arbitrary collection $\{F_1, \ldots, F_m\}$ of subsets of V there is a (unique) smallest simplicial complex, denoted by $\langle F_1, \ldots, F_m \rangle$, which contains all F_i. This simplicial complex is said to be generated by $F_1, \ldots, F_m$. It consists of all subsets $G \subset V$ which are contained in some F_i. A simplicial complex generated by one face is called a *simplex*.

Each simplicial complex has a geometric realization as a certain subset (composed of simplices) of a finite dimensional affine space. This explains the geometric terminology introduced above. Geometric realizations will be discussed in the next section. As an example consider the *octahedron* with vertex set $\{v_1, \ldots, v_6\}$ (Figure 5.1). Its facets are the sets $\{v_1, v_3, v_4\}$,

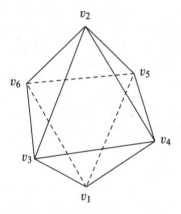

Figure 5.1

$\{v_1, v_4, v_5\}$, $\{v_1, v_5, v_6\}$, $\{v_1, v_3, v_6\}$, $\{v_2, v_3, v_4\}$, $\{v_2, v_4, v_5\}$, $\{v_2, v_5, v_6\}$, and $\{v_2, v_3, v_6\}$.

An important class of simplicial complexes arises from finite sets with partial order $\leq$, called *posets* for short. The *order complex* $\Delta(\Pi)$ of a poset Π is the set of chains of Π. Recall that a subset C of Π is a *chain* if any two elements of C are comparable. Obviously, $\Delta(\Pi)$ is a simplicial complex.

For example, if we order the elements of the set $\{v_1, \ldots, v_5\}$ according to Figure 5.2, then the order complex of the corresponding poset has the facets $\{v_1, v_2, v_3, v_5\}$ and $\{v_1, v_2, v_4, v_5\}$.

Stanley–Reisner rings and f-vectors. Now let Δ be an arbitrary simplicial complex of dimension $d - 1 \geq 0$ on a vertex set V. We denote by f_i the number of i-dimensional faces of Δ. We have $f_0 = |V|$, and $f_{-1} = 1$ since $\emptyset \in \Delta$. The d-tuple

$$f(\Delta) = (f_0, f_1, \ldots, f_{d-1})$$

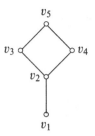

Figure 5.2

is called the f-*vector* of $\varDelta$. For example, the octahedron has the f-vector $(6, 12, 8)$, while the above order complex has the f-vector $(5, 9, 7, 2)$.

The possible f-vectors of simplicial complexes have been determined by Kruskal [243] and Katona [232]. Given two integers $a, d > 0$, let

$$a = \binom{k(d)}{d} + \binom{k(d-1)}{d-1} + \cdots + \binom{k(j)}{j},$$

$k(d) > k(d-1) > \cdots > k(j) \geq j \geq 1$, be the unique d-th Macaulay representation of a; see 4.2.6 and the definition following 4.2.7. We set

$$a^{(d)} = \binom{k(d)}{d+1} + \binom{k(d-1)}{d} + \cdots + \binom{k(j)}{j+1}.$$

Then $(f_0, f_1, \ldots, f_{d-1}) \in \mathbb{Z}^d$ is the f-vector of some $(d-1)$-dimensional simplicial complex if and only if

$$0 < f_{i+1} \leq f_i^{(i+1)}, \qquad 0 \leq i \leq d-2.$$

However, if we consider more restricted classes of simplicial complexes, for instance those simplicial complexes whose geometric realization is a sphere, new constraints appear; this will be the topic of the next sections. It turns out that the Stanley–Reisner rings are the appropriate tool to attack these problems.

Definition 5.1.2. Let $\varDelta$ be a simplicial complex on the vertex set $V = \{v_1, \ldots, v_n\}$, and k a ring. The *Stanley–Reisner ring* (or *face ring*) of the complex $\varDelta$ (with respect to k) is the homogeneous k-algebra

$$k[\varDelta] = k[X_1, \ldots, X_n]/I_\varDelta,$$

where $I_\varDelta$ is the ideal generated by all monomials $X_{i_1} X_{i_2} \cdots X_{i_s}$ such that $\{v_{i_1}, v_{i_2}, \ldots, v_{i_s}\} \notin \varDelta$.

The choice of the letter k in the definition indicates that, with a few exceptions, we usually have in mind a field for the coefficient ring of a Stanley–Reisner ring.

Note that I_Δ is generated by squarefree monomials. On the other hand, if $I \subset (X_1,\ldots,X_n)^2$ is any ideal which is generated by squarefree monomials, then $k[X_1,\ldots,X_n]/I \cong k[\Delta]$ for some simplicial complex Δ.

The correspondence between simplicial complexes and squarefree ideals is inclusion reversing: if Δ and Δ' are simplicial complexes on the same vertex set, then $\Delta \subset \Delta' \iff I_{\Delta'} \subset I_\Delta$.

Throughout this chapter we will assume, unless otherwise stated, that $V = \{v_1,\ldots,v_n\}$ is the vertex set of the simplicial complex Δ.

Example 5.1.3. Let $P = \{v_1,\ldots,v_n\}$ be a poset, and Δ the order complex of P. Then I_Δ is generated by all monomials X_iX_j for which v_i and v_j are incomparable. In the above example, $I_\Delta = (X_3X_4)$.

The dimension of a Stanley–Reisner ring can be easily determined.

Theorem 5.1.4. *Let Δ be a simplicial complex, and k a field. Then*

$$I_\Delta = \bigcap_F \mathfrak{P}_F,$$

where the intersection is taken over all facets F of Δ, and $\mathfrak{P}_F$ denotes the (prime) ideal generated by all X_i such that $v_i \notin F$. In particular,

$$\dim k[\Delta] = \dim \Delta + 1.$$

PROOF. By Exercise 4.4.17, $k[\Delta]$ is reduced, and hence I_Δ is the intersection of its minimal prime ideals; by Exercise 4.4.15, all these ideals are generated by subsets of $\{X_1,\ldots,X_n\}$. Let $\mathfrak{P} = (X_{i_1},\ldots,X_{i_s})$; notice that $I_\Delta \subset \mathfrak{P}$ if and only if $\{v_1,\ldots,v_n\} \setminus \{v_{i_1},\ldots,v_{i_s}\}$ is a face of Δ, and that $\mathfrak{P}$ is a minimal prime ideal of I_Δ if and only if $\{v_1,\ldots,v_n\} \setminus \{v_{i_1},\ldots,v_{i_s}\}$ is a facet. □

A simplicial complex Δ is *pure* if all its facets are of the same dimension, namely $\dim \Delta$, and Δ is called a *Cohen–Macaulay complex over k* if $k[\Delta]$ is a Cohen–Macaulay ring. We say that Δ is a *Cohen–Macaulay complex* if Δ is Cohen–Macaulay over some field. According to Exercise 5.1.25, Δ is a Cohen–Macaulay complex over every Cohen–Macaulay ring k if and only if $\mathbb{Z}[\Delta]$ is Cohen–Macaulay.

As a consequence of 2.1.2 and the previous theorem we obtain

Corollary 5.1.5. *A Cohen–Macaulay complex is pure.*

We are going to relate the f-vector of a simplicial complex to the Hilbert series of $k[\Delta]$. To this end we introduce a $\mathbb{Z}^n$-*grading* or *fine grading* on $k[\Delta]$.

More generally, let $(G,+)$ be an Abelian group. A *G-graded ring* is a ring R together with a decomposition $R = \bigoplus_{a\in G} R_a$ (as a $\mathbb{Z}$-module) such that $R_aR_b \subset R_{a+b}$ for all $a,b \in G$.

Similarly one defines a G-graded R-module, the category of G-graded R-modules, G-graded ideals etc. simply by mimicking the corresponding definitions for graded rings and modules with $G = \mathbb{Z}$; see Section 1.5. If M is a G-graded R-module, then $x \in M$ is *homogeneous* (*of degree* $a \in G$) if $x \in M_a$, and we set $\deg x = a$.

Example 5.1.6. The polynomial ring $R = k[X_1, \ldots, X_n]$ has a natural $\mathbb{Z}^n$-grading: for $a = (a_1, \ldots, a_n) \in \mathbb{Z}^n$, $a_i \geq 0$ for $i = 1, \ldots, n$, we let $R_a = \{cX^a : c \in k\}$ be the a-th homogeneous component of R, and set $R_a = 0$ if $a_i < 0$ for some i. Here, $X^a = X_1^{a_1} \cdots X_n^{a_n}$ for $a = (a_1, \ldots, a_n)$. Note that the $\mathbb{Z}^n$-graded ideals in R are just the ideals generated by monomials, and the $\mathbb{Z}^n$-graded prime ideals are just the finitely many ideals which are generated by subsets of $\{X_1, \ldots, X_n\}$.

Let $I \subset R$ be an ideal generated by monomials. Since I is $\mathbb{Z}^n$-graded, the factor ring R/I inherits the natural $\mathbb{Z}^n$-grading given by $(R/I)_a = R_a/I_a$ for all $a \in \mathbb{Z}^n$. In particular, Stanley–Reisner rings are $\mathbb{Z}^n$-graded in this way.

Now let R be an arbitrary $\mathbb{Z}^n$-graded ring, and M a $\mathbb{Z}^n$-graded R-module. Each homogeneous component M_a of M is an R_0-module. Just as for $\mathbb{Z}$-graded modules we define the *Hilbert function* $H(M, \_): \mathbb{Z}^n \to \mathbb{Z}$ by $H(M, a) = \ell(M_a)$, provided all homogeneous components of M have finite length, and call $H_M(t) = \sum_{a \in \mathbb{Z}^n} H(M, a) t^a$ the *Hilbert series* of M. Here $t = (t_1, \ldots, t_n)$ where the t_i are indeterminates, and $t^a = t_1^{a_1} \cdots t_n^{a_n}$ for $a = (a_1, \ldots, a_n)$.

For example, the $\mathbb{Z}^n$-graded polynomial ring $R = k[X_1, \ldots, X_n]$ has the Hilbert series

$$H_R(t) = \sum_{a \in \mathbb{N}^n} t^a = \prod_{i=1}^n (1 - t_i)^{-1}.$$

Let us return to Stanley–Reisner rings. Given a simplicial complex Δ, we denote by x_i the residue classes of the indeterminates X_i in $k[\Delta]$; then $k[\Delta] = k[x_1, \ldots, x_n]$.

We define the *support* of an element $a \in \mathbb{Z}^n$ to be the set $\operatorname{supp} a = \{v_i : a_i > 0\}$. If x^a and x^b are non-zero monomials (with non-negative exponents) in $x_1, \ldots, x_n$, then $x^a = x^b$ if and only if $a = b$. Therefore, without ambiguity, we may set $\operatorname{supp} x^a = \operatorname{supp} a$ for any non-zero monomial.

Note that $x^a \neq 0$ if and only if $\operatorname{supp} a \in \Delta$, and that the non-zero monomials x^a form a k-basis of $k[\Delta]$. Therefore,

$$H_{k[\Delta]}(t) = \sum_{\substack{a \in \mathbb{N}^n \\ \operatorname{supp} a \in \Delta}} t^a = \sum_{F \in \Delta} \sum_{\substack{a \in \mathbb{N}^n \\ \operatorname{supp} a = F}} t^a.$$

If $F = \emptyset$, then $\sum_{\text{supp } a = F} t^a = 1$, and if $F \neq \emptyset$, then $\sum_{\text{supp } a = F} t^a = \prod_{v_i \in F} t_i/(1 - t_i)$. Thus, if we understand that the product over an empty index set is 1, we get

$$
(1) \qquad\qquad H_{k[\Delta]}(t) = \sum_{F \in \Delta} \prod_{v_i \in F} \frac{t_i}{1 - t_i}.
$$

We are actually interested in the Hilbert series of $k[\Delta]$ as a homogeneous $\mathbb{Z}$-graded algebra. Note that for all $i \in \mathbb{Z}$ we have

$$
k[\Delta]_i = \bigoplus_{a \in \mathbb{Z}^n, \ |a| = i} k[\Delta]_a,
$$

where $|a| = a_1 + \cdots + a_n$ for $a = (a_1, \ldots, a_n)$. (This relation explains the alternative terminology 'fine grading' for '$\mathbb{Z}^n$-grading'.)

It follows that the Hilbert series of $k[\Delta]$ with respect to the $\mathbb{Z}$-grading is obtained from (1) by replacing all t_i by t. Thus we have shown

Theorem 5.1.7. *Let Δ be a simplicial complex with f-vector $(f_0, \ldots, f_{d-1})$. Then*

$$
H_{k[\Delta]}(t) = \sum_{i=-1}^{d-1} \frac{f_i t^{i+1}}{(1 - t)^{i+1}}.
$$

From the Hilbert series of $k[\Delta]$ we can read off its Hilbert function:

$$
H(k[\Delta], n) = \begin{cases} 1 & \text{if } n = 0, \\ \sum_{i=0}^{d-1} f_i \binom{n-1}{i} & \text{if } n > 0. \end{cases}
$$

We note the following interesting fact: $H(k[\Delta], n)$ is a polynomial function for $n > 0$, and hence coincides with the Hilbert polynomial for all $n \geq 0$ except possibly for $n = 0$. Evaluating $\sum_{i=0}^{d-1} f_i \binom{n-1}{i}$ at $n = 0$ gives

$$
\chi(\Delta) = \sum_{i=0}^{d-1} (-1)^i f_i,
$$

the so-called *Euler characteristic* of Δ. Thus the Hilbert function and the Hilbert polynomial of Δ agree for all $n \geq 0$ if and only if $\chi(\Delta) = 1$. The geometric significance of the Euler characteristic will become clear in 5.2.17.

Two other conclusions can be drawn from 5.1.7. First, we recover that $\dim k[\Delta] = d$ since the degree of the Hilbert polynomial $P_{k[\Delta]}(t)$ is $d - 1$; secondly, we see that the multiplicity of $k[\Delta]$ equals f_{d-1}, the number of $(d - 1)$-dimensional facets of Δ.

The h-vector. Recall from 4.1.8 that a homogeneous k-algebra R of dimension d has a Hilbert series of the form $H_R(t) = Q_R(t)/(1-t)^d$ where

$Q_R(t)$ is a polynomial with integer coefficients. Let Δ be a simplicial complex, and write

$$H_{k[\Delta]}(t) = \frac{h_0 + h_1 t + \cdots}{(1-t)^d}.$$

The finite sequence of integers $h(\Delta) = (h_0, h_1, \ldots)$ is called the *h-vector* of Δ.

A comparison with 5.1.7 yields

Lemma 5.1.8. *The f-vector and h-vector of a $(d-1)$-dimensional simplicial complex Δ are related by*

$$\sum_i h_i t^i = \sum_{i=0}^{d} f_{i-1} t^i (1-t)^{d-i}.$$

In particular, the h-vector has length at most d, and for $j = 0, \ldots, d$,

$$h_j = \sum_{i=0}^{j} (-1)^{j-i} \binom{d-i}{j-i} f_{i-1} \quad and \quad f_{j-1} = \sum_{i=0}^{j} \binom{d-i}{j-i} h_i.$$

PROOF. Comparing the coefficients in the polynomial identity gives the formula for the h_j in terms of the f_i. In order to prove the inverse relation replace t by $s/(1+s)$. Then the above polynomial identity transforms into

$$\sum_{i=0}^{d} h_i s^i (1+s)^{d-i} = \sum_{i=0}^{d} f_{i-1} s^i$$

from which one obtains the last set of equations. □

The octahedron has f-vector $(6, 12, 8)$. Applying 5.1.8 we see that its h-vector is $(1, 3, 3, 1)$.

We single out some special cases of the above equations:

Corollary 5.1.9. *With the assumptions of 5.1.8 one has*

$$h_0 = 1, \quad h_1 = f_0 - d, \quad h_d = (-1)^{d-1}(\chi(\Delta) - 1) \quad and \quad \sum_{i=0}^{d} h_i = f_{d-1}.$$

Since the f-vector and h-vector of a simplicial complex determine each other, bounds for the h-vector implicitly contain certain constraints for the f-vector. We treat an important case:

Theorem 5.1.10. *Let Δ be a $(d-1)$-dimensional Cohen–Macaulay complex with n vertices and h-vector $(h_0, \ldots, h_d)$. Then*

$$0 \leq h_i \leq \binom{n-d+i-1}{i}, \quad 0 \leq i \leq d.$$

PROOF. Let $R = k[\Delta]$, where k is a field for which $k[\Delta]$ is Cohen–Macaulay. We may assume that k is infinite. Then, since R is Cohen–Macaulay, there exists an R-sequence x of elements of degree 1 such that $\bar{R} = \dim R/xR$ is of dimension 0; see 1.5.12. Now it follows from 4.1.11 that $h_i = H(\bar{R}, i)$ for all i. This implies already that $h_i \geq 0$ for all i.

Notice that $\bar{R}$ is generated over k by $n - d$ elements of degree 1. Therefore, the Hilbert function of $\bar{R}$ is bounded by the Hilbert function of a polynomial ring in just as many variables. This yields the second inequality.   □

To illustrate the theorem consider the simplicial complex Δ in Figure 5.3 with facets $F_1 = \{v_1, v_2, v_3\}$ and $F_2 = \{v_1, v_4, v_5\}$. We have $f(\Delta) =$

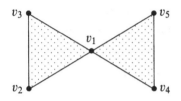

Figure 5.3

$(5, 6, 2)$, and so $h(\Delta) = (1, 2, -1)$. It follows that Δ is not a Cohen–Macaulay complex.

Shellable simplicial complexes. The previous theorem will be of real use only when we are able to exhibit interesting classes of Cohen–Macaulay complexes. Such a class is given in the following definition.

Definition 5.1.11. A pure simplicial complex Δ is called *shellable* if one of the following equivalent conditions is satisfied: the facets of Δ can be given a linear order $F_1, \ldots, F_m$ in such a way that
(a) $\langle F_i \rangle \cap \langle F_1, \ldots, F_{i-1} \rangle$ is generated by a non-empty set of maximal proper faces of $\langle F_i \rangle$ for all i, $2 \leq i \leq m$, or
(b) the set $\{F : F \in \langle F_1, \ldots, F_i \rangle, F \notin \langle F_1, \ldots, F_{i-1} \rangle\}$ has a unique minimal element for all i, $2 \leq i \leq m$, or
(c) for all i, j, $1 \leq j < i \leq m$, there exist some $v \in F_i \setminus F_j$ and some $k \in \{1, 2, \ldots, i - 1\}$ with $F_i \setminus F_k = \{v\}$.
A linear order of the facets satisfying the equivalent conditions (a), (b), and (c) is called a *shelling* of Δ.

Let us check that these conditions are indeed equivalent:
(a) $\Rightarrow$ (b): We may assume that $F_i = \{v_1, \ldots, v_m\}$, and that $\langle F_i \rangle \cap \langle F_1, \ldots, F_{i-1} \rangle$ is generated by the faces $\{v_1, \ldots, v_{j-1}, v_{j+1}, \ldots, v_m\}$, $1 \leq j \leq r \leq m$. The unique minimal element in the set $S_i = \{F : F \in \langle F_1, \ldots, F_i \rangle, F \notin \langle F_1, \ldots, F_{i-1} \rangle\}$ is $\{v_1, \ldots, v_r\}$.

(b) $\Rightarrow$ (c): Let G be the unique minimal element in S_i. Since $G \not\subseteq F_j$, there exists $v \in G \setminus F_j$. Then $v \in F_i \setminus F_j$, and it follows from the definition of G that there exists a k, $1 \le k \le i - 1$, such that $F_i \setminus F_k = \{v\}$.

(c) $\Rightarrow$ (a): Let $F \in \langle F_i \rangle \cap \langle F_1, \ldots, F_{i-1} \rangle$. Then $F \subset F_j$ for some $j < i$. Let $v \in F_i \setminus F_j$ as in (c). Then $F_i \setminus \{v\}$ is a maximal proper face of $\langle F_i \rangle$ belonging to $\langle F_i \rangle \cap \langle F_1, \ldots, F_{i-1} \rangle$ and containing F. This proves (a).

In Figure 5.4 the first simplicial complex is shellable, the second is not. Shellable simplicial complexes arise naturally in geometry; see Section

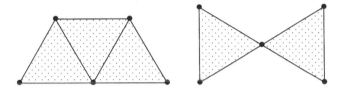

Figure 5.4

5.2. Other interesting classes arise from order complexes of certain posets. Here we discuss one important case. For this we need to introduce some more terminology: a (finite) poset is said to be *bounded* if it has a least and a greatest element, denoted $\hat{0}$ and $\hat{1}$. The poset is *pure* if all maximal chains have the same length, and *graded* if it is bounded and pure. In this case all unrefinable chains between two comparable elements have the same length (Exercise 5.1.18).

Let Π be a poset, and $v \in \Pi$. The *rank* of v, rank v, is defined to be the maximal length of all chains descending from v. The *length* of Π is the maximal rank of an element of Π. Let $u, v \in \Pi$; we say that v *covers* u, written $u \prec v$, if $u < v$, and if there is no $w \in \Pi$ such that $u < w < v$.

The poset Π is *locally upper semimodular* if whenever v_1 and v_2 cover u, and $v_1, v_2 < v$ for some $v \in \Pi$, then there is $t \in \Pi$, $t \le v$, which covers each of v_1 and v_2; see Figure 5.5.

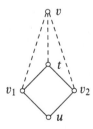

Figure 5.5

Theorem 5.1.12 (Björner). *The order complex of a bounded, locally upper semimodular poset is shellable.*

PROOF. In a first step we shall prove that for every graded poset $(\Pi, \leq)$ the set $\mathfrak{M}$ of maximal chains can be given a linear order Ω such that for any two chains $m, m' \in \mathfrak{M}$, $m : \hat{0} = x_0 \prec x_1 \prec \cdots \prec x_n = \hat{1}$ and $m' : \hat{0} = y_0 \prec y_1 \prec \cdots \prec y_n = \hat{1}$ with $x_i = y_i$ for $i = 0, 1, \ldots, e$, and $x_{e+1} \neq y_{e+1}$ the following conditions are satisfied:

(i) If $\{y_0, y_1, \ldots, y_{e+1}\}$ is contained in a maximal chain m'' and if $m' <^{\Omega} m$, then $m'' <^{\Omega} m$;

(ii) If $m' \setminus \{x_e\}$ is contained in some maximal chain m'' with $m'' <^{\Omega} m$ but $m \setminus \{x_e\}$ is contained in no maximal chain m''' with $m''' <^{\Omega} m$, then $m' <^{\Omega} m$.

In a second step we shall see that such a linear order is a shelling of the order complex $\Delta(\Pi)$, provided Π satisfies the hypothesis of the theorem.

First we prove by induction on the length n of Π the existence of a linear order on $\mathfrak{M}$ satisfying (i) and (ii). The assertion being trivial for $n = 2$, we may assume that $n \geq 3$. We denote by Π' the subposet of Π consisting of the elements $x \in \Pi$ with rank $x \neq n - 1$. (The order $<'$ of Π' is induced by the order $<$ of Π, that is, for all $x, y \in \Pi'$, $x <' y$ if and only if $x < y$.) Since Π' is graded of length $n - 1$, the induction hypothesis implies that there exists a linear order Ω' of the set of maximal chains $\mathfrak{M}'$ of Π' satisfying (i) and (ii). Let $m'_1, m'_2, \ldots, m'_s$ be the elements of $\mathfrak{M}'$ in their linear order. For $m'_i : \hat{0} = x_0 \prec x_1 \prec \cdots \prec x_{n-2} < \hat{1}$ we define the set $A_i = \{z \in \Pi : x_{n-2} \prec z \prec \hat{1}\}$, and the set B_i of all $z \in A_i$ for which there is an element $y \in \Pi$ such that $x_{n-3} \prec y \prec z$ and $(m'_i \setminus \{x_{n-2}\}) \cup \{y\} <^{\Omega} m'_i$. Finally we let $C_i = A_i \setminus B_i$. Now we order the elements of A_i linearly in such a way that all elements of B_i are less than all elements of C_i. We label the elements of A_i, $z_{i1}, z_{i2}, \ldots, z_{ia_i}$, $a_i = |A_i|$, according to their order, and set $m_{ij} = m'_i \cup \{z_{ij}\}$ for $i = 1, 2, \ldots, s$ and $j = 1, 2, \ldots, a_i$. The lexicographic order of the indices determines a linear order Ω of the set $\mathfrak{M} = \{m_{ij} : 1 \leq i \leq s, \ 1 \leq j \leq a_i\}$ of maximal chains of Π.

We claim that Ω satisfies (i) and (ii). Indeed, let $m, m' \in \mathfrak{M}$, $m : \hat{0} = x_0 \prec x_1 \prec \cdots \prec x_n = \hat{1}$ and $m' : \hat{0} = y_0 \prec y_1 \prec \cdots \prec y_n = \hat{1}$ with $x_i = y_i$ for $i = 0, 1, \ldots, e$, and $x_{e+1} \neq y_{e+1}$. We distinguish two cases.

In the first case suppose $e + 1 = n - 1$; then $m' = m_{ij}$ and $m = m_{ik}$ for some i, j, k with $1 \leq i \leq s$ and $1 \leq j, k \leq a_i$. Condition (i) is trivially satisfied since necessarily $m'' = m'$. The hypothesis of condition (ii) implies $y_{n-1} \in B_i$ and $x_{n-1} \in C_i$. Therefore, $j < k$, and hence $m' <^{\Omega} m$.

In the second case we assume that $e + 1 < n - 1$. If $\{y_0, y_1, \ldots, y_{e+1}\} \subset m'' \in \mathfrak{M}$, then $\{y_0, y_1, \ldots, y_{e+1}\} \subset (m'' \setminus \{z\}) \in \mathfrak{M}'$ where $z \in m''$, rank $z = n - 1$. Suppose now that $m' <^{\Omega} m$; then $(m' \setminus \{y_{n-1}\}) <^{\Omega'} (m \setminus \{x_{n-1}\})$, and thus, by the induction hypothesis, $(m'' \setminus \{z\}) <^{\Omega'} (m \setminus \{x_{n-1}\})$. But then $m'' <^{\Omega} m$, and this proves condition (i). In a similar manner one checks condition (ii).

Now suppose that Π is a bounded, locally upper semimodular poset. By 5.1.18, Π is pure and hence graded. Therefore, as we have just seen, the set $\mathfrak{M}$ of maximal chains of Π admits a linear order Ω satisfying the conditions (i) and (ii). In order to prove that this is a shelling of $\Delta(\Pi)$, we consider $m, m' \in \mathfrak{M}$, $m: \hat{0} = x_0 \prec x_1 \prec \cdots \prec x_n = \hat{1}$ and $m': \hat{0} = y_0 \prec y_1 \prec \cdots \prec y_n = \hat{1}$ with $m' <^\Omega m$. Let d be the greatest integer such that $x_i = y_i$ for $i \leq d$, and let g be the least integer for which $y_{d+1} \prec x_g$.

Since Π is locally upper semimodular there exists an element z_{d+2} which covers both y_{d+1} and x_{d+1} and such that $z_{d+2} \prec x_g$. If $g > d + 2$, we find again an element z_{d+3} which covers z_{d+2} and x_{d+2} and such that $z_{d+2} \prec x_g$. This process ends with $z_g = x_g$. Setting $z_{d+1} = y_{d+1}$ we obtain Figure 5.6. By the choice of g we have $y_e \neq x_e$ and $z_e \neq x_e$ for all e,

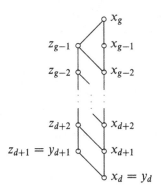

Figure 5.6

$d + 1 \leq e \leq g - 1$. It remains to show that for some e in this range there is a maximal chain m'' with $m'' <^\Omega m$ such that $m \setminus \{x_e\} \subset m''$.

For $i = d + 1, d + 2, \ldots, g - 1$ we let m_i be the maximal chain $\hat{0} = x_0 \prec x_1 \prec \cdots \prec x_{i-1} \prec z_i \prec z_{i+1} \prec \cdots \prec z_{g-1} \prec x_g \prec x_{g+1} \prec \cdots \prec x_n = \hat{1}$. As we assume that $m' <^\Omega m$, property (i) of Ω implies that $m_{d+1} <^\Omega m$. Now, either $m \setminus \{x_{d+1}\} \subset m''$ for some maximal chain $m'' <^\Omega m$ as we want, or, otherwise, property (ii) of Ω implies that $m_{d+2} <^\Omega m$. Again, if $m \setminus \{x_{d+2}\} \subset m''$ for some maximal chain $m'' <^\Omega m$ the proof is completed; otherwise $m_{d+3} <^\Omega m$. Continuing this argument we conclude that either $m \setminus \{x_e\}$ is contained in some earlier maximal chain for some e, or $m_{g-1} <^\Omega m$. In the latter case however, $m \setminus \{x_{g-1}\} \subset m_{g-1}$ with $m_{g-1} <^\Omega m$. $\square$

After this purely combinatorial result we return to algebra.

Theorem 5.1.13. *A shellable simplicial complex is Cohen–Macaulay over every field.*

PROOF. The proof is based on the following simple observation: let I_1 and I_2 be two ideals of a ring R. Then there exists an exact sequence of R-modules

$$(2) \qquad 0 \longrightarrow R/(I_1 \cap I_2) \overset{\alpha}{\longrightarrow} R/I_1 \oplus R/I_2 \overset{\beta}{\longrightarrow} R/(I_1 + I_2) \longrightarrow 0$$

with $\alpha(a+I_1 \cap I_2) = (a+I_1, -a+I_2)$, and $\beta(a+I_1, b+I_2) = (a+b)+I_1+I_2$. Suppose moreover that R is a polynomial ring over a field, that I_1 and I_2 are graded ideals such that R/I_1 and R/I_2 are d-dimensional Cohen–Macaulay rings, and that $R/(I_1 + I_2)$ is a $(d-1)$-dimensional Cohen–Macaulay ring; then $R/(I_1 \cap I_2)$ is a d-dimensional Cohen–Macaulay ring. The proof of these statements is left to the reader.

Let Δ be a shellable complex of dimension $d-1$ (on the vertex set $\{v_1, \ldots, v_n\}$), and $F_1, \ldots, F_m$ a shelling of Δ. By 5.1.4, we have $I_\Delta = \bigcap_{i=1}^{m} \mathfrak{P}_{F_i}$ where $\mathfrak{P}_{F_i}$ is the ideal generated by all X_j such that $v_j \notin F_i$. Set $\Delta_j = \langle F_1, \ldots, F_j \rangle$, $1 \le j \le m$; then $k[\Delta_j] \cong k[X_1, \ldots, X_n]/\bigcap_{i=1}^{j} \mathfrak{P}_{F_i}$. In fact, we may suppose that $\{v_1, \ldots, v_r\}$, $r \le n$, is the vertex set of $\langle F_1, \ldots, F_j \rangle$. For all $i \le j$, let $\mathfrak{p}_{F_i}$ be the ideal in $k[X_1, \ldots, X_r]$ generated by all X_s, $s \le r$, such that $X_s \in \mathfrak{P}_{F_i}$. Then $\mathfrak{P}_{F_i} = \mathfrak{p}_{F_i} + (X_{r+1}, \ldots, X_n)$ for all $i \le j$, and it follows from 4.4.16 that

$$\bigcap_{i=1}^{j} \mathfrak{P}_{F_i} = \left(\bigcap_{i=1}^{j} \mathfrak{p}_{F_i} \right) + (X_{r+1}, \ldots, X_n).$$

Therefore, $k[\Delta_j] \cong k[X_1, \ldots, X_r]/\bigcap_{i=1}^{j} \mathfrak{p}_{F_i} \cong k[X_1, \ldots, X_n]/\bigcap_{i=1}^{j} \mathfrak{P}_{F_i}$.

We show by induction on j that Δ_j is Cohen–Macaulay. If $j = 1$, then $k[\Delta_1]$ is a polynomial ring, and there is nothing to prove. Now suppose that $j > 1$. The sequence (2) with $I_1 = \bigcap_{i=1}^{j-1} \mathfrak{P}_{F_i}$ and $I_2 = \mathfrak{P}_{F_j}$ yields the exact sequence

$$(3) \qquad 0 \longrightarrow k[\Delta_j] \longrightarrow k[\Delta_{j-1}] \oplus k[F_j] \longrightarrow k[\langle F_j \rangle \cap \Delta_{j-1}] \longrightarrow 0.$$

It follows easily from Definition 5.1.11(a) and 5.1.19 that $k[\langle F_j \rangle \cap \Delta_{j-1}]$ is isomorphic to a residue class ring of a polynomial ring in d variables modulo a single equation, and therefore is a Cohen–Macaulay ring of dimension $d-1$. By our induction hypothesis $k[\Delta_{j-1}]$ is a d-dimensional Cohen–Macaulay ring. Finally, since $k[F_j]$ is a d-dimensional polynomial ring (Δ is pure), it follows from the general properties of the sequence (2) that $k[\Delta_j]$ is Cohen–Macaulay. $\qquad \square$

From the exact sequence (3) we easily derive a combinatorial interpretation of the h-vector of a shellable simplicial complex, due to McMullen and Walkup [273].

Corollary 5.1.14. *Let Δ be a $(d-1)$-dimensional shellable simplicial complex with shelling $F_1, \ldots, F_m$. For $j = 2, \ldots, m$, let r_j be the number of facets of*

$\langle F_j \rangle \cap \langle F_1, \ldots, F_{j-1} \rangle$, *and set* $r_1 = 0$. *Then*

$$h_i = |\{j : r_j = i\}| \qquad \text{for} \quad i = 0, \ldots, d.$$

In particular, up to their order, the numbers r_j *do not depend on the particular shelling.*

PROOF. Set $\Delta_j = \langle F_1, \ldots, F_j \rangle$, and write $H_{k[\Delta_j]}(t) = Q_j(t)/(1-t)^d$. Then it follows from the sequence (3) that

$$\frac{Q_j(t)}{(1-t)^d} = \frac{Q_{j-1}(t)}{(1-t)^d} + \frac{1}{(1-t)^d} - \frac{P_j(t)}{(1-t)^{d-1}},$$

where $P_j(t)/(1-t)^{d-1}$ is the Hilbert series of $\langle F_j \rangle \cap \Delta_{j-1}$. According to Exercise 5.1.19 one has $P_j(t) = 1 + t + \cdots + t^{r_j-1}$; therefore, $Q_j(t) = Q_{j-1}(t) + t^{r_j}$. As $Q_1(t) = 1$, it follows that $Q_m(t) = \sum_{j=1}^{m} t^{r_j}$. This implies the assertion since the coefficient vector of $Q_m(t)$ is just the h-vector of Δ. □

Analyzing the proof of 5.1.13 we see that we did not use all the properties of shellability. It suffices to require that Δ is *constructible* which means that Δ can be obtained by the following recursive procedure: (i) any simplex is constructible, (ii) if Δ' and Δ'' are constructible of the same dimension d, and if $\Delta' \cap \Delta''$ is constructible of dimension $d-1$, then $\Delta' \cup \Delta''$ is constructible.

It is clear that the following implications hold for simplicial complexes:

shellable $\Rightarrow$ constructible $\Rightarrow$ Cohen–Macaulay.

Nevertheless the h-vectors of all these types of complexes are characterized by the same condition. For the next theorem recall the definition of $a^{\langle n \rangle}$ given before 4.2.8.

Theorem 5.1.15 (Stanley). *Let* $s = (h_0, \ldots, h_d)$ *be a sequence of integers. The following conditions are equivalent:*
(a) $h_0 = 1$ *and* $0 \le h_{i+1} \le h_i^{\langle i \rangle}$ *for all* i, $1 \le i \le d-1$;
(b) s *is the* h-*vector of a shellable complex;*
(c) s *is the* h-*vector of a constructible complex;*
(d) s *is the* h-*vector of a Cohen–Macaulay complex.*

PROOF. The implications (b) $\Rightarrow$ (c) $\Rightarrow$ (d) are obvious, while (d) $\Longleftrightarrow$ (a) follows from 4.2.15. Following [39] we indicate the proof of the implication (a) $\Rightarrow$ (b), which is a purely combinatorial result. A full proof can be found in [46]. Given a vector s satisfying (a), let $n = h_1 + d$ and $V = \{1, 2, \ldots, n\}$. Let $\mathscr{F}$ be the collection of all subsets of V with d elements, and $\mathscr{F}_i$ the set of those members F of $\mathscr{F}$ such that $d + 1 - i$ is the smallest element of V not in F. The elements of $\mathscr{F}$ are ordered in such a way that $F < G$ if the largest element in their symmetric difference lies in G. For each i, $0 \le i \le d$, choose the first h_i members of $\mathscr{F}_i$. The resulting collection $\mathscr{C}$ consists of the facets of the required shellable complex, and the given order on $\mathscr{F}$ induces the shelling order. $\quad\square$

Systems of parameters. Let Δ be a simplicial complex. Given two faces $G \subset F$, the set of faces $[G, F] = \{H : G \subset H \subset F\}$ is called the *interval between G and F.* Now assume Δ is shellable with shelling $F_1, \ldots, F_m$. By definition, there is a unique minimal element $G_i \in \langle F_i \rangle \setminus \langle F_1, \ldots, F_{i-1} \rangle$, and it is clear that Δ is the disjoint union of the intervals $[G_i, F_i]$, $i = 1, \ldots, m$. In the following we use that $k[F]$ is a residue class ring of $k[\Delta]$ in a natural way.

Theorem 5.1.16. *Let k be a field, Δ a $(d-1)$-dimensional simplicial complex, and $y = y_1, \ldots, y_d$ a sequence of elements of degree 1 in $k[\Delta]$.*
(a) *The following conditions are equivalent:*
   (i) *y is a homogeneous system of parameters of $k[\Delta]$;*
   (ii) *for all facets F of Δ, the $k[\Delta]$-module $k[F]/(y)k[F]$ is isomorphic to k.*
(b) *Suppose the equivalent conditions in (a) hold. Then the images of the monomials $x^F = \prod_{v_i \in F} x_i$ in $S = k[\Delta]/(y)$, $F \in \Delta$, form a system of generators of the (finite) k-vector space S.*
(c) *(Kind–Kleinschmidt) Assume in addition that Δ is shellable with decomposition $\Delta = \bigcup_{i=1}^{m}[G_i, F_i]$, as described above. If y is a homogeneous system of parameters of $k[\Delta]$ and $S = k[\Delta]/(y)$, then the images of the monomials x^{G_i} in S form a k-basis of S. In particular, $k[\Delta]$ is a free $k[y_1, \ldots, y_d]$-module with basis $x^{G_1}, x^{G_2}, \ldots, x^{G_m}$.*

PROOF. (a)(i) $\Rightarrow$ (ii): Since $k[F]$ is a homomorphic image of $k[\Delta]$ it follows that $k[F]/yk[F]$ has finite length. Note $k[F]$ is a polynomial ring and y a sequence of elements of degree 1. Therefore $k[F]/yk[F]$ is also a polynomial ring. If it has finite length, it must be isomorphic to k.

(ii) $\Rightarrow$ (i): Let $\varphi : k[X_1, \ldots, X_n] \to \bigoplus_F(k[X_1, \ldots, X_n]/\mathfrak{P}_F) = \bigoplus_F k[F]$ be the homomorphism which on each component is the canonical epimorphism. The direct sum is taken over all facets of Δ. Since $\operatorname{Ker} \varphi = \bigcap_F \mathfrak{P}_F = I_\Delta$, we obtain an induced homomorphism $k[\Delta] \to \bigoplus_F k[F]$ of finite $k[\Delta]$-modules which actually is a monomorphism. As we did for local rings (see 4.7.1) one shows that $k[\Delta]/(y)$ has finite length if the module $(\bigoplus_F k[F])/y(\bigoplus_F k[F])$ has finite length. But this follows from assumption (ii).

(b) Let F be a facet of Δ. Since $\mathfrak{P}_F k[\Delta]$ is the annihilator of x^F in $k[\Delta]$ it follows that $x^F S$ is a $(k[F]/yk[F])$-module. Therefore, by (a)(ii), $x_i x^F = 0$ in S for all $i = 1, \ldots, n$ which clearly implies that the elements $x^{F'}$, $F' \in \Delta$, form a system of generators of the k-vector space S.

(c) First note that S is generated as an algebra over k by the monomials x^{G_i}, $i = 0, \ldots, m$. This follows from (b) simply because any other monomial

x^F is a multiple of some x^{G_i}.

Set $\Delta_j = \langle F_1, \ldots, F_j \rangle$; then $k[\Delta_j]$ is a residue class ring of $k[\Delta]$ of the same dimension. Therefore y is a homogeneous system of parameters of $k[\Delta_j]$, too. We set $S_j = k[\Delta_j]/(y)$, and show by induction on j that $S_j = \bigoplus_{i=1}^{j} kx^{G_i}$. For $j = m$ this is the desired assertion.

Since $\Delta_1 = \langle F_1 \rangle$ it follows from (a)(ii) that $S_1 \cong k$. Since $G_1 = \emptyset$, we have $1 = x^{G_1}$ which is a basis of S_1. Now suppose $j > 1$; then we have $S_{j-1} = \bigoplus_{i=1}^{j-1} kx^{G_i}$ by the induction hypothesis. Further we know that $S_j/(x^{G_j}) = S_{j-1}$ since S_j is generated as a k-algebra by the monomials x^{G_i}, $i \leq j$. Note that $k[\Delta_j]$ is Cohen–Macaulay. Thus y is a $k[\Delta_j]$-sequence, and so Remark 4.1.11 implies that $\dim_k S_j = \sum_i h_i(\Delta_j)$. By 5.1.9, this sum equals the number of facets of Δ_j, which is j. But this is only possible if $S_j = \bigoplus_{i=1}^{j} kx^{G_i}$.

It remains to show that $k[\Delta]$ is a free $k[y_1, \ldots, y_d]$-module with basis $x^{G_1}, \ldots, x^{G_m}$: from Nakayama's lemma for graded modules (see 1.5.24) it follows that $x^{G_1}, \ldots, x^{G_m}$ is a minimal set of generators of the $k[y_1, \ldots, y_d]$-module $k[\Delta]$. Let $\mathfrak{n}$ be the graded maximal ideal of $k[y_1, \ldots, y_d]$. Then $k[\Delta]_{\mathfrak{n}}$ is a maximal Cohen–Macaulay module over $k[y_1, \ldots, y_d]_{\mathfrak{n}}$. Hence by 2.2.11, $k[\Delta]_{\mathfrak{n}}$ is free over $k[y_1, \ldots, y_d]_{\mathfrak{n}}$. But then $k[\Delta]$ is a free $k[y_1, \ldots, y_d]$-module; see 1.5.15. In particular, $x^{G_1}, \ldots, x^{G_m}$ is a basis of $k[\Delta]$ over $k[y_1, \ldots, y_d]$. □

Exercises

5.1.17. Let k be a field, and $I \subset k[X_1, \ldots, X_n]$ an ideal generated by squarefree monomials of degree 2. Does there exist a poset Π such that $k[X_1, \ldots, X_n]/I \cong k[\Delta]$ with $\Delta = \Delta(\Pi)$?

5.1.18. (a) Show that in a graded poset all unrefinable chains between two comparable elements have the same length.
(b) Show that a bounded, locally upper semimodular poset is pure.

5.1.19. Let Δ be a simplicial complex which is generated by m maximal proper faces F_i of the simplex with vertex set $\{v_1, \ldots, v_n\}$, say $F_i = \{v_1, \ldots, v_n\} \setminus \{v_i\}$. Show
(a) $k[\Delta] = k[X_1, \ldots, X_n]/(X_1 \cdots X_m)$,
(b) $h(\Delta)$ is the vector $(1, 1, \ldots, 1)$ with m components.

5.1.20. Let Γ and Δ be simplicial complexes on disjoint vertex sets V and W, respectively. The *join* $\Gamma * \Delta$ is the simplicial complex on the vertex set $V \cup W$ with faces $F \cup G$ where $F \in \Gamma$ and $G \in \Delta$. Compute $h(\Gamma * \Delta)$ in terms of $h(\Gamma)$ and $h(\Delta)$.
Hint: first show that $k[\Gamma * \Delta] \cong k[\Gamma] \otimes_k k[\Delta]$ (as graded k-algebras).

5.1.21. Let Γ and Δ be simplicial complexes. Prove that Γ and Δ are Cohen–Macaulay if and only if their join $\Gamma * \Delta$ is Cohen–Macaulay.

5.1.22. Let Δ be a $(d-1)$-dimensional simplicial complex. For r, $0 \leq r \leq d-1$, one defines the *r-skeleton* of Δ to be $\Delta_r = \{F \in \Delta : \dim F \leq r\}$. Compute $h(\Delta_{d-2})$ in terms of $h(\Delta)$.

5.1.23. Let Δ be a simplicial complex with r-skeleton Δ_r. Show:
(a) depth $k[\Delta] = \max\{r : \Delta_r$ is Cohen–Macaulay over $k\} + 1$. Hint: Use induction on the number of faces.
(b) If Δ is Cohen–Macaulay, then Δ_r is Cohen–Macaulay.

5.1.24. Prove all skeletons of a shellable complex are shellable.

5.1.25. Let Δ be a simplicial complex. Show:
(a) The following conditions are equivalent:
   (i) $\mathbb{Z}[\Delta]$ is Cohen–Macaulay;
   (ii) $k[\Delta]$ is Cohen–Macaulay for all fields k;
   (iii) $R[\Delta]$ is Cohen–Macaulay for all Cohen–Macaulay rings R.
(The Cohen–Macaulay property of $k[\Delta]$ may well depend upon k; see Reisner's example at the end of Section 5.3.)
Hint: It is crucial that $R[\Delta]$ is a free R-module for an arbitrary ring R. For (i) $\Rightarrow$ (ii) one uses 2.1.10; for (ii) $\Rightarrow$ (iii) note that (ii) applies to $k(\mathfrak{p}) \otimes R[\Delta]$, $\mathfrak{p} \in \operatorname{Spec} R$, so that 2.1.23 becomes applicable.
(b) The following conditions are equivalent:
   (i) $\mathbb{Q}[\Delta]$ is Cohen–Macaulay;
   (ii) there exist prime numbers $p_1, \ldots, p_n$ such that $k[\Delta]$ is Cohen–Macaulay for any field k whose characteristic is different from p_i, $i = 1, \ldots, n$.
   (iii) there exists a prime number p such that $k[\Delta]$ is Cohen–Macaulay for any field k whose characteristic is p.
Hint: 2.1.29.

5.1.26. Let Δ be a simplicial complex. Δ is called *disconnected* if the vertex set V of Δ is a disjoint union $V = V_1 \cup V_2$ such that no face of Δ has vertices in both V_1 and V_2. Otherwise Δ is *connected*. Show:
(a) If $\dim \Delta = 0$, then Δ is Cohen–Macaulay.
(b) depth $k[\Delta] = 1$ if Δ is disconnected. In particular, all Cohen–Macaulay complexes of positive dimension are connected.
Hint: let Δ_i, $i = 1, 2$, be the subcomplex of Δ consisting of all faces of Δ whose vertices belong to V_i, and represent $k[\Delta]$ as the kernel of a suitable map $k[\Delta_1] \oplus k[\Delta_2] \to k$.
(c) Suppose $\dim \Delta = 1$. The following conditions are equivalent: (i) Δ is connected; (ii) Δ is shellable; (iii) Δ is Cohen–Macaulay.

5.1.27. Let Δ be a $(d-1)$-dimensional simplicial complex with h-vector $h_0, \ldots, h_d$. We define the *a-invariant* $a(\Delta)$ *of* Δ to be $a(k[\Delta])$ where k is an arbitrary field. Show:
(a) $a(\Delta) \leq 0$.
(b) The following conditions are equivalent: (i) $a(\Delta) = 0$; (ii) $\chi(\Delta) = 1$.
If moreover Δ is shellable with shelling $F_1, \ldots, F_m$, then (i) and (ii) are equivalent to (iii) There exists an integer $i \leq m$ such that $\langle F_1, \ldots, F_{i-1} \rangle \cap \langle F_i \rangle$ consists of all maximal proper faces of $\langle F_i \rangle$.

5.1.28. Let k be a finite field. Find a simplicial complex Δ for which $k[\Delta]$ does not have a homogeneous system of parameters consisting of linear forms.

5.2 Polytopes

We briefly discuss combinatorial properties of polytopes, and give an outline of McMullen's proof of the upper bound theorem for polytopes. Stanley's far-reaching generalization to simplicial spheres will be proved in the next sections. The topic as well as the methods employed in this section are non-algebraic. Therefore most of the statements will be given without proof. Though many of them seem obvious from our geometric intuition, they need a rigorous proof. We refer the interested reader to the standard work on polytopes by Grünbaum [144], and to the excellent monograph [272] by McMullen and Shephard of which large parts of this section are an abstract. Another very good reference is the recent book by Ziegler [399].

We consider $\mathbb{R}^d$ as a d-dimensional Euclidean space whose points are d-tuples $x = (\xi_1, \ldots, \xi_d)$ of real numbers, and whose scalar product is given by

$$\langle x, y \rangle = \sum_{i=1}^{d} \xi_i \eta_i, \qquad x = (\xi_1, \ldots, \xi_d), \quad y = (\eta_1, \ldots, \eta_d).$$

A subset K of $\mathbb{R}^d$ is *convex* if for any two points $x_0, x_1 \in K$ the *line segment* with end points x_0 and x_1, that is, the set of points $x = (1 - \lambda)x_0 + \lambda x_1$, $\lambda \in \mathbb{R}$, $0 \le \lambda \le 1$, belongs to K. The intersection of any non-empty family of convex sets is again convex. This allows us to define the *convex hull*, conv X, of a subset $X \subset \mathbb{R}^d$ to be the intersection of all convex sets $K \subset \mathbb{R}^d$ which contain X. The convex hull of X can also be described as the set of all *convex combinations* of finite subsets of X, that is, as the set of linear combinations

$$\lambda_1 x_1 + \cdots + \lambda_r x_r \qquad \text{with} \quad x_i \in X, \quad \lambda_i \ge 0, \quad \sum_{i=1}^{r} \lambda_i = 1.$$

Definition 5.2.1. A *polytope* is the convex hull of a finite set of points in $\mathbb{R}^d$.

There is an alternative description of a polytope as the intersection of a finite number of (closed) half-spaces: let $a \in \mathbb{R}^d$, $a \ne 0$, and $\beta \in \mathbb{R}$; the set

$$H = \{x \in \mathbb{R}^d : \langle a, x \rangle = \beta\}$$

is a *hyperplane* with *normal vector a*. The set of points lying on one side of a hyperplane (including the hyperplane) is a *closed half-space*. Thus H determines two half-spaces

$$H^+ = \{x \in \mathbb{R}^d : \langle a, x \rangle \ge \beta\} \quad \text{and} \quad H^- = \{x \in \mathbb{R}^d : \langle a, x \rangle \le \beta\}.$$

Definition 5.2.2. A *polyhedral set* or *polyhedron* is the intersection of a finite number of closed half-spaces.

Obviously polyhedra are convex sets, but of course need not be bounded.

Theorem 5.2.3. *A subset of $\mathbb{R}^d$ is a polytope if and only if it is a bounded polyhedron.*

Let P be a polyhedron, and H a hyperplane. Then H is called a *supporting hyperplane* if $H \cap P \neq \emptyset$ and P is contained in one of the closed half-spaces determined by H. If H is a supporting hyperplane of P, then $H \cap P$ is called a *face* of P.

It is convenient to consider the empty set and P as faces, the *improper faces*. All the other faces of P are called *proper faces*. The faces of a polyhedron (polytope) are again polyhedra (polytopes).

The *dimension*, $\dim P$, of a polyhedron P is the dimension of its affine hull; a *d-polyhedron* is a polyhedron of dimension d. Recall that for an arbitrary set $X \subset \mathbb{R}^d$ there is a smallest (under inclusion) affine space A containing X, namely just the intersection of all affine subspaces of $\mathbb{R}^d$ containing X. This affine space A is called the *affine hull*, denoted by aff X. A *j-face* is a face whose dimension as a polyhedron is j, and we set $\dim \emptyset = -1$. If $\dim P = t$, faces of dimension $0, 1, t - 2, t - 1$ are called *vertices, edges, subfacets* and *facets*, respectively.

In the following theorem we collect a few facts about the facial structure of a polyhedron.

Theorem 5.2.4. *Let P be a polyhedron.*
(a) *P has only a finite number of faces.*
(b) *Let F be a face of P and F' a face of F. Then F' is a face of P.*
(c) *Any proper face of P is a face of some facet of P.*
(d) *The set of faces of P, ordered by inclusion, is a lattice.*

The lattice in 5.2.4(d), denoted by $\mathscr{F}(P)$, is called the *face lattice* or *boundary complex* of P. Two polyhedra are called *combinatorially equivalent* if their face lattices are isomorphic. An invariant under combinatorial equivalence is the *f-vector* $(f_0, f_1, \ldots, f_{d-1})$ of a d-polyhedron P. Here $f_j = f_j(P)$ is the number of j-faces of P.

Simplicial polytopes. Let $A \subset \mathbb{R}^d$ be a k-dimensional affine subspace of $\mathbb{R}^d$. We pick $x \in A$; then there exists a linear subspace U of $\mathbb{R}^d$ (not depending on x) such that $A = x + U$. The vector space U is called the *associated linear space of A*. Let $u_1, \ldots, u_k$ be a basis of U. Then each element $y \in A$ has a (unique) presentation

$$y = x + \lambda_1 u_1 + \cdots + \lambda_k u_k$$

with $\lambda_i \in \mathbb{R}$. Set $x_0 = x$, $\mu_0 = (1 - \sum_i \lambda_i)$, $x_i = x + u_i$ and $\mu_i = \lambda_i$ for $i = 1, \ldots, k$. Then

(4) $y = \mu_0 x_0 + \mu_1 x_1 + \cdots + \mu_k x_k$ and $\mu_0 + \cdots + \mu_k = 1$.

This suggests defining y to be *affinely dependent on* $x_0, \ldots, x_k$ if there exists an equation as in (4). It is clear that the set of elements which are affinely dependent on $x_0, \ldots, x_k$ is just the affine hull of $\{x_0, \ldots, x_k\}$. The elements $x_0, \ldots, x_k$ are called *affinely independent* if each element $y \in \text{aff}\{x_0, \ldots, x_k\}$ has a unique presentation as in (4), or equivalently, if the elements $x_1 - x_0, x_2 - x_0, \ldots, x_k - x_0$ form a basis of the associated linear space of $\text{aff}\{x_0, \ldots, x_k\}$.

Definition 5.2.5. A d-*simplex* is the convex hull of $d+1$ affinely independent points. A polytope is called *simplicial* if all its proper faces are simplices.

Let P be a simplex defined by $d+1$ affinely independent points $x_0, x_1, \ldots, x_d$, and let X be a subset of $\{x_0, x_1, \ldots, x_d\}$ consisting of d points. For the following argument we may assume that $P \subset \mathbb{R}^d$. Then $\text{aff}\, X$ is a hyperplane which supports P, and thus $\text{conv}\, X = P \cap \text{aff}\, X$ is a facet of P. Since any subset of a set of affinely independent points is again affinely independent it follows that $\text{conv}\, X$ is a $(d-1)$-simplex. Thus induction on the dimension yields

Proposition 5.2.6. *Every j-face of a d-simplex P is a j-simplex, and every $j+1$ vertices of P are the vertices of a j-face of P.*

Corollary 5.2.7. *Let P be a simplicial polytope with vertex set V, and let $\Delta(P)$ be the collection of subsets of V consisting of the empty set and the vertices of the proper faces of P. Then $\Delta(P)$ is a simplicial complex.*

We call $\Delta(P)$ the *vertex scheme* of P. It is clear that not every simplicial complex is the vertex scheme of some simplicial polytope P. Nevertheless, to any simplicial complex Δ we may associate a geometric object whose construction is in a sense inverse to the one given in 5.2.7. Let X be an arbitrary subset in $\mathbb{R}^d$. We define the *relative interior* of X, denoted $\text{relint}\, X$, as the interior of X relative to $\text{aff}\, X$. For example, it is not difficult to see that the relative interior of the convex hull of $\{x_1, \ldots, x_r\}$, $x_i \in \mathbb{R}^d$, is the set of points

$$\lambda_1 x_1 + \cdots + \lambda_r x_r, \qquad \lambda_i > 0, \qquad \sum_{i=1}^{r} \lambda_i = 1.$$

Definition 5.2.8. Let Δ be a simplicial complex on the vertex set V. Suppose the map $\rho : V \to \mathbb{R}^d$ satisfies the following conditions:
(a) ρ is injective,
(b) the elements of $\rho(F)$ are affinely independent for all $F \in \Delta$,
(c) $\text{relint}(\text{conv}\, \rho(F)) \cap \text{relint}(\text{conv}\, \rho(G)) = \emptyset$ for all $F, G \in \Delta$, $F \neq G$.
Then $\bigcup_{F \in \Delta} \text{relint}(\text{conv}\, \rho(F))$ is called a *geometric realization* of Δ.

Giving a geometric realization of Δ its natural topology as a subspace of $\mathbb{R}^d$, we note that any two geometric realizations of Δ are homeomorphic, and we denote the underlying topological space by $|\Delta|$.

A geometric realization always exists. Indeed, if $V = \{v_1,\ldots,v_n\}$ is the vertex set of Δ, and $x_1,\ldots,x_n$ are affinely independent elements in $\mathbb{R}^d$, then $\rho: V \to \mathbb{R}^d$ with $\rho(v_i) = x_i$ for $i = 1,\ldots,n$ defines a geometric realization of Δ.

Cyclic polytopes. Consider the algebraic curve $M \subset \mathbb{R}^d$, defined parametrically by

$$x(\tau) = (\tau,\tau^2,\ldots,\tau^d), \qquad \tau \in \mathbb{R};$$

M is called the *moment curve*. It is a curve of degree d which implies that a hyperplane not containing M intersects it in at most d points.

Definition 5.2.9. Let $n \geq d + 1$ be an integer. A *cyclic polytope*, denoted $C(n,d)$, is the convex hull of any n distinct points on M.

The notation $C(n,d)$ is justified since, as we shall see in a moment, its face lattice depends only on n and d. We first observe

Proposition 5.2.10. *Any $d+1$ distinct points on M are affinely independent. In particular, $C(n,d)$ is a simplicial d-polytope.*

PROOF. Let $\tau_0,\ldots,\tau_d$ be the distinct parameters of these points. We need to show that the vectors $x(\tau_1) - x(\tau_0),\ldots,x(\tau_d) - x(\tau_0)$ are linearly independent, or, equivalently, that the corresponding matrix with these row vectors is non-singular. Clearly, this is the case if and only if the Vandermonde matrix

$$A = \begin{pmatrix} 1 & \tau_0 & \tau_0^2 & \cdots & \tau_0^d \\ 1 & \tau_1 & \tau_1^2 & \cdots & \tau_1^d \\ \vdots & & & & \vdots \\ 1 & \tau_d & \tau_d^2 & \cdots & \tau_d^d \end{pmatrix}$$

is non-singular. The determinant of A is known to be $\prod_{0 \leq i < j \leq d}(\tau_i - \tau_j)$, and this expression is non-zero since the τ_i are pairwise distinct. $\square$

Next we determine the vertex scheme $\Delta(C(n,d))$ which encodes the combinatorial properties of $C(n,d)$. Let $C(n,d)$ be the convex hull of the points $x_i = x(\tau_i)$, $\tau_1 < \tau_2 < \cdots < \tau_n$, $n \geq d + 1$. A subset X of $V = \{x_1,\ldots,x_n\}$ will be called an *end set* if there exists an integer i, $1 \leq i \leq n$, such that either $X = \{x_1,\ldots,x_i\}$ or $X = \{x_i,\ldots,x_n\}$. The set X will be called *contiguous* if there exist integers $1 < i \leq j < n$ such that $X = \{x_i,\ldots,x_j\}$, and an *odd* (*even*) *contiguous* set if it is contiguous and $|X|$ is odd (even). It is clear that any proper subset $W \subset V$ has a unique decomposition

$$W = Y_1 \cup X_1 \cup X_2 \cup \cdots \cup X_t \cup Y_2,$$

where the X_i are contiguous, and Y_1 and Y_2 are end sets or empty. The set W is of *type* (r,s) if $|W| = r$, and if there are exactly s odd contiguous subsets X_i of W.

Theorem 5.2.11. *Let j be an integer with $0 \leq j \leq d-1$. A subset $W \subset V$ is a j-face of $C(n,d)$ if and only if W is of type $(j+1, s)$ for some s with $0 \leq s \leq d-j-1$.*

PROOF. We first show the assertion for $j = d-1$. Since $C(n,d)$ is simplicial, any $(d-1)$-face has d vertices. Thus we have to show that if $W \subset V$ is of type (d, s), then conv W is a facet if and only if $s = 0$. By 5.2.10, the points of W are affinely independent, and hence define a hyperplane $H \subset \mathbb{R}^d$. It is clear that $W \subset H \cap M$. But actually, $W = H \cap M$ since M is a curve of degree d, and it follows that the points of W divide M into $d+1$ arcs lying alternately on each side of H. Now conv W is a facet of $C(n,d)$ if and only if H supports $C(n,d)$, or in other words, if and only if all points of $V \setminus W$ lie on one side of H. Obviously this happens exactly when every two points of $V \setminus W$ are separated by an even number of points of W, that is, when $s = 0$.

Let us now treat the general case, and assume that $|W| = j+1$. Suppose that W has at most $d-j-1$ odd contiguous subsets. Then it is possible to find a subset T of $d-j-1$ points of M such that $V \cap T = \emptyset$, and $W \cup T$ has only even contiguous subsets. Since $|W \cup T| = d$ it follows from the first part of the proof that conv$(W \cup T)$ is a facet of $C(n+d-j-1, d)$ supported by the hyperplane $H = \text{aff}(W \cup T)$. As $W = H \cap V$ we conclude that conv $W = H \cap C(n,d)$ is a face of $C(n,d)$.

Conversely, if conv W is a j-face of $C(n,d)$, then there exists some facet conv W' of $C(n,d)$ with $W \subset W'$. Since W' has no odd contiguous subsets, W can have at most $d-j-1$ odd contiguous subsets. $\square$

Corollary 5.2.12. *The combinatorial type of a cyclic polytope $C(n,d)$ depends only upon n and d, and not on the particular vertex set $V \subset M$.*

A polytope P has the highest possible number of j-faces when every subset of $j+1$ elements of the vertex set of P is the set of vertices of a proper face of P. In this case we say that P is $(j+1)$-*neighbourly*.

Corollary 5.2.13. *$C(n,d)$ is $[d/2]$-neighbourly.*

The upper bound theorem. In 1957 Motzkin [278] made the following conjecture. Let P be a d-polytope with n vertices; then $f_j(P) \leq f_j(C(n,d))$ for all j, $1 \leq j \leq d$. This conjecture was proved in 1970 by McMullen [271]. We indicate the ideas of his proof: given a d-polytope P with n vertices, one applies in a first step a process, known as 'pulling the vertices', with the effect of transforming P into a simplicial polytope with the same number of vertices as P, and at least as many faces of higher dimension. Thus one may assume from the beginning that P is a simplicial polytope.

Just as for simplicial complexes one defines the h-vector $(h_0, \ldots, h_d)$ of P by the equation $\sum_{i=0}^{d} h_i t^i = \sum_{i=0}^{d} f_{i-1} t^i (1-t)^{d-i}$, $f_{-1} = 1$. Then owing to the fact that $C(n, d)$ is $[d/2]$-neighbourly we have

(a) $h_i(C(n, d)) = \binom{n-d+i-1}{i}$ for all i, $0 \le i \le [d/2]$.

Moreover, the existence of a line shelling of P (see below) yields

(b) $0 \le h_i(P) \le \binom{n-d+i-1}{i}$, and

(c) $h_i(P) = h_{d-i}(P)$ for all i, $0 \le i \le d$.

The identities in (c) are the famous *Dehn–Sommerville equations*.

Now (a), (b) and (c) imply $h_i(P) \le h_i(C(n, d))$ for all i, $0 \le i \le d$. Finally, since the $f_j(P)$ are non-negative linear combinations of the $h_i(P)$ (see 5.1.8), the proof of the upper bound theorem is completed.

Shellings. A *shelling* of the boundary complex of a d-polytope P (or simply a shelling of P) is an order of its facets $F_1, \ldots, F_m$ such that $F_i \cap \bigcup_{i=0}^{j-1} F_i$ is homeomorphic to a $(d-2)$-dimensional ball or sphere for all j, $2 \le j \le m$.

Theorem 5.2.14 (Bruggesser–Mani). *Every polytope is shellable.*

We give a sketch of the proof. Present P as an intersection of closed half-spaces. One may assume without loss of generality that $P = \{x \in \mathbb{R}^d : \langle a_i, x \rangle \le 1\}$, $0 \le i \le m$, where a_i is a normal vector for the face F_i. Choose a vector c such that $\langle a_i, c \rangle \ne 0$, and order the faces in such a way that $\langle a_1, c \rangle > \langle a_2, c \rangle > \cdots > \langle a_m, c \rangle$. Then $F_1 \ldots, F_m$ is a shelling of P. Such a shelling is called a *line shelling* of P. It can be imagined as follows: moving along the line L in direction c starting from the origin, one lists the facets of P as they become 'visible'. (This happens exactly when one meets the corresponding supporting hyperplane.) Coming back from the opposite side one lists the remaining facets in the order they 'disappear'.

Corollary 5.2.15. *Let $F_1, \ldots, F_m$ be a line shelling of the polytope P. Then $F_m, F_{m-1}, \ldots, F_1$ is a line shelling of P, too.*

PROOF. Let $F_1, \ldots, F_m$ be the line shelling induced by c. Then $F_m, \ldots, F_1$ is the line shelling induced by $-c$.   □

Suppose now that P is a simplicial polytope. Then the h-vector of P and the h-vector of the vertex scheme $\Delta(P)$ coincide. Furthermore, it is clear that a shelling of P induces a shelling of $\Delta(P)$ (in the sense of Section 5.1). Thus we may apply 5.1.14 to compute the h-vector from a line shelling of P. In particular, it follows from 5.1.14 that $h_i \ge 0$ for all i. Moreover, in view of 5.2.15, we obtain the Dehn–Sommerville equations,

Theorem 5.2.16 (Sommerville). *Let* $(h_0, \ldots, h_d)$ *be the h-vector of a simplicial polytope. Then* $h_i = h_{d-i}$ *for* $0 \leq i \leq d$.

These formulas imply in particular that $h_d = 1$. Thus 5.1.9 yields

Corollary 5.2.17. *Let* P *be a simplicial d-polytope with* f*-vector* $(f_0, \ldots, f_{d-1})$. *Then*

$$\sum_{i=0}^{d-1} (-1)^i f_i = 1 - (-1)^d.$$

This formula is valid not only for simplicial polytopes, but more generally for all polytopes, and is known as the *Euler relation*.

For the proof of the inequalities $h_i \leq \binom{n-d+i-1}{i}$ one again uses line shellings. We refer the reader to McMullen's original paper [271] or [272].

Exercise

5.2.18. Let $\Delta(n, d)$ denote the boundary complex of the cyclic polytope $C(n, d)$.
(a) Show that the cyclic permutation $x_i \mapsto x_{i+1 \bmod n}$ induces an automorphism of $\Delta(n, d)$ for d even.
(b) Show that the substitution $X_1 \mapsto X_1 X_{n+1}$, $X_i \mapsto X_i$ for $i = 2, \ldots, n$, maps the monomial generators of $I_{\Delta(n,d)}$, d even, to those of $I_{\Delta(n+1,d+1)}$.

5.3 Local cohomology of Stanley–Reisner rings

We will compute the local cohomology of a Stanley–Reisner ring $k[\Delta]$ in terms of the modified Čech complex $C^\bullet$ introduced in Section 3.5. It is not surprising that $C^\bullet$, just like $k[\Delta]$, is equipped with a fine grading. This allows us to decompose the local cohomology groups of $k[\Delta]$. As it turns out, their homogeneous pieces can be interpreted as the reduced simplicial homology of certain subcomplexes of Δ. This basic result of Hochster is the main content of this section. As a corollary one obtains Reisner's Cohen–Macaulay criterion for simplicial complexes.

For the reader's convenience we recall the notion of reduced simplicial homology. Let Δ be a simplicial complex with vertex set V. An *orientation* on Δ is a linear order on V. A simplicial complex together with an orientation is an *oriented* simplicial complex.

Suppose Δ is an oriented simplicial complex of dimension $d - 1$, and $F \in \Delta$ an i-face. We write $F = [v_0, \ldots, v_i]$ if $F = \{v_0, \ldots, v_i\}$ and $v_0 < v_1 < \cdots < v_i$, and $F = [\,]$ if $F = \emptyset$. Having introduced this notation, we define the *augmented oriented chain complex of* Δ,

$$\tilde{\mathscr{C}}(\Delta): 0 \longrightarrow \mathscr{C}_{d-1} \overset{\partial}{\longrightarrow} \mathscr{C}_{d-2} \longrightarrow \cdots \longrightarrow \mathscr{C}_0 \overset{\partial}{\longrightarrow} \mathscr{C}_{-1} \longrightarrow 0$$

by setting

$$\mathscr{C}_i = \bigoplus_{\substack{F \in \Delta \\ \dim F = i}} \mathbb{Z}F \quad \text{and} \quad \partial F = \sum_{j=0}^{i} (-1)^j F_j$$

for all $F \in \Delta$; here $F_j = [v_0, \dots, \hat{v}_j, \dots, v_i]$ for $F = [v_0, \dots, v_i]$. A straight-forward computation shows that $\partial \circ \partial = 0$. Let G be an Abelian group. We set

$$\tilde{H}_i(\Delta; G) = H_i(\mathscr{C}(\Delta) \otimes G), \qquad i = -1, \dots, d-1,$$

and call $\tilde{H}_i(\Delta; G)$ the *i-th reduced simplicial homology of Δ with values in* G. It follows from the next lemma that a reference to the orientation is superfluous.

Lemma 5.3.1. *Define $\tilde{\mathscr{C}}'(\Delta)$ in the same way as $\tilde{\mathscr{C}}(\Delta)$, but with respect to a different orientation of Δ. Then there exists an isomorphism of complexes $\tilde{\mathscr{C}}(\Delta) \cong \tilde{\mathscr{C}}'(\Delta)$.*

PROOF. Let $<$ and $\prec$ be the different linear orders on the vertex set of V. Given $F = \{v_0, \dots, v_i\}$, $v_0 < v_1 < \cdots < v_i$, there exists a permutation $\pi = \pi_F$ of the vertices of F such that $v_{\pi(0)} \prec v_{\pi(1)} \prec \cdots \prec v_{\pi(i)}$. We leave it to the reader to verify that $\alpha \colon \tilde{\mathscr{C}}(\Delta) \to \tilde{\mathscr{C}}(\Delta)$ with $\alpha(F) = \sigma(\pi_F)F$ is the desired isomorphism. $\qquad\square$

The *i-th reduced simplicial cohomology of Δ with values in* G is defined to be

$$\tilde{H}^i(\Delta; G) = H^i(\operatorname{Hom}_{\mathbb{Z}}(\tilde{\mathscr{C}}(\Delta), G)), \qquad i = -1, \dots, d-1.$$

We set $\tilde{H}_i(\Delta) = \tilde{H}_i(\Delta; \mathbb{Z})$ and $\tilde{H}^i(\Delta) = \tilde{H}^i(\Delta; \mathbb{Z})$ for all i. The simplicial complex Δ is called *acyclic* if $\tilde{H}_\bullet(\Delta) = 0$. In this case, $\tilde{\mathscr{C}}(\Delta)$ is split exact, and so $\tilde{H}_\bullet(\Delta; G) = 0$ and $\tilde{H}^\bullet(\Delta; G) = 0$ for all Abelian groups G. Examples of acyclic simplicial complexes are the cones: the *cone* $\operatorname{cn}(\Delta)$ of Δ is the join (see 5.1.20) of a point $\Pi = \{v_0\}$ with Δ. The reader is referred to Exercise 5.3.10 for further details.

The cone construction can be iterated. We set $\operatorname{cn}^j(\Delta) = \operatorname{cn}(\operatorname{cn}^{j-1}(\Delta))$ for all $j > 1$. It is immediate that $\operatorname{cn}^j(\Delta)$ is the join of Δ with a j-simplex, and it follows that

$$(5) \qquad\qquad \tilde{H}_\bullet(j\text{-simplex} * \Delta) = 0.$$

If $G = k$ is a field, then the reduced simplicial homology and cohomology groups are k-vector spaces, and there are canonical isomorphisms

$$\tilde{H}^i(\Delta; k) \cong \operatorname{Hom}_k(\tilde{H}_i(\Delta; k), k), \qquad \tilde{H}_i(\Delta; k) \cong \operatorname{Hom}_k(\tilde{H}^i(\Delta; k), k)$$

for all i; see Exercise 5.3.11. In particular it follows that $\dim \tilde{H}_i(\Delta; k) = \dim \tilde{H}^i(\Delta; k)$ for all i.

Since $\mathscr{C}_i \otimes k$ is a vector space of dimension f_i, elementary linear algebra yields

$$\sum_{i=-1}^{d-1} (-1)^i \dim \widetilde{H}_i(\varDelta;k) = \sum_{i=-1}^{d-1} (-1)^i f_i.$$

This sum, denoted by $\widetilde{\chi}(\varDelta)$ is called the *reduced Euler characteristic of $\varDelta$*. A comparison with the Euler characteristic $\chi(\varDelta)$ introduced in Section 5.1 shows that $\widetilde{\chi}(\varDelta) = \chi(\varDelta) - 1$, and we can rewrite the Euler relation 5.2.17 as $\widetilde{\chi}(\varDelta) = (-1)^{d-1}$.

A geometric realization of $\varDelta$ in $\mathbb{R}^n$ inherits the structure of a topological space (with the subspace topology). In Section 5.2 we denoted this space by $|\varDelta|$ and remarked that it is unique up to homeomorphism. Let X be a topological space, and $\varphi: |\varDelta| \to X$ a homeomorphism. The pair $(\varDelta, \varphi)$ is called a *triangulation* of X. Less precisely, we often say in this situation that $\varDelta$ is a triangulation of X.

It is a fundamental theorem in topology (see [280], Theorem 34.3) that the reduced singular homology $\widetilde{H}_i(X;k)$ of a topological space X with triangulation $\varDelta$ can be computed by means of the reduced simplicial homology of $\varDelta$.

Theorem 5.3.2. *Let X be a topological space with triangulation $\varDelta$. Then*

$$\widetilde{H}_i(X;k) \cong \widetilde{H}_i(\varDelta;k) \qquad \text{for all } i.$$

Examples 5.3.3. (a) Let $\varDelta$ be the d-simplex with vertices $V = \{v_0, \ldots, v_d\}$. Then $|\varDelta|$ is homeomorphic to the d-dimensional closed ball B^d, whose reduced singular homology is trivial since B^d is contractible to a point. Thus 5.3.2 implies that $\widetilde{H}_\bullet(\varDelta;k) = 0$. That the reduced simplicial homology of $\varDelta$ is trivial can be seen directly: one immediately identifies $\mathscr{C}_\bullet \otimes k$ with the Koszul complex $K_\bullet(f)$ associated with $f: k^{d+1} \to k$ where f maps the canonical basis elements of k^{d+1} to 1. It follows from 1.6.5(b) that this Koszul complex is exact.

(b) Consider the subcomplex $\varGamma \subset \varDelta$ obtained from $\varDelta$ by deleting the face $F = \{v_0, \ldots, v_d\}$; then $|\varGamma|$ is homeomorphic to the $(d-1)$-dimensional sphere S^{d-1}. It is clear that the quotient $U_\bullet = \mathscr{C}(\varDelta)/\mathscr{C}(\varGamma)$ has $U_i = 0$ for $i \neq d$ and $U_d \cong \mathbb{Z} \cdot [v_0, \ldots, v_d]$. Therefore

$$\widetilde{H}_i(S^{d-1};k) \cong \widetilde{H}_i(\varGamma;k) \cong \begin{cases} k & \text{if } i = d - 1, \\ 0 & \text{if } i \neq d - 1. \end{cases}$$

(c) Let $\varDelta$ be the vertex scheme of a simplicial $(d-1)$-polytope P. Then $|\varDelta|$ is homeomorphic to the $(d-1)$-sphere. Therefore, by (b),

$$\widetilde{\chi}(\varDelta) = \sum_{i=-1}^{d-1} (-1)^i \dim \widetilde{H}_i(S^{d-1};k) = (-1)^{d-1}.$$

Thus we have recovered the Euler relation.

The following notions will be crucial in the analysis of the local cohomology of a Stanley–Reisner ring.

Definition 5.3.4. Let Δ be a simplicial complex, and F a subset of the vertex set of Δ. The *star of* F is the set $\mathrm{st}_\Delta F = \{G \in \Delta : F \cup G \in \Delta\}$, and the *link of* F is the set $\mathrm{lk}_\Delta F = \{G : F \cup G \in \Delta, \ F \cap G = \emptyset\}$.

To simplify notation we occasionally omit the index Δ in st_Δ or lk_Δ. It is clear that $\mathrm{st}\, F$ is a subcomplex of Δ, $\mathrm{lk}\, F$ a subcomplex of $\mathrm{st}\, F$, and that $\mathrm{st}\, F = \mathrm{lk}\, F = \emptyset$ if $F \notin \Delta$.

In Figure 5.7 let v be the vertex in the centre of the hexagon. Then $\mathrm{st}\, v$ is the full simplicial complex, while $\mathrm{lk}\, v$ is the subcomplex constituting the boundary of the hexagon.

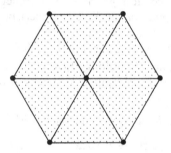

Figure 5.7

Lemma 5.3.5. *Let F be a face of the simplicial complex Δ, and $G \in \mathrm{lk}\, F$. Then*
(a) $F \in \mathrm{lk}\, G$ *and* $\mathrm{lk}_{\mathrm{st}\, G} F = \langle G \rangle * \mathrm{lk}_{\mathrm{lk}\, G} F$;
(b) $\mathrm{lk}_{\mathrm{st}\, G} F$ *is acyclic, if* $G \neq \emptyset$.

PROOF. Statement (a) is trivial, and (b) follows from equation (5). $\square$

Local cohomology. Let Δ be a simplicial complex, k a field, and $R = k[X_1, \ldots, X_n]/I_\Delta$ the Stanley–Reisner ring of Δ. Let $\mathfrak{m}$ be the maximal ideal generated by the residue classes x_i of the indeterminates X_i. Note that $(R, \mathfrak{m})$ is a *local ring, and hence by 2.1.27, R is Cohen–Macaulay if and only if $R_\mathfrak{m}$ is Cohen–Macaulay. Thus, in order to determine when Δ is Cohen–Macaulay, we are led to compute the local cohomology $H^\bullet_{\mathfrak{m}R_\mathfrak{m}}(R_\mathfrak{m})$ of $R_\mathfrak{m}$. To simplify notation we will write $H^\bullet_\mathfrak{m}(R)$ for $H^\bullet_{\mathfrak{m}R_\mathfrak{m}}(R_\mathfrak{m})$. Let $x = x_1, \ldots, x_n$; as in Section 3.4 we consider the complex $\varinjlim K^\bullet(x^k)$ which is isomorphic to

$$C^\bullet : 0 \longrightarrow C^0 \longrightarrow C^1 \longrightarrow \cdots \longrightarrow C^n \longrightarrow 0,$$

$$C^t = \bigoplus_{1 \le i_1 < i_2 < \cdots < i_t \le n} R_{x_{i_1} x_{i_2} \cdots x_{i_t}},$$

and whose differential is composed of the maps

$$(-1)^{s-1} \operatorname{nat} : R_{x_{i_1} \cdots x_{i_t}} \longrightarrow R_{x_{j_1} \cdots x_{j_{t+1}}}$$

if $\{i_1, \ldots, i_t\} = \{j_1, \ldots, \widehat{j_s}, \ldots, j_{t+1}\}$, and 0 otherwise. It follows from 3.5.6 that $H_{\mathfrak{m}}^i(R) \cong H^i(C_{\mathfrak{m}}^{\boldsymbol{\cdot}}) \cong H^i(C^{\boldsymbol{\cdot}})_{\mathfrak{m}}$. We claim that $\operatorname{Supp} H^i(C^{\boldsymbol{\cdot}}) \subset \{\mathfrak{m}\}$ for all i. This finally implies

$$H_{\mathfrak{m}}^i(R) \cong H^i(C^{\boldsymbol{\cdot}})$$

for all i. Indeed, $C_{x_j}^{\boldsymbol{\cdot}}$ is exact for $j = 1, \ldots, n$ because the identity and the zero-map of $C_{x_j}^{\boldsymbol{\cdot}}$ are homotopic via $\sigma^{\boldsymbol{\cdot}}$, where $\sigma^k : C^k \to C^{k-1}$ is defined on the component $(R_{x_{i_1} \cdots x_{i_k}})_{x_j} \to (R_{x_{j_1} \cdots x_{j_{k-1}}})_{x_j}$ to be $(-1)^{s-1} \operatorname{id}$ if $\{i_1, \ldots, i_k\} = \{j_1, \ldots, j_{k-1}, j\}$ and $i_s = j$, and 0 otherwise.

Next note that $C^{\boldsymbol{\cdot}}$ is a $\mathbb{Z}^n$-graded complex: recall from Section 5.1 that R itself is $\mathbb{Z}^n$-graded. Let $a \in \mathbb{Z}^n$, $a = (a_1, \ldots, a_n)$; then $R_a \cong k$ if $a \in \mathbb{N}^n$ and $\{v_i : a_i > 0\} \in \Delta$, and $R_a = 0$ otherwise. The components of C^i are of the form R_x for some element $x \in R$ which is homogeneous in the fine grading of R. One defines a $\mathbb{Z}^n$-grading on R_x by setting

$$(R_x)_a = \{ \frac{r}{x^m} : r \text{ homogeneous, } \deg r - m \deg x = a \}.$$

Of course the terms 'homogeneous' and 'deg' refer to the fine grading of R.

We extend this grading on the components to C^i. Then it is clear that $C^{\boldsymbol{\cdot}}$ becomes a $\mathbb{Z}^n$-graded complex, and we may equip the homology of $C^{\boldsymbol{\cdot}}$ with the induced $\mathbb{Z}^n$-graded structure. In other words, the local cohomology modules $H_{\mathfrak{m}}^i(R)$ are in a natural way $\mathbb{Z}^n$-graded modules.

As R is a homogeneous k-algebra (in the sense of Section 1.5), we may as well consider the graded local cohomology modules ${}^{\boldsymbol{\cdot}}H_{\mathfrak{m}}^i(R)$ of R introduced in Section 3.6. Then

$${}^{\boldsymbol{\cdot}}H_{\mathfrak{m}}^i(R)_j \cong \bigoplus_{a \in \mathbb{Z}^n, \ |a|=j} H_{\mathfrak{m}}^i(R)_a$$

for all i and j; see Exercise 5.3.12.

Given $x = x_{i_1} \cdots x_{i_r}$, $i_1 < \cdots < i_r$, we set $F = \{v_{i_1}, \ldots, v_{i_r}\}$. In order to analyze when $(R_x)_a \neq 0$ for $a \in \mathbb{Z}^n$ we introduce some more notation, and put

$$G_a = \{v_i : a_i < 0\} \quad \text{and} \quad H_a = \{v_i : a_i > 0\}.$$

Lemma 5.3.6. (a) $\dim_k(R_x)_a \le 1$ *for all* $a \in \mathbb{Z}^n$.
(b) $(R_x)_a \cong k$ *if and only if* $F \supset G_a$ *and* $F \cup H_a \in \Delta$.

PROOF. (a) Let r_i/x^{n_i}, $i = 1, 2$, be non-zero elements in $(R_x)_a$. Then $x^{n_2} r_1$ and $x^{n_1} r_2$ are homogeneous of the same degree, and hence are linearly dependent over k. We may assume that $\kappa(x^{n_2}r_1) = x^{n_1}r_2$ for some $\kappa \in k$; then $\kappa(r_1/x^{n_1}) = r_2/x^{n_2}$.

(b) We have $(R_x)_a \neq 0$ if and only if there exist a monomial v in R and an integer l such that

(i) $x^m v \neq 0$ for all $m \in \mathbb{N}$,   and   (ii) $\deg v/x^l = a$.

Condition (i) is equivalent to (i') $v/x^l \neq 0$.

Now (i) implies $F \cup \operatorname{supp} v \in \Delta$, and (ii) implies $F \supset G_a$ and $H_a \subset \operatorname{supp} v$. In particular, $F \cup H_a \in \Delta$.

Conversely, suppose $F \supset G_a$ and $F \cup H_a \in \Delta$. Set $v = \prod_{a_i > 0} x_i^{a_i}$, $w = \prod_{a_i < 0} x_i^{-a_i}$. Since $F \supset G_a$ there exists an integer l such that $x^l = wu$ where u is a monomial (with non-negative exponents) in the x_i. Since $F \cup H_a \in \Delta$, we have $vu/x^l \neq 0$, and it follows that $\deg vu/x^l = a$.   □

Let $a \in \mathbb{Z}^n$; as a consequence of the lemma we see that $(C^i)_a$ has a basis

$$\{b_F : \ F \supset G_a, \ F \cup H_a \in \Delta, \ |F| = i\}.$$

Restricting the differentiation of $C^{\bullet}$ to the a-th graded piece we obtain a complex $(C^{\bullet})_a$ of finite dimensional vector spaces with differentiation $\partial : (C^i)_a \to (C^{i+1})_a$ given by $\partial(b_F) = \sum (-1)^{\sigma(F,F')} b_{F'}$ where the sum is taken over all F' such that $F' \supset F$, $F' \cup H_a \in \Delta$ and $|F'| = i + 1$, and where $\sigma(F,F') = s$ for $F' = [v_0, \ldots, v_i]$ and $F = [v_0, \ldots, \hat{v}_s, \ldots, v_i]$.

Lemma 5.3.7. *For all $a \in \mathbb{Z}^n$ there exists an isomorphism of complexes*

$$\alpha^{\bullet} : (C^{\bullet})_a \to \operatorname{Hom}_{\mathbb{Z}}(\widetilde{\mathscr{C}}(\operatorname{lk}_{\operatorname{st} H_a} G_a)[-j-1], k), \qquad j = |G_a|.$$

PROOF. The assignment $F \mapsto F' = F \setminus G_a$ establishes a bijection between the set

$$\mathscr{B} = \{F \in \Delta : \ F \supset G_a, \ F \cup H_a \in \Delta, \ |F| = i\}$$

and the set $\mathscr{B}' = \{F' \in \Delta : \ F' \in \operatorname{lk}_{\operatorname{st} H_a} G_a, \ |F'| = i - j\}$. Therefore it is clear that

$$\alpha^i : (C^i)_a \to \operatorname{Hom}_{\mathbb{Z}}(\widetilde{\mathscr{C}}(\operatorname{lk}_{\operatorname{st} H_a} G_a)_{i-j-1}, k), \qquad b_F \mapsto \varphi_{F \setminus G_a}$$

is an isomorphism of vector spaces. Here $\varphi_{F'}$ is defined by

$$\varphi_{F'}(F'') = \begin{cases} 1 & \text{if } F' = F'', \\ 0 & \text{otherwise.} \end{cases}$$

By 5.3.1, we have the possibility of adjusting the orientation of Δ suitably. We choose it in such a way that the elements in G_a are latest in the linear order of the vertex set of Δ. Furthermore we give the subcomplex $\operatorname{lk}_{\operatorname{st} H_a} G_a$ the induced orientation. With this standardization, $\alpha^{\bullet}$ becomes a complex homomorphism.   □

We are ready to prove the main result of this section.

Theorem 5.3.8 (Hochster). *Let Δ be a simplicial complex, and k a field. Then the Hilbert series of the local cohomology modules of $k[\Delta]$ with respect to the fine grading is given by*

$$H_{H_{\mathfrak{m}}^i(k[\Delta])}(t) = \sum_{F \in \Delta} \dim_k \widetilde{H}_{i-|F|-1}(\mathrm{lk}\, F; k) \prod_{v_j \in F} \frac{t_j^{-1}}{1 - t_j^{-1}}.$$

PROOF. By the previous lemma we have

$$H_{\mathfrak{m}}^i(k[\Delta])_a \cong \widetilde{H}^{i-|G_a|-1}(\mathrm{lk}_{\mathrm{st}\, H_a} G_a; k),$$

and therefore $\dim_k H_{\mathfrak{m}}^i(k[\Delta])_a = \dim_k \widetilde{H}_{i-|G_a|-1}(\mathrm{lk}_{\mathrm{st}\, H_a} G_a; k)$; see Exercise 5.3.11.

If $H_a \neq \emptyset$, then, by 5.3.5, $\mathrm{lk}_{\mathrm{st}\, H_a} G_a$ is acyclic, and if $H_a = \emptyset$, then $\mathrm{st}\, H_a = \Delta$, and so $\mathrm{lk}_{\mathrm{st}\, H_a} G_a = \mathrm{lk}\, G_a$.

Let $\mathbb{Z}_-^n = \{a \in \mathbb{Z}^n : a_i \leq 0 \text{ for } i = 1, \dots, n\}$; then $H_a = \emptyset$ if and only if $a \in \mathbb{Z}_-^n$, and it follows that

$$H_{H_{\mathfrak{m}}^i(k[\Delta])}(t) = \sum_{F \in \Delta} \sum_{a \in \mathbb{Z}_-^n,\, G_a = F} \dim_k \widetilde{H}_{i-|F|-1}(\mathrm{lk}\, F; k) t^a$$

$$= \sum_{F \in \Delta} \dim_k \widetilde{H}_{i-|F|-1}(\mathrm{lk}\, F; k) \prod_{v_j \in F} \frac{t_j^{-1}}{1 - t_j^{-1}}. \qquad \square$$

Hochster's theorem yields an important Cohen–Macaulay criterion for simplicial complexes:

Corollary 5.3.9 (Reisner). *Let Δ be a simplicial complex, and k a field. The following conditions are equivalent:*
(a) *Δ is Cohen–Macaulay over k;*
(b) *$\widetilde{H}_i(\mathrm{lk}\, F; k) = 0$ for all $F \in \Delta$ and all $i < \dim \mathrm{lk}\, F$.*

PROOF. Let $\dim \Delta = d - 1$. Then Δ is Cohen–Macaulay over k if and only if $H^i(C^{\bullet}) = 0$ for $i < d$. The latter is equivalent to

(6) $\widetilde{H}_{i-|F|-1}(\mathrm{lk}\, F; k) = 0$ for all $F \in \Delta$ and all $i < d$.

(a) $\Rightarrow$ (b): If Δ is Cohen–Macaulay over k, then Δ is pure (see 5.1.5), and so $\dim \mathrm{lk}\, F = d - |F| - 1$. Therefore (6) implies $\widetilde{H}_i(\mathrm{lk}\, F; k) = 0$ for $i < \dim \mathrm{lk}\, F$.

(b) $\Rightarrow$ (a): Let $F \in \Delta$ and $G \in \mathrm{lk}\, F$. Then $\mathrm{lk}_{\mathrm{lk}\, F} G = \mathrm{lk}(G \cup F)$. Therefore, by induction on the dimension of the simplicial complex we may assume that all proper links of Δ are Cohen–Macaulay over k. In particular, the links of the vertices are pure. Now since $\widetilde{H}_0(\Delta; k) = \widetilde{H}_0(\mathrm{lk}\, \emptyset; k) = 0$ if $\dim \Delta \geq 1$, we conclude that Δ itself is pure. Then obviously (b) implies (6). $\qquad \square$

As a first application we consider an example of Reisner: let $\varDelta$ be a triangulation of the real projective plane $\mathbb{P}^2$. Figure 5.8 indicates such a triangulation. For reasons of readability the triangles in the figure have

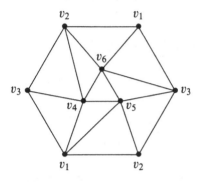

Figure 5.8

not been shadowed. Also note that edges with the same vertices have to be identified according to their orientations.

Let k be a field. If $\varDelta$ is Cohen–Macaulay over k, then 5.3.9 implies $\widetilde{H}_i(\varDelta;k) = 0$ for $i < 2$. Since $\mathbb{P}^2$ is connected, we have $\widetilde{H}_0(\mathbb{P}^2;k) = 0$. But

$$\widetilde{H}_1(\mathbb{P}^2;k) = \begin{cases} k & \text{if char } k = 2, \\ 0 & \text{otherwise;} \end{cases}$$

see [280], Theorem 40.6 and Theorem 55.1. In particular, $\varDelta$ is not Cohen–Macaulay over a field k of characteristic 2. On the other hand, it follows from Exercise 5.3.14 that $\varDelta$ is Cohen–Macaulay over k if char $k \neq 2$.

Exercises

5.3.10. (a) Let $\varGamma * \varDelta$ be the join of the simplicial complexes $\varGamma$ and $\varDelta$. Show that $\widetilde{\mathscr{C}}(\varGamma * \varDelta) \cong \widetilde{\mathscr{C}}(\varGamma) \otimes \widetilde{\mathscr{C}}(\varDelta)$
(b) Let $\varPi = \{v_0\}$ be the simplicial complex consisting of one point, and form the cone $\mathrm{cn}(\varDelta) = \varPi * \varDelta$. Show there exists an exact sequence

$$0 \longrightarrow \widetilde{\mathscr{C}}(\varDelta) \longrightarrow \widetilde{\mathscr{C}}(\mathrm{cn}(\varDelta)) \longrightarrow \widetilde{\mathscr{C}}(\varDelta)(-1) \longrightarrow 0$$

where $\widetilde{\mathscr{C}}(\varDelta) \to \widetilde{\mathscr{C}}(\mathrm{cn}(\varDelta)) = \widetilde{\mathscr{C}}(\varPi) \otimes \widetilde{\mathscr{C}}(\varDelta)$ is the natural complex homomorphism $a \mapsto 1 \otimes a$.
(c) Prove that the connecting homomorphisms in the associated long exact homology sequence are isomorphisms, and conclude that $\widetilde{H}_*(\mathrm{cn}(\varDelta)) = 0$. (This conclusion can also be drawn directly from 1.6.12 with $x = 1$.)
5.3.11. Let $\varDelta$ be a simplicial complex, and k a field. Show there exist natural isomorphisms

$$\widetilde{H}^i(\varDelta;k) \cong \mathrm{Hom}_k(\widetilde{H}_i(\varDelta;k),k), \qquad \widetilde{H}_i(\varDelta;k) \cong \mathrm{Hom}_k(\widetilde{H}^i(\varDelta;k),k).$$

5.3.12. Let Δ be a simplicial complex, and k a field.

(a) Consider $k[\Delta]$ as a homogeneous k-algebra, and give the modules C^i the structure of $\mathbb{Z}$-graded $k[\Delta]$-modules by setting $(C^i)_j = \bigoplus_{a \in \mathbb{Z}^n, \ |a|=j}(C^i)_a$. Show $C^{\bullet}$ is a complex of $\mathbb{Z}$-graded modules.

(b) Give $H^i(C^{\bullet})$ the induced $\mathbb{Z}$-graded structure, and deduce $^{\bullet}H^i_{\mathfrak{m}}(k[\Delta]) \cong H^i(C^{\bullet})$ as graded $k[\Delta]$-modules.

(c) Conclude from (a) that $^{\bullet}H^i_{\mathfrak{m}}(k[\Delta])_j \cong \bigoplus_{a \in \mathbb{Z}^n, |a|=j} H^i(C^{\bullet})_a$ for all i and j.

5.3.13. Use Reisner's criterion to give an alternative proof of the equivalence (i) $\Longleftrightarrow$ (iii) of Exercise 5.1.26(c).

5.3.14. Let Δ be a simplicial complex of dimension 2. Show that Δ is Cohen–Macaulay over k if and only if the following conditions are satisfied:

(a) Δ is connected;

(b) $\tilde{H}_1(\Delta; k) = 0$;

(c) each point of $|\Delta|$ has arbitrarily small connected punctured neighborhoods. Hint: (c) is equivalent to the condition that the links of the vertices of Δ be connected.

5.3.15. Let Δ be a simplicial complex of dimension $d - 1$, k a field, and $\mathfrak{m}$ the graded maximal ideal of $k[\Delta]$.

(a) Show the following conditions are equivalent:

(i) Δ is pure and $k[\Delta]_{\mathfrak{p}}$ is Cohen–Macaulay for all prime ideals $\mathfrak{p} \neq \mathfrak{m}$;

(ii) $H^i_{\mathfrak{m}}(k[\Delta])$ has finite length for all $i < d$;

(iii) $H^i_{\mathfrak{m}}(k[\Delta])_a = 0$ for all $a \neq 0$ and $i < d$;

(iv) $\tilde{H}_i(\operatorname{lk} F; k) = 0$ for all $F \in \Delta$, $F \neq \emptyset$, and all $i < \dim \operatorname{lk} F$;

(v) $H^i_{\mathfrak{m}}(k[\Delta]) \cong \tilde{H}_{i-1}(\Delta; k)$ for all $i < d$.

(b) (Reisner) Show the following conditions are equivalent:

(i) Δ is Cohen–Macaulay;

(ii) $\tilde{H}_i(\Delta; k) = 0$ for all $i < d - 1$, and the links of all vertices of Δ are Cohen–Macaulay.

Hint: (a)(i) is equivalent to the condition that $k[\Delta]_{x_i}$ be Cohen–Macaulay for all $i = 1, \ldots, n$. Further, observe that $k[\Delta]_{x_i} \cong k[x_i, x_i^{-1}][\operatorname{lk}\{x_i\}]$.

5.4 The upper bound theorem

This section is devoted to the proof of the upper bound theorem for simplicial spheres, that is, simplicial complexes whose geometric realization is topologically a sphere. It follows from a result of Kalai [229] that there are many more simplicial spheres than polytopes. Therefore, the upper bound theorem for simplicial spheres properly generalizes McMullen's theorem for polytopes whose proof we sketched in Section 5.2.

The upper bound theorem for simplicial spheres was conjectured by Klee in 1964 and proved by Stanley [356] in 1975.

The proof is carried out in three steps: first we show that Euler complexes satisfy the Dehn–Sommerville equations; secondly we prove the

upper bound theorem for Cohen–Macaulay Euler complexes; and finally
we show that simplicial spheres are Cohen–Macaulay Euler complexes.

Definition 5.4.1.   The simplicial complex Δ is an *Euler complex* if Δ is
pure, and $\tilde{\chi}(\operatorname{lk} F) = (-1)^{\dim \operatorname{lk} F}$ for all $F \in \Delta$.

Theorem 5.4.2 (Dehn, Sommerville, Klee).   *Let Δ be an Euler complex of
dimension $d - 1$ with h-vector $(h_0, \ldots, h_d)$. Then $h_i = h_{d-i}$ for $i = 0, \ldots, d$.*

The proof will easily follow from

Lemma 5.4.3.   *Let Δ be a simplicial complex on $V = \{v_1, \ldots, v_n\}$. Then*

$$H_{k[\Delta]}(t_1^{-1}, \ldots, t_n^{-1}) = \sum_{F \in \Delta} (-1)^{\dim F} \tilde{\chi}(\operatorname{lk} F) \prod_{v_i \in F} \frac{t_i}{1 - t_i}.$$

PROOF. We have $H_{k[\Delta]}(t) = \sum_{F \in \Delta} \prod_{v_i \in F} t_i/(1 - t_i)$; see Section 5.1. The
substitution $t_i \mapsto t_i^{-1}$ transforms $t_i/(1 - t_i)$ into $-(1 + t_i/(1 - t_i))$. It follows
that $\prod_{v_i \in F} t_i/(1 - t_i)$ is transformed into

$$(-1)^{\dim F + 1} \prod_{v_i \in F} \left(1 + \frac{t_i}{1 - t_i}\right) = (-1)^{\dim F + 1} \sum_{G \subset F} \prod_{v_i \in G} \frac{t_i}{1 - t_i},$$

so that

$$H_{k[\Delta]}(t_1^{-1}, \ldots, t_n^{-1}) = \sum_{F \in \Delta} (-1)^{\dim F + 1} \sum_{G \subset F} \prod_{v_i \in G} \frac{t_i}{1 - t_i}$$

$$= \sum_{G \in \Delta} \left(\sum_{\substack{F \in \Delta \\ G \subset F}} (-1)^{\dim F + 1} \right) \prod_{v_i \in G} \frac{t_i}{1 - t_i}.$$

Since $\sum_{F \in \Delta, \, G \subset F} (-1)^{\dim F + 1} = \sum_{F \in \operatorname{lk} G} (-1)^{\dim F - \dim G} = (-1)^{\dim G} \tilde{\chi}(\operatorname{lk} G)$,
the assertion follows. $\square$

PROOF OF 5.4.2.   If Δ is an Euler complex of dimension $d - 1$, then
$\tilde{\chi}(\operatorname{lk} F) = (-1)^{\dim \operatorname{lk} F} = (-1)^{d - \dim F}$. The latter equality holds since Δ is
pure. Now 5.4.3 yields

$$H_{k[\Delta]}(t_1, \ldots, t_n) = (-1)^d H_{k[\Delta]}(t_1^{-1}, \ldots, t_n^{-1}).$$

Replacing the t_i by t we obtain the identity $H_{k[\Delta]}(t) = (-1)^d H_{k[\Delta]}(t^{-1})$ for
the Hilbert function of $k[\Delta]$. It is clear that this yields the desired result;
see also 4.4.7(a). $\square$

Let Δ be an Euler complex, and k a field. It follows from Reisner's
criterion 5.3.9 that Δ is Cohen–Macaulay over k if and only if for all
$F \in \Delta$

$$(7) \qquad\qquad \tilde{H}_i(\operatorname{lk} F; k) \cong \begin{cases} k & \text{if } i = \dim \operatorname{lk} F, \\ 0 & \text{otherwise.} \end{cases}$$

In view of the results in the previous sections it is now an easy matter to show that the upper bound theorem holds for any simplicial complex whose faces satisfy condition (7). In other words we have

Theorem 5.4.4. *Let Δ be an Euler complex of dimension $d - 1$ with n vertices which is Cohen–Macaulay over a field k. Then $f_i(\Delta) \le f_i(C(n, d))$ for $i = 1, \ldots, d - 1$.*

PROOF. Just as in the proof of the upper bound theorem for polytopes it suffices to show (a) $h_i(\Delta) \le \binom{n-d+i-1}{i}$ for $i = 0, \ldots, d$, and (b) Δ satisfies the Dehn–Sommerville equations. But (a) follows from 5.1.10, and (b) from 5.4.2. $\qquad\square$

The final step in the proof of the upper bound theorem for simplicial spheres is to show that the faces of a simplicial complex satisfy (7) if the geometric realization is homeomorphic to a sphere.

In the next lemma which is a reformulation of [280], Lemma 63.1 we refer to the notation used in 5.2.8, and denote, as usual, by $H_*(X, Y; k)$ the relative singular homology of the pair (X, Y) where X is a topological space and Y a subspace of X.

Lemma 5.4.5. *Let Δ be a simplicial complex on the vertex set V, and k be a field. Suppose that X is a geometric realization of Δ given by $\rho : V \to \mathbb{R}^d$, that $F \in \Delta$ is a face of dimension j, and that $p \in \operatorname{relint}(\operatorname{conv}(\rho(F)))$. If $\operatorname{lk} F \ne \emptyset$, then*

$$H_i(X, X \setminus \{p\}; k) \cong \widetilde{H}_{i-j-1}(\operatorname{lk} F; k) \qquad \text{for all } i,$$

and if $\operatorname{lk} F = \emptyset$, then

$$H_i(X, X \setminus \{p\}; k) \cong \begin{cases} k & \text{for } i = j, \\ 0 & \text{otherwise.} \end{cases}$$

As a consequence of this lemma and Reisner's criterion we see that the Cohen–Macaulay property of Δ only depends on the topology of $|\Delta|$.

Corollary 5.4.6 (Munkres, Stanley). *Let Δ be a $(d - 1)$-dimensional simplicial complex, $X = |\Delta|$, and k a field. The following conditions are equivalent:*
(a) Δ is Cohen–Macaulay over k;
(b) for all $p \in X$ and all $i < \dim X$ one has

$$\widetilde{H}_i(X; k) = H_i(X, X \setminus \{p\}; k) = 0.$$

Moreover, if the equivalent conditions are satisfied, then Δ is an Euler complex if and only if

$$\widetilde{H}_{d-1}(X; k) \cong H_{d-1}(X, X \setminus \{p\}; k) \cong k \qquad \text{for all } p \in X.$$

PROOF. We only prove the implication (b) $\Rightarrow$ (a); the converse implication is proved similarly. We have $\operatorname{lk} F = \emptyset$ if and only if F is a facet. Thus assumption (b) and 5.4.5 imply that all facets have dimension $d - 1$, that is, Δ is pure.

Now suppose that $\operatorname{lk} F \neq \emptyset$. Since Δ is pure we have $\dim \operatorname{lk} F = d - 2 - \dim F = d - 2 - j$. Therefore, by 5.4.5 and assumption (b), if $F \neq \emptyset$ and $i < \dim \operatorname{lk} F$, then $\tilde{H}_i(\operatorname{lk} F; k) \cong H_{i+j+1}(X, X \setminus \{p\}; k) = 0$ since $i + j + 1 < d - 1$. Finally, if $F = \emptyset$, then $\operatorname{lk} F = \Delta$, and (b) implies that $\tilde{H}_i(\operatorname{lk} F; k) \cong \tilde{H}_i(X; k) = 0$ for $i < \dim \operatorname{lk} F$. By Reisner's criterion it follows that Δ is Cohen–Macaulay over k.

The supplement concerning the Euler property is obvious. $\square$

Corollary 5.4.7 (The upper bound theorem for simplicial spheres). *Let Δ be a simplicial complex with n vertices and $|\Delta| \cong S^{d-1}$. Then $f_i(\Delta) \leq f_i(C(n, d))$ for $i = 1, \ldots, d - 1$.*

PROOF. The assertion is clear in view of 5.4.4 and 5.4.6. $\square$

Exercises

5.4.8. (a) Give an example of a simplicial complex which does not satisfy the Dehn–Sommerville equations.
(b) Give an example of a simplicial complex Δ which for some i fails the condition $h_i \leq \binom{n-d+i-1}{i}$, $d - 1 = \dim \Delta$, $n = f_0(\Delta)$.

5.4.9. Let k be a field, and Δ a Cohen–Macaulay complex over k. Δ is called *level over k* if $k[\Delta]$ is a level ring, that is, if all generators in a minimal set of generators of the ˙canonical module $\omega_{k[\Delta]}$ have the same degree. The *type of Δ over k*, denoted by $r_k(\Delta)$, is the type of $k[\Delta]$. Let $s = \max\{i : h_i(\Delta) \neq 0\}$. Show that $h_s \leq r_k(\Delta)$, and that equality holds if and only if Δ is level over k.

5.5 Betti numbers of Stanley-Reisner rings

Let k be a field, and Δ a simplicial complex on a vertex set V with $|V| = n$. We write $k[\Delta] \cong R/I_\Delta$ with $R = k[X_1, \ldots, X_n]$. Since $k[\Delta]$ is a $\mathbb{Z}^n$-graded R-module, it has a minimal $\mathbb{Z}^n$-graded resolution

$$F_\bullet : 0 \longrightarrow F_p \xrightarrow{\varphi_p} F_{p-1} \longrightarrow \cdots \longrightarrow F_1 \xrightarrow{\varphi_1} F_0 \longrightarrow 0,$$

where $F_i = \bigoplus_{j=1}^{\beta_i} R(-a_{ij})$ for $i = 0, \ldots, p$ with certain $a_{ij} \in \mathbb{N}^n$, and where the maps φ_i are homogeneous of degree 0; see 1.5.16 where a similar result has been established for $\mathbb{Z}$-graded resolutions. Minimality of the resolution means that $\varphi_i(F_i) \subset (X_1, \ldots, X_n)F_{i-1}$ for all i. The numbers $\beta_{ia} = |\{j : a_{ij} = a\}|$, $a \in \mathbb{Z}^n$, are called the *fine Betti numbers of $k[\Delta]$*. It is easily seen that the minimal $\mathbb{Z}^n$-graded resolution is uniquely determined up to isomorphism.

In order to compute the shifts a_{ij} in the resolution $F_\bullet$, we consider the k-vector spaces $T_i = \text{Tor}_i^R(k, k[\Delta])$ and notice that $T_i \cong F_i/(X_1, \ldots, X_n)F_i$ as a $\mathbb{Z}^n$-graded vector space. Obviously, $\beta_{ia} = \dim_k(T_i)_a$.

Let $W \subset V$; we set $\Delta_W = \{F \in \Delta : F \subset W\}$, and call Δ_W the *restriction of Δ to W*. It is clear that Δ_W is again a simplicial complex.

The following theorem gives a combinatorial interpretation of the fine Betti numbers of $k[\Delta]$.

Theorem 5.5.1 (Hochster). *Let* $H_{T_i}(t) = \sum_{a \in \mathbb{Z}^n} \beta_{ia} t^a$ *be the fine Hilbert series of the module* $T_i = \text{Tor}_i^R(k, k[\Delta])$. *Then*

$$H_{T_i}(t) = \sum_{W \subset V} \left(\dim_k \tilde{H}_{|W|-i-1}(\Delta_W; k) \right) \prod_{v_j \in W} t_j.$$

We say that $a \in \mathbb{Z}^n$ is *squarefree* if each of its entries is either 0 or 1. One remarkable consequence of Hochster's theorem is

Corollary 5.5.2. *The shifts in the minimal $\mathbb{Z}^n$-graded R-resolution of $k[\Delta]$ are squarefree.*

For the proof of 5.5.1 we shall need *Alexander duality*. It involves the *dual complex* of Δ which is given by

$$\bar{\Delta} = \{G \in \Sigma : \bar{G} \notin \Delta\}.$$

Here $\bar{G}$ denotes the complement of G in V, and Σ the simplex on the vertex set V. It is easy to see that $\bar{\Delta}$ is again a simplicial complex, and that $\bar{\bar{\Delta}} = \Delta$.

Let $\Gamma \subset \Delta$ be a simplicial subcomplex of Δ; then $\tilde{\mathscr{C}}(\Gamma)$ is a subcomplex $\tilde{\mathscr{C}}(\Delta)$, and we may form the quotient complex $\tilde{\mathscr{C}}(\Delta)/\tilde{\mathscr{C}}(\Gamma)$. For an Abelian group G we set

$$\tilde{H}_i(\Delta, \Gamma; G) = H_i\big(\tilde{\mathscr{C}}(\Delta)/\tilde{\mathscr{C}}(\Gamma) \otimes G\big)$$

and

$$\tilde{H}^i(\Delta, \Gamma; G) = H_i\big(\text{Hom}_{\mathbb{Z}}(\tilde{\mathscr{C}}(\Delta)/\tilde{\mathscr{C}}(\Gamma), G)\big).$$

These groups are called the *reduced relative simplicial homology* and *cohomology* of the pair (Δ, Γ) (with values in G). The following lemma is the relative version of *Alexander duality*.

Lemma 5.5.3. *Let k be a field, and let $\Gamma \subset \Delta \subset \Sigma$ be simplicial complexes, where Σ is the simplex on the vertex set V, $|V| = n$. Then*

$$\tilde{H}_i(\Delta, \Gamma; k) \cong \tilde{H}^{n-2-i}(\bar{\Gamma}, \bar{\Delta}; k) \cong \tilde{H}_{n-2-i}(\bar{\Gamma}, \bar{\Delta}; k).$$

for all i. In particular, one has $\tilde{H}_i(\Delta; k) \cong \tilde{H}^{n-3-i}(\bar{\Delta}; k) \cong \tilde{H}_{n-3-i}(\bar{\Delta}; k)$.

PROOF. Let $e_1, \ldots, e_n$ be a basis of the free $\mathbb{Z}$-module $L = \mathbb{Z}^n$. The exterior products $e_F = \bigwedge_{j \in F} e_j$, $F \subset V$, $|F| = j$, are a basis of $\bigwedge^j L$, and $\bigwedge L$ together with the differential

$$\partial_j : \overset{j}{\bigwedge} L \to \overset{j-1}{\bigwedge} L, \quad e_{i_1} \wedge \ldots \wedge e_{i_j} \mapsto \sum_{k=1}^{j} (-1)^{k+1} e_{i_1} \wedge \ldots e_{i_{k-1}} \wedge e_{i_{k+1}} \wedge \ldots e_{i_j}$$

is an exact complex; in fact, it is just the Koszul complex $K_\bullet(\varepsilon)$ of the linear form $\varepsilon : L \to \mathbb{Z}$ with $\varepsilon(e_i) = 1$ for all i. Evidently $\widetilde{\mathscr{C}}(\Delta)(-1)$ may be identified with the subcomplex of $\bigwedge L$ spanned by the basis elements e_F, $F \in \Delta$.

In 1.6.10 we have exhibited an isomorphism $\tau : K_\bullet(\varepsilon) \to K^\bullet(\varepsilon)$, which is induced by the multiplication on $\bigwedge L$ and the orientation $\omega_n : \bigwedge^n L \to \mathbb{Z}$, $\omega_n(e_1 \wedge \cdots \wedge e_n) = 1$. The restriction of τ to $\widetilde{\mathscr{C}}(\Delta)(-1)$ yields an isomorphism $\widetilde{\mathscr{C}}(\Delta)/\widetilde{\mathscr{C}}(\Gamma) \cong \mathrm{Hom}_{\mathbb{Z}}(\widetilde{\mathscr{C}}(\Delta)/\widetilde{\mathscr{C}}(\Gamma), \mathbb{Z})$. Upon tensoring with k one gets the first of our isomorphisms whereas the second holds because we are taking coefficients in a field.

In the special case in which Γ is the empty set one has $\widetilde{H}_i(\Delta; k) \cong \widetilde{H}^{n-2-i}(\Sigma, \bar{\Delta}; k)$. On the other hand, $\widetilde{H}^{n-2-i}(\Sigma, \bar{\Delta}; k) \cong \widetilde{H}^{n-3-i}(\bar{\Delta}; k)$, as follows from the long exact cohomology sequence

$$\widetilde{H}^{j-1}(\Sigma; k) \longrightarrow \widetilde{H}^{j-1}(\bar{\Delta}; k) \longrightarrow \widetilde{H}^j(\Sigma, \bar{\Delta}; k) \longrightarrow \widetilde{H}^j(\Sigma; k)$$

and the fact that $\widetilde{H}^\bullet(\Sigma; k) = 0$.   □

PROOF OF 5.5.1. The Koszul complex $K_\bullet(x; R)$ of the sequence $x = X_1, \ldots, X_n$ is a minimal graded free resolution of the R-module $k = R/(x)$ (see 1.6.14). Thus for each $i \geq 0$, and each $a \in \mathbb{Z}^n$

$$H_i(x; k[\Delta])_a \cong \mathrm{Tor}_i^R(k, k[\Delta])_a.$$

We will compute the graded components of $\mathrm{Tor}_i^R(k, k[\Delta])$ by means of these isomorphisms.

With a subset $F \subset \{1, \ldots, n\}$ we associate the vector $\varepsilon(F) = \sum_{i \in F} e_i$, where e_i is the i-th canonical unit vector in $\mathbb{Z}^n$. Now it is straightforward to verify that $K_i(x; I_\Delta)_a$ is a k-vector space with basis

$$x^b e_F, \quad b + \varepsilon(F) = a, \quad |F| = i, \quad \text{and} \quad \mathrm{supp}(b) \notin \Delta.$$

(As above, $e_F = \bigwedge_{j \in F} e_j$.) Thus, if Δ_a is the simplicial complex consisting of those faces $F \in \Sigma$, $F \subset \mathrm{supp}\, a$, for which $\mathrm{supp}(a \setminus \varepsilon(F)) \notin \Delta$, then the map

$$\alpha_i : \widetilde{C}_{i-1}(\Delta_a) \longrightarrow K_i(x, I_\Delta)_a, \quad F \mapsto x^{a-\varepsilon(F)} e_F,$$

is an isomorphism of vector spaces.

One easily checks that $\alpha_{\bullet}$ is a chain map, so that we actually have an isomorphism of complexes $\alpha_{\bullet} \colon \widetilde{C}_{\bullet}(\Delta_a)(-1) \longrightarrow K_{\bullet}(x, I_{\Delta})_a$. Therefore the exact sequence of complexes

$$0 \longrightarrow K_{\bullet}(x, I_{\Delta}) \longrightarrow K_{\bullet}(x, R) \longrightarrow K_{\bullet}(x, k[\Delta]) \longrightarrow 0$$

yields the isomorphisms

$$\operatorname{Tor}_i^R(k, k[\Delta])_a \cong H_i(x, k[\Delta])_a \cong H_{i-1}(x, I_{\Delta})_a \cong \widetilde{H}_{i-2}(\Delta_a; k)$$

for $i > 0$. The case $i = 0$ is trivial: $\dim_k \widetilde{H}_{|W|-1}(\Delta_W; k) \neq 0$ if and only if $W = \emptyset$ and, equivalently, $\Delta_W = \emptyset$; furthermore $\dim_k \widetilde{H}_{-1}(\Delta_{\emptyset}; k) = 1$.

Suppose first that $a = (a_1, \ldots, a_n)$ is not squarefree. We pick j such that $a_j \geq 2$, and consider the element $a(r) = (a_1, \ldots, a_j + r, \ldots, a_n)$. Then $\Delta_a = \Delta_{a(r)}$ for all $r \geq 0$. Hence it follows that $\operatorname{Tor}_i^R(k, k[\Delta])_a = \operatorname{Tor}_i^R(k, k[\Delta])_{a(r)}$ for all $r \geq 0$. This is only possible if $\operatorname{Tor}_i^R(k, k[\Delta])_a = 0$, because otherwise there would exist infinitely many shifts in the finite resolution $F_{\bullet}$ of $k[\Delta]$.

Now we assume that a is squarefree. Let $W = \operatorname{supp} a$; then $F \in \Delta_a$ if and only if $W \setminus F \notin \Delta_W$. Therefore, $\Delta_a = \overline{\Delta_W}$ with respect to the vertex set W, and the assertion follows from Alexander duality. $\qquad\square$

Exercises

5.5.4. Let k be field, and Δ a simplicial complex with n vertices.
(a) Show that the Betti numbers β_{ia} of $k[\Delta]$ are independent of k for $i = 0, 1, 2, n-1$, and n.
Hint: Use Alexander duality and the fact that $\widetilde{H}_i(\Gamma; k)$ does not depend on k for $i = 0, -1$.
(b) Prove that all Betti numbers of $k[\Delta]$ are independent of k if $n \leq 5$.
(c) Give an example of a simplicial complex Δ with 6 vertices for which the Betti numbers of $k[\Delta]$ depend on k.

5.5.5. Let k be field, and Δ a simplicial complex on a vertex set V with n elements.
(a) Let F be a face of the dual simplicial complex $\overline{\Delta}$, and set $W = V \setminus F$. Show that

$$\widetilde{H}^{i-2}(\operatorname{lk}_{\overline{\Delta}} F; k) \cong \widetilde{H}_{|W|-i-1}(\Delta_W; k).$$

(b) Use (a) and 5.5.1 to prove the following theorem of Eagon and Reiner [88]: the Stanley-Reisner ring $k[\Delta]$ has an m-linear resolution (see 4.1.17) if and only if $\overline{\Delta}$ is Cohen-Macaulay over k. Determine m.

5.6 Gorenstein complexes

Let Δ be a simplicial complex on the vertex set V, and k a field. The complex Δ is called *Gorenstein over k* if $k[\Delta]$ is Gorenstein. Our main concern in this section is to characterize the Gorenstein complexes.

We define core Δ to be $\Delta_{\text{core } V}$ where core $V = \{v \in V : \text{st } v \neq \Delta\}$. Notice that $\Delta = (\text{core } \Delta) * \Delta_{V \setminus \text{core } V}$. Therefore,

$$k[\Delta] \cong k[\text{core } \Delta] \otimes k[\Delta_{V \setminus \text{core } V}] \cong k[\text{core } \Delta][X_i : v_i \in V \setminus \text{core } V].$$

It follows that Δ is Gorenstein if and only if core Δ is Gorenstein.

Theorem 5.6.1. *Let Δ be a simplicial complex, $\Gamma = \text{core } \Delta$, and k a field. The following conditions are equivalent:*
(a) *Δ is Gorenstein over k;*
(b) *for all $F \in \Gamma$ one has*

$$\widetilde{H}_i(\text{lk}_\Gamma F; k) \cong \begin{cases} k & \text{if } i = \dim \text{lk}_\Gamma F, \\ 0 & \text{if } i < \dim \text{lk}_\Gamma F; \end{cases}$$

(c) *for $X = |\Gamma|$ and $p \in X$ one has*

$$\widetilde{H}_i(X; k) \cong H_i(X, X \setminus \{p\}; k) \cong \begin{cases} k & \text{if } i = \dim X, \\ 0 & \text{if } i < \dim X. \end{cases}$$

PROOF. The equivalence of (b) and (c) follows from 5.4.5, and that of (a) and (b) from the next theorem. □

Theorem 5.6.2. *Let Δ be a simplicial complex with $\Delta = \text{core } \Delta$. Then Δ is Gorenstein over k if and only if Δ is an Euler complex which is Cohen–Macaulay over k.*

For the proof of 5.6.2 we need the following two lemmas.

Lemma 5.6.3. *Let Δ be a simplicial complex, and k a field. Let M be a $\mathbb{Z}^n$-graded $k[\Delta]$-module whose fine Hilbert series coincides with that of $k[\Delta]$. Suppose M is indecomposable. Then $k[\Delta]$ and M are isomorphic as $\mathbb{Z}^n$-graded modules.*

PROOF. We set $R = k[\Delta]$. There exists a non-zero $\mathbb{Z}^n$-graded homomorphism $\varphi : R \to M$ of degree 0. We want to show that φ is an isomorphism. Consider the exact sequence

$$0 \longrightarrow K \longrightarrow R \overset{\varphi}{\longrightarrow} M \overset{\varepsilon}{\longrightarrow} N \longrightarrow 0,$$

where $K = \text{Ker } \varphi$ and $N = \text{Coker } \varphi$. Since R and M have the same Hilbert series (with respect to the fine grading), this is true for K and N as well. We choose homogeneous generators $x_0, \ldots, x_n$ of M with $\varphi(1) = x_0$, and such that $\varepsilon(x_1), \ldots, \varepsilon(x_n)$ form a minimal system of generators of N. Consider the sets $A = \{a \in \mathbb{Z}^n : (Rx_0)_a \neq 0\}$, $B = \{a \in \mathbb{Z}^n : K_a \neq 0\}$ and $C = \{a \in \mathbb{Z}^n : N_a \neq 0\}$. Then $A \cap B = \emptyset$, $B = C$, and $A \cup B = D$ where $D = \{a \in \mathbb{Z}^n : R_a \neq 0\} = \{a \in \mathbb{N}^n : \text{supp } a \in \Delta\}$; see Section 5.1 for the last equality and the definition of supp.

We want to show that $M = Rx_0 \oplus (Rx_1 + \cdots + Rx_n)$. As, by assumption, M is indecomposable, the assertion of the lemma will follow.

Suppose $Rx_0 \cap (Rx_1 + \cdots + Rx_n) \neq 0$. Then there exists a homogeneous element $y \in Rx_0 \cap (Rx_1 + \cdots + Rx_n)$, $y \neq 0$. It follows that $a = \deg y \in A$. On the other hand, there exist a homogeneous element $r \in R$ and some x_i, $i \geq 1$, such that $y = rx_i$. Therefore, $a = \deg r + \deg x_i$. Note that $\deg x_i \in C = B$. Hence there exists a homogeneous element $z \in K$, $z \neq 0$, with $\deg z = \deg x_i$. Consider $w = rz$; since $\operatorname{supp}(\deg r) \cup \operatorname{supp}(\deg z) = \operatorname{supp}(\deg r) \cup \operatorname{supp}(\deg x_i) = \operatorname{supp} a \in \Delta$, it follows that $w \neq 0$. Therefore, $a = \deg w \in B$, a contradiction. $\qquad\square$

Lemma 5.6.4. *Let Δ be a $(d-1)$-dimensional Gorenstein complex over a field k with $\Delta = \operatorname{core} \Delta$. Then $h_d(\Delta) = 1$.*

PROOF. It is enough to show that $h_d(\Delta) \neq 0$. We write $k[\Delta] \cong R/I_\Delta$, $R = k[X_1, \ldots, X_n]$, and consider the minimal $\mathbb{Z}^n$-graded resolution

$$F_{\bullet} : 0 \longrightarrow F_p \xrightarrow{\varphi_p} F_{p-1} \longrightarrow \cdots \longrightarrow F_1 \xrightarrow{\varphi_1} F_0 \longrightarrow 0,$$

where $F_i = \bigoplus_{j=1}^{\beta_i} R(-a_{ij})$ for $i = 0, \ldots, p$ with certain $a_{ij} \in \mathbb{N}^n$.

It is obvious that $(F_{\bullet})_{\mathfrak{m}}$ is a minimal $R_{\mathfrak{m}}$-resolution of the Gorenstein ring $k[\Delta]_{\mathfrak{m}}$ where $\mathfrak{m} = (X_1, \ldots, X_n)$. It follows from 1.3.3 that $p = n - d$, and 3.3.9 implies that $F_{n-d} = R(-a)$ for some squarefree $a \in \mathbb{N}^n$.

Notice that $F_{\bullet}$ is also a minimal $\mathbb{Z}$-graded resolution: simply replace the shifts a_{ij} by $|a_{ij}|$, where $|b|$ denotes the sum of the components of a vector b. Thus we may apply 4.1.14(a), and conclude that $|a| - (n - d)$ is the largest integer s for which $h_s(\Delta) \neq 0$. The assertion of the lemma follows once we have shown that $|a| \geq n$.

We claim that $a = (1, \ldots, 1)$ (which implies $|a| = n$). Indeed, Hochster's theorem 5.5.1 shows that 0 or 1 are the only possible entries of a. By 3.3.9, the R-dual $(F_{\bullet})^*$ of $F_{\bullet}$ (suitably shifted) is a minimal free resolution of $k[\Delta]$. Thus, if an entry of a was zero, then the corresponding variable would not divide any of the generators of I_Δ, a contradiction to our hypothesis that $\Delta = \operatorname{core} \Delta$. $\qquad\square$

PROOF OF 5.6.2. According to Exercise 5.6.6 the *canonical module $\omega_{k[\Delta]}$ has a natural $\mathbb{Z}^n$-grading.

Suppose that Δ is an Euler complex which is Cohen–Macaulay over k. Then the formula given in Exercise 5.6.6 implies that $k[\Delta]$ and $\omega_{k[\Delta]}$ have the same Hilbert series with respect to the fine grading. By 5.6.3, this implies that $\omega_{k[\Delta]} \cong k[\Delta]$ which in turn implies that $k[\Delta]$ is Gorenstein; see 3.6.11.

Conversely, suppose that Δ is Gorenstein over k. Then, just as in 3.6.11, one sees that $\omega_{k[\Delta]} \cong k[\Delta](c)$, $c = (c_1, \ldots, c_n) \in \mathbb{Z}^n$, where $|c|$ is the a-invariant of $k[\Delta]$. As we assume that $\Delta = \operatorname{core} \Delta$, it follows from 5.6.4

that $|c| = 0$. Since by Exercise 5.6.6, $c_i \leq 0$ for $i = 0, \ldots, n$, this implies $c = 0$. Therefore, again by Exercise 5.6.6,

$$\sum_{F \in \Delta} \dim_k \widetilde{H}_{\dim \operatorname{lk} F}(\operatorname{lk} F; k) \prod_{v_i \in F} \frac{t_i}{1 - t_i} = \sum_{F \in \Delta} \prod_{v_i \in F} \frac{t_i}{1 - t_i}.$$

Comparing coefficients we see that $\dim_k \widetilde{H}_{\dim \operatorname{lk} F}(\operatorname{lk} F; k) = 1$ for all $F \in \Delta$. This together with the fact that Δ is Cohen–Macaulay over k implies that Δ is an Euler complex. $\qquad\square$

Corollary 5.6.5. *Simplicial spheres are Gorenstein over every field.*

Exercises

5.6.6. Let k be a field, and Δ a $(d-1)$-dimensional Cohen–Macaulay complex over k. According to 3.6.19, the ˙canonical module $\omega_{k[\Delta]}$ of $k[\Delta]$ is isomorphic to ˙$\operatorname{Hom}_k(\text{˙}H_m^d(k[\Delta]); k)$. Conclude that $\omega_{k[\Delta]}$ has a natural $\mathbb{Z}^n$-grading, and show that

$$H_{\omega_{k[\Delta]}}(t) = \sum_{F \in \Delta} \dim_k \widetilde{H}_{\dim \operatorname{lk} F}(\operatorname{lk} F; k) \prod_{X_i \in F} \frac{t_i}{1 - t_i}$$

$$= (-1)^d H_{k[\Delta]}(t_1^{-1}, \ldots, t_n^{-1}).$$

Hint: 5.3.8 and 5.4.3.

5.6.7. Let k be a field, and Δ be a $(d-1)$-dimensional Cohen–Macaulay complex over k. Prove the following conditions are equivalent:
(a) Δ is an Euler complex;
(b) $H_{k[\Delta]}(t_1, \ldots, t_n) = (-1)^d H_{k[\Delta]}(t_1^{-1}, \ldots, t_n^{-1})$;
(c) $\omega_{k[\Delta]} \cong k[\Delta]$ as a $\mathbb{Z}^n$-graded $k[\Delta]$-module.

5.6.8. With the assumptions of 5.6.7 show the following conditions are equivalent:
(a) Δ is Gorenstein over k;
(b) $t^a H_{k[\Delta]}(t_1, \ldots, t_n) = (-1)^d H_{k[\Delta]}(t_1^{-1}, \ldots, t_n^{-1})$.
Suppose the equivalent conditions hold. Show t^a is a squarefree monomial in $t_1, \ldots, t_n$ of degree $|V \setminus \operatorname{core} V|$. Conclude that $a(k[\Delta]) = -|V \setminus \operatorname{core} V|$.

5.6.9. Determine all 1-dimensional Gorenstein complexes.

5.6.10. Let k be a field and Δ a Gorenstein complex over k of even dimension d such that $\Delta = \operatorname{core} \Delta$. Show Δ is $d/2$-neighbourly if and only if $k[\Delta]$ is an extreme Gorenstein ring.

5.7 The canonical module of a Stanley–Reisner ring

Let k be a field, and Δ a Cohen–Macaulay complex over k. In the previous section we have already considered the ˙canonical module $\omega_{k[\Delta]}$ of $k[\Delta]$. By Exercise 5.6.6, it has a natural fine grading with Hilbert series

$$(8) \qquad H_{\omega_{k[\Delta]}}(t) = \sum_{F \in \Delta} \dim_k \widetilde{H}_{\dim \operatorname{lk} F}(\operatorname{lk} F; k) \prod_{v_i \in F} \frac{t_i}{1 - t_i}.$$

In 3.3.16 we defined the canonical module of a non-local ring R to be
a finite module which is locally isomorphic to the canonical modules of
the corresponding local rings, and observed in 3.3.17 that a canonical
module, if it exists, is only unique up to tensor products with locally
free R-modules of rank 1. Hochster ([182], Theorem 6.1) showed that all
locally free $k[\Delta]$-modules of rank 1 are actually free. Hence for $k[\Delta]$ we
do not have to distinguish between the canonical and $^{\bullet}$canonical module.
The reader may recover the proof of Hochster's theorem in Exercise 5.7.8
where we indicate the steps.

As $k[\Delta]$ is reduced, it follows from 3.3.18 that $\omega_{k[\Delta]}$ can be identified
with an ideal I of $k[\Delta]$. Unfortunately we cannot expect that I be $\mathbb{Z}^n$-
graded (n the number of vertices of Δ), simply because it may happen that
$\dim_k(\omega_{k[\Delta]})_a > 1$ for some $a \in \mathbb{Z}^n$. Indeed, consider the 1-dimensional
simplex in Figure 5.9. By (8), the fine Hilbert series of its canonical

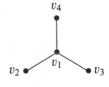

Figure 5.9

module is

$$2\frac{t_1}{1 - t_1} + \sum_{i=2}^{4} \frac{t_1 t_i}{(1 - t_1)(1 - t_i)}.$$

Thus for $a = (1, 0, 0, 0)$ we have $\dim_k(\omega_{k[\Delta]})_a = 2$.

Theorem 5.7.1. *Let Δ be a $(d - 1)$-dimensional Cohen–Macaulay complex
over a field k. Then the following conditions are equivalent:*
(a) *Δ is not an Euler complex, and there exists an embedding $\omega_{k[\Delta]} \to k[\Delta]$
of $\mathbb{Z}^n$-graded $k[\Delta]$-modules;*
(b) *there exists a $(d - 2)$-dimensional subcomplex Σ of Δ which is Euler
and Cohen–Macaulay over k such that for all $F \in \Delta$*

$$\widetilde{H}_{\dim \mathrm{lk}\, F}(\mathrm{lk}\, F; k) \cong \begin{cases} 0 & \text{if } F \in \Sigma, \\ k & \text{if } F \notin \Sigma. \end{cases}$$

*If the equivalent conditions hold, then as a $\mathbb{Z}^n$-graded $k[\Delta]$-module, $\omega_{k[\Delta]}$
is isomorphic to the ideal J in $k[\Delta]$ which is generated by the monomials
$x^F = \prod_{v_i \in F} x_i$, $F \in \Delta \setminus \Sigma$.*

PROOF. (a) $\Rightarrow$ (b): Let I be the $\mathbb{Z}^n$-graded ideal in $k[\Delta]$ which is isomorphic
to the image of $\omega_{k[\Delta]} \to k[\Delta]$. As we assume that Δ is not an Euler
complex, Exercise 5.6.7 implies that $I \neq k[\Delta]$. Further note that if $x^a \in I$,

then $\prod_{a_i>0} x_i \in I$. This can be deduced from the Hilbert series of $\omega_{k[\Delta]}$; see (11). Thus if we set $\Sigma = \{\operatorname{supp} u : u \notin I\}$, then

$$k[\Delta]/\omega_{k[\Delta]} \cong k[\Delta]/I \cong k[\Sigma].$$

It follows that Σ is a $(d-2)$-dimensional Cohen–Macaulay complex over k. Since $H_{k[\Sigma]}(t) = H_{k[\Delta]}(t) - H_{\omega_{k[\Delta]}}(t)$ we conclude from Exercise 5.6.6 that

$$H_{k[\Sigma]}(t_1^{-1},\ldots,t_n^{-1}) = (-1)^d H_{\omega_{k[\Delta]}}(t) - (-1)^d H_{k[\Delta]}(t)$$

$$= (-1)^{d-1}(H_{k[\Delta]}(t) - H_{\omega_{k[\Delta]}}(t)) = (-1)^{d-1} H_{k[\Sigma]}(t).$$

By 5.6.7, this implies that Σ is an Euler complex. Once again applying (11) we obtain

$$\sum_{F \in \Delta} \dim_k \widetilde{H}_{\dim \operatorname{lk} F}(\operatorname{lk} F; k) \prod_{v_i \in F} \frac{t_i}{1-t_i} = H_{\omega_{k[\Delta]}}(t) = H_{k[\Delta]}(t) - H_{k[\Sigma]}(t)$$

$$= \sum_{F \in \Delta \setminus \Sigma} \prod_{v_i \in F} \frac{t_i}{1-t_i}.$$

A comparison of the coefficients on both sides yields the assertion concerning the links of the faces of Δ. Moreover it follows that I equals the ideal J described in the theorem since both ideals have the same Hilbert series.

(b) $\Rightarrow$ (a): First observe that Δ is not an Euler complex, since the links of the faces which belong to Σ are acyclic.

In order to obtain the desired embedding of the canonical module we add a vertex w, form the cone $\operatorname{cn}(\Sigma) = \{w\} * \Sigma$, and let $\Gamma = \operatorname{cn}(\Sigma) \cup \Delta$. Then $\dim \Gamma = \dim \Delta = d - 1$, $k[\Gamma] = k[X_1, \ldots, X_n, Y]/I_\Gamma$ (Y corresponding to the vertex w), and $k[\Gamma]/(y) \cong k[\Delta]$ where y denotes the residue of Y modulo I_Γ.

We will show that Γ is an Euler complex which is Cohen–Macaulay over k. In particular Γ will be Gorenstein. Then 3.6.12 implies

$$\omega_{k[\Delta]} \cong \operatorname{Hom}_{k[\Gamma]}(k[\Delta], k[\Gamma]) \cong \operatorname{Ann}(y) = Jk[\Gamma] = J.$$

Since these isomorphisms are obviously $\mathbb{Z}^n$-graded, the desired conclusion follows.

It remains to be shown that the links of the faces $F \in \Gamma$ are homology spheres, that is, satisfy condition (7) of Section 5.4. We distinguish several cases:

(i) $F \in \Delta \setminus \Sigma$; then $\operatorname{lk}_\Gamma F = \operatorname{lk}_\Delta F$, and (7) is satisfied by assumption.

(ii) $F \in \Sigma$; then $\operatorname{lk}_\Gamma F = \operatorname{cn}(\operatorname{lk}_\Sigma F) \cup \operatorname{lk}_\Delta F$. Since $\operatorname{cn}(\operatorname{lk}_\Sigma F) \cap \operatorname{lk}_\Delta F = \operatorname{lk}_\Sigma F$, the Mayer–Vietoris sequence ([280], Theorem 25.1) applied to this situation yields the long exact sequence

$$\cdots \longrightarrow \widetilde{H}_i(\operatorname{lk}_\Sigma F; k) \longrightarrow \widetilde{H}_i(\operatorname{cn}(\operatorname{lk}_\Sigma F); k) \oplus \widetilde{H}_i(\operatorname{lk}_\Delta F; k)$$

$$\longrightarrow \widetilde{H}_i(\operatorname{lk}_\Gamma F; k) \longrightarrow \widetilde{H}_{i-1}(\operatorname{lk}_\Sigma F; k) \longrightarrow \cdots,$$

provided $\mathrm{lk}_\Sigma F \neq \emptyset$. Note that $\tilde{H}_.(\mathrm{lk}_\Delta F;k) = 0$ by assumption, and $\tilde{H}_.(\mathrm{cn}(\mathrm{lk}_\Sigma F);k) = 0$ by Exercise 5.3.10, so that

$$\tilde{H}_{i-1}(\mathrm{lk}_\Sigma F;k) \cong \tilde{H}_i(\mathrm{lk}_\Gamma F;k) \qquad \text{for all } i.$$

As Σ is an Euler complex which is Cohen–Macaulay over k, it follows that $\mathrm{lk}_\Gamma F$ is a homology sphere.

If $\mathrm{lk}_\Sigma F = \emptyset$, then $\mathrm{lk}_\Gamma F = \{w\} \cup \mathrm{lk}_\Delta F$. Note that F is a facet of Σ, so that $\dim \mathrm{lk}_\Delta F = 0$. Hence assumption (b) implies that $\mathrm{lk}_\Delta F$ consists of one vertex. Therefore $\mathrm{lk}_\Gamma F = \{w,v\}$ where v is a vertex of Δ, and thus it is a sphere.

(iii) $w \in F$; then $F = \{w\} \cup G$ where $G \in \Sigma$, and $\mathrm{lk}_\Gamma F = \mathrm{lk}_\Sigma G$. Again we derive the desired conclusion. □

Let Δ be a simplicial complex whose geometric realization $X = |\Delta|$ is a manifold with boundary ∂X. Then $\partial X = |\Sigma|$ where Σ is the subcomplex of Δ which is characterized by the property that its facets are faces of precisely one facet of Δ ([280], §35 and Exercise 4).

As an application of 5.7.1 we obtain

Theorem 5.7.2 (Hochster). *Let k be a field, and Δ a $(d-1)$-dimensional Cohen–Macaulay complex over k whose geometric realization $X = |\Delta|$ is a manifold with a non-empty boundary ∂X. Further let Σ be the subcomplex of Δ with $\partial X = |\Sigma|$, and J the ideal in $k[\Delta]$ generated by the monomials x^F, $F \in \Delta \setminus \Sigma$. Then the following conditions are equivalent:*

(a) *$\omega_{k[\Delta]} \cong J$ as a $\mathbb{Z}^n$-graded $k[\Delta]$-module;*

(b) *Σ is a Gorenstein complex over k;*

(c) *Σ is an Euler complex which is Cohen–Macaulay over k.*

PROOF. (a) $\Rightarrow$ (b): Suppose J is the canonical module of $k[\Delta]$. Then 3.3.18(b) in conjunction with 3.6.20(c) shows that $k[\Sigma] \cong k[\Delta]/J$ is Gorenstein.

(b) $\Rightarrow$ (c): By 5.6.2, it suffices to show that $\Sigma = \mathrm{core}\,\Sigma$. Suppose this is not the case. Then there exists a vertex $v \in \Sigma$ such that $\mathrm{st}\,v = \Sigma$, and so $\Sigma = \{v\} * \Gamma$ for some subcomplex Γ of Σ. But then $\partial(\partial X) = \partial|\Sigma| \supset |\Gamma| \neq \emptyset$, a contradiction since the boundary of a manifold is a manifold without boundary.

(c) $\Rightarrow$ (a): We have to check the conditions 5.7.1(b) for the links of the faces of Δ. Let $\rho : \Delta \to \mathbb{R}^n$ be the map defining the geometric realization of Δ.

Suppose $F \in \Delta$, $F \neq \emptyset$, and $p \in \mathrm{relint}(\mathrm{conv}\,\rho(F))$. If $\mathrm{lk}\,F \neq \emptyset$, then 5.4.5 yields

$$\tilde{H}_{\dim \mathrm{lk} F}(\mathrm{lk}\,F;k) \cong H_{d-1}(X, X \setminus \{p\};k) \cong \begin{cases} 0 & \text{if } p \in \partial X, \\ k & \text{if } p \notin \partial X. \end{cases}$$

The first case happens when $F \in \Sigma$, the second when $F \notin \Sigma$. If $\operatorname{lk} F = \emptyset$, then $F \notin \Sigma$ and again $\tilde{H}_{\dim \operatorname{lk} F}(\operatorname{lk} F; k) = \tilde{H}_{-1}(\emptyset; k) \cong k$.

Now suppose $F = \emptyset$. Then $\operatorname{lk} F = \Delta$, and we need to show that $\tilde{H}_{d-1}(X; k) \cong 0$, or equivalently, that any $(d-1)$-cycle $z = \sum a_F F$ of the chain complex $\tilde{\mathscr{C}}(\Delta)$ is trivial. As z is a cycle we have

$$\sum_{\substack{F \supset F' \\ \dim F = d-1}} \pm a_F = 0$$

for all $F' \in \Delta$ with $\dim F' = d - 2$. Now since X is a manifold with boundary, each $(d-2)$-face $F' \in \Delta$ is a face of precisely one facet of Δ when $F' \in \Sigma$, and of precisely two facets of Δ when $F' \notin \Sigma$. Hence (i) $a_F = 0$ if F contains a facet $F' \in \Sigma$, and (ii) $a_{F_1} \pm a_{F_2} = 0$ if $F' \notin \Sigma$ and F_1 and F_2 are the facets of Δ containing F'. Since by assumption $\Sigma \neq \emptyset$, we conclude from (i) that $a_F = 0$ for at least one facet of Δ. Now let $G \in \Delta$ be any other facet. Notice that Δ is connected since it is Cohen–Macaulay of positive dimension; see Exercise 5.1.26. Therefore we can find a chain of faces

$$F = F_0 \supset F_1 \subset F_2 \supset \cdots \supset F_{2m-1} \subset F_{2m} = G$$

with alternating inclusions where $\dim F_{2i} = d - 1$ and $\dim F_{2i-1} = d - 2$ for $i = 0, \ldots, m$. Thus it follows from (ii) and by induction on i that $a_{F_{2i}} = 0$ for $i = 0, \ldots, m$; in particular, $a_G = 0$. $\qquad\square$

A $\mathbb{Z}^n$-*graded embedding of* $\omega_{k[\Delta]}$. Though the canonical module of a Stanley–Reisner ring $k[\Delta]$ cannot always be identified with a $\mathbb{Z}^n$-graded ideal, it may be realized as a kernel of a certain $\mathbb{Z}^n$-graded homomorphism. In order to derive such a presentation we first observe that the homology of the complex $C^\bullet$ is concentrated in 'negative degrees'; see Theorem 5.3.8. To be precise, we have $H^\bullet(C_a^\bullet) = 0$ if some component a_i of a is positive. Thus if we set

$$D^i = \bigoplus_{a \in \mathbb{Z}^n_-} C_a^i$$

where $\mathbb{Z}^n_- = \{a \in \mathbb{Z}^n : a_i \leq 0 \text{ for } i = 1, \ldots, n\}$, then

$$H^\bullet(D^\bullet) \cong H^\bullet(C^\bullet) \cong H^\bullet_{\mathfrak{m}}(k[\Delta]),$$

and these are isomorphisms of $\mathbb{Z}$-graded modules. Write $k[\Delta] = k[X]/I_\Delta$ with $k[X] = k[X_1, \ldots, X_n]$. Then by virtue of the local duality theorem for graded modules we obtain the isomorphisms

$$H^i(D^\bullet)^\vee \cong {}^*\operatorname{Ext}^{n-i}_{k[X]}(k[\Delta], k[X]), \qquad i \geq 0,$$

where $H^i(D^\bullet)^\vee \cong {}^*\operatorname{Hom}_k(H^i(D^\bullet), k) \cong H_i({}^*\operatorname{Hom}_k(D^\bullet, k))$.

Let us more closely inspect the complex $G_{\bullet} = {}^*\mathrm{Hom}_k(D^{\bullet},k)$. Recall that C^t is a direct sum of modules $R_{x_{i_1}\cdots x_{i_t}}$ where $R = k[\varDelta]$. Let $F = \{v_{i_1},\ldots,v_{i_t}\}$, $X = X_{i_1}\cdots X_{i_t}$ and $x = x_{i_1}\cdots x_{i_t}$; then $R_x \cong k[X_i, X_i^{-1} : v_i \in F][X_i : v_i \notin F]/(I_\varDelta)_X$ where $(I_\varDelta)_X$ is an ideal generated by certain squarefree monomials in the variables X_i for which $v_i \notin F$. It is clear that $1 \in (I_\varDelta)_X$ if and only if $F \notin \varDelta$. Thus we see that

$$\bigoplus_{a\in\mathbb{Z}_-^n}(R_x)_a \cong \begin{cases} 0 & \text{if } F \notin \varDelta, \\ k[X_i^{-1} : v_i \in F] & \text{if } F \in \varDelta, \end{cases}$$

so that ${}^*\mathrm{Hom}_k(\bigoplus_{a\in\mathbb{Z}_-^n}(R_x)_a,k) \cong k[X_i : v_i \in F] \cong k[X_1,\ldots,X_n]/\mathfrak{P}_F$ if $F \in \varDelta$. By definition, G_t is a direct sum of such modules. Thus we have

Theorem 5.7.3. *Let $\varDelta$ be a $(d-1)$-dimensional simplicial complex, and k a field. For each $i = 0,\ldots,d$ let G_i be the direct sum of the $k[\varDelta]$-modules $k[X_1,\ldots,X_n]/\mathfrak{P}_F$ where $F \in \varDelta$ and $|F| = i$. Consider the complex*

$$G_{\bullet} : 0 \longrightarrow G_d \longrightarrow G_{d-1} \longrightarrow \cdots \longrightarrow G_1 \longrightarrow G_0 = k \longrightarrow 0$$

of $k[\varDelta]$-modules whose differentiation is composed of the maps

$$(-1)^{j-1}\,\mathrm{nat} : k[X_1,\ldots,X_n]/\mathfrak{P}_F \longrightarrow k[X_1,\ldots,X_n]/\mathfrak{P}_{F'}$$

if $F = \{v_{i_1},\ldots,v_{i_r}\}$ and $F' = \{v_{i_1},\ldots,\widehat{v}_{i_j},\ldots,v_{i_r}\}$, and zero otherwise. Then for $i = 0,\ldots,d$,

$$H_i(G_{\bullet}) \cong \mathrm{Ext}_{k[X]}^{n-i}(k[\varDelta],k[X]).$$

In particular, if $\varDelta$ is Cohen–Macaulay, then one obtains the exact sequence of $\mathbb{Z}^n$-graded $k[\varDelta]$-modules

$$0 \longrightarrow \omega_{k[\varDelta]} \longrightarrow G_d \longrightarrow G_{d-1} \longrightarrow \cdots \longrightarrow G_1 \longrightarrow G_0 \longrightarrow 0.$$

As a consequence of 5.7.3 we derive a result of Gräbe [137].

Corollary 5.7.4. *Let k be a field, and $\varDelta$ a $(d-1)$-dimensional simplicial complex which is Cohen–Macaulay over k. Then there is a $\mathbb{Z}$-graded embedding*

$$\omega_{k[\varDelta]}(-d) \longrightarrow k[\varDelta].$$

PROOF. Let $\varepsilon : k[X_1,\ldots,X_n] \to G_d = \bigoplus_{|F|=d}(k[X_1,\ldots,X_n]/\mathfrak{P}_F)$ be the homomorphism which on each component is just the canonical epimorphism. Then $\mathrm{Ker}\,\varepsilon = \bigcap_{|F|=d}\mathfrak{P}_F$, and so ε induces an isomorphism $\bar\varepsilon : k[\varDelta] \to \mathrm{Im}\,\varepsilon$.

Let $x = \sum_{|F|=d} x^F$; then x is homogeneous of degree d. Moreover, x is G_d-regular and $xG_d \subset \mathrm{Im}\,\varepsilon$. To see this, note that if $a = (a_F) \in G_d$, then $xa = (x^F a_F)$. From this it follows immediately that x is indeed G_d-regular, and it also follows that $\bar\varepsilon(x^F) = xe_F$ for all facets F. Here e_F denotes the element of G_d whose projection to $k[X_1, \ldots, X_n]/\mathfrak{P}_{F'}$ is 1 if $F = F'$, and 0 otherwise. Since these elements generate G_d, the element x multiplies G_d into the submodule $\operatorname{Im} \varepsilon$, as asserted.

In conclusion we have

$$\omega_{k[\Delta]}(-d) \cong x\omega_{k[\Delta]} \subset xG_d \subset \operatorname{Im} \varepsilon \cong k[\Delta]. \qquad \square$$

We illustrate 5.7.4 by means of the simplicial complex Δ illustrated in Figure 5.9. Theorem 5.7.3 yields the exact sequence

$$0 \longrightarrow \omega_{k[\Delta]} \longrightarrow \bigoplus_{i=2}^{4} k[X_1, X_i] \longrightarrow \bigoplus_{i=1}^{4} k[X_i] \longrightarrow k \longrightarrow 0,$$

and it is readily seen that $\omega_{k[\Delta]}$ is generated by the elements $(X_1, -X_1, 0)$ and $(X_1, 0, -X_1)$ in $\bigoplus_{i=2}^{4} k[X_1, X_i]$. Then $(\sum_{|F|=2} x^F)\omega_{k[\Delta]}$ has the generators $(X_1^2 X_2, -X_1^2 X_3, 0)$ and $(X_1^2 X_2, 0, -X_1^2 X_4)$. Thus we see that the ideal in $k[\Delta]$ corresponding to $x\omega_{k[\Delta]}$ via $\bar{\varepsilon}^{-1}$ is generated by $x_1^2 x_2 - x_1^2 x_3$ and $x_1^2 x_2 - x_1^2 x_4$.

Doubly Cohen–Macaulay complexes. Let k be a field. In Exercise 5.4.9 we noticed that the type $r_k(\Delta)$ of a Cohen–Macaulay complex Δ over k is at least h_s, the last non-vanishing component of the h-vector of Δ. Unfortunately, we may have $r_k(\Delta) > h_s$; see Exercise 5.7.10. By 5.4.9 equality holds exactly when Δ is level over k. The situation is particularly simple when Δ is level and $s = d = \dim \Delta + 1$. Then 5.1.9 implies that $r_k(\Delta) = (-1)^{d-1}\tilde{\chi}(\Delta)$, and this number is reasonably accessible.

Definition 5.7.5. Let k be a field. A simplicial complex Δ on the vertex set V is *doubly Cohen–Macaulay over* k if Δ is Cohen–Macaulay over k, and for all $v \in V$ the subcomplex $\Delta_{V \setminus \{v\}}$ is Cohen–Macaulay over k of the same dimension as Δ.

Concluding this chapter we present two results of Baclawski [35] on doubly Cohen–Macaulay complexes.

Theorem 5.7.6 (Baclawski). *Let k be a field, and Δ a $(d-1)$-dimensional doubly Cohen–Macaulay complex over k. Then Δ is level and*

$$r_k(\Delta) = (-1)^{d-1}\tilde{\chi}(\Delta).$$

PROOF. We make use of Hochster's formula 5.5.1 which gives the Hilbert series of $\operatorname{Tor}_i^R(k, k[\Delta])$ where $R = k[X_1, \ldots, X_n]$. Note Δ is Cohen–Macaulay over k if and only if $\operatorname{Tor}_i^R(k, k[\Delta]) = 0$ for $i > n - d$. Thus we have the following result:

Δ is Cohen–Macaulay over k $\iff$

$\tilde{H}_j(\Delta_W; k) = 0$ for all $W \subset V$ and $j < |W| - (n - d) - 1.$

We claim that $\mathrm{Tor}^R_{n-d}(k, k[\Delta])_a = 0$ for $a \neq (1, \ldots, 1)$. Suppose this is not the case. Then from 5.5.1 we deduce that there exists a proper subset W of V such that $\tilde{H}_j(\Delta_W; k) \neq 0$ for $j = |W| - (n-d) - 1$. Choose i such that $W \subset V' = V \setminus \{v_i\}$; then $(\Delta_{V'})_W = \Delta_W$. Since by assumption $\Delta_{V'}$ is Cohen–Macaulay it follows from (8) (applied to $\Delta_{V'}$) that $|W| - (n-d) - 1 = j \geq |W| - (n-1-d) - 1$, a contradiction.

We leave it to the reader to complete the proof. Simply observe that the degrees of the non-zero components of $\mathrm{Tor}^R_{n-d}(k, k[\Delta])$ determine the degrees of the generators of $\omega_{k[\Delta]}$, and that $h_d = \dim_k \mathrm{Tor}^R_{n-d}(k, k[\Delta])_a$ for $a = (1, \ldots, 1)$. $\qquad\square$

We may view $\tilde{\mathscr{C}}(\Delta) \otimes k$ as a graded k-vector subspace of $k[\Delta]$ simply by identifying the elements $F \otimes 1$ with x^F for all $F \in \Delta$. Then $\tilde{H}_{d-1}(\Delta; k)$ is identified with a k-vector subspace of $k[\Delta]$.

Corollary 5.7.7 (Baclawski). *If Δ is doubly Cohen–Macaulay over k, then as a $\mathbb{Z}$-graded module, $\omega_{k[\Delta]}(-d)$ is isomorphic to the ideal generated by $\tilde{H}_{d-1}(\Delta; k)$.*

PROOF. We view $\omega_{k[\Delta]}$ as a submodule of G_d. By Exercise 5.7.9, $\omega_{k[\Delta]}$ is generated by elements of degree 0, that is, by elements of $\mathrm{Ker}((G_d)_0 \to (G_{d-1})_0)$. Let $x = \sum_{|F|=d} x^F$ be as in the proof of 5.7.4. Then $x\omega_{k[\Delta]}$ is the ideal in $k[\Delta]$ which is generated by $\mathrm{Ker}((xG_d)_d \to (xG_{d-1})_d)$, and this yields the desired conclusion since $(xG_d)_d \to (xG_{d-1})_d$ can be identified with $\mathscr{C}_{d-1}(\Delta) \otimes k \to \mathscr{C}_{d-2}(\Delta) \otimes k$. $\qquad\square$

Exercises

5.7.8. Let k be a field, Δ a simplicial complex, and P a finite $k[\Delta]$-module of rank 1 which is locally free. Show P is free. The proof can be accomplished in the following steps.
(a) Let R be a Noetherian ring, P a finite module, and I_1, I_2 two ideals in R such that P/I_jP is a free (R/I_j)-module of rank 1 for $j = 1, 2$. Assume that the group of units of $R/(I_1 \cap I_2)$ is mapped surjectively onto that of $R/(I_1 + I_2)$. Show $P/(I_1 \cap I_2)P$ is a free $R/(I_1 \cap I_2)$-module of rank 1.
(b) Use (a) and induction on the number of facets of Δ. To start the induction observe that a finite, locally free $k[X_1, \ldots, X_n]$-module of rank 1 is actually free. Indeed such a module is isomorphic to a projective ideal, and since $k[X_1, \ldots, X_n]$ is factorial, projective ideals are principal; see [270], Theorem 20.7.

5.7.9. Let k be a field, and Δ a $(d-1)$-dimensional doubly Cohen–Macaulay complex over k. Show $\mathrm{Ker}(G_d \to G_{d-1})$ is generated by elements of degree 0 in G_d.

5.7.10. Show a 1-dimensional simplicial complex Δ on V satisfies $r(\Delta) = \tilde{\chi}(\Delta)$ if and only if for all $v \in V$ the subcomplex $\Delta_{V \setminus \{v\}}$ is connected. (Reference to a field k is not needed in dimension 1. Why?)

5.7.11. Give an example of a Cohen–Macaulay complex whose type depends on the field k.

5.7.12. Prove the converse of 5.7.6: if Δ is a Cohen–Macaulay complex and k a field such that $r_k(\Delta) = (-1)^{d-1}\tilde{\chi}(\Delta)$, then Δ is doubly Cohen–Macaulay.

5.7.13. Characterize the 1-dimensional simplicial complexes Δ for which there exists a $\mathbb{Z}^n$-graded embedding $\omega_{k[\Delta]} \to k[\Delta]$.

Notes

Simplicial complexes have been considered in topology since Poincaré [300] who computed homology groups of topological spaces via triangulations.

Another motivation comes from polytope theory where simplicial complexes appear as boundary complexes of simplicial polytopes. The question of how the number of the faces in various dimensions are related to each other has attracted combinatorialists and geometers since Euler who discovered the familiar equation $f_0 - f_1 + f_2 = 2$ for 3-polytopes in 1752.

A new technique in studying simplicial complexes was introduced by Stanley [356]. His proof of the upper bound theorem for simplicial spheres depends heavily on methods from commutative algebra whose foundations were laid by Hochster [182] and Reisner [306]. Naturally our exposition concentrates on the algebraic aspects of the theory. It is very much influenced by Stanley's monograph [363] and the lectures by McMullen and Stanley held at the DMV-Seminar in Blaubeuren, July 1991. The reader interested in a general, up-to-date survey on convex polytopes is referred to the excellent article [39] by Bayer and Lee. Hibi's book [169] offers an attractive introduction to algebraic combinatorics.

The results of Kruskal–Katona mentioned in Section 5.1 can be understood as a theorem on Hilbert functions of residue class rings of an exterior algebra; see Aramova, Herzog, and Hibi [12] for this approach.

Hochster's formula 5.5.1 appeared in [182]. Our treatment is taken from Bruns and Herzog [59] where a more general result for monomial ideals of semigroup rings has been given.

There are other notable results in the direction of Baclawski's theorem. For example, Miyazaki [275] proved that the barycentric subdivision of a level complex is again level, and Hibi [166] showed that the proper skeletons of a Cohen–Macaulay complex are all level.

There are several aspects in the algebraic theory of simplicial complexes not considered in this book or only discussed in passing: for instance, a careful account of order complexes of posets, or Schenzel's characterization of Buchsbaum complexes; see [276], [329], and [365]. It should be mentioned that the statements (i)–(v) in Exercise 5.3.15(a) are all equivalent to the Buchsbaum property. Fröberg and Hoa [120] investigated Segre products of Stanley-Reisner rings. For an excellent and up-to-date overview see Stanley [363].

A thorough study of order complexes of posets can be found in Björner's paper [45]. Theorem 5.1.12 on the shellability of bounded, semimodular posets is taken from [45]. Garsia's paper [124] is another source of information on this topic. In [165] Hibi classifies those order complexes of distributive lattices which are Gorenstein.

In the notes of Chapter 2 we have mentioned the problem as to whether the Poincaré series of a local ring is a rational function. For a positively graded ring R over a field k one defines its Poincaré series with respect to a minimal free graded resolution of k. Fröberg [119] showed that if R is defined by monomial relations of degree 2, then k has a linear resolution over R; in particular the Poincaré series of R is rational. Backelin [34] proved the rationality of the Poincaré series for graded algebras defined by monomial relations of arbitrary degree.

Another important result left out is the g-theorem whose existing proof goes beyond the scope of this book. A vector $h = (h_0, \ldots, h_d) \in \mathbb{N}^{d+1}$ satisfies the g-condition if $h_0 = 1$, $h_i = h_{d-i}$ for all i, and if $(h_0, h_1 - h_0, \ldots, h_{[d/2]} - h_{[d/2]-1})$ is the h-vector of a homogeneous k-algebra. According to 4.2.10, the latter condition is satisfied if and only if $h_0 \leq h_1 \leq \cdots \leq h_{[d/2]}$, and $h_{i+1} - h_i \leq (h_i - h_{i-1})^{\langle i \rangle}$ for all $i \leq d/2 - 1$. The name g-condition stems from the fact that one commonly denotes by g_i the differences $h_i - h_{i-1}$.

It was conjectured by McMullen in 1971 that $(h_0, \ldots, h_d) \in \mathbb{N}^{d+1}$ is the h-vector of a simplicial polytope if and only if it satisfies the g-condition. The 'sufficiency' was proved by Billera and Lee [43], while the 'necessity' was shown by Stanley [359] who exhibited a homogeneous system of parameters $\Theta_1, \ldots, \Theta_d$ of $k[\Delta]$ such that $\deg \Theta_i = 1$ and $A = k[\Delta]/(\Theta_1, \ldots, \Theta_d)$ has a Lefschetz element, that is, an element $\omega \in A_1$ for which multiplication by ω induces linear maps $A_{i-1} \to A_i$ of maximal rank.

6 Semigroup rings and invariant theory

This chapter opens with the study of affine semigroup rings, i.e. sub-algebras of Laurent polynomial rings generated by a finite number of monomials. We relate the structure of such a ring R to that of the semigroup C formed by the exponent vectors of the monomials in R, and to the cone D spanned by C. From the face lattice of D we then construct a complex for the local cohomology of R.

The connection between R and D is strongest if R is normal: this is the case if and only if R contains all monomials which correspond to the integral points in D. By a theorem of Hochster normal semigroup rings are Cohen–Macaulay. Moreover, we shall determine their canonical modules and, as a combinatorial application, derive the reciprocity laws of Ehrhart and Stanley.

We are led to the second topic of this chapter by the fact that rings of invariants of torus actions are normal semigroup rings. We also treat finite groups, covering Watanabe's characterization of Gorenstein invariants and the famous Shephard–Todd theorem on invariants of reflection groups. The discussion of invariant theory culminates in the Hochster–Roberts theorem which warrants the Cohen–Macaulay property for rings of invariants of all linearly reductive groups.

6.1 Affine semigroup rings

An *affine semigroup* C is a finitely generated semigroup which for some n is isomorphic to a subsemigroup of $\mathbb{Z}^n$ containing 0. Let k be a field. We write $k[C]$ for the vector space $k^{(C)}$, and denote the basis element of $k[C]$ which corresponds to $c \in C$ by X^c. This 'monomial' notation is suggested by the fact that $k[C]$ carries a natural multiplication whose table is given by $X^c X^{c'} = X^{c+c'}$ (we use $+$ to denote the semigroup operation). For example, $k[\mathbb{Z}^n]$ is isomorphic to the Laurent polynomial ring $k[X_1, X_1^{-1}, \ldots, X_n, X_n^{-1}]$ if we let X_i correspond to the i-th element of the canonical basis of $\mathbb{Z}^n$; similarly $k[\mathbb{N}^n]$ is isomorphic to $k[X_1, \ldots, X_n]$. The rings $k[C]$ where k is a field and C is an affine semigroup are called *affine semigroup rings*.

There is a 'smallest' group G containing C, characterized by the fact that every homomorphism from C to a group factors in a unique way through G. We write $\mathbb{Z}C$ for G, for if $C \subset \mathbb{Z}^n$, then G is just the

$\mathbb{Z}$-submodule of $\mathbb{Z}^n$ generated by C. Since an affine semigroup can be embedded into $\mathbb{Z}^n$ for some n, we see that $\mathbb{Z}C \cong \mathbb{Z}^d$ for some $d \in \mathbb{N}$ which we call the *rank* of C. We set $\mathbb{Q}C = \mathbb{Q} \otimes_\mathbb{Z} \mathbb{Z}C$ and $\mathbb{R}C = \mathbb{R} \otimes_\mathbb{Z} \mathbb{Z}C$. In the following we will consider $\mathbb{Z}C$ as a subgroup of $\mathbb{Q}C$ and $\mathbb{Q}C$ as a $\mathbb{Q}$-vector subspace of $\mathbb{R}C$ where the inclusions are the map $z \mapsto 1 \otimes z$ and the one induced by the embedding $\mathbb{Q} \to \mathbb{R}$.

An embedding $C \hookrightarrow \mathbb{Z}^n$ of semigroups induces an embedding $k[C] \hookrightarrow k[\mathbb{Z}^n]$ of k-algebras. Therefore $k[C]$ is a domain; it is Noetherian since C is finitely generated. Obviously $k[C]$ and $k[\mathbb{Z}C]$ have the same field of fractions if we regard $k[C]$ as a subalgebra of $k[\mathbb{Z}C]$ in a natural way. It follows that $\mathbb{Z}C \cong \mathbb{Z}^d$ where $d = \dim k[C]$; so $\dim k[C] = \operatorname{rank} C$.

The ring $k[C]$ is a k-subalgebra of $k[\mathbb{Z}C]$; it is in fact a graded subring of the $\mathbb{Z}C$-graded ring $k[\mathbb{Z}C]$ (see Section 5.1 for this notion), and without further specification the attributes 'graded' and 'homogeneous' always refer to the $\mathbb{Z}C$-graduation of $k[C]$. The graded ideals of $k[C]$ are those generated by homogeneous elements. Each homogeneous component of $k[C]$ is a one dimensional k-vector space, and therefore the graded ideals correspond to certain subsets of C which will be identified below. In order to switch from the ring $k[C]$ to the semigroup C we introduce the operator

$$\log I = \{c : X^c \in I\} \qquad \text{for a subset} \quad I \subset k[C].$$

It is clear that log establishes a bijection between the set of graded vector subspaces of $k[C]$ and the set of subsets of C.

In a semigroup C we may define ideals, and even radical, prime, or primary ideals: $S \subset C$ is an *ideal* if $c + s \in S$ for all $c \in C$, $s \in S$ (so $\emptyset$ is an ideal). The *radical* of an ideal S is $\operatorname{Rad} S = \{s : ms \in S$ for some $m \in \mathbb{N}\}$; $\operatorname{Rad} S$ is itself an ideal, and S is a *radical ideal* if $S = \operatorname{Rad} S$. An ideal $S \neq C$ is *prime* if $c + c' \in S$ implies $c \in S$ or $c' \in S$, and it is *primary* if $c + c' \in S$, $c \notin S$ implies $c' \in \operatorname{Rad} S$. It is easy to check that the radical of a primary ideal is prime. The following proposition whose proof is left for the reader (Exercise 6.1.9) establishes the correspondence of the ideal theory of C and that of the graded ideals of $k[C]$.

Proposition 6.1.1. *Let C be an affine semigroup, and I, $I' \subset k[C]$ graded k-vector subspaces. Then*
(a) *$I \subset I' \iff \log I \subset \log I'$, $\log(I_1 \cap I_2) = \log I_1 \cap \log I_2$, $\log I_1 + I_2 = \log I_1 \cup \log I_2$,*
(b) *I is a (radical, prime, primary) ideal if and only if $\log I$ is a (radical, prime, primary) ideal; furthermore $\log \operatorname{Rad} I = \operatorname{Rad} \log I$,*
(c) *the minimal prime overideals of I are graded.*

Normal semigroup rings. An affine semigroup C is called *normal* if it satisfies the following condition: if $mz \in C$ for some $z \in \mathbb{Z}C$ and $m \in \mathbb{N}$,

$m > 0$, then $z \in C$. One sees immediately that C must be normal if $k[C]$ is a normal domain: X^z is an element of the field of fractions of $k[C]$, and if $(X^z)^m \in k[C]$ and $k[C]$ is normal, then $X^z \in k[C]$. That the converse is also true will be shown below. First we explore the geometric significance of normal semigroups.

A non-empty subset D of an $\mathbb{R}$-vector space V is called a *cone* if it is closed under linear combinations with non-negative coefficients in $\mathbb{R}$. For $S \subset V$ the set

$$\mathbb{R}_+ S = \{ \sum_{i=1}^{n} a_i v_i : a_i \in \mathbb{R}_+, \ v_i \in S, \ n \in \mathbb{N} \}$$

is obviously the smallest cone containing S; it is the *cone generated by S*. Finitely generated cones can be characterized in complete analogy with convex polytopes: a subset D of a finite dimensional $\mathbb{R}$-vector space V is a finitely generated cone if and only if there exist finitely many vector half-spaces

$$H_i^+ = \{ v \in V : \langle a_i, v \rangle \geq 0 \}, \qquad a_i \in V, \quad a_i \neq 0, \quad i = 1, \dots, m,$$

such that $D = H_1^+ \cap \cdots \cap H_m^+$.

In the following it will be necessary to consider rational polytopes and cones. Let V be an $\mathbb{R}$-vector space of finite dimension, and U a $\mathbb{Q}$-vector subspace of V such that $\dim_{\mathbb{Q}} U = \dim_{\mathbb{R}} V$. A polytope $P \subset V$ is *rational* (with respect to U) if its vertices lie in U, and a cone is *rational* if it is generated by a subset of U. We choose a scalar product which has an orthonormal basis in U, and define a *rational* half-space to be a set

$$H^+ = \{ v \in V : \langle a, v \rangle \geq \beta \}$$

with $a \in U$, $a \neq 0$ and $\beta \in \mathbb{Q}$. Of course, the notion of rationality makes sense only with respect to a fixed $\mathbb{Q}$-subspace U (and, for a half-space, is independent of the choice of the scalar product, provided it has an orthonormal basis in U). If $V = \mathbb{R}^n$, then it is tacitly understood that $U = \mathbb{Q}^n$, and when $V = \mathbb{R}C$ for an affine semigroup C, $U = \mathbb{Q}C$.

We need some results about rational polytopes and cones:

(i) A subset $P \subset V$ is a rational polytope if and only if it is bounded and the intersection of finitely many rational half-spaces.

(ii) A subset $D \subset V$ is a finitely generated rational cone if and only if it is the intersection of finitely many rational vector half-spaces.

(iii) Let $v_1, \dots, v_m \in U$. Then $u \in U \cap \operatorname{conv}\{v_1, \dots, v_m\}$ if and only if there exist $r_1, \dots, r_m \in \mathbb{Q}_+$ with $\sum_{i=1}^{m} r_i = 1$ such that $u = \sum_{i=1}^{m} r_i v_i$; in other words

$$U \cap \operatorname{conv}\{v_1, \dots, v_m\} = \operatorname{conv}_{\mathbb{Q}}\{v_1, \dots, v_m\}.$$

(iv) Let $v_1, \dots, v_m \in U$. Then $u \in U \cap \mathbb{R}_+\{v_1, \dots, v_m\}$ if and only if there exist $r_1, \dots, r_m \in \mathbb{Q}_+$ such that $u = \sum_{i=1}^{m} r_i v_i$; in other words

$$U \cap \mathbb{R}_+\{v_1, \dots, v_m\} = \mathbb{Q}_+\{v_1, \dots, v_m\}.$$

It is a good exercise for the reader to prove (i)–(iv). An essential argument is that a linear system of equations with rational coefficients is soluble over $\mathbb{Q}$ if and only if it has a solution over $\mathbb{R}$.

Proposition 6.1.2 (Gordan's lemma). (a) *If C is a normal semigroup, then $C = \mathbb{Z}C \cap \mathbb{R}_+ C$ (within $\mathbb{R}C$).*
(b) *Let G be a finitely generated subgroup of $\mathbb{Q}^n$ and D a finitely generated rational cone in $\mathbb{R}^n$. Then $C = G \cap D$ is a normal semigroup.*

PROOF. (a) It follows from (iv) above that $\mathbb{Z}C \cap \mathbb{R}_+ C = \mathbb{Z}C \cap \mathbb{Q}_+ C$, and that $C = \mathbb{Z}C \cap \mathbb{Q}_+ C$ is (almost) the definition of a normal semigroup.

(b) The essential point to prove is that $G \cap D$ is a finitely generated rational semigroup; the rest is again elementary.

We claim that $D \cap \mathbb{R}C$ is a finitely generated rational cone in $\mathbb{R}C$. In fact, let $D = \bigcap H_i^+$ be given as the intersection of finitely many rational half-spaces of $\mathbb{R}^n$. Then $D \cap \mathbb{R}C = \bigcap(H_i^+ \cap \mathbb{R}C)$, and because of $\mathbb{Q}C = \mathbb{R}C \cap \mathbb{Q}^n$, each $H_i^+ \cap \mathbb{R}C$ is a rational half-space of $\mathbb{R}C$ or equal to $\mathbb{R}C$.

Replacing G by $\mathbb{Z}C$ and $\mathbb{R}^n$ by $\mathbb{R}C$ we may now assume that $G = \mathbb{Z}^n$. By hypothesis there exist $q_1, \ldots, q_v \in \mathbb{Q}^n$ with $D = \{\sum_{i=1}^v a_i q_i : a_i \in \mathbb{R}, a_i \geq 0\}$. Multiplying by a suitable common denominator we may assume that $q_1, \ldots, q_v \in \mathbb{Z}^n$.

Choose $c \in C$. Then $c = \sum_{i=1}^v a_i q_i$ with $a_i \in \mathbb{Q}_+$, and therefore

$$c = \sum_{i=1}^v a_i' q_i + \sum_{i=1}^v a_i'' q_i$$

with $a_i' \in \mathbb{N}$ and $a_i'' \in \mathbb{Q}$, $0 \leq a_i'' < 1$. Since $C = \mathbb{Z}^n \cap D$, we have $c'' = \sum_{i=1}^v a_i'' q_i \in C$. But c'' lies in the bounded set $B = \{\sum_{i=1}^v a_i'' q_i : 0 \leq a_i'' < 1\}$ so that $\mathbb{Z}^n \cap B$ is finite. The finite set $(B \cap \mathbb{Z}^n) \cup \{q_1, \ldots, q_v\}$ generates C. □

The invertible elements in a semigroup C form a group C_0, the largest group contained in C. If $C_0 = 0$, we say that C is *positive*. If C is normal, then C splits into a direct sum of C_0 and a positive normal semigroup:

Proposition 6.1.3. *Let C be a normal semigroup, and C_0 the group of its invertible elements.*
(a) *Then $C \cong C_0 \oplus C'$ with a positive normal semigroup C'. Furthermore $C_0 \cong \mathbb{Z}^u$ for some $u \geq 0$.*
(b) *One has $k[C] \cong k[C_0] \otimes_k k[C'] \cong k[\mathbb{Z}^u] \otimes_k k[C']$ for every field k.*

PROOF. It follows immediately from the normality of C that the group $\mathbb{Z}C/C_0$ is torsion-free. Therefore C_0 is a direct summand of $\mathbb{Z}C$, and hence of C itself. The rest of (a) is quite obvious. Part (b) is a special case of the general fact that $k[C_1 \oplus C_2] \cong k[C_1] \otimes_k k[C_2]$. □

With the notation of the previous proposition, all essential ring-theoretic properties are shared by $k[C]$ and $k[C']$: the ring $k[C]$ arises from $k[C']$ by a polynomial extension followed by the inversion of the indeterminates, and is a free, thus faithfully flat, $k[C']$-module.

Theorem 6.1.4. *Let C be an affine semigroup, and k a field. Then the following are equivalent:*
(a) *C is a normal semigroup;*
(b) *$k[C]$ is normal.*

PROOF. The implication (b) $\Rightarrow$ (a) has already been observed.

For (a) $\Rightarrow$ (b) we note that C is the intersection of finitely many rational half-spaces $H_i^+ = \{q \in \mathbb{R}C : \langle a_i, q \rangle \geq 0\}$ of $\mathbb{R}C$ with $\mathbb{Z}C$, $a_i \in \mathbb{Q}C$; see 6.1.2. Set $C_i = \mathbb{Z}C \cap H_i^+$. One has $(C_i)_0 = \{z \in C_i : \langle a_i, z \rangle = 0\}$. It follows that $(C_i)_0 \cong \mathbb{Z}^{d-1}$ where $d = \operatorname{rank} C$. Thus the semigroup C_i' in the splitting $C_i = (C_i)_0 \oplus C_i'$ has rank 1.

Since C_i is normal, C_i' is also normal. Being a normal subsemigroup of $\mathbb{Z}$, and not a group, C_i' is isomorphic to $\mathbb{N}$. Therefore $k[C_i] \cong k[\mathbb{Z}^{d-1} \oplus \mathbb{N}]$ is even regular. As $k[C]$ is the intersection of the normal rings $k[C_i]$, it is normal itself. $\square$

In order to use the results on $\mathbb{Z}$-graded rings and modules for affine semigroup rings we say that a decomposition

$$k[C] = \bigoplus_{i \in \mathbb{N}} k[C]_i$$

of the k-vector space $k[C]$ is an *admissible grading* if $k[C]$ is a positively graded k-algebra with respect to this decomposition, and furthermore each component $k[C]_i$ is a direct sum of *finitely* many $\mathbb{Z}C$-graded components. It follows that X^c is homogeneous for each $c \in C$, and that the ˙maximal ideal $\mathfrak{m}$ of $k[C]$ is generated by the monomials X^c, $c \neq 0$. Thus $k[C]$ has an admissible grading only if C is positive. That the converse is also true, will be very important in the following.

Proposition 6.1.5. *Let C be a positive affine semigroup. Then C is isomorphic with a subsemigroup of $\mathbb{N}^m$ for some m. In particular $k[C]$ is isomorphic with a graded k-subalgebra of $k[X_1, \ldots, X_m]$, and has an admissible grading.*

PROOF. We choose a scalar product that has a $\mathbb{Z}$-basis of $\mathbb{Z}C$ as an orthonormal basis. The cone $\mathbb{R}_+ C$ is the intersection of half-spaces

$$H_i^+ = \{v \in \mathbb{R}C : \langle a_i, v \rangle \geq 0\}, \qquad a_i \in \mathbb{Q}C, \quad a_i \neq 0, \quad i = 1, \ldots, m.$$

Multiplying by a suitable common denominator we may assume that $a_i \in \mathbb{Z}C$. Then $\langle a_i, c \rangle \in \mathbb{Z}$ for all $c \in \mathbb{Z}C$, and $\varphi : \mathbb{Z}C \to \mathbb{Z}^m$, $\varphi(c) =$

$(\langle a_1, c \rangle, \ldots, \langle a_m, c \rangle)$ is a group homomorphism with $\varphi(C) \subset \mathbb{N}^m$. The kernel of φ is the intersection of the hyperplanes $H_i = \{ v \in \mathbb{R}C : \langle a_i, v \rangle = 0 \}$; therefore the group $\operatorname{Ker} \varphi \cap \mathbb{Z}C$ is contained in C. Since C is positive, $\varphi|_C$ is injective. The rest is obvious. □

The graded prime ideals of an affine semigroup ring. The results of Sections 6.2 and 6.3 depend crucially on the fact that one can determine the graded prime ideals of $k[C]$ from the geometry of the cone $\mathbb{R}_+C$. Let us first show that the set of non-zero graded radical ideals in $k[C]$ has a unique minimal element. For an affine semigroup C we set

$$\operatorname{relint} C = C \cap \operatorname{relint} \mathbb{R}_+C.$$

Lemma 6.1.6. *Let C be an affine semigroup. Then the ideal generated by the elements X^c, $c \in \operatorname{relint} C$ is a radical ideal, and is contained in every non-zero graded radical ideal of $k[C]$.*

PROOF. In view of 6.1.1 we may equivalently prove that $\operatorname{relint} C$ is the smallest non-empty radical ideal of C.

Set $I = \operatorname{relint} C$. It is obvious that I is a radical ideal of C. Let $J \subset C$ be an arbitrary non-empty radical ideal, $c \in I$, and $s \in J$. We must show that $c \in J$, for which there is only something to prove if $c \neq s$. As $c \in \operatorname{relint} \mathbb{R}_+C$, the intersection of $\operatorname{relint} \mathbb{R}_+C$ with the line L through s and c is a neighbourhood of c in L. Since L is rational, there exist rational points on both sides of c in L arbitrarily close to c. So there exists $t \in L \cap (\operatorname{relint} \mathbb{R}_+C) \cap \mathbb{Q}C$ such that c lies in the line segment $[s, t]$. Therefore we have an equation

$$c = \lambda s + (1 - \lambda)t \qquad \text{with} \quad \lambda \in \mathbb{Q}, \quad 0 < \lambda < 1.$$

Multiplication with a suitable common denominator yields an equation

$$mc = ns + t' \qquad \text{with} \quad m, n \in \mathbb{N} \setminus \{0\}$$

and $t' \in C$. It follows that $c \in J$ because J is a radical ideal and $s \in J$. □

We shall see in Theorem 6.3.5 that the ideal considered in 6.1.6 is the canonical module of $k[C]$ if C is a normal semigroup.

Let C be an affine semigroup, and suppose that F is a face of $\mathbb{R}_+C$. The set $C \setminus F$ is immediately seen to be a prime ideal of C. By 6.1.1 it follows that the ideal $\mathfrak{P}(F)$ of $k[C]$ generated by the elements X^c, $c \in C \setminus F$, is a graded prime ideal of $k[C]$. In fact, all homogeneous prime ideals can be represented in this way:

Theorem 6.1.7. *Let C be an affine semigroup, and k a field. Then the assignment $F \mapsto \mathfrak{P}(F)$ is a bijection between the set of non-empty faces of $\mathbb{R}_+C$ and the set of graded prime ideals of $k[C]$.*

PROOF. In view of 6.1.1 we may equivalently show that the assignment $F \mapsto \Pi(F) = C \setminus F$ is a bijection between the set of non-empty faces of $\mathbb{R}_+ C$ and the set of prime ideals of C.

It is easy to see that Π is injective; in fact, $F = \mathbb{R}_+(C \cap F) = \mathbb{R}_+(C \setminus \Pi(F))$ for every face F of $\mathbb{R}_+ C$.

Surjectivity of Π is proved by induction on rank C, the case rank $C = 0$ being trivial. Let rank $C > 0$ and $P \subset C$ be a prime ideal. If $P = \emptyset$, then $P = \Pi(\mathbb{R}_+ C)$. So suppose $P \neq \emptyset$. By 6.1.6 we have $P \supset$ relint C. As relint $C = \Pi(F_1) \cap \cdots \cap \Pi(F_m)$ where $F_1, \ldots, F_m$ are the maximal proper faces of $\mathbb{R}_+ C$, it follows that $P \supset \Pi(F_i)$ for at least one i, say $P \supset \Pi(F_1)$.

The intersection $C \cap F_1$ is an affine semigroup with rank $C \cap F_1 <$ rank C. As $P \cap F_1$ is a prime ideal in $C \cap F_1$, there exists a face G of $\mathbb{R}_+ F_1$ with $P \cap F_1 = (C \cap F_1) \setminus G$. Being a face of a face of $\mathbb{R}_+ F_1$, G is a face of $\mathbb{R}_+ C$, and elementary set theory shows that $P = \Pi(G)$. $\qquad \square$

In the next section the homogeneous localizations $k[C]_{(\mathfrak{p})}$ will play a crucial role. Since we shall argue rather geometrically, it is more suggestive to denote them by

$$k[C]_F$$

where F is the face of $\mathbb{R}_+ C$ with $\mathfrak{p} = \mathfrak{P}(F)$. This notation is also justified by the fact that $k[C]_F$ is the ring of fractions of $k[C]$ with respect to the multiplicatively closed set $\{X^c : c \in C \cap F\}$.

Finally we want to relate the faces of the cone $\mathbb{R}_+ C$ to those of a suitably chosen polytope. For simplicity we restrict ourselves to the case in which C is positive. More generally, let

$$D = \{x \in \mathbb{R}^n : \langle a_i, x \rangle \geq 0 \text{ for } i = 1, \ldots, m\}$$

be a cone in $\mathbb{R}^n$ given as the intersection of vector half-spaces defined by $a_i \in \mathbb{R}^n$, $i = 1, \ldots, m$. Let us say that D is *positive* if 0 is the only element $v \in D$ with $-v \in D$.

This is the case if and only if $a_1, \ldots, a_m$ generate $\mathbb{R}^n$. Set $b = a_1 + \cdots + a_m$ and define

$$T = \{x \in D : \langle b, x \rangle = 1\}.$$

It follows easily that T is bounded. Being the intersection of finitely many affine half-spaces, it is a convex polytope. We say that the hyperplane $\{x : \langle b, x \rangle = 1\}$ is *transversal* to D, and call T a *cross-section* of D. Cross-sections are introduced because their combinatorial structure will lead us to a complex by which one can compute the local cohomology of an affine semigroup ring.

A non-empty face of D is given by D itself or by $H \cap D$ where H is a supporting hyperplane of D. Since D is a cone, H must contain 0. Therefore there is a unique minimal non-empty face of D, namely $\{0\}$, and we choose $\mathscr{F}(D)$ to be the set of non-empty faces of D.

Proposition 6.1.8. *Let D be a positive cone, and T a cross-section of D. Then the assignment $F \mapsto F \cap T$ induces an isomorphism $\mathscr{F}(D) \cong \mathscr{F}(T)$ of partially ordered sets. Its inverse is given by $G \mapsto \mathbb{R}_+ G$.*

The proof is easy and left as an exercise for the reader.

At several places below we will have to use the correspondence between the faces of $\mathbb{R}_+ C$ and those of a cross-section T of $\mathbb{R}_+ C$, as given by 6.1.8. In order to avoid cumbersome notation we agree on denoting corresponding faces by corresponding capital and small letters. So, if F is a face of $\mathbb{R}_+ C$, then $f = (\mathbb{R}_+ C) \cap T$.

Exercises

6.1.9. Prove Proposition 6.1.1.
Hint: For the implication '$\Leftarrow$' in (b) and for (c) one uses that $\mathbb{Z}C \cong \mathbb{Z}^d$, $d = \operatorname{rank} C$, can be given a linear order under which it becomes an ordered group. (For example one may choose the reverse degree-lexicographical order introduced in Section 4.2.) Then the homogeneous components of an element are linearly ordered, and one argues similarly as in the proof of Lemma 1.5.6.

6.1.10. Let S, T be affine semigroups, $S \subset T$. One says that S is a *full* subsemigroup of T if $S = T \cap \mathbb{Z}S$. Show
(a) a full subsemigroup of a normal semigroup is again normal,
(b) a positive affine semigroup is normal if and only if it is isomorphic to a full subsemigroup of $\mathbb{N}^n$ for some $n \geq 0$,
(c) if S is full in T, then $k[S]$ is a direct $k[S]$-summand of $k[T]$.

6.1.11. Let C be an affine semigroup. Then $k[C]$ is regular if and only if C is of the form $\mathbb{Z}^u \oplus \mathbb{N}^v$.
Hint: The implication '$\Leftarrow$' is easy. For the implication '$\Rightarrow$' one uses 6.1.5 and 2.2.25, noting that a minimal set of generators of the •maximal ideal of $k[C]$ can be chosen of the form $X^{c_1}, \ldots, X^{c_v}$.

6.1.12. Let C be an affine semigroup, and F a face of $\mathbb{R}_+ C$. Show
(a) the composition $k[C \cap F] \to k[C] \to k[C]/\mathfrak{P}(F)$ of natural maps is an isomorphism of affine semigroup rings,
(b) if C is normal, then $k[C \cap F]$ is also normal,
(c) $k[C]_F$ is an affine semigroup ring.

6.1.13. Let $D \subset \mathbb{R}^n$ be a positive cone, and $z \in \mathbb{R}^n$. Show that $z \notin -D$ if and only if there exists a hyperplane H which is transversal to D and contains z.

6.2 Local cohomology of affine semigroup rings

In this section we shall define a complex by which we can compute the local cohomology of an affine semigroup ring; it is based on a construction of algebraic topology, namely the oriented augmented chain complex associated with a finite regular cell complex.

Cell complexes.   Regular cell complexes generalize the simplicial complexes of Chapter 5. Massey [266] gives an introduction to the theory of cell complexes which is very well suited for our purpose. We introduce the chain complex associated with a cell complex axiomatically, borrowing the existence and uniqueness theorems from algebraic topology.

A *finite regular cell complex* is a non-empty topological space X together with a finite set Γ of subsets of X such that the following conditions are satisfied:

(i) $X = \bigcup_{e \in \Gamma} e$;

(ii) the subsets $e \in \Gamma$ are pairwise disjoint;

(iii) for each $e \in \Gamma$, $e \neq \emptyset$ there exists a homeomorphism from a closed i-dimensional ball $B^i = \{x \in \mathbb{R}^i : \|x\| \leq 1\}$ onto the closure $\bar{e}$ of e which maps the open ball $U^i = \{x \in \mathbb{R}^i : \|x\| < 1\}$ onto e;

(iv) $\emptyset \in \Gamma$.

By the invariance of dimension the number i in (iii) is uniquely determined by e, and e is called an *open i-cell*; $\emptyset$ is a (-1)-cell. By Γ^i we denote the set of the i-cells in Γ. The dimension of Γ is given by $\dim \Gamma = \max\{i : \Gamma^i \neq \emptyset\}$. It is finite since Γ is finite. One sets $|\Gamma| = X$.

Finite regular cell complexes are special cases of a more general topological structure, namely that of a CW-complex. Since all our CW-complexes are finite and regular, we shall simply call them *cell complexes*.

A cell e' is a *face* of the cell $e \neq e'$ if $e' \subset \bar{e}$, and a subset Σ of Γ is a *subcomplex* if for each $e \in \Sigma$ all the faces of e are contained in Σ.

The classical examples of cell complexes are convex polytopes P together with their decomposition $P = \bigcup_{f \in \mathcal{F}(P)} \operatorname{relint} f$. For them the following property, which follows from (i)–(iv), is an elementary theorem:

(v) if $e \in \Gamma^i$ and $e' \in \Gamma^{i-2}$ is a face of e, then there exist exactly two cells $e_1, e_2 \in \Gamma^{i-1}$ such that e_j is a face of e and e' is a face of e_j.

Each simplicial complex Δ may be identified with a cell complex, namely the cell complex it defines in a natural way on a geometric realization and whose open cells correspond to the faces of Δ. It is convenient to denote this cell complex simply by Δ, and an open cell by the corresponding face of Δ. Let $\{v_1, \dots, v_n\}$ be the vertex set of Δ. For $e \in \Delta^i$ and $e' \in \Delta^{i-1}$ we set $\varepsilon(e, e') = 0$ if e' is not a face of e, and $\varepsilon(e, e') = (-1)^{k+1}$ if e corresponds to $\{v_{i_1}, \dots, v_{i_m}\}$ and e' to $\{v_{i_1}, \dots, \widehat{v}_{i_k}, \dots, v_{i_m}\}$, $i_1 < \dots < i_m$. Then the augmented oriented chain complex of Δ, which has been introduced in Chapter 5, is a complex of free $\mathbb{Z}$-modules $\mathscr{C}^i(\Delta) = \bigoplus_{e \in \Delta^i} \mathbb{Z}e$ whose differential is given by $\partial(e) = \sum_{e' \in \Delta^{i-1}} \varepsilon(e, e')e'$. The crucial point in constructing a similar complex for an arbitrary cell complex is to find a suitable function ε.

Let us say that ε is an *incidence function* on Γ if the following conditions are satisfied:

(a) to each pair (e, e') such that $e \in \Gamma^i$ and $e' \in \Gamma^{i-1}$ for some $i \geq 0$, ε assigns a number $\varepsilon(e, e') \in \{0, \pm 1\}$;

(b) $\varepsilon(e, e') \neq 0 \Longleftrightarrow e'$ is a face of e;

(c) $\varepsilon(e, \emptyset) = 1$ for all 0-cells e;

(d) if $e \in \Gamma^i$ and $e' \in \Gamma^{i-2}$ is aface of e, then

$$\varepsilon(e, e_1)\varepsilon(e_1, e') + \varepsilon(e, e_2)\varepsilon(e_2, e') = 0$$

where e_1 and e_2 are those $(i-1)$-cells such that e_j is a face of e and e' is a face of e_j (see (v) above).

Lemma 6.2.1. *Let Γ be a cell complex. Then there exists an incidence function on Γ.*

For a proof see Lemma IV.7.1 in [266] where the incidence numbers $\varepsilon(e, e')$ appear as topological data determined by orientations of the cells. Figure 6.1 indicates two incidence functions on the solid rectangle and how they are induced by orientations.

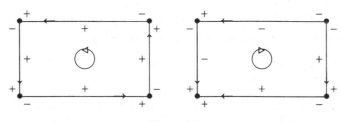

Figure 6.1

Let $\delta : \Gamma \to \{\pm 1\}$ be a function with $\delta(\emptyset) = 1$ and $\delta(e) = 1$ for all 0-cells e. Then the function

$$\varepsilon'(e, e') = \delta(e')\varepsilon(e, e')\delta(e)$$

is also an incidence function. On the other hand, all pairs ε, ε' of incidence functions differ only by a 'sign' δ:

Theorem 6.2.2. *Let Γ be a cell complex with incidence functions ε and ε'. Then there exists $\delta : \Gamma \to \{\pm 1\}$ such that $\delta(\emptyset) = 1$ and $\varepsilon'(e, e') = \delta(e')\varepsilon(e, e')\delta(e)$ for all $e \in \Gamma^i$, $e' \in \Gamma^{i-1}$, $i = 0, \ldots, \dim \Gamma$.*

This is Theorem IV.7.2 of [266] (in a different formulation). Its proof shows that incidence functions can be constructed in a completely naive manner. (i) One starts with Γ^0 on which there is no choice according to property (c) of incidence functions. (ii) If one has constructed an incidence function ε^i on $\Gamma^0 \cup \cdots \cup \Gamma^i$, then there exists an incidence function ε^{i+1} on $\Gamma^0 \cup \cdots \cup \Gamma^{i+1}$ whose restriction to $\Gamma^0 \cup \cdots \cup \Gamma^i$ is just ε^i. The reader is advised to construct incidence functions for some three dimensional polytopes.

Let Γ be a cell complex of dimension $d-1$, and ε an incidence function on Γ (as in Chapter 5 it is convenient to denote dimension by $d-1$). We define the *augmented oriented chain complex* of Γ by the complex

$$\widetilde{\mathscr{C}}(\Gamma): 0 \longrightarrow \mathscr{C}_{d-1} \xrightarrow{\ \partial\ } \mathscr{C}_{d-2} \longrightarrow \cdots \longrightarrow \mathscr{C}_0 \xrightarrow{\ \partial\ } \mathscr{C}_{-1} \longrightarrow 0$$

where we set

$$\mathscr{C}_i = \bigoplus_{e \in \Gamma^i} \mathbb{Z}e \quad \text{and} \quad \partial(e) = \sum_{e' \in \Gamma^{i-1}} \varepsilon(e, e')e' \quad \text{for } e \in \Gamma^i,$$

$i = 0, \ldots, d-1$. That $\partial^2 = 0$ follows from the definition of an incidence function and property (v) of cell complexes. The notation $\widetilde{\mathscr{C}}(\Gamma)$ is justified since the dependence of $\widetilde{\mathscr{C}}(\Gamma)$ on ε is inessential; Theorem 6.2.2 guarantees that we obtain an isomorphic complex upon replacing ε by another incidence function ε'. (The isomorphism is given by $e \mapsto \delta(e)e$.) For simplicity of notation we set $\widetilde{H}_i(\Gamma) = H_i(\widetilde{\mathscr{C}}(\Gamma))$.

The fundamental importance of $\widetilde{\mathscr{C}}(\Gamma)$ in algebraic topology relies on the fact that it computes reduced singular homology :

Theorem 6.2.3. *Let Γ be a cell complex. Then $\widetilde{H}_i(\Gamma) = \widetilde{H}_i(|\Gamma|)$ for all $i \geq 0$ (and $\widetilde{H}_{-1}(\Gamma) = 0$).*

Theorem IV.4.2 of [266] states that $H_i(\mathscr{C}(\Gamma)) \cong H_i(|\Gamma|)$ for the non-augmented complex $\mathscr{C}(\Gamma)$ which arises from $\widetilde{\mathscr{C}}(\Gamma)$ if we replace $\widetilde{\mathscr{C}}^{-1}$ by 0. It follows easily that $\widetilde{H}_0(\Gamma) \cong \widetilde{H}_0(|\Gamma|)$ as well.

We use 6.2.3 via the following corollary:

Corollary 6.2.4. *Let Γ be a cell complex such that $|\Gamma|$ is homeomorphic to a closed ball B^n. Then $\widetilde{H}_i(\Gamma) = 0$ for all $i \geq -1$.*

Local cohomology. Let C be a positive affine semigroup, and k a field. The ideal $\mathfrak{m}$ in $R = k[C]$ generated by the elements X^c, $c \in C \setminus \{0\}$, is maximal. For an R-module M we denote by $H^i_{\mathfrak{m}}(M)$ the i-th right derived functor of

$$\Gamma_{\mathfrak{m}}(M) = \{x \in M : \mathfrak{m}^i x = 0 \text{ for } i \gg 0\}.$$

As in 3.5.3 one has a natural isomorphism

$$H^i_{\mathfrak{m}}(M) \cong \varinjlim \operatorname{Ext}^i_R(R/\mathfrak{m}^j, M) \quad \text{for all} \quad i \geq 0.$$

The natural map $\operatorname{Ext}^i_R(R/\mathfrak{m}^j, M) \to \operatorname{Ext}^i_{R_{\mathfrak{m}}}(R_{\mathfrak{m}}/(\mathfrak{m}R_{\mathfrak{m}})^j, M_{\mathfrak{m}})$ is an isomorphism. Therefore $H^i_{\mathfrak{m}}(M) \cong H^i_{\mathfrak{m}R_{\mathfrak{m}}}(M_{\mathfrak{m}})$, and we are justified in calling $H^i_{\mathfrak{m}}(M)$ a local cohomology module. We now want to construct a complex 'computing' $H_{\mathfrak{m}}^i(M)$ which resembles the combinatorial structure of C as closely as possible.

Suppose for the moment that $C = \mathbb{N}^n$ so that $R = k[C] \cong k[X_1, \ldots, X_n]$ and $\mathfrak{m} = (X_1, \ldots, X_n)$. As we saw in 3.5.6, the modified Čech complex

$$C^{\bullet} : 0 \longrightarrow C^0 \longrightarrow C^1 \longrightarrow \cdots \longrightarrow C^n \longrightarrow 0$$

with

$$C_t = \bigoplus R_{x_{i_1} \cdots x_{i_t}}$$

computes $H_{\mathfrak{m}}^{\bullet}(M)$ in the sense that $H_{\mathfrak{m}}^i(M) \cong H^i(M \otimes C^{\bullet})$ for all $i \geq 0$. The components of C^t are of the form R_F where F is a face of $\mathbb{R}_+^n = \mathbb{R}_+\mathbb{N}^n$, and the differential is composed of maps

$$\varepsilon \cdot \mathrm{nat} : R_{x_{i_1} \cdots x_{i_t}} \longrightarrow R_{x_{i_1} \cdots x_{i_t} x_j}$$

whose signs ε are just the values of an incidence function on the pair $(\mathrm{conv}\{e_{i_1}, \ldots, e_{i_t}, e_j\}, \mathrm{conv}\{e_{i_1}, \ldots, e_{i_t}\})$ of faces of the simplex spanned by the canonical basis $e_1, \ldots, e_n$ of $\mathbb{R}^n$. This simplex is a cross-section of the cone $\mathbb{R}_+^n$.

It is easy to generalize this construction. Let C be a positive affine semigroup of rank d, $R = k[C]$, T a cross-section of the cone $\mathbb{R}_+C$, and $\mathscr{F} = \mathscr{F}(T)$ its face lattice. (We remind the reader of our convention of denoting corresponding faces of $\mathbb{R}_+C$ and T by F and f respectively.) Let

$$L^t = \bigoplus_{f \in \mathscr{F}^{t-1}} R_F, \qquad t = 0, \ldots, d,$$

and define $\partial : L^{t-1} \to L^t$ by specifying its component

$$\partial_{f',f} : R_{F'} \to R_F \qquad \text{to be} \qquad \begin{cases} 0 & \text{if } F' \not\subset F, \\ \varepsilon(f, f')\,\mathrm{nat} & \text{if } F' \subset F; \end{cases}$$

here ε is an incidence function on $\mathscr{F}$. It is clear that

$$L^{\bullet} : 0 \longrightarrow L^0 \xrightarrow{\ \partial\ } L^1 \longrightarrow \cdots \longrightarrow L^{d-1} \xrightarrow{\ \partial\ } L^d \longrightarrow 0$$

is a complex.

Theorem 6.2.5. *Let C be a positive affine semigroup, and k a field. Let $\mathfrak{m}$ be the maximal ideal generated by the elements X^c, $c \in C \setminus \{0\}$. Then for every $k[C]$-module M, and all $i \geq 0$,*

$$H_{\mathfrak{m}}^i(M) \cong H^i(L^{\bullet} \otimes M).$$

PROOF. We follow the pattern of the proof of 3.5.6. Let I be the ideal generated by the elements X^c, $c \neq 0$ for which there exists a one dimensional face F of $\mathbb{R}_+C$ with $c \in F$. In order to show $H^0(L^0 \otimes M) = H_{\mathfrak{m}}^0(M)$ for all $k[C]$-modules M, we must verify that $\operatorname{Rad} I = \mathfrak{m}$. Let $c_1, \ldots, c_m \in C$ be a minimal set of generators of $\mathbb{Q}_+ C$. Then each one dimensional face F of $\mathbb{R}_+ C$ contains exactly one of the c_i, and it is enough to show that $\operatorname{Rad} J = \mathfrak{m}$ where J is the ideal generated by $X^{c_1}, \ldots, X^{c_m}$. Let $c \in C$, $c \neq 0$. There exist $q_1, \ldots, q_m \in \mathbb{Q}_+$ with $c = q_1 c_1 + \cdots + q_m c_m$. Multiplication by a common denominator yields $rc = s_1 c_1 + \cdots + s_m c_m$ with r, $s_i \in \mathbb{N}$. Since $s_i \neq 0$ for at least one i, it follows that $(X^c)^r \in J$.

Now let $0 \to M_1 \to M_2 \to M_3 \to 0$ be an exact sequence of $k[C]$-modules. Since all the summands of $L^{\bullet}$ are flat $k[C]$-modules, this yields an exact sequence

$$0 \to L^{\bullet} \otimes M_1 \to L^{\bullet} \otimes M_2 \to L^{\bullet} \otimes M_3 \to 0.$$

As desired we have a long exact sequence

$$\cdots \to H^i(L^{\bullet} \otimes M_1) \to H^i(L^{\bullet} \otimes M_2) \to H^i(L^{\bullet} \otimes M_3) \to H^{i+1}(L^{\bullet} \otimes M_1) \to \cdots$$

Finally we must show that $H^i(L^{\bullet} \otimes M) = 0$ for all i if M is an injective $k[C]$-module. It suffices to consider the indecomposable modules $E(R/\mathfrak{p})$ where $\mathfrak{p}$ is a prime ideal of $R = k[C]$. Then, as shown in the proof of 3.5.6, there are only two possibilities for an element x of R: either every element of $E(R/\mathfrak{p})$ is annihilated by some power of x, namely if $x \in \mathfrak{p}$, or multiplication by x is bijective on $E(R/\mathfrak{p})$. So

$$E(R/\mathfrak{p}) \otimes R_F = \begin{cases} 0 & \text{if } F \cap \mathfrak{p} \neq \emptyset, \\ E(R/\mathfrak{p}) & \text{if } F \cap \mathfrak{p} = \emptyset. \end{cases}$$

Set $P = \log \mathfrak{p}$. Then P is a prime ideal in the semigroup C, and by 6.1.7 there is a face G of $\mathbb{R}_+ C$ with $P = C \setminus G$. Thus

$$E(R/\mathfrak{p}) \otimes R_F = \begin{cases} 0 & \text{if } F \not\subset G, \\ E(R/\mathfrak{p}) & \text{if } F \subset G. \end{cases}$$

Let $\mathscr{G} = \mathscr{F}(g)$ denote the face lattice of the face $g = G \cap T$ of a cross-section T of $\mathbb{R}_+ C$. It follows that

$$L^t \otimes E(R/\mathfrak{p}) = \bigoplus_{f \in \mathscr{G}^{t-1}} E(R/\mathfrak{p})$$

for all $t \geq 0$. Of course $\mathscr{G}$ is a subcomplex of $\mathscr{F} = \mathscr{F}(T)$, and the restriction of an incidence function on $\mathscr{F}$ to $\mathscr{G}$ is an incidence function on $\mathscr{G}$. Therefore we have

$$L^{\bullet} \otimes E(R/\mathfrak{p}) \cong \operatorname{Hom}_{\mathbb{Z}}\big(\widetilde{\mathscr{C}}(\mathscr{G})(-1), \, E(R/\mathfrak{p})\big).$$

(This statement is the heart of the proof; the reader should verify it carefully.) Since g is a convex polytope, it is homeomorphic to a closed ball. So $\widetilde{\mathscr{C}}(\mathscr{G})$ is an exact complex; see 6.2.4. Since $\widetilde{\mathscr{C}}(\mathscr{G})$ is a complex of free $\mathbb{Z}$-modules, exactness is preserved in $\operatorname{Hom}_{\mathbb{Z}}(\widetilde{\mathscr{C}}(\mathscr{G})(-1), E(R/\mathfrak{p}))$. $\square$

Corollary 6.2.6. *Let C be a positive affine semigroup of rank d, and k be a field. Then $k[C]$ is Cohen–Macaulay if and only if $H^i(L^{\bullet}) = 0$ for $i = 0, \ldots, d - 1$.*

PROOF. Set $R = k[C]$, and note that $d = \dim R_{\mathfrak{m}}$ (why?). If R is Cohen–Macaulay, then $R_{\mathfrak{m}}$ is Cohen–Macaulay. Thus it follows from 3.5.7 that $H^i(L^{\bullet}) = 0$ for $i = 0, \ldots, d - 1$. Conversely, 3.5.7 also implies that $R_{\mathfrak{m}}$ is Cohen–Macaulay if $H^i(L^{\bullet}) = 0$ for $i = 0, \ldots, d - 1$. By virtue of 6.1.5 R is a *local ring with *maximal ideal $\mathfrak{m}$. Now 2.1.27 yields that R is Cohen–Macaulay. $\square$

Exercises

6.2.7. We will see in the next section that a normal semigroup ring is Cohen–Macaulay. This exercise presents an example (due to Hochster [174]) of an affine semigroup ring showing that Serre's condition (S_2) alone is not sufficient for the Cohen–Macaulay property. Let k be a field, and Y_1, Y_2, Z_1, Z_2 be indeterminates over k. Prove:
(a) The semigroup C generated by the monomials $x_{ij} = Y_i Z_j$, $i, j = 1, 2$, is normal; $S = k[C]$ is a normal domain of dimension 3.
(b) The substitution $X_{ij} \mapsto x_{ij}$ induces an isomorphism

$$k[X_{11}, X_{12}, X_{21}, X_{22}]/(X_{11}X_{22} - X_{21}X_{12}) \cong S;$$

S is a Cohen–Macaulay ring.
(c) The subsemigroup C' of C generated by all monomials f with $\deg_{Y_1} f > 1$ and $\deg_{Y_2} f > 1$ is finitely generated.
(d) The elements x_{11}^2, x_{22}^2, $x_{12}^2 + x_{21}^2$ form a homogeneous system of parameters of $R = k[C']$, but not an R-sequence.
(e) The ideals generated by x_{11}^2 and x_{22}^2 in R are unmixed. (Hint: Use that the associated primes of a $\mathbb{Z}C'$-graded module are $\mathbb{Z}C'$-graded; this follows as in 1.5.6.)
(f) $R[x_{11}^{-2}, x_{22}^{-2}] = S[x_{11}^{-2}, x_{22}^{-2}]$.
(g) R satisfies Serre's condition (S_2), but is not Cohen–Macaulay.

6.2.8. One says that an n-dimensional positive cone D is *simplicial* if it is generated by n elements, and a positive affine semigroup C is *simplicial* if the cone $\mathbb{R}_+ C$ is simplicial. Let k be a field.
(a) Let C be an arbitrary positive affine semigroup. Prove that $X^{c_1}, \ldots, X^{c_n}$ with $c_1 \ldots, c_n \in C$ form an $k[C]$-sequence if and only if X^{c_i}, X^{c_j} is an $k[C]$-sequence for all $i \neq j$, equivalently, $c_i + s = c_j + t$ for $s, t \in C$ implies $s \in c_i + C$.
(b) Show that C is simplicial if and only if $k[C]$ has a homogeneous system of parameters $X^{c_1}, \ldots, X^{c_n}$ with $c_1 \ldots, c_n \in C$.
(c) Let C be simplicial. Deduce from (a) and (b) that $k[C]$ is Cohen–Macaulay if and only if it satisfies Serre's condition (S_2), and that this property is independent of k (Goto, Watanabe, and Suzuki [133]).
(d) Formulate a Gorenstein criterion for $k[C]$ with C simplicial, using the socle of $k[C]/(X^{c_1}, \ldots, X^{c_n})$, and show that this property is also independent of k.

6.3   Normal semigroup rings

In this section we want to show that a normal semigroup ring is a
Cohen–Macaulay ring and to determine its canonical module.

The complex $L^{\bullet}$ constructed in the previous section is $\mathbb{Z}C$-graded in
a natural way, and in order to compute its cohomology we analyze its
graded components just as in the proof of 5.3.8. Given $z \in \mathbb{Z}C$, the main
point is to determine those faces F of C for which $(R_F)_z \neq 0$. As we shall
see, this is the case if and only if the face F is not 'visible' from z.

Let P be a polyhedron in a $\mathbb{R}$-vector space V. Let $x, y \in V$. We
say that y is *visible* from x if $y \neq x$ and the line segment $[x, y]$ does not
contain a point $y' \in P$, $y' \neq y$. A subset $S \subset V$ is *visible* if each $v \in S$ is
visible.

Proposition 6.3.1. *Let P be a polytope in $\mathbb{R}^n$ with face lattice $\mathscr{F}$, and
$x \in \mathbb{R}^n$ a point outside P. Set $\mathscr{S} = \{F \in \mathscr{F} : F \text{ visible from } x\}$. Then $\mathscr{S}$ is
a subcomplex of $\mathscr{F}$; its underlying space $S = \bigcup_{F \in \mathscr{S}} F$ is the set of points
$y \in P$ which are visible from x, and is homeomorphic to a closed ball.*

PROOF. Let $y \in P$ be visible from x. There exists a (unique) face F with
$y \in \text{relint } F$, and one concludes easily (for example by 6.3.2 below) that
the whole of F is visible from x. Therefore $S = \{y \in P : y \text{ visible from } x\}$,
and it follows easily that S is homeomorphic to a closed ball. That $\mathscr{S}$ is
a subcomplex is obvious. $\square$

Let P be a polyhedron in an $\mathbb{R}$-vector space V, $\dim V < \infty$. Suppose
that P is given as the intersection of finitely many half-spaces

$$H_i^+ = \{x \in V : \langle a_i, x \rangle \geq \beta_i\}, \qquad i = 1, \dots, m.$$

We set

$$x^0 = \{i : \langle a_i, x \rangle = \beta_i\}, \quad x^+ = \{i : \langle a_i, x \rangle > \beta_i\}, \quad x^- = \{i : \langle a_i, x \rangle < \beta_i\}.$$

Lemma 6.3.2. *With the notation introduced, a point $y \in P$ is visible from
$x \in V \setminus P$ if and only if $y^0 \cap x^- \neq \emptyset$.*

The elementary proof is left for the reader. Figure 6.2 illustrates the
following lemma. Let $C = \mathbb{N}^2 \subset \mathbb{R}^2$, and F be the positive X-axis, G
the positive Y-axis. Then $k[C]_F = k[X, Y, X^{-1}]$, and $(k[C]_F)_z \neq 0$ for
$z \notin C$ exactly when z is in the second quadrant (including the negative
X-axis). Thus $(k[C]_F)_z \neq 0$ if and only if F is not visible from z. Similar
arguments work for the faces $\{0\}$, G, and C.

Lemma 6.3.3. *Let C be a normal semigroup, k a field, and $R = k[C]$. Let F
be a face of $\mathbb{R}_+C$ and $z \in \mathbb{Z}C$. Then $(R_F)_z \neq 0$ (and therefore $(R_F)_z \cong k$)
if and only if F is not visible from z.*

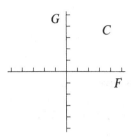

Figure 6.2

PROOF. Suppose first that F is not visible from z. Then there exists $c \in C \cap F$ which is not visible from z. We have $c^+ \supset z^-$ (note that $c^- = \emptyset$), and it follows that $(mc + z)^- = \emptyset$ for $m \gg 0$, whence $mc + z \in C$. (Of course z^+, c^- etc. are defined with respect to a representation of $\mathbb{R}_+ C$ as an intersection of vector half-spaces.) That $mc + z \in C$ is equivalent with $(X^c)^m X^z \in R$ so that $X^z \in R_F$.

Conversely suppose that $(R_F)_z \neq 0$. Then there exists $c \in C \cap F$ with $X^c X^z \in R$. Consequently $c + z \in C$, and $(c + z)^- = \emptyset$, which is only possible if c is not visible from z. $\square$

Now we can compute the local cohomology of normal semigroup rings. In the sequel $-C$ is the affine semigroup $\{-c : c \in C\}$.

Theorem 6.3.4. *Let C be a positive normal semigroup of rank d, k a field, and $z \in \mathbb{Z}C$.*
(a) *If $z \in \text{relint}(-C)$, then $(L^\bullet)_z$ is isomorphic to $0 \to k \to 0$ with k in degree d. Consequently $H^i(L^\bullet)_z = 0$ for $i \neq d$, and $H^d(L^\bullet)_z \cong k \cong (L^\bullet)_z$.*
(b) *Suppose that $z \notin \text{relint}(-C)$. Let T be a cross-section of $\mathbb{R}_+ C$ with face lattice $\mathscr{F}$, and $\mathscr{S} = \{F \cap T : F \in \mathscr{F}(\mathbb{R}_+ C) \text{ visible from } z\}$. Then*
(i) $(L^\bullet)_z \cong \text{Hom}_{\mathbb{Z}}\big((\widetilde{\mathscr{C}}(\mathscr{F})/\widetilde{\mathscr{C}}(\mathscr{S}))(-1), k\big)$,
(ii) $\widetilde{H}_i(\mathscr{F}) = \widetilde{H}_i(\mathscr{S}) = 0$ *for all i,*
(iii) $(H^i(L^\bullet))_z = 0$ *for all i.*

PROOF. (a) For $z \in \text{relint}(-C)$ one has $z \in R_F$ if and only if $F = \mathbb{R}_+ C$.

(b)(i) The complex $\widetilde{\mathscr{C}}(\mathscr{F})$ consists of direct summands $\mathbb{Z}f$, $f \in \mathscr{F}$. As $\mathscr{S}$ is a cell subcomplex of $\mathscr{F}$, $\widetilde{\mathscr{C}}(\mathscr{S})$ is a chain subcomplex of $\widetilde{\mathscr{C}}(\mathscr{F})$, and we obtain $\widetilde{\mathscr{C}}(\mathscr{F})/\widetilde{\mathscr{C}}(\mathscr{S})$ if we replace all the direct summands $\mathbb{Z}f$ with $f \in \mathscr{S}$ by 0. The complex $\text{Hom}_{\mathbb{Z}}((\widetilde{\mathscr{C}}(\mathscr{F})/\widetilde{\mathscr{C}}(\mathscr{S}))(-1), k)$ is therefore isomorphic to the complex

$$D^\bullet : 0 \longrightarrow D^0 \overset{\partial}{\longrightarrow} \cdots \overset{\partial}{\longrightarrow} D^d \to 0$$

with

$$D^t = \bigoplus_{f \in \mathscr{F}^{t-1} \setminus \mathscr{S}} kf^* \quad \text{and} \quad \partial((f')^*) = \sum \varepsilon(f, f')f^*.$$

According to 6.3.3, $(L^\bullet)_z$ is given by $D^\bullet$.

(ii) Note that the combinatorial structures of $\mathscr{F}$ and $\mathscr{S}$ do not depend on the chosen cross-section T. (This follows from 6.1.8.) Therefore we may vary T. Furthermore it was observed above that $\tilde{H}_i(\mathscr{F}) = 0$ for all i.

If $z \in C$, then $\mathscr{S} = \emptyset$, and $\tilde{\mathscr{C}}(\mathscr{S})$ is the zero complex. So suppose that $z \notin C$ in the following.

If $z \notin -C$, then, by virtue of 6.1.13, there exists a hyperplane E through z which is transversal to $\mathbb{R}_+ C$. Choose $T = E \cap \mathbb{R}_+ C$. Then $\mathscr{S}$ is the set of faces of T which are visible from z, and we invoke 6.3.1 in conjunction with 6.2.4 to conclude that $\tilde{H}_i(\mathscr{S}) = 0$ for all i.

If $z \in -C$, then there exists a point $z' \in \mathbb{R}C \setminus (-C)$ with $(z')^- = z^-$. (By hypothesis $z \notin \text{relint}(-C)$; consider a sufficiently small neighbourhood of z.) Because of 6.3.2 we may replace z by z' in defining $\mathscr{S}$ and argue as in the case $z \notin -C$.

(iii) We have a long exact sequence

$$\cdots \longrightarrow \tilde{H}_i(\mathscr{S}) \longrightarrow \tilde{H}_i(\mathscr{F}) \longrightarrow \tilde{H}_i(\tilde{\mathscr{C}}(\mathscr{F})/\tilde{\mathscr{C}}(\mathscr{S})) \longrightarrow \tilde{H}_{i+1}(\mathscr{S}) \longrightarrow \cdots$$

Thus it follows from (ii) that $\tilde{\mathscr{C}}(\mathscr{F})/\tilde{\mathscr{C}}(\mathscr{S})$ is exact. As it is a complex of free $\mathbb{Z}$-modules, the dual (of a shifted copy) with respect to an arbitrary $\mathbb{Z}$-module is also exact.   □

The previous theorem allows us not only to show that normal semigroup rings are Cohen–Macaulay, but also to determine their canonical modules.

Theorem 6.3.5. *Let C be a normal semigroup, and k a field. Then*
(a) (Hochster) $k[C]$ *is a Cohen–Macaulay ring,*
(b) (Danilov, Stanley) *the ideal I generated by the monomials X^c with $c \in \text{relint}\, C$ is the canonical module of $k[C]$.*

PROOF. (a) We write $k[C]$ in the form $k[C] \cong k[C_0] \otimes k[C']$ as in 6.1.3; then $k[C] \cong k[C'][X_1, X_1^{-1}, \ldots, X_u, X_u^{-1}]$ for some $u \geq 0$. In view of 2.1.9 it is therefore enough to show that $k[C']$ is Cohen–Macaulay. But this follows immediately from 6.2.6 and 6.3.4, the latter of which in particular says that $H^i(L^\bullet) = 0$ for $i = 0, \ldots, d-1$.

(b) Suppose first that C is positive. As $L^d = k[\mathbb{Z}C]$, we have an exact sequence

(1) $$0 \longrightarrow U \longrightarrow k[\mathbb{Z}C] \longrightarrow H^d(L^\bullet) \longrightarrow 0$$

of $\mathbb{Z}C$-graded $k[C]$-modules. The functor $^*\mathrm{Hom}_k(\_, k)$ in the category of $\mathbb{Z}C$-graded $k[C]$-modules assigns each module the k-vector space

$$\bigoplus_{z \in \mathbb{Z}C} \mathrm{Hom}_k(M_{-z}, k),$$

which is a $\mathbb{Z}C$-graded $k[C]$-module in a natural way. Applying this functor to the exact sequence above we obtain an exact sequence

$$0 \longrightarrow {}^*\mathrm{Hom}_k(H^d(L^*), k) \longrightarrow k[\mathbb{Z}C] \longrightarrow U' \longrightarrow 0.$$

It follows from 6.3.4 that $^*\mathrm{Hom}_k(H^d(L^*), k)$ consists exactly of those graded components $k[\mathbb{Z}C]_z$ with $z \in \mathrm{relint}\, C$. Therefore

$$I \cong {}^*\mathrm{Hom}_k(H^d(L^*), k) \qquad \text{as } \mathbb{Z}C\text{-graded modules.}$$

As in the proof of 6.2.6 we use that $k[C]$ has an admissible grading. Thus it is a *local $\mathbb{Z}$-graded ring whose *maximal ideal $\mathfrak{m}$ is generated by the monomials X^c, $c \in C$, $c \neq 0$. Furthermore each $\mathbb{Z}$-homogeneous component of $k[C]$ is the direct sum of *finitely* many $\mathbb{Z}C$-graded components. The same holds for $H^d(L^*)$. As Hom_k commutes with finite direct sums, we conclude that

$$I \cong {}^*\mathrm{Hom}_k(H^d(L^*), k) \qquad \text{as } \mathbb{Z}\text{-graded modules.}$$

In Section 3.5 we defined the *local cohomology functors $^*H^i_{\mathfrak{m}}(\_)$ in the category of $\mathbb{Z}$-graded $k[C]$-modules. If M is a $\mathbb{Z}$-graded $k[C]$-module, then $L^* \otimes M$ is a complex of $\mathbb{Z}$-graded modules, and virtually the same arguments as in the proof of 6.2.5 show that $^*H^i_{\mathfrak{m}}(M) \cong H^i(L^* \otimes M)$ for all i. Finally we deduce from 3.6.19 and 3.6.9 that I is the canonical module of $k[C]$.

The general case of (b) in which C is not necessarily positive follows as in (a) if we use 3.3.21 to compute a canonical module of a polynomial extension. $\qquad\qquad\square$

Corollary 6.3.6. *Suppose, in addition to the hypothesis of 6.3.5, that C is positive. Then I is the (unique) *canonical module of $k[C]$ with respect to an arbitrary admissible grading.*

Remark 6.3.7. The formulation 'the canonical module' of 6.3.5 needs justification beyond 6.3.6. First, if we had developed the theory of $\mathbb{Z}^n$-graded rings to the same extent as that of $\mathbb{Z}$-graded rings, it would be immediate that I is the unique $\mathbb{Z}C$-graded canonical module of $k[C]$ (up to an isomorphism of $\mathbb{Z}C$-graded modules). Second, and even more, a canonical module of $k[C]$ is unique in the category of all $k[C]$-modules. We briefly indicate the argument; it exploits the theory of class groups (Fossum [108]), and will be explained in detail in Section 7.3 where it is more essential. With our usual notation, the extension $k[C'] \to k[C]$

induces an isomorphism of class groups $\text{Cl}(k[C']) \cong \text{Cl}(k[C])$. Because of this isomorphism a canonical module ω of $k[C]$ is of the form $\omega' \otimes k[C]$ for some $k[C']$-module ω'. The extension $k[C'] \to k[C]$ is faithfully flat. Applying 3.3.30 one concludes that ω' is a canonical module of $k[C']$. Thus it is enough to consider positive semigroups C. For those one has an isomorphism $\text{Cl}(k[C]) \cong \text{Cl}(k[C]_{\mathfrak{m}})$ ([108], 10.3). Finally one uses that the canonical module of a local ring is unique.

The preceding argument amounts to the fact that a projective rank 1 module over $k[C]$ is free. This was shown for arbitrary projective $k[C]$-modules by Gubeladze [145].

Corollary 6.3.8. *Let C be a normal semigroup, and k a field. Then $k[C]$ is Gorenstein if and only if there exists $c \in \text{relint}\, C$ with $\text{relint}\, C = c + C$.*

PROOF. If $\text{relint}\, C = c + C$, then the ideal I of 6.3.5 is principal, and $k[C]$ is Gorenstein by 3.3.7. For the converse implication we decompose C in the form $C = C_0 \oplus C'$ where C_0 is a group and C' is positive. If $k[C]$ is Gorenstein, then $k[C']$ is Gorenstein: the extension $k[C'] \to k[C]$ is faithfully flat, and the Gorenstein property descends from $k[C']$ to $k[C]$ by 3.3.30. For $k[C']$ we can apply 3.6.11 (with respect to an admissible grading), and thus $I \cong k[C']$. It follows that I is generated by an element X^c. Therefore $\text{relint}\, C' = c + C'$, and it is easy to verify that $\text{relint}\, C = c + C$ as well. $\square$

Combinatorial applications. Let $\mathscr{S}$ be a system of homogeneous linear Diophantine equations in n variables. It follows directly from 6.1.2 that the set C of solutions $c \in \mathbb{N}^n$ of $\mathscr{S}$ is a positive normal semigroup. This fact enables us to apply results on Hilbert functions to the combinatorial object C.

The set C can be represented by the power series

$$C(t) = \sum_{c \in C} t^c$$

in n variables $t = t_1, \ldots, t_n$. Obviously $C(t)$ is the $\mathbb{Z}^n$-graded Hilbert series of $k[C]$ if we consider the $\mathbb{Z}^n$-grading on $k[C]$ it inherits from $k[\mathbb{N}^n] = k[X_1, \ldots, X_n]$. As we have not developed the theory of $\mathbb{Z}^n$-graded modules to the necessary extent, we restrict ourselves to considering the specialization

$$c(t) = \sum_{c \in C} t^{|c|}.$$

It is the Hilbert series of $k[C]$ for the $\mathbb{Z}$-grading induced by the total degree of a monomial. Under this grading $k[C]$ is a positively graded k-algebra.

Let C^+ be the set of strictly positive integral solutions of $\mathscr{S}$, i.e. solutions $c \in \mathbb{N}^n$ with $c_i > 0$ for $i = 1, \ldots, n$. It may of course happen that $C^+ = \emptyset$, but otherwise we have $C^+ = \text{relint}\, C$ (Exercise 6.3.14). Therefore, and by 6.3.6, the power series

$$c^+(t) = \sum_{c \in C^+} t^{|c|}$$

is the Hilbert series of the °canonical module ω of $k[C]$. Hence 4.4.6 immediately yields the following reciprocity law .

Theorem 6.3.9 (Stanley). *With the notation introduced, suppose that C^+ is non-empty. Then*

$$c^+(t) = (-1)^d c(t^{-1}), \qquad d = \text{rank}\, C.$$

Of all the results of Section 4.4 only 4.4.6 has been applied to $k[C]$. We could extend 6.3.9 by furthermore considering the Hilbert function $H(k[C], m) = |\{c \in C : |c| = m\}|$. Below, such an extension is carried out for the Ehrhart function of a rational polytope.

Remark 6.3.10. In [361], Theorem 4.6.14, Stanley proves the 'fine' version

$$C^+(t) = (-1)^d C(t^{-1})$$

of the previous theorem by combinatorial methods. In order to obtain it by ring-theoretic arguments one needs the $\mathbb{Z}^n$-graded variant of 4.4.6 which was also given by Stanley; see [357], Theorem 6.1. (Exercise 5.6.6 is the $\mathbb{Z}^n$-graded variant of 4.4.6 for Stanley–Reisner rings.)

Conversely, the computation of the canonical module of a normal semigroup ring in [357] uses the fine reciprocity law: similarly as in 5.6.3 one shows that the ideal generated by the monomials X^c, $c \in C^+$, is the canonical module of $k[C]$ once the equation $C^+(t) = (-1)^d C(t^{-1})$ has been established.

Let $P \subset \mathbb{R}^n$ be a polytope of dimension d. Since P is bounded, we may define its *Ehrhart function* by

$$E(P, m) = |\{z \in \mathbb{Z}^n : \frac{z}{m} \in P\}|, \quad m \in \mathbb{N},\ m > 0, \quad \text{and} \quad E(P, 0) = 1.$$

and its *Ehrhart series* by

$$E_P(t) = \sum_{m \in \mathbb{N}} E(P, m) t^m.$$

It is clear that $E(P, m) = |\{z \in \mathbb{Z}^n : z \in mP\}|$ where $mP = \{mp : p \in P\}$. Similarly as above we set

$$E^+(P, m) = |\{z \in \mathbb{Z}^n : \frac{z}{m} \in \text{relint}\, P\}| \quad \text{for } m > 0, \quad E^+(P, 0) = 0,$$

and

$$E_P^+(t) = \sum_{m \in \mathbb{N}} E^+(P, m) t^m.$$

Note that $E^+(P, m) = |\{z \in \mathbb{Z}^n : z \in \text{relint}\, mP\}|$ for $m > 0$.

We define the cone $D \subset \mathbb{R}^{n+1}$ by $D = \mathbb{R}_+\{(p,1): p \in P\}$. Then $C = D \cap \mathbb{Z}^{n+1}$ is a subsemigroup of $\mathbb{Z}^{n+1}$. Therefore one may consider the k-algebra $k[C]$. Suppose P is a rational polytope; then D is a rational cone, and C is a positive normal semigroup. Let us fix a grading on $k[C]$ by assigning to $c = (c_1, \ldots, c_{d+1})$ the degree c_{d+1}. For this grading the Hilbert functions of $k[C]$ and of the ideal I generated by the monomials X^c, $c \in \operatorname{relint} C$, are given by

$$H(k[C], m) = E(P, m) \quad \text{and} \quad H(I, m) = E^+(P, m).$$

The grading under consideration is admissible for $k[C]$, and therefore we may apply the theory of Chapter 4 to $k[C]$. Part (b) of the following theorem is Ehrhart's remarkable reciprocity law for rational polytopes.

Theorem 6.3.11 (Ehrhart). *Let $P \subset \mathbb{R}^n$ be a d-dimensional rational polytope, $d > 0$. Then*

(a) $E_P(t)$ *is a rational function, and there exists a quasi-polynomial q with $E(P, m) = q(m)$ for all $m \geq 0$;*

(b) $E_P^+(t) = (-1)^{d+1} E_P(t^{-1})$, *equivalently*

$$E^+(P, m) = (-1)^d E(P, -m) \qquad \text{for all} \quad m \geq 1$$

where $E(P, -m) = q(-m)$ is the natural extension of $E(P, \_)$.

PROOF. (a) Since $E_P(t)$ is the Hilbert series of a positively graded Noetherian k-algebra, it is a rational function. According to 4.4.3 we must show for the second statement in (a) that $E_P(t)$ has negative degree, or, equivalently, that the a-invariant of $k[C]$ is negative. By 6.3.5 the ring $k[C]$ is Cohen–Macaulay, and by 6.3.6 its ˙canonical module is generated by the elements X^c, $c \in \operatorname{relint} C$. These have positive degrees under the grading of $k[C]$, and hence $a(k[C]) < 0$.

(b) By what has just been said, $E_P^+(t)$ is the Hilbert series of the ˙canonical module of $k[C]$. Furthermore, $\dim k[C] = d + 1$. Thus the first equation is a special case of 4.4.6. The second equation results from $\sum_{m \geq 1} E(P, -m)t^m = -E_P(t^{-1})$. The reader may prove this identity as an exercise, or look up [361], 4.2.3. $\qquad \square$

The quasi-polynomial q in 6.3.11 is called the *Ehrhart quasi-polynomial of P*.

Suppose that P is even an *integral* polytope, that is, a polytope whose vertex set V is contained in $\mathbb{Z}^n$. Then, in addition to $k[C]$, we may also consider its subalgebra

$$k[V] = k[X^{(v,1)} : v \in V].$$

Obviously $k[V]$ is a homogeneous k-algebra. Let $c \in C$; then there exist $q_v \in \mathbb{Q}_+$ such that $c = \sum_{v \in V} q_v v$. If we multiply this equation by a suitable common denominator e and interpret the result in terms of monomials, then we see that $(X^c)^e \in k[V]$. Thus $k[C]$ is integral over $k[V]$. Since it is also a finitely generated $k[V]$-algebra, it is even a finite $k[V]$-module. In particular, by Hilbert's theorem 4.1.3, the Ehrhart quasi-polynomial of P is a polynomial and therefore called the *Ehrhart polynomial*. Furthermore $k[C]$ has a well defined multiplicity. In concluding this section we want to illuminate the beautiful relation between the volume $\operatorname{vol} P$ of an n-dimensional integral polytope $P \subset \mathbb{R}^n$ and the multiplicity of $k[C]$.

Theorem 6.3.12. *Let* $P \subset \mathbb{R}^n$ *be an n-dimensional integral polytope, and let* $k[C]$ *the normal semigroup ring constructed above. Then*

$$e(k[C]) = n! \operatorname{vol} P.$$

PROOF. Elementary arguments of measure theory show that the volume of P is

$$\operatorname{vol} P = \lim_{m \to \infty} \frac{E(P,m)}{m^n}.$$

Being the Hilbert polynomial of a $(n + 1)$-dimensional $k[V]$-module, $E(P,m)$ has degree n. Thus its leading coefficient is given by $\operatorname{vol} P$. On the other hand, it is also given by $e(k[C])/n!$. $\qquad\square$

The restriction to n-dimensional polytopes $P \subset \mathbb{R}^n$ is only for simplicity; see [361], Section 4.6, for the general case. Using the fact that the volume of P is the leading coefficient of its Ehrhart polynomial one can derive classical formulas for $\operatorname{vol} P$. Exercise 6.3.17 presents the cases $n = 2$ and $n = 3$.

Exercises

6.3.13. Let C be a positive normal semigroup. For each $i = 0,\dots,d$ let G_i be the direct sum of the residue class rings $k[C]/\mathfrak{P}(F)$ where F is an i-dimensional face of $\mathbb{R}_+ C$. Define the map $k[C]/\mathfrak{P}(F) \to k[C]/\mathfrak{P}(F')$ to be $\varepsilon(f,f')$ nat if $F' \subset F$, $\dim F' = \dim F - 1$, or 0 otherwise. Show that the induced sequence

$$0 \longrightarrow I \overset{\mathrm{nat}}{\longrightarrow} k[C] = G_d \longrightarrow G_{d-1} \longrightarrow \cdots \longrightarrow G_1 \longrightarrow G_0 = k \longrightarrow 0$$

is exact (of course I is defined as in 6.3.5). Hint: The proof is similar to that of 5.7.3.

6.3.14. Let C be the semigroup of solutions $c \in \mathbb{N}^n$ of a system of homogeneous linear Diophantine equations, and $C^+ = \{c \in C : c_i > 0 \text{ for all } i\}$. Show that if $C^+ \neq \emptyset$, then $C^+ = \operatorname{relint} C$.

6.3.15. Let P be an integral polytope of dimension n, and define the semigroup C and the grading of $k[C]$ as above. It is customary to call $(h_0,\dots,h_n)$ the h-vector of P where h_i is the i-th coefficient of the (Laurent) polynomial $Q(t)$ in the numerator of the Ehrhart series of $k[C]$; it follows from 6.3.11 that $h_i = 0$ for $i > n$. Prove the following inequalities due to Stanley [362] and Hibi [168]:

(a) $h_i \geq 0$ for all i;

(b) $\sum_{i=0}^{j} h_i \leq \sum_{i=0}^{j} h_{s-i}$ for all $j = 0, \ldots, s$ where $s = \max\{i : h_i \neq 0\}$;

(c) $\sum_{i=n-j}^{n} h_i \leq \sum_{i=0}^{j+1} h_i$ for all $j = 0, \ldots, n$.

Hint: For (b) and (c) study (again) the proof of 4.4.9. For $R = k[C]$ we also have an exact sequence $0 \to \omega \to R \to R/\omega \to 0$.

6.3.16. With P, C, and $k[C]$ as in 6.3.15, set $L = \{(p, 1) : p \in P \cap \mathbb{Z}^n\}$ and let $k[P]$ be the k-algebra generated by the elements X^w, $w \in L$.

(a) Show the following are equivalent:
    (i) $k[C] = k[P]$;
    (ii) $k[C]$ is homogeneous;
    (iii) $k[P]$ is normal and $\mathbb{Z}L = \mathbb{Z}^{n+1}$.

(b) Discuss the conditions of (a)(iii) for the polytopes $P_1, P_2 \subset \mathbb{R}^3$ spanned by (1) $v_0 = 0$, $v_1 = (1, 0, 0)$, $v_2 = (0, 1, 0)$, and $v_3 = (1, 1, 3)$, and (2) $v_0 = 0$, $v_1 = (2, 0, 0)$, $v_2 = (0, 3, 0)$, and $v_3 = (0, 0, 5)$.

6.3.17. Prove that the volume of an n-dimensional integral polytope P in $\mathbb{R}^n$ is

$$\mathrm{vol}\, P = \frac{1}{2}(E(P, 1) + E^+(P, 1) - 2) \qquad \text{for } n = 2, \text{ and}$$

$$\mathrm{vol}\, P = \frac{1}{6}(E(P, 2) - 3E(P, 1) - E^+(P, 1) + 3) \qquad \text{for } n = 3.$$

Hint: The coefficients of a polynomial can be determined by interpolation.

6.4 Invariants of tori and finite groups

In the following we use some elementary notions and results from the theory of linear algebraic groups for which we refer the reader to Humphreys [208], Kraft [241], or Mumford and Fogarty [279].

Let k be an algebraically closed field, and V a k-vector space of finite dimension. Each $\varphi \in GL(V)$ yields a k-algebra automorphism α_φ of the symmetric algebra $R = S(V)$. In concrete terms, if $e_1, \ldots, e_n$ is a basis of V, then $S(V) \cong k[X_1, \ldots, X_n]$, the isomorphism being induced by the linear map which sends e_i to X_i, $i = 1, \ldots, n$. If we identify V and $kX_1 + \cdots + kX_n$ via this map, then α_φ is just the k-algebra automorphism of $R = k[X_1, \ldots, X_n]$ given by the substitution $X_i \mapsto \varphi(X_i)$. (From a categorical point of view it would be better to consider the action of $GL(V)$ on $S(V^*)$, the ring of polynomial functions on V.)

Suppose that G is a linear algebraic group over k; such a group is always isomorphic to a Zariski closed subgroup of $GL(W)$ where W is a suitable finite dimensional k-vector space. A morphism $\Phi : G \to GL(V)$ (in the category of algebraic groups) is called a *representation* of G. It assigns the automorphism $\alpha_{\Phi(g)}$ of R to each $g \in G$, so that we say that G *acts linearly* on R. It is the classical problem of (algebraic) invariant theory to determine the structure of the *ring of invariants*

$$R^G = \{f \in R : g(f) = f \text{ for all } g \in G\},$$

where we have set $g(f) = \alpha_{\Phi(g)}(f)$ for simplicity of notation. If f is homogeneous of total degree d, then so is $g(f)$. Therefore R^G is a positively graded k-algebra inheriting its grading from R.

A *character* of G is a representation $\chi: G \to \mathrm{GL}(k)$. To each character χ we associate the set

$$R^\chi = \{f \in R: g(f) = \chi(g)f \text{ for all } g \in G\}$$

of *semi-invariants of weight* χ. It is easily verified that R^χ is a graded R^G-submodule of R. Especially important in the following is the *inverse determinant character* $g \mapsto \det^{-1}(g) = \det \Phi(g)^{-1}$ associated with Φ.

The ring R^G of invariants only depends on $\Phi(G) \subset \mathrm{GL}(V)$; thus we shall often simplify the situation by directly considering a subgroup of $\mathrm{GL}(V)$. Furthermore, for concrete groups the requirement that k be algebraically closed can sometimes be relaxed.

More generally, one may always form the ring R^G when R is a ring and G is a subgroup of $\mathrm{Aut}\, R$. Clearly R^G inherits all properties of R which descend to subrings, and is a normal domain along with R:

Proposition 6.4.1. *Let R be a normal domain, and G a subgroup of $\mathrm{Aut}\, R$. Then R^G is a normal domain.*

PROOF. It is easy to see that R^G is the intersection of its field of fractions $Q(R^G)$ with R (within $Q(R)$). □

Invariants of diagonalizable groups. Let k be an algebraically closed field. For each $m \in \mathbb{N}$ the group $\mathrm{GL}(k)^m$ is called a *torus*; it is isomorphic to the group of $m \times m$ diagonal matrices of rank m over k. Slightly more generally we want to consider *diagonalizable* groups over k, i.e. direct products

$$D = T \times H$$

where T is a torus and H is a finite Abelian group whose order is not divisible by $\mathrm{char}\, k$. Since k contains a primitive q-th root of unity for each q not divisible by $\mathrm{char}\, k$, H may be written in the form

$$H = \langle \zeta_1 \rangle \times \cdots \times \langle \zeta_w \rangle$$

where $\langle \zeta_j \rangle$ is the cyclic subgroup of $\mathrm{GL}(k)$ generated by a root of unity ζ_j. Thus we may write each element in D in the form $(d_1, \ldots, d_m, \zeta_1^{s_1}, \ldots, \zeta_w^{s_w})$ with $s_j \in \mathbb{N}$.

Suppose now that we are given a representation of D, that is, a homomorphism $\Phi: D \to \mathrm{GL}(V)$. Then Φ can be diagonalized: there exists a basis $e_1, \ldots, e_n$ of V such that each e_i is an eigenvector of $\Phi(d)$ for every $d \in D$. Thus the vector subspace $ke_i \cong k$ is stable under the action of D, and therefore Φ induces a character χ_i of D; χ_i associates to each element $d \in D$ its eigenvalue with respect to e_i. It is sufficient to determine the characters of the direct factors GL(k) and $\langle \zeta_j \rangle$ of D. One sees easily that in both cases the characters are the powers $a \mapsto a^s$, $s \in \mathbb{Z}$. Thus there exist $t_{1i}, \ldots, t_{mi} \in \mathbb{Z}$ and $u_{1i}, \ldots, u_{wi} \in \mathbb{N}$, $i = 1, \ldots, n$, such that

$$\Phi(d_1, \ldots, d_m, \zeta_1^{s_1}, \ldots, \zeta_w^{s_w})(e_i) = d_1^{t_{1i}} \cdots d_m^{t_{mi}} \zeta_1^{s_1 u_{1i}} \cdots \zeta_w^{s_w u_{wi}} e_i$$

for $i = 1, \ldots, n$.

Theorem 6.4.2. *Let k be an algebraically closed field, and D a diagonalizable group over k acting linearly on a polynomial ring $R = k[X_1, \ldots, X_n]$. Then*

(a) *(Hochster) the ring R^D of invariants is a graded Cohen–Macaulay ring,*

(b) *(Danilov, Stanley) $R^{\det^{-1}}(-n)$ is the *canonical module of R^D, provided $R^{\det^{-1}} \neq 0$.*

PROOF. We may assume right away that D acts diagonally as just described. It follows that each monomial $X_1^{a_1} \cdots X_n^{a_n}$ is mapped to a multiple of itself by every $d \in D$. Therefore $f \in R$ is invariant if and only if all its monomials are invariant, so that $R^D = k[C]$ for some semigroup $C \subset \mathbb{N}^n$. Extending the formula for the action of D to monomials, we see that $X_1^{a_1} \cdots X_n^{a_n}$ is an invariant if and only if $(a_1, \ldots, a_n)$ satisfies the system

$$t_{j1} a_1 + \cdots + t_{jn} a_n = 0, \qquad j = 1, \ldots, m,$$

of homogeneous linear equations with integral coefficients, and simultaneously the system

$$s_j(u_{j1} a_1 + \cdots + u_{jn} a_n) \equiv 0 \bmod (\text{ord}\, \zeta_j), \qquad j = 1, \ldots, w,$$

of homogeneous congruences (of course $\text{ord}\, \zeta_j$ denotes the order of the root of unity ζ_j).

It follows easily that C is the intersection of $\mathbb{R}_+^n$ with a finitely generated group $G \subset \mathbb{Q}^n$. Therefore C is normal, and part (a) is an immediate consequence of 6.3.5(a).

Similarly, part (b) can be derived rather quickly from 6.3.5(b). Set $S = R^D$, $M = R^{\det^{-1}}$, and $P = X_1 \cdots X_n$. Then $d(P) = \det(d)P$ for all $d \in D$. Hence, for every $f \in R$, $f \in M$ if and only if $Pf \in S$. Obviously M is an S-module generated by monomials (even as a k-vector space), and therefore a graded S-module.

Let I be the ideal generated by the monomials $X_1^{c_1} \cdots X_n^{c_n}$ with $(c_1, \ldots, c_n) \in \text{relint}\, C$. We know from 6.3.5 in conjunction with 6.3.7 that I is the graded canonical module of S (up to an isomorphism of graded modules). Evidently it is enough to show that $PM \subset I$ and $P^{-1}I \subset M$, provided $M \neq 0$.

The representation $C = \mathbb{R}_+^n \cap G$ readily yields that

$$\{(c_1, \ldots, c_n) \in C : c_i > 0 \text{ for all } i\} \subset \text{relint}\, C.$$

Therefore $PM \subset I$. Conversely, suppose $M \neq 0$, and let $X_1^{b_1} \cdots X_n^{b_n} \in M$. Then $P X_1^{b_1} \cdots X_n^{b_n} \in S$, and so C contains an element $(c_1, \ldots, c_n)$ with $c_i > 0$ for all i. Hence C is not contained in a coordinate hyperplane, and consequently no relative interior point of C lies in such a hyperplane. It follows that $P^{-1}I \subset k[X_1, \ldots, X_n]$, and thus $P^{-1}I \subset M$. □

The preceding proof shows that the degenerate case $R^{\det^{-1}} = 0$ occurs precisely when, after diagonalization, R^G is contained in one of the subrings $k[X_1, \ldots, \hat{X}_i, \ldots, X_n]$. Furthermore, the condition that k be algebraically closed is dispensable once the action of D is *a priori* diagonal. If k is infinite, then the proof of 6.4.2 remains valid without modification; if k is finite, then one must set $T = \{\mathrm{id}\}$ and $m = 0$. This generalization can also be extended to the following corollary.

Corollary 6.4.3. *Under the hypothesis of* 6.4.2 *suppose additionally that* $\det d = 1$ *for all* $d \in D$. *Then* R^D *is a Gorenstein ring.*

The proof of 6.4.2 suggests that 6.4.2 is just a special case of 6.3.5; however, Exercise 6.4.16 shows that these theorems are actually equivalent.

Finite groups. Theorem 6.4.2 in particular covers the case in which a finite Abelian group G acts linearly on a polynomial ring $k[X_1, \ldots, X_n]$, provided the order $|G|$ of G is invertible in k. With the same proviso, we now want to treat the case of an arbitrary finite group. It is convenient to restrict oneself to subgroups G of $\mathrm{GL}(V)$.

More generally let us first consider a ring R and a finite group G of automorphisms of R such that $|G|$ is invertible in R. Let S be the ring R^G of invariants, and set

$$\rho(r) = |G|^{-1} \sum_{g \in G} g(r)$$

for every $r \in R$. It is straightforward to verify that ρ is an S-linear map from R to S with $\rho|_S = \mathrm{id}_S$. A map satisfying these conditions is called a *Reynolds operator* (for the pair (R, S).) The existence of a Reynolds operator is obviously equivalent to the fact that S is a direct summand of R as an S-module.

Proposition 6.4.4. *Let* R *be a ring,* S *a subring of* R, *and suppose that there exists a Reynolds operator for* (R, S). *Then the following hold:*
(a) *for every ideal* I *of* S *one has* $IR \cap S = I$;
(b) *if* R *is Noetherian, then so is* S;
(c) *if* x *is an* R-sequence in S, then it is also an S-sequence.

PROOF. (a) For $s_1, \ldots, s_n \in S$, $r_1, \ldots, r_n \in R$ with $r = \sum s_i r_i \in S$ one has $r = \rho(r) = \sum s_i \rho(r_i)$.

(b) If $I_0 \subset I_1 \subset \cdots$ is an ascending sequence of ideals in S, then the sequence $I_0 R \subset I_1 R \subset \cdots$ is stationary in R. Therefore, and by (a), the sequence $I_0 \subset I_1 \subset \cdots$ is also stationary.

(c) This follows easily from (a). $\square$

In the case of a group action considered above, each $r \in R$ is a solution of the equation

$$\prod_{g \in G}(X - g(r)) = 0.$$

The left hand side is a monic polynomial in X whose coefficients are elementary symmetric functions in the elements $g(r)$, $g \in G$. Therefore all the coefficients belong to the ring S of invariants, and we see that R is integral over S.

Theorem 6.4.5 (Hochster–Eagon). *Suppose R is a Cohen–Macaulay ring and S is a subring such that there exists a Reynolds operator ρ, and R is integral over S. Then, if R is Cohen–Macaulay, so is S.*

PROOF. We must show that the localizations $S_{\mathfrak{n}}$ of S with respect to its maximal ideals $\mathfrak{n}$ are Cohen–Macaulay. Given a maximal ideal $\mathfrak{n}$ of S, we replace S by $S_{\mathfrak{n}}$ and R by $R \otimes S_{\mathfrak{n}}$. Therefore we may assume that S is a local ring with maximal ideal $\mathfrak{n}$. Since R is integral over S, it is a semi-local ring. (This follows easily from A.6.)

We argue by induction on the length of a maximal R-sequence in $\mathfrak{n}$. Suppose first that $\mathfrak{n}$ consists entirely of zero-divisors of R. Then each $s \in \mathfrak{n}$ is contained in one of the associated prime ideals $\mathfrak{p}_1, \ldots, \mathfrak{p}_m$ of R. So $\mathfrak{n} = \bigcup \mathfrak{p}_i \cap S$, and there exists a j with $\mathfrak{n} = \mathfrak{p}_j \cap S$. As R is Cohen–Macaulay, all the $\mathfrak{p}_i$ are minimal prime ideals of R. On the other hand, since R is integral over S, $\mathfrak{p}_j$ is also a maximal ideal of R.

If $\mathfrak{p}_j$ is the only maximal ideal of R, then it follows immediately that $\dim S = \dim R = 0$ so that S is Cohen–Macaulay as desired. Otherwise the zero ideal of R can be written $\mathfrak{q} \cap \mathfrak{r}$ where $\mathfrak{q}$ is $\mathfrak{p}_j$-primary and $\mathfrak{r} \not\subset \mathfrak{p}_j$. As $\mathfrak{q} + \mathfrak{r} = R$, the Chinese remainder theorem implies that R splits into the direct product of subrings R_1, R_2. If we can replace R by one of them, then we can finish the case under consideration by induction on the number of maximal ideals of R.

Let π_1 and π_2 be the projections of R onto R_1 and R_2, and ι the embedding of S into R. Then both $\rho \circ \pi_1 \circ \iota$ and $\rho \circ \pi_2 \circ \iota$ are endomorphisms of the S-module S. Hence there exist s_1 and s_2 such that $\rho \circ \pi_i \circ \iota$ is multiplication by s_i. It follows that

$$1 = \rho \circ \iota(1) = \rho \circ (\pi_1 + \pi_2) \circ \iota(1) = s_1 + s_2$$

so that at least one of s_1 and s_2 is a unit in S, say s_1. Then $\pi_1 \circ \iota$ is an embedding of S into R_1, and one easily checks that all the hypotheses pass on to the pair (R_1, S). This finishes the case in which $\mathfrak{n}$ consists entirely of zero-divisors of R.

Now suppose that $s \in \mathfrak{n}$ is R-regular. Then 6.4.4 implies that S/sS is in a natural way a subring of R/sR, and it is again easily verified that the remaining hypotheses hold for the subring S/sS of R/sR. As R/sR is Cohen–Macaulay, we conclude that S/sS and, hence, S are Cohen–Macaulay. $\square$

Corollary 6.4.6. *Let R be a Cohen–Macaulay ring, and G a finite group of automorphisms of R whose order is invertible in R. Then the ring R^G of invariants is Cohen–Macaulay.*

Remark 6.4.7. In our derivation of 6.4.6 we have used that $|G|$ is invertible in R in order to show that R^G is Noetherian. However, for this property of R^G the hypothesis on $|G|$ is quite inessential: if R is a finitely generated algebra over a Noetherian ring k such that G acts trivially on k, then, by a famous theorem of E. Noether, R^G is a finitely generated k-algebra. We saw above that R is integral over R^G. Therefore R is already integral over the k-subalgebra A generated by the coefficients of the equations $f_i(x) = 0$, $f_i \in k[T]$, which establish that the finitely many generators x_i of R are integral over R^G. It follows that R and, hence, R^G are finite A-modules.

On the other hand, that $|G|$ is invertible in R is essential for the Cohen-Macaulay property of R^G. In fact, if k is a field of characteristic 2, then $k[X_1, \ldots, X_4]^G$ is a non–Cohen–Macaulay factorial domain for the group G of cyclic permutations of $X_1, \ldots, X_4$; see Bertin [41].

Similarly to 6.4.2 one can determine the canonical module of R^G from invariant theoretic data if G acts linearly on a polynomial ring R. Let V be again a vector space of finite dimension over a field k which we now assume to be of characteristic 0 (see Remark 6.4.11 for the more general case in which $|G|$ is not divisible by $\operatorname{char} k$). As above we extend the action of G to the symmetric algebra $R = S(V)$ which we may identify with the polynomial ring $k[X_1, \ldots, X_n]$, $n = \dim V$, whenever it is appropriate.

Let $S = R^G$. Since the action of G can be restricted to the graded components R_i of R, S is a positively graded k-algebra. Being a finitely generated integral extension of S, R is a finite graded S-module, and in fact a maximal Cohen–Macaulay S-module: according to 1.5.17 there exists a homogeneous system of parameters x in S; it follows that $\operatorname{height} xR = n$, and thus x is an R-sequence. (In conjunction with 6.4.4 this observation yields a quick proof of the previous corollary in the special case under consideration.)

It is customary to call the Hilbert series of S the *Molien series* of G:

$$M_G(t) = H_S(t) = \sum_{i=0}^{\infty} \dim S_i t^i.$$

We also need the Molien series and the Reynolds operator for the semi-invariants of G. For a character χ of G we set

$$M_\chi(t) = H_{M_\chi}(t) = \sum_{i=0}^{\infty} \dim R_i^\chi t^i \quad \text{and} \quad \rho^\chi(r) = |G|^{-1} \sum_{g \in G} \chi(g)^{-1} g(r).$$

It is easy to check that $\rho^\chi(R) = R^\chi$ and $\rho^\chi(r) = r$ for $r \in R^\chi$. The operator ρ^χ is a k-endomorphism of the graded k-vector space R. Let ρ_i^χ denote its restriction to R_i; then $\rho_i^\chi = (\rho_i^\chi)^2$, and therefore

$$\dim R_i^\chi = \dim \mathrm{Im}\, \rho_i^\chi = \mathrm{Tr}\, \rho_i^\chi = |G|^{-1} \sum_{g \in G} \chi(g)^{-1} \mathrm{Tr}\, g|_{R_i}.$$

Here Tr denotes the trace, and we use its linearity. Combining the formulas yields

$$M_\chi(t) = |G|^{-1} \sum_{g \in G} \chi(g)^{-1} \sum_{i=0}^{\infty} (\mathrm{Tr}\, g|_{R_i}) t^i.$$

Theorem 6.4.8 (Molien's formula). *Let k be a field of characteristic 0, V a finite dimensional k-vector space, and G a finite subgroup of $\mathrm{GL}(V)$. Then the Molien series of a character χ of G is given by*

$$M_\chi(t) = |G|^{-1} \sum_{g \in G} \frac{\chi(g)^{-1}}{\det(\mathrm{id} - tg)}.$$

PROOF. We need to show that

$$\frac{1}{\det(\mathrm{id} - tg)} = \sum_{i=0}^{\infty} (\mathrm{Tr}\, g|_{R_i}) t^i$$

for each $g \in G$. In fact, this equation holds for an arbitrary element $g \in \mathrm{GL}(V)$. In order to prove it we may extend k to an algebraically closed field. Then, for a suitable basis $X_1, \ldots, X_n$ of V, g is given by an upper triangular matrix whose diagonal entries are the eigenvalues $\lambda_1, \ldots, \lambda_n$ of g (as an element of $\mathrm{GL}(V)$).

The monomials of total degree i in $X_1, \ldots, X_n$ form a basis of the vector space R_i. If these monomials are ordered lexicographically, then $g|_{R_i}$ is again represented by an upper triangular matrix whose diagonal entry corresponding to the monomial $X^a = X_1^{a_1} \cdots X_n^{a_n}$ is $\lambda^a = \lambda_1^{a_1} \cdots \lambda_n^{a_n}$. Therefore

$$\mathrm{Tr}\, g|_{R_i} = \sum_{|a|=i} \lambda^a,$$

and the expansion of the product of the geometric series $1/(1 - \lambda_j t)$, $j = 1, \ldots, n$, gives us

$$\sum_{i=0}^{\infty} (\text{Tr } g|_{R_i}) t^i = \sum_{i=0}^{\infty} \sum_{|a|=i} \lambda^a t^i = \prod_{j=1}^{n} \frac{1}{1 - \lambda_j t}.$$

Using that $\lambda_1^{-1}, \ldots, \lambda_n^{-1}$ are the eigenvalues of g^{-1}, we finally get

$$\prod_{j=1}^{n} \frac{1}{1 - \lambda_j t} = \prod_{j=1}^{n} \frac{\lambda_j^{-1}}{\lambda_j^{-1} - t} = \frac{\det g^{-1}}{\det(g^{-1} - t \, \text{id})} = \frac{1}{\det(\text{id} - gt)}. \qquad \square$$

We can now easily prove the analogues of 6.4.2 and 6.4.3 for linear actions of finite groups:

Theorem 6.4.9 (Watanabe). *Let k be a field of characteristic 0, V a k-vector space of dimension n, $R = S(V)$, and G a finite subgroup of $\text{GL}(V)$.*
(a) *Then $R^{\det^{-1}}(-n)$ is the *canonical module of R^G.*
(b) *In particular R^G is Gorenstein if $G \subset \text{SL}(V)$.*

PROOF. Set $S = R^G$ and $\chi = \det^{-1}$. Since ρ^χ is an S-linear map from R onto $N = R^\chi$, we see that N is a direct S-summand of R. It was observed above that R is a maximal Cohen–Macaulay module over S; therefore N is also a maximal Cohen–Macaulay S-module. Furthermore

$$M_\chi(t) = |G|^{-1} \sum_{g \in G} \frac{\det g}{\det(\text{id} - tg)} = |G|^{-1} \sum_{g \in G} \frac{1}{\det(g^{-1} - t \, \text{id})}$$

$$= |G|^{-1} \sum_{g \in G} \frac{1}{\det(g - t \, \text{id})} = |G|^{-1} \sum_{g \in G} \frac{(-1)^n t^{-n}}{\det(\text{id} - t^{-1} g)}$$

$$= (-1)^n t^{-n} M_G(t^{-1}).$$

As the Molien series are Hilbert series, we may apply 4.4.5 to conclude that $N(-n)$ is the *canonical module of S. This proves (a).

If $G \subset \text{SL}(V)$, then, by (a), S is isomorphic to the *canonical module of S. As a *canonical module is canonical, S is Gorenstein. $\qquad \square$

Very easy examples show that 6.4.9(b) cannot be reversed. The obstruction is the presence of pseudo-reflexions in G: $g \in \text{GL}(V)$ is called a *pseudo-reflexion* if it has finite order and its eigenspace for the eigenvalue 1 has dimension $\dim V - 1$. (Thus the remaining eigenvalue is the determinant.)

Theorem 6.4.10. *With the notation of 6.4.9 the following hold.*
(a) (Stanley) *R^G is Gorenstein if and only if*

$$\sum_{g \in G} \frac{1}{\det(\text{id} - tg)} = t^{-m} \sum_{g \in G} \frac{\det g}{\det(\text{id} - tg)}$$

where m is the number of pseudo-reflexions in G.

(b) (Watanabe) *Suppose G contains no pseudo-reflexions. Then R^G is Gorenstein if and only if $G \subset SL(V)$.*

PROOF. (a) If we apply 4.4.6 to the Molien series of R^G, then it follows easily that R^G is Gorenstein if and only if the equation in (a) holds for some $m \in \mathbb{Z}$. It remains to determine m. To this end we expand both sides in a Laurent series at $t = 1$. Let $n = \dim V$, and Σ denote the set of pseudo-reflexions in G.

The pole order of $1/\det(\mathrm{id} - tg)$ at $t = 1$ is the multiplicity of 1 as an eigenvalue of g. Thus the only summand with a pole of order n is $1/\det(\mathrm{id} - t\,\mathrm{id}) = 1/(1 - t)^n$, and those with a pole of order $n - 1$ are exactly the summands

$$\frac{1}{\det(\mathrm{id} - t\sigma)} = \frac{1}{(1 - t)^{n-1}} \frac{1}{1 - \det \sigma} + \cdots, \qquad \sigma \in \Sigma,$$

where $\cdots$ denotes terms of higher order in $(1 - t)$. Thus the left hand side is

$$\frac{1}{(1 - t)^n} + \frac{1}{(1 - t)^{n-1}} \sum_{\sigma \in \Sigma} \frac{1}{1 - \det \sigma} + \cdots$$

whereas the right hand side is

$$(1 + m(1 - t) + \cdots) \left(\frac{1}{(1 - t)^n} + \frac{1}{(1 - t)^{n-1}} \sum_{\sigma \in \Sigma} \frac{\det \sigma}{1 - \det \sigma} + \cdots \right)$$

so that a comparison of coefficients yields $m = |\Sigma|$ as required.

(b) Evaluating the formula in (a) for $t = 0$ gives $|G| = \sum_{g \in G} \det g$. Since the eigenvalues of the elements of G are roots of unity, we must have $\det g = 1$ for all $g \in G$. (Note that the elements of k which are algebraic over k may be considered complex numbers.) $\qquad \square$

Remark 6.4.11. Theorems 6.4.9 and 6.4.10 were proved by Watanabe [387], [388] under the weaker assumption that $|G|$ is not divisible by char k. His proofs use divisorial methods. Hinič [172] extended Watanabe's results to invariant subrings of Gorenstein rings.

Finite groups generated by pseudo-reflexions. That the pseudo-reflexions in a finite group $G \subset GL(V)$ play a special role has already been demonstrated by 6.4.10. However, the most ostensive indication of this fact is the celebrated theorem which characterizes the regular ones among the rings of invariants of finite groups:

Theorem 6.4.12 (Shephard–Todd, Chevalley, Serre). *Let k be a field of characteristic 0, V a k-vector space of dimension n, $R = S(V)$, and G a finite subgroup of $GL(V)$. Then the following are equivalent:*
(a) *G is generated by pseudo-reflexions;*

(b) R is a free R^G-module;
(c) the k-algebra R^G is generated by (necessarily n) algebraically independent elements.

That (c) is equivalent with the regularity of R^G follows from Exercise
2.2.25, which also shows that the algebraically independent elements
can be chosen homogeneous; their number is n, as $\dim R^G = \dim R = \dim V = n$.
The remainder of this section is devoted to a proof of 6.4.12. The
next lemma covers the equivalence (b) $\iff$ (c).

Lemma 6.4.13. *Let R be a positively graded, finitely generated algebra
over an arbitrary field k, and S a graded k-subalgebra such that R is a
finite S-module.*
(a) *Then S is a finitely generated k-algebra.*
(b) *If R is Cohen–Macaulay and S is generated by algebraically independent
elements over k, then R is a free S-module. Moreover, it has a basis of
homogeneous elements.*
(c) *If R is generated by algebraically independent elements over k and a
free S-module, then S is generated by algebraically independent elements.*

PROOF. (a) This is a special case of E. Noether's theorem proved in 6.4.7.
   (b) By hypothesis S is a regular ring; a minimal homogeneous system
$x_1, \ldots, x_n$ of generators of its *maximal ideal is algebraically independent,
and furthermore generates S as a k-algebra (see Exercise 2.2.25). Since R
is a finite S-module, $x_1, \ldots, x_n$ is also a homogeneous system of parameters
of R, and thus an R-sequence by hypothesis on R. Consequently R is a
maximal Cohen–Macaulay S-module. It follows from 2.2.11 that R is a
projective S-module, and then 1.5.15 implies that R is a free S-module,
and that every minimal homogeneous system of generators of R over S
is a basis.
   (c) Let $\mathfrak{m}$ and $\mathfrak{n}$ be the *maximal ideals of R and S. The hypothesis
implies that $R_{\mathfrak{m}}$ is a regular local ring and a flat local extension of $S_{\mathfrak{n}}$.
Thus $S_{\mathfrak{n}}$ is regular according to 2.2.12. Again we apply 2.2.25 to conclude
that S is generated by algebraically independent elements. $\square$

We now show that (a) $\Rightarrow$ (b) in 6.4.12. For R to be a free S-module,
$S = R^G$, it is sufficient that $M = \operatorname{Tor}_1^S(R, S/\mathfrak{n}) = 0$. In fact this implies
that $R_{\mathfrak{n}}$ is a free $S_{\mathfrak{n}}$-module, whence R is free over S by 1.5.15. The module
M is the kernel of the homomorphism $\varphi : R \otimes_S \mathfrak{n} \to R \otimes_S S$ induced by
the embedding $\mathfrak{n} \subset S$. Given a minimal homogeneous system $x_1, \ldots, x_m$
of generators of $\mathfrak{n}$, M consists of all the elements $\sum y_i \otimes x_i$, $y_i \in R$ with
$\sum y_i x_i = 0$.
   Evidently M is a graded submodule of the graded S-module $R \otimes \mathfrak{n}$
with $\deg a \otimes b = \deg a + \deg b$ for homogeneous elements $a \in R$, $b \in \mathfrak{n}$. We
assume that $M \neq 0$. In order to derive a contradiction choose a non-zero

homogeneous element $\sum y_i \otimes x_i$ of minimal degree in M. Replacing y_i by a suitable homogeneous component, we may suppose that each y_i is itself homogeneous.

We claim that $\sum y_i \otimes x_i = \sum g(y_i) \otimes x_i$ for all $g \in G$. Since G is generated by pseudo-reflexions, it is enough to show this for a pseudo-reflexion σ. We choose a basis $X_1, \ldots, X_n$ of V such that $\sigma(X_i) = X_i$ for $i = 2, \ldots, n$ and $\sigma(X_1) = \zeta X_1$ where ζ is a root of unity. For each monomial f in $X_1, \ldots, X_n$ it follows easily that X_1 divides $\sigma(f) - f$. Therefore X_1 divides $\sigma(f) - f$ for every element f of R. Let $\sigma(y_i) - y_i = X_1 y_i'$ for $i = 1, \ldots, n$. Then $\sum y_i' x_i = 0$ so that $\sum y_i' \otimes x_i \in M$. From the assumption on $\sum y_i \otimes x_i$ we conclude $\sum y_i' \otimes x_i = 0$ and, hence, $\sum y_i \otimes x_i = \sum \sigma(y_i) \otimes x_i$.

The Reynolds operator ρ, viewed as an S-endomorphism of R, induces an S-linear map $\rho' = \rho \otimes \mathrm{id} : R \otimes \mathfrak{n} \to R \otimes \mathfrak{n}$. By what has just been proved, $\rho'(\sum y_i \otimes x_i) = \sum y_i \otimes x_i$. On the other hand $\mathrm{Im}\, \rho = S$ so that ρ' factors as

$$R \otimes \mathfrak{n} \xrightarrow{\rho''} S \otimes \mathfrak{n} \xrightarrow{\iota} R \otimes \mathfrak{n}$$

where ι is induced by the embedding $S \to R$. It is immediate that $\rho''(M)$ is mapped to the kernel $\mathrm{Tor}_1^S(S, S/\mathfrak{n}) = 0$ of the natural map $S \otimes \mathfrak{n} \to S \otimes S = S$. Thus $\rho'(\sum y_i \otimes x_i) = 0$, and therefore $\sum y_i \otimes x_i = 0$, which is the required contradiction.

It remains to prove the implication (c) $\Rightarrow$ (a) for which we use a combinatorial argument based on the following lemma.

Lemma 6.4.14. *Let k be a field of characteristic 0, V a k-vector space of finite dimension, $R = S(V)$, and G a finite subgroup of $\mathrm{GL}(V)$. Let $x_1, \ldots, x_n$, $n = \dim V$, be a homogeneous system of parameters of R^G.*

(a) *Then $x_1, \ldots, x_n$ are algebraically independent over k, and R^G is a free $k[x_1, \ldots, x_n]$-module; it has a basis of homogeneous elements $h_1, \ldots, h_m$.*

(b) *Let $d_i = \deg x_i$, $i = 1, \ldots, n$, and $e_j = \deg h_j$, $j = 1, \ldots, m$, and let Σ denote the set of pseudo-reflexions in G. Then*

$$m|G| = d_1 \cdots d_n, \quad \text{and} \quad m|\Sigma| + 2(e_1 + \cdots + e_m) = m(d_1 + \cdots + d_n - n).$$

PROOF. (a) According to 1.5.17 R^G is a finite $k[x_1, \ldots, x_n]$-module. Thus we have $\dim k[x_1, \ldots, x_n] = n$ so that $x_1, \ldots, x_n$ are algebraically independent over k. One now applies 6.4.13.

(b) The Hilbert series of $k[x_1, \ldots, x_n]$ is $1/\prod_{i=1}^n (1 - t^{d_i})$. Thus the Hilbert series of the $k[x_1, \ldots, x_n]$-module $R^G = \bigoplus h_i k[x_1, \ldots, x_n]$ is

$$M_G(t) = \frac{t^{e_1} + \cdots + t^{e_m}}{\prod_{i=1}^n (1 - t^{d_i})} = \frac{1}{(1-t)^n} f(t)$$

where $f(t) = (t^{e_1} + \cdots + t^{e_m})/\prod_{i=1}^n \sum_{j=0}^{d_i-1} t^j$ does not have a pole at $t = 1$. Expansion in a Laurent series at $t = 1$ yields

$$M_G(t) = \frac{1}{(1-t)^n}(f(1) - f'(1)(1-t) + \cdots)$$

with $f(1) = m/(d_1 \cdots d_n)$ and

$$f'(1) = \frac{e_1 + \cdots + e_m - (m/2)\sum_{i=1}^n (d_i - 1)}{d_1 \cdots d_n}.$$

As we saw in the proof of 6.4.10, we also have

$$M_G(t) = |G|^{-1}\left(\frac{1}{(1-t)^n} + \frac{1}{(1-t)^{n-1}} \sum_{\sigma \in \Sigma} \frac{1}{1 - \det \sigma} + \cdots\right).$$

Observe that

$$\sum_{\sigma \in \Sigma} \frac{1}{1 - \det \sigma} = \frac{1}{2}\left(\sum_{\sigma \in \Sigma} \frac{1}{1 - \det \sigma} + \sum_{\sigma \in \Sigma} \frac{1}{1 - \det \sigma^{-1}}\right) = \frac{1}{2} \sum_{\sigma \in \Sigma} 1 = \frac{1}{2}|\Sigma|.$$

Comparing coefficients in the Laurent expansions gives the required formulas. □

We now complete the proof of 6.4.12 with the implication (c) ⇒ (a). Let H be the subgroup of G generated by the pseudo-reflexions in G. Using the implication (a) ⇒ (c), we see that R^H is generated by algebraically independent homogeneous elements $y_1, \ldots, y_n$. Since R^G is, by hypothesis, also generated by algebraically independent homogeneous elements $x_1, \ldots, x_n$, we have an inclusion $k[x_1, \ldots, x_n] \subset k[y_1, \ldots, y_n]$. We want to show that there exists a permutation π of $\{1, \ldots, n\}$ such that $\deg x_i \geq \deg y_{\pi(i)}$ for all i.

To this end we define P_i to be the smallest subset of $\{1, \ldots, n\}$ such that $x_i \in k[y_j : j \in P_i]$. For each subset I of $\{1, \ldots, n\}$ the set $\bigcup_{i \in I} P_i$ must have at least $|I|$ elements since the x_i, $i \in I$, are algebraically independent. Thus the marriage theorem of elementary combinatorics guarantees an injective map π with $\pi(i) \in P_i$ for all i. By definition of P_i we have $\deg x_i \geq \deg y_{\pi(i)}$.

Arranging $y_1, \ldots, y_n$ in the order prescribed by π we may assume that $d_i = \deg x_i \geq z_i = \deg y_i$ for all i. Lemma 6.4.14 applied to $R^G = k[x_1, \ldots, x_n]$ yields

$$|\Sigma| = d_1 + \cdots + d_n - n,$$

since we have $m = 1$, $h_1 = 1$, and $d_1 = 0$. But 6.4.14 also applies to $R^H = k[y_1, \ldots, y_n]$, and since H contains all the pseudo-reflexions of G, we similarly obtain

$$|\Sigma| = z_1 + \cdots + z_n - n.$$

Summing up, we must have $d_i = z_i$ for all i. Therefore $|G| = |H|$ by the first equation in 6.4.14, and $G = H$.

Remarks 6.4.15. (a) The equivalence of (b) and (c) in 6.4.12 is independent of any assumption on the characteristic of k or the order of G, as is clearly exhibited by 6.4.13. Furthermore, the proof of the implication (a) $\Rightarrow$ (a) only uses that $|G|$ is not divisible by $\operatorname{char} k$. It is due to Serre [335], as well as a proof of (c) $\Rightarrow$ (a) (based on ramification theory) which does not require any assumption on k or G; see also Bourbaki [49], Ch. 5. For (c) $\Rightarrow$ (a) we have reproduced the original argument of Shephard and Todd [345] which exploits the fact that $\operatorname{char} k = 0$ in an essential way.

(b) Within the hierarchy 'regular, complete intersection, Gorenstein, Cohen–Macaulay', the property of being a complete intersection is the most difficult for rings of invariants of linear actions of finite groups. A necessary condition for R^G to be a complete intersection was given by Kac and Watanabe [228]: if R^G is a complete intersection, then G is generated by elements g with $\operatorname{rank}(g - \operatorname{id}) \leq 2$. The proof uses geometric methods. Exercise 6.4.21 presents an example showing that this condition is not sufficient for R^G to be a complete intersection. See Gordeev [129] and Nakajima and Watanabe [287] for a classification of the groups G for which R^G is a complete intersection. Nakajima [286] has classified the hypersurface rings R^G.

Exercises

6.4.16. Let S, T be affine semigroups, $S \subset T$. One says that S is an *expanded* subsemigroup of T if $S = T \cap \mathbb{Q}S$ (in $\mathbb{Q}T$). Prove:

(a) An expanded subsemigroup is a full subsemigroup.

(b) The following are equivalent for a subsemigroup S of $\mathbb{N}^n$:
   (i) S is expanded;
   (ii) there exists a vector subspace U of $\mathbb{Q}^n$ with $S = U \cap \mathbb{Q}^n$;
   (iii) there exists a homogeneous system of linear equations with integral coefficients such that S is the set of its non-negative solutions;
   (iv) $k[S]$ is the ring of invariants of a linear torus action on $k[X_1, \ldots, X_n]$.

(c) Every positive normal semigroup C is isomorphic to an expanded subsemigroup of $\mathbb{N}^n$ for some $n \geq 0$.

Hint for (c) (communicated by Hochster): By 6.1.10 we may assume that C is a full subsemigroup of $\mathbb{N}^m$ for some $m \geq 0$, thus $C = \mathbb{N}^m \cap \mathbb{Z}C$. Set $\hat{C} = \mathbb{N}^m \cap \mathbb{Q}C$. Then $\mathbb{Z}\hat{C}/\mathbb{Z}C$ is a torsion group, so that there exist a basis $e_1, \ldots, e_r$ of $\mathbb{Z}\hat{C}$ and positive integers q_i for which $q_1 e_1, \ldots, q_r e_r$ is a basis of $\mathbb{Z}C$. Extend $e_1, \ldots, e_r$ to a basis of $\mathbb{Q}^m$, and let φ_i, $i = 1, \ldots, r$, be the linear form on $\mathbb{Q}^m$ which assigns each vector its i-th coordinate with respect to this basis. Note that $\varphi_i(a) \in \mathbb{Z}$ for all $a \in \mathbb{Z}\hat{C}$. Then C is the set of elements of $c \in \mathbb{N}^m$ satisfying (i) a system of homogeneous linear equations with rational coefficients whose set of solutions is $\mathbb{Q}C$, and secondly the congruence conditions $\varphi_i(c) \equiv 0 \bmod q_i$. Adding positive integral multiples of q_i to the coefficients of φ_i (with respect to the dual canonical basis of $\mathbb{Q}^m$), we may replace the φ_i by linear forms which are non-negative on $\mathbb{N}^m$. Then $\varphi_i(c) \equiv 0 \bmod q_i$ if and only if the linear equation $\varphi_i(c) = y_i q_i$

has a non-negative integral solution $y_i(c)$. Consider the map $C \to \mathbb{N}^m \times \mathbb{N}^r$, $c \mapsto (c, y_1(c), \ldots, y_r(c))$.

6.4.17. Let k be an infinite field, and $R = [Y_1, Y_2, Z_1, Z_2]$. Suppose that $\mathrm{GL}(1, k) = k \setminus \{0\}$ acts on R by the substitutions $Y_i \mapsto aY_i$, $Z_i \mapsto a^{-1}Z_i$.
(a) Show that $S = R^G$ is generated by the elements $x_{ij} = Y_iZ_j$, $i, j = 1, 2$.
In Exercise 6.2.7 we have seen that $S \cong k[X_{11}, X_{12}, X_{21}, X_{22}]/(X_{11}X_{22} - X_{21}X_{12})$, the isomorphism being induced by the substitution $X_{ij} \mapsto x_{ij}$.
(b) Let $\mathfrak{p} = (x_{11}, x_{12})$ and $\mathfrak{q} = (x_{11}, x_{21})$. Show (i) $\mathfrak{p}$ and $\mathfrak{q}$ are prime ideals in S and maximal Cohen–Macaulay S-modules, (ii) $\mathfrak{p}^j$ and $\mathfrak{q}^j$ are not Cohen–Macaulay for $j \geq 2$. (4.7.11 is helpful for (ii); use a system of parameters consisting of 1-forms. Or use the Hilbert–Burch theorem.)
(c) The characters of $\mathrm{GL}(1, k)$ are given by the maps $a \mapsto a^j$, $j \in \mathbb{Z}$. Compute the semi-invariants for each of the characters, and find out which of them are Cohen–Macaulay S-modules.

6.4.18. Let V be a finite dimensional vector space over a field k, $R = S(V)$, and G a finite subgroup of $\mathrm{GL}(V)$ such that $|G|$ is invertible in k. Show that for each character χ of G the R^G-module M^χ is a direct R^G-summand of R and a rank 1 maximal Cohen–Macaulay R^G-module.

6.4.19. Let G be the cyclic subgroup of $\mathrm{GL}(2, \mathbb{C})$ generated by the matrix

$$\begin{pmatrix} 0 & -1 \\ 1 & 0 \end{pmatrix}$$

and $R = S(\mathbb{C}^2) = \mathbb{C}[X_1, X_2]$. Compute the Molien series of G, and show that R^G is a complete intersection. (In order to determine the generators of R^G one should draw as much information as possible from the Molien series.)

6.4.20. Show that the subgroup of $\mathrm{GL}(2, \mathbb{C})$ generated by the matrices

$$\begin{pmatrix} 0 & -1 \\ 1 & -1 \end{pmatrix} \quad \text{and} \quad \begin{pmatrix} 0 & 1 \\ 1 & 0 \end{pmatrix}$$

is isomorphic to S_3, the permutation group of three letters. Prove that R^G (with $R = S(\mathbb{C}^2) = \mathbb{C}[X_1, X_2]$) is generated by algebraically independent elements x_1, x_2, and determine their degrees.

6.4.21. (a) Let k be a field and S a graded k-algebra generated by elements $x_1, \ldots, x_n$ of positive degrees $d_1, \ldots, d_n$. If S is a complete intersection, then there exist positive integers $e_1, \ldots, e_r$ with $r = n - \dim S$ such that $H_S(t) = \prod_{i=1}^{r}(1 - t^{e_i})/\prod_{j=1}^{n}(1 - t^{d_j})$.
(b) Embed the group G of the previous problem into $\mathrm{GL}(4, \mathbb{C})$ by sending each matrix $A \in G$ to the matrix $\begin{pmatrix} A & 0 \\ 0 & A \end{pmatrix}$, and let $R = S(\mathbb{C}^4) = \mathbb{C}[X_1, \ldots, X_4]$. Show that R^G is not a complete intersection. Is R^G Gorenstein?

6.4.22. Let k be a field, $R = k[X_1, \ldots, X_n]$, and $\sigma_1, \ldots, \sigma_n$ be the elementary symmetric polynomials in $X_1, \ldots, X_n$.
(a) Show that $\mathrm{height}(\sigma_1, \ldots, \sigma_n)R = n$.
(b) Let G be the subgroup of $\mathrm{GL}(n, k)$ formed by the permutation matrices. Noting that G is generated by pseudo-reflexions, give a fast proof of the main theorem on symmetric functions in the case in which $\mathrm{char} k = 0$.

6.5   Invariants of linearly reductive groups

Let k be an algebraically closed field. A linear algebraic group G over k is called *linearly reductive* if for every finite dimensional representation $G \to \mathrm{GL}(V)$ the k-vector space V splits into the direct sum of irreducible G-subspaces U. Here U is a *G-subspace* if $g(u) \in U$ for all $g \in G$ and $u \in U$; it is *irreducible* if it has no G-subspaces other than $\{0\}$ and U.

The main objective of this section is the proof of the following fundamental result.

Theorem 6.5.1 (Hochster–Roberts). *Let k be an algebraically closed field and G be a linearly reductive group over k acting linearly on a polynomial ring $R = k[X_1, \ldots, X_n]$. Then the ring R^G is Cohen–Macaulay.*

The most classical examples of linearly reductive groups are finite groups G whose order is not divisible by $\operatorname{char} k$; this is Maschke's theorem. The tori $\mathrm{GL}(1, k)^m$ are linearly reductive independently of $\operatorname{char} k$, as follows easily from the fact that a torus action can be diagonalized. Thus the results of the previous section about the Cohen–Macaulay property of rings of invariants of tori or finite groups are special cases of 6.5.1.

In characteristic 0 the groups $\mathrm{GL}(n, k)$ and $\mathrm{SL}(n, k)$ are linearly reductive, and so are the orthogonal and symplectic groups. However, in characteristic $p > 0$ there exist only a few linearly reductive groups so that 6.5.1 has its main applications in characteristic 0.

Let $G \to \mathrm{GL}(V)$ be a finite dimensional representation of a linearly reductive group G. Then the set V^G of invariants is the maximal trivial G-subspace of V where *trivial* means for a G-subspace U that $g(u) = u$ for all $g \in G$ and $u \in U$. Let W be the sum of all non-trivial irreducible G-subspaces of V. Then W is in fact a direct sum $W_1 \oplus \cdots \oplus W_t$ of non-trivial irreducible subspaces W_i, and it follows easily that $V^G \cap W = W_1^G \oplus \cdots \oplus W_t^G = 0$. Thus $V = V^G \oplus W$, and W is the unique complementary G-subspace of V^G, as every irreducible subspace U with $U \cap V^G = 0$ is contained in W. The projection $\rho : V \to V^G$ with kernel W satisfies the condition $\rho(g(v)) = \rho(v)$ for all $g \in G$ and $v \in V$. It is called the *Reynolds operator*.

As in the previous section let now $R = k[X_1, \ldots, X_n]$ be a polynomial ring over k whose space of 1-forms is identified with V. Then G acts linearly on the graded components R_i of R for each of which we have a Reynolds operator ρ_i. The direct sum of the ρ_i is a surjective map $\rho : R \to R^G$ which is easily seen to be R^G-linear. In fact, the R^G-linearity of ρ is equivalent to $r \operatorname{Ker} \rho \subset \operatorname{Ker} \rho$ for all $r \in R^G$. It is enough to show that $rU \subset \operatorname{Ker} \rho$ for a homogeneous element $r \in R^G$ of degree i and a non-trivial G-subspace $U \subset R_j$. As multiplication by an invariant is G-linear, rU is either 0 or G-isomorphic to U. Therefore $rU \subset \operatorname{Ker} \rho_{i+j}$. In the general context of linearly reductive groups we have thus recovered the existence of a Reynolds operator $R \to R^G$ which was first encountered above 6.4.4 and which is the crucial fact in the proof of 6.5.1.

By 6.4.4 the existence of a Reynolds operator $R \to R^G$ implies that $IR \cap R^G = I$ for every ideal I of R^G, and furthermore that R^G is a Noetherian k-algebra. Being positively graded, it is even finitely generated over k. So 6.5.1 follows from the more general, purely ring-theoretic

Theorem 6.5.2. *Let k be a field, and $R = k[X_1, \ldots, X_n]$. Suppose S is a finitely generated graded k-subalgebra of R such that $IR \cap S = I$ for all ideals I of S. Then S is Cohen–Macaulay.*

Remarks 6.5.3. (a) The original Hochster–Roberts theorem [201] is more general than stated in 6.5.1. It says: *let G be a linearly reductive group over a field k acting rationally on a regular Noetherian k-algebra R; then R^G is Cohen–Macaulay.*

(b) Let $S \subset R$ be rings; S is called a *pure subring* (or R a *pure extension* of S) if for every S-module M the natural homomorphism $M = M \otimes_S S \to M \otimes_S R$ is injective. The reader may prove that S is a pure subring of R if one of the following conditions holds: (i) there exists a Reynolds operator $R \to S$; (ii) R is faithfully flat over S; (iii) R/S is flat over S. Thus, under the conditions of 6.5.1, R^G is a pure subring of R. The choice $M = S/I$ yields that $IR \cap S = I$ for every ideal I of S if S is a pure subring of R. See [270], §7 or Hochster and Roberts [202], Section 5 for a discussion of purity.

Using the notion just introduced we can formulate the following even more general theorem of Hochster and Huneke [197]: *let R be a regular ring, and S a pure subring of R containing a field; then S is Cohen–Macaulay.* We will prove this theorem under the slightly weaker hypothesis that S is a direct S-summand of R; see 10.4.1. The case in which R contains a field of characteristic $p > 0$ was already given by Hochster and Roberts [201], and the case in which R and S are finitely generated algebras over a field was established by Kempf [235].

(c) An important variant of the theorem of Hochster and Roberts is due to Boutot [50]: *Let R be a finitely generated algebra over an algebraically closed field of characteristic 0, and S a pure subring of R; if R has rational singularities, then so has S.* It is remarkable that Boutot's theorem (for which there is also an analytic version) weakens the hypothesis of the Hochster–Roberts theorem while strengthening its conclusion. For the notion of rational singularity we refer the reader to [236] and [44].

(d) The hypotheses of the theorems presented in (b) and (c) cannot be weakened essentially. In particular it is not true that R^G is Cohen–Macaulay whenever a linearly reductive group G acts linearly on a Cohen–Macaulay ring R. See 6.5.7 for a simple counterexample.

(e) By 6.1.10 a positive normal semigroup ring $S = k[C]$ can be embedded as a graded subring into a polynomial ring R over k such that there exists a Reynolds operator $R \rightarrow S$. In conjunction with 6.5.2 this argument is the fastest and most elementary proof of the Cohen–Macaulay property of normal semigroup rings, especially if one uses the simple reduction to characteristic p indicated in Exercise 2.1.29. Nevertheless the proof given in 6.3 retains is value since it gave us insight into the combinatorial structure of S, and, above all, allowed us to compute the canonical module.

The following proof of 6.5.2 has been drawn from Knop [239]. Its characteristic p part is an argument from tight closure theory, which we will study systematically in Chapter 10.

The first step is the reduction of 6.5.2 to Theorem 6.5.4 below. The k-algebra S is positively graded. By 1.5.17 it has a homogeneous system of parameters $f_1, \ldots, f_s$. Suppose that $g_{r+1} f_{r+1} = g_1 f_1 + \cdots + g_r f_r$ for some r, $0 \leq r \leq s - 1$. If we can show that $g_{r+1} \in (f_1, \ldots, f_r)$, then $f_1, \ldots, f_s$ is an S-sequence, and the theorem is proved. Suppose on the contrary that $g_{r+1} \notin (f_1, \ldots, f_r)$. As $(f_1, \ldots, f_r)R \cap S = (f_1, \ldots, f_r)$ by hypothesis, one even has $g_{r+1} \notin (f_1, \ldots, f_r)R$. The elements $f_1, \ldots, f_s$ are algebraically independent over k. Moreover, S is a finite $k[f_1, \ldots, f_s]$-module. Therefore it is enough to prove the following theorem.

Theorem 6.5.4. *Let k be a field, and $f_1, \ldots, f_s$ algebraically independent homogeneous elements of positive degree in $R = k[X_1, \ldots, X_n]$. Suppose that S is a module-finite graded $k[f_1, \ldots, f_s]$-algebra such that there exists a homogeneous homomorphism $\psi : S \rightarrow R$ of $k[f_1, \ldots, f_s]$-algebras. If $g_{r+1} f_{r+1} = g_1 f_1 + \cdots + g_r f_r$ with $g_1, \ldots, g_{r+1} \in S$ for some r, $0 \leq r \leq s - 1$, then $\psi(g_{r+1}) \in (f_1, \ldots, f_r)R$.*

PROOF. Without restriction we may assume that the g_i are homogeneous elements of S. Let $r_1, \ldots, r_m$ be a system of generators of S as a $k[f_1, \ldots, f_s]$-module. Suppose A is a finitely generated $\mathbb{Z}$-subalgebra of R containing all the elements of k which appear as coefficients in

(i) $\psi(g_i)$ as a polynomial in $X_1, \ldots, X_n$, $i = 1, \ldots, r + 1$,

(ii) the polynomials $p_{iju} \in k[Y_1, \ldots, Y_s]$ with $r_i r_j = \sum_{u=1}^{m} p_{iju}(f_1, \ldots, f_s) r_u$, and

(iii) the analogous representations $g_i = \sum_{u=1}^{m} q_{iu}(f_1, \ldots, f_s) r_u$.

Let $B = A[f_1, \ldots, f_s]$, $C = A[X_1, \ldots, X_n]$, and $T = B[r_1, \ldots, r_m] \subset S$. Then

$$A[f_1, \ldots, f_s] \subset C, \quad \psi(T) \subset C, \quad g_i \in T, \quad T = Br_1 + \cdots Br_m.$$

Thus, if we replace k by the finitely generated $\mathbb{Z}$-subalgebra A, then all the assumptions of the theorem (except that on k) remain valid. It is enough to show $\psi(g_{r+1}) \in (f_1, \ldots, f_r)C$ for one such A. As $\psi(g_{r+1}), f_1, \ldots, f_r$ are homogeneous this is equivalent to the solubility of a system $\mathscr{S}$ of linear equations with coefficients in A. (The system $\mathscr{S}$ arises from comparing coefficients in $C = A[X_1,\ldots,X_n]$.)

If $\mathscr{S}$ has a solution over the field of fractions of A, then we enlarge A by adjoining the reciprocals of the finitely many denominators of a solution, and obtain a solution in the new A.

So suppose that $\mathscr{S}$ is insoluble over the field of fractions of A. Then there is a non-zero element $d \in A$ such that the reduction of $\mathscr{S}$ modulo a maximal ideal $\mathfrak{m}$ of A does not have a solution whenever $d \notin \mathfrak{m}$. (Simply take d as a suitable subdeterminant of the matrix of $\mathscr{S}$ including the right hand side.) Adjoining $d^{-1} \in k$ to A, we may assume that the reduction of $\mathscr{S}$ modulo any maximal ideal $\mathfrak{m}$ of A is insoluble.

We want to pass to such a reduction. It may happen however that the induced map $B/\mathfrak{m}B \to C/\mathfrak{m}C$ is not injective. Therefore an extra condition must first be satisfied. In fact, by the theorem on generic flatness, which we will prove below, there exists $t \in B$ such that C_t is a free B_t-algebra. As A and B are finitely generated $\mathbb{Z}$-algebras, they are Hilbert rings. This implies that (1) there exists a maximal ideal $\mathfrak{n}$ of B with $t \notin \mathfrak{n}$, (2) $\mathfrak{m} = \mathfrak{n} \cap A$ is a maximal ideal of A, and furthermore (3) $A/\mathfrak{m}$ is a finite field; see A.17, A.18. One has a commutative diagram

$$
\begin{array}{ccc}
B/\mathfrak{m}B & \xrightarrow{\;\iota\;} & C/\mathfrak{m}C \\
\kappa \downarrow & & \downarrow \\
B_t/\mathfrak{m}B_t & \xrightarrow{\;\lambda\;} & C_t/\mathfrak{m}C_t
\end{array}
$$

Since $\mathfrak{m}B$ is a prime ideal with $t \notin \mathfrak{m}B$, κ is injective. Next λ is injective because the extension $B_t \to C_t$ is faithfully flat, and so ι is injective as desired.

One now replaces all objects by their residue classes modulo $\mathfrak{m}$. Since the field $A/\mathfrak{m}$ is finite, the theorem has been reduced to the case in which k is a finite field! Let p be its characteristic.

The finite $k[f_1,\ldots,f_s]$-module S has a rank (just because $k[f_1,\ldots,f_s]$ is a domain). Let F be a free submodule of S such that $\operatorname{rank} F = \operatorname{rank} S$. There exists a non-zero element $c \in k[f_1,\ldots,f_s]$ such that $cS \subset F$. We set $q = p^e$, and take the q-th power of the equation $g_{r+1}f_{r+1} = g_1f_1 + \cdots + g_rf_r$ and multiply by c to obtain

$$
(cg_{r+1}^q)f_{r+1}^q = \sum_{i=1}^{r}(cg_i^q)f_i^q.
$$

The elements cg_i^q, $i = 1,\ldots,r+1$ are in the free $k[f_1,\ldots,f_s]$-module F. Then an elementary argument yields $h_{iq} \in F$ with $cg_i^q = h_{iq}f_{r+1}^q$ for $i = 1,\ldots,r$. By substituting these expressions into the previous equation and applying $\psi : S \to k[X_1, \ldots, X_n]$ one has

$$cf_{r+1}^q \psi(g_{r+1})^q = \sum_{i=1}^r f_i^q f_{r+1}^q \psi(h_{iq}), \quad \text{hence} \quad c\psi(g_{r+1})^q = \sum_{i=1}^r f_i^q \psi(h_{iq}).$$

Let M be the set of monomials $\mu = X_1^{\mu_1} \cdots X_1^{\mu_n}$ with $\mu_i < q$ for $i = 1, \ldots, n$. Taking q-th powers in k is bijective since k is finite. Therefore every element $h \in k[X_1, \ldots, X_n]$ has a necessarily *unique* representation $h = \sum_{\mu \in M} (h_\mu)^q \mu$; in particular

$$\psi(h_{iq}) = \sum_{\mu \in M} (h_{iq\mu})^q \mu.$$

Thus

$$\sum_{i=1}^r f_i^q \psi(h_{iq}) = \sum_{\mu \in M} (\sum_{i=1}^r h_{iq\mu} f_i)^q \mu = \sum_{\mu \in M} (h_{q\mu})^q \mu$$

with $h_{q\mu} \in (f_1, \ldots, f_r)R$.

The crucial point is that c does not depend on q. We choose q so large that $c = \sum_{\mu \in M} c'_\mu \mu$ with $c'_\mu \in k$. Let $c'_\mu = (c_\mu)^q$. Then

$$\sum_{\mu \in M} (c_\mu \psi(g_{r+1}))^q \mu = \sum_{\mu \in M} (h_{q\mu})^q \mu.$$

Since $c \neq 0$ there exists μ with $c_\mu \neq 0$, and so

$$\psi(g_{r+1}) = \frac{1}{c_\mu} h_{q\mu} \in (f_1, \ldots, f_r)R. \qquad \square$$

A remarkable feature of the preceding proof is that a theorem which has its main applications in characteristic 0 has been reduced to its characteristic p case. Such a reduction will also be fundamental for the results of Chapters 8, 9, and 10.

Remark 6.5.5. In view of 6.4.2 and 6.4.9 it is tempting to conjecture that $R^{\det^{-1}}$, if non-zero, is the canonical module of R^G under the hypothesis of 6.5.1. Then, in particular, R^G would be a Gorenstein ring if $\det g = 1$ for all $g \in G$. This was however disproved by Knop [238]: in fact, every ring of invariants R^G can be written in the form $(R')^{G'}$ where $\det g' = 1$ for all $g' \in G'$; see Exercise 6.5.8. On the other hand, Knop showed that over an algebraically closed field of characteristic zero $R^{\det^{-1}}$ is indeed the canonical module of R^G if the action of G on the vector space V (of 1-forms of R) satisfies a mild non-degeneracy condition; the proof uses methods of geometric invariant theory beyond the scope of this book. Knop also proved estimates for the a-invariant $a(R^G)$; in particular one always has $a(R^G) \leq -\dim R^G$ (compare this with 6.4.2 and 6.4.9).

However, for one class of groups the ring of invariants is always Gorenstein in characteristic 0: if G is semisimple and connected (for example $G = \mathrm{SL}(n, k)$), then R^G is factorial, and therefore Gorenstein by 6.5.1 and 3.3.19. In fact, let $f = \pi_1 \cdots \pi_m$ be the prime decomposition of an invariant f. Then the action of an element $g \in G$ permutes the prime ideals $R\pi_i$. Since $\{g \in G : g(R\pi_i) = R\pi_i\}$ is a non-empty Zariski closed subset of the connected variety G, this set equals G. So $g(\pi_i) = \chi_i(g)\pi_i$ with $\chi_i(g) \in k \setminus \{0\}$; χ_i is a character of G. But a semisimple group has no non-trivial characters. Thus $g(\pi_i) = \pi_i$ for all $g \in G$, and therefore f has a prime decomposition in R^G.

Generic flatness. In the proof of 6.5.4 we used the following theorem on 'generic flatness':

Theorem 6.5.6. *Let R be a Noetherian domain, S a finitely generated R-algebra, and M a finite S-module. Then there exists $f \in R$ such that $M \otimes R_f$ is a free (in particular flat) R_f-module.*

PROOF. There is nothing to prove for $M = 0$. So suppose that $M \neq 0$. Then there exists in M a chain $0 = M_0 \subset M_1 \subset \cdots \subset M_m = M$ of submodules such that $M_{i+1}/M_i \cong S/\mathfrak{p}_i$ for some prime ideal $\mathfrak{p}_i$ of S. (One only needs that $\mathrm{Ass}\, N \neq \emptyset$ for an S-module $N \neq 0$.) It is enough to prove that the theorem holds for each quotient M_{i+1}/M_i, since $N = U \oplus N/U$ if U is a submodule of N for which N/U is free. That is to say, we may suppose that $M = S$, and, furthermore, that S is a domain.

If the natural homomorphism $R \to S$ is not injective, we simply take f from its kernel. Thus R may be considered as a subring of S. Let Q be the field of fractions of R. Then $S \otimes Q = S_{R \setminus \{0\}}$ is a domain contained in the field of fractions of S. It is a finitely generated Q-algebra, and therefore has finite Krull dimension, say d. We go by induction on d.

By the Noether normalization theorem A.14 the Q-algebra $S \otimes Q$ contains $y_1, \ldots, y_d$ such that $S \otimes Q$ is integral over $Q[y_1, \ldots, y_d]$; moreover, $y_1, \ldots, y_d$ are algebraically independent over Q. Multiplying by a suitable common denominator, we may assume that $y_i \in S$ for all i. Let z be an element of S. As z is integral over $Q[y_1, \ldots, y_d]$, it is easy to find $g \in R$ such that z is already integral over $R_g[y_1, \ldots, y_d]$. Since R is a finitely generated R-algebra, one therefore has that S_g is integral over $R_g[y_1, \ldots, y_d]$ for some element $g \in R$. In view of what is to be proved we may replace R by R_g and S by $S \otimes R_g$.

Thus we have reached a situation in which S is a finite module over the ring $T = R[f_1, \ldots, f_d] \subset S$ which in turn is isomorphic to a polynomial ring over R, and therefore a free R-module. Let F be a free T-submodule of S such that S/T is a torsion module. Then F is a free R-module. It remains to show that the theorem holds for S/T as a finite

T-module. As above, S/T has a finite filtration with successive quotients of type $T/\mathfrak{p}$ where $\mathfrak{p} \in \operatorname{Spec} T$. Since S/T is a torsion module, $\mathfrak{p} \neq 0$. Therefore $T/\mathfrak{p} \otimes Q$, if non-zero, is a proper residue class ring of $T \otimes Q$, and so has dimension $< \dim T \otimes Q = d$. Thus we may repeatedly apply the induction hypothesis in order to complete the proof. $\square$

Exercises

6.5.7. Let k be an infinite field, and let $R = k[Y_1, Y_2, Z_1, Z_2]/(Y_1^2 - Y_2^2)$. Obviously R is a Cohen–Macaulay ring, and reduced if $\operatorname{char} k \neq 2$. Let $G = \operatorname{GL}(1, k)$ act on $k[Y_1, Y_2, Z_1, Z_2]$ by the substitutions $Y_i \mapsto aY_i$, $Z_i \mapsto a^{-1}Z_i$, $a \in k$, $a \neq 0$. Prove:
(a) The action of G induces an action of G on R, and R^G is the k-subalgebra generated by the products $y_i z_j$, $i, j = 1, 2$. (Small letters denote residue classes in R.)
(b) The substitution $X_{ij} \mapsto Y_i Z_j$ induces a surjective k-algebra homomorphism $k[X_{11}, X_{12}, X_{21}, X_{22}] \to R^G$. Its kernel is generated by the elements $X_{11}X_{22} - X_{12}X_{21}$, $X_{11}^2 - X_{12}^2$, $X_{11}X_{21} - X_{12}X_{22}$, $X_{21}^2 - X_{22}^2$.
(c) R^G is not Cohen–Macaulay.
By increasing the number of variables Y_i (and the degree of the equation defining R) one can even produce examples of factorial hypersurface rings R such that R^G is not Cohen–Macaulay. (A *hypersurface ring* is a residue class ring of a regular ring with respect to a principal ideal.)

6.5.8. Suppose G be a subgroup of $\operatorname{GL}(V)$ where V is a finite dimensional vector space over an infinite field k. Set $V' = k^2 \oplus k^3 \oplus V$ and let $G' = \operatorname{SL}(2, k) \times \operatorname{SL}(3, k) \times G$ act on V' by

$$(f, h, g)(u, w, v) = \Big((\det g)f(u), (\det g)^{-1}h(w), g(v) \Big).$$

Then obviously $\det g' = 1$ for all $g' \in G'$. Let $R = S(V)$ and $R' = S(V')$. Show $R^G \cong (R')^{G'}$.

Notes

Hochster [174] proved the Cohen–Macaulay property of normal semigroup rings using the shellability of convex polytopes (see Section 5.2 for the notion of shellability.) A purely algebraic proof was provided by Goto and Watanabe [135]; they computed local cohomology from a complex similar to $L^{\bullet}$. Such complexes, or their graded k-duals (which are dualizing complexes) have been constructed by several authors. See Trung and Hoa [371] or Schäfer and Schenzel [321]; these articles give Cohen–Macaulay criteria for general affine semigroup rings. For a general affine semigroup C the Cohen–Macaulay and the Gorenstein property of $k[C]$ may depend on the field k; see [371] and Hoa [173]. Gilmer [126] treats semigroup rings from a more general point of view.

Our approach is close to that of Ishida [227]; from Danilov [76] we borrowed the idea of proving the vanishing of certain cohomology groups by a topological argument. See also Stanley [360] and [363] where the method is also applied to certain modules over normal semigroup rings, namely those which in the invariant-theoretic situation arise as semi-invariants. (See 6.4.17 for a non–Cohen–Macaulay such module.) More recently the Cohen–Macaulay property of modules of semi-invariants was investigated by Van den Bergh [378]. The complete intersection normal semigroup rings were classified by Nakajima [285].

Stanley computed the canonical module of a normal semigroup ring by a combinatorial argument outlined in 6.3.10, whereas Danilov [76] applied differentials. Local cohomology as in 6.3.5 was used by Goto and Watanabe.

For the theory of Ehrhart polynomials and related combinatorial functions we refer the reader to Ehrhart [89], Stanley [361], and to Hibi's survey [168] where numerous references are given. See also Danilov [76].

Normal semigroup rings, or rather their spectra, are the most special cases of toric varieties which connect combinatorics and algebraic geometry. We must confine ourselves to a list of references: Kempf, Knudsen, Mumford, and Saint–Donat [236], Danilov [76], Oda [293], and Ewald [99]. The recent book by Sturmfels [366] treats the combinatorial aspects of Gröbner bases for the defining ideals of semigroup rings.

The invariant theory of finite groups is a classical subject whose literature we cannot cover adequately; instead we refer the reader to Springer [355], Stanley [358], Benson [40], and Smith [353]. While 6.4.5 is due to Hochster and Eagon [189], the Cohen–Macaulay property of rings of invariants of finite groups seems to have been realized by several authors. The characterization of Gorenstein invariants is the work of Watanabe as pointed out in 6.4.11; Stanley gave the combinatorial proof reproduced by us. The determination of the canonical module is only implicit in Watanabe's papers; according to Stanley [358] it was made explicit by Eisenbud.

References for the Hochster–Roberts theorem, its variants and extensions have been indicated in 6.5.3. Hochster [187] contains an extensive discussion of the problem of determining the canonical module of a ring of invariants. As pointed out in 6.5.5, this problem was satisfactorily solved by Knop [238].

The example in 6.5.7 is a simplification of that of Hochster and Eagon [189], p. 1056. It is a very special instance of the Segre product of graded rings. The Cohen–Macaulay property of Segre products was explored by Chow [69] and Goto and Watanabe [134].

Hochster [185] is a survey of the invariant theory of commutative rings.

The theorem of generic flatness is due to Grothendieck [142], IV.6.9. A more refined version was given by Hochster and Roberts [201]; see also [270], §24.

7 Determinantal rings

Determinantal rings occur in algebraic geometry as coordinate rings of classical algebraic varieties. From the algebraic point of view they are graded algebras with straightening law which themselves form a subclass of the class of graded Hodge algebras. The special feature of such an algebra is that it is free over the ground ring with a monomial basis whose multiplication table is compatible with a partial order on the algebra generators.

The results on filtered rings in Section 4.5 will be applied to 'trivialize' a graded Hodge algebra: by repeatedly passing to a suitable associated graded ring one eventually gets a discrete Hodge algebra, which is nothing but the residue class ring of a polynomial ring modulo an ideal generated by monomials. A discrete algebra with straightening law may be considered the Stanley–Reisner ring of the order complex of a certain poset, and as an application we will thus obtain a Cohen–Macaulay criterion.

The remaining sections of the chapter are devoted to the most important examples of algebras with straightening law, the determinantal rings. It will be shown that these rings are normal Cohen–Macaulay domains. The class group and the canonical module will be identified, and we will characterize the Gorenstein determinantal rings.

7.1 Graded Hodge algebras

In this section we introduce graded Hodge algebras and study their basic properties.

Let A be a ring, H a finite subset of A, and $c \in \mathbb{N}^H$, $c = (c_\xi)$. An element $u = \prod_{\xi \in H} \xi^{c_\xi}$ is called a *monomial on H with exponent c*. Its *support* is the set $\operatorname{supp} u = \{\xi \in H : c_\xi \neq 0\}$. Let u and u' be monomials on H with exponents c and c', respectively. We say u *divides* u' or u *is a factor of* u' if $c' - c \in \mathbb{N}^H$. Finally, if $\Sigma \subset \mathbb{N}^H$ is a semigroup ideal, we call $c \in \Sigma$ a *generator* of Σ if $c - c' \notin \mathbb{N}^H$ for all $c' \in \Sigma$, $c' \neq c$.

Definition 7.1.1. Let A be a B-algebra, $H \subset A$ a finite subset with partial order $\leq$, and $\Sigma \subset \mathbb{N}^H$ a semigroup ideal. A is a *graded Hodge algebra on H over B governed by Σ* if the following conditions hold.
(H_0) $A = \bigoplus_{i \geq 0} A_i$ is a graded B algebra with $A_0 = B$, and H consists of elements of positive degree and generates A over B.

(H$_1$) The monomials on H with exponent in $\mathbb{N}^H \setminus \Sigma$ are linearly independent over B. They are called *standard monomials*.

(H$_2$) (*Straightening law*) If v is a monomial on H whose exponent is a generator of Σ, then v has a presentation

$$v = \sum b_u u, \qquad b_u \in B, \quad b_u \neq 0, \quad u \text{ a standard monomial,}$$

such that for each $\xi \in H$ which divides v there exists for every u a factor ζ_u with $\zeta_u < \xi$.

The right hand side of a straightening relation may of course be the empty sum, i.e. equal to zero. If this happens for all straightening relations, the graded Hodge algebra is called *discrete*. In this case $A \cong B[X_\xi : \xi \in H]/I$ where I is generated by the monomials $\prod_{\xi \in H} X_\xi^{c_\xi}$, $c = (c_\xi)$, $c \in \Sigma$. In particular, Stanley–Reisner rings are discrete Hodge algebras.

The graded Hodge algebra A is called a *graded algebra with straightening law* (*on H over B*), abbreviated *graded ASL*, if Σ is generated by the exponents of monomials ξv where ξ and v are incomparable elements in H. It follows that a monomial u is standard if and only if all factors of u are comparable with each other, and for all incomparable $\xi, v \in H$ we have a straightening relation

$$\xi v = \sum b_u u, \qquad b_u \in B, \quad b_u \neq 0, \quad u \text{ a standard monomial,}$$

satisfying the condition: every u contains a factor $\zeta \in H$ such that $\zeta < \xi$, $\zeta < v$. In fact, by (H$_2$) there exist factors ζ_u and ζ_u' of u such that $\zeta_u < \xi$ and $\zeta_u' < v$. Since all factors of u are comparable with each other we may choose for ζ the minimum of ζ_u and ζ_u'.

ASLs are the most important graded Hodge algebras. Significant examples will be treated in the next sections.

Proposition 7.1.2. *Let A be a graded Hodge algebra on H over B governed by Σ. Then the standard monomials form a B-basis of A.*

PROOF. Let $\xi \in H$; we define $\dim \xi$ to be the maximal length of chains $\xi = \zeta_0 < \zeta_1 < \cdots$ in H, and define the *weight of a monomial* $u = \prod_{\xi \in H} \xi^{c_\xi}$ to be $\sum_{\xi \in H} c_\xi (d+1)^{\dim \xi}$, where d is the maximum of the numbers $\sum_{\xi \in H} c_\xi$ of generators $c = (c_\xi)$ of Σ.

It suffices to show that all non-standard monomials are linear combinations of standard monomials. Let v' be a monomial with exponent $c' \in \Sigma$, and let c be a generator of Σ such that $c' - c \in \mathbb{N}^H$. Then $v = \prod_{\xi \in H} \xi^{c_\xi}$ divides v', and so $v' = vw$ where w is a monomial on H. Applying the straightening law for v, we obtain the equation $v' = \sum b_u uw$, $b_u \in B$, $b_u \neq 0$, u standard. We claim that all monomials on the right hand side of this equation are of strictly greater weight than v'. In fact, if $\alpha = \max\{\dim \xi : c_\xi \neq 0\}$, then for any u in the straightening equation for v there exists a factor ξ_u with $\alpha < \dim \xi_u$, so that weight $v \leq \sum_{\xi \in H} c_\xi (d + 1)^\alpha < (d + 1)^{\alpha+1} \leq (d + 1)^{\dim \xi_u} \leq$ weight u. Since the weight of a product of monomials is the sum of the weights of the factors, the claim follows.

On the other hand, the monomials on the right hand side of the equation for v' have the same degree as v'. Therefore descending induction concludes the proof. $\square$

The previous proposition guarantees that every element of A has a unique presentation as a B-linear combination of standard monomials, which we call its *standard representation*.

Proposition 7.1.3. *Let A be a graded Hodge algebra on H over B governed by Σ, and T_ξ, $\xi \in H$, a set of indeterminates over B. For each monomial $u = \xi_1 \xi_2 \cdots \xi_n$ on H we set $T_u = T_{\xi_1} \cdots T_{\xi_n}$. Then the kernel of the B-algebra epimorphism*

$$\varphi : B[T_\xi : \xi \in H] \longrightarrow A, \qquad T_\xi \mapsto \xi,$$

is generated by the elements $T_v - \sum b_u T_u$ corresponding to the straightening relations.

PROOF. Let I be the ideal in $B[T_\xi : \xi \in H]$ generated by the elements $T_v - \sum b_u T_u$ corresponding to the straightening relations. It is clear that $I \subset \operatorname{Ker} \varphi$. Conversely, let $f \in \operatorname{Ker} \varphi$; then the proof of 7.1.2 shows that there exists $g \in I$ such that $f - g = \sum b_u T_u$, u standard. It follows that $0 = \varphi(f - g) = \sum b_u u$. According to (H$_1$) all $b_u = 0$, and hence $f \in I$. $\square$

Among the graded Hodge algebras on H over B governed by Σ, the discrete Hodge algebra is in a sense the simplest. Its ring-theoretic properties are determined only by the ground ring B and the combinatorial properties of H and Σ. Surprisingly this is true in part for a general graded Hodge algebra as well. The set $\operatorname{Ind} A$ of elements $\xi \in H$ which appear as factors in the monomials on the right-hand side of the straightening relations is called the *indiscrete part of A*. It serves as a measure of how much A differs from a discrete Hodge algebra.

The following theorem allows the stepwise approach from a general graded Hodge algebra to a discrete one by forming suitable associated graded rings. The results of 4.4 permit us to control ring-theoretic properties of the algebras involved in this operation.

Suppose $\operatorname{Ind} A \neq \emptyset$, choose a minimal element $\xi_0 \in \operatorname{Ind} A$, and set $I = (\xi_0)$. We will first prove a refinement of 7.1.2.

Lemma 7.1.4. *The ideal I^j has a B-basis consisting of all standard monomials $u = \prod_{\xi \in H} \xi^{c_\xi}$ such that $c_{\xi_0} \geq j$.*

PROOF. Certainly the elements $\xi_0^j u$, u a standard monomial, generate I^j as a B-module. We claim that $\xi_0^j u$ either is a standard monomial or is zero. This, in view of 7.1.2, will prove the lemma.

Suppose $\xi_0^j u$ is not standard; then it is a multiple of a monomial v whose exponent is a generator of Σ. Since u is standard, the element ξ_0 is a factor of v, and thus for each monomial on the right hand side of the straightening relation for v there exists a factor less than ξ_0. Since ξ_0 is a minimal element among such factors, the straightening relation must be trivial. It follows that $\xi_0^j u = 0$. □

For an element $a \in A$ we define $\operatorname{ord} a$ as the supremum of integers j for which $a \in I^j$, and call $a^\star = a + I^{\operatorname{ord} a + 1}$ the initial form of a in $\operatorname{gr}_I(A)$. We have $\operatorname{ord} \xi_0 = 1$, and $\operatorname{ord} \xi = 0$ for all $\xi \in H$, $\xi \neq \xi_0$.

Let $a, b \in A$; then $\operatorname{ord} ab \geq \operatorname{ord} a + \operatorname{ord} b$, and $(ab)^\star = a^\star b^\star$ if $\operatorname{ord} ab = \operatorname{ord} a + \operatorname{ord} b$. By induction one proves a similar formula for more than just two factors. Thus, if $u = \prod_{\xi \in H} \xi^{c_\xi}$ is a standard monomial in A, it follows from the previous lemma that $u^\star = \prod_{\xi \in H} (\xi^\star)^{c_\xi}$. In conclusion we see that $\operatorname{gr}_I(A)$ is generated over B by the elements $\xi^\star$, $\xi \in H$, and that $\operatorname{gr}_I(A)$ is a free B-module with basis $\{u^\star : u \text{ is a standard monomial of } A\}$. Moreover, $\operatorname{gr}_I(A)$ may be viewed as a positively graded B-algebra, if, for all $j \geq 0$, the set of homogeneous elements of degree j of $\operatorname{gr}_I(A)$ is defined to be $\{a^\star : a \in A \text{ is homogeneous of degree } j\}$.

Now it is easy to give $\operatorname{gr}_I(A)$ the structure of a graded Hodge algebra: we let $H^\star = \{\xi^\star : \xi \in H\}$. The map $H \to H^\star$, $\xi \mapsto \xi^\star$, induces a bijection $\varphi : \mathbb{N}^H \to \mathbb{N}^{H^\star}$, and we set $\Sigma^\star = \varphi(\Sigma)$. The partial order defined on $H^\star$ will of course be given by $\xi^\star < \eta^\star \iff \xi < \eta$.

Theorem 7.1.5. $\operatorname{gr}_I(A)$ *is a graded Hodge algebra on* $H^\star$ *over* B *governed by* $\Sigma^\star$, *and* $\operatorname{Ind} \operatorname{gr}_I(A) \subset \{\xi^\star : \xi \in \operatorname{Ind} A\} \setminus \{\xi_0^\star\}$.

PROOF. It remains to check (H2): let $d = (d_{\xi^\star})$ be a generator of $\Sigma^\star$, and $w = \prod_{\xi^\star \in H^\star} (\xi^\star)^{d_{\xi^\star}}$. Then $c = (c_\xi)$, $c_\xi = d_{\xi^\star}$ for all $\xi \in H$, is a generator of Σ, and for $v = \prod_{\xi \in H} \xi^{c_\xi}$ we have the straightening relation $v = \sum b_u u$.

We want to define the straightening relation for w. There are two cases to consider. In the first case, ξ_0 is a factor of v. Then $v = \xi_0 v'$ where v' is a standard monomial, and it follows that $v = 0$ as we saw in the proof of 7.1.4. Therefore $w = 0$ is the desired straightening relation in this case. In the second case, ξ_0 is not a factor of v. Then $w = \sum_{\operatorname{ord} u = 0} b_u u^\star$ is the straightening relation for w. In particular it follows that $\xi_0^\star \notin \operatorname{Ind} \operatorname{gr}_I(A)$. □

Let A (respectively A') be a graded Hodge algebra on H (respectively H') over B governed by Σ (respectively Σ'). We say that A *and* A' *are Hodge algebras with the same data*, if there is an isomorphism $H \to H'$ of posets for which the corresponding map $\mathbb{N}^H \to \mathbb{N}^{H'}$ induces a bijection

$\Sigma \to \Sigma'$. We have just seen that A and $\mathrm{gr}_I(A)$ are graded Hodge algebras with the same data, the only difference being that the indiscrete part has become smaller. Thus after finitely many such steps we arrive at a discrete Hodge algebra with the same data as A.

Corollary 7.1.6. *Let A be a graded Hodge algebra on H over the Noetherian ring B governed by Σ, and let A_0 be the discrete Hodge algebra over B with the same data. Then:*
(a) $\dim A = \dim A_0$;
(b) *A is reduced, Cohen–Macaulay, or Gorenstein if A_0 is too.*

PROOF. A basic observation for (a) and (b) is that both A and A_0 are free and, hence, faithfully flat B-algebras. In conjunction with A.11 it implies

$$\dim A = \max(\dim B_{\mathfrak{p}} + \dim A \otimes k(\mathfrak{p}))$$

where $\mathfrak{p}$ ranges over $\operatorname{Spec} B$. It is clear that $A \otimes k(\mathfrak{p})$ is a Hodge algebra over $k(\mathfrak{p})$ with the same data as A; therefore $A_0 \otimes k(\mathfrak{p}) \cong (A \otimes k(\mathfrak{p}))_0$. Thus it is enough to consider the case in which $B = k$ is a field. Next we may replace A_0 by $\mathrm{gr}_I(A)$. Both these rings are positively graded so that $\dim A = \dim A_{\mathfrak{m}}$ and $\dim A_0 = \dim A_{\mathfrak{m}_0}$ where $\mathfrak{m}$ and $\mathfrak{m}_0$ are the *maximal ideals. Furthermore I is generated by homogeneous elements of positive degrees. Therefore 4.5.6 yields the desired equality of dimensions.

We show (b) for the Cohen–Macaulay property. The Cohen–Macaulay property of A_0 implies that of B and that of $A_0 \otimes k(\mathfrak{p})$ for all prime ideals $\mathfrak{p}$ of B (see 2.1.7). Let $\mathfrak{P}$ be a prime ideal of A and set $\mathfrak{p} = B \cap \mathfrak{P}$. Then

$$\operatorname{depth} A_{\mathfrak{P}} = \operatorname{depth} B_{\mathfrak{p}} + \operatorname{depth} A_{\mathfrak{P}} \otimes k(\mathfrak{p})$$
$$= \dim B_{\mathfrak{p}} + \operatorname{depth}(A \otimes k(\mathfrak{p}))_{\mathfrak{q}}$$

where $\mathfrak{q}$ is the image of $\mathfrak{P}$ in $A \otimes k(\mathfrak{p})$ (see 1.2.16). It follows that $A_{\mathfrak{P}}$ is Cohen–Macaulay if $\operatorname{depth}(A \otimes k(\mathfrak{p}))_{\mathfrak{q}} = \dim(A \otimes k(\mathfrak{p}))_{\mathfrak{q}}$. Thus the isomorphism $A_0 \otimes k(\mathfrak{p}) \cong (A \otimes k(\mathfrak{p}))_0$ reduces the contention once more to the case in which $B = k$ is a field.

It is enough to derive the Cohen–Macaulay property of A from that of $\mathrm{gr}_I(A)$. Let $\mathfrak{m}$ be the *maximal ideal of A. Since $I \subset \mathfrak{m}$, 4.5.7 implies that $A_{\mathfrak{m}}$ is Cohen–Macaulay, and then A is, too, by Exercise 2.1.27.

For the Gorenstein property one argues similarly, using 3.3.15, 4.5.7, and Exercise 3.6.20. The assertion about A being reduced follows immediately from 4.5.8. □

In case A is an ASL on H over B, the discrete ASL with these data is the Stanley–Reisner ring of the order complex $\Delta(H)$ (see Section 5.1) over B. Thus we may use the results of Chapter 5 in order to conclude that certain ASLs are Cohen–Macaulay. In the following corollary we extend the poset H by adding absolutely minimal and maximal elements $\hat{0}$ and $\hat{1}$.

Corollary 7.1.7. *Let A be an ASL on H over B. If B is Cohen–Macaulay and $H \cup \{\hat{0}, \hat{1}\}$ is a locally upper semimodular poset, then A is Cohen–Macaulay .*

PROOF. By virtue of 5.1.12 and 5.1.14 the discrete ASL A_0' on $H \cup \{\hat{0}, \hat{1}\}$ is Cohen–Macaulay, provided B is a field. Since A_0' is obviously a polynomial ring over the discrete ASL A_0 on H, it follows that A_0 is Cohen–Macaulay. According to Exercise 5.1.25, A_0 is Cohen–Macaulay for every Cohen–Macaulay ring B. Now one applies 7.1.6. $\square$

There is a simple proof of 7.1.7 which avoids the combinatorially difficult theorem 5.1.12; see [90] or [61], (5.14).

Exercises

7.1.8. Let A be a graded Hodge algebra on H over a ring B governed by Σ.
(a) Let H' be a subset of H such that the ideal $H'A$ is generated as a B-module by the standard monomials it contains. Show that $A/H'A$ is in a natural way a Hodge algebra on $H \setminus H'$ governed by Σ' where $H \setminus H'$ is considered as a subposet of H and Σ' consists of all elements of Σ which are exponents of monomials on $H \setminus H'$.
(b) Show that (a) in particular applies if H' is an ideal in H. (An ideal in H' is a subset satisfying the following condition: $h' \leq h \in H' \Rightarrow h' \in H'$.)
(c) Let H' and H'' be ideals in H. Then $H'A \cap H''A$ is the ideal of A generated by $H' \cap H''$.
(d) Specialize (a), (b), and (c) to the case of an ASL A.

7.1.9. With the notation of 7.1.8 assume that Σ is generated by squarefree monomials. Show that A is reduced if (and only if) B is reduced. In particular a graded ASL over a reduced ring B is reduced.

7.1.10. Let A be a graded ASL on the poset H over a Noetherian ring B. Show that $\dim A = \dim B + \operatorname{rank} H + 1$.
Hint: First prove the formula for a field B. Next deduce $\dim A = \max\{\dim B_\mathfrak{p} + \dim k(\mathfrak{p}) \otimes A : \mathfrak{p} \in \operatorname{Spec} B\}$ from A.11 and the fact that A is a free B-module, and note that $k(\mathfrak{p}) \otimes A$ is an ASL on H over the field $k(\mathfrak{p})$.

7.1.11. (Hibi) Let k be a field, C a positive affine semigroup generated by $c_1, \ldots, c_n$, and $A = k[C]$. We order $\{c_1, \ldots, c_n\}$ by setting $c_1 < \cdots < c_n$, and let Σ be the set of exponents $(a_1, \ldots, a_n)$ such that there exists $(b_1, \ldots, b_n)$ which is lexicographically greater than $(a_1, \ldots, a_n)$ and satisfies the condition $(X^{c_1})^{a_1} \cdots (X^{c_n})^{a_n} = (X^{c_1})^{b_1} \cdots (X^{c_n})^{b_n}$.
(a) Show $k[C]$ is a graded Hodge algebra over k with these data.
(b) Let C be the subsemigroup of $\mathbb{N}^2$ generated by $(2,0)$, $(2,1)$, $(1,2)$, and $(0,2)$. Determine the sets Σ for the orders (i) $(2,0) < (2,1) < (1,2) < (0,2)$ and (ii) $(2,1) < (1,2) < (2,0) < (0,2)$, and show that the discrete Hodge algebra A_0 is Gorenstein in case (i), and not even Cohen–Macaulay in case (ii).

7.2 Straightening laws on posets of minors

The most important examples of ASLs are rings related to matrices and determinants, and the prototype of such a ring is

$$R_{r+1} = R_{r+1}(X) = B[X]/I_{r+1}(X)$$

where $B[X]$ is the polynomial ring in the entries of an $m \times n$ matrix of indeterminates X_{ij} over some ring B of coefficients, and $I_{r+1}(X)$ denotes the ideal generated by the $(r+1)$-minors of X. We always suppose that $0 \le r \le \min(m, n)$, the trivial cases $r = 0$ and $r = \min(m, n)$ being included for reasons of systematics.

What makes the analysis of R_{r+1} difficult is the fact that the generators of $I_{r+1}(X)$ are very complicated expressions in terms of the X_{ij}. Therefore one enlarges the set of generators of the B-algebra $B[X]$ by considering each minor as a generator. Of course, apart from trivial cases, we lose the algebraic independence of the generating set, but only to the extent that $B[X]$ is an ASL on the set of minors of X.

The minor corresponding to the submatrix of X with rows $a_1, \dots, a_u$ and columns $b_1, \dots, b_u$ is denoted by

$$[a_1 \dots a_u \,|\, b_1 \dots b_u].$$

The set Δ consists of those minors $[a_1 \dots a_u \,|\, b_1 \dots b_u]$ which satisfy the condition $a_1 < \cdots < a_u$, $b_1 < \cdots < b_u$. It is partially ordered by the rule

$$[a_1 \dots a_u \,|\, b_1 \dots b_u] \le [c_1 \dots c_v \,|\, d_1 \dots d_v]$$
$$\iff \quad u \ge v \text{ and } a_i \le c_i, \ b_i \le d_i, \ i = 1, \dots, v.$$

It is easy to see that Δ is a distributive lattice under this partial order.

Rather than proving directly that $B[X]$ is a graded ASL on Δ we take a detour which leads to a substantial simplification of the combinatorial details, and introduces another interesting and important class of rings. We suppose that $m \le n$. Then the maximal minors of X are the m-minors. An m-minor of X is simply denoted by

$$[a_1 \dots a_m]$$

where $a_1, \dots a_m$ are the column indices of the submatrix whose determinant is taken. The subset of Δ consisting of all m-minors in Δ is called Γ. Obviously Γ is a sublattice of Δ. We write

$$G(X),$$

or, if appropriate, $G_B(X)$ for the B-subalgebra of $B[X]$ generated by Γ. The letter G has been chosen since (over a field B) $G(X)$ is the coordinate ring of the Grassmannian of m-dimensional vector subspaces of B^n.

Theorem 7.2.1 (Hodge). *Let B be a ring, and X an $m \times n$ matrix of indeterminates over B with $m \leq n$. Then $G(X)$ is a graded ASL on Γ.*

Condition (H_0) is evidently satisfied, and the validity of (H_1) is stated in the following lemma:

Lemma 7.2.2. *The standard monomials in Γ are linearly independent.*

PROOF. For $\gamma = [a_1 \ldots a_m]$ let U_γ be the following $m \times n$ matrix whose entries U_{ij} are indeterminates over B:

$$\begin{pmatrix} 0 & \cdots & 0 & U_{1a_1} & \cdots & U_{1a_2-1} & U_{1a_2} & \cdots & U_{1a_3-1} & \cdots & U_{1a_m} & \cdots & U_{1n} \\ & & 0 & \cdots & 0 & U_{2a_2} & \cdots & U_{2a_3-1} & & & & & \\ & & & & 0 & \cdots & 0 & \cdots & & \vdots & & & \vdots \\ \vdots & & \vdots & \vdots & & \vdots & \vdots & & \vdots & & & & \\ 0 & \cdots & 0 & 0 & \cdots & 0 & 0 & \cdots & 0 & \cdots & U_{ma_m} & \cdots & U_{mn} \end{pmatrix}.$$

The substitution which maps X_{ij} to the corresponding entry of U_γ induces a B-algebra homomorphism $\varphi_\gamma : G(X) \to G(U_\gamma)$ where $G(U_\gamma)$ denotes the B-subalgebra of $B[U_{ij} : j \geq a_i]$ generated by the maximal minors of U_γ. Observe that for $\delta \geq \gamma$ the matrix U_γ has indeterminate entries where U_δ has non-zero entries. Therefore the analogous substitution yields a B-algebra homomorphism $\psi_{\delta\gamma} : G(U_\gamma) \to G(U_\delta)$ with $\varphi_\delta = \psi_{\delta\gamma} \circ \varphi_\gamma$.

The matrix U_γ is chosen in such a way that the submatrix of its columns $1, \ldots, a_i - 1$ has rank $i - 1$ for $i = 1, \ldots, m + 1$ (where we let $a_{m+1} = n + 1$). The reader may carefully check that this implies $\varphi_\gamma(\beta) = 0$ for all $\beta \not\geq \gamma$. So the application of φ_γ to a linear combination of standard monomials strips off all terms which contain a factor $\beta \not\geq \gamma$.

The lemma follows immediately from the following claim (with $\gamma = [1 \ldots m]$): let $\Sigma(\gamma)$ be the set of standard monomials all of whose factors are $\geq \gamma$; then $\varphi_\gamma(\Sigma(\gamma))$ is a linearly independent subset of $G(U_\gamma)$.

We prove this claim by descending induction over the poset Γ. Let $\sum_{u \in U} b_u \varphi_\gamma(u) = 0$ be a linear combination with $U \subset \Sigma(\gamma)$, $U \neq \emptyset$, and $b_u \neq 0$ for all $u \in U$. The element $\varphi_\gamma(\gamma)$ is a product of indeterminates, and therefore $G(U_\gamma)$-regular. Thus, cancelling $\varphi_\gamma(\gamma)$ if necessary, we may suppose that γ does not occur as a factor of at least one of the standard monomials in the sum, say u_0. Let γ' be the smallest factor of u_0. Then $\gamma' > \gamma$, and $0 = \sum_{u \in U} b_u \psi_{\gamma'\gamma}(\varphi_\gamma(u)) = \sum_{u' \in U'} b_{u'} \varphi_{\gamma'}(u')$ where $U' = U \cap \Sigma(\gamma')$. Since $u_0 \in U'$, we obtain a contradiction to the induction hypothesis. □

Next we want to show that every product $\gamma\delta$ of (incomparable) minors $\gamma, \delta \in \Gamma$ can be written as a linear combination of standard monomials. This 'straightening' will be performed by iterated applications of the *Plücker relations* given in the following lemma. (We use $\sigma(i_1 \ldots i_s)$ to denote the sign of the permutation of $\{1, \ldots, s\}$ represented by the sequence $i_1, \ldots, i_s$.)

Lemma 7.2.3. *For every $m \times n$-matrix, $m \leq n$, with elements in a ring A and all indices $a_1, \ldots, a_p$, $b_q, \ldots, b_m$, $c_1, \ldots, c_s \in \{1, \ldots, n\}$ such that $s = m - p + q - 1 > m$ and $t = m - p > 0$ one has*

$$\sum_{\substack{i_1 < \cdots < i_t \\ i_{t+1} < \cdots < i_s \\ \{i_1, \ldots, i_s\} = \{1, \ldots, s\}}} \sigma(i_1 \ldots i_s)[a_1 \ldots a_p c_{i_1} \ldots c_{i_t}][c_{i_{t+1}} \ldots c_{i_s} b_q \ldots b_m] = 0.$$

PROOF. It suffices to prove the lemma for a matrix X of indeterminates over $\mathbb{Z}$. Next we may replace $\mathbb{Z}$ by $\mathbb{Q}$, and finally the ring $\mathbb{Q}[X]$ by its field of fractions $\mathbb{Q}(X)$. We consider the $\mathbb{Q}(X)$-vector space V generated by the columns of X. Let $\mathscr{S}$ be the group of permutations of $\{1, \ldots, s\}$, and let X_j denote the j-th column of X. We define $\alpha \colon V^s \to \mathbb{Q}(X)$ by

$$\alpha(y_1, \ldots, y_s) =$$

$$\sum_{\pi \in \mathscr{S}} \sigma(\pi) \det(X_{a_1}, \ldots, X_{a_p}, y_{\pi(1)}, \ldots, y_{\pi(t)}) \det(y_{\pi(t+1)}, \ldots, y_{\pi(s)}, X_{b_q}, \ldots, X_{b_m}).$$

It is straightforward to check that α is a multilinear form on V^s. When two of the vectors y_i coincide, every term in the expansion of α which does not vanish anyway is cancelled by a term of the opposite sign. Thus α is alternating. Since $s > \dim V = m$, one has $\alpha = 0$.

Let us fix a subset $\{i_1, \ldots, i_t\}$, $i_1 < \cdots < i_t$, of $\{1, \ldots, s\}$. Then, for all π such that $\pi(\{1, \ldots, t\}) = \{i_1, \ldots, i_t\}$, the summand corresponding to π in the expansion of α equals

$$\sigma(i_1 \ldots i_s) \det(X_{a_1}, \ldots, X_{a_p}, y_{i_1}, \ldots, y_{i_t}) \det(y_{i_{t+1}}, \ldots, y_{i_s}, X_{b_q}, \ldots, X_{b_m})$$

where $i_{t+1}, \ldots, i_s$ are chosen as above. Therefore each of these terms appears with multiplicity $t! \, (s - t)!$ in the expansion of α, and cancelling this factor we obtain the desired formula. $\square$

Let us 'straighten' the product $[1\ 4\ 6][2\ 3\ 5]$. For the data $p = 1$, $a_1 = 1$, $q = 3$, $b_3 = 5$, $(c_1, \ldots, c_4) = (4, 6, 2, 3)$ one has the following Plücker relation:

$$[1\ 4\ 6][2\ 3\ 5] + [1\ 2\ 4][3\ 5\ 6] - [1\ 3\ 4][2\ 5\ 6]$$
$$+ [1\ 2\ 6][3\ 4\ 5] - [1\ 3\ 6][2\ 4\ 5] + [1\ 2\ 3][4\ 5\ 6] = 0.$$

When solved for $[1\ 4\ 6][2\ 3\ 5]$, it is not yet a linear combination of standard monomials. However, $[1\ 4\ 6][2\ 3\ 5]$ is the 'worst' term: for it incomparability occurs already in the second position, whereas in the remaining terms at least positions 1 and 2 are ordered. Using the Plücker relations

$$[1\ 2\ 6][3\ 4\ 5] - [1\ 2\ 3][4\ 5\ 6] + [1\ 2\ 4][3\ 5\ 6] - [1\ 2\ 5][3\ 4\ 6] = 0,$$
$$[1\ 3\ 6][2\ 4\ 5] + [1\ 2\ 3][4\ 5\ 6] + [1\ 3\ 4][2\ 5\ 6] - [1\ 3\ 5][2\ 4\ 6] = 0,$$

one finally gets

$$[1\ 4\ 6][2\ 3\ 5] = -3[1\ 2\ 3][4\ 5\ 6] - [1\ 2\ 5][3\ 4\ 6] + [1\ 3\ 5][2\ 4\ 6].$$

We now describe how to apply the Plücker relations in general.

Lemma 7.2.4. *Let* $[a_1 \ldots a_m]$, $[b_1 \ldots b_m]$ *be elements of* Γ *such that* $a_i \leq b_i$ *for* $i = 1, \ldots, p$ *and* $a_{p+1} > b_{p+1}$. *We put* $q = p + 2$, $s = m + 1$, $(c_1, \ldots, c_s) = (a_{p+1}, \ldots, a_m, b_1, \ldots, b_{p+1})$. *Then, in the Plücker relation 7.2.3 with these data, all the non-zero terms* $[d_1 \ldots d_m][e_1 \ldots e_m] \neq [a_1 \ldots a_m][b_1 \ldots b_m]$ *have the following properties (after their column indices have been arranged in ascending order):*

(a) $[d_1 \ldots d_m] \leq [a_1 \ldots a_m]$;   (b) $d_1 \leq e_1, \ldots, d_{p+1} \leq e_{p+1}$.

PROOF. Since $b_1 < \cdots < b_{p+1} < a_{p+1} < \cdots < a_m$, $[d_1 \ldots d_m]$ arises from $[a_1 \ldots a_m]$ by a replacement of some of the a_i by smaller indices. This implies (a) and $d_i \leq e_i$ for $i = 1, \ldots, p$. Furthermore $d_{p+1} \in \{a_1, \ldots, a_p, b_1, \ldots, b_{p+1}\}$, so $d_{p+1} \leq b_{p+1}$, and $e_{p+1} \in \{a_{p+1}, \ldots, a_m, b_{p+1}, \ldots, b_m\}$, so $b_{p+1} \leq e_{p+1}$. □

By induction on p it follows immediately from 7.2.4 that a product $\gamma\delta$, $\gamma, \delta \in \Gamma$, is a B-linear combination of standard monomials $\alpha\beta$, $\alpha \leq \beta$, such that $\alpha \leq \gamma$. This however does not yet imply (H₂). (In general the straightening procedure based on 7.2.4 produces intermediate results violating (H₂).) In order to see that (H₂) is indeed satisfied we must also straighten the product in the order $\delta\gamma$. The standard monomials $\alpha\beta$ obtained now satisfy $\alpha \leq \delta$. By linear independence of the standard monomials, both representations of $\gamma\delta = \delta\gamma$ coincide, and (H₂) follows. This completes the proof of 7.2.1.

Before we turn to the discussion of the polynomial ring $B[X]$, we state a useful corollary of 7.2.1.

Corollary 7.2.5. (a) *Let* T_γ, $\gamma \in \Gamma$, *be a set of indeterminates over* B. *Then the kernel of the surjective homomorphism* $B[T_\gamma : \gamma \in \Gamma] \to G(X)$ *is generated by the elements representing the Plücker relations.*
(b) *One has* $G_B(X) = G_{\mathbb{Z}}(X) \otimes B$.
(c) *Suppose* B *is a Noetherian ring. Then* $\dim G(X) = \dim B + m(n - m) + 1$.

PROOF. By 7.1.3, $G_B(X)$ is the residue class ring of $B[T_\gamma : \gamma \in \Gamma]$ modulo the ideal generated by the elements representing the straightening relations. As seen above, the straightening relations are linear combinations of the Plücker relations. This proves (a), and (b) is a simple consequence of (a) if one notes that the Plücker relations are defined over $\mathbb{Z}$.

By virtue of 7.1.10 one has $\dim G(X) = \dim B + \operatorname{rank} \Gamma + 1$. Each cover of $[a_1 \ldots a_m] \in \Gamma$ is obtained by replacing one of the indices a_i by $a_i + 1$ (of course, this is only feasible if $a_i + 1 < a_{i+1}$). Furthermore $\operatorname{rank}[1 \ldots m] = 0$. Therefore $\operatorname{rank}[a_1 \ldots a_m] = \sum_{i=1}^{m} a_i - i = \sum_{i=1}^{m} a_i - m(m + 1)/2$. This immediately yields $\operatorname{rank} \Gamma = \operatorname{rank}[n - m + 1 \ldots n] = m(n - m)$. □

As before let X be a matrix of indeterminates, and Δ the poset of its minors; the condition $m \le n$ is no longer required. We extend X by m columns of further indeterminates, obtaining

$$X' = \begin{pmatrix} X_{11} & \cdots & X_{1n} & X_{1,n+1} & \cdots & X_{1,n+m} \\ \vdots & & \vdots & \vdots & & \vdots \\ X_{m1} & \cdots & X_{mn} & X_{m,n+1} & \cdots & X_{m,n+m} \end{pmatrix}.$$

Then $B[X']$ is mapped onto $B[X]$ by substituting for each entry of X' the corresponding entry of the matrix

$$\begin{pmatrix} X_{11} & \cdots & X_{1n} & 0 & \cdots & \cdots & 0 & 1 \\ & & \vdots & & \iddots & \iddots & & 0 \\ \vdots & & \vdots & \vdots & \iddots & \iddots & \iddots & \vdots \\ & & & 0 & \iddots & \iddots & & \vdots \\ X_{m1} & \cdots & X_{mn} & 1 & 0 & \cdots & \cdots & 0 \end{pmatrix}.$$

Let $\varphi : G(X') \to B[X]$ be the induced homomorphism. Then

(1) $$\varphi([b_1 \ldots b_m]) = \pm[a_1 \ldots a_t \mid b_1 \ldots b_t]$$

where $t = \max\{i : b_i \le n\}$ and $a_1, \ldots, a_t$ have been chosen such that

$$\{a_1, \ldots a_t, n + m + 1 - b_m, \ldots, n + m + 1 - b_{t+1}\} = \{1, \ldots, m\}.$$

Equation (1) shows that φ is surjective, and furthermore sets up a bijective correspondence between the set Γ' of m-minors of X' and $\Delta \cup \{\pm 1\}$. Note that the maximal element $\tilde{\mu} = [n + 1 \ldots n + m]$ of Γ' is mapped to ± 1, and that the restriction of φ to $\Gamma' \setminus \{\tilde{\mu}\}$ is an isomorphism of posets. (We leave the verification of this fact to the reader; the details can also be found in [61], (4.9).)

Lemma 7.2.6. *The kernel of* $\varphi : G(X') \to B[X]$ *is generated by* $\tilde{\mu} \pm 1$.

PROOF. Note that $G(X')$, $B[X]$, and φ are obtained from the corresponding objects over $\mathbb{Z}$ by taking tensor products of the latter with B. (This is non-trivial only for $G(X')$ for which it has been stated in 7.2.5.) Therefore it is sufficient to consider the case $B = \mathbb{Z}$. Then $G(X')$ is an integral domain, and it follows from the properties of dehomogenization (see 1.5.18) that $\tilde{\mu} \pm 1$ generates a prime ideal $\mathfrak{p}$ of height 1. By virtue of 7.2.5 one has

$$\dim G(X') = \dim \mathbb{Z} + mn + 1 = \dim \mathbb{Z}[X] + 1.$$

As $\mathfrak{p} \subset \operatorname{Ker} \varphi$, we in fact have $\mathfrak{p} = \operatorname{Ker} \varphi$. $\square$

Theorem 7.2.7 (Doubilet–Rota–Stein). *Let B be a ring, and X an $m \times n$ matrix of indeterminates. Then $B[X]$ is a graded ASL on the poset Δ of minors of X.*

PROOF. From our previous arguments it is obvious that φ maps the standard monomials in $\Gamma' \setminus \{\tilde{\mu}\}$ to standard monomials in Δ. Since φ is surjective and $G(X')$ is a graded ASL on Γ', the standard monomials in Δ generate $B[X]$ as a B-module.

The smallest factor of a standard monomial on the right hand side of a straightening relation in an arbitrary ASL is never a maximal element of the underlying poset. Therefore the validity of (H₂) cannot be destroyed by a substitution which takes such an element to an element of B.

It only remains to observe the linear independence of the standard monomials in Δ, or, equivalently, that there is no non-trivial relation

$$\sum a_u u = (\tilde{\mu} \pm 1) \sum b_v v$$

where u and v represent pairwise distinct standard monomials in $G(X')$, and none of the u contains $\tilde{\mu}$ as a factor. $\quad\square$

Exercises

7.2.8. Let $u = \delta_1 \cdots \delta_r$ be a product of minors of X. The *content* of u is the vector of length $m + n$ which for each row and each column lists the multiplicity with which the row or column appears in u. Let v be a standard monomial in the standard representation of u. Show that v has the same content as u, and has at most r factors.

7.2.9. Let $m \leq n$, and X be an $m \times n$ matrix of indeterminates over a ring B. Let $1 \leq r_1 < \cdots < r_s \leq m$ be integers, and consider the subposet $\Gamma(r_1, \ldots, r_s)$ of Δ formed by all minors which are of the form $[1 \ldots r_i | a_1 \ldots a_{r_i}]$ for some i. Show that $B[\Gamma(r_1, \ldots, r_s)]$ is a graded ASL on $\Gamma(r_1, \ldots, r_s)$. (Note that this class of rings generalizes $G(X)$.)

(For a field $B = k$, $k[\Gamma(r_1, \ldots, r_s)]$ is the multihomogeneous coordinate ring of the variety of flags $0 \subset U_1 \subset \cdots \subset U_r \subset k^n$ of linear subspaces such that $\dim U_i = r_i$.)

7.2.10. Let H be a finite poset with partial order $\leq$, and H^- the poset with the reverse partial order $\preceq$: $h \preceq h' \iff h' \leq h$. A graded ASL R on H is called *symmetric* if it is also an ASL on H^- (with respect to the same embedding $H \to R$).

(a) Show that $G(X)$ is a symmetric ASL.

(b) Show that the graded ASLs $B[\Gamma(r_1, \ldots, r_s)]$ of the previous exercise are symmetric.

7.3 Properties of determinantal rings

In this section we shall assume that the ground ring B is a field, and therefore replace the letter B by k throughout. The transfer of the results to more general ground rings is indicated in Exercise 7.3.8.

As in the previous section let X be an $m \times n$ matrix of indeterminates over k. The determinantal ring R_{r+1} is the residue class ring of $k[X]$ with respect to the ideal generated by the $(r + 1)$-minors of X. In view of the

ASL structure of $k[X]$ it is useful to extend this system of generators by including all t-minors with $t \geq r+1$. The enlarged system of generators is an ideal in the poset Δ of all minors of X. By Exercise 7.1.8, R_{r+1} inherits the ASL property of $k[X]$; its underlying poset is the coideal Δ_{r+1} of Δ which consists of the u-minors of X with $u < r + 1$. (A *coideal* is the complement of an ideal.) Evidently Δ_{r+1} has a single minimal element, namely $[1 \dots r \,|\, 1 \dots r]$.

More generally, for $\delta \in \Delta$ we want to investigate the residue class rings R_δ of $k[X]$ modulo the ideal I_δ generated by all minors $\varepsilon \not\geq \delta$. As in the special case $\delta = [1 \dots r \,|\, 1 \dots r]$ above, R_δ is an ASL on the coideal $\Delta_\delta = \{\zeta \in \Delta : \zeta \geq \delta\}$. In a distributive lattice a coideal with a single minimal element is again a distributive lattice, and it follows directly from 7.1.7 and 7.1.9 that R_δ is a reduced Cohen–Macaulay ring. Moreover we can easily compute its dimension.

Similarly we may consider the residue class rings G_γ of $G(X)$. These are the residue class rings of $G(X)$ with respect to the ideal J_γ generated by all $\beta \in \Gamma$, $\beta \not\geq \gamma$. The corresponding coideal in Γ is denoted Γ_γ.

Theorem 7.3.1. *Let k be a field, and X an $m \times n$ matrix of indeterminates over k.*

(a) (Hochster, Laksov, Musili) *Suppose $m \leq n$, and let $\gamma = [a_1 \dots a_m] \in \Gamma$. Then G_γ is a normal Cohen–Macaulay domain of dimension*

$$m(n - m) + \frac{m(m + 1)}{2} - \sum_{i=1}^{m} a_i + 1.$$

(b) *Let $\delta = [a_1 \dots a_r \,|\, b_1 \dots b_r] \in \Delta$. Then R_δ is a normal Cohen–Macaulay domain of dimension*

$$(m + n + 1)r - \sum_{i=1}^{r}(a_i + b_i).$$

(c) (Hochster–Eagon) *In particular, R_{r+1} is a normal Cohen–Macaulay domain of dimension $(m + n - r)r$.*

PROOF. The Cohen–Macaulay property of R_δ and G_γ follows from the fact that the posets Δ_δ and Γ_γ are distributive lattices, as explained above. In order to compute their dimensions one must determine the ranks of the posets Δ_δ and Γ_γ; see 7.1.10.

Since all maximal chains in a distributive lattice have the same length, one has

$$\operatorname{rank} \Gamma_\gamma = \operatorname{rank} \Gamma - (\operatorname{rank} \gamma + 1).$$

Both $\operatorname{rank} \Gamma$ and $\operatorname{rank} \gamma$ were computed in the proof of 7.2.5.

For the computation of $\dim R_\delta$ and for the proof of normality it is convenient to relate R_δ to a ring of type $G_{\gamma'}$ in the same way as $B[X]$

was related to $G(X')$ for the proof of 7.2.7. We choose

$$\gamma' = [b_1 \dots b_r \ (n+m+1) - a'_1 \ \dots \ (n+m+1) - a'_{m-r}]$$

with $\{a'_1, \dots, a'_{m-r}\}$ being the complement of $\{a_1, \dots, a_r\}$ in $\{1, \dots, m\}$. Then $\varphi : G(X') \to B[X]$ as defined before 7.2.6 maps γ' to $\pm\delta$ and the generators of $J_{\gamma'}$ to a generating set of I_δ. It follows from 7.2.6 that the induced homomorphism $\varphi_\delta : G_{\gamma'} \to R_\delta$ is surjective, and that its kernel is generated by $\tilde{\mu} \pm 1$ where the maximal element $\tilde{\mu}$ of Γ' is considered as an element of $G_{\gamma'}$. Now an easy computation yields the dimension of R_δ.

As we just saw, R_δ is a dehomogenization of a ring of type Γ_γ, and therefore it is sufficient to prove that G_γ is a normal domain (see Exercise 2.2.34). Note first that G_γ is indeed a domain: the surjective homomorphism $\varphi_\gamma : G(X) \to G(U_\gamma)$ constructed in the proof of 7.2.2 induces a homomorphism $\bar{\varphi}_\gamma : G_\gamma \to G(U_\gamma)$, and $\bar{\varphi}_\gamma$ maps the standard basis of G_γ onto a linearly independent subset of $G(U_\gamma)$ as was shown there. So G_γ is isomorphic to the integral domain $G(U_\gamma)$.

To prove normality we apply the criterion in Exercise 2.2.33 with $x = \gamma$: being the single minimal element of Γ_γ, γ is evidently Γ_γ-regular; moreover, $G_\gamma/(\gamma)$ is an ASL on $\Gamma_\gamma \setminus \{\gamma\}$, and therefore reduced. Finally 7.3.2 shows that $G_\gamma[\gamma^{-1}]$ is a normal domain. $\qquad\square$

Theorem 7.3.1(c) entails the classical formula

$$\text{height } I_{r+1}(X) = (m-r)(n-r).$$

Thus $I_{r+1}(X)$ has maximal height: by a theorem of Eagon and Northcott one has height $I_{r+1}(x) \leq (m-r)(n-r)$ for an arbitrary $m \times n$ matrix x over a Noetherian ring S, provided $I_{r+1}(x) \neq S$; see [61], (2.1) or [270], §13.

Lemma 7.3.2. *With the notation of* 7.3.1 *let*

$$\Psi = \big\{ \, [d_1 \dots d_m] \in \Gamma_\gamma : a_i \notin [d_1 \dots d_m] \text{ for at most one index } i \, \big\}.$$

Then

$$G_\gamma[\gamma^{-1}] = k[\Psi, \gamma^{-1}],$$

and the elements of Ψ *are algebraically independent over* k. *In particular* $G_\gamma[\gamma^{-1}]$ *is a regular domain.*

PROOF. We show that $[e_1 \dots e_m] \in k[\Psi, \gamma^{-1}]$ for all $[e_1 \dots e_m] \in \Gamma_\gamma$ by induction on the number w of indices i such that $e_i \notin [a_1 \dots a_m]$. For $w = 0$ and $w = 1$, $[e_1 \dots e_m] \in \Psi$ by definition. Let $w > 1$ and choose an index j such that $e_j \notin [a_1 \dots a_m]$. We use the Plücker relation 7.2.3, the data of 7.2.3 corresponding to the present ones in the following manner: $p = 0$, $q = 2$, $s = m+1$, $(b_2, \dots, b_m) = (e_1, \dots, e_{j-1}, e_{j+1}, \dots, e_m)$, and $(c_1, \dots, c_s) = (a_1, \dots, a_m, e_j)$. In this relation all the terms different from

$$[a_1 \dots a_m][e_j \, e_1 \dots e_{j-1} \, e_{j+1} \dots e_m] = (-1)^{j-1}[a_1 \dots a_m][e_1 \dots e_m]$$

and non-zero in G_γ have the form $\delta\varepsilon$ such that $\delta \in \Psi$ and ε has only $w-1$ indices not occurring in γ. Solving for $[a_1 \ldots a_m][e_1 \ldots e_m]$ and dividing by γ, one gets $[e_1 \ldots e_m] \in k[\Psi, \gamma^{-1}]$.

For the proof of the algebraic independence of Ψ we first note that $\dim G_\gamma[\gamma^{-1}] = \dim G_\gamma$. This follows easily from A.16 if one uses that G_γ is an affine domain over k as was demonstrated in the proof of 7.3.1. Now it is enough to verify that $|\Psi| = \dim G_\gamma$, a combinatorial exercise which we leave for the reader. $\quad\square$

Lemma 7.3.2 enables one to compute the singular locus of the rings G_γ, and, again by dehomogenization, that of R_δ; see [61], 6.B. We confine ourselves to the rings R_{r+1}, for which there is a simpler approach.

Suppose that $x = (x_{ij})$ is an $m \times n$ matrix over a ring R such that x_{11} is a unit. Then we may transform x by elementary row and column operations into the matrix

$$\begin{pmatrix} x_{11} & 0 & \cdots & 0 \\ 0 & y_{11} & \cdots & y_{1,n-1} \\ \vdots & \vdots & & \vdots \\ 0 & y_{m-1,1} & \cdots & y_{m-1,n-1} \end{pmatrix}, \quad y_{ij} = x_{i+1,j+1} - x_{1,j+1}x_{i+1,1}x_{11}^{-1},$$

and clearly $I_{r+1}(x) = I_r(y)$. The equation $y_{ij} = x_{i+1j+1} - x_{1j+1}x_{i+1,1}x_{11}^{-1}$, read as a substitution of indeterminates, suggests the following elementary lemma.

Lemma 7.3.3. *Let $X = (X_{ij})$ and $Y = (Y_{ij})$ be matrices of indeterminates over a ring B of sizes $m \times n$ and $(m-1) \times (n-1)$. Then the substitution $Y_{ij} \mapsto X_{i+1j+1} - X_{1j+1}X_{i+1,1}X_{11}^{-1}$ yields an isomorphism*

$$B[Y, X_{11}, \ldots, X_{m1}, X_{12}, \ldots, X_{1n}, X_{11}^{-1}] \cong B[X, X_{11}^{-1}]$$

which for every $t > 0$ maps the extension of $I_{t-1}(Y)$ to the extension of $I_t(X)$. In particular it induces an isomorphism

$$(2) \qquad R_{t-1}(Y)[X_{11}, \ldots, X_{m1}, X_{12}, \ldots, X_{1n}, X_{11}^{-1}] \cong R_t(X)[x_{11}^{-1}]$$

where x_{11} denotes the residue class of X_{11} in $R_t(X)$.

Proposition 7.3.4. *For a prime ideal $\mathfrak{p} \in \operatorname{Spec} R_{r+1}$ the localization is regular if and only if $\mathfrak{p} \not\supset I_r(X)/I_{r+1}(X)$.*

PROOF. We use induction on r, starting from the trivial case $r = 0$ (note that $I_0(X) = k[X]$). Suppose now that $r > 0$. If $\mathfrak{p}$ is the maximal ideal of $R = R_{r+1}$ generated by the x_{ij}, then $R_\mathfrak{p}$ is evidently non-regular and $\mathfrak{p}$ contains $I_r(X)/I_{r+1}(X)$. Otherwise $\mathfrak{p}$ does not contain one of the residue classes x_{ij}, and by symmetry we may assume that $x_{11} \notin \mathfrak{p}$. Then $R_\mathfrak{p}$ is a localization of $R[x_{11}^{-1}]$, and contracting the extension of $\mathfrak{p}$ via the isomorphism (2) to $S = R_r(Y)$ we obtain a prime ideal $\mathfrak{q} \in \operatorname{Spec} S$. As the extension from S to $R[x_{11}^{-1}]$ is an adjunction of indeterminates followed by the inversion of one of them, $R_\mathfrak{p}$ is regular if and only if $S_\mathfrak{q}$ is regular. Furthermore $\mathfrak{p} \supset I_r(X)/I_{r+1}(X)$ if and only if $\mathfrak{q} \supset I_{r-1}(Y)/I_r(Y)$. $\square$

The rings R_{r+1} satisfy Serre's condition (R_2) since

$$\text{height}(I_r(X)/I_{r+1}(X)) = \text{height}\, I_r(X) - \text{height}\, I_{r+1}(X)$$
$$= m + n - 2r + 1 \geq 3.$$

By Serre's normality criterion this argument, together with the Cohen–Macaulay property, proves independently of 7.3.2 that R_{r+1} is a normal domain. (In fact, all the rings R_δ and G_γ satisfy (R_2); see [61], (6.12).) The example $m = n = 2$, $r = 1$ shows that (R_3) fails in general.

Finally we want to determine which of the rings R_{r+1} are Gorenstein. The easiest way to solve this problem is to determine the canonical module, or rather the divisor class it represents (see 3.3.19; the canonical module of R_{r+1} is unique by (vi) below). In the following we use elementary facts from the theory of class groups of Noetherian normal domains R; see [47], Ch. 7, or [108].

(i) The elements of $\text{Cl}(R)$ are the isomorphism classes $[I]$ of fractionary divisorial ideals of R; a fractionary ideal is divisorial if and only if it is a reflexive R-module, and $\mathfrak{p} \in \text{Spec}\, R$ is divisorial if and only if height $\mathfrak{p} = 1$. One has $[I] = 0$ if and only if I is principal. In particular, R is factorial if and only if $\text{Cl}(R) = 0$.

(ii) The addition in $\text{Cl}(R)$ is given by $[I] + [J] = [(IJ)^{**}]$ where * denotes the R-dual $\text{Hom}_R(\_, R)$.

(iii) ('Gauss' lemma') The extension $[I] \mapsto [IR[T]]$ yields an isomorphism of class groups $\text{Cl}(R) \cong \text{Cl}(R[T])$ (here T denotes an indeterminate over R).

(iv) (Nagata's theorem) If $S \subset R$ is multiplicatively closed, then the assignment $[I] \mapsto [IR_S]$ maps $\text{Cl}(R)$ surjectively onto $\text{Cl}(R_S)$; the kernel of this map is generated by the classes $[\mathfrak{p}]$ of the divisorial prime ideals $\mathfrak{p}$ with $\mathfrak{p} \cap S \neq \emptyset$.

(v) An ideal $I \subset R$ is divisorial if and only if $I = \bigcap_{i=1}^r \mathfrak{p}_i^{(e_i)}$ with divisorial prime ideals $\mathfrak{p}_i$; one then has $[I] = \sum_{i=1}^r e_i[\mathfrak{p}_i]$ ($\mathfrak{p}^{(e)}$ is the e-th symbolic power $R \cap \mathfrak{p}^e R_\mathfrak{p}$.)

(vi) If R is a positively graded k-algebra with maximal ideal $\mathfrak{m}$, then one has a natural isomorphism $\text{Cl}(R) \cong \text{Cl}(R_\mathfrak{m})$. It follows that the canonical module of R is unique (up to isomorphism) since this holds for $R_\mathfrak{m}$.

Theorem 7.3.5. *Suppose that $0 < r < \min(m, n)$, and let $\mathfrak{p}$ be the ideal of R_{r+1} generated by the r-minors of the first r rows of the residue class x of X, and $\mathfrak{q}$ the corresponding ideal for the first r columns. Then*
(a) $\mathfrak{p}$ and $\mathfrak{q}$ are prime ideals of height 1,
(b) $\text{Cl}(R_{r+1})$ is isomorphic to $\mathbb{Z}$, and is generated by $[\mathfrak{p}] = -[\mathfrak{q}]$.

PROOF. (a) follows from the isomorphism $R_{r+1}/\mathfrak{p} \cong R_\varepsilon$, $\varepsilon = [1 \ldots r-1\,r+1\,|$ $1 \ldots r]$, together with 7.3.1 and the analogous isomorphism for $R_{r+1}/\mathfrak{q}$.

(b) Let $\delta = [1 \ldots r\,|\,1 \ldots r]$, and $\Psi = \{x_{ij} : i \leq r \text{ or } j \leq r\}$. We claim that Ψ is algebraically independent over k and that $R_{r+1}[\delta^{-1}] = k[\Psi, \delta^{-1}]$. In fact, one sees that $x_{uv} \in k[\Psi, \delta^{-1}]$ by expanding the minor $[1 \ldots r\,u\,|\,1 \ldots r\,v]$ along row u (or column v); in R_{r+1} one has $[1 \ldots r\,u\,|\,1 \ldots r\,v] = 0$. The algebraic independence of Ψ follows as in the proof of 7.3.2. By (iii) and (iv) above, $Cl(k[\Psi, \delta^{-1}]) = 0$ so that, again by (iv), $Cl(R_{r+1})$ is generated by the classes of those divisorial prime ideals which contain δ.

The systems of generators of $\mathfrak{p}$ and $\mathfrak{q}$ specified in the theorem are ideals in the poset Δ_{r+1} and their intersection is exactly $\{\delta\}$. In conjunction with Exercise 7.1.8 this shows $(\delta) = \mathfrak{p} \cap \mathfrak{q}$. We conclude that $[\mathfrak{p}] = -[\mathfrak{q}]$ and that $[\mathfrak{p}]$ generates $Cl(R_{r+1})$.

It remains to be shown that $u[\mathfrak{p}] \neq 0$ for all $u > 0$. Suppose that $u[\mathfrak{p}] = 0$, or, equivalently, that $\mathfrak{p}^{(u)}$ is a principal ideal (a), $a \in R$. Since $\mathfrak{p}^{(u)}$ contains δ^u, the extension $\mathfrak{p}^{(u)}k[\Psi, \delta^{-1}]$ equals $k[\Psi, \delta^{-1}]$. Hence a is a unit in $k[\Psi, \delta^{-1}]$. In $k[\Psi]$ the element δ is the determinant of a matrix of indeterminates, and therefore a prime element according to 7.3.1. Thus $a = e\delta^v$ with $e \in k$ and $v \geq 0$. In the case where $u > 0$ we would have $v > 0$, and $\mathfrak{p}$ and $\mathfrak{q}$ would be minimal prime ideals of $\mathfrak{p}^{(u)}$. $\square$

It is now easy to reduce the computation of the canonical class of R_{r+1} to the case $r = 1$; fortunately R_2 is a normal semigroup ring, and we can draw on the results of Chapter 6. The hypothesis $0 < r < n \leq m$ in the following theorem has been inserted in order to exclude the trivial cases $r = 0$ or $r = \min(m, n)$. The condition $n \leq m$ is no restriction since we may replace X by its transpose if necessary.

Theorem 7.3.6. *With the notation of 7.3.5 suppose that $0 < r < n \leq m$. Then*

(a) $\mathfrak{p}^{(m-n)}$ *is the canonical module of R_{r+1},*

(b) *(Svanes) R_{r+1} is Gorenstein if and only if $m = n$.*

PROOF. The canonical module of R_{r+1} is uniquely determined as was observed above. In particular (a) implies (b).

In proving (a) we first suppose that $r \geq 2$. The isomorphism (2) in 7.3.3, and (iii) and (iv) above induce a homomorphism

$$Cl(R_{r+1}) \to Cl(R_{r+1}[x_{11}^{-1}]) \cong Cl(R_r(Y)[X_{11}, \ldots, X_{m1}, X_{12}, \ldots, X_{1n}, X_{11}^{-1}])$$
$$\cong Cl(R_r(Y))$$

which maps the generator $[\mathfrak{p}]$ of $Cl(R_{r+1})$ to the analogous generator $[\mathfrak{p}']$ of $Cl(R_r(Y))$; in particular it is an isomorphism.

We set $S = R_{r+1}[x_{11}^{-1}]$, and identify R_{r+1} and $R_r(Y)$ with subrings of S. Let ω be the canonical module of R_{r+1}. Since the formation of the canonical module commutes with localization, $\omega \otimes_{R_{r+1}} S$ is a canonical module of S. Let ω' be a divisorial ideal of $R_r(Y)$ which under the above isomorphism has the same class as ω. Then $\omega \otimes_{R_{r+1}} S \cong \omega' \otimes_{R_r(Y)} S$. As the extension $R_r(Y) \to S$ is faithfully flat, 3.3.30 implies that ω' is a canonical module of $R_r(Y)$.

Summing up, we conclude that $u[\mathfrak{p}]$ is the class of the canonical module of R_{r+1} if and only if $u[\mathfrak{p}']$ is the class of the canonical module of $R_r(Y)$. An iteration of the argument reduces (a) to the case $r = 1$.

Suppose that $r = 1$. Let $U_1, \ldots, U_m$ and $V_1, \ldots, V_n$ be indeterminates over k. The 2-minors of the matrix $x = (U_i V_j)$ vanish so that the substitution $X_{ij} \mapsto x_{ij} = U_i V_j$ induces a surjective homomorphism from R_2 onto the normal semigroup ring $k[C]$ generated by the monomials x_{ij}. An easy calculation of $\dim k[C]$ yields that we may in fact identify R_2 and $k[C]$. Let I be the ideal generated by relint C in $k[C]$. By virtue of 6.3.5 I is the canonical module of $k[C]$.

Let $\mathfrak{p}_i$ be the prime ideal generated by the entries in the i-th row of $x = (x_{ij})$, and $\mathfrak{q}_j$ the corresponding ideal for the j-th column. Then the ideals $\mathfrak{p}_i$ and $\mathfrak{q}_j$ are exactly the height 1 $\mathbb{Z}C$-graded prime ideals. In fact, the $\mathbb{Z}C$-graded prime ideals are those prime ideals which are generated by some of the elements x_{ij}, and thus are the prime ideals generated by all the x_{ij} in the union of a set of columns and a set of rows. It follows from 6.1.1 and 6.1.6 (or direct arguments) that $I = \mathfrak{p}_1 \cap \cdots \cap \mathfrak{p}_m \cap \mathfrak{q}_1 \cap \cdots \cap \mathfrak{q}_n$. Therefore

$$[I] = \sum_{i=1}^{m} [\mathfrak{p}_i] + \sum_{j=1}^{n} [\mathfrak{q}_j] = m[\mathfrak{p}] + n[\mathfrak{q}] = (m - n)[\mathfrak{p}]. \qquad \square$$

Remarks 7.3.7. (a) One can show that the symbolic and ordinary powers of the prime ideals $\mathfrak{p}$ and $\mathfrak{q}$ in 7.3.5 coincide so that $\mathfrak{p}^{m-n}$ is the canonical module of R_{r+1}. (The case $r = 1$ is indicated in Exercise 7.3.9.) Furthermore 7.3.5 and 7.3.6 can be extended to all the rings R_δ and G_γ; see [61].

(b) With the notation of 7.3.6 and its proof, I is the *canonical module of R_2. But it is impossible to preserve the grading under the divisorial arguments by which we computed the canonical module of R_{r+1} from that of R_2. In Bruns and Herzog [57] it has been shown that the a-invariant of R_{r+1} is $-rm$. As $\mathfrak{p}^{m-n}$ is generated by elements of degree $(m - n)r$, it follows that $\mathfrak{p}^{m-n}(-rn)$ is the *canonical module of R_{r+1}.

(c) Let X be a symmetric $n \times n$ matrix of indeterminates; more precisely, the entries X_{ij} of X with $i \leq j$ are algebraically independent, and $X_{ij} = X_{ji}$ for $i > j$. The residue class rings $S_{r+1} = k[X]/I_{r+1}(X)$ are as well understood as the rings R_{r+1} constructed from 'generic' $m \times n$ matrices. In particular S_{r+1} is a normal Cohen–Macaulay domain of dimension $r(r + 1)/2 + (n - r)r$ (Kutz [255]), its divisor class group is

$\mathbb{Z}/(2)$, and it is Gorenstein if and only if $r + 1 \equiv n \mod 2$ (Goto [130], [131]); see Exercise 7.3.10. There is also a 'standard monomial' approach to the structure of S_{r+1}, in which 'doset algebras' replace ASLs (see De Concini, Eisenbud, and Procesi [73]).

(d) Let X be a alternating $n \times n$ matrix of indeterminates; this of course means that the entries X_{ij} of X with $i < j$ are algebraically independent, $X_{ii} = 0$, and $X_{ij} = -X_{ji}$ for $i > j$. The residue class ring P_{r+2} of $k[X]$ with respect to the ideal $\mathrm{Pf}_{r+2}(X)$, r even, is a normal Cohen–Macaulay domain of dimension $r(r-1)/2 + (n-r)r$ (Kleppe and Laksov [237]); it is factorial (Avramov [24]), and therefore Gorenstein by 3.3.19. The rings P_{r+2} carry a 'natural' ASL structure [73].

Exercises

7.3.8. Let R be a finitely generated faithfully flat $\mathbb{Z}$-algebra, and let $\mathscr{P}$ be one of the properties 'Cohen–Macaulay', 'Gorenstein', 'reduced', 'normal', 'integral domain', (S_n), (R_n).
(a) Show that the following are equivalent: (i) $R \otimes k$ has $\mathscr{P}$ for every field k; (ii) $R \otimes B$ has $\mathscr{P}$ for every Noetherian ring B which satisfies $\mathscr{P}$.
(b) Suppose that $K \otimes_k L$ has $\mathscr{P}$ whenever K and L are extension fields of k one of which is finitely generated (for which of the listed properties is this true?). Show that (a)(ii) already follows from the fact that R has $\mathscr{P}$.
Hint: Exercise 5.1.25 is a similar problem.

7.3.9. With the notation of 7.3.5 assume $r = 1$. Show $\mathfrak{p}^{(i)} = \mathfrak{p}^i$ for all i. Hint: 6.1.1.

7.3.10. With the notation of 7.3.7(c) let $\mathfrak{p}$ be the ideal generated by the r-minors of the first r rows of X, and $\delta = [1 \ldots r \mid 1 \ldots r]$. We use the fact that S_{r+1} is a normal Cohen–Macaulay domain, and that $\mathfrak{p}$ is a prime ideal in S_{r+1}. Show
(a) $\mathfrak{p}^2 \subset (\delta)$, $\mathfrak{p}^{(2)} = (\delta)$, and $\mathrm{Cl}(S_{r+1}) = \mathbb{Z}/(2)$,
(b) the canonical module of S_{r+1} is S_{r+1} if $r + 1 \equiv n \mod 2$, and $\mathfrak{p}$ otherwise,
(c) S_{r+1} is Gorenstein if and only if $r + 1 \equiv n \mod 2$.
Hint: S_2 can be considered as the second Veronese subring of a polynomial ring $k[Y_1, \ldots, Y_n]$, or as a normal semigroup ring.

7.3.11. Let X be an $m \times n$ matrix of indeterminates over a field k with $m \leq n$. Show that $[1, \ldots, m]$ is a prime element in $G(X)$, and deduce $G(X)$ is factorial.

Notes

The notion of an algebra with straightening law is due to Eisenbud [90]. It was generalized to that of a (not necessarily graded) Hodge algebra in De Concini, Eisenbud, and Procesi [73]. This monograph contains all the theory developed in Section 7.1 as well as numerous examples of Hodge algebras. A significant class of non-ASL Hodge algebras are the coordinate rings of the varieties of complexes (De Concini and Strickland [75]). That the notion of a graded Hodge algebra is very general is illustrated by a theorem of Hibi [164]: every positively graded affine algebra over a field is a Hodge algebra. A non-graded Hodge algebra may behave pathologically as was shown by Trung [370].

The term 'Hodge algebra' reflects the fact that the first standard monomial theory was created by Hodge [203] as a method for establishing the 'postulation formula' for the Grassmannian and its Schubert subvarieties. In algebraic language this amounts to the computation of the Hilbert function of the rings $G(X)$ and G_γ, and therefore is 'only' a matter of counting the standard monomials of a fixed degree. See also Hodge and Pedoe [204]. More recent accounts are due to Laksov [256] and Musili [282], whom we follow in proving the linear independence of the standard monomials. The straightening law on the polynomial ring is due to Doubilet, Rota, and Stein [77]. We follow De Concini, Eisenbud, and Procesi [72] in deriving it from that of $G(X)$.

For a detailed account of the history of determinantal ideals we refer the reader to Bruns and Vetter [61], 2.E. It begins with Macaulay [262] who proved (in a special case) that the ideals $I_{r+1}(X)$ are unmixed for $r + 1 = \min(m, n)$. In the influential paper [86] Eagon and Northcott constructed a finite free resolution of these ideals and proved their perfection (which (over a field of coefficients) is equivalent to the Cohen–Macaulay property of the rings R_{r+1}; see 2.2.15). This resolution, the so-called Eagon–Northcott complex, has served as a model for several related constructions. In this connection one should mention the theory of generic perfection which was also developed by Eagon and Northcott [87]; also see [61], Section 3. Its main result is that a 'generic' acyclic complex remains acyclic under extensions of the ring of coefficients.

The Cohen–Macaulay property of the rings R_{r+1} and their normality for general r are due to Hochster and Eagon [189]. They used an inductive scheme based on a 'principal radical system'. That the rings G_γ are Cohen–Macaulay seems to have been realized independently by Hochster [177], Laksov [256], and Musili [282]. The Gorenstein determinantal rings were determined by Svanes [367] whereas the divisor class group and the canonical module were computed by Bruns [52], [55].

A driving force in the investigation of determinantal ideals was their relation to invariant theory: the rings R_{r+1} and $G(X)$ appear as ring of invariants of 'natural' linear group actions. In order to prove this fact (in arbitrary characteristic) De Concini and Procesi [74] established the straightening laws on which the ASL structures are built; also see [61].

The Rees and associated graded rings of $k[X]$ with respect to the ideals $I_{r+1}(X)$ are ASLs in a natural way and Cohen–Macaulay when $r + 1 = \min(m, n)$; see Bruns, Simis, and Trung [60]. For $r + 1 < \min(m, n)$ the Cohen–Macaulay property holds at least in characteristic zero, but fails in general [56].

The Hilbert function of R_{r+1} and the numerical invariants derived from it are the subject of a monograph by Abhyankar [7]. See Herzog and Trung [163] for an approach using Gröbner bases.

The homological properties of R_{r+1} discussed in this chapter were proved by inductive methods. It would be much more satisfactory to derive them from a minimal free resolution of R_{r+1} over $k[X]$. As pointed out above, in the case $r + 1 = \min(m, n)$ the Eagon–Northcott complex is such a resolution, and for $r + 1 = \min(m, n) - 1$ a suitable complex was constructed by Akin, Buchsbaum, and Weyman [8]. Both these complexes are characteristic-free; they are defined over $\mathbb{Z}$ and specialize to a minimal free resolution under base change from $\mathbb{Z}$ to an arbitrary field. Recently Hashimoto [154] showed that such a resolution also exists for $r + 1 = \min(m, n) - 2$. In characteristic zero Lascoux [257] described a minimal resolution of R_{r+1} for arbitrary r. However, the construction of such resolutions seems to be exceedingly difficult in positive characteristics as is indicated by a result of Hashimoto [153]: for $1 < r + 1 < \min(m, n) - 2$ the Betti numbers of R_{r+1} depend on the characteristic of k.

The theory of determinantal rings has many aspects not considered in this chapter. For these, as well as for an extensive bibliography, we refer the reader to [61].

Part III

Characteristic p methods

8 Big Cohen–Macaulay modules

In this chapter we prove Hochster's theorem on the existence of big Cohen–Macaulay modules M for Noetherian local rings R containing a field. An R-module is called a *big Cohen–Macaulay module* if there is a system of parameters x for which M is x-regular. Note that one does not require M to be finite, thus the attribute 'big'. The importance of big Cohen–Macaulay modules stems from the fact that one can deduce many fundamental homological theorems from their existence (as we shall see in Chapter 9).

Their construction is a paradigm for the application of characteristic p methods: one first shows that big Cohen–Macaulay modules exist in characteristic p; then the result is transferred to characteristic zero via a rather abstract principle. It asserts that certain 'generic' systems of equations are soluble over some local ring of characteristic p if there is a solution in characteristic zero.

Rings of characteristic p are endowed with a canonical endomorphism, the Frobenius homomorphism $a \mapsto a^p$. Its homological power seems to have first been realized by Peskine and Szpiro. They also introduced M. Artin's approximation theorem to commutative algebra. The approximation theorem guarantees the descent from complete, 'analytic' local rings to 'algebraic' ones.

8.1 The annihilators of local cohomology

Let $(R, \mathfrak{m})$ be a Noetherian local ring and let

$$\mathfrak{a}_i = \operatorname{Ann}_R H^i_{\mathfrak{m}}(R)$$

be the annihilator of the i-th local cohomology. This notation is kept throughout the section. As we shall see, the products $\mathfrak{a}_0 \cdots \mathfrak{a}_j$ annihilate the homology of certain complexes, and furthermore $\mathfrak{a}_0 \cdots \mathfrak{a}_{n-1}$ annihilates the ideals $(x_1, \ldots, x_j) : x_{j+1}$ modulo $(x_1, \ldots, x_j)$ for all systems $x_1, \ldots, x_n$ of parameters and $j = 0, \ldots, n-1$. This will be important in the construction of big Cohen–Macaulay modules in characteristic p.

Theorem 8.1.1. *Let R be a Noetherian local ring of dimension n which is a residue class ring of a Gorenstein local ring S, $\dim S = d$. Then the following hold for $i = 0, \ldots, n$:*

(a) $\mathfrak{a}_i = \mathrm{Ann}_R \, \mathrm{Ext}_S^{d-i}(R, S)$;

(b) $\dim R/\mathfrak{a}_i \leq i$;

(c) $\mathfrak{a}_0 \cdots \mathfrak{a}_{n-1}$ contains a non-nilpotent element;

(d) for $\mathfrak{p} \in \mathrm{Spec}\, R$, $\dim R/\mathfrak{p} = i$, one has $\mathfrak{p} \in \mathrm{Ass}\, R/\mathfrak{a}_i \iff \mathfrak{p} \in \mathrm{Ass}\, R$.

PROOF. (a) We want to show first that both R and S can be replaced by their completions $\hat{R}$ and $\hat{S}$ for the proof of (a). Of course one has $\hat{R} \cong R \otimes_S \hat{S}$, and the formation of local cohomology commutes with completion: by 3.5.4,

$$H_{\mathfrak{m}}^i(R) \cong H_{\mathfrak{m}}^i(R) \otimes_R \hat{R} \cong H_{\hat{\mathfrak{m}}}^i(\hat{R}).$$

The same holds for Ext since $\hat{S}$ is a flat S-module and R has a resolution by finite free S-modules. Finally, for every R-module N, one has $\mathrm{Ann}_R N = R \cap (\mathrm{Ann}_{\hat{R}}(N \otimes_R \hat{R}))$ because $\hat{R}$ is faithfully flat. So we may assume that S and R are complete.

We saw in the proof of 3.5.7 that $H_{\mathfrak{m}}^i(R) \cong H_{\mathfrak{n}}^i(R)$ for all i (as an S- or R-module), $\mathfrak{n}$ denoting the maximal ideal of S. Let E be the injective hull of $S/\mathfrak{n}$ over S, and $'$ the functor $\mathrm{Hom}_S(\_, E)$. Since $H_{\mathfrak{n}}^i(R) = H_{\mathfrak{n}}^i(R)''$ we have

$$\mathrm{Ann}_R H_{\mathfrak{n}}^i(R) \subset \mathrm{Ann}_R H_{\mathfrak{n}}^i(R)' \subset \mathrm{Ann}_R H_{\mathfrak{n}}^i(R)'' = \mathrm{Ann}_R H_{\mathfrak{n}}^i(R),$$

and so the local duality theorem 3.5.8, applied to the S-module R, yields

$$\mathrm{Ann}_R H_{\mathfrak{n}}^i(R) = \mathrm{Ann}_R H_{\mathfrak{n}}^i(R)' = \mathrm{Ann}_R \mathrm{Ext}_S^{d-i}(R, S).$$

(b) This inequality is 3.5.11(c).

(c) By (b) one has $\dim(R/(\mathfrak{a}_0 \cdots \mathfrak{a}_{n-1})) \leq n - 1$. Therefore $\mathfrak{a}_0 \cdots \mathfrak{a}_{n-1}$ is not contained in any minimal prime ideal $\mathfrak{p}$ with $\dim R/\mathfrak{p} = n$.

(d) Consider the preimage q of $\mathfrak{p}$ in S. Since $\dim S_\mathfrak{q} = d - i$, one has $\mathrm{depth}\, R_\mathfrak{p} = \mathrm{depth}\, R_\mathfrak{q} = 0$ if and only if $\mathrm{Ext}_S^{d-i}(R, S)_\mathfrak{p} = \mathrm{Ext}_S^{d-i}(R, S)_\mathfrak{q} \neq 0$. Consequently $\mathfrak{p} \in \mathrm{Ass}\, R$ if and only if $\mathfrak{p} \in \mathrm{Supp}\, \mathrm{Ext}_S^{d-i}(R, S)$. Because of (a) and (b) the latter is equivalent to $\mathfrak{p} \in \mathrm{Ass}\, R/\mathfrak{a}_i$ for prime ideals $\mathfrak{p}$ such that $\dim R/\mathfrak{p} = i$. $\square$

A very important property of the ideals $\mathfrak{a}_i$ is expressed by the following theorem.

Theorem 8.1.2. Let R be a Noetherian local ring, and

$$F_\bullet : 0 \longrightarrow F_m \longrightarrow F_{m-1} \longrightarrow \cdots \longrightarrow F_0 \longrightarrow 0$$

a complex of finitely generated free R-modules such that all the homology modules $H_i(F_\bullet)$ have finite length. Then $\mathfrak{a}_0 \cdots \mathfrak{a}_{m-i}$ annihilates $H_i(F_\bullet)$ for $i = 0, \ldots, m$.

PROOF. First we construct an object which connects the local cohomology of R and the homology of $F_\bullet$. Let $x_1, \ldots, x_n$ be a system of parameters, and let $K_\bullet$ denote the complex

$$0 \longrightarrow K_n \longrightarrow \cdots \longrightarrow K_1 \longrightarrow K_0 \longrightarrow 0,$$

$$K_j = \bigoplus_{1 \le i_1 < \cdots < i_{n-j} \le n} R_{x_{i_1} \cdots x_{i_{n-j}}}.$$

Then $H_{\mathfrak{m}}^i(R) = H_{n-i}(K_\bullet)$ (see 3.5.5 where we write C^i for K_{n-i}). Now we form the tensor product $K_\bullet \otimes_R F_\bullet$, a first quadrant bicomplex. The crucial point is that the homology of the associated total complex $T_\bullet$ can be computed from two spectral sequences.

First we consider the spectral sequence whose E_{pq}^1-term is given by $H_q(K_p \otimes_R F_\bullet)$, the homology of the columns of $K_\bullet \otimes_R F_\bullet$ (see [318], Theorem 11.18 where the E^2-terms are described). This spectral sequence converges to the homology of the total complex. The modules K_p are flat R-modules, thus

$$H_q(K_p \otimes_R F_\bullet) = K_p \otimes_R H_q(F_\bullet).$$

Being an R-module of finite length, $H_q(F_\bullet)$ is annihilated by a power of each of the elements x_i. Hence

$$H_q(K_p \otimes F_\bullet) = \begin{cases} 0 & \text{for } p < n, \\ H_q(F_\bullet) & \text{for } p = n. \end{cases}$$

Since the E^1-terms are concentrated in a single column, it follows immediately that

$$E_{p,q}^\infty = E_{p,q}^1 = \begin{cases} 0 & \text{for } p < n, \\ H_q(F_\bullet) & \text{for } p = n, \end{cases}$$

and therefore $H_{i+n}(T_\bullet) = H_i(F_\bullet)$ for all i.

Secondly, one determines the homology of the total complex by first computing the homology of its rows:

$$E_{p,q}^1 = H_q(K_\bullet \otimes F_p) = H_{\mathfrak{m}}^{n-q}(F_p) = (H_{\mathfrak{m}}^{n-q}(R))^r, \qquad r = \operatorname{rank} F_p.$$

By the definition of the ideals $\mathfrak{a}_i$ one has $\mathfrak{a}_{n-q} E_{p,q}^1 = 0$. Since all the terms $E_{p,q}^r$ are subquotients of $E_{p,q}^1$, they are equally annihilated by $\mathfrak{a}_{n-q}$:

$$\mathfrak{a}_{n-q} E_{p,q}^\infty = 0.$$

This spectral sequence also converges to the total homology of $T_\bullet$ ([318], Theorem 11.19). For every t one therefore has a filtration

$$0 \subset U^{-1} \subset U^0 \subset \cdots \subset U^u = H_t(T_\bullet)$$

where $U^p/U^{p-1} = E_{p,t-p}^\infty$. Observe that $E_{p,t-p}^\infty = 0$ for $p > m$ or $t - p > n$. Thus the filtration is already given by

$$0 = U^{t-n-1} \subset U^{t-n} \subset \cdots \subset U^m = H_t(T_\bullet),$$

and $H_t(T_\bullet)$ is annihilated by $\mathfrak{a}_{n-(t-(t-n))} \cdots \mathfrak{a}_{n-(t-m)} = \mathfrak{a}_0 \cdots \mathfrak{a}_{n-t+m}$. Taking into account that $H_i(F_\bullet) = H_{i-n}(T_\bullet)$ we get the desired result. $\square$

As a consequence we derive another 'annihilation theorem' whose second part is crucial in the construction of big Cohen–Macaulay modules in characteristic p.

Corollary 8.1.3.   *Let R be a Noetherian local ring of dimension n. Then, given a sequence $x = x_1, \ldots, x_m \in R$ such that $\operatorname{codim}(x_1, \ldots, x_m) = m$, the following hold:*
(a) *$\mathfrak{a}_0 \cdots \mathfrak{a}_{n-i}$ annihilates the Koszul homology $H_i(x)$, $i = 0, \ldots, m$;*
(b) *$\mathfrak{a}_0 \cdots \mathfrak{a}_{n-1}$ annihilates $((x_1, \ldots, x_{m-1}) : x_m)/(x_1, \ldots, x_{m-1})$.*

PROOF. (a) The sequence x can be extended to a system of parameters $x_1, \ldots, x_n$ (recall that $\operatorname{codim} I = \dim R - \dim R/I$). We start a descending induction at $m = n$ for which the assertion is obviously a special case of 8.1.2.

Suppose now that $m < n$ and put $x' = x_1, \ldots, x_m, x_{m+1}^t$, $t \geq 1$. By 1.6.13 we have an exact sequence

$$H_i(x) \xrightarrow{\pm x_{m+1}^t} H_i(x) \xrightarrow{\varphi} H_i(x').$$

By induction the submodule $\operatorname{Im} \varphi \cong H_i(x)/x_{m+1}^t H_i(x)$ of $H_i(x')$ is annihilated by $\mathfrak{a}_0 \cdots \mathfrak{a}_{n-i}$. Since $\bigcap x_{m+1}^t H_i(x) = 0$, we are done.

(b) We use another segment of the long exact sequence of Koszul homology, now relating x and $x'' = x_1, \ldots, x_{m-1}$:

$$H_1(x) \xrightarrow{\psi} H_0(x'') \xrightarrow{x_m} H_0(x'').$$

Since $\mathfrak{a}_0 \cdots \mathfrak{a}_{n-1} H_1(x) = 0$, we also have $\mathfrak{a}_0 \cdots \mathfrak{a}_{n-1}(\operatorname{Im} \psi) = 0$. Im ψ consists of exactly those elements in $H_0(x'') = R/(x_1, \ldots, x_{m-1})$ annihilated by x_m, that is $\operatorname{Im} \psi = ((x_1, \ldots, x_{m-1}) : x_m)/(x_1, \ldots, x_{m-1})$. $\square$

The preceding corollary is completely vacuous if R happens to be a Cohen–Macaulay ring, but in connection with 8.1.2 it shows that certain local rings, among them the complete ones, preserve a faint trace of the Cohen–Macaulay property: the modules

$$((x_1, \ldots, x_{j-1}) : x_j)/(x_1, \ldots, x_{j-1}),$$

which are zero for R Cohen–Macaulay, cannot be arbitrarily 'big'.

Corollary 8.1.4.   *Let R be a Noetherian local ring which is a residue class ring of a Gorenstein local ring S. Then there exists a non-nilpotent element $c \in R$ such that $c \cdot ((x_1, \ldots, x_j) : x_{j+1})/(x_1, \ldots, x_j) = 0$ for all systems of parameters $x_1, \ldots, x_n$ and all $j = 0, \ldots, n-1$.*

PROOF. According to 8.1.1 there exists a non-nilpotent $c \in \mathfrak{a}_0 \cdots \mathfrak{a}_{n-1}$. By 8.1.3 such a c satisfies our needs. $\square$

Remark 8.1.5. Parts (b) of 8.1.1 and 8.1.4 are not true for arbitrary Noetherian local rings. One of the most used counterexamples of commutative algebra (constructed by Nagata [284], Example 2, p. 203) works here, too: let R be a 2-dimensional local domain such that its completion $\hat{R}$ has an associated prime ideal $\mathfrak{p}$ with $\dim \hat{R}/\mathfrak{p} = 1$. Put $\mathfrak{a}_1 = \operatorname{Ann}_R H^1_\mathfrak{m}(R)$ and $\mathfrak{b}_1 = \operatorname{Ann}_{\hat{R}} H^1_{\hat{\mathfrak{m}}}(\hat{R})$. Since $H^1_\mathfrak{m}(R) \cong H^1_{\hat{\mathfrak{m}}}(\hat{R})$ as R-modules, $\mathfrak{a}_1 = \mathfrak{b}_1 \cap R$. Of course 8.1.1 applies to $\hat{R}$, and by its third part, $\mathfrak{p} \in \operatorname{Ass} \hat{R}/\mathfrak{b}_1$. So $\mathfrak{a}_1 \subset \mathfrak{p} \cap R = 0$ and $\dim R/\mathfrak{a}_1 = 2$. (Note that a regular element of R stays regular in $\hat{R}$.)

Let $x = x, y$ be a system of parameters of R and put $x^t = x^t, y^t$. Then $H^1_\mathfrak{m}(R)$ is the direct limit of the Koszul cohomology modules $H^1(x^t)$. By 1.6.10 one has $H^1(x^t) \cong H_1(x^t)$. Consider the long exact sequence (which appeared already in the proof of 8.1.3):

$$H_1(y^t) \longrightarrow H_1(x^t) \longrightarrow H_0(y^t) \xrightarrow{x^t} H_0(y^t).$$

R is a domain, so $H_1(y^t) = 0$, and $H_1(x^t) = (y^t : x^t)/(y^t)$. An element c annihilating all the modules $(y^t : x^t)/(y^t)$ must annihilate $H^1_\mathfrak{m}(R)$, hence $c = 0$ as seen above.

Exercises

8.1.6. Let M be an arbitrary module over a Noetherian local ring $(R, \mathfrak{m})$, and $\mathfrak{a}_i(M) = \operatorname{Ann} H^i_\mathfrak{m}(M)$. Let $F_\bullet$ be a complex of finite free R-modules with homology of finite length as in 8.1.2. Prove that $\mathfrak{a}_0(M) \cdots \mathfrak{a}_{m-i}(M)$ annihilates $H_i(F_\bullet \otimes M)$ for $i = 0, \ldots, m$.

8.1.7. With R and M as in 8.1.6 assume that $H^i_\mathfrak{m}(M) = 0$ for $i = 0, \ldots, n-1$ where $n = \dim R$, and $M/\mathfrak{m}M \neq 0$. Let x be a system of parameters of R. Show $H_i(x, M) = 0$ for $i = 1, \ldots, n$, and that x is M-quasi-regular. (See 1.6.20.)

8.2 The Frobenius functor

Let R be a ring of characteristic p, i.e. a ring with a monomorphism $\mathbb{Z}/p\mathbb{Z} \longrightarrow R$ where p is a prime number. The *Frobenius homomorphism* is the map $F: R \longrightarrow R$, $F(a) = a^p$. Via F one may consider R as an R-algebra in a non-trivial way. The crucial point in the construction of the Frobenius functor is to work simultaneously with two essentially different module structures of R itself. This is an unusual idea in commutative algebra and has to be kept firmly in mind. Let R^F denote the $(R-R)$-bimodule with additive group R and left and right scalar multiplication given by

$$a \cdot r \circ b = arF(b) = arb^p, \qquad a, b \in R, \quad r \in R^F.$$

(The standard associative laws are obviously satisfied.)

Let M be a left R-module. Then we take the tensor product $R^F \otimes_R M$ with R^F as a right R-module, i.e.

$$a \otimes bx = a \circ b \otimes x = ab^p \otimes x, \qquad a \in R^F, \quad b \in R, \quad x \in M.$$

The left R-module structure of R^F endows $R^F \otimes_R M$ with a like structure such that $c(a \otimes x) = ca \otimes x$. (This tensor product is merely biadditive; bilinearity is lost because in general $a \cdot r \neq r \circ a$ for $a \in R$, $r \in R^F$.) The *Frobenius functor* $\mathscr{F}$ acts on a left R-module M by assigning to it the left R-module $R^F \otimes_R M$. For an R-linear map $\varphi : M \to N$ one consequently considers $\mathscr{F}(\varphi)$ to be the R-linear map $\mathrm{id}_{R^F} \otimes_R \varphi$. The following properties of $\mathscr{F}$ are just the fundamental ones of tensor products.

Proposition 8.2.1. *Let R be a ring of characteristic p. Then $\mathscr{F}$ is a covariant, additive, and right exact functor from the category of left R-modules to itself.*

We want to compute some specific values of $\mathscr{F}$. First we see that $\mathscr{F}(R) = R^F \otimes_R R = R^F$ as a left R-module, so $\mathscr{F}(R) = R$; then additivity implies $\mathscr{F}(R^n) = R^n$. For a cyclic R-module R/I one gets $\mathscr{F}(R/I) = R^F \otimes_R R/I = R^F/(R^F \circ I)$. Now $r \circ a = ra^p$ for $r \in R^F$, $a \in I$, and $R^F \circ I$ turns out to be *the ideal $I^{[p]}$ generated by the p-th powers of the elements of I*. Hence $R^F/(R^F \circ I) = R/I^{[p]}$ with its ordinary left scalar multiplication.

Proposition 8.2.2. *Let R be a ring of characteristic p. Then*

(a) $\mathscr{F}(R^n) = R^n$ *for all n (as left R-modules), and if $e_1, \ldots, e_n$ is a basis of R^n, $1 \otimes e_1, \ldots, 1 \otimes e_n$ is a basis of $\mathscr{F}(R^n)$,*

(b) $\mathscr{F}(R/I) = R/I^{[p]}$ *for all ideals I of R.*

More generally, we denote by $I^{[q]}$, $q = p^e$, the ideal generated by the q-th power of the elements of I; $I^{[q]}$ is called the *q-th Frobenius power* of I.

The Frobenius functor owes its power to its non-linearity, again something remarkable. It is straightforward to verify the following.

Proposition 8.2.3. *Let R be a ring of characteristic p, M and N be R-modules, and $\varphi : M \to N$ an R-linear map. Then*

(a) $\mathscr{F}(a\varphi) = a^p \mathscr{F}(\varphi)$ *for all $a \in R$,*

(b) *if $\varphi(x) = \sum a_i y_i$ for $x \in M$, $a_i \in R$, $y_i \in N$, then $\mathscr{F}(\varphi)(1 \otimes x) = \sum a_i^p(1 \otimes y_i)$,*

(c) *the map $M \to \mathscr{F}(M)$, $x \mapsto 1 \otimes x$, is not R-linear in general: instead one has $(ax) \mapsto a^p(1 \otimes x)$.*

We can now give a concrete description of $\mathscr{F}$ in terms of 'generators and relations':

Proposition 8.2.4. *Let R be a ring of characteristic p and M an R-module with a presentation $R^m \xrightarrow{\varphi} R^n \longrightarrow M \longrightarrow 0$.*
(a) Then $\mathscr{F}(M)$ has the presentation $R^m \xrightarrow{\mathscr{F}(\varphi)} R^n \longrightarrow \mathscr{F}(M) \longrightarrow 0$;
(b) furthermore, if φ is given by a matrix (a_{ij}), then $\mathscr{F}(\varphi)$ is given by the matrix (a_{ij}^p).

Part (a) follows from the right exactness of $\mathscr{F}$ and the fact that $\mathscr{F}$ leaves R^n untouched. Part (b) follows from 8.2.3.

We conclude the list of basic properties of the Frobenius functor with its behaviour under localization:

Proposition 8.2.5. *Let R be a ring of characteristic p. The Frobenius functor commutes with rings of fractions: $R_S \otimes_R \mathscr{F}(M) = \mathscr{F}(R_S \otimes_R M)$ for all R-modules M, and analogously for R-linear maps.*

PROOF. We have $R_S \otimes_R \mathscr{F}(M) = R_S \otimes_R R^F \otimes_R M$ and

$$\mathscr{F}(R_S \otimes_R M) = R_S^F \otimes_{R_S} R_S \otimes_R M = R_S^F \otimes_R M.$$

As left R_S-modules, $R_S \otimes_R R^F \cong R_S \otimes_R R \cong R_S$ and R_S^F are naturally isomorphic; this isomorphism is also an isomorphism of right R-modules. $\square$

We cannot resist trying the strength of the Frobenius functor by proving the 'new intersection theorem' in characteristic p. (This nomenclature will be explained in Chapter 9.) The elegant argument, including 8.1.2, is due to Roberts.

Theorem 8.2.6. *Let $(R, \mathfrak{m})$ be a Noetherian local ring of characteristic p, and*

$$F_\bullet : 0 \longrightarrow F_s \xrightarrow{\varphi_s} F_{s-1} \longrightarrow \cdots \longrightarrow F_1 \xrightarrow{\varphi_1} F_0 \longrightarrow 0$$

a complex of finite free R-modules such that each homology $H_i(F_\bullet)$ has finite length. If $s < \dim R$, the complex $F_\bullet$ is exact.

PROOF. Note that it is enough to cover the case of a complete ring R: if R is not complete, we simply tensor all our objects by $\hat{R}$, a faithfully flat extension of the same dimension.

Assume that $F_\bullet$ is not exact. If $H_0(F_\bullet) = 0$, the map φ_1 is a split epimorphism, and $F_\bullet$ decomposes into the direct sum of two shorter complexes of the same type. So we may suppose $H_0(F_\bullet) \neq 0$. Furthermore, if $\varphi_1(F_1) \not\subseteq \mathfrak{m}F_0$, $F_\bullet$ splits off an isomorphism $\varphi_1' : F_1' \to F_0'$ of direct summands of F_1 and F_0. This leaves the essential case $\varphi_1(F_1) \subset \mathfrak{m}F_0$.

Apply the Frobenius functor $\mathscr{F}$ to $F_\bullet$. The modules appearing in $\mathscr{F}(F_\bullet)$ are the same as in $F_\bullet$. Furthermore $\mathscr{F}(F_\bullet)$ has finite length homology: by hypothesis $F_\bullet \otimes R_\mathfrak{p}$ is split exact for all prime ideals $\mathfrak{p} \in \operatorname{Spec} R$, $\mathfrak{p} \neq \mathfrak{m}$. Since the Frobenius functor commutes with localization, this also holds for $\mathscr{F}(F_{\bullet})$. Something has changed however, namely we have $\mathscr{F}(\varphi_1)(F_1) \subset \mathfrak{m}^p F_0$ by 8.2.4(b).

Now one iterates this procedure: all the complexes $\mathscr{F}^e(F_{\bullet}), e \geq 0$, have finite length homology, and $\mathscr{F}^e(\varphi_1)(F_1) \subset \mathfrak{m}^{p^e} F_0$. On the other hand $H_0(\mathscr{F}^e(F_{\bullet}))$ is annihilated by the ideal $\mathfrak{a}_0 \cdots \mathfrak{a}_s$ where $\mathfrak{a}_i = \operatorname{Ann} H_{\mathfrak{m}}^i(R)$; see 8.1.2. This forces $\mathfrak{a}_0 \cdots \mathfrak{a}_s$ to be contained in $\bigcap \mathfrak{m}^{p^e} = 0$, contradicting 8.1.1 for $s < \dim R$ since R, a complete local ring, is a residue class ring of a Gorenstein ring. $\square$

A crucial point in the preceding proof is that for a finite free complex $F_{\bullet}$ the Frobenius functor $\mathscr{F}$ preserves the property of having finite length homology. It also preserves acyclicity:

Theorem 8.2.7 (Peskine–Szpiro). *Let R be a Noetherian ring of character-istic p, and $F_{\bullet} : 0 \longrightarrow F_s \xrightarrow{\varphi_s} \cdots \xrightarrow{\varphi_1} F_0$ a complex of finite free R-modules. Then $F_{\bullet}$ is acyclic if and only if $\mathscr{F}(F_{\bullet})$ is acyclic.*

PROOF. Set $r_j = \sum_{i=j}^{s}(-1)^{i-j} \operatorname{rank} F_i$. By the acyclicity criterion 1.4.13 it depends only on the grades of the ideals $I_{r_i}(\varphi_i)$ whether $F_{\bullet}$ is acyclic or not: it is acyclic if and only if $\operatorname{grade} I_{r_i}(\varphi_i) \geq i$ for $i = 1, \ldots, s$.

By virtue of 8.2.2 and 8.2.4, first $\mathscr{F}(F_j) = F_j$, and next $I_{r_i}(\mathscr{F}(\varphi_i)) = I_{r_i}(\varphi_i)^{[p]}$. The two ideals have the same radical, hence the same grade. $\square$

The following corollary will play an important rôle in Chapter 10.

Corollary 8.2.8 (Kunz). *Let R be a regular ring of characteristic p. Then R^F is a flat R-algebra; equivalently, $\mathscr{F}$ is an exact functor.*

PROOF. Since flatness is a local property and $\mathscr{F}$ commutes with localiza-tion, we may assume that R is a regular local ring. By a standard flatness criterion (for example, see [270], 7.8) it is enough that $\operatorname{Tor}_1^R(R^F, R/I) = 0$ for every ideal I of R. This follows from 8.2.7: the finite free resolution of R/I stays acyclic when tensored with R^F. $\square$

The assertion of 8.2.8 is usually called the *flatness of the Frobenius*. Kunz [245] also showed the converse of 8.2.8; see Exercise 8.2.11 for the case in which R is Cohen–Macaulay.

Exercises

8.2.9. Let R be a Noetherian ring, M a finite R-module of finite projective dimension with finite free resolution $F_{\bullet}$, and $e \geq 1$ an integer. Prove that $\mathscr{F}^e(F_{\bullet})$ is acyclic, and $\operatorname{proj dim} \mathscr{F}^e(M) = \operatorname{proj dim} M$.

8.2.10. Let R be a regular local ring of characteristic p. Show $\ell(\mathscr{F}^e(M)) = p^{e \dim R} \ell(M)$ for every finite length module M. (Use induction on $\ell(M)$.)

8.2.11. Herzog [157] proved that 8.2.9 characterizes modules of finite projective dimension; then it follows immediately from 2.2.7 that the exactness of $\mathscr{F}$ characterizes the regular ones among the Noetherian local rings. For simplicity we restrict ourselves to Cohen–Macaulay rings. So suppose R is a Cohen–Macaulay local ring of characteristic p.

(a) Let $F_{\bullet}$ be a minimal free resolution of a finite R-module M and x a maximal R-sequence. If $\mathscr{F}^e(F_{\bullet})$ is acyclic for all $e \geq 1$, show $\mathrm{Tor}^R_i(R/(x), \mathscr{F}^e(M)) \cong (R/(x))^{b_i(M)}$ for $i > 0$ and $e \gg 0$.

(b) Conclude that $\operatorname{proj\,dim} M < \infty$.

8.3 Modifications and non-degeneracy

In this section we show that for a system of parameters $x = x_1, \ldots, x_n$ of a Noetherian local ring of characteristic p there exists an x-regular R-module. The conditions to satisfy are: (i) x_{s+1} is a regular element of $M/(x_1, \ldots, x_s)M$, $s = 0, \ldots, n-1$, (ii) $M \neq xM$. Since the trivial choice $M = 0$ satisfies (i), we see that (i) is completely useless without (ii), and we need results from the preceding sections in order to show that the construction below does not degenerate by violating condition (ii).

Suppose M is an R-module such that x_{s+1} is not $(M/(x_1, \ldots, x_s)M)$-regular. Then there exists a $y \in M, y \notin (x_1, \ldots, x_s)M$, for which $x_{s+1}y \in (x_1, \ldots, x_s)M$. Equivalent to $y \notin (x_1, \ldots, x_s)M$ is the non-existence of a solution $z_1, \ldots, z_s \in M$ of the equation $y = x_1 z_1 + \cdots + x_s z_s$. The *deus ex machina* by which algebraists force equations to be soluble is to extend the given object by some 'free' variables and to introduce the as yet insoluble equation as a relation on them. In our case we pass to $M' = (M \oplus R^s)/Rw$, where $w = y - (x_1 e_1 + \cdots + x_s e_s)$, and $e_1, \ldots, e_s$ a basis of R^s. The element y', the image of y under the natural map $M \to M'$, no longer keeps x_{s+1} from being regular on $M'/(x_1, \ldots, x_s)M'$. It is quite obvious that a well organized iteration of this construction in the limit yields a module $\widetilde{M}$ satisfying condition (i) of x-regularity.

It is however equally obvious that we may lose condition (ii) for $\widetilde{M}$. One attempt to control (ii), successful in characteristic p, is to keep track of a fixed element $f \in M$ on its way to the limit and to make sure that $f \notin xM_i$ for all approximations M_i.

For a pair $(M, f), f \in M$, let M' be constructed as above, and f' be the image of f under the natural map $M \to M'$. Then (M', f') is called an x-*modification of* (M, f) (*of type s*). More generally, if there is a sequence

$$(M, f) = (M_0, f_0) \longrightarrow (M_1, f_1) \longrightarrow \cdots \longrightarrow (M_r, f_r) = (N, g),$$

with (M_{i+1}, f_{i+1}) an x-modification of (M_i, f_i) (of type s_{i+1}), then (N, g) is an x-*modification of* (M, f) (*of type* $(s_1, \cdots, s_r)$). As soon as x is fixed, we may simply speak of a *modification*. If $g \notin xN$, then (N, g) is *non-degenerate*.

Proposition 8.3.1. *Let R be a Noetherian ring, and $x = x_1, \ldots, x_n \in R$. Then the following are equivalent:*
(a) *there exists an x-regular R-module M;*
(b) *every x-modification (N, g) of $(R, 1)$ is non-degenerate.*

PROOF. We start with the more important implication (b) $\Rightarrow$ (a). Our goal is to construct a direct system of modules (M_i, φ_{ij}), $i \in \mathbb{N}$, starting from $M_0 = R$ such that $M = \varinjlim M_i$ is x-regular. Each $(M_i, \varphi_{0i}(1))$ will be a modification of $(R, 1)$. Therefore our hypothesis (b) forces $\varinjlim \varphi_{0i}(1) \notin xM$.

Suppose that $M_1, \ldots, M_j$ (together with the natural maps in a sequence of modifications) have been determined. Now choose first i, then s, minimal such that there exists a $y \in M_i$ with $x_{s+1}\varphi_{ij}(y) \in (x_1, \ldots, x_s)M_j$ while $\varphi_{ij}(y) \notin (x_1, \ldots, x_s)M_j$. Then put $M_{j+1} = (M_j \oplus R^s)/Rw$, $w = y - (x_1 e_1 + \cdots + x_s e_s)$ as above, the maps $\varphi_{i,j+1}$ being the natural ones. Let us say that *step $j + 1$ has index (i, s).*

We claim that for each pair (i, s) there are only finitely many steps of index (i, s). For, if the sequence $j + 1, \ldots$ of steps of index (i, s) does not stop, one finds a non-stationary ascending chain of submodules of M_i by taking the preimages of $(x_1, \ldots, x_s)M_{j+1}, (x_1, \ldots, x_s)M_{j+2}, \ldots$ in M_i. But R is Noetherian, and all the modules M_i are finite.

If there is an equation $x_{s+1}y = x_1 z_1 + \cdots + x_s z_s$ for elements $y, z_1, \ldots, z_s$ of the limit M, it has to hold in an approximation M_i as well. According to the claim, $\varphi_{ij}(y) \in (x_1, \ldots, x_s)M_j$ for $j \gg i$, hence $y \in (x_1, \ldots, x_s)M$.

The validity of the implication (a) $\Rightarrow$ (b) is forced by our choice of a *free* direct summand in the construction of a modification. Let $f \in M$ be any element $\notin xM$. Trivially there is a homomorphism $R \to M$, $1 \mapsto f$. So it is enough to show that if (N', g') is a type s modification of (N, g) and there is a map $\varphi : N \to M$, $\varphi(g) = f$, then this map can be extended to $\varphi' : N' \to M$, $\varphi'(g') = f$. Suppose that $N' = (N \oplus R^s)/Rw$, $w = y - (x_1 e_1 + \cdots + x_s e_s)$. Since $x_{s+1}\varphi(y) \in (x_1, \ldots, x_s)M$ and M is x-regular, there are elements $e_1', \ldots, e_s' \in M$ such that $\varphi(y) = x_1 e_1' + \cdots + x_s e_s'$. Thus take φ' to be the map induced by φ and the assignment $e_i \mapsto e_i'$. $\square$

Of course, the implication (a) $\Rightarrow$ (b) of the preceding proposition does not help in the construction of a big Cohen–Macaulay module. However, the idea in its proof, namely to compare a sequence of modifications to some 'universal' object, is very useful:

Lemma 8.3.2. *Let R be a Noetherian local ring which is a residue class ring of a Gorenstein ring. Then there exists a non-nilpotent element $c \in R$ such that for every system of parameters x and every sequence $(R, 1) = (M_0, f_0) \to \cdots \to (M_r, f_r)$ of x-modifications one has a commutative dia-*

gram

$$(M_0, f_0) \longrightarrow (M_1, f_1) \longrightarrow \cdots \longrightarrow (M_{r-1}, f_{r-1}) \longrightarrow (M_r, f_r)$$

$$\downarrow \varphi_0 \qquad \downarrow \varphi_1 \qquad \qquad \downarrow \varphi_{r-1} \qquad \downarrow \varphi_r$$

$$(R, 1) \xrightarrow{\ c\ } (R, c) \xrightarrow{\ c\ } \cdots \xrightarrow{\ c\ } (R, c^{r-1}) \xrightarrow{\ c\ } (R, c^r)$$

the commutativity including $\varphi_i(f_i) = c^i$, $i = 0, \ldots, r$.

PROOF. Take c as in 8.1.4, i.e. non-nilpotent and

$$c \cdot ((x_1, \ldots, x_s) : x_{s+1})/(x_1, \ldots, x_s) = 0$$

for every system of parameters x and $s = 0, \ldots, \dim R - 1$.
Naturally $\varphi_0 = \mathrm{id}$. Suppose φ_i has been chosen. If

$$M_{i+1} = (M_i \oplus R^s)/Rw,$$

$$w = y - (x_1 e_1 + \cdots + x_s e_s),$$

$$x_{s+1} y = x_1 z_1 + \cdots + x_s z_s, \qquad z_j \in M_i,$$

then $\varphi_i(y) \in (x_1, \ldots, x_s) : x_{s+1}$, and there are elements $e'_1, \ldots, e'_s \in R$ for which $c\varphi_i(y) = x_1 e'_1 + \cdots + x_s e'_s$. The homomorphism $\varphi' : M_i \oplus R^s \to R$, $\varphi'(g) = c\varphi_i(g)$ for $g \in M_i$, $\varphi'(e_i) = e'_i$, factors through M_{i+1}, yielding the desired map φ_{i+1}. $\qquad \square$

Suppose R has characteristic p, and let $\mathscr{F}$ denote the Frobenius functor. Given an R-module M and $f \in M$, we write $\mathscr{F}(f)$ for $1 \otimes f \in \mathscr{F}(M) = R^F \otimes M$. We want to investigate how modifications behave under $\mathscr{F}$. With the standard meanings of y, z_i, w, M' we have

$$x_{s+1}^p \mathscr{F}(y) = x_1^p \mathscr{F}(z_1) + \cdots + x_s^p \mathscr{F}(z_s),$$

$$\mathscr{F}(w) = \mathscr{F}(y) - (x_1^p \mathscr{F}(e_1) + \cdots + x_s^p \mathscr{F}(e_s)),$$

$$\mathscr{F}(M') = (\mathscr{F}(M) \oplus \mathscr{F}(R^s))/R\mathscr{F}(w),$$

and $\mathscr{F}(e_1), \ldots, \mathscr{F}(e_s)$ form a basis of $\mathscr{F}(R^s) \cong R^s$; see 8.2.1 – 8.2.3. This shows that if (M', f') is an x-modification of (M, f), then $(\mathscr{F}(M'), \mathscr{F}(f'))$ is an x^p-modification of $(\mathscr{F}(M), \mathscr{F}(f))$.

All the arguments necessary to prove the existence of big Cohen–Macaulay modules in characteristic p have now been collected. Let R be a Noetherian local ring of characteristic p. Note that a system of parameters x of R is a system of parameters of $\hat{R}$, and every x-regular $\hat{R}$-module is also an x-regular R-module. Therefore we may assume that R is complete. According to 8.3.1 the existence of a degenerate x-modification (N, g) of $(R, 1)$ must be excluded. Suppose (N, g) is degenerate and of type $(s_1, \ldots, s_r)$. Let us now iterate the application of the Frobenius functor to the given data. After the e-th iteration we have reached an x^{p^e}-modification $(N^{(e)}, g^{(e)})$ of $(R, 1)$. Since (N, g) is degenerate, i.e. $g \in xN$, $(N^{(e)}, g^{(e)})$ is degenerate, too, i.e. $g^{(e)} \in x^{p^e} N^{(e)}$. Since R is complete, 8.3.2

can be invoked: there exists a homomorphism $\varphi_r : N^{(e)} \to R$ such that $c^r = \varphi_r(g^{(e)})$. However, $g^{(e)} \in x^{p^e} N^{(e)}$, so $c^r \in (x^{p^e})$ for all e. Since $\bigcap(x^{p^e}) = 0$, c must be nilpotent – a contradiction.

Theorem 8.3.3. *Let R be a Noetherian local ring of characteristic $p > 0$, and x a system of parameters. Then there exists an x-regular module M. In particular R has a big Cohen–Macaulay module.*

An equational criterion for degeneracy of modifications. In the coming section we want to derive the existence of big Cohen–Macaulay modules in characteristic zero from their existence in characteristic p. The key argument will be Hochster's finiteness theorem which guarantees the solubility of certain systems of polynomial equations over some local ring of characteristic p provided there is a solution in characteristic zero. The following proposition gives a sufficiently detailed description of the equations to be used. Combined with 8.3.1 it is a criterion for the existence of x-regular modules, in particular the existence of big Cohen–Macaulay modules.

Proposition 8.3.4. *Let $n \in \mathbb{Z}$, $n \geq 1$, and let $s_1, \ldots, s_r \in \mathbb{Z}$ with $0 \leq s_1, \ldots, s_r \leq n - 1$. Then there exists a set $\mathcal{S}(s_1, \ldots, s_r)$ of polynomials $p \in \mathbb{Z}[X_1, \ldots, X_n, Y_1, \ldots, Y_m]$, m determined by $s_1, \ldots, s_r$, such that for every ring R and every sequence $x = x_1, \ldots, x_n \in R$ the following are equivalent:*

(a) *there is a degenerate x-modification (N, g) of $(R, 1)$ of type $(s_1, \ldots, s_r)$;*

(b) *there exist $y_1, \ldots, y_m \in R$ such that $p(x_1, \ldots, x_n, y_1, \ldots, y_m) = 0$ for all $p \in \mathcal{S}(s_1, \ldots, s_r)$.*

PROOF. Consider a sequence

$$(R, 1) = (M_0, f_0) \longrightarrow \cdots \longrightarrow (M_r, f_r) = (N, g),$$

(M_i, f_i) being a modification of (M_{i-1}, f_{i-1}) of type s_i. M_i is constructed by adding generators $e_1^i, \ldots, e_{s_i}^i$ and a relation $w_i = y_i - (x_1 e_1^i + \cdots + x_{s_i} e_{s_i}^i)$ to M_{i-1}, so

$$N = M_r = (\bigoplus_{j=0}^{r} F_j) / \sum_{v=1}^{r} R w_v.$$

The module $M_0 = R$ is simply generated by $e_0^1 = 1$, and F_i has the basis $e_1^i, \ldots, e_{s_i}^i$. Writing y_i as a linear combination of the basis elements one obtains $a_{ij}^l \in R$ such that

(1)
$$
\left.
\begin{aligned}
y_i &= \sum_{j=0}^{i-1} \sum_{l=1}^{s_j} a_{il}^j e_l^j, \\
w_i &= \sum_{j=0}^{i-1} \sum_{l=1}^{s_j} a_{il}^j e_l^j - (x_1 e_1^i + \cdots + x_{s_i} e_{s_i}^i),
\end{aligned}
\right\} \quad i = 1, \ldots, r.
$$

The condition $x_{s_i+1}y_i \in (x_1, \ldots, x_{s_i})M_{i-1}$ can be formulated in $\bigoplus_{j=0}^{i-1} F_j$:

$$x_{s_i+1}y_i = \sum_{u=1}^{s_i} x_u g_i^u + \sum_{v=1}^{i-1} b_i^v w_v$$

with $g_i^u \in \bigoplus_{j=0}^{i-1} F_j$, $b_i^v \in R$. Expressing the g_i^u in the given basis of $\bigoplus_{j=0}^{i-1} F_{i-1}$ and substituting the right sides of (1) for y_i and w_i yields

$$(2) \quad \sum_{j=0}^{i-1} \sum_{l=1}^{s_j} x_{s_i+1} a_{il}^j e_l^j$$
$$= \sum_{u=1}^{s_i} \sum_{j=0}^{i-1} \sum_{l=1}^{s_j} x_u c_{il}^{uj} e_l^j + \sum_{v=1}^{i-1} \Big(\sum_{j=0}^{v-1} \sum_{l=1}^{s_j} b_i^v a_{vl}^j e_l^j - \sum_{l=1}^{s_v} b_i^v x_l e_l^v \Big), \quad i = 1, \ldots, r.$$

Each of these equations relating elements of the free module $\bigoplus_{j=0}^{i-1} F_i$ splits into its components with respect to the elements of the given basis. Replacing the coefficients $a_{il}^j, b_i^v, c_{il}^{uj}, x_u$ by algebraically independent elements $A_{il}^j, B_i^v, C_{il}^{uj}, X_u$ over the ring $\mathbb{Z}$ and collecting all the terms in the components of (2) on one side, one obtains a set $\mathscr{S}_0(s_1, \ldots, s_r)$ of polynomials over $\mathbb{Z}$ which depends only on $(s_1, \ldots, s_r)$.

We have seen that an *x*-modification of type $(s_1, \ldots, s_r)$ leads to a solution of the system $\mathscr{S}_0(s_1, \ldots, s_r)$ in which the variables $A_{il}^j, B_i^v, C_{il}^{uj}$ take values in R whereas $x_1, \ldots, x_n$ are substituted for $X_1, \ldots, X_n$. Conversely, given such a solution, one defines the elements y_i and w_i by their representations in (1). The validity of (2) then guarantees that one has constructed a chain of modifications.

Next we write down what it means for (M_r, f_r) to be degenerate. The element f_r is just the residue class of $e_0^1 = 1$ in M_r. Therefore (M_r, f_r) is degenerate if and only if

$$e_0^1 \in x\Big(\bigoplus_{j=0}^{r} F_j\Big) + \sum_{v=1}^{r} Rw_v.$$

This adds an $(r+1)$-th relation to the system (2):

$$e_0^1 = \sum_{u=1}^{n} \sum_{j=0}^{r} \sum_{l=1}^{s_j} x_u c_{r+1,l}^{uj} e_l^j + \sum_{v=1}^{r} \Big(\sum_{j=0}^{v-1} \sum_{l=1}^{s_j} b_{r+1}^v a_{vl}^j e_l^j - \sum_{l=1}^{s_v} b_{r+1}^v x_l e_l^v \Big).$$

Accordingly we enlarge our sets of indeterminates and of equations. In view of what must be proved, there is no need to distinguish the variables $A_{il}^j, B_i^v, C_{il}^{uj}$. We order them in some sequence and rename them $Y_1, \ldots, Y_m$. □

Exercise

8.3.5. Let $(R, \mathfrak{m}, k)$ be a Noetherian local ring of dimension n, M a big Cohen–Macaulay module, and y a system of parameters. Prove

(a) $H_{\mathfrak{m}}^i(M) = 0$ for $i = 0, \ldots, n - 1$ and $H_{\mathfrak{m}}^n(M) \neq 0$,

(b) $H_i(y, M) = 0$ for $i = 1, \ldots, n$ and $H_0(y, M) \neq 0$,

(c) y is M-quasi-regular,

(d) $\mathrm{Ext}_R^i(k, M) = 0$ for $i = 0, \ldots, n - 1$ and $\mathrm{Ext}_k^n(k, M) \neq 0$.

(Use 8.1.7.)

8.4 Hochster's finiteness theorem

The following theorem is fundamental for the application of characteristic p methods to (local) rings containing a field of characteristic zero. Let $X = X_1, \ldots, X_n$, $Y = Y_1, \ldots, Y_m$ be families of independent indeterminates over $\mathbb{Z}$. Then we call a subset $\mathscr{E}$ of $\mathbb{Z}[X, Y]$ simply *a system of equations* (*over* $\mathbb{Z}$). It has *a solution of height* n in a Noetherian ring R if there are families $x = x_1, \ldots, x_n$, $y = y_1, \ldots, y_m$ in R such that

$$\text{(i)} \quad p(x, y) = 0 \text{ for all } p \in \mathscr{E}, \qquad \text{(ii)} \quad \text{height } xR = n.$$

Note that condition (ii) implies $xR \neq R$ (by definition, height $R = \infty$).

Theorem 8.4.1 (Hochster). (a) *Suppose that the system $\mathscr{E}$ of equations has a solution of height n in a Noetherian ring R containing a field. Then $\mathscr{E}$ has a solution x', y' in a local ring R' of characteristic $p > 0$ such that x' is a system of parameters for R'.*

Moreover, R' can be chosen as a localization of an affine domain over a finite field with respect to a maximal ideal.

(b) *If, in addition, R is a regular local ring such that x is a regular system of parameters, then the ring R' in (a) can be chosen as a regular local ring with regular system of parameters x'.*

The theorem suggests the following strategy for proving a statement $\mathscr{S}$ about Noetherian rings containing a field:

(i) prove $\mathscr{S}$ for local rings of characteristic $p > 0$,

(ii) show that there exists a family $(\mathscr{E}_i)_{i \in I}$ of systems of equations with the property that $\mathscr{S}$ holds for R if and only if none of the systems $\mathscr{E}_i$ has a solution of the appropriate height in R.

Both steps (i) and (ii) have been carried out for the statement 'If R is local and x a system of parameters, then there exists an x-regular R-module'; see 8.3.1, 8.3.4, and 8.3.3. Thus one obtains

Theorem 8.4.2 (Hochster). *Let R be a Noetherian local ring containing a field, and x a system of parameters for R. Then there exists an x-regular R-module M. In particular, R has a big Cohen–Macaulay module.*

The proof of 8.4.1 falls into three parts: (i) the reduction to its part (b), (ii) the reduction to the case in which R is the localization of an affine algebra, and (iii) the final step.

Before we set out for the proof of 8.4.1, we show that certain conditions of linear algebra over a regular local ring can be formulated by stating that the ring elements involved and some auxiliary elements satisfy a suitable system of equations.

The equational presentation of acyclicity. Some conditions for elements in a ring R and vectors and matrices formed by them are evidently of an equational nature: for example, the membership of a vector in a finite submodule of a free module of finite rank (especially that of an element in a finite ideal) and the assertion that a sequence of matrices forms a complex. The crucial fact for regular local rings is that the acyclicity of a complex can also be cast into equational conditions:

Lemma 8.4.3. *Suppose R is a regular local ring with regular system of parameters $x_1, \ldots, x_n$, and let the matrices $\varphi_p, \ldots, \varphi_1$ represent the linear maps in a finite free resolution over R. Denote the family of the entries of all the φ_i by $z = (z_{jk}^{(i)})$.*

Then there is a set $\mathcal{A}$ of polynomials over $\mathbb{Z}$ in the indeterminates $X = X_1, \ldots, X_n$, the family of indeterminates $Z = (Z_{jk}^{(i)})$ representing the entries of all the matrices φ_i, and auxiliary indeterminates $W = W_1, \ldots, W_u$ such that the following holds:

(a) there are $w = w_1, \ldots, w_u \in R$ for which x, z, w is a solution of $\mathcal{A}$;

(b) whenever x', z', w' are specializations of X, Z, W in a regular local ring R' satisfying the systems of equations $\mathcal{A}$ and such that x' is a regular system of parameters, then the complex formed by the matrices φ_i' with entries $z_{jk}^{(i)'}$ is acyclic.

PROOF. We use the Buchsbaum–Eisenbud acyclicity criterion 1.4.13. Let r_i be the (expected) rank of φ_i. Then $\operatorname{grade} I_{r_i}(\varphi_i) \geq i$. For $i \leq \dim R$ this is equivalent to the existence of elements $a_{ij} \in (x)$, $j = 1, \ldots, n - i$, for which $I_{r_i}(\varphi_i) + (a_{i1}, \ldots, a_{i,n-i})$ contains a power of each of the x_l. For $i > \dim R$, the condition $\operatorname{grade} I_{r_i}(\varphi_i) \geq i$ just says $1 \in I_{r_i}(\varphi_i)$. Thus the grade condition of the acyclicity criterion can be interpreted equationally. In R' we use the converse direction of the acyclicity criterion. $\square$

This lemma describes very precisely which indeterminates and equations between them must be introduced in order to transfer objects of linear algebra represented by matrices and some of their properties $\mathcal{P}$ from the regular local ring R to another regular local ring R' of the same dimension via a generic presentation as in 8.4.3. Let us simply say that $\mathcal{P}$ has a *regular equational presentation*. The following corollary lists some properties with regular equational presentations.

Corollary 8.4.4. *Let* $(R, \mathfrak{m})$ *be a regular local ring,* $U \subset V$ *submodules of* R^r, $v \in R^r$, W *a submodule of* R^s, *and* $\psi : R^r \to R^t$ *a linear map. Then the following properties have regular equational presentations:*
(a) $U = \operatorname{Ker} \psi$;
(b) $V/U \cong R^s/W$;
(c) $v \notin U$;
(d) $\operatorname{Ext}_R^i(R^r/U, R) \cong R^s/W$ *for some fixed* i;
(e) $\dim R^r/U = d$ *for some integer* d.

PROOF. (a) We choose a system $u_1, \ldots, u_q$ of generators of U and define the linear map $\rho : R^q \to R^r$ by sending the i-th basis vector to u_i. Then the sequence $R^q \xrightarrow{\rho} R^r \xrightarrow{\psi} R^t$ can be extended to a finite free resolution $F_{\bullet}$ of $\operatorname{Coker} \psi$. By virtue of the previous lemma the acyclicity of $F_{\bullet}$ has a regular equational presentation, and the acyclicity includes the condition $U = \operatorname{Ker} \psi$.

(b) The given isomorphism $V/U \cong R^s/W$ and the choice of a system of generators of U as above induce a commutative diagram

$$
\begin{array}{ccccccccc}
0 & \longrightarrow & R^q & \longrightarrow & R^{q+s} & \xrightarrow{\ \pi\ } & R^s & \longrightarrow & 0 \\
 & & \downarrow{\scriptstyle \rho} & & \downarrow{\scriptstyle \sigma} & & \downarrow{\scriptstyle \tau} & & \\
0 & \longrightarrow & U & \longrightarrow & V & \longrightarrow & V/U & \longrightarrow & 0
\end{array}
$$

with exact rows and epimorphisms ρ, σ, τ. Conversely, given such a diagram, one has $\operatorname{Ker} \tau = \pi(\operatorname{Ker} \sigma)$. In other words, a system of generators of W is obtained by projecting a system of generators of $\operatorname{Ker} \sigma$ onto the last s coordinates. Let $T = \operatorname{Ker} \sigma$. After the specification of matrices for ρ, σ, τ it is sufficient that the condition $T = \operatorname{Ker} \sigma$ has a regular equational presentation, and this is warranted by (a) since σ can also be considered as a linear map with target R^r.

(c) Set $V = U + Rv$; then $V/U \neq 0$, and so $V/U \cong R^s/W$ with $s > 0$ and $W \subset \mathfrak{m}R^s$. The condition $V/U \cong R^s/W$ has a regular equational presentation by (b), and the same evidently holds for $W \subset \mathfrak{m}R^s$.

(d) Again we choose an epimorphism $\rho : R^q \to U$ and extend it to a free resolution $F_{\bullet}$ of R^r/U. Let φ_j, $j = 1, \ldots, p$ be the maps in $F_{\bullet}$. Then $\operatorname{Ext}_R^i(R^r/U, R) \cong \operatorname{Im} \varphi_i^* / \operatorname{Ker} \varphi_{i+1}^*$ (here $*$ denotes the R-dual). Set $M = \operatorname{Im} \varphi_i^*$ and $N = \operatorname{Ker} \varphi_{i+1}^*$. Since the acyclicity of the resolution and the condition $N = \operatorname{Ker} \varphi_{i+1}^*$ have a regular equational presentation (for $M = \operatorname{Im} \varphi_i^*$ this is trivial), $\operatorname{Ext}_R^i(R^r/U, R) \cong N/M$ also has such a presentation, and an application of (b) concludes the argument.

(e) One has $\dim R - \dim R^r/U = \operatorname{grade} R^r/U$ since R is a regular local ring. However, $\operatorname{grade} R^r/U = \min\{i : \operatorname{Ext}_R^i(R^r/U, R) \neq 0\}$, and the vanishing and non-vanishing of Ext can be captured by equations according to (d). $\qquad\square$

The reduction to the affine case. The reader should note that we are free to extend the set of indeterminates appearing in $\mathscr{E}$ and the system $\mathscr{E}$ itself. Moreover, we can also change the family X of distinguished variables that guarantees the height of the solution. We only have to make sure that the elements to which the variables X will finally specialize generate an ideal of height n.

The very first step is a routine matter: we choose a prime ideal $\mathfrak{p}$ minimal over (x) such that height $\mathfrak{p} = n$, then we complete $R_{\mathfrak{p}}$ with respect to the $\mathfrak{p}R_{\mathfrak{p}}$-adic topology, and finally replace R by $\hat{R}_{\mathfrak{p}}$. Because of Cohen's structure theorem A.21 we can write R as a residue class ring of a regular local ring (containing a field), say $R = S/I$ where I is generated by elements $\boldsymbol{b} = b_1, \ldots, b_s$ and S has a regular system of parameters $\boldsymbol{a} = a_1, \ldots, a_r$. Extend the set of indeterminates by $A = A_1, \ldots, A_r$ and $B = B_1 \ldots, B_s$, and modify the system $\mathscr{E}$ to a system $\widetilde{\mathscr{E}}$ as follows: each equation $p(X, Y) = 0$ is replaced by the equation

$$p(X, Y) = C_{p1}B_1 + \cdots + C_{ps}B_s$$

where the C_{pj} are new indeterminates.

Next we enlarge $\widetilde{\mathscr{E}}$ by further equations expressing (i) the condition that $\dim S/(b_1, \ldots, b_s) = n$ and (ii) the fact that a power of each a_i lies in the ideal generated by $\boldsymbol{b}$ and preimages of the x_j. While the equations for (ii) simply exist because x is a system of parameters of R, we must invoke 8.4.4(e) for (i).

Next suppose part (b) of the theorem has been proved. Then we can find a solution $\widetilde{x}, \widetilde{y}$ to the system $\widetilde{\mathscr{E}}$ in a regular local ring $\widetilde{S}$ in which A specializes to a regular system of parameters $\widetilde{a}$. We simply set $R' = S'/(b')$. The original system $\mathscr{E}$ is solved by the residue classes x' and y' in R' of the families $\widetilde{x}$ and $\widetilde{y}$. Moreover, $\dim R' = n$ because of the extra equations for (i) above, and x' is a system of parameters by the additional equations for (ii).

The reduction to the affine case. For the next reduction step Artin's approximation theorem will be crucial. In order to explain it we need the theory of Henselian local rings. We must content ourselves with a very brief sketch, referring the reader to Grothendieck [142], IV, §18, Nagata [284], or Raynaud [301] for a full treatment. A local ring $(R, \mathfrak{m})$ is Henselian if it has the following property: suppose $f \in R[X]$ is a monic polynomial such that its residue class $\bar{f}$ modulo $\mathfrak{m}R[X]$ has a factorization $\bar{f} = g'h'$ with monic polynomials $g', h' \in (R/\mathfrak{m})[X]$ for which $(g', h') = (R/\mathfrak{m})[X]$; then there exist monic polynomials $f, g \in R[X]$ such that $f = gh$ and $\bar{g} = g', \bar{h} = h'$. A more abstract characterization is that R is Henselian if and only if every R-algebra S which is a finite R-module is a product of local rings.

Hensel's lemma says that a complete local ring is Henselian. Moreover, for each local ring $(R, \mathfrak{m})$ there exists a Henselization $(R^h, \mathfrak{m}^h)$ which, in a sense, is the smallest Henselian local ring containing R. One has local embeddings $(R, \mathfrak{m}) \subset (R^h, \mathfrak{m}^h) \subset (\hat{R}, \hat{\mathfrak{m}})$, and $\mathfrak{m} R^h = \mathfrak{m}^h$. The ring R^h is a direct limit of subrings S each of which is the localization of a module-finite extension of R with respect to a maximal ideal lying over $\mathfrak{m}$. More precisely, S has the form $(R[X]/f)_{(\mathfrak{m}, X)}$ where $f = X^n + c_{n-1} X^{n-1} + \cdots + c_0$ is a monic polynomial with $c_0 \in \mathfrak{m}$, $c_1 \notin \mathfrak{m}$. (It follows that R^h is a flat extension of R.)

We can now formulate (a special case of) Artin's approximation theorem [13]:

Theorem 8.4.5. *Let R be a local ring which is a localization of an affine algebra over a field k. Let $\mathcal{E}$ be a system of polynomial equations over R in n variables. If $\mathcal{E}$ has a solution $x \in \hat{R}^n$, then it has a solution $x' \in (R^h)^n$. Furthermore, given t, the solution x' can be chosen such that it approximates x to order t, that is, $x \equiv x' \bmod \mathfrak{m}^t \hat{R}^n$.*

We will not need the statement about approximation to order t; it has only been included for completeness.

Proposition 8.4.6. *Let $\mathcal{E} \subset \mathbb{Z}[X, Y]$ be a system of equations with $X = X_1, \ldots, X_n$ and $Y = Y_1, \ldots, Y_m$. Suppose that $\mathcal{E}$ has a solution x, y in a regular local ring R that contains a field K and such that x is a regular system of parameters. Then there exist an algebraically closed field L of like characteristic, an affine domain A over L, and a maximal ideal $\mathfrak{m}$ of A with $A_{\mathfrak{m}}$ regular such that $\mathcal{E}$ has a solution x', y' in $A_{\mathfrak{m}}$ for which x' is a regular system of parameters of $A_{\mathfrak{m}}$.*

PROOF. We may assume R is complete. By Cohen's structure theorem A.21, R is just a formal power series ring $K[\![x]\!]$ in which the elements of x are the indeterminates. It is obviously harmless to replace K by an algebraic closure L.

Next the indeterminates X in the system $\mathcal{E}$ are replaced by the elements of x so that we obtain a system of polynomial equations $\mathcal{E}'$ in the unknowns Y over $L[x]$. By hypothesis it has a solution in the completion $L[\![x]\!]$ of $A' = L[x]_{(x)}$ with respect to its maximal ideal $\mathfrak{m}' = x A'$. Thus the approximation theorem yields a solution in the Henselization of A', and therefore in an extension $A'' = (A'[X]/(f))_{(\mathfrak{m}', X)}$ where $f = X^n + c_{n-1} X^{n-1} + \cdots + c_0$ is a monic polynomial with $c_0 \in \mathfrak{m}'$, $c_1 \notin \mathfrak{m}'$. It is easily verified that $\dim A'' = n$ and that the image of x generates the maximal ideal of A''. It follows that A'' is a regular local ring for which x is a regular system of parameters. In order to arrive at an integral domain A we replace $A'[X]/(f)$ by the residue class ring with respect to its unique minimal prime ideal contained in $(\mathfrak{m}', X) A'[X]/(f)$. $\qquad\square$

The final step. Let L, A, $\mathfrak{m}$, x', and y' be as in 8.4.6. We rename x' and y' by setting $x = x'$, $y = y'$. Since L is algebraically closed, the injection $L \to A$ induces an isomorphism $L \cong A/\mathfrak{m}$ by Hilbert's Nullstellensatz in its algebraic version; see A.15. In other words, $A \cong L \oplus \mathfrak{m}$ as an L-vector space. This implies that $A = L[z_1, \ldots, z_r]$ where $z_1, \ldots, z_r$ generates the ideal $\mathfrak{m}$. We write $A \cong L[\mathbf{Z}]/I$, $\mathbf{Z} = Z_1, \ldots, Z_r$, being indeterminates over L. The ideal I is finitely generated by polynomials $f_1, \ldots, f_s$. Let $C \subset L$ be the finite set of coefficients appearing in

(1) the polynomials $f_1, \ldots, f_s$,

(2) $n + m$ polynomials expressing $x_1, \ldots, x_n$ and $y_1, \ldots, y_m$ in terms of $z_1, \ldots, z_r$,

(3) a polynomial $g \notin \mathfrak{m}$ and polynomials h_{ij} such that $g z_i = \sum h_{ij} x_j$ for all i. (The polynomials h_{ij} can be found since x generates $\mathfrak{m} A_\mathfrak{m}$.)

Let L_0 be the subfield generated by C over the prime field of L and set $R' = L_0[z_1, \ldots, z_r]$.

The first point to be observed is that $I \cap L_0[\mathbf{Z}]$ is generated by $f_1, \ldots, f_s$, since the extension $L_0[\mathbf{Z}] \to L[\mathbf{Z}]$ is faithfully flat. This implies $R' \cong L_0[\mathbf{Z}]/(f_1, \ldots, f_s)L_0[\mathbf{Z}]$; therefore $A = L \otimes_{L_0} R'$. Obviously $z_1, \ldots, z_r$ generate a maximal ideal $\mathfrak{m}'$ of R', and $R'/\mathfrak{m}' \cong L_0$; hence $R' = L_0 \oplus (z_1, \ldots, z_r)R'$. The extension $R'_{\mathfrak{m}'} \to A_\mathfrak{m}$ is flat, and so $\dim R'_{\mathfrak{m}'} = \dim A_\mathfrak{m} = n$ by A.11. The solution x, y is contained in R', and $\mathfrak{m}' R'_{\mathfrak{m}'}$ is generated by x because of (3) above.

Let us first treat the case of characteristic 0; after all, the descent from characteristic 0 to positive characteristic is the main point of 8.4.1. (In positive characteristic it only remains to replace L_0 by a finite field of the same characteristic.) With the notation just introduced, L_0 is the field of fractions of the finitely generated $\mathbb{Z}$-algebra $B = \mathbb{Z}[C]$, and apart from the fact that the coefficients no longer form a field, almost nothing is lost if we replace L_0 by B.

(i) $R'' = B[z_1, \ldots, z_r]$ is a domain containing x, y. This is obvious.

(ii) $\mathfrak{p} = (z_1, \ldots, z_r)R' \cap R''$ is a prime ideal of height n: evidently $\mathfrak{p}$ is a prime ideal, and it has the same height as its extension in R', a ring of fractions.

(iii) $\mathfrak{p} = (z_1, \ldots, z_r)R''$ and $R'' = B \oplus \mathfrak{p}$. This is an immediate consequence of $R' = L_0 \oplus (z_1, \ldots, z_r)R'$.

(iv) $\mathfrak{p} R''_\mathfrak{p}$ is generated by x because C contains all the coefficients appearing in (3) above and $g \notin \mathfrak{p}$.

For the very last step we choose a maximal ideal $\mathfrak{q}$ of B not containing the constant coefficient of the polynomial g appearing in (3) above. This is possible since $\mathbb{Z}$ and, hence, B are Hilbert rings, and for the same reason $B/\mathfrak{q}$ is a finite field (see A.18). We claim that replacing all the data by their residue classes mod $\mathfrak{q}$ gives us the desired solution of $\mathscr{E}$ in a local ring of characteristic p.

First note that x generates $\bar{\mathfrak{p}}$ since $\bar{\mathfrak{p}}$ is generated by the residue classes $\bar{z}_i$ of the z_i, and these in turn can be written as linear combinations of the $\bar{x}_j$ because the polynomial g of (3) above is non-zero modulo $\bar{\mathfrak{p}}$.

Second, height $\bar{\mathfrak{p}} = n$ as will now be shown. Let $\mathfrak{P} = \mathfrak{q} \oplus \mathfrak{p}$. Then height $\mathfrak{P} \leq$ height $\mathfrak{q} + \dim(R''_{\mathfrak{P}}/\mathfrak{q}R''_{\mathfrak{P}})$ by A.5, applied to the homomorphism $B_{\mathfrak{q}} \to R''_{\mathfrak{P}}$, and therefore

$$\text{height } \bar{\mathfrak{p}} = \dim(R''_{\mathfrak{P}}/\mathfrak{q}R''_{\mathfrak{P}}) \geq \text{height } \mathfrak{P} - \text{height } \mathfrak{q}.$$

The ring R'', a residue class ring of a polynomial ring over $\mathbb{Z}$, and thus of a Cohen–Macaulay ring, is catenary; see 2.1.12. Since R'' is also a domain,

$$\text{height } \mathfrak{P} = \text{height } \mathfrak{p} + \text{height } \mathfrak{P}/\mathfrak{p} = \text{height } \mathfrak{p} + \text{height } \mathfrak{q},$$

or height $\bar{\mathfrak{p}} \geq$ height $\mathfrak{p} = n$, as desired.

In positive characteristic the argument is essentially the same: one only has to replace $\mathbb{Z}$ by the prime field of L_0.

Exercise

8.4.7. Let R be a regular local ring and U, M, N finite R-modules, given as quotients of finite free R-modules by submodules. Show that both the acyclicity and the non-acyclicity of a complex $U \to M \to N$ have regular equational presentations. (Describe the maps by matrices.)

8.5 Balanced big Cohen–Macaulay modules

Big Cohen–Macaulay modules M lack many of the properties of finite Cohen–Macaulay modules. For example, let $R = K[[X, Y]]$ and $M = R \oplus Q$, where Q is the field of fractions of $R/(Y)$. Then X is obviously regular on M, and $M/XM \cong R/(X)$. Thus M is (X, Y)-regular, but not (Y, X)-regular. However, it is important for the applications in Chapter 9 and an interesting fact in itself that every local ring R possessing a big Cohen–Macaulay module even has a *balanced* big Cohen–Macaulay module, i.e. a module M such that every system of parameters is an M-sequence. More precisely, we want to prove that the $\mathfrak{m}$-adic completion of any big Cohen–Macaulay module is balanced. Our main argument will be that 1.1.8 has a converse for complete modules:

Theorem 8.5.1. Let R be a ring, $x = x_1, \ldots, x_n$ a sequence of elements of R, and M an R-module. Let $I = xR$, and denote the I-adic completion of M by $\hat{M}$. Then the following are equivalent:
(a) x is M-quasi-regular;
(b) x is $\hat{M}$-quasi-regular;
(c) x is $\hat{M}$-regular.

PROOF. By definition quasi-regularity includes the requirement $IM \neq M$. Since $I^j \hat{M}/I^{j+1} \hat{M}$ is naturally isomorphic with $I^j M/I^{j+1} M$, one has a commutative diagram

$$
\begin{array}{ccc}
M \otimes (R/I)[X_1,\ldots,X_n] & \longrightarrow & \mathrm{gr}_I\, M \\
\| & & \| \\
\hat{M} \otimes (R/I)[X_1,\ldots,X_n] & \longrightarrow & \mathrm{gr}_I\, \hat{M}
\end{array}
$$

Together with the description of quasi-regularity by the conclusion of 1.1.8 this diagram immediately yields the equivalence of (a) and (b).

Theorem 1.1.8 says that (c) implies (b). We want to prove the crucial implication (b) $\Rightarrow$ (c) by induction on n, and recall the results of Exercise 1.1.15, namely
(i) if $x_1 z \in I^i M$ for $z \in M$, then $z \in I^{i-1} M$,
(ii) the sequence $x_2,\ldots,x_n$ is $(M/x_1 M)$-quasi-regular.
Let $z \in M$ such that $x_1 z = 0$. Then, by (i), $z \in \bigcap I^j \hat{M} = 0$, and hence x_1 is $\hat{M}$-regular. Because of (ii) it remains to prove that $\hat{M}/x_1 \hat{M} \cong (\hat{M}/x_1 \hat{M})\hat{\;}$. There is a natural exact sequence

$$
0 \to (x_1 \hat{M})' \longrightarrow \hat{M} \longrightarrow (\hat{M}/x_1 \hat{M})\hat{\;} \longrightarrow 0,
$$

in which $(x_1 \hat{M})'$ is the completion of $x_1 \hat{M}$ with respect to its subspace topology (see [270], Theorem 8.1; note that the quotient topology on $\hat{M}/x_1 \hat{M}$ is just the I-adic topology). The subspace topology is given by the filtration $(x_1 \hat{M} \cap I^j \hat{M})$. Of course $x_1 \hat{M} \cong \hat{M}$ is complete in its own I-adic topology, and we are left to verify the following claim of Artin–Rees type: if x is M-quasi-regular for some R-module M, then $x_1 M \cap I^j M \subset I^{j-1} x_1 M$. But this follows immediately from (i). □

Since quasi-regularity of a sequence is invariant under permutations of its elements, one can permute $\hat{M}$-regular sequences:

Corollary 8.5.2. *With the notation of 8.5.1 assume that* $x = x_1,\ldots,x_n$ *is* $\hat{M}$-regular. *Then for every permutation* σ *of* $\{1,\ldots,n\}$ *the sequence* $x_\sigma = x_{\sigma(1)},\ldots,x_{\sigma(n)}$ *is* $\hat{M}$-regular.

Another consequence is the existence of balanced big Cohen–Macaulay modules:

Corollary 8.5.3. *Let* $(R,\mathfrak{m})$ *be a Noetherian local ring, and* M *a big Cohen–Macaulay* R-module. *Then the* $\mathfrak{m}$-adic completion $\hat{M}$ *is a balanced big Cohen–Macaulay module. In particular, if* R *contains a field, it has a balanced big Cohen–Macaulay module.*

PROOF. Note that for any system of parameters x the $\mathfrak{m}$-adic and (x)-adic topologies on M coincide. Therefore we can apply 8.5.1 to the $\mathfrak{m}$-adic completion.

Suppose that the system of parameters $x = x_1, \ldots, x_n$ is an M-sequence, and let $y = y_1, \ldots, y_n$ be an arbitrary system of parameters. By the standard prime avoidance argument there exists an element $w \in \mathfrak{m}$ not contained in any minimal prime ideal of $(x_1, \ldots, x_{n-1})$ or $(y_1, \ldots, y_{n-1})$. Hence $x_1, \ldots, x_{n-1}, w$ and $y_1, \ldots, y_{n-1}, w$ are systems of parameters.

Note that $x_1, \ldots, x_{n-1}, w$ is an M-sequence: a power of x_n being a multiple of w modulo $(x_1, \ldots, x_{n-1})$, the element w must be regular on $M/(x_1, \ldots, x_{n-1})M$. Then $w, x_1, \ldots, x_{n-1}$ is M-quasi-regular and therefore $\hat{M}$-regular. Furthermore $\hat{M}/w\hat{M}$ is an $(x_1, \ldots, x_{n-1})$-regular module for the local ring $\bar{R} = R/Rw$. By induction on n one may assume that $(\bar{y}_1, \ldots, \bar{y}_{n-1})$ is $(\hat{M}/w\hat{M})$-regular, too. Then $w, y_1, \ldots, y_{n-1}$ is $\hat{M}$-regular, and applying the preceding arguments in reverse order we get that y is an $\hat{M}$-sequence.

Now the second part of the corollary follows immediately from the existence of big Cohen–Macaulay modules for local rings containing a field; see 8.4.2.   □

Remark 8.5.4. A different construction of balanced big Cohen–Macaulay modules was given by Griffith in [139] and [140]. Let R be a Noetherian complete local domain containing a field K. As in the proof of 8.4.6, R is a module-finite extension of a formal power series ring $K[[x]]$ where x is an arbitrary system of parameters. By [139], Theorem 3.1, there exists an R-module which is a free A-module (with countable basis). Such a module is a balanced big Cohen–Macaulay module ([140], Proposition 1.4).

Balanced big Cohen–Macaulay modules are much closer to finite modules than is apparent from their definition. Sharp [342] developed the theory of grade for balanced big Cohen–Macaulay modules, similar to that for finite modules, using Theorem 8.5.6 below as a prime avoidance argument. Unfortunately, however, the property of being a balanced big Cohen–Macaulay module is not stable under localization. In Chapter 9 we shall introduce a general notion of grade that overcomes this obstacle.

Proposition 8.5.5. *Let R be a Noetherian local ring, and M a balanced big Cohen–Macaulay module.*
(a) *One has $\dim R/\mathfrak{p} = \dim R$ for all $\mathfrak{p} \in \operatorname{Ass} M$;*
(b) *in particular $\operatorname{Ass} M$ is finite;*
(c) *$\operatorname{Ass} M$ consists of the minimal prime ideals of $\operatorname{Ann} M$, and so $\operatorname{Supp} M = V(\operatorname{Ann} M)$.*

PROOF. (a) Let $\mathfrak{q} \in \operatorname{Spec} R$ with $\dim R/\mathfrak{q} < \dim R$. Then $\mathfrak{q}$ is not contained in a prime ideal $\mathfrak{p}$ such that $\dim R/\mathfrak{p} = \dim R$, hence not contained in the union of these prime ideals. So there exists $x \in \mathfrak{q}$ with $\dim R/(x) < \dim R$. The element x can be extended to a system of parameters. Thus it is regular on M, and $\mathfrak{q} \notin \operatorname{Ass} M$.

Part (b) follows immediately from (a).

(c) Let $\operatorname{Ass} M = \{\mathfrak{p}_1, \ldots, \mathfrak{p}_r\}$. Since $\operatorname{Ann} M \subset \mathfrak{p}_i$ for all i, it remains to be shown for the first assertion in (c) that $f^j \in \operatorname{Ann} M$ for all $f \in \bigcap_{i=1}^r \mathfrak{p}_i$ and $j \gg 0$. Given such an element f, there exists j with $f^j/1 = 0$ in $R_{\mathfrak{p}_i}$ for all i. It follows that $f^j M_{\mathfrak{p}_i} = 0$, and so $\mathfrak{p}_i \notin \operatorname{Ass} f M$. On the other hand, $\operatorname{Ass} f^j M \subset \operatorname{Ass} M$. Therefore $\operatorname{Ass} f^j M = \emptyset$, which is only possible if $f^j M = 0$.

The second assertion in (c) follows from the first since, over a Noetherian ring, every prime ideal $\mathfrak{q} \in \operatorname{Supp} M$ contains a $\mathfrak{p} \in \operatorname{Ass} M$. $\quad\square$

Theorem 8.5.6. *Let R be a Noetherian local ring, and M a balanced big Cohen–Macaulay module. Suppose that $x = x_1, \ldots, x_r$ is an M-sequence. Then $\operatorname{Ass} M/xM$ is finite, and $\dim R/\mathfrak{p} = \dim R - r$ for all $\mathfrak{p} \in \operatorname{Ass} M/xM$.*

PROOF. Let $\mathfrak{q}$ be prime ideal such that $\dim R/\mathfrak{q} = \dim R$. Then $Rx_1 + \operatorname{Ann} M \not\subset \mathfrak{q}$: if $\mathfrak{q} \in \operatorname{Ass} M$, then $x_1 \notin \mathfrak{q}$; if $\mathfrak{q} \notin \operatorname{Ass} M$, then $\operatorname{Ann} M \not\subset \mathfrak{q}$ by 8.5.5. As the number of such prime ideals is finite, $Rx_1 + \operatorname{Ann} M$ is not contained in their union. By 1.2.2 one therefore finds an element $y \in \operatorname{Ann} M$ with $x_1 + y \notin \mathfrak{q}$ for all $\mathfrak{q}$ such that $\dim R/\mathfrak{q} = \dim R$. The following facts are now obvious:

(i) $\dim(R/(x_1 + y)) = \dim R - 1$, and $M/(x_1 + y)M$ is a balanced big Cohen–Macaulay module over $R/(x_1 + y)$;

(ii) $x_1 + y, x_2, \ldots, x_r$ is an M-sequence;

(iii) $M/xM \cong M/(x_1 + y, x_2, \ldots, x_r)M$.

Set $\bar{R} = R/(x_1 + y)$ and $\bar{M} = M/(x_1 + y)M$. Because of (i) and (ii) we can apply an inductive argument to the $\bar{M}$-sequence $\bar{x}_2, \ldots, \bar{x}_r$. By (iii) the associated primes of M/xM are exactly the preimages of the associated primes of $\bar{M}/(\bar{x}_2, \ldots, \bar{x}_r)M$ over $\bar{R}$. $\quad\square$

Exercises

8.5.7. Prove that each of the conditions (a), (b), and (c) of 8.3.5 is equivalent to $\hat{M}$ being a (balanced) big Cohen–Macaulay module.

8.5.8. Let R be a Noetherian local ring, and M a balanced big Cohen–Macaulay module over R. One sets $\operatorname{supp} M = \{\mathfrak{p} \in \operatorname{Spec} R : M_{\mathfrak{p}}/\mathfrak{p}M_{\mathfrak{p}} \neq 0\}$. Show
(a) $M_{\mathfrak{p}}$ is a big Cohen–Macaulay module for $R_{\mathfrak{p}}$ if and only if $\mathfrak{p} \in \operatorname{supp} M$,
(b) one has $\operatorname{height} \mathfrak{p} + \dim R_{\mathfrak{p}} = \dim R$ for every $\mathfrak{p} \in \operatorname{supp} M$.
(For general M one uses 8.5.9(c) to define $\operatorname{supp} M$; see Foxby [116].)

8.5.9. Let R and M be as in 8.5.8. Then verify that the following are equivalent for $\mathfrak{p} \in \operatorname{Spec} R$:
(a) $\mathfrak{p} \in \operatorname{supp} M$;
(b) there exists an M-sequence $a_1, \ldots, a_r$ with $\mathfrak{p} \in \operatorname{Ass}(M/(a_1, \ldots, a_r)M)$;
(c) there exists i with $\mu_i(\mathfrak{p}, M) \neq 0$;
(d) there exists i with $H^i_{\mathfrak{p}R_{\mathfrak{p}}}(M_{\mathfrak{p}}) \neq 0$;
(e) $H^i_{\mathfrak{p}R_{\mathfrak{p}}}(M_{\mathfrak{p}}) = 0$ for $i = 0, \ldots, \operatorname{height} \mathfrak{p} - 1$ and $H^h_{\mathfrak{p}R_{\mathfrak{p}}}(M_{\mathfrak{p}}) \neq 0$ for $h = \operatorname{height} \mathfrak{p}$.

Hint: (b) $\Rightarrow$ (c): localize and consider $i = r$; (c) $\Rightarrow$ (d): use 3.5.12; (d) $\Rightarrow$ (e): by hypothesis on M there exists an M-sequence $a_1, \ldots, a_h$ in $\mathfrak{p}$; (e) $\Rightarrow$ (a): this holds for arbitrary M by 3.5.6.

Notes

The results in this chapter on the existence of big Cohen–Macaulay modules are entirely due to Hochster, as well as the method of their construction. With one exception we have followed closely Hochster's original treatment in [178] and his influential lecture notes [181]. The exception is the existence of 'amiable' systems of parameters, for which Hochster avoids local cohomology. The results 8.1.3, 8.1.4, and 8.1.5 are due to Schenzel [326], [329]. The proof of Roberts' theorem 8.1.2 is a slight variation of his original argument [309] introduced by Schenzel [328] in order to obtain a somewhat more general result.

There have been suggestions for modifying Hochster's methods. Bartijn and Strooker [364] constructed a 'pre–Cohen–Macaulay' module by 'monomial modifications', and showed that the $\mathfrak{m}$-adic completion of such a module is a balanced big Cohen–Macaulay module. Our proof of the existence of such modules is a variant of their arguments, whereas the first construction was given by Hochster in [180] (based on an extension of the 'modification method'). Griffith's work was mentioned in 8.5.4. There are several articles which deal with the properties of balanced big Cohen–Macaulay modules: see Duncan [78], Sharp [342], [343], Zarzuela [398] and the literature quoted in these papers. Theorem 8.5.6 and Exercises 8.5.8 and 8.5.9 have been taken from Sharp [342], who coined the notion 'balanced' in that article.

A very interesting revision of Hochster's arguments is due to van den Dries who introduced methods of model theory to our subject; see Chapter 12 of Strooker [364]. A completely different construction in characteristic zero was given by Roberts [310] who derived the existence of a 'Cohen–Macaulay complex' from resolution of singularities and the Grauert–Riemenschneider vanishing theorem. We refer the reader to Hochster and Huneke [197] for a more extensive list of properties that have a regular equational presentation.

It is still open whether there exist big Cohen–Macaulay modules for local rings of mixed characteristic. The most intensive attempts towards their construction can be found in Hochster's article [180].

We saw in 2.1.14 that one cannot expect a finite maximal Cohen–Macaulay module for every local ring. Of course a local ring of dimension 1 has such a module, and also a complete local ring of dimension 2 (2.1.20 and 2.2.31). It is an open question whether there exist finite maximal Cohen–Macaulay modules for all complete local rings. A very special positive result in dimensions > 2 is due to Hartshorne, Peskine, and Szpiro; see [180], (5.12).

Hochster's article [180] also contains a discussion of the question whether there exist big Cohen–Macaulay algebras. For positive characteristics Hochster and Huneke [193] have answered this question: if R is excellent, then the integral closure of $R/\mathfrak{p}$ where $\mathfrak{p}$ is a prime ideal with $\dim R/\mathfrak{p} = \dim R$ in an algebraic closure of its field of fractions is a big Cohen–Macaulay module for R (see also Remark 10.1.6).

The material on the Frobenius functor has been taken from Peskine and Szpiro's ingenious thesis [297]. Theorem 8.2.7 is just the first in their long series of surprising results, many of which will be dealt with in Chapter 9. Kunz's characterization [245] of regular local rings of characteristic p and Herzog's converse [157] of 8.2.7 were mentioned in 8.2.11. More recently, the Frobenius functor was investigated by Dutta [79], [83] and Seibert [331].

9 Homological theorems

This chapter is devoted to the consequences of the existence of big Cohen–Macaulay modules for local rings containing a field. Among the theorems covered, the reader will find Hochster's direct summand theorem for regular local rings, his canonical element theorem, the Peskine–Szpiro intersection theorem and its extensions, the theorem of Evans and Griffith on ranks of syzygy modules, and, finally, bounds for the Bass numbers of modules. These bounds entail surprising characterizations of Cohen–Macaulay and Gorenstein local rings.

There exist derivations of all the theorems in this chapter avoiding big Cohen–Macaulay modules; most of them will only be outlined briefly. They were found in attempts to prove the theorems in mixed characteristic. With the main exception of Roberts' new intersection theorem (whose proof in mixed characteristic requires methods beyond the scope of this book) these efforts have not yet succeeded.

9.1 Grade and acyclicity

The fundamental argument in Sections 9.4–9.6 is that certain finite free complexes become exact when tensored with a balanced big Cohen–Macaulay module. This section contains the acyclicity criterion on which our treatment is based.

Let R be a Noetherian ring,

$$F_{\bullet} : 0 \longrightarrow F_s \overset{\varphi_s}{\longrightarrow} F_{s-1} \longrightarrow \cdots \longrightarrow F_1 \overset{\varphi_1}{\longrightarrow} F_0 \longrightarrow 0$$

a complex of finite free R-modules, and M an R-module. We saw in 1.4.13 that $F_{\bullet}$ is acyclic if and only if $\operatorname{grade} I_{r_i}(\varphi_i) \geq i$ for $i = 1, \dots, s$. Here $r_i = \sum_{j=i}^{s} (-1)^{j-i} \operatorname{rank} F_j$ is the expected rank of φ_i. Now we want to develop a more general criterion by which one can decide whether $F_{\bullet} \otimes M$ is acyclic for a given R-module M. As we shall see, the condition $\operatorname{grade} I_{r_i}(\varphi_i) \geq i$ is just to be replaced by $\operatorname{grade}(I_{r_i}(\varphi_i), M) \geq i$. It will be crucial that we can use the general criterion for a balanced big Cohen–Macaulay module M. Therefore we must first introduce a concept of grade which does not exclude non-finite modules.

Definition 9.1.1. Let R be a ring, I an ideal generated by $x = x_1, \dots, x_n$, and M an R-module. If all the Koszul homology modules $H_i(x, M)$

348

vanish, then we set $\mathrm{grade}(I, M) = \infty$; otherwise $\mathrm{grade}(I, M) = n - h$ where $h = \sup\{i : H_i(x, M) \neq 0\}$.

Note that by 1.6.22 $\mathrm{grade}(I, M)$ is well defined: it does not depend on the choice of x. Furthermore, 1.6.17 shows that for a finite module M over a Noetherian ring R the definition of grade is consistent with that in Chapter 1. There is not much point in considering non-finite ideals I; for completeness let us define $\mathrm{grade}(I, M)$ to be the supremum of $\mathrm{grade}(I', M)$ where I' ranges over the finitely generated subideals of I. (This makes sense because grade is monotone with respect to inclusion of (finite) ideals; see 9.1.2.) On the other hand, there is no reason to restrict ourselves to Noetherian rings, as we shall see below.

Proposition 9.1.2. *Let R be a ring, I a finite ideal, and M an R-module. Then*

(a) $\mathrm{grade}(I, M) = 0 \iff \mathrm{Hom}_R(R/I, M) \neq 0 \iff \{z \in M : Iz = 0\} \neq 0$;

(b) *if* $y = y_1, \ldots, y_m$ *is a weak M-sequence in I, then* $\mathrm{grade}(I, M) \geq m$, *and* $\mathrm{grade}(I, M/yM) = \mathrm{grade}(I/(y), M/yM) = \mathrm{grade}(I, M) - m$;

(c) *if* $R \to S$ *is a flat ring homomorphism, then* $\mathrm{grade}(IS, M \otimes S) \geq \mathrm{grade}(I, M)$; *in particular* $\mathrm{grade}(I_\mathfrak{p}, M_\mathfrak{p}) \geq \mathrm{grade}(I, M)$ *for* $\mathfrak{p} \in \mathrm{Spec}\, R$;

(d) *if* $R \to S$ *is faithfully flat, then* $\mathrm{grade}(IS, M \otimes S) = \mathrm{grade}(I, M)$;

(e) *if* $0 \to U \to M \to N \to 0$ *is an exact sequence of R-modules, then*

$$\mathrm{grade}(I, M) \geq \min\{\mathrm{grade}(I, U), \mathrm{grade}(I, N)\},$$

$$\mathrm{grade}(I, U) \geq \min\{\mathrm{grade}(I, M), \mathrm{grade}(I, N) + 1\},$$

$$\mathrm{grade}(I, N) \geq \min\{\mathrm{grade}(I, U) - 1, \mathrm{grade}(I, M)\};$$

(f) *if* $J \supset I$ *is finite, then* $\mathrm{grade}(J, M) \geq \mathrm{grade}(I, M)$;

(g) *if* S *is a subring of R containing a system of generators x of I, then* $\mathrm{grade}(xS, M) = \mathrm{grade}(I, M)$.

PROOF. (a) By virtue of 1.6.16 one has $H_n(x, M) = \mathrm{Hom}_R(R/I, M)$ for every system of generators $x = x_1, \ldots, x_n$ of I.

(b) The inequality $\mathrm{grade}(I, M) \geq m$ follows immediately from 1.6.16.

Let $^-$ denote residue classes modulo (y). We have an isomorphism $K_\bullet(x) \otimes_R \bar{M} \cong K_\bullet(x) \otimes_R \bar{R} \otimes_{\bar{R}} \bar{M} \cong K_\bullet(\bar{x}) \otimes_{\bar{R}} \bar{M}$; see 1.6.7. This shows $\mathrm{grade}(I, \bar{M}) = \mathrm{grade}(\bar{I}, \bar{M})$. Now we extend y by a sequence z to a system of generators of I. Then $\mathrm{grade}((z), \bar{M}) = \mathrm{grade}(I, M) - m$ by 1.6.13, and $\mathrm{grade}((z), \bar{M}) = \mathrm{grade}((\bar{z}), \bar{M}) = \mathrm{grade}(\bar{I}, \bar{M})$ follows as above.

(c) and (d) are immediate consequences of 1.6.7.

(e) One argues as in the proof of 1.2.9, but uses the exact sequence 1.6.11 of Koszul homology rather than that of Ext. (To carry the analogy one step further, one could work with Koszul cohomology; see 1.6.10.)

(f) It is enough to consider the case in which $J = I + (y)$. Let $I = (x)$ and compare $H_\bullet(x, M)$ and $H_\bullet((x, y), M)$ via 1.6.13.

(g) Let x_S denote x as a sequence in S. By 1.6.7 one has $K_{\bullet}(x_S) \otimes_S M \cong$ $K_{\bullet}(x_S) \otimes_S R \otimes_R M \cong K_{\bullet}(x) \otimes_R M$. □

Part (g) of 9.1.2 explains why the computation of grade can always be reduced to a situation in which R is Noetherian: one simply replaces R by the $\mathbb{Z}$-subalgebra generated by a system of generators of I.

For inductive proofs one must be able to decrease grade by passing to residue classes modulo a regular element. In general, one cannot find such an element in an ideal of positive grade, but one need not go very far. For simplicity we write $I[X]$ for $IR[X]$ and $M[X]$ for $M \otimes R[X]$.

Proposition 9.1.3. *Let R be a ring, I and J finite ideals, and M an R-module.*
(a) *Suppose* grade$(I, M) \geq 1$. *Then $I[X]$ contains an $M[X]$-regular element.*
(b) *One has* grade$(IJ, M) = \min(\text{grade}(I, M), \text{grade}(J, M))$.

PROOF. (a) We may replace R by a Noetherian subring. Let $x_1, \ldots, x_n$ generate I, and set $y = x_1 + x_2 X + \cdots + x_n X^{n-1}$. If y is a zero-divisor, then it is contained in an associated prime of $M[X]$, whether M is finite or not. But $M[X]$ is a graded module over the graded ring $R[X]$, so by 1.5.6 y annihilates a non-zero homogeneous element of $M[X]$ which necessarily has the form $X^p z$, $z \in M$, $z \neq 0$. It follows that $Iz = 0$ which contradicts our hypothesis.

(b) We go by induction on grade(IJ, M). If grade$(IJ, M) = \infty$, then the assertion follows from 9.1.2(f). If grade$(IJ, M) = 0$, then grade$(I, M) = 0$ or grade$(J, M) = 0$ by 9.1.2(a). In the other case we may first adjoin an indeterminate because of 9.1.2(d). Then IJ contains an M-regular element y by (a); now one replaces all data by their residue classes modulo y, and applies the induction hypothesis in conjunction with 9.1.2(b). □

It remains to add a proposition which describes the special properties of grade over Noetherian rings R; for these it makes sense to introduce the notation

$$\text{depth } M_{\mathfrak{p}} = \text{grade}(\mathfrak{p}R_{\mathfrak{p}}, M_{\mathfrak{p}}).$$

Proposition 9.1.4. *Let R be a Noetherian ring, I an ideal in R, and M an R-module. Then*
(a) grade$(I, M) = 0$ *if and only if there exists $\mathfrak{p} \in \text{Ass } M$ with $I \subset \mathfrak{p}$,*
(b) $\mathfrak{p} \in \text{Ass } M \Longleftrightarrow \text{depth } M_{\mathfrak{p}} = 0$,
(c) grade$(I, M) = \inf\{\text{depth } R_{\mathfrak{p}} : \mathfrak{p} \in V(I)\}$.

PROOF. (a) One has grade$(I, M) = 0$ exactly when there exists a non-zero $x \in M$ such that $Ix = 0$. Over a Noetherian ring, I must be contained in an associated prime ideal of M.

(b) Because of (a), depth $R_{\mathfrak{p}} = 0$ is equivalent with $\mathfrak{p}R_{\mathfrak{p}} \in \text{Ass } M_{\mathfrak{p}}$, and this holds if and only if $\mathfrak{p} \in \text{Ass } M$. (One only needs that $\mathfrak{p}$ is finite.)

(c) One has grade$(I, M) \leq \text{depth } M_{\mathfrak{p}}$ for all $\mathfrak{p} \in V(I)$. So (c) is trivial when grade$(I, M) = \infty$. Suppose that in the case of finite grade we have found $\mathfrak{P} \in V(I[X])$ such that depth $M[X]_{\mathfrak{P}} = \text{grade}(I[X], M[X])$, and set $\mathfrak{p} = R \cap \mathfrak{P}$. Then

$$\text{depth } M_{\mathfrak{p}} = \text{grade}(\mathfrak{p}R_{\mathfrak{p}}, M_{\mathfrak{p}}) = \text{grade}(\mathfrak{p}R[X]_{\mathfrak{P}}, M[X]_{\mathfrak{P}}) \leq \text{depth } M[X]_{\mathfrak{P}}.$$

Together with $\text{grade}(I[X], M[X]) = \text{grade}(I, M)$ this yields $\text{grade}(I, M) = \text{depth } R_{\mathfrak{p}}$, and thus the assertion.

In order to find $\mathfrak{P}$ one proceeds by induction, using the fact that $I[X]$ contains an M-regular element if $\text{grade}(I, M) > 0$; the case of grade zero is covered by (a) and (b). $\qquad\square$

Let $\varphi : F \to G$ be a homomorphism of finite free R-modules, and $M \neq 0$ an R-module. We say the φ has rank r with respect to M, if $\text{grade}(I_r(\varphi), M) \geq 1$, whereas $I_{r+1}(\varphi)M = 0$. We write $\text{rank}(\varphi, M) = r$. Note that $\text{rank}(\varphi, M)$ may not be defined. Furthermore, $\text{rank}(\varphi, R) = \text{rank } \varphi$ by 1.4.11. For systematic reasons one sets $\text{rank}(\varphi, M) = 0$ when M is the zero module.

Proposition 9.1.5. *Let R be a ring, $M \neq 0$ an R-module, and $F_{\bullet} : 0 \to F_s \to F_{s-1} \to \cdots \to F_1 \to F_0 \to 0$ a complex of finite free R-modules such that $F_{\bullet} \otimes M$ is acyclic. Let φ_i denote the map $F_i \to F_{i-1}$. Then $\text{rank}(\varphi_i, M)$ is the expected rank r_i of φ_i for $i = 1, \ldots, s$: $\text{rank}(\varphi_i, M) = \sum_{j=i}^{s}(-1)^{j-i} \text{rank } F_j$.*

PROOF. We choose bases of the free modules, and matrices A_i representing the homomorphisms φ_i. Let S be the $\mathbb{Z}$-subalgebra generated by the entries of all these matrices. They define a complex $F'_{\bullet}$ of finite free S-modules such that $F'_{\bullet} \otimes_S R = F_{\bullet}$. Therefore $F'_{\bullet} \otimes_S M = F_{\bullet} \otimes_R M$ is acyclic. The ring S is Noetherian. For $\mathfrak{p} \in \text{Ass}_S M$ the complex $F'_{\bullet} \otimes S_{\mathfrak{p}}$ is split exact by 1.4.12 which furthermore implies that $I_{r_i}(A_i)_{\mathfrak{p}} = S_{\mathfrak{p}}$ and $I_{r_i+1}(A_i)_{\mathfrak{p}} = 0$.

Since S is Noetherian, one has $\text{grade}(I_{r_i}(A_i), M) \geq 1$ by 9.1.4. Let $I = I_{r_i+1}(A_i)$. Assume $IM \neq 0$, and choose $z \in M$ such that $Iz \neq 0$. Then $\text{Ass } Iz \neq \emptyset$. For $\mathfrak{p} \in \text{Ass } Iz$ one has $(Iz)_{\mathfrak{p}} \neq 0$, and hence $IM_{\mathfrak{p}} \neq 0$, which is a contradiction: since $\text{Ass } Iz \subset \text{Ass } M$, one even has $I_{\mathfrak{p}} = 0$ as seen above. It follows that $\text{rank}(A_i, M) = r_i$.

By the definition of rank and 9.1.2 it is irrelevant for $\text{rank}(A_i, M)$ whether one considers A_i as a matrix over S or R. $\qquad\square$

Theorem 9.1.6 (Buchsbaum–Eisenbud; Northcott). *Let R be a ring, M an R-module, and*

$$F_{\bullet} : 0 \longrightarrow F_s \xrightarrow{\varphi_s} F_{s-1} \longrightarrow \cdots \longrightarrow F_1 \xrightarrow{\varphi_1} F_0 \longrightarrow 0$$

a complex of finite free R-modules. Let r_i be the expected rank of φ_i. Then the following are equivalent:
(a) $F_{\bullet} \otimes M$ is acyclic;
(b) $\text{grade}(I_{r_i}(\varphi_i), M) \geq i$ for $i = 1, \ldots, s$.

The remark about r_i that follows 1.4.13 applies here, too: each of (a) and (b) implies that $r_i \geq 0$ for $i = 1, \ldots, s$.

PROOF. As in the proof of 9.1.5 one reduces the theorem to the case in which R is Noetherian. Then the proof is *mutatis mutandis* the same as that of 1.4.13. We indicate some of the modifications. There is nothing to prove if $M = 0$, so assume that M is non-zero.

For (a) $\Rightarrow$ (b) one uses 9.1.5 to get $\text{grade}(I_{r_i}(\varphi_i), M) \geq 1$. Then one adjoins an indeterminate, which affects neither the acyclicity of the complex nor the grades under consideration. By virtue of 9.1.3 one finds an M-regular element in the intersection of the ideals $I_{r_i}(\varphi_i)$, and completes the proof of (a) $\Rightarrow$ (b) as in the case of 1.4.13. It is not necessary to pass from R to $R/(x)$; instead one substitutes M/xM for M in order to apply the induction hypothesis. If $xM = M$, then $\text{grade}((x), M) = \text{grade}(I_{r_i}(\varphi_i), M) = \infty$ for all i.

For (b) $\Rightarrow$ (a) one sets $M_i = \text{Coker}\,\varphi_{i+1} \otimes M$, and replaces depth $R_\mathfrak{p}$ by depth $M_\mathfrak{p}$, and F_i by $F_i \otimes M$. That $(M_i)_\mathfrak{p}$ is free for depth $R_\mathfrak{p} \leq i$, must be modified to '$(M_i)_\mathfrak{p}$ is a direct sum of finitely many copies of $M_\mathfrak{p}$ if depth $M_\mathfrak{p} \leq i$'. $\qquad\square$

We introduce a new invariant of a complex and provide a lemma which is fundamental for the results in Sections 9.4–9.6. (Recall that $\text{codim}\, I = \dim R - \dim R/I$ for an ideal I in a local ring R.)

Definition 9.1.7. Let $(R, \mathfrak{m})$ be a Noetherian local ring,

$$F_\bullet : 0 \longrightarrow F_s \xrightarrow{\varphi_s} F_{s-1} \longrightarrow \cdots \longrightarrow F_1 \xrightarrow{\varphi_1} F_0 \longrightarrow 0$$

a complex of finite free R-modules, and r_i the expected rank of φ_i. We define the *codimension of $F_\bullet$* by

$$\text{codim}\, F_\bullet = \inf\{\text{codim}\, I_{r_i}(\varphi_i) - i : \ i = 1, \ldots, s\}.$$

If $F_\bullet$ is acyclic, then $\text{codim}\, F_\bullet \geq 0$ by the Buchsbaum–Eisenbud acyclicity criterion 1.4.13 (or 9.1.6) since $\text{grade}\, I \leq \text{codim}\, I$ for all ideals. Conversely, if $\text{codim}\, F_\bullet \geq 0$, then $F_\bullet$ need not be acyclic, but $F_\bullet \otimes M$ is acyclic for a balanced big Cohen–Macaulay module M:

Lemma 9.1.8. *Let* $(R, \mathfrak{m})$ *be a Noetherian local ring, and* $F_\bullet$ *a complex of finite free R-modules as above. Suppose that* $\text{codim}\, F_\bullet \geq 0$. *Then* $F_\bullet \otimes M$ *is acyclic for every balanced big Cohen–Macaulay module M.*

PROOF. In view of 9.1.6 it is enough that $\text{grade}(I_{r_i}(\varphi_i), M) \geq i$ for $i = 1, \ldots, s$. In fact, if I is an ideal with $\text{codim}\, I \geq i$, then it contains a sequence $x_1, \ldots, x_i$ which is part of a system of parameters, as is easily shown by induction on i. Such a sequence is M-regular. $\qquad\square$

In Section 9.5 we shall investigate lower bounds for the numbers r_i. A first result in this direction can be recorded already. Let $(R, \mathfrak{m})$ be a local ring. We say that a complex of finite free R-modules $F.$ as in 9.1.7 is *minimal of length* s if $F_s \neq 0$ and $\varphi_i(F_i) \subset \mathfrak{m}F_{i-1}$ for all i. Considering minimal complexes only is not a severe restriction since every complex of finite free modules over a local ring decomposes into a direct sum of a minimal such complex and a split exact one.

Proposition 9.1.9. *Let* $(R, \mathfrak{m}, k)$ *be a local ring, and* $F.$ *a length* s *minimal complex of finite free R-modules as above. Suppose there exists an R-module M such that $M \neq \mathfrak{m}M$ and $F. \otimes M$ is acyclic. Let r_i denote the expected rank of φ_i. Then $r_i \geq 1$ for $i = 1, \ldots, s$.*

PROOF. One has $r_s = \operatorname{rank} F_s \geq 1$ by hypothesis, and it follows from Proposition 9.1.5 that $r_i = \operatorname{rank}(\varphi_i, M) \geq 0$ for all i. Arguing inductively, we must only show $r_1 = 0$ implies $r_2 = 0$.

If $r_1 = \operatorname{rank}(\varphi_1, M) = 0$, then $I_1(\varphi_1)M = 0$, and so $\varphi_1 \otimes M = 0$. Therefore we have an exact sequence

$$F_2 \otimes M \xrightarrow{\varphi_2 \otimes M} F_1 \otimes M \longrightarrow 0.$$

Consequently $F_2 \otimes M \otimes k \to F_1 \otimes M \otimes k \to 0$ is also exact. By hypothesis $M \neq \mathfrak{m}M$, equivalently, $M \otimes k$ is a non-zero k-vector space. Thus the sequence $F_2 \otimes k \to F_1 \otimes k \to 0$ must be exact. On the other hand, $\varphi_2 \otimes k = 0$ since $\varphi_2(F_2) \subset \mathfrak{m}F_1$. Hence we get $F_1 = 0$, and $r_2 = \operatorname{rank} F_1 - r_1 = 0$. $\square$

Exercises

9.1.10. Let R be a ring, I a finitely generated ideal, and M an R-module. Furthermore let R_∞ be a polynomial ring over R in an infinite number of indeterminates, $I_\infty = IR_\infty$, and $M_\infty = M \otimes R_\infty$. Prove the following:
(a) If $\operatorname{grade}(I, M) < \infty$, then every maximal weak M_∞-sequence in I_∞ has length equal to $\operatorname{grade}(I, M)$.
(b) One has $\operatorname{grade}(I, M) = \infty$ if and only if I_∞ contains an infinite weak M_∞-sequence.
(c) One has $\operatorname{grade}(I, M) = \inf\{i : \operatorname{Ext}^i_{R_\infty}(R_\infty/I_\infty, M_\infty) \neq 0\}$.
(d) For Noetherian R one has $\operatorname{grade}(I, M) = \inf\{i : \operatorname{Ext}^i_R(R/I, M) \neq 0\}$.
(e) Suppose that the number of associated prime ideals of $M/(x)M$ is finite for every weak M-sequence x. (For example, this holds when M is a balanced big Cohen–Macaulay module over a Noetherian local ring; see 8.5.6). Then one can drop the subscript ∞ in (a), (b), and (c).

9.1.11. For a finite module M over a Noetherian ring R we have $\operatorname{grade}(I, M) = \infty$ if $IM = M$. For non-finite M this may be false. Find an example.

9.1.12. Let $(R, \mathfrak{m})$ be a Noetherian local ring, and M an R-module. Prove
(a) if $M \neq \mathfrak{m}M$, then depth M is finite,

(b) if depth M is finite, then depth $M \leq \dim R$,

(c) depth $M = \inf\{i : H^i_m(M) \neq 0\}$,

(d) depth $M = \dim R \Longleftrightarrow \hat{M}$ is a (balanced) big Cohen–Macaulay module.

9.1.13. Sometimes it may be more natural to work with homology modules rather than the ideals $I_{r_i}(\varphi_i)$. Therefore it is worth while reformulating the crucial condition for acyclicity. One must however use the homology of $F_\bullet^* = \operatorname{Hom}_R(F_\bullet, R)$. With the notation of 9.1.6, show the following are equivalent:

(a) $\operatorname{grade}(I_{r_i}(\varphi_i), M) \geq i$ for $i = 1, \ldots, s$;

(b) $\operatorname{grade}(\operatorname{Ann} H^i(F_\bullet^*), M) \geq i$ for $i = 1, \ldots, s$.

9.1.14. Generalize the 'lemme d'acyclicité' 1.4.24 to the case of arbitrary R-modules L_i.

9.1.15. Let R be a Noetherian local ring and M a balanced big Cohen–Macaulay module. Prove $\operatorname{Tor}^R_i(N, M) = 0$ for all finite R-modules N and $i > 0$. In particular, M is faithfully flat if R is regular.

9.2 Regular rings as direct summands

Let R and S be Noetherian local rings such that $R \subset S$ and S is a finite R-module. Suppose that R is regular and, for the moment, S is a Cohen–Macaulay ring. Since every system of parameters of R is a system of parameters of S, the R-module S, having finite projective dimension, must be free. Furthermore the element $1 \in R$ is part of an R-basis of S, and it follows that R is a direct summand of S as an R-module. Quite surprisingly this holds true regardless of the Cohen–Macaulay property of S, at least when S contains a field.

The argument above uses the fact that a system of parameters of R is an S-sequence. As we shall see, a much weaker property suffices. It is given by the following 'monomial theorem':

Theorem 9.2.1. *Let S be a Noetherian local ring containing a field. Then for every system $x = x_1, \ldots, x_n$ of parameters and all $t \geq 0$ one has $x_1^t \cdots x_n^t \notin (x^{t+1})$.*

PROOF. By 8.4.2 there exists an x-regular module M. Suppose that $x_1^t \cdots x_n^t \in (x^{t+1})$. Then $x_1^t \cdots x_n^t M \subset x^{t+1} M$. The associated graded module $\operatorname{gr}_{(x)} M$ is an $(R/(x)[X_1, \ldots, X_n])$-module in a natural way, and, *a fortiori*, $X_1^t \cdots X_n^t \operatorname{gr}_{(x)} M \subset (X_1^{t+1}, \ldots, X_n^{t+1}) \operatorname{gr}_{(x)} M$.

On the other hand, since x is an M-sequence, the associated graded module $\operatorname{gr}_{(x)} M$ is isomorphic to $M \otimes R/(x)[X_1, \ldots, X_n]$ (see 1.1.8). Therefore

$$(\operatorname{gr}_{(x)} M)/(X_1^{t+1}, \ldots, X_n^{t+1}) \operatorname{gr}_{(x)} M \cong \bigoplus X_1^{e_1} \cdots X_n^{e_n}(M/xM)$$

as an R-module where the direct sum is taken over all monomials

$$X_1^{e_1} \cdots X_n^{e_n} \notin (X_1^{t+1}, \ldots, X_n^{t+1}).$$

This is a contradiction since $X_1^t \cdots X_n^t \notin (X_1^{t+1}, \ldots, X_n^{t+1})$. $\square$

The proof of 9.2.1 shows much more than stated in the theorem: let I_0, J_0 be ideals in $\mathbb{Z}[X_1, \ldots, X_n]$ generated by monomials, and I_1, J_1 the ideals generated by the corresponding monomials in $x_1, \ldots, x_n$; then $I_1 \subset J_1 \Longleftrightarrow I_0 \subset J_0$.

Suppose that $S = R \oplus C$ as an R-module. Then for every ideal $I \subset R$ one has $IS = I \oplus IC$, hence $IS \cap R = I$. Let $x = x_1, \ldots, x_n$ be a system of parameters of R. If R is regular, then, as the proof of 9.2.1 shows, $x_1^t \cdots x_n^t \notin (x^{t+1})$; so $x_1^t \cdots x_n^t \notin x^{t+1}S$, for otherwise $x_1^t \cdots x_n^t \in (x^{t+1})S \cap R = (x^{t+1})$. This simple observation proves the easy part of the following lemma.

Lemma 9.2.2. *Let* $(R, \mathfrak{m})$ *be a regular local ring and* $x = x_1, \ldots, x_n$ *a regular system of parameters. Suppose that* $S \supset R$ *is an* R-*algebra which is finite as an* R-*module. Then* R *is a direct* R-*summand of* S *if and only if* $x_1^t \cdots x_n^t \notin x^{t+1}S$ *for every* $t \geq 0$.

PROOF. Since the $\mathfrak{m}$-adic completion $\hat{R}$ is a faithfully flat extension of R, the same holds true for the extension $S \otimes \hat{R}$ of S. Thus $x_1^t \cdots x_n^t \notin x^{t+1}S$ implies that $x_1^t \cdots x_n^t \notin x^{t+1}S \otimes \hat{R}$.

Suppose that the implication still open holds under the additional assumption that the regular local ring is complete. Then $\hat{R}$ is a direct $\hat{R}$-summand of $S \otimes \hat{R}$ and the natural homomorphism (given by restriction of maps)

$$\hat{\rho}: \operatorname{Hom}_{\hat{R}}(S \otimes \hat{R}, \hat{R}) \longrightarrow \operatorname{Hom}_{\hat{R}}(\hat{R}, \hat{R})$$

is surjective. Since S is a finitely presented R-module, one has a natural commutative diagram

$$
\begin{array}{ccc}
\operatorname{Hom}_R(S, R) \otimes \hat{R} & \xrightarrow{\ \rho \otimes \hat{R}\ } & \operatorname{Hom}_R(R, R) \otimes \hat{R} \\
\| & & \| \\
\operatorname{Hom}_{\hat{R}}(S \otimes \hat{R}, \hat{R}) & \xrightarrow{\ \hat{\rho}\ } & \operatorname{Hom}_{\hat{R}}(\hat{R}, \hat{R})
\end{array}
$$

where $\rho: \operatorname{Hom}_R(S, R) \to \operatorname{Hom}_R(R, R)$ is again given by restriction. Since $\rho \otimes \hat{R}$ is surjective, ρ itself must be surjective: the identity map on R can be extended to an R-homomorphism $S \to R$, so R is a direct R-summand of S.

After these preparations we may assume that R is complete. Let $R_t = R/(x^t)$, and $S_t = S/x^t S$; R_t is a Gorenstein ring of dimension zero. Since $x_1^{t-1} \cdots x_n^{t-1} \notin x^t$, but $\mathfrak{m}x_1^{t-1} \cdots x_n^{t-1} \subset x^t$, the residue class of $x_1^{t-1} \cdots x_n^{t-1}$ generates Soc R_t. Therefore each of the induced maps $\varphi_t: R_t \to S_t$ is injective; otherwise its kernel would contain Soc R_t, whence $x_1^{t-1} \cdots x_n^{t-1} \in x^t S$, contradicting the hypothesis of the lemma. Furthermore R_t is an injective R_t-module. Thus each of the maps φ_t splits: there is an R_t-homomorphism $\psi_t: S_t \to R_t$ such that $\psi_t \circ \varphi_t = \operatorname{id}_{R_t}$.

The ideals (x^t) form a system cofinal with that of the powers of $\mathfrak{m}$. Since R is $\mathfrak{m}$-adically complete, one has

$$\operatorname{Hom}_R(S, R) = \operatorname{Hom}_R(S, \varprojlim R_t) = \varprojlim \operatorname{Hom}_R(S, R_t) = \varprojlim \operatorname{Hom}_{R_t}(S_t, R_t).$$

In the latter inverse system the map $\pi_{ij} : \operatorname{Hom}_{R_i}(S_i, R_i) \to \operatorname{Hom}_{R_j}(S_j, R_j)$ associates to each homomorphism $S_i \to R_i$ the induced map $S_j \to R_j$.

We have to find homomorphisms $\psi_t : S_t \to R_t$ such that (i) $\pi_{ij}(\psi_i) = \psi_j$, and (ii) $\psi_t \circ \varphi_t = \operatorname{id}_{R_t}$. Because of (i) we then obtain a homomorphism $\varprojlim \psi_t = \psi : S \to R$, which by (ii) satisfies $\psi|R = \operatorname{id}_R$: if $\psi(y) \neq y$ for some $y \in R$, then $\psi_t \circ \varphi_t \neq \operatorname{id}_{R_t}$ for every t such that $\psi(y) - y \notin (x^t)$.

Let D_t be the set of homomorphisms $\rho : S_t \to R_t$ for which $\rho \circ \varphi_t = \operatorname{id}_{R_t}$. Obviously the sets $D_t \subset \operatorname{Hom}_{R_t}(S_t, R_t)$ are non-empty and form an inverse system. However, since the maps $\pi_{ij}|_{D_i} : D_i \to D_j$ may not be surjective, we cannot immediately conclude that $\varprojlim D_t \neq \emptyset$, as desired. Instead we define subsets

$$E_t = \bigcap_{i \geq t} \pi_{it}(D_i).$$

Then $E_t = \pi_{it}(E_i)$ for all i with $i \geq t$, and it is enough to show that $E_t \neq \emptyset$ for some, equivalently all, t.

Every D_i is an affine subspace of $\operatorname{Hom}_{R_i}(S_i, R_i)$; that is, it is of the form $\gamma_i + U_i$ with a submodule U_i. Therefore

$$\pi_{tt}(D_t) \supset \pi_{t+1,t}(D_{t+1}) \supset \cdots \supset \pi_{it}(D_i) \supset \cdots$$

is a decreasing chain of non-empty affine subspaces A_j of $\operatorname{Hom}_{R_t}(S_t, R_t)$. Consequently the submodules $M_j = \{\rho - \sigma : \rho, \sigma \in A_j\}$ are non-zero and form a decreasing chain, too. This chain stabilizes in the Artinian module $\operatorname{Hom}_{R_t}(S_t, R_t)$, and so does the chain of affine subspaces A_j. □

A consequence of 9.2.1 and 9.2.2 is the 'direct summand theorem' for regular local rings:

Theorem 9.2.3 (Hochster). *Let R be a regular local ring containing a field, and $S \supset R$ an R-algebra which is a finite R-module. Then R is a direct summand of the R-module S.*

PROOF. As in the proof of the lemma, we may assume that R is complete. Let $\mathfrak{p}$ be a prime ideal of S lying over the zero-ideal of R. If R is a direct R-summand of $S/\mathfrak{p}$, then it is a direct R-summand of S: compose a section of the natural embedding $R \to S/\mathfrak{p}$ with the natural epimorphism $S \to S/\mathfrak{p}$. Being an integral domain which is module-finite over a complete local ring, $S/\mathfrak{p}$ is local itself ([284], (30.5)), and we can invoke 9.2.1 and 9.2.2. □

Remarks 9.2.4. (a) In characteristic zero a much weaker property than regularity is sufficient for the direct summand property of R as described by 9.2.3. Let R be a Noetherian *normal* domain containing a field of characteristic zero, and S a module-finite extension ring. In showing that R is a direct R-summand of S, it is harmless to replace S by any S-algebra T (see the proof of 9.2.3). So we first factor out a prime ideal $\mathfrak{p}$ of S lying over the zero-ideal of R, and may assume that S is a domain. Then we extend the field of fractions of S to a finite normal extension L of the field K of fractions of R, and replace S by the integral closure T of R in L. Let $d = \dim_K L$ and $\mathrm{Tr}: L \to K$ denote the trace map. Then for every $x \in K$ one has $(1/d)\,\mathrm{Tr}\,x = x$, and $\mathrm{Tr}\,y \in R$ for every $y \in T$, since the trace of an integral element is integral and R is integrally closed in K. (We refer the reader to [397], Chapter II for the field theory involved.)

As a consequence one obtains a proof of 9.2.1 avoiding big Cohen–Macaulay modules: if $x_1, \ldots, x_n$ is a system of parameters of S, then there is a regular subring R of $\hat{S}$ in which $x_1, \ldots, x_n$ generate the maximal ideal and over which $\hat{S}$ is finite (see A.22). Since the conclusion of 9.2.1 is invariant under completion, one obtains 9.2.1 in the same way as the implication '$\Rightarrow$' of 9.2.2.

(b) In characteristic p the situation is just inverted: there is a direct proof of 9.2.1. Let $\mathfrak{n}$ be the maximal ideal of S. By 3.5.6,

$$H_{\mathfrak{n}}^n(S) \cong \varinjlim H^n(x^t) \neq 0.$$

One has $H^n(x^t) \cong S/(x^t)$, and the map $S/(x^t) \to S/(x^{t+i})$ is induced by the multiplication by $x_1^i \cdots x_n^i$. Since $H_{\mathfrak{n}}^n(S) \neq 0$, this map must be non-zero for t sufficiently large. Equivalently, $x_1^i \cdots x_n^i \notin (x_1^{t+i}, \ldots, x_n^{t+i})$ for t sufficiently large and $i \geq t$. On the other hand, if $x_1^t \cdots x_n^t \in (x_1^{t+1}, \ldots, x_n^{t+1})$, then one applies the Frobenius homomorphism repeatedly to obtain

$$x_1^{tp^e} \cdots x_n^{tp^e} \in (x_1^{p^e + tp^e}, \ldots, x_n^{p^e + tp^e}),$$

which is a contradiction for e large.

Via 9.2.2 this argument yields an 'elementary' proof of 9.2.3. For still another proof of 9.2.1 in characteristic p see [176], as well as for a counterexample showing that normality is not sufficient in characteristic p for R to have the direct summand property.

9.3 Canonical elements in local cohomology modules

Independently of characteristic, the discussion in 9.2.4(b) shows that 9.2.1 and, hence, 9.2.3 are equivalent to the non-vanishing of certain elements in the local cohomology module $H_{\mathfrak{n}}^n(S)$ (notation as in 9.2.1): one has $x_1^k \cdots x_n^k \notin (x^{k+1})$ for all k if and only if the image of 1 under the map $S \to S/(x) \to \varinjlim S/(x^k) = H^n_{\mathfrak{m}}(S)$ is non-zero. As the example $S = K[[X]]$ and $x = X$ or $x = X^2$ shows, the element thus obtained depends heavily on the choice of the system of parameters; for example, its annihilator varies with x. In the following we shall discuss a theorem which involves a 'canonical element' in a local cohomology module although local cohomology does not appear explicitly:

Theorem 9.3.1. *Let $(R, \mathfrak{m}, k)$ be a Noetherian local ring of dimension n containing a field. Let $F.$ be a free resolution of the residue class field k, and x a system of parameters. If $\gamma : K.(x) \to F.$ is a complex homomorphism extending the natural epimorphism $R/(x) \to k$, then the homomorphism $\gamma_n : K_n(x) \to F_n$ is non-zero.*

PROOF. In order to derive a contradiction we assume that there exists a complex homomorphism γ with $\gamma_n = 0$.

There exists an x-regular module M by 8.4.2. Since $(x) \supset \mathfrak{m}^i$ for i large and $M/xM \neq 0$, we can pick an element $y \in M$ such that $y \notin xM$, but $\mathfrak{m}y \in xM$. The assignment $1 \mapsto y$ then defines a homomorphism $R/\mathfrak{m} \to M/xM$. This homomorphism can be lifted to a complex homomorphism $\beta : F. \to K(x, M)$ since the Koszul complex $K(x, M)$ is acyclic; see 1.6.12. Composition with γ gives a homomorphism $\alpha = \beta \circ \gamma : K(x) \to K(x, M)$ with $\alpha_n = 0$.

The complex homomorphism α extends the homomorphism $\alpha_0 : R \to M$ with $\alpha_0(1) = y$. As $K.(x, M) = K.(x) \otimes M$, one obtains a second such extension by $\alpha' = \mathrm{id}_{K.(x)} \otimes \alpha_0$. The complex $K.(x)$ is projective and $K.(x, M)$ is acyclic; therefore α and α' differ only by a homotopy σ. In particular $\alpha'_n = \alpha'_n - \alpha_n = \sigma_{n-1} \circ \partial_n$:

$$
\begin{array}{ccc}
0 \longrightarrow & K_n(x) & \xrightarrow{\ \partial_n\ } K_{n-1}(x) \\
 & \Big\downarrow{\scriptstyle \alpha'_n} & \\
0 \longrightarrow & K_n(x, M). &
\end{array}
$$

We may identify $K_n(x)$ with R, $K_n(x, M) = K_n(x) \otimes M$ with M and $K_{n-1}(x)$ with R^n. Then $\partial_n(R) \subset xR^n$, and so $y = \alpha'_n(1) = \sigma_{n-1} \circ \partial_n(1) \in xM$, which is a contradiction. $\qquad\square$

Let us fix the data x and $F.$ of the theorem. Complex homomorphisms γ and γ' both extending the epimorphism $R/(x) \to k$ differ by a homotopy σ:

$$
\begin{array}{ccccccc}
0 & \longrightarrow & K_n(x) & \xrightarrow{\ \partial_n\ } & K_{n-1}(x) & \longrightarrow & \cdots \\
 & {\scriptstyle \sigma_n}\nearrow & {\scriptstyle \gamma_n}\Big\downarrow{\scriptstyle \gamma'_n} & {\scriptstyle \sigma_{n-1}}\nearrow & \Big\downarrow & & \\
\cdots \longrightarrow & F_{n+1} & \xrightarrow[\varphi_{n+1}]{} & F_n & \xrightarrow[\varphi_n]{} & F_{n-1} & \longrightarrow \cdots
\end{array}
$$

As above we identify $K_n(x)$ with R; furthermore we consider $N = \text{Ker}\,\varphi_{n-1} = \text{Im}\,\varphi_n$ as the target of φ_n. (The module N is the n-th syzygy of k with respect to the resolution $F_\bullet$.) Then

$$\varphi_n \circ \gamma_n(1) - \varphi_n \circ \gamma_n'(1) = \varphi_n \circ \sigma_{n-1} \circ \partial_n(1).$$

This element belongs to xN, since $\partial_n(1) \subset xK_{n-1}(x)$. So different choices of the complex homomorphism yield the same residue class $(\varphi_n \circ \gamma_n(1))^- \in N/xN$. On the other hand, given a complex homomorphism γ, we may freely choose σ to define γ' by $\gamma' = \gamma + \sigma \circ \partial + \varphi \circ \sigma$. For the possible choices of σ_{n-1}, the elements $\sigma_{n-1} \circ \partial_n(1)$ exhaust xF_n; note that $\partial_n(1) = \pm x_1 e_1 \pm \cdots \pm x_n e_n$ with respect to a suitable basis of R^n. In sum, $\gamma_n \neq 0$ for every choice of γ if and only if $\varphi_n \circ \gamma_n(1) \notin xN$ for a specific choice.

Now consider the systems of parameters x^t, $t > 0$. There is a natural map $K_\bullet(x^t) \to K_\bullet(x)$; it sends $e_{i_1} \wedge \cdots \wedge e_{i_u}$ to $x_{i_1}^{t-1} \cdots x_{i_u}^{t-1} e_{i_1} \wedge \cdots \wedge e_{i_u}$. Composition with $\gamma : K_\bullet(x) \to F_\bullet$ gives a complex homomorphism γ^t with $\gamma_n^t(1) = x_1^{t-1} \cdots x_n^{t-1} \gamma_n(1)$. If all the homomorphisms $\delta : K_\bullet(x^t) \to F_\bullet$ which lift the epimorphism $R/(x^t) \to k$ have $\delta_n \neq 0$, then the arguments above imply

(1) $x_1^{t-1} \cdots x_n^{t-1} \varphi_n \circ \gamma_n(1) \notin x^t N$ for all $t > 0$.

Observe that $H_{\mathfrak{m}}^n(N) = \varinjlim N/x^t N$. So condition (1) is equivalent to the following: the image of $\varphi_n \circ \gamma_n(1)$ under the map $N \to N/xN \to H_{\mathfrak{m}}^n(N)$ is non-zero.

The module $H_{\mathfrak{m}}^n(N)$ can also be represented as $\varinjlim \text{Ext}_R^n(R/\mathfrak{m}^t, N)$ (see 3.5.3). Hence there is a natural homomorphism $\text{Ext}_R^n(k, N) \to H_{\mathfrak{m}}^n(N)$. Moreover, the exact sequence $0 \to N \to F_{n-1} \to \cdots \to F_0 \to k \to 0$ represents an element $\varepsilon(F_\bullet) \in \text{Ext}_R^n(k, N)$ and thus an element $\eta(F_\bullet) \in H_{\mathfrak{m}}^n(N)$. The connection between 'extensions' like the previous exact sequence and Ext^n is discussed in [264], pp. 82–87, or [48], Ch. X, §7; if one writes

$$\text{Ext}_R^n(k, N) \cong \text{Hom}_R(N, N)/\iota^*(\text{Hom}_R(F_{n-1}, N))$$

where $\iota : N \to F_{n-1}$ is the natural embedding, then $\varepsilon(F_\bullet)$ is the residue class of $\pm \text{id}_N$. The elements $\varepsilon(F_\bullet)$ and $\eta(F_\bullet)$ may be called 'canonical' since they depend functorially on $F_\bullet$. In particular the vanishing of $\eta(F_\bullet)$ is independent of $F_\bullet$. Hochster [184] contains a detailed discussion of these facts and a proof of the following crucial statement: the element of $H_{\mathfrak{m}}^n(N)$ constructed from a complex homomorphism $\gamma : K_\bullet(x) \to F_\bullet$ as above can be identified with $\eta(F_\bullet)$. The conclusion of 9.3.1 is therefore equivalent to $\eta(F_\bullet) \neq 0$, which justifies the name 'canonical element theorem' for 9.3.1.

As an application of 9.3.1 we prove a generalization of Krull's principal ideal theorem. (Another one will be given in 9.4.4.) Let M be a module over a commutative ring R, and $x \in M$. Then

$$\mathcal{O}(x) = \{\alpha(x) : \alpha \in \operatorname{Hom}_R(M, R)\}$$

is called the *order ideal* of x.

Every finitely generated ideal is an order ideal: given $x_1, \ldots, x_n \in R$, one sets $M = R^n$ and $x = (x_1, \ldots, x_n)$. Obviously $\mathcal{O}(x) = \sum R x_i$. By Krull's principal ideal theorem $\mathcal{O}(x)$ has height $\leq n$ if R is Noetherian, provided $\mathcal{O}(x)$ is a proper ideal. For $x \in M = R^n$ this condition is equivalent to the existence of a maximal ideal $\mathfrak{m}$ such that $x \in \mathfrak{m}M$. The following theorem generalizes Krull's bound on height $\mathcal{O}(x)$ to arbitrary finite modules. In the general version the number $n = \operatorname{rank} R^n$ must be replaced by

$$\operatorname{big\,rank} M = \max\{\mu(M_\mathfrak{p}) : \mathfrak{p} \in \operatorname{Spec} R \text{ minimal}\}.$$

If M has a rank, then $\operatorname{big\,rank} M = \operatorname{rank} M$.

Theorem 9.3.2 (Eisenbud–Evans). *Let* $(R, \mathfrak{m}, k)$ *be a Noetherian local ring containing a field, and* M *a finite* R-*module. Then* height $\mathcal{O}(x) \leq \operatorname{big\,rank} M$ *for all elements* $x \in \mathfrak{m}M$.

PROOF. There is a prime ideal $\mathfrak{p}$ with height $\mathcal{O}(x) = \operatorname{height}((\mathcal{O}(x) + \mathfrak{p})/\mathfrak{p})$. Let $^-$ denote taking residue classes modulo $\mathfrak{p}$. Every linear form $M \to R$ induces an $\bar{R}$-linear form $\bar{M} \to \bar{R}$; therefore $\mathcal{O}(x)^- \subset \mathcal{O}(\bar{x})$. Suppose the theorem has been proved for $\bar{R}$ and $\bar{M}$. Then

$$\operatorname{height} \mathcal{O}(x) = \operatorname{height} \mathcal{O}(x)^- \leq \operatorname{height} \mathcal{O}(\bar{x}) \leq \operatorname{big\,rank} \bar{M} \leq \operatorname{big\,rank} M.$$

Furthermore note that $\bar{x} \in \bar{\mathfrak{m}}\bar{M}$ if and only if $x \in \mathfrak{m}M$.

As these arguments show, it suffices to treat the case of an integral domain R. Then $\operatorname{big\,rank} M = \operatorname{rank} M$. Let $h = \operatorname{height} \mathcal{O}(x)$ and $n = \dim R$. There exists a system of parameters $x_1, \ldots, x_n$ with $x_1, \ldots, x_h \in \mathcal{O}(x)$. Replacing M by $M \oplus R^{n-h}$ and x by $x \oplus (x_{h+1}, \ldots, x_n)$, we may assume that $\mathcal{O}(x)$ is $\mathfrak{m}$-primary.

As usual, * denotes $\operatorname{Hom}_R(\_, R)$. Choose $\alpha_i \in M^*$ such that $a_1 = \alpha_1(x)$, $\ldots, a_n = \alpha_n(x)$ is a system of parameters. The collection $\alpha_1, \ldots, \alpha_n$ defines a map $\alpha : M \to R^n$ through $\alpha(z) = (\alpha_1(z), \ldots, \alpha_n(z))$. Let $y = y_1, \ldots, y_m$ generate $\mathfrak{m}$. Since $x \in \mathfrak{m}M$, there is a homomorphism $\pi : R^m \to M$ with $\pi(y_1, \ldots, y_m) = x$. Let us put $F = (R^n)^*$ and $G = (R^m)^*$, and define $f : F \to G$ by $f = \pi^* \circ \alpha^*$. Then f 'writes' $a = a_1, \ldots, a_n$ in terms of y; i.e. f makes the diagram

$$
\begin{array}{ccc}
F & \xrightarrow{\ a\ } & R \\
{\scriptstyle f}\downarrow & & \parallel \\
G & \xrightarrow{\ y\ } & R
\end{array}
$$

commute. By 1.6.8 the exterior powers of f yield a complex homomorphism

$$0 \longrightarrow \textstyle\bigwedge^n F \longrightarrow \cdots \longrightarrow \textstyle\bigwedge^2 F \longrightarrow F \overset{a}{\longrightarrow} R \longrightarrow 0$$

$$\Big\downarrow \textstyle\bigwedge^n f \qquad\qquad \Big\downarrow \textstyle\bigwedge^2 f \quad \Big\downarrow f \qquad \Big\|$$

$$\textstyle\bigwedge^{n+1} G \longrightarrow \textstyle\bigwedge^n G \longrightarrow \cdots \longrightarrow \textstyle\bigwedge^2 G \longrightarrow G \overset{y}{\longrightarrow} R \longrightarrow 0$$

of Koszul complexes. By definition f factors through M^*. So rank $f \leq$ rank $M^* =$ rank M. On the other hand, $\bigwedge^n f \neq 0$ because there exists a complex homomorphism β from $K_{\bullet}(y)$ to a free resolution $F_{\bullet}$ of k which extends the identity on k; 9.3.1 guarantees that $\beta_n \circ \bigwedge^n f \neq 0$. Therefore rank $f \geq n$ and, hence, rank $M \geq n$. $\qquad\square$

Remarks 9.3.3. (a) Bruns [54] gave a more elementary proof of 9.3.2 which works for arbitrary local rings.

(b) Formula (1) above shows that the canonical element theorem 9.3.1 implies 9.2.1 and thus the direct summand theorem 9.2.3. Surprisingly one can conversely derive 9.3.1 from 9.2.3 if the residue class field of the local ring under consideration has characteristic $p > 0$; see Hochster [184]. (There seems to be no such derivation in characteristic zero.) Furthermore the main homological theorems like 9.4.3 and 9.5.6 can be derived from 9.3.1. (See 9.4.8 and 9.5.7.)

(c) It is not difficult to reduce 9.3.1 to characteristic p via 8.4.1; see [184]. (Such a reduction will be carried out in detail for 9.4.3.) In connection with (b) and 9.2.4 that yields a proof of 9.3.1 which does not use big Cohen–Macaulay modules.

9.4 Intersection theorems

We have already met an intersection theorem in Section 8.2, the 'new intersection theorem' (for local rings of characteristic p). We now want to prove a very powerful generalization and to derive several consequences, one of which will eventually explain why the results in this section are called 'intersection theorems': it generalizes a variant of Serre's intersection theorem for the spectrum of a regular local ring.

Theorem 9.4.1. *Let $(R, \mathfrak{m}, k)$ be a Noetherian local ring containing a field, and*

$$F_{\bullet} : 0 \longrightarrow F_s \overset{\varphi_s}{\longrightarrow} F_{s-1} \longrightarrow \cdots \longrightarrow F_1 \overset{\varphi_1}{\longrightarrow} F_0 \longrightarrow 0$$

a complex of finite free R-modules such that codim $F_{\bullet} \geq 0$. *Let $C =$ Coker φ_1 and $e \in C$, $e \notin \mathfrak{m}C$. Then* codim$(\text{Ann } e) \leq s$.

PROOF. We use induction on dim$(R/(\text{Ann } e))$. Suppose dim$(R/(\text{Ann } e)) = 0$ first. By 8.5.3 there exists a balanced big Cohen–Macaulay R-module M.

Lemma 9.1.8 implies that $F_\bullet \otimes M$ is acyclic. One has depth $M = \dim R$, and 9.1.2 yields depth$(C \otimes M) \geq \dim R - s$; note that $C \otimes M \cong \text{Coker}(\varphi_1 \otimes M)$. On the other hand, the natural surjection $C \otimes M \to C/\mathfrak{m}C \otimes M/\mathfrak{m}M$ maps $e \otimes M$ onto a module isomorphic to $M/\mathfrak{m}M$. In particular, $e \otimes M \neq 0$. Since $\dim(R/(\text{Ann } e)) = 0$, one has $\mathfrak{m}^p(e \otimes M) = 0$ for some p, whence $\{\mathfrak{m}\} = \text{Ass}(e \otimes M) \subset \text{Ass } C \otimes M$. Therefore depth $C \otimes M = 0$, and so $s \geq \dim R$.

Now suppose that $\dim(R/(\text{Ann } e)) > 0$. There is nothing to prove if $s = \dim R$. So we may assume that $s < \dim R$. Let P be the finite set of prime ideals $\mathfrak{p}$ such that (i) Ann $e \subset \mathfrak{p}$ and codim $\mathfrak{p} = \text{codim}(\text{Ann } e)$, or (ii) there exists i with $I_{r_i}(\varphi_i) \subset \mathfrak{p}$ and codim $\mathfrak{p} = i$. Then $\mathfrak{m} \notin P$; so we can choose $x \in \mathfrak{m}$ such that $x \notin \mathfrak{p}$ for any $\mathfrak{p} \in P$. Let $^-$ denote residue classes modulo x. It is a routine matter to verify that codim $\bar{F}_\bullet \geq 0$. Furthermore $\bar{e} \notin \bar{\mathfrak{m}}\bar{C}$, and $\dim(\bar{R}/(\text{Ann } \bar{e})) < \dim(R/(\text{Ann } e))$. The inductive hypothesis yields codim$(\text{Ann } \bar{e}) \leq s$. Since $\dim R \leq \dim \bar{R} + 1$, we have, as desired,

$$\text{codim}(\text{Ann } e) = \dim R - \dim(R/(\text{Ann } e))$$
$$\leq \dim \bar{R} + 1 - (\dim(\bar{R}/(\text{Ann } \bar{e})) + 1) \leq s. \qquad \square$$

The following corollary is usually called the 'improved new intersection theorem':

Corollary 9.4.2 (Evans–Griffith). *With the notation of* 9.4.1 *suppose that* $F_\bullet \otimes R_\mathfrak{p}$ *is acyclic for all* $\mathfrak{p} \in \text{Spec } R$, $\mathfrak{p} \neq \mathfrak{m}$. *If* $\ell(Re) < \infty$, *then* $s \geq \dim R$.

PROOF. Assume that $s < \dim R$. In order to apply the theorem one must show that codim $I_{r_i}(\varphi_i) \geq i$ for $i = 1, \dots, s$.

We even claim that height $I_{r_i}(\varphi_i) \geq i$. If this is false, then there exist j and a prime ideal $\mathfrak{p} \supset I_{r_j}(\varphi_j)$ with height $\mathfrak{p} = \text{height } I_{r_j}(\varphi_j) < j$. Since $j \leq s < \dim R$, one has $\mathfrak{p} \neq \mathfrak{m}$. On the other hand, the acyclicity of $F_\bullet \otimes R_\mathfrak{p}$ implies that $\text{grade}(I_{r_j}(\varphi_j))_\mathfrak{p} = \text{grade } I_{r_j}(\varphi_j \otimes R_\mathfrak{p}) \geq j$ by virtue of 1.4.13, which is a contradiction.

Now that we can apply 9.4.1, we get a contradiction to our initial assumption $s < \dim R$: $\ell(Re) < \infty$ is equivalent with codim$(\text{Ann } e) \geq \dim R$. $\qquad \square$

The next level of specialization is the 'new intersection theorem' which for local rings of characteristic p was already proved in 8.2.6:

Corollary 9.4.3 (Peskine–Szpiro, Roberts). *With the notation of* 9.4.1 *suppose that* $F_\bullet \otimes R_\mathfrak{p}$ *is exact for all* $\mathfrak{p} \in \text{Spec } R$, $\mathfrak{p} \neq \mathfrak{m}$. *If* $s < \dim R$, *then the complex* $F_\bullet$ *is exact.*

PROOF. The complex $F_\bullet$ satisfies the hypothesis of 9.4.2, and furthermore $\ell(Re) < \infty$ for all $e \in \text{Coker } \varphi_1$. So $\text{Coker } \varphi_1 = 0$ by Nakayama's lemma. The map φ_1 is a split epimorphism, and we obtain a shorter complex which also satisfies the hypothesis of the corollary. Induction on s yields the assertion. $\qquad \square$

At least once in this chapter we want to give a complete proof of a theorem by direct reduction to characteristic p via Hochster's finiteness theorem 8.4.1. Since we have a proof of 9.4.3 in characteristic p which is independent of big Cohen–Macaulay modules, 9.4.3 is the best candidate for such a demonstration.

SECOND PROOF OF 9.4.3. Suppose that 9.4.3 is violated for a local ring containing a field of characteristic zero. Arguing as in the proof of 8.2.6, we may assume that $F_0 \neq 0$ and $\operatorname{Im} \varphi_1 \subset \mathfrak{m} F_0$. Choose a basis for F_i, $i = 0, \ldots, s$. Then each homomorphism φ_i is represented by a matrix $A_i = (a_{jl}^{(i)})$. Since $F_{\scriptscriptstyle \bullet}$ is a complex, one has

$$(2) \qquad\qquad A_{i-1}A_i = 0, \qquad i = 1, \ldots, s.$$

Let x be a system of parameters for R. That $\operatorname{Im} \varphi_1 \subset \mathfrak{m} F_0$, can be expressed by the relation

$$(3) \qquad\qquad (a_{jl}^{(i)})^t \in xR$$

for some $t > 0$ and all i, j, l. That $F_{\scriptscriptstyle \bullet} \otimes R_\mathfrak{p}$ is exact for all $\mathfrak{p} \neq \mathfrak{m}$, is described by the following two conditions:

(i) $F_{\scriptscriptstyle \bullet} \otimes R_\mathfrak{p}$ is split acyclic for each $\mathfrak{p} \in \operatorname{Spec} R$, $\mathfrak{p} \neq \mathfrak{m}$;

(ii) $\sum_{i=0}^s (-1)^i \operatorname{rank} F_i = 0$.

Let r_i be the expected rank of φ_i. Via 1.4.12 condition (i) can be translated into the non-vanishing of $I_{r_i}(A_i)$ modulo $\mathfrak{p}$ for all prime ideals $\mathfrak{p} \neq \mathfrak{m}$, equivalently

$$(4) \qquad\qquad (xR)^u \subset I_{r_i}(A_i)$$

for some $u > 0$ and all $i = 1, \ldots, s$.

It is mechanical to express (2) – (4) in terms of polynomial equations over $\mathbb{Z}$ satisfied by the entries of the matrices, the elements of x, and the coefficients in the linear combinations involved. These equations only depend on the numerical parameters s, rank F_i, t, and u. Conversely, given a solution to one of the systems of equations thus obtained, one immediately constructs a counterexample to 9.4.3, interpreting the matrices as homomorphisms. $\square$

The reader is invited to try similar reductions for 9.4.1, 9.5.2, and 9.5.5.

The next member of the chain of corollaries is the 'homological height theorem'. It belongs to the class of 'superheight theorems'. For a proper ideal I of a Noetherian ring R let us define its *superheight* as the supremum of height IS where S is any Noetherian ring to which there exists a ring homomorphism $R \to S$, with $IS \neq S$. The fundamental superheight theorem is Krull's principal ideal theorem; it says that superheight I is bounded above by the minimal number of generators of I.

Theorem 9.4.4 (Hochster). *Let R be a Noetherian ring containing a field, and $M \neq 0$ a finite R-module. Then* superheight Ann $M \leq$ proj dim M.

Before proving this theorem one should note that it is a far-reaching generalization of Krull's principal ideal theorem for Noetherian rings containing a field k: take $R = k[X_1, \ldots, X_n]$ and $M = k \cong R/(X_1, \ldots, X_n)$. Then proj dim $M = n$, and therefore height$(x_1, \ldots, x_n) \leq n$ for elements $x_1, \ldots, x_n$ of a K-algebra S with $(x_1, \ldots, x_n) \neq S$. (Simply consider the extension $R \to S$ induced by the substitution $X_i \to x_i$.)

PROOF OF 9.4.4. The theorem is trivial if proj dim $M = \infty$, so assume it is finite, and let $R \to S$ be a Noetherian extension of R such that $(\text{Ann } M)S \neq S$. Replacing S by a localization S_q for a minimal prime ideal of $(\text{Ann } M)S$ and R by $R_{q \cap R}$, one may assume that $R \to S$ is a local extension, and $(\text{Ann } M)S$ is not contained in any prime ideal $\mathfrak{p}$ of S different from the maximal ideal $\mathfrak{q}$ of S.

Let $F_{\bullet}$ be a minimal free resolution of M over R. Then $\mathfrak{p} \cap R \not\supset \text{Ann } M$ for every $\mathfrak{p} \in \text{Spec } S$ with $\mathfrak{p} \neq \mathfrak{q}$. Hence $M \otimes R_{\mathfrak{p} \cap R} = 0$, and $F_{\bullet} \otimes R_{\mathfrak{p} \cap R}$ is split exact. Split exactness is preserved under ring extensions, and so $F_{\bullet} \otimes S_{\mathfrak{p}}$ is split exact: $F_{\bullet} \otimes S$ satisfies the hypotheses of 9.4.3, whence proj dim $M \geq \dim S$. $\square$

Let k be an algebraically closed field, and Y, Z subvarieties of the affine space $\mathbb{A}^n(k)$ (or the projective space $\mathbb{P}^n(k)$). Then a classical theorem of algebraic geometry asserts that

$$\dim W \geq \dim Y + \dim Z - n$$

for every irreducible component of $Y \cap Z$ ([152], Prop. 7.1). If $\mathfrak{p}, \mathfrak{q}, \mathfrak{r}$ are the prime ideals defining the varieties Y, Z, and W respectively, then this inequality can be written

(5)   height $\mathfrak{r} \leq$ height $\mathfrak{p} +$ height $\mathfrak{q}$.

Note that $\mathfrak{r}$ is a minimal prime ideal of $\mathfrak{p} + \mathfrak{q}$. Serre showed in [334], Théorème 3, p. V-18 that the inequality (5) holds for prime ideals $\mathfrak{p}, \mathfrak{q}, \mathfrak{r}$ of a regular local ring such that $\mathfrak{r}$ is a minimal prime ideal of $\mathfrak{p} + \mathfrak{q}$. Suppose that $\mathfrak{r} = \mathfrak{m}$ is the maximal ideal of R. Then $\mathfrak{r}$ contains all the minimal prime ideals of any ideal $I \subset R$, and we can replace $\mathfrak{p}$ and $\mathfrak{q}$ by arbitrary ideals I and J to obtain the following version of Serre's theorem: let I, J be ideals of a regular local ring $(R, \mathfrak{m})$ such that $I + J$ is $\mathfrak{m}$-primary; then height $I +$ height $J \geq \dim R$, or, returning to dimensions,

(6) $\dim R/I \leq \dim R - \dim R/J$.

The example $R = k[[X_1, X_2, Y_1, Y_2]]/(X_1 Y_2 - X_2 Y_1)$, $I = (x_1, x_2)$, $J = (y_1, y_2)$, shows that the last inequality is false in non-regular local rings. However, one can hope that in the presence of their characteristic property namely finite projective dimension of finite modules, one can generalize the inequality above, reading I and J as the annihilators of modules M and N. The best possible result to be expected is the direct generalization of (6):

(7) $$\dim N \leq \dim R - \dim M$$

for all modules M, N over a local ring $(R, \mathfrak{m})$ such that M has finite projective dimension and $\operatorname{Supp} M \cap \operatorname{Supp} N = \{\mathfrak{m}\}$. It seems to be unknown whether (7) holds, but (7) turns into a valid inequality if we replace its right side by $\operatorname{depth} R - \operatorname{depth} M = \operatorname{proj\,dim} M$ (the Auslander–Buchsbaum formula; see 1.3.3). It should now be clear why the following corollary is named the 'intersection theorem'. (It is customary in this context to express the condition $\operatorname{Supp} M \cap \operatorname{Supp} N = \{\mathfrak{m}\}$ by $\ell(M \otimes N) < \infty$, which is an equivalent requirement if $M, N \neq 0$.)

Theorem 9.4.5 (Peskine–Szpiro). *Let R be a Noetherian local ring containing a field, and $M, N \neq 0$ finite R-modules such that $\ell(M \otimes N) < \infty$. Then $\dim N \leq \operatorname{proj\,dim} M$.*

PROOF. There is nothing to prove if $\operatorname{proj\,dim} M = \infty$. So assume it is finite. Neither the condition $\ell(M \otimes N) < \infty$, nor the number $\dim N$, can change if we replace N by another finite module with the same support. In particular we may replace N by $S = R/\operatorname{Ann} N$. Then $(\operatorname{Ann} M)S$ is primary to the maximal ideal of S, and the desired inequality proves to be a special case of 9.4.4. □

It is easy to generalize 9.4.5 to situations in which $\ell(M \otimes N)$ is not necessarily finite. Suppose that $\dim(M \otimes N) > 0$. Then $\dim N > 0$, and none of the finitely many minimal prime ideals of $M \otimes N$ or N equals $\mathfrak{m}$. Therefore there exists $x \in \mathfrak{m}$ such that $\dim N/xN = \dim N - 1$ and $\dim(M \otimes (N/xN)) = \dim(M \otimes N) - 1$. Applied inductively, this argument proves the following corollary:

Corollary 9.4.6. *Let R be a Noetherian local ring containing a field, and $M, N \neq 0$ finite R-modules. Then $\dim N \leq \operatorname{proj\,dim} M + \dim(M \otimes N)$.*

One of the reasons for which we have stated the corollary, is that it explains why 9.4.5 is easier to prove than inequality (7) above: 9.4.6 is equivalent to $\dim R - \dim M \leq \operatorname{depth} R - \operatorname{depth} M$ for $N = R$.

The following theorem, often called 'Auslander's conjecture', does not strictly fall under the title of this section, but its proof is short and an elegant application of the intersection theorem 9.4.5.

Theorem 9.4.7 (Peskine–Szpiro). *Let $(R, \mathfrak{m})$ be a Noetherian local ring containing a field, and $M \neq 0$ a finite module of finite projective dimension. Then every M-sequence is an R-sequence; in particular every M-regular element is R-regular.*

PROOF. If $x \in R$ is regular on M and on R, then $\operatorname{proj\,dim}_{R/(x)} M/xM = \operatorname{proj\,dim} M$ by 1.3.5. Thus it is enough to prove the second statement; the first follows by induction.

One has to show that every $\mathfrak{p} \in \operatorname{Ass} R$ is contained in some $\mathfrak{q} \in \operatorname{Ass} M$. We proceed by induction on $\dim M$. If $\dim M = 0$, then $\mathfrak{m} \in \operatorname{Ass} M$, and certainly $\mathfrak{p} \subset \mathfrak{m}$. Assume that $\dim M > 0$.

If there is a prime ideal $\mathfrak{q} \in \operatorname{Supp} M$ such that $\mathfrak{m} \neq \mathfrak{q} \supset \mathfrak{p}$, one can apply the inductive hypothesis to $M_\mathfrak{q}$: there exists $\mathfrak{q}' \in \operatorname{Spec} R_\mathfrak{q}$ with $\mathfrak{q}' \in \operatorname{Ass} M_\mathfrak{q}$ and $\mathfrak{q}' \supset \mathfrak{p} R_\mathfrak{q}$; hence $\mathfrak{q}' \cap R$ satisfies our needs.

Otherwise $V(\mathfrak{p}) \cap \operatorname{Supp} M = \{\mathfrak{m}\}$. So $\dim R/\mathfrak{p} \leq \operatorname{proj\,dim} M$ by 9.4.5. On the other hand $\operatorname{depth} R \leq \dim R/\mathfrak{p}$ according to 1.2.13, and furthermore the Auslander–Buchsbaum formula says that $\operatorname{proj\,dim} M + \operatorname{depth} M = \operatorname{depth} R$. Therefore $\operatorname{depth} M = 0$, whence $\mathfrak{m} \in \operatorname{Ass} M$.   □

Remarks 9.4.8. (a) The new intersection theorem 9.4.3 was proved for all local rings by Roberts [314]. Consequently 9.4.4–9.4.7 and 9.4.9–9.4.10 hold without the hypothesis that R contains a field. In particular 9.4.4 is a true generalization of Krull's principal ideal theorem (take $R = \mathbb{Z}[X_1, \ldots, X_n]$).

(b) It is possible to avoid the use of big Cohen–Macaulay modules in the proof of the improved new intersection theorem 9.4.2. In fact, 9.4.2 is on a par with the canonical element theorem 9.3.1. Hochster [184] derived 9.4.2 from 9.3.1, and Dutta [82] found the converse. As pointed out in 9.3.3, the canonical element theorem can be proved independently of the existence of big Cohen–Macaulay modules.

(c) The intersection theorem 9.4.5 can be improved to the best conceivable result if M is perfect; see [297], p. 94, Théorème 4.2:

(i) $\operatorname{grade} M + \dim M = \dim R$;

(ii) if N is a finite R-module such that $l(M \otimes N) < \infty$, then $\dim M + \dim N \leq \dim R$.

Furthermore both (i) and (ii) hold if $R = \bigoplus_{i=0}^{\infty} R_i$ is a graded ring with R_0 an Artinian local ring, M is a finite graded R-module of finite projective dimension, and N is a finite graded R-module; see Peskine and Szpiro [299]. Equation (i), sometimes called the 'codimension conjecture', was proved by Foxby [116] for modules M of finite projective dimension over a large class of equicharacteristic local rings.

(d) Let R be a Noetherian ring, and M, N finite R-modules such that $\operatorname{proj\,dim} M < \infty$ and $\ell(M \otimes N) < \infty$. Then the modules $\operatorname{Tor}_i^R(M, N)$ have finite length, and only finitely many are non-zero. Thus one can define the *intersection multiplicity* of M and N by

$$e(M, N) = \sum_{i=0}^{\infty} (-1)^i \ell(\operatorname{Tor}_i^R(M, N)).$$

This notion was introduced by Serre [334]. He proved that the following hold if $(R, \mathfrak{m})$ is an unramified regular local ring (see [270] for this notion):

(i) if $\dim M + \dim N < \dim R$, then $e(M, N) = 0$;

(ii) if $\dim M + \dim N = \dim R$, then $e(M, N) > 0$.

(Note that $\dim M + \dim N \leq \dim R$, as discussed above.) Recently Gabber [121] showed that over an arbitrary regular local ring one has $\chi(M, N) \geq 0$. However, both (i) and (ii) fail if R is allowed to be an arbitrary local ring: Dutta, Hochster, and McLaughlin [85] constructed counterexamples over the hypersurface ring $k[X_1, X_2, Y_1, Y_2]/(X_1 Y_2 - X_2 Y_1)$. However, (i) was shown to hold if both M and N have finite projective dimension and R is a complete intersection (Roberts [313], [315], Gillet and Soulé [125]) or $\dim \operatorname{Sing} R \leq 1$ ([315]).

Exercises

9.4.9. A Noetherian local ring $(R, \mathfrak{m})$ containing a field is Cohen–Macaulay if (and only if) there exists an R-module of finite length and finite projective dimension. Prove this.

9.4.10. Let $\varphi : R \to S$ be a surjective homomorphism of Noetherian local rings containing a field such that $\operatorname{proj\,dim}_R S < \infty$. Show the following are equivalent:

(i) R is Cohen–Macaulay and S is a perfect R-module (of type 1);

(ii) S is Cohen–Macaulay (Gorenstein).

Hint: 9.4.9 is essential for the difficult implication (ii) $\Rightarrow$ (i).

9.4.11. Prove the assertions on perfect R-modules in 9.4.8(c) for Noetherian local rings R containing a field.

Hint: It suffices to prove that $\operatorname{grade} M + \dim M \leq \dim R$ which is quite evident.

9.4.12. Let R be a Cohen–Macaulay local ring, and x a system of parameters for R. Show that $e(x, N) = e(R/(x), N)$ for all finite R-modules N.

9.5 Ranks of syzygies

Let R be a local ring, and M a finite R-module of finite projective dimension. Then $\operatorname{proj\,dim} M \leq \operatorname{depth} R$: the length of a minimal free resolution is bounded by $\operatorname{depth} R$. Moreover, each of the values $s = 0, \ldots, \operatorname{depth} R$ occurs if we choose $M = R/(x)$ with an R-sequence $x = x_1, \ldots, x_s$. In this section we shall discuss the possible values for the Betti numbers of M and the ranks of its syzygy modules. For systematic reasons and in view of an application to Bass numbers below, it is useful to consider a larger class of complexes than just minimal free resolutions, namely minimal complexes of codimension ≥ 0.

Let M be a module over a commutative ring R, and $x \in M$. The notion of order ideal, which was introduced in connection with 9.3.2, plays an important role in the following. The next lemma describes a property of x which is controlled by $\mathcal{O}(x)$.

Lemma 9.5.1. *Let R be a Noetherian ring, M a finite R-module, $x \in M$, and $\mathfrak{p}$ a prime ideal. Then x generates a non-zero free direct summand of $M_\mathfrak{p}$ if and only if $\mathfrak{p} \not\supseteq \mathcal{O}(x)$.*

PROOF. Since $\mathrm{Hom}_{R_\mathfrak{p}}(M_\mathfrak{p}, R_\mathfrak{p})$ is naturally isomorphic to $\mathrm{Hom}(M, R)_\mathfrak{p}$, the formation of order ideals commutes with localization. We may therefore assume that $(R, \mathfrak{p})$ is a local ring. If $M = Rx \oplus N$ and $Rx \cong R$, then there obviously exists $\alpha \in \mathrm{Hom}_R(M, R)$ such that $\alpha(x) = 1$. Conversely, if $\alpha(x) = 1$, then $M = Rx \oplus \mathrm{Ker}\,\alpha$.   □

Suppose now that $\varphi : F \to G$ is a map of finite free modules. Let $e \in F$. Given a basis $g_1, \ldots, g_n$ of G, there are uniquely determined elements $a_1, \ldots, a_n \in R$ such that $\varphi(e) = a_1 g_1 + \cdots + a_n g_n$. The elements $g_1^*, \ldots, g_n^*$ of the dual basis of $\mathrm{Hom}_R(G, R)$ yield the values $g_j^*(\varphi(e)) = a_j$. Therefore $\mathcal{O}_G(\varphi(e)) = (a_1, \ldots, a_n)$.

Theorem 9.5.2. *Let $(R, \mathfrak{m})$ be a local ring containing a field, and*

$$F_\bullet : 0 \longrightarrow F_s \xrightarrow{\varphi_s} F_{s-1} \longrightarrow \cdots \longrightarrow F_1 \xrightarrow{\varphi_1} F_0 \longrightarrow 0$$

a complex of finite free R-modules. Then, for $j = 1, \ldots, s$ and every $e \in F_j$ with $e \notin \mathfrak{m}F_j + \mathrm{Im}\,\varphi_{j+1}$, one has $\mathrm{codim}\,\mathcal{O}(\varphi_j(e)) \geq \mathrm{codim}\,F_\bullet + j$.

PROOF. Let $t = \mathrm{codim}\,F_\bullet$. For given j we truncate the complex at F_{j-1}, and adjust the indices by setting $F_i' = F_{i+j-1}$ and $t' = t + j - 1$. Replacing the given data by those just defined, we may assume that $j = 1$. Let $J = \mathcal{O}(\varphi_1(e))$. There is something to prove only if $J \subset \mathfrak{m}$ and $\mathrm{codim}\,F_\bullet \geq 0$. We put $\bar{R} = R/J$ and $\bar{F} = F_\bullet \otimes \bar{R}$. From the description of J preceding the theorem one sees that $\bar{\varphi}_1(\bar{e}) = 0$. In order to derive a contradiction, we assume that $\mathrm{codim}\,J \leq t$. Note that $I_{r_i}(\bar{\varphi}_i) = (I_{r_i}(\varphi_i) + J)/J$. Hence

$$\dim(\bar{R}/I_{r_i}(\bar{\varphi}_i)) \leq \dim(R/I_{r_i}(\varphi_i)) \leq \dim R - i - t \leq \dim \bar{R} - i.$$

This inequality shows that $\mathrm{codim}\,\bar{F}_\bullet \geq 0$. Let M be a balanced big Cohen–Macaulay module for $\bar{R}$. By virtue of 9.1.8, $\bar{F}_\bullet \otimes M$ is acyclic. Since $\bar{\varphi}_1(\bar{e}) = 0$, we have $(\bar{\varphi}_1 \otimes M)(\bar{e} \otimes M) = 0$. Let $C = \mathrm{Coker}\,\varphi_2$, and $\pi : F_1 \to C$ be the natural epimorphism. Since $\bar{F}_\bullet \otimes M$ is acyclic, $\bar{\varphi}_1 \otimes M$ induces an isomorphism $\bar{C} \otimes M \to \mathrm{Im}(\bar{\varphi}_1 \otimes M)$. So $\bar{\pi}(\bar{e}) \otimes M = 0$.

On the other hand, the hypothesis $e \notin \mathfrak{m}F_1 + \mathrm{Im}\,\varphi_2$ implies that $\bar{\pi}(\bar{e}) \notin \mathfrak{m}\bar{C}$. Thus the image of $\bar{\pi}(\bar{e}) \otimes M$ under the natural epimorphism $\bar{C} \otimes M \to (\bar{C}/\mathfrak{m}\bar{C}) \otimes (M/\mathfrak{m}M)$ is isomorphic to $M/\mathfrak{m}M \neq 0$. This is a contradiction.   □

An application of the following corollary was anticipated in the proof of 2.3.14.

Corollary 9.5.3. *Let $(R, \mathfrak{m}, k)$ be a regular local ring containing a field, and $I \subset \mathfrak{m}$ an ideal generated by a sequence x. Then the natural homomorphism from $H_i(x, k) = K_\bullet(x) \otimes k$ to $\mathrm{Tor}_i^R(R/I, k)$ is zero for $i > \mathrm{grade}\,I$.*

PROOF. The natural homomorphism $H_i(x,k) \to \mathrm{Tor}_i^R(R/I,k)$ is induced by a complex homomorphism γ from $K_\bullet(x)$ to a free resolution $F_\bullet$ of R/I; see 1.6.9. It only depends on I and x, so that we may assume that $F_\bullet$ is a minimal free resolution. Since R is regular, $F_\bullet$ has finite length by 2.2.7. That $H_\bullet(x,k) = K_\bullet(x) \otimes k$ and $\mathrm{Tor}_\bullet^R(R/I,k) \cong F_\bullet \otimes k$, follows from the minimality of the complexes $K_\bullet(x)$ and $F_\bullet$. Thus the map $H_\bullet(x,k) \to \mathrm{Tor}_\bullet^R(R/I,k)$ is just $\gamma \otimes k$.

The assertion amounts to $\gamma(K_i(x)) \subset \mathfrak{m}F_i$ for $i > \mathrm{grade}\,I$. Let $z \in K_i(x)$, and ∂ and φ denote differentiation in $K_\bullet(x)$ and $F_\bullet$. If $\gamma(z) \notin \mathfrak{m}F_i$, then

$$\mathrm{grade}\,\mathcal{O}(\gamma(\partial(z))) = \mathrm{codim}\,\mathcal{O}(\varphi(\gamma(z))) \geq i$$

by 9.5.2: an acyclic complex has non-negative codimension as observed above. On the other hand, $\mathcal{O}(\varphi(\gamma(z))) = \mathcal{O}(\gamma(\partial(z))) \subset I$ since $\mathrm{Im}\,\partial \subset IK_\bullet(x)$. □

As indicated above, we aim at a bound for the expected ranks r_i of the maps in a free complex $F_\bullet$. Reasoning inductively, we will have to pass to a complex $0 \to F_s \to F_{s-1} \to \cdots \to F_2 \to F_1' \to F_0' \to 0$ in which rank $F_1' = \mathrm{rank}\,F_1 - 1$. Theorem 9.5.2 enables us to find F_1', whereas the following lemma contains the construction of F_0'.

Lemma 9.5.4. *Let R be a Noetherian ring and M a finite R-module. Then there is a finite free R-module F and a homomorphism $\varphi: M \to F$ with the following property: If $\mathfrak{p}$ is a prime ideal, and $N \subset M_\mathfrak{p}$ is a free direct $R_\mathfrak{p}$-summand of rank r, then $(\varphi \otimes R_\mathfrak{p})(N)$ is a free direct $R_\mathfrak{p}$-summand of $F_\mathfrak{p}$ with $\mathrm{rank}(\varphi \otimes R_\mathfrak{p})(N) = r$.*

PROOF. Let * denote the functor $\mathrm{Hom}_R(\_,R)$. There is a finite free R-module G with an epimorphism $\pi: G \to M^*$. Let $h: M \to M^{**}$ be the canonical homomorphism, and choose $\varphi = \pi^* \circ h$, $F = G^*$. Then $\varphi: M \to F$ has the property that every linear form $\alpha \in M^*$ can be extended to F along φ. Since R is Noetherian and the modules involved are finite, the preceding construction commutes with every localization of R. Thus assume $R = R_\mathfrak{p}$.

Now the hypothesis on N is equivalent to the existence of $g_1, \ldots, g_r \in N$ and $\alpha_1, \ldots, \alpha_r \in M^*$ such that $N = Rg_1 + \cdots + Rg_r$ and $\alpha_i(g_j) = \delta_{ij}$. Since the α_i can be extended to F, the elements $\varphi(g_1), \ldots, \varphi(g_r)$ generate a free direct summand of rank r. □

As pointed out before 9.1.9 every complex of finite free modules over a local ring decomposes into a split exact direct summand and a direct summand which is minimal. For the ranks of the maps in a split exact complex one can only say that they are non-negative, but for those of a minimal complex there exists a non-trivial lower bound. (It was essentially given by Evans and Griffith in the form of Corollary 9.5.6.)

Theorem 9.5.5. Let $(R, \mathfrak{m})$ be a local ring containing a field, and

$$F_\bullet : 0 \longrightarrow F_s \xrightarrow{\varphi_s} F_{s-1} \longrightarrow \cdots \longrightarrow F_1 \xrightarrow{\varphi_1} F_0 \longrightarrow 0$$

a length s minimal complex of finite free R-modules. Let r_i denote the expected rank of φ_i. If $\operatorname{codim} F_\bullet \geq 0$, then $r_i \geq \operatorname{codim} F_\bullet + i$ for $i = 1, \ldots, s - 1$.

PROOF. The same manipulation as in the proof of 9.5.2 reduces the theorem to a statement about r_1. Since the theorem makes an assertion only on $r_1, \ldots, r_{s-1}$, the complex which remains after the truncation has length ≥ 2; there is nothing to prove if $s = 1$. We introduce an auxiliary variable t, and use induction on t to show that $\operatorname{codim} F_\bullet \geq t$ implies $r_1 \geq t + 1$.

Since $\operatorname{codim} F_\bullet \geq 0$, Lemma 9.1.8 yields acyclicity of $F_\bullet \otimes M$ for a balanced big Cohen–Macaulay module M of R; such a module exists by 8.5.3. Therefore 9.1.9 implies $r_i \geq 1$ for $i = 1, \ldots, s$. This inequality covers the case $t = 0$, and shows furthermore that $F_1 \neq 0$: one has $\operatorname{rank} F_1 = r_1 + r_2 \geq 2$. So there exists $e \in F_1$ with $e \notin \mathfrak{m} F_1$. Since $F_\bullet$ is minimal, $e \notin \mathfrak{m} F_1 + \operatorname{Im} \varphi_2$.

Let $t \geq 1$. Put $F_1' = F_1/Re$, and choose φ_2' as the induced map $F_2 \to F_1'$. Applying 9.5.4 to $\operatorname{Coker} \varphi_2'$ one obtains a homomorphism $\operatorname{Coker} \varphi_2' \to F = F_0'$. Its composition with the natural epimorphism $F_1' \to \operatorname{Coker} \varphi_2'$ then yields $\varphi_1' : F_1' \to F_0'$. For the complex

$$F_\bullet' : 0 \longrightarrow F_s \longrightarrow F_{s-1} \longrightarrow \cdots \longrightarrow F_2 \xrightarrow{\varphi_2'} F_1' \xrightarrow{\varphi_1'} F_0' \longrightarrow 0$$

one has $r_2' = r_2$, $r_1' = r_1 - 1$. In order to show that $\operatorname{codim} F_\bullet' \geq t - 1$ we must verify the following inequalities: (i) $\operatorname{codim} I_{r_2'}(\varphi_2') \geq t + 1$, and (ii) $\operatorname{codim} I_{r_1'}(\varphi_1') \geq t$.

For (i) we choose a prime ideal $\mathfrak{p}$ with $\operatorname{codim} \mathfrak{p} \leq t$. Certainly $I_{r_i}(\varphi_i) \not\subset \mathfrak{p}$ for $i = 1, \ldots, s$. Therefore $F_\bullet \otimes R_\mathfrak{p}$ is split acyclic by 1.4.12. In particular we have a decomposition

$$(F_1)_\mathfrak{p} \cong (\operatorname{Im} \varphi_2)_\mathfrak{p} \oplus (\operatorname{Coker} \varphi_2)_\mathfrak{p}$$

with $\operatorname{rank}(\operatorname{Im} \varphi_2)_\mathfrak{p} = r_2$ and $\operatorname{rank}(\operatorname{Coker} \varphi_2)_\mathfrak{p} = r_1$. Moreover – and this is the crucial argument – $\operatorname{codim} \mathcal{O}(\varphi_1(e)) \geq t + 1$ by 9.5.2. Therefore $\varphi_1(e)$ generates a free direct summand of $(F_0)_\mathfrak{p}$ by 9.5.1. A fortiori the residue class $\bar{e}$ of e generates a non-zero free direct summand of $(\operatorname{Coker} \varphi_2)_\mathfrak{p}$. So $(\operatorname{Coker} \varphi_2')_\mathfrak{p} \cong (\operatorname{Coker} \varphi_2)_\mathfrak{p}/R_\mathfrak{p}\bar{e}$ is free of rank $r_1' = r_1 - 1$, and the exact sequence

$$0 \longrightarrow (\operatorname{Im} \varphi_2')_\mathfrak{p} \longrightarrow (F_1')_\mathfrak{p} \longrightarrow (\operatorname{Coker} \varphi_2')_\mathfrak{p} \longrightarrow 0$$

splits. Also this shows that $(\operatorname{Im} \varphi_2')_\mathfrak{p}$ is a free direct summand of rank $r_2 = r_2'$ of $(F_1')_\mathfrak{p}$. By 1.4.8 we get $I_{r_2'}(\varphi_2') \not\subset \mathfrak{p}$. Since $\mathfrak{p}$ is an arbitrary prime ideal with $\operatorname{codim} \mathfrak{p} \leq t$, the inequality (i) has been proved.

Slightly more than required for (ii) we show that $\operatorname{codim}(R/I_{r'_1}(\varphi'_1)) \geq t + 1$. Pick $\mathfrak{p}$ as before. We saw that $(\operatorname{Coker}\varphi'_2)_\mathfrak{p}$ is free of rank r'_1. Since φ'_1 was constructed as prescribed by 9.5.4, $(\operatorname{Coker}\varphi'_2)_\mathfrak{p}$ is mapped isomorphically onto a free direct summand of F'_0. As desired, $I_{r'_1}(\varphi'_1) \not\subset \mathfrak{p}$.

If it should happen that $F'_{\bullet}$ is not minimal, then one splits off a direct summand id : $R^u \to R^u$ from $F'_1 \to F'_0$. This does not affect the codimension, and even improves the desired inequality $r'_1 \geq t$ which holds by induction. (Because of $s \geq 2$ the construction of $F'_{\bullet}$ does not touch F_s, so that $F'_{\bullet}$ also has length s.) $\Box$

Corollary 9.5.6 (Evans–Griffith). *Let R be a Noetherian local ring containing a field, and $M \neq 0$ a module of projective dimension $s < \infty$. Then*
(a) *for $i = 1, \ldots, s - 1$ the i-th syzygy M_i of M has rank $\geq i$,*
(b)

$$\beta_i(M) \geq \begin{cases} 2i + 1, & i = 0, \ldots, s - 2, \\ s, & i = s - 1, \\ 1, & i = s. \end{cases}$$

PROOF. A minimal free resolution $F_{\bullet}$ of M is acyclic, and thus has codimension ≥ 0, as was observed above. Theorem 9.5.5 says that for $i = 1, \ldots, s - 1$ the i-th map φ_i has expected rank $r_i \geq i$. Since $F_{\bullet}$ is acyclic, $r_i = \operatorname{rank}\varphi_i = \operatorname{rank}M_i$; see 1.4.6. This proves (a) from which (b) follows with $\beta_i(M) = r_i + r_{i+1}$. (Note that $\beta_s(M) > 0$ because of $\operatorname{proj\,dim} M = s$.) $\Box$

It is of course not difficult to give a non-local version of the corollary, which we leave to the reader.

Remarks 9.5.7. (a) Corollary 9.5.6 is the best possible result. In fact, if R is a Noetherian local ring, and M the m-th syzygy module of a module of finite projective dimension, then M contains a free submodule L such that M/L inherits this property and $\operatorname{rank} M/L \leq m$; see Bruns [53]. Similarly one can find modules M for all preassigned values of $\operatorname{proj\,dim} M = s \leq \operatorname{depth} R$ and $\beta_i(M)$, $i = 0, \ldots, s$, which are consistent with 9.5.6.

(b) It is not necessary to use big Cohen–Macaulay modules in the proof of 9.5.6. Ogoma [295] derived it from the improved new intersection theorem 9.4.2.

(c) Theorem 9.5.2 and its consequences admit conclusions even for local rings *not* containing a field. Let $p = \operatorname{char} R/\mathfrak{m}$. Then one passes from a given complex $F_{\bullet}$ over R to $F_{\bullet} \otimes R/(p)$, and $R/(p)$ contains a field. The reader may verify that $\operatorname{codim} \mathcal{O}(\varphi_j(e)) \geq \operatorname{codim} F_{\bullet} + j - 1$ in 9.5.2, regardless of whether $\dim R/(p) = \dim R$, or $\dim R/(p) = \dim R - 1$. Similarly the bounds in 9.5.3, 9.5.5, and 9.5.6(a) become worse by at most 1.

9.6 Bass numbers

Let R be a Noetherian ring, and M a finite R-module. The Bass numbers

$$\mu_i(\mathfrak{p}, M) = \dim_{k(\mathfrak{p})} \operatorname{Ext}^i_{R_\mathfrak{p}}(k(\mathfrak{p}), M_\mathfrak{p}), \qquad \mathfrak{p} \in \operatorname{Spec} R,$$

determine the modules in a minimal injective resolution

$$I^\bullet : 0 \longrightarrow E^0(M) \longrightarrow E^1(M) \longrightarrow \cdots \longrightarrow E^i(M) \longrightarrow \cdots$$

of M; by 3.2.9 one has $E^i(M) = \bigoplus_{\mathfrak{p} \in \operatorname{Spec} R} E(R/\mathfrak{p})^{\mu_i(\mathfrak{p},M)}$ for all $i \geq 0$. In this section we want to derive inequalities satisfied by the numbers $\mu_i(\mathfrak{m}, M)$ when $(R, \mathfrak{m})$ is a local ring; since the Bass numbers are local data by definition, such inequalities can be translated into assertions about the $\mu_i(\mathfrak{p}, M)$ in general.

Suppose that $(R, \mathfrak{m}, k)$ is a local ring with $\mathfrak{m}$-adic completion $(\hat{R}, \hat{\mathfrak{m}}, k)$. Since $\operatorname{Ext}^i_R(k, M) \cong \operatorname{Ext}^i_R(k, M) \otimes \hat{R} \cong \operatorname{Ext}^i_{\hat{R}}(k, \hat{M})$ for all $i \geq 0$, one has $\mu_i(\mathfrak{m}, M) = \mu_i(\hat{\mathfrak{m}}, \hat{M})$. Therefore it is no restriction to assume R is complete. For simplicity of notation we set $\mu_i = \mu_i(\mathfrak{m}, M)$.

By their very definition the local cohomology modules of M are given as $H^i_{\mathfrak{m}}(M) = H^i(\Gamma_{\mathfrak{m}}(I^\bullet))$, see Section 3.4. It is easy to determine $\Gamma_{\mathfrak{m}}(I^\bullet)$ since a non-zero element of $E(R/\mathfrak{p})$ cannot be annihilated by a power of $\mathfrak{m}$ if $\mathfrak{p} \neq \mathfrak{m}$; see 3.2.7. Thus $\Gamma_{\mathfrak{m}}(I^\bullet)$ is the subcomplex

$$J^\bullet : 0 \longrightarrow E(k)^{\mu_0} \xrightarrow{\sigma_0} \cdots \xrightarrow{\sigma_{i-1}} E(k)^{\mu_i} \xrightarrow{\sigma_i} \cdots$$

By Grothendieck's theorem 3.5.7 we have $H^i_{\mathfrak{m}}(M) \neq 0$ for $i = \operatorname{depth} M$ and $i = \dim M$, in particular $\mu_i \neq 0$ for these values of i. On the other hand, $\mu_i = 0$ for $i < \operatorname{depth} M$.

By assumption R is complete. So Theorem 3.2.13 yields

$$\operatorname{Hom}_R(E(k), E(k)) = R,$$

and one obtains a complex of finite free modules from an application of the functor $\operatorname{Hom}_R(E(k), \_)$ to $J^\bullet$:

$$G^\bullet = \operatorname{Hom}_R(E(k), J^\bullet) : 0 \longrightarrow R^{\mu_0} \xrightarrow{\psi_0} R^{\mu_1} \longrightarrow \cdots \xrightarrow{\psi_{i-1}} R^{\mu_i} \xrightarrow{\psi_i} \cdots$$

Moreover there is some information on the maps σ_i and ψ_i. The endomorphisms of $E(k)$ are just given by multiplication by elements of R; therefore the maps σ_i can naturally be considered as matrices over R, and ψ_i is given by the same matrix as σ_i. Since $I^\bullet$ is a minimal injective resolution, the entries of these matrices are in $\mathfrak{m}$.

Also, one obtains a complex of finite free R-modules if one applies $\operatorname{Hom}_R(\_, E(k))$ to $J^\bullet$:

$$L_\bullet = \operatorname{Hom}_R(J^\bullet, E(k)) : \cdots \xrightarrow{\chi_i} R^{\mu_i} \xrightarrow{\chi_{i-1}} \cdots \xrightarrow{\chi_1} R^{\mu_1} \xrightarrow{\chi_0} R^{\mu_0} \longrightarrow 0;$$

the matrix representing χ_i is the transpose of σ_i. Let * denote the functor $\text{Hom}_R(_-, R)$. As just seen,

$$(G^{\cdot})^* = L_{\cdot}, \qquad (L_{\cdot})^* = G^{\cdot}.$$

The advantage of $L_{\cdot}$ over $G^{\cdot}$ is that we know its homology. By the exactness of $\text{Hom}_R(_-, E(k))$,

$$H_i(L_{\cdot}) \cong \text{Hom}_R(H^i(J^{\cdot}), E(k)) \cong \text{Hom}_R(H^i_{\mathfrak{m}}(M), E(k)).$$

We claim that $\dim H_i(L_{\cdot}) \leq i$: for this to hold it is surely sufficient that $\dim\big(R/(\text{Ann}\, H^i_{\mathfrak{m}}(M))\big) \leq i$, and the latter inequality has already been proved in 8.1.1.

In order to adapt the present notation to that in the previous section we set

$$d = \dim R, \qquad v_i = \mu_{d-i}, \qquad \varphi_i = \psi_{d-i},$$

and define the complex $F_{\cdot}$ by

$$F_{\cdot} : 0 \longrightarrow R^{v_d} \xrightarrow{\varphi_d} R^{v_{d-1}} \longrightarrow \cdots \longrightarrow R^{v_1} \xrightarrow{\varphi_1} R^{v_0} \longrightarrow 0.$$

We want to show that $\text{codim}\, F_{\cdot} \geq 0$. We consider the truncation

$$(L_{\cdot}|d - i + 1) : R^{\mu_{d-i+1}} \xrightarrow{\chi_{d-i}} R^{\mu_{d-i}} \longrightarrow \cdots \longrightarrow R^{\mu_1} \xrightarrow{\chi_0} R^{\mu_0} \longrightarrow 0.$$

Since $\dim H_v(L_{\cdot}) \leq v$, the complex $(L_{\cdot}|d - i + 1) \otimes R_{\mathfrak{p}}$ is exact, and thus split exact for prime ideals $\mathfrak{p}$ satisfying $\text{codim}\, \mathfrak{p} \leq i - 1$. We dualize to get that

$$0 \longrightarrow R^{v_d} \xrightarrow{\varphi_d} R^{v_{d-1}} \longrightarrow \cdots \longrightarrow R^{v_{i-1}} \longrightarrow 0$$

is split acyclic. Thus 1.4.12 gives $I_{r_i}(\varphi_i) \not\subseteq \mathfrak{p}$, and $\text{codim}\, I_{r_i}(\varphi_i) \geq i$ as desired.

Let $t = \text{depth}\, M$. As noticed above, $R^{v_{d-i+j}} = 0$ for $j \geq 1$, $R^{v_{d-t}} \neq 0$, and $F_{\cdot}$ is a minimal complex of length $d - t$. Now we have reached our goal; 9.5.5 yields

$$v_i = r_{i+1} + r_i \geq \begin{cases} 1, & i = d - t, \\ d - t, & i = d - t - 1, \\ 2i + 1, & i = 0, \dots, d - t - 2. \end{cases}$$

Returning to the previous notation we get part (a) of

Theorem 9.6.1. *Let R be a Noetherian local ring containing a field, $\dim R = d$, and M a finite R-module of depth t.*
(a) *Then*

$$\mu_i(\mathfrak{m}, M) \geq \begin{cases} 1, & i = t, \\ d - t, & i = t + 1, \\ 2(d - i) + 1, & i = t + 2, \dots, d. \end{cases}$$

(b) *If $t < \dim M = d$, then $\mu_d(\mathfrak{m}, M) \geq 2$.*

PROOF. (a) was proved above. In its proof we exploited results on the vanishing of local cohomology and its non-vanishing at the depth of a module. Part (b) relies on its non-vanishing at the dimension, as will be seen now.

Consider the interval $R^{\mu_{d+1}} \xrightarrow{\chi_d} R^{\mu_d} \xrightarrow{\chi_{d-1}} R^{\mu_{d-1}}$ of $L_{\bullet}$. Its homology at R^{μ_d} is $H_d(L_{\bullet}) = \operatorname{Hom}_R(H^d_{\mathfrak{m}}(M), E(k))$, and the transpose of χ_{d-1} is the map φ_1 in

$$F_{\bullet} : 0 \longrightarrow R^{\nu_d} \longrightarrow \cdots \longrightarrow R^{\nu_1} \xrightarrow{\varphi_1} R^{\nu_0} \longrightarrow 0.$$

Suppose that $\nu_0 = \mu_d = 1$. Since depth $M < d$, the arguments that proved (a) yield that $r_1 \geq 1$ (with respect to $F_{\bullet}$), hence $r_1 = 1$; note that $r_1 \leq \nu_0$. Furthermore $\dim(R/I_{r_1}(\varphi_1)) \leq d-1$, as stated above. This implies $\varphi_1 \otimes R_{\mathfrak{p}}$ is surjective for prime ideals $\mathfrak{p}$ with $\dim R/\mathfrak{p} = \dim R$. Therefore $\chi_{d-1} \otimes R_{\mathfrak{p}}$ is injective, and $\dim H_d(L_{\bullet}) < d$.

We choose a Gorenstein ring S with an epimorphism $S \to R$. By the variant 3.5.14 of the local duality theorem

$$H_d(L_{\bullet}) \cong \operatorname{Hom}_R(H^d_{\mathfrak{m}}(M), E(k)) \cong \operatorname{Ext}^{n-d}_S(M, S), \qquad n = \dim S.$$

Let $\mathfrak{q} \in \operatorname{Supp}_S M$ with $\dim S/\mathfrak{q} = d$. Then $\operatorname{Ext}^{n-d}_{S_{\mathfrak{q}}}(M_{\mathfrak{q}}, S_{\mathfrak{q}}) = 0$, since $\dim H_d(L_{\bullet}) < d$; that however contradicts 3.5.11 (note that $\dim M_{\mathfrak{q}} = 0$, $\dim S_{\mathfrak{q}} = n - d$). $\qquad \square$

Two corollaries are immediate. The first of them is usually called 'Bass' conjecture'; the second was conjectured by Vasconcelos.

Corollary 9.6.2 (Peskine–Szpiro). *Let R be a Noetherian local ring containing a field. If R has a finite module $M \neq 0$ of finite injective dimension, then R is a Cohen–Macaulay ring.*

In fact, if $\operatorname{inj} \dim M$ is finite, then it equals depth R by 3.1.17. The theorem yields $\operatorname{inj} \dim M \geq \dim R$. The converse could already have been proved in Chapter 3. Let $(R, \mathfrak{m}, k)$ be a local Cohen–Macaulay ring, x a system of parameters, and E the injective hull of k over R. Then $\operatorname{Hom}_R(R/(x), E)$ has finite length by 3.2.12. The Koszul complex $K_{\bullet}(x)$ is a projective resolution of $R/(x)$. Therefore the acyclic complex $K^{\bullet}(x, E) = \operatorname{Hom}_R(K_{\bullet}(x), E)$ is an injective resolution of $K^0(x, E) \cong \operatorname{Hom}_R(R/(x), E)$.

Corollary 9.6.3 (Foxby). *Let R be a Noetherian local ring containing a field, and $d = \dim R$. If $\mu_d(\mathfrak{m}, R) = 1$, then R is a Cohen–Macaulay ring, hence Gorenstein.*

Remarks 9.6.4. (a) Both the corollaries hold for all local rings:
  (i) Roberts [312] gave a characteristic-free proof of 9.6.3. It exploits the properties of dualizing complexes. Kawasaki [233] generalized 9.6.3 using the methods of this section: a complete local ring of type n

satisfying Serre's condition (S_{n-1}) is Cohen–Macaulay (for $n = 2$ one has additionally to assume that R is unmixed).

(ii) For a large class of local rings, 9.6.2 was first proved by Peskine and Szpiro. Their argument rests mainly on the intersection theorem 9.4.5 and the following fact which is interesting in itself: *let $(R, \mathfrak{m})$ be a Noetherian complete local ring, and $M \neq 0$ a finite R-module of finite injective dimension; then there exists a finite R-module N such that* $\operatorname{proj\,dim} N = \operatorname{depth} R - \operatorname{depth} M$ *and* $\operatorname{Supp} N = \operatorname{Supp} M$. Since Roberts [314] proved the intersection theorem for all local rings, 9.6.2 holds without any restriction.

The theorem of Peskine–Szpiro just mentioned can be proved by the method we used for 9.6.1, independently of the hypothesis that R contains a field. (One constructs the complex $F_{\bullet}$ as in the proof of 9.6.1 and chooses $N = \operatorname{Coker} \varphi_{d-u+1}$ where $d = \dim R$, $u = \operatorname{depth} R = \operatorname{inj\,dim} M$.) On the other hand, it can also be obtained as a consequence of 9.6.2 in conjunction with Exercise 9.6.5. In fact, if R contains a field and has a finite module of finite injective dimension, then it is Cohen–Macaulay by 9.6.2. Furthermore it has a canonical module since it is complete, and thus it satisfies the hypothesis of 9.6.5.

(b) Using 9.5.7(c) one can derive slightly weaker bounds for Bass numbers over an arbitrary Noetherian local ring.

(c) If R is a Cohen–Macaulay ring, then the complex $F_{\bullet}$ above is acyclic, and already 9.1.9 gives

$$\mu_i(\mathfrak{m}, M) \geq \begin{cases} 1, & i = \operatorname{depth} M \text{ and } i = \dim R, \\ 2, & \operatorname{depth} M < i < \dim R. \end{cases}$$

This inequality and 9.6.1(b) were first obtained by Foxby [115] for Cohen–Macaulay rings and local rings containing a field.

(d) Whenever $\mu_d(\mathfrak{m}, M) > 0$, $d = \dim R$, and $\operatorname{inj\,dim} M = \infty$, then $\mu_i(\mathfrak{m}, M) > 0$ for all $i \geq \dim R$; see 3.5.12.

Exercise

9.6.5. Let R be a Cohen–Macaulay local ring with canonical module ω. Recall from Exercise 3.3.28 that a finite R-module of finite injective dimension has a minimal augmented ω-resolution $\Omega_{\bullet} : 0 \to \omega^{r_p} \to \cdots \to \omega^{r_0} \to M \to 0$ with $p = \dim R - \operatorname{depth} M$. The following assertions (due to Sharp [339]) set up a bijective correspondence between finite modules M of finite injective dimension and those of finite projective dimension that is given by the assignment $M \mapsto \operatorname{Hom}_R(\omega, M)$ and its inverse $N \mapsto N \otimes \omega$.

(a) Let N be a finite R-module of finite projective dimension with minimal free resolution $F_{\bullet}$. Show that $F_{\bullet} \otimes \omega$ is a minimal ω-resolution of $M = N \otimes \omega$; in particular $\dim R - \operatorname{depth} M = \operatorname{proj\,dim} N$ and $\operatorname{Supp} M = \operatorname{Supp} N$.

(b) Conversely, let M be a finite R-module of finite injective dimension with minimal ω-resolution $\Omega_{\bullet}$. Show that $\mathrm{Hom}_R(\omega, \Omega_{\bullet})$ is a minimal free resolution of $N = \mathrm{Hom}_R(\omega, M)$.

(c) Using 9.5.7(a) show that 9.6.1 gives the best possible lower bounds for the Bass numbers of an R-module.

Hint: use 9.1.6 for (a) and (b), noting that ω is a Cohen–Macaulay module with $\mathrm{Supp}\,\omega = \mathrm{Spec}\,R$ and that $\mathrm{End}(\omega) = R$.

Notes

The acyclicity criterion 9.1.6 is essentially due to Buchsbaum and Eisenbud [63]. The general version without any finiteness condition on the ring R or the module M was given by Northcott [290]. The concept of grade on which it is based goes back to Hochster [179]. To us it seemed most convenient to use Koszul homology in the definition of grade.

Section 9.2 is based on Hochster's article [176]. We outlined the fact that essentially all the homological theorems can be derived from the direct summand theorem 9.2.3 or its equivalent, the monomial theorem 9.2.1. One of the rare results in mixed characteristic is due to Hochster and McLaughlin [199]; it says that a regular local ring is a direct summand of a finite extension domain if the extension of the fields of fractions has degree two. As a surprising spin-off of an investigation of the monomial theorem in mixed characteristic, Roberts [317] obtained a counterexample for Hilbert's fourteenth problem, and furthermore a prime ideal in a formal power series ring whose symbolic Rees algebra is not finitely generated.

The material on the canonical element theorem in Section 9.3 is taken from Hochster's comprehensive treatise [184]. It seems however that the idea to compare a Koszul complex for a system of parameters with a free resolution of the residue class field, was first used by Eisenbud and Evans [92] in the demonstration of their generalized principal ideal theorem 9.3.2. Hochster [184] contains many more results than indicated in 9.3.3 and 9.4.8. In particular we would like to mention a connection between canonical elements and canonical modules. The canonical element theorem has also been studied by Dutta [82], [84], and Huneke and Koh [216].

A 'tremendous breakthrough' (Hochster [186], p. 496) was made by Peskine and Szpiro in [297]. As mentioned already in Chapter 8, they were the first to apply the Frobenius morphism in the context of homological questions and to reduce such questions from characteristic zero to characteristic p through Artin approximation. They proved the intersection theorem 9.4.5 in characteristic p and for local rings which can be obtained as inductive limits of local étale extensions of localizations of affine algebras over a field of characteristic zero. Furthermore, for the same class of local rings they were able to deduce Auslander's conjecture 9.4.7 and Bass' conjecture 9.6.2 from the intersection theorem.

An equally fundamental achievement is Hochster's construction of big Cohen–Macaulay modules. It enabled him to extend Peskine and Szpiro's results to all local rings containing a field, and had the side-effect of a considerable technical simplification. See [175], [178], [181].

The new intersection theorem is due independently to Peskine and Szpiro [299] and Roberts [309]. It seems that Foxby [115] published the first complete proof valid for all equicharacteristic local rings; using big Cohen–Macaulay modules he gave an even more general theorem than 9.4.3. As pointed out above, Roberts [314], [316] proved the new intersection theorem in full generality; it has been noted which of the theorems therefore become valid without a restriction on the characteristic. The improved new intersection theorem 9.4.2 is implicitly contained in Evans and Griffith [97]; it was explicitly formulated (and given its name) by Hochster [184]. Still another extension of the intersection theorem must be mentioned, namely Foxby's version for complexes in [116].

In 9.4.8 we commented on generalizations of Serre's theorem for intersection multiplicities. It should be added here that some positive results were obtained by Foxby [117] and Dutta [80], [81].

The original argument of Evans and Griffith's remarkable syzygy theorem 9.5.6 is found in [97]. It requires a weak condition on the underlying ring. Such conditions were removed by Ogoma [295], as pointed out in 9.5.7. Our proof of the more general result 9.5.5 is a direct generalization of the argument in [98]. This monograph of Evans and Griffith contains an extensive discussion of questions related to the syzygy theorem; its bibliography gives an overview of the pertinent literature.

Successively better inequalities for Bass numbers were obtained by Foxby [112], Fossum, Foxby, Griffith, and Reiten [110], and again Foxby [115]; the last two articles make use of big Cohen–Macaulay modules. The relationship of injective resolutions to finite free complexes was realized by Peskine and Szpiro in their proof of Bass' conjecture 9.6.2. Their arguments were extended by Foxby [115]. In particular 9.6.1(b) and 9.6.3 (even a more general version for modules) are due to him. As pointed out already, Roberts gave a characteristic free version of 9.6.3. The investigation of 9.6.3 originated from Vasconcelos [380] who proved it for certain one dimensional local rings.

10 Tight closure

The final chapter extends the characteristic p methods by introducing the tight closure of an ideal, a concept that, via the comparison to a regular subring or overring, conveys the flatness of the Frobenius to non-regular rings. It was invented by Hochster and Huneke about ten years ago and is still in rapid development.

The principal classes of rings whose definition is suggested by tight closure theory consist of the F-regular and F-rational rings; they are characterized by the condition that all ideals or, in the case of F-rationality, the ideals of the principal class are tightly closed. Under a mild extra hypothesis F-rationality implies the Cohen–Macaulay property. More is true: F-rational rings are the characteristic p counterparts of rings with rational singularities; we will at least indicate this connection – a full treatment would require methods of algebraic geometry beyond our scope.

Tight closure theory has many powerful applications. Among them we have selected the Briançon–Skoda theorem, whose proof is based on the relationship of tight closure and integral closure, and the theorem of Hochster and Huneke that equicharacteristic direct summands of regular rings are Cohen–Macaulay.

10.1 The tight closure of an ideal

Throughout this section we suppose that all rings are Noetherian and of prime characteristic p, unless stated otherwise. Recall from Section 8.2 that $I^{[q]}$, $q = p^e$, denotes the q-th Frobenius power of an ideal I, that is, $I^{[q]}$ is the ideal generated by the q-th powers of the elements of I; equivalently, $I^{[q]}$ is the ideal generated by the image of I under the e-fold iteration F^e of the Frobenius homomorphism $F: R \to R$, $F(a) = a^p$. We reserve the letter q for powers of p; for example, we will say 'for $q \gg 0$' when we mean 'for $q = p^e$ with $e \gg 0$'.

In the following the set R° of elements of R that are not contained in a minimal prime ideal of R will play an important rôle. Note that R° is multiplicatively closed.

Definition 10.1.1. Let $I \subset R$ be an ideal. The *tight closure I^* of I* is the set of all elements $x \in R$ for which there exists $c \in R^\circ$ with $cx^q \in I^{[q]}$ for $q \gg 0$. One says *I is tightly closed* if $I = I^*$.

In previous chapters I^* has denoted the ideal generated by the homogeneous elements in I where I is an ideal in a graded ring. Since there is no danger of confusion, we keep the 'traditional' notation for tight closure.

The next proposition lists some basic properties of tight closure; in particular it behaves as expected for a closure operation.

Proposition 10.1.2. *Let I and J be ideals in R. Then the following hold:*
(a) *I^* is an ideal and $I \subset J \Rightarrow I^* \subset J^*$;*
(b) *there exists $c \in R^\circ$ with $c(I^*)^{[q]} \subset I^{[q]}$ for $q \gg 0$;*
(c) *$I \subset I^* = I^{**}$;*
(d) *if I is tightly closed, then so is $I : J$;*
(e) *$x \in I^*$ if and only if the residue class of x lies in $((I + \mathfrak{p})/\mathfrak{p})^*$ for all minimal prime ideals $\mathfrak{p}$ of R;*
(f) *if R is reduced or height $I > 0$, then $x \in I^*$ implies that there exists $c \in R^\circ$ with $cx^q \in I^{[q]}$ for all q.*

PROOF. (a) is obvious.

(b) We choose a system $y_1, \ldots, y_m$ of generators of I^*. For each i there exist $c_i \in R^\circ$ such that $c_i y_i^q \in I^{[q]}$ for $q \gg 0$, and therefore $c(I^*)^{[q]} \subset I^{[q]}$ for $c = c_1 \cdots c_m$ and $q \gg 0$.

(c) Suppose $dx^q \in (I^*)^{[q]}$ for $q \gg 0$ with $d \in R^\circ$. With c as in (b) one then has $(cd)x^q \in I^{[q]}$ for $q \gg 0$. Since $cd \in R^\circ$, it follows that $x \in I^*$.

(d) Note that $(I : J)^{[q]} \subset I^{[q]} : J^{[q]}$. Thus $cx^q \in (I : J)^{[q]}$ for $q \gg 0$ implies $c(xy)^q \in I^{[q]}$ for all $y \in J$ and $q \gg 0$. Hence $xy \in I^* = I$ for all $y \in J$, and therefore $x \in I : J$.

(e) If $x \in I^*$, then the residue class $\bar{x}$ belongs to $((I + \mathfrak{p})/\mathfrak{p})^*$ since $R^\circ \cap \mathfrak{p} = \emptyset$.

Conversely, let $\mathfrak{p}_1, \ldots, \mathfrak{p}_n$ be the minimal prime ideals of R, and suppose $\bar{x} \in ((I + \mathfrak{p}_i)/\mathfrak{p}_i)^*$ for all i. Then there exist $c_i \in R \setminus \mathfrak{p}_i$ with $c_i x^q \in I^{[q]} + \mathfrak{p}_i$ for $q \gg 0$. We may assume that $c_i \in R^\circ$: replace c_i by $c_i + c_i'$ where $c_i' \in \mathfrak{p}_j$ if and only if $c_i \notin \mathfrak{p}_j$. (Such c_i' exist since the intersection of some minimal prime ideals is not contained in the union of the remaining ones.) In the next step we take $d = \sum_i c_i d_i$ where $d_i \notin \mathfrak{p}_i$, but $d_i \in \prod_{j \neq i} \mathfrak{p}_j$.

Now pick $r = p^f$ so large that $(\mathfrak{p}_1 \cdots \mathfrak{p}_m)^{[r]} = 0$. Then we have

$$(d_i c_i)^r x^{rq} \in (I^{[rq]} + \mathfrak{p}_i^{[r]}) \prod_{j \neq i} \mathfrak{p}_j^{[r]} \subset I^{[rq]} \qquad \text{for all } i.$$

This implies $d^r x^q \in I^{[q]}$ for $q \gg 0$. Since $d \in R^\circ$, we conclude $x \in I^*$.

(f) If height $I > 0$, then $R^\circ \cap I \neq \emptyset$, so that $c \in R^\circ$ with $cx^q \in I^{[q]}$ for $q \gg 0$ can be replaced by ca^r with $a \in I$ and r sufficiently large.

Suppose now that R is reduced. Applying the previous argument for the case of positive height to the residue class rings $R/\mathfrak{p}_i$, where $\mathfrak{p}_1, \ldots, \mathfrak{p}_m$

are again the minimal prime ideals of R, we find $c_i \in R^\circ$ such that $c_i x^q \in I^{[q]} + \mathfrak{p}_i$ for all $q \geq 0$. Now we choose d as in the proof of (e) and find that $dx^q \in I^{[q]} + \mathfrak{p}_1 \cdots \mathfrak{p}_m = I^{[q]}$.   □

Usually the computation of I^* is very difficult. We give two examples.

Examples 10.1.3. (a) Let $R_1 = k[X, Y, Z]/(X^2 - Y^3 - Z^7)$ where k is a field of arbitrary characteristic $p > 0$. Evidently R_1 is an integral domain, a complete intersection, and therefore Cohen–Macaulay. Furthermore, the ideal generated by the residue classes of the partial derivatives of $X^2 - Y^3 - Z^7$ is primary to the maximal ideal $\mathfrak{m} = (x, y, z)$ (small letters denote residue classes). The Jacobian criterion (for example, see [270], (30.4)) shows that $(R_1)_\mathfrak{p}$ is a regular local ring for $\mathfrak{p} \neq \mathfrak{m}$. Especially, R_1 is a normal ring by Serre's criterion 2.2.22. Setting $\deg X = 21$, $\deg Y = 14$, and $\deg Z = 6$ makes R_1 a positively graded k-algebra with *maximal ideal $\mathfrak{m}$. (All these assertions hold over an arbitrary field k.)

We claim that $x \in (y, z)^*$. If $p = 2$, then obviously $x^q \in (y, z)^{[q]}$ for all $q = p^e$. For $p > 2$ one has $cx^q \in (y, z)^{[q]}$ for $c = x$. In fact, set $u = (q + 1)/2$. Then x^{q+1} is a k-linear combination of monomials $y^{3v} z^{7w}$ with $v + w = u$. It is an elementary exercise that $3v \geq q$ or $7w \geq q$.

(b) Let $R_2 = k[X, Y, Z]/(X^2 - Y^3 - Z^5)$. Then, as in (a), R_2 is a normal complete intersection domain. The graduation is now given by $\deg X = 15$, $\deg Y = 10$, and $\deg Z = 6$. We claim that $x \notin (y, z)^*$ if and only if $\operatorname{char} k > 7$. In this case (y, z) is tightly closed because the only proper ideal of $R/(y, z)$ is generated by the residue class of x.

Evidently $S = k[y, z]$ is isomorphic to the polynomial ring in two indeterminates over k and R_1 is a free S-module with basis $1, x$. Therefore every element $f \in R_2$ has a unique presentation of the form $f_0 + x f_1$ with $f_0, f_1 \in k[y, z]$.

The case $p = 2$ is trivial. So suppose p is an odd prime. As above, set $u = (q + 1)/2$, choose $c \in R_2^\circ$, and let s and t denote the highest exponents with which y and z respectively appear in c_0 and c_1 where $c = c_0 + x c_1$ with $c_0, c_1 \in k[y, z]$. One has

$$x^{q+1} = c \sum_{v+w=u} \binom{u}{v} y^{3v} z^{5w}.$$

Therefore $cx^q \in (y^q, z^q)$ only if all the binomial coefficients $\binom{u}{v}$ for which $3v + s < q$ and $5w + t < q$ vanish modulo p.

First let $p > 7$. Then at least one (for $p > 30$ each) of the following inequalities has an integral solution:

$$(\text{i}) \quad \frac{3}{10}p < \alpha_p < \frac{p}{3}, \qquad (\text{ii}) \quad \frac{1}{6}p < \beta_p < \frac{1}{5}p.$$

If (i) has a solution α_p, then $v = p^{e-1}\alpha_p$ and $w = u - v$ satisfy the inequalities $3v + s < q$ and $5w + t < q$ for $e \gg 0$, $q = p^e$. Since none of the factors in the 'numerator' of $\binom{u}{v} = u(u-1)\cdots(u-v+1)/v!$ is divisible by p^e, one sees easily that $\binom{u}{v}$ is non-zero modulo p. If (ii) has an integral solution, the argument is analogous. This shows $cx^q \in (y^q, z^q)$ for $q \gg 0$ is impossible.

Second, for $p = 7$ neither (i) nor (ii) has an integral solution. Nevertheless, there appears exactly one multiple of 7^{e-1} in the 'numerator' as well as in the 'denominator' of $\binom{u}{w} = \binom{u}{v}$. Therefore it is enough if we can choose w as an integral multiple of 7^{e-2} in the critical range, and this is possible since $49/6 < 9 < 49/5$.

The argument showing that $x \in (y, z)^*$ for $p = 3$ and $p = 5$ is left to the reader.

Though R_1 and R_2 have a very similar structure, there is an invariant distinguishing them: the a-invariant of R_1 is non-negative, namely $a(R_1) = 1$, whereas $a(R_2) = -1$ (see 3.6.14 and 3.6.15 for the computation of a-invariants). Therefore, if k is a field of characteristic 0, R_2 has a rational singularity by the criterion of Flenner [107] and Watanabe [389] whereas the singularity of R_1 is non-rational. The connection between rational singularities and tight closure will be discussed in Section 10.3, and we will see that the different behaviour of R_1 and R_2 with respect to tight closure is by no means accidental.

Remark 10.1.4. While it is usually not difficult to show that homologically defined invariants commute with localization or, in the case of a local ring $(R, \mathfrak{m})$, with $\mathfrak{m}$-adic completion, tight closure so far has resisted all efforts to establish these properties for it. It is obvious that $(I^*)_\mathfrak{p} \subset (I_\mathfrak{p})^*$ and $I^*\hat{R} \subset (I\hat{R})^*$, but the converse inclusions are only known in special cases, some of which will be discussed below. The best result available for localization is due to Aberbach, Hochster, and Huneke [2]: under some mild conditions on R one has $(I^*)_\mathfrak{p} = (I_\mathfrak{p})^*$ for ideals I of finite phantom projective dimension; this includes all ideals of finite projective dimension. The definition of finite phantom projective dimension requires the introduction of tight closure for submodules $U \subset M$ (see Hochster and Huneke [192] and Aberbach [1]).

The following proposition indicates how elements in the tight closure of an ideal may arise in a non-trivial way.

Proposition 10.1.5. *Let $S \supset R$ be a module-finite R-algebra. Then one has $(IS)^* \cap R \subset I^*$ for all ideals I of R.*

PROOF. Assume first that R and S are integral domains. Then there are a free R-submodule F of S and an element $e \in R$, $e \neq 0$, with $eS \subset F$, and for each element $u \in F$, $u \neq 0$, there exists an R-linear map $f : F \to R$ with $f(u) \neq 0$. Therefore, given $c \in S^\circ$, one can find an R-linear map $g : S \to R$ with $g(c) \neq 0$.

Now pick $x \in (IS)^* \cap R$. Then there is a $c \in S^\circ$ with $cx^q \in (IS)^{[q]} = I^{[q]}S$ for all $q \gg 0$, and applying an R-linear map g one gets $g(c)x^q \in I^{[q]}$. Choosing g as above, one concludes $x \in I^*$.

In the general case let $\mathfrak{p}_1, \ldots, \mathfrak{p}_m$ be the minimal prime ideals of R, and pick minimal prime ideals $\mathfrak{q}_1, \ldots, \mathfrak{q}_m$ with $\mathfrak{p}_i = \mathfrak{q}_i \cap R$. (This is possible by A.6.) Let $\pi_i : R \to R/\mathfrak{p}_i$ be the natural map and φ_i the composition $R \to R/\mathfrak{p}_i \to S/\mathfrak{q}_i$. Then

$$(IS)^* \cap R \subset \bigcap_i \varphi_i^{-1}\big((IS)^*/\mathfrak{q}_i\big) \subset \bigcap_i \varphi_i^{-1}(IS/\mathfrak{q}_i)^* \subset \pi_i^{-1}(IR/\mathfrak{p}_i)^*$$

where the last inclusion is given by the first part of the proof. By virtue of 10.1.2(e) it follows that $(IS)^* \cap R \subset I^*$.   □

In the next remark and in Section 10.3 we will need the notion of excellence for rings. A Noetherian ring R is called *excellent* if it satisfies the following conditions:

(i) R is universally catenary;

(ii) for all prime ideals $\mathfrak{p}$ of R, all prime ideals $\mathfrak{q}$ of $R_\mathfrak{p}$, and all finite field extensions $L \supset k(\mathfrak{q})$ the ring $(R_\mathfrak{p})\widehat{} \otimes L$ is regular $((R_\mathfrak{p})\widehat{}$ is the $\mathfrak{p}R_\mathfrak{p}$-adic completion of $R_\mathfrak{p}$);

(iii) for every finitely generated R-algebra S the *singular locus* $\operatorname{Sing} S = \{\mathfrak{q} \in \operatorname{Spec} S : S_\mathfrak{q}$ non-regular$\}$ is closed in $\operatorname{Spec} S$.

Property (ii) is called the *geometric regularity of the formal fibres of all localizations of R*. Complete local rings, and in particular fields are excellent. Furthermore the localizations of an excellent ring R and the finitely generated R-algebras are excellent as well. We refer the reader to [270], §32 or [142], IV.7.8 for a systematic development of this concept.

Remarks 10.1.6. (a) Suppose R is a domain and S a module-finite extension domain. Then the field of fractions of S is an algebraic extension of R and can therefore be embedded into a fixed algebraic closure L of the field of fractions of R. Through this embedding, S is contained in the integral closure R^+ of R in L; one calls R^+ the *absolute integral closure* of R. Conversely, R^+ is the union of module-finite extension domains of R. Thus 10.1.5 implies $IR^+ \cap R \subset I^*$. It is not known whether equality holds in general, but Smith [348] has proved that $IR^+ \cap R = I^*$ for ideals I of the principal class in domains R such that $R_\mathfrak{p}$ is excellent for all $\mathfrak{p} \in \operatorname{Spec} R$.

(b) By a remarkable theorem of Hochster and Huneke [193], the ring R^+ is a big Cohen–Macaulay algebra for R if R is an excellent local domain of characteristic p. This allows one to construct big Cohen–Macaulay algebras for all Noetherian local rings containing a field; moreover, the construction is 'functorial' in the best possible way. See

Hochster and Huneke [197] for the numerous applications of the existence and functoriality of big Cohen–Macaulay algebras.

The next theorem gives a crucial property of tight closure. It also shows that the attribute 'tight' is well chosen.

Theorem 10.1.7. *Let R be a regular ring. Then*
(a) $I^{[q]} : J^{[q]} = (I : J)^{[q]}$ *for all ideals I and J of R, and*
(b) *every ideal of R is tightly closed.*

PROOF. (a) By induction it is enough to show $I^{[p]} : J^{[p]} = (I : J)^{[p]}$. One has $I^{[p]} = IR^F$ where R^F is R viewed as an R-algebra via the Frobenius endomorphism F. For a regular ring R, the R-algebra R^F is flat by Kunz's theorem 8.2.8, and we show more generally that $IS : JS = (I : J)S$ if S is a flat algebra over R.

The ideal $I : J$ is the annihilator of the R-module $(J + I)/I$. Since S is flat, one has natural isomorphisms $(I : J) \otimes S \cong (I : J)S$ and

$$((J + I)/I) \otimes S \cong ((J + I) \otimes S)/(I \otimes S) \cong (J + I)S/IS.$$

Therefore it is enough to show $(\mathrm{Ann}_R M)S = \mathrm{Ann}_S(M \otimes S)$ for a finite R-module M. This follows by tensoring the exact sequence $0 \to \mathrm{Ann}_R M \to R \to \mathrm{End}_R(M)$ with S and using the natural isomorphism $\mathrm{End}_R(M) \otimes S \cong \mathrm{End}_S(M \otimes S)$.

(b) Let I be an ideal of R and suppose that $cx^q \in I^{[q]}$ for $x \in R$, $x \notin I$, $c \in R^\circ$, and $q \gg 0$. Then $I : x \neq R$, and all the conditions remain true after localization at a prime ideal containing $I : x$. In order to derive a contradiction we may therefore assume that R is local with maximal ideal $\mathfrak{m}$.

By (a) one has $(I : x)^{[q]} = I^{[q]} : x^q$ for all $q \geq 0$. Therefore, if $c \in I^{[q]} : x^q$ for $q \gg 0$, then $c \in (I : x)^{[q]} \subset \mathfrak{m}^{[q]} \subset \mathfrak{m}^q$ for $q \gg 0$. This implies $c = 0$, the desired contradiction. □

For several theorems below it will be essential that R is *equidimensional*: this means $\dim R/\mathfrak{p} = \dim R < \infty$ for all minimal prime ideals $\mathfrak{p}$ of R.

Corollary 10.1.8. *Suppose R is equidimensional and a finite module over a regular domain A. Then $IR :_R JR \subset ((I :_A J)R)^*$ and $IR \cap JR \subset ((I \cap J)R)^*$ for all ideals I and J of A.*

PROOF. There exist $c \in A$, $c \neq 0$, and a free A-submodule F of R such that $cR \subset F$. Choose $x \in IR :_R JR$. Then $x^q J^{[q]} \subset I^{[q]}R$ for all q. Multiplication with c yields $J^{[q]}(cx^q) \subset I^{[q]}F$. Since F is a free A-module, this implies $cx^q \in (J^{[q]} : I^{[q]})F$. By 10.1.7(a) one has $(J^{[q]} : I^{[q]})F = (I : J)^{[q]}F$, and so $cx^q \in (I : J)^{[q]}F \subset (I : J)^{[q]}R$. The argument for $IR \cap JR$ is similar.

It remains to show $c \in R^{\circ}$ for which we need the hypothesis that R is equidimensional. Let $\mathfrak{p}$ be a minimal prime ideal of R. Then $\dim R/\mathfrak{p} = \dim A/(\mathfrak{p} \cap A)$ by the corollary A.8 of the going-up theorem, and there exists such a prime ideal $\mathfrak{p}_0$ with $\mathfrak{p}_0 \cap A = 0$. Especially, $\dim A = \dim R = \dim A/(\mathfrak{p} \cap A)$ and, hence, $\mathfrak{p} \cap A = 0$ for all minimal prime ideals $\mathfrak{p}$ of R. (Conversely, this fact implies that R is equidimensional.) $\square$

If, in the situation of 10.1.8, R (and therefore A) is a local ring, then every system of parameters $x_1 \ldots, x_d$ of A is also a system of parameters of R and an A-sequence. The last condition is equivalent to

$$(x_1, \ldots, x_j) :_A x_{j+1} = (x_1, \ldots, x_j).$$

Since A is regular, $(x_1, \ldots, x_j)^* = (x_1, \ldots, x_j)$ for $j = 0, \ldots, d-1$ by 10.1.7, and so

$$(x_1, \ldots, x_j) :_R x_{j+1} \subset (x_1, \ldots, x_j)^*, \qquad j = 0, \ldots, d-1.$$

If R is an equidimensional complete local ring, then we can always find a suitable regular 'Noether normalization' A (see A.22). Roughly speaking one may therefore say that R is 'Cohen–Macaulay up to tight closure'. This holds for a larger class of local rings.

Theorem 10.1.9 (Hochster–Huneke). *Let R be an equidimensional residue class ring of a Cohen–Macaulay local ring A, and $x_1 \ldots, x_d$ a system of parameters of R. Then*

$$(x_1, \ldots, x_j) :_R x_{j+1} \subset (x_1, \ldots, x_j)^*, \qquad j = 0, \ldots, d-1.$$

PROOF. We write $R = A/I$. Lemma 10.1.10 below shows that there exists a system of parameters $z_1, \ldots, z_g, y_1, \ldots, y_d$ in A with $g = \operatorname{codim} I$ such that $z_1, \ldots, z_g \in I$ and x_i is the residue class of y_i. Since A is Cohen–Macaulay, $z_1, \ldots, z_g, y_1, \ldots, y_d$ is an A-sequence.

Set $J = (z_1, \ldots, z_g)$. Since R is equidimensional, all the minimal prime ideals $\mathfrak{p}_1, \ldots, \mathfrak{p}_m$ of I have height g, and are therefore minimal prime ideals of J. Let $\mathfrak{p}_{m+1}, \ldots, \mathfrak{p}_n$ be the remaining minimal prime ideals of J. If we now choose $c \in (\mathfrak{p}_{m+1} \cap \cdots \cap \mathfrak{p}_n)^s \setminus (\mathfrak{p}_1 \cup \cdots \cup \mathfrak{p}_m)$ for s sufficiently large, then $cI^r \subset J$ for some $r > 0$. Furthermore the residue class d of c in R belongs to R°.

Suppose that $b x_{j+1} \in (x_1, \ldots, x_j)$ for some $b \in R$. Then we pick a preimage a of b in A, obtaining a relation $a y_{j+1} - (a_1 y_1 + \cdots + a_j y_j) \in I$. For $q = p^e \geq r$ this entails

$$c a^q y_{j+1} - (a_1^q y_1^q + \cdots + a_j^q y_j^q) = b_{q1} z_1 + \cdots + b_{qg} z_g \qquad \text{with } b_{uv} \in A.$$

However $z_1, \ldots, z_g, y_1, \ldots, y_d$ is an A-sequence, and so is $z_1, \ldots, z_g, y_1^q, \ldots, y_d^q$ (see 1.1.10). Therefore $c a^q \in (y_1, \ldots, y_j)^{[q]} + I$ for all $q \geq r$, and taking residue classes we get the desired result. $\square$

Lemma 10.1.10. *Let* $(A, \mathfrak{m})$ *be a Noetherian local ring (not necessarily of characteristic p),* I *a proper ideal of* A, *and* $x_1, \ldots, x_d$ *a system of parameters of* A/I. *Then one can find representatives* $y_1, \ldots, y_d$ *of* $x_1, \ldots, x_d$ *in* A *and* $z_1, \ldots, z_g \in I$, $g = \operatorname{codim} I$, *such that* $z_1, \ldots, z_g, y_1, \ldots, y_d$ *is a system of parameters for* A.

PROOF. Note that $g + d = \dim A$. Suppose we have constructed representatives $y_1, \ldots, y_d$ of $x_1, \ldots, x_d$ such that $\operatorname{codim} J = d$ for $J = (y_1, \ldots, y_d)$. Since $\dim A/J = g$ and since $(I + J)/J$ is $\mathfrak{m}/J$-primary, we can then find $z_1, \ldots, z_g \in I$ that complement $y_1, \ldots, y_d$ to a system of parameters.

The elements $y_1, \ldots, y_d$ are constructed inductively. Assume that $y_1, \ldots, y_{j-1}$ have been found such that $\operatorname{codim}(y_1, \ldots, y_{j-1}) = j - 1$. Choose a representative y_j' of x_j. Then

$$\dim A/(y_1, \ldots, y_{j-1}, I, y_j') = d - j < g + d - j + 1 = \dim A/(y_1, \ldots, y_{j-1}).$$

Thus $I + (y_j')$ is not contained in any of the finitely many prime ideals $\mathfrak{p}_1, \ldots, \mathfrak{p}_m \supset (y_1, \ldots, y_{j-1})$ with $\dim A/\mathfrak{p}_i = g + d - j + 1$. Now Lemma 1.2.2 (with $M = A$ and $N = I + (y_j')$) yields a representative y_j of x_j such that $y_j \notin \mathfrak{p}_i$ for $i = 1, \ldots, m$. $\qquad\square$

In the case in which R is a residue class ring of a Gorenstein local ring, one can give a shorter proof of 10.1.9, using 8.1.3. This technique will be applied in the proof of 10.4.4.

F-regularity. Theorem 10.1.9 shows that rings in which every ideal is tightly closed have special properties. They deserve a special name.

Definition 10.1.11. One says R is *weakly F-regular* if every ideal of R is tightly closed. If all rings R_T of fractions of R are weakly F-regular, then R is *F-regular*.

The distinction between weak F-regularity and F-regularity is undesirable but hard to avoid as long as the localization of tight closure has not been proved. However, it is enough to require F-regularity for the localizations $R_\mathfrak{p}$, $\mathfrak{p} \in \operatorname{Spec} R$:

Proposition 10.1.12. (a) *Let* I *be an ideal primary to a maximal ideal* $\mathfrak{m}$. *Then* $(I R_\mathfrak{m})^* = I^* R_\mathfrak{m}$.
(b) *If every ideal primary to a maximal ideal is tightly closed, then* R *is weakly F-regular.*
(c) R *is weakly F-regular if and only if* $R_\mathfrak{m}$ *is weakly F-regular for all maximal ideals* $\mathfrak{m}$.
(d) *A weakly F-regular ring is normal.*
(d) *If* R *is a weakly F-regular residue class ring of a Cohen–Macaulay ring, then* R *is Cohen–Macaulay.*

PROOF. (a) We only need to show the inclusion $(IR_\mathfrak{m})^* \subset I^*R_\mathfrak{m}$, and it holds if $(IR_\mathfrak{m})^* \cap R \subset I^*$. By virtue of 10.1.2 it is enough that

$$((IR_\mathfrak{m})^* \cap R + \mathfrak{p})/\mathfrak{p} \subset ((I + \mathfrak{p})/\mathfrak{p})^*$$

for all minimal prime ideals $\mathfrak{p}$ of R. If $\mathfrak{m} \not\subset \mathfrak{p}$, equivalently $I \not\subset \mathfrak{p}$, then both sides equal $R/\mathfrak{p}$. So suppose $\mathfrak{p} \subset \mathfrak{m}$. Then the image of $x \in (IR_\mathfrak{m})^* \cap R$ under the natural map $R \to R_\mathfrak{m}/\mathfrak{p}R_\mathfrak{m}$ certainly belongs to the tight closure of $(I + \mathfrak{p})R_\mathfrak{m}/\mathfrak{p}R_\mathfrak{m}$. This observation reduces (a) to the case of an integral domain R in which we have $R_\mathfrak{m}^\circ \cap R = R^\circ$. (So far we have only used that $R_\mathfrak{m}$ is a localization of R.)

Suppose that $cx^q \in I_\mathfrak{m}^{[q]}$ for $x \in R$, $c \in R_\mathfrak{m}^\circ$, and $q \gg 0$. Then we can obviously assume $c \in R$. It follows that $c \in R^\circ$. Furthermore $cx^q \in I_\mathfrak{m}^{[q]} \cap R = I^{[q]}$ where for the last equation we have used that $I^{[q]}$ is $\mathfrak{m}$-primary because $\mathrm{Rad}\, I^{[q]} = \mathfrak{m}$ and $\mathfrak{m}$ is a maximal ideal.

(b) By Krull's intersection theorem, every ideal I of R is the intersection of the ideals $I + \mathfrak{m}^n$ where $\mathfrak{m}$ is a maximal ideal containing I and $n \in \mathbb{N}$. Furthermore the intersection of tightly closed ideals is tightly closed.

(c) is an immediate consequence of (a) and (b).

(d) will be proved after 10.2.7.

(e) We must show that $R_\mathfrak{m}$ is Cohen–Macaulay for all maximal ideals $\mathfrak{m}$. Part (c) implies that $R_\mathfrak{m}$ is weakly F-regular. Thus $R_\mathfrak{m}$ is a normal domain by (d) and, therefore, equidimensional. Now the Cohen–Macaulay property results from 10.1.9.   □

The following proposition yields the most important examples of F-regular rings.

Proposition 10.1.13. *Let $S \supset R$ be a (weakly) F-regular R-algebra such that $IS \cap R = I$ for all ideals I of R. If $R^\circ \subset S^\circ$, then R is (weakly) F-regular.*

PROOF. The hypothesis $\varphi(R^\circ) \subset S^\circ$ implies that $(I^*)S \subset (IS)^*$, whence the assertion about weak F-regularity is obvious. Furthermore it is inherited by every localization, as is the condition $IS \cap R = I$: if the induced homomorphism $R/I \to S/IS$ is injective, then so is $R_T/IR_T \to S_T/IS_T$ for all multiplicatively closed subsets T of R, and every ideal of R_T has the form IR_T for an ideal I of R.   □

The hypothesis $IS \cap R = I$ is satisfied if R is a direct summand of S as an R-module or, more generally, if S is pure over R (see 6.5.3(b) for the notion of purity). An immediate corollary is the characteristic p version of the Hochster–Roberts theorem.

Corollary 10.1.14. *Let the ring R be a direct summand of the regular ring S. If R is a residue class ring of a Cohen–Macaulay ring, then R is Cohen–Macaulay.*

Exercises

10.1.15. (a) Let I and J be ideals of R. Show $(I \cap J)^* \subset I^* \cap J^*$, $(I+J)^* = (I^*+J^*)^*$, and $(IJ)^* = (I^*J^*)^*$; furthermore $(0)^* = \text{Rad}(0)$.
(b) Let $\bar{R} = R/\text{Rad}(0)$. Prove I^* is the preimage of $(I\bar{R})^*$ under the natural homomorphism $R \to \bar{R}$.

10.1.16. (a) Let $x_1, \ldots, x_n, y, z$ be elements of R such that the ideals $(x_1, \ldots, x_n, y)$ and $(x_1, \ldots, x_n, z)$ are tightly closed and $\text{grade}(x_1, \ldots, x_n, y) = n + 1$. Show $(x_1, \ldots, x_n, yz)$ is tightly closed. (Use 1.6.17.)
(b) Suppose that $\text{grade}(x_1, \ldots, x_n) = n$ and $(x_1, \ldots, x_n)$ is tightly closed. Show $(x_1^{a_1}, \ldots, x_n^{a_n})$ is tightly closed for all integers $a_1, \ldots, a_n \geq 1$.

10.1.17. Find a tight closure proof of the 'monomial theorem' 9.2.1 in characteristic p.

10.1.18. (a) Show that a primary component $\mathfrak{q}$ of a tightly closed ideal I that belongs to a minimal prime ideal of I is tightly closed. (Hint: $\mathfrak{q} = I : x$ for a suitable x.)
(b) Let I be a tightly closed ideal such that the maximal ideal $\mathfrak{m}$ is a minimal prime ideal of I. Show $I_\mathfrak{m}$ is tightly closed.
(c) Let I be an ideal all of whose minimal prime ideals are maximal ideals. Show I is tightly closed if and only if all the localizations $I_\mathfrak{m}$ with respect to maximal ideals $\mathfrak{m}$ are tightly closed.

10.1.19. (Smith) Prove the following assertions:
(a) If tight closure commutes with localization in $R/\mathfrak{p}$ for each minimal prime of $\mathfrak{p}$ of R, then tight closure commutes with localization in R.
(b) Let R be a domain that has an F-regular module-finite extension. Then tight closure commutes with localization in R.
(c) Tight closure commutes with localization in rings $R = k[X_1, \ldots, X_n]/I$ where I is generated by monomials and binomials. (Hint: the minimal prime ideals of I are again generated by such elements, and if I is prime, then R is an affine semigroup ring. See Eisenbud and Sturmfels [95] for the theory of binomial ideals.)

10.2 The Briançon–Skoda theorem

This section is devoted to the relationship between the tight closure and the integral closure of an ideal. Our major objective is a proof of the Briançon–Skoda theorem for regular rings containing a field. It will be derived from its tight closure variant by reduction to characteristic p.

Integral dependence on an ideal. We first discuss the basic notion of integral dependence on an ideal I and introduce the integral closure of I.

Definition 10.2.1. Let R be a ring and $I \subset R$ an ideal. Then $x \in R$ is *integrally dependent on* I or *integral over* I if and only if there exists an equation

$$x^m + a_1 x^{m-1} + \cdots + a_m = 0 \qquad \text{with } a_i \in I^i, \ i = 1, \ldots, m.$$

The elements $x \in R$ that are integral over I form the *integral closure* $\bar{I}$ *of* I.

It is evident that $I \subset \bar{I} \subset \mathrm{Rad}\, I$, that $I_1 \subset I_2 \Rightarrow \bar{I}_1 \subset \bar{I}_2$, and that integral dependence is preserved under ring homomorphisms. The following proposition lists less obvious properties of integral dependence.

Proposition 10.2.2. (a) *The following are equivalent:*
(i) $x \in \bar{I}$;
(ii) *there exists* $m \geq 1$ *with* $x^m \in I(I + Rx)^{m-1}$;
(iii) *there exists* $m \geq 1$ *with* $(I + Rx)^{m+k} = I^{k+1}(I + Rx)^{m-1}$ *for all* $k \in \mathbb{N}$;
(iv) *there exists a finite ideal* $J \subset R$ *such that* $xJ \subset IJ$ *and* $\mathrm{Ann}\, J$ *annihilates a power of* x.
(b) $\bar{I}$ *is an integrally closed ideal.*
(c) *Suppose that R is Noetherian. Then $x \in \bar{I}$ if and only if the residue class of x is integral over* $(I + \mathfrak{p})/\mathfrak{p}$ *for all minimal prime ideals $\mathfrak{p}$ of R.*

PROOF. (a) The equivalence of (i) and (ii) is evident, and (ii) results from (iii) with $k = 0$. Conversely, $x^m \in I(I + Rx)^{m-1}$ implies $(I + Rx)^m = I(I + Rx)^{m-1}$ from which (iii) follows by induction on k.

For (i) $\Rightarrow$ (iv) pick $x \in \bar{I}$. Then there exists a finite subideal I' of I over which x is integral. Therefore we may assume I to be finite and choose $J = Rx^{m-1} + Ix^{m-2} + \cdots + I^{m-1}$.

For (iv) $\Rightarrow$ (i) let J be generated by $y_1, \ldots, y_n$. Then there exists an $n \times n$ matrix $A = (a_{ij})$ with $a_{ij} \in I$ such that $(xE_n - A)y = 0$ where y is the column vector with components y_j and E_n is the $n \times n$ unit matrix. It follows that $\det(xE_n - A)J = 0$, and so $\det(xE_n - A) \in \mathrm{Ann}\, J$. Upon multiplication by a power of x we obtain an equation showing $x \in \bar{I}$.

(b) It is obvious that $ax \in \bar{I}$ for all $x \in \bar{I}$ and $a \in R$. Suppose $x_1, x_2 \in \bar{I}$. Again we may assume that I is finitely generated and we choose J_1 for x_1 and J_2 for x_2 as we have chosen J for x above; especially, both J_1 and J_2 contain a power of I. It follows immediately that $(x_1 + x_2)J_1J_2 \subset IJ_1J_2$; furthermore $\mathrm{Ann}\, J_1J_2$ annihilates a power of I and therefore annihilates $(x_1 + x_2)^n$ for $n \gg 0$.

The argument showing that $\bar{I}$ is integrally closed is similar and can be left to the reader.

(c) The 'only if' part is obvious. For the 'if' part let $\mathfrak{p}_1, \ldots, \mathfrak{p}_r$ be the minimal prime ideals of R. We lift an integral dependence equation of the residue class of x with respect to $(I + \mathfrak{p}_i)/\mathfrak{p}_i$ to a relation $F_i(x) \in \mathfrak{p}_i$ such that the coefficients of the powers of x satisfy the requirements of Definition 10.2.1. Then $F(x) = F_1(x) \cdots F_r(x) \in \mathfrak{p}_1 \cdots \mathfrak{p}_r$, and a suitable power of $F(x)$ vanishes.   □

For ideals $J \subset I$ of a Noetherian ring R one has $\bar{I} = \bar{J}$ if and only if J is a reduction ideal of I (see Exercise 10.2.10).

We note a useful criterion for normality.

Proposition 10.2.3. *A Noetherian ring R is normal if and only if it satisfies the following conditions:*
(i) $R_\mathfrak{p}$ *is a field for each prime ideal* $\mathfrak{p}$ *that is both minimal and maximal;*
(ii) *the principal ideals* (x), $x \in R^\circ$, *are integrally closed.*

PROOF. The essential observation relating normality and condition (ii) is the following: let x be a regular element of R and suppose we have an integral dependence relation

$$y^m + a_1 x y^{m-1} + \cdots + a_{m-1} x^{m-1} y + a_m x^m = 0, \qquad a_i \in R.$$

Then the element y/x of the total ring of fractions Q of R is integral over R. Now, if R is integrally closed in Q, it follows that $y/x \in R$ and, hence, $y \in (x)$. Conversely, if f/g ($f, g \in R$, g a regular element) is integral over R, one sees immediately that f is integral over the ideal (g).

Suppose now that R is normal. Then it is the direct product of finitely many integrally closed domains. Therefore it obviously satisfies condition (i). Furthermore every element $x \in R^\circ$ is a regular element of R so that the previous observation immediately yields that (x) is integrally closed.

For the converse we first split R into a direct product $R_1 \times \cdots \times R_r$ such that $\operatorname{Spec} R_i$ is irreducible for each of the rings R_i. It suffices to show that each R_i is a normal domain. Note that condition (ii) is inherited by R_i. Furthermore condition (i) implies that R_i is a field if R_i has a prime ideal that is both minimal and maximal. So we can assume that $\operatorname{Spec} R$ is irreducible and R has no such prime ideal.

The first (and crucial) step is to show that R is reduced. For each minimal prime ideal $\mathfrak{p}_i$, $i = 1, \ldots, s$, there is a non-minimal prime ideal $\mathfrak{q}_i \supset \mathfrak{p}_i$. Choose $a \in (\bigcap_i \mathfrak{q}_i) \setminus (\bigcup_i \mathfrak{p}_i)$. The nilradical N of R is contained in every integrally closed ideal, and therefore it is contained in $\bigcap_j (a^j)$ since $a \in R^\circ$. There exists an element $c \in R$ such that $b = 1 - ca$ annihilates $\bigcap_j (a^j)$ (this is the usual argument from which Krull's intersection theorem is derived). A fortiori, $bN = 0$. The choice of a ensures that $b \in R^\circ$ as well, and, by the same token, we have $(1 - db)N = 0$ for some $d \in R$. This shows $N = 0$.

Since R is reduced, the total ring of fractions Q of R is the direct product of fields Q_i. The idempotents e_i representing the unit elements of Q_i satisfy the equation $e_i^2 - e_i = 0$. Write $e_i = f_i/g_i$ with $f_i \in R$ and a regular element $g_i \in R$. The initial observation yields $e_i \in R$. By the assumption on R this is only possible if Q is a field and, hence, R is a domain. Now we apply the initial observation once more to conclude that R is integrally closed. $\square$

For the connection with tight closure it is important that in a Noetherian ring integral dependence can be characterized by homomorphisms to valuation rings. Let K be a field. Recall that a proper subring V of K is a *valuation ring of* K if $x \in V$ or $x^{-1} \in V$ for all $x \in K$. It follows that the set of ideals of V is linearly ordered by inclusion; in particular V is local and every finite ideal of V is principal. If V is Noetherian, then the maximal ideal $\mathfrak{m}_V$ of V is principal, and conversely; a Noetherian valuation ring V is a regular local ring of dimension 1 and is termed a discrete valuation ring.

The following theorem will be crucial: let K be a field, A a subring of K, and $\mathfrak{p} \subset A$ a prime ideal of A; then there exists a valuation ring V of K such that $A \subset V$ and $\mathfrak{m}_V \cap A = \mathfrak{p}$. Furthermore it is easily proved that a valuation ring is normal. (See [270], §10, [47], Ch. VI, or [397], Vol. II, Ch. VI for proofs and more information on valuation rings.)

Proposition 10.2.4. (a) *Let R be an integral domain with field of fractions K and I an ideal of R. Then $\bar{I}$ is the intersection of all ideals IV where V ranges over the valuation rings of K containing R.*
(b) *Suppose R is a Noetherian ring. Then there exist a finite number of homomorphisms φ_i from R to discrete valuation rings V_i such that $\mathrm{Ker}\,\varphi_i$ is a minimal prime ideal of R and $\bar{I}$ is the intersection of the preimages $\varphi^{-1}(IV_i)$.*

PROOF. (a) Let J be the intersection of the ideals IV. For $\bar{I} \subset J$ it is enough that all the ideals IV are integrally closed. As observed above, IV is a principal ideal of V. Since V is a normal domain, principal ideals of V are integrally closed (see 10.2.3 – for this implication the Noetherian property is irrelevant).

For the converse inclusion choose $x \in J$. Let L be the set of all quotients a/x with $a \in I$ and consider the ideal $LR[L]$ in the subring $R[L]$ of K. If $LR[L]$ were a proper ideal of $R[L]$, then there would exist a valuation ring V of K with $LR[L] \subset \mathfrak{m}_V$. In particular we would have $a/x \in \mathfrak{m}_V$ for all $a \in I$; this implies $x/a \notin V$, and so $x \notin IV$, a contradiction. Thus $LR[L] = R[L]$, and there exists a representation $1 = f(a_1/x, \ldots, a_m/x)$ where f is a polynomial with coefficients in R and $a_1, \ldots, a_m \in I$. Multiplication by a sufficiently high power of x yields an integral dependence relation for x on I.

(b) In view of 10.2.2 we can restrict ourselves to the case of a domain R. Choose a system of generators $x_1, \ldots, x_n$ of I and let R_i be the integral closure of $R[x_j/x_i : j = 1, \ldots, n]$ in K. We claim that $\bar{I} = \bigcap_i R_i x_i$. The inclusion '$\subset$' holds because the principal ideal $R_i x_i$ is integrally closed in the normal domain R_i. For the converse we use part (a). Let V be a valuation ring of K containing R, and pick an index i such that x_i generates IV. Then $x_j/x_i \in V$ for all j and therefore $R_i x_i \subset IV$. It follows that the intersection $\bigcap_i R_i x_i$ is contained in every ideal IV where V ranges over the valuation rings of K.

Though the ring R_i need not be Noetherian, it is a Krull ring (see

[270], §12). A divisorial ideal $\mathfrak{a}$ of a Krull ring, and especially a principal ideal, has the primary decomposition $\mathfrak{a} = \bigcap_j (\mathfrak{a} R_{\mathfrak{p}_j} \cap R)$ where the $\mathfrak{p}_j$ are the finitely many divisorial prime ideals containing $\mathfrak{a}$, and furthermore $R_{\mathfrak{p}_j}$ is a discrete valuation ring. □

Tight closure and integral closure. After these preparations we can easily show that tight closure is tighter than integral closure. In the sequel we shall again assume that R is a Noetherian ring of characteristic p.

Proposition 10.2.5. *One has* $I^* \subset \bar{I}$ *for all ideals I of R.*

PROOF. Let φ be a homomorphism from R to a discrete valuation ring V such that Ker φ is a minimal prime ideal of R. Then $\varphi(I^*)V \subset (IV)^*$ since $\varphi(R^\circ) \subset V^\circ$. Moreover, V is a regular local ring and, thus, $(IV)^* = IV$. So 10.2.4 implies $I^* \subset \bar{I}$. □

It is easy to give examples of tightly closed ideals that are not integrally closed. For example, every ideal in a polynomial ring R over a field is tightly closed, but not every ideal of R is integrally closed if $\dim R > 1$ (see Exercise 10.2.12).

The tight closure version of the Briançon–Skoda theorem is an 'asymptotic' converse of the previous proposition.

Theorem 10.2.6 (Hochster–Huneke). *Let I be an ideal of R generated by elements $f_1, \ldots, f_n$.*
(a) *Then $\overline{I^{n+w}} \subset (I^{w+1})^*$ for all $w \in \mathbb{N}$.*
(b) *If R is regular or just weakly F-regular, then $\overline{I^{n+w}} \subset I^{w+1}$, and in particular $\overline{I^n} \subset I$.*

PROOF. We must relate Frobenius powers and ordinary powers of I. This is possible through the equation

$$I^{k(n+w)} = (f_1^k, \ldots, f_n^k)^{w+1} I^{k(n-1)},$$

whose elementary verification is left to the reader.

In view of 10.1.2 and 10.2.2 we may assume that R is an integral domain. Set $J = I^{n+w}$ and pick $x \in \bar{J}$. By 10.2.2 there exists $m \geq 1$ with $(J + Rx)^{m+k} = J^{k+1}(J + Rx)^{m-1}$ for all $k \in \mathbb{N}$; in particular

$$x^m x^k \in J^k = I^{k(n+w)} \subset (f_1^k, \ldots, f_n^k)^{w+1} I^{k(n-1)}$$

for all $k \in \mathbb{N}$. Setting $c = x^m$ and $k = q = p^e$ we obtain

$$cx^q \in (f_1^q, \ldots, f_n^q)^{w+1} I^{q(n-1)} \subset (I^{w+1})^{[q]},$$

as desired.

Part (b) results immediately from (a). □

The following corollary is crucial for issues of normality.

Corollary 10.2.7. *Let $I = (x)$ be a principal ideal. Then $\bar{I} = I^*$.*

PROOF. The inclusion $I^* \subset \bar{I}$ is Proposition 10.2.5, and the converse inclusion is contained in the theorem.   □

As a consequence of 10.2.3 and 10.2.5 we derive the normality of a weakly F-regular ring R, which has already been stated in 10.1.12. Since $\text{Rad}(0) = (0)^*$, an F-regular ring is reduced, whence it satisfies condition (i) of 10.2.3. By definition it also fulfills condition (ii) so that normality follows immediately.

The original Briançon–Skoda theorem [347] is essentially the assertion of 10.2.6 for $R = \mathbb{C}\langle X_1, \ldots, X_d \rangle$, the ring of convergent power series in d indeterminates. It was motivated by the following problem: given $f \in (X_1, \ldots, X_d)$, what is the smallest number m such that $f^m \in I = (X_1 \partial_1 f, \ldots, X_d \partial_d f)$? (Here ∂_i is the partial derivative with respect to X_i.) Answering this question obviously generalizes the well-known rule $uf = X_1 \partial_1 f + \cdots + X_d \partial_d f$ for a homogeneous polynomial f of degree u. The connection with integral closure is given by the fact that $f \in \bar{I}$; this results easily from the criterion 10.2.4.

We derive a generalization of the original Briançon–Skoda theorem from 10.2.6 by reduction to characteristic p.

Theorem 10.2.8 (Lipman–Sathaye). *Let R be a regular ring containing a field of arbitrary characteristic and I be an ideal of R generated by elements $f_1, \ldots, f_n$. Then $\overline{I^{n+w}} \subset I^{w+1}$ for all $w \in \mathbb{N}$, and in particular $\overline{I^n} \subset I$.*

PROOF. The theorem has already been proved in characteristic p. So suppose that there exists a counterexample $(R, f_1, \ldots, f_n)$ in characteristic 0. Suppose that $y \in \overline{I^{n+w}}$ but, $y \notin I^{w+1}$. Then there is a maximal ideal $\mathfrak{m}$ of R such that $x \notin I_{\mathfrak{m}}^{w+1}$, and since integral closure commutes with localization (see Exercise 10.2.10), we may assume R is local.

For the application of the 'regular' variant (b) of Theorem 8.4.1 we must show that our data have a regular equational presentation. That $y \in \overline{I^{n+w}}$ can easily be expressed in terms of a single equation: we simply choose indeterminates representing y, the generators of I, and the coefficients in an integral dependence relation for y on I^{n+w}. The difficult part of the problem, namely to express the condition $y \notin I^{w+1}$, has fortunately been solved in Corollary 8.4.4. (Observe that the generators of I^{w+1} are polynomials in $f_1, \ldots, f_n$.)

It follows that there exists a counterexample to the theorem in which R is a regular local ring of characteristic $p > 0$, a contradiction to 10.2.6.   □

Remarks 10.2.9. (a) If $(R, \mathfrak{m})$ is a local ring with an infinite residue class field, then every ideal $I \subset \mathfrak{m}$ has a reduction ideal $J \subset I$ generated by at most $d = \dim R$ elements; see 4.6.8. Since J^w is a reduction of I^w for all $w \in \mathbb{N}$, one can replace n by the minimum of n and d in 10.2.6 and 10.2.8 if the hypothesis on R is satisfied.

(b) Theorem 10.2.8 was proved by Lipman and Sathaye [260] for arbitrary regular rings. A variant, valid for an ideal I generated by a regular sequence in a pseudo-rational ring (see 10.3.23 below), was given by Lipman and Teissier [261].

(c) Since it seems impossible to derive the mixed characteristic cases of these theorems from characteristic p results, the tight closure approach does not supersede the proofs given by Lipman–Sathaye and Lipman–Teissier. However it offers a refinement we have neglected so far, namely the extra factor $I^{q(n-1)}$ that appears in the proof of 10.2.6. Taking care of it leads one to the Briançon–Skoda theorems with coefficients of Aberbach and Huneke [4].

Exercises

10.2.10. (a) Let I and J be ideals in a ring R and $x \in R$ integral over I, $y \in R$ integral over J. Deduce xy is integral over IJ.

(b) Show that x is integral over the ideal I if and only if $xt \in R[t]$ is integral over the Rees algebra $\mathscr{R}(I) = R[It]$. (Thus integral dependence on ideals can be considered a special case of integral dependence on rings.)

(c) Let $J \subset I$ be ideals and suppose I is finitely generated. Prove that $\bar{I} = \bar{J}$ if and only if there exists $r \in \mathbb{N}$ with $JI^r = I^{r+1}$.

(d) Let $T \subset R$ be a multiplicatively closed set and $S = T^{-1}R$. Show $\overline{IS} = \bar{I}S$.

10.2.11. The definition of integral dependence can be extended as follows: let $R \subset S$ be rings and $I \subset R$ an ideal; then $x \in S$ is integral over I if it satisfies an equation as in 10.2.1. Extend 10.2.2 and 10.2.10 to this situation.

10.2.12. Let K be an arbitrary field and $I \subset K[X_1, \ldots, X_n]$ an ideal generated by monomials. Show that the integral closure of I is the ideal generated by all monomials whose exponent vector belongs to the convex hull (in $\mathbb{R}^n$ or $\mathbb{Q}^n$) of the set of exponent vectors of the monomials in I.

10.2.13. Given a regular local ring of dimension n, find an n-generated ideal I of R with $\overline{I^{n-1}} \not\subset I$.

10.3 F-rational rings

Throughout this section we suppose that all rings are Noetherian and of characteristic p, unless stated otherwise. Recall that in a weakly F-regular ring every ideal is tightly closed by definition. Now we discuss a weaker condition for a ring R:

Definition 10.3.1. One says R is *F-rational* if the ideals of the principal class, that is, ideals I generated by height I elements, are tightly closed.

Note that for an equidimensional, universally catenary local ring $(R, \mathfrak{m})$ an ideal $I = (x_1, \ldots, x_i)$ is of the principal class if and only if $x_1, \ldots, x_i$ are part of a system of parameters of R.

The name 'F-rational' indicates that such rings are the characteristic p analogues of rings with rational singularities. The results of Smith [349] and Hara [149] discussed at the end of this section justify this comparison.

In the following we present some basic properties of F-rational rings. Just as for weakly F-regular rings it results from 10.2.3 and 10.2.7 that

Proposition 10.3.2. F-rational rings are normal.

The following lemma is essential in the study of F-rational local rings.

Proposition 10.3.3. Let $(R, \mathfrak{m})$ be an equidimensional local ring that is a homomorphic image of a Cohen–Macaulay ring, and $(x_1, \ldots, x_d)$ a system of parameters of R. Then

(a) $(x_1, \ldots, x_{i-1})^* : x_i = (x_1, \ldots, x_{i-1})^*$ for all $i = 1, \ldots, d$;

(b) If $(x_1, \ldots, x_d)$ is tightly closed, then so is $(x_1, \ldots, x_i)$ for all $i = 1, \ldots, d$.

PROOF. Set $J_i = (x_1, \ldots, x_i)$ and pick $r \in J_{i-1}^* : x_i$. Then $rx_i \in J_{i-1}^*$, and hence there exists $c \in R^\circ$ such that $c(rx_i)^q \in J_{i-1}^{[q]}$ for q large; see 10.1.2(b). Thus from 10.1.9 we conclude that $cr^q \in J_{i-1}^{[q]} : x_i^q \subset J_{i-1}^*$, which yields $r \in J_{i-1}^*$. This proves (a).

We derive (b) by descending induction on i. Suppose it is already known that J_i is tightly closed. Let $r \in J_{i-1}^*$; then $r \in J_i^*$, and hence $r \in J_i$ by the induction hypothesis. So $r = a + x_i b$ with $a \in J_{i-1}$ and $b \in R$. Then $r - a \in J_{i-1}^*$, whence $b \in J_{i-1}^* : x_i = J_{i-1}^*$ by (a). This shows $J_{i-1}^* = J_{i-1} + x_i J_{i-1}^*$, and the conclusion follows from Nakayama's lemma. □

Corollary 10.3.4. An F-rational ring R is Cohen–Macaulay if it is a homomorphic image of a Cohen–Macaulay ring.

PROOF. Let $\mathfrak{m}$ be a maximal ideal of R. We choose elements $x_1, \ldots, x_d \in \mathfrak{m}$, $d = \dim R_\mathfrak{m}$ that generate an ideal I of the principal class. Especially, $x_1, \ldots, x_d$ form a system of parameters in $R_\mathfrak{m}$. By hypothesis I is tightly closed. As $\mathfrak{m}$ is a minimal prime ideal of I, we conclude from 10.1.18 that $I_\mathfrak{m}$ is tightly closed.

Notice that $R_\mathfrak{m}$ is a domain since it is normal by 10.3.2. Hence 10.3.3(b) entails that the ideals $(x_1, \ldots, x_i)R_\mathfrak{m}$ are tightly closed for all i. Now 10.3.3(a) and 10.1.9 imply that $x_1, \ldots, x_d$ is an $R_\mathfrak{m}$-sequence. Thus $R_\mathfrak{m}$ is Cohen–Macaulay. □

For local rings, F-rationality is easier to control. In fact one has

Proposition 10.3.5. *Let* $(R, \mathfrak{m})$ *be a local ring that is a homomorphic image of a Cohen–Macaulay ring. Then R is F-rational if and only if it is equidimensional and one ideal generated by a system of parameters is tightly closed.*

PROOF. Let $x_1, \ldots, x_d$ be a system of parameters of R generating a tightly closed ideal. By 10.3.3(b), R is F-rational if any other system of parameters $y_1, \ldots, y_d$ of R generates a tightly closed ideal as well. Choose $t \in \mathbb{N}$ such that $y_1, \ldots, y_d \in (x_1^t, \ldots, x_d^t)$, and write $y_i = \sum_{j=1}^d a_{ij} x_j^t$. Then $(y_1, \ldots, y_n) = (x_1^t, \ldots, x_d^t) : a$ with $a = \det(a_{ij})$. This follows from 2.3.10, since, by 10.3.4, R is Cohen–Macaulay, so that every system of parameters is R-regular. Now Exercise 10.1.16(b) tells us that $(x_1^t, \ldots, x_d^t)$ is tightly closed. Finally 10.1.2 implies that $(y_1, \ldots, y_d)$ is tightly closed, too.   □

Proposition 10.3.6. *Let R be a homomorphic image of a Cohen–Macaulay ring. Then R is F-rational if and only if $R_\mathfrak{m}$ is F-rational for every maximal ideal $\mathfrak{m}$ of R.*

PROOF. '⇐': Let $I \subset R$ be an ideal of the principal class. Suppose that I is strictly contained in I^*. Then for some maximal ideal $\mathfrak{m}$ of R we have again a strict inclusion $I_\mathfrak{m} \subset (I^*)_\mathfrak{m}$. It follows that $I_\mathfrak{m}$ is not tightly closed as $(I^*)_\mathfrak{m} \subset (I_\mathfrak{m})^*$. This is a contradiction since $I_\mathfrak{m}$ is of the principal class, and $R_\mathfrak{m}$ is F-rational.

'⇒': Let $\mathfrak{m}$ be a maximal ideal of R. As in the proof of 10.3.4 we conclude that some ideal in $R_\mathfrak{m}$ generated by a system of parameters is tightly closed. Hence the assertion results from 10.3.5.   □

Now we can easily show that for a Gorenstein ring 'F-rational' is a condition as strong as 'F-regular'.

Proposition 10.3.7. *A Gorenstein ring is F-regular if and only if it is F-rational.*

PROOF. In view of 10.3.6 and 10.1.12(c) we only need to show that an F-rational Gorenstein local ring is F-regular. In order to apply 10.1.12(b), we choose an ideal I which is primary to the maximal ideal $\mathfrak{m}$ of R and show it is tightly closed. There exists an ideal $J \subset I$ generated by a system of parameters. Since R is Gorenstein, we have $I = J : (J : I)$. (This follows immediately from Exercise 3.2.15 applied to the Artinian ring R/J). The ideal J is tightly closed since R is F-rational, and by 10.1.2(d), I is tightly closed as well.   □

The previous proposition cannot be generalized essentially; there exist F-rational, but not weakly F-regular rings of dimension 2; see Watanabe [390] or Hochster and Huneke [196], (7.15), (7.16).

F-rationality has good permanence properties; for example, it localizes as will easily follow from

Proposition 10.3.8. *Suppose I is an ideal generated by an R-sequence. Then $(IR_S)^* = I^*R_S$ for every multiplicatively closed set $S \subset R$.*

The proof of the proposition uses

Lemma 10.3.9. *Let R be an arbitrary Noetherian ring, $I \subset R$ an ideal, and $S \subset R$ a multiplicatively closed set.*
(a) *Then there exists an element $s \in S$ such that $\bigcup_{w \in S} I^m : w = I^m : s^m$ for all $m \in \mathbb{N}$.*
(b) *Suppose in addition that char $R = p > 0$, and that I is generated by an R-sequence $x_1, \ldots, x_n$. Then, with $s \in S$ as in part (a), we have*

$$\bigcup_{w \in S} I^{[q]} : w = I^{[q]} : s^{(n+1)q} \qquad \text{for all } q.$$

PROOF. (a) It is enough to show the inclusion '$\subset$'. Let T be the associated graded ring $\mathrm{gr}_I(R)$. Since T is Noetherian, there exists $s \in S$ such that $\mathrm{Ann}_T(s) = \mathrm{Ann}_T(ws)$ for all $w \in S$. (One chooses an element s for which $\mathrm{Ann}(s)$ is maximal.)

Now suppose $u \in I^m : w$, $u \in I^r \setminus I^{r+1}$. We may assume that $r < m$, since otherwise the assertion is trivial. We claim that $us^{m-r} \in I^m$. Indeed, $uw \in I^m \subset I^{r+1}$, and so $uws \in I^{r+1}$. By the choice of s this implies $us \in I^{r+1}$. Induction on r concludes the proof of (a).

(b) Again only the inclusion '$\subset$' needs proof. Given $w \in S$ and q, fix $u \in I^{[q]} : w$. We shall prove by induction on $h \in \mathbb{N}$ that the element $d_h = s^{q+h}u$ belongs to $I^{[q]} + I^{q+h}$. Once we know this, it follows for $h = qn$ that $s^{(n+1)q}u \in I^{[q]} + I^{q(n+1)} = I^{[q]}$. (The last equality holds, since $I^{q(n+1)} \subset I^{[q]}$.)

We start the induction with $h = 0$. Then $I^{[q]} + I^q = I^q$, and the assertion follows from (a).

Now suppose that $d_h \in I^{[q]} + I^{q+h}$ for some $h > 0$. Say,

$$d_h = \sum_i r_i' x_i^q + \sum_a r_a x^a, \qquad a = (a_1, \ldots, a_n) \in \mathbb{N}^n,$$

with $\sum_i a_i = q + h$ and $a_i < q$ for every i. As $wu \in I^{[q]}$, we get an equation

$$\sum_i r_i'' x_i^q = \sum_i w r_i' x_i^q + \sum_a w r_a x^a$$

with certain $r_i'' \in R$. This implies

$$\sum_i (w r_i'' - r_i') x_i^q + \sum_a w r_a x^a = 0.$$

Since $x_1, \ldots, x_n$ is R-regular, all the $w r_a$ are in I. Therefore $s r_a \in I$ for all a, and we conclude that $d_{h+1} = s d_h = d_h = \sum_i s r_i' x_i^q + \sum_a s r_a x^a$ lies in $I^{[q]} + I^{q+(h+1)}$. Indeed, the first sum belongs to $I^{[q]}$, the second to $I^{q+(h+1)}$. $\square$

PROOF OF 10.3.8. Let $u/1 \in (IR_S)^*$; then there exists an element $c \in R^\circ$ such that for all $q \gg 0$ one finds $s(q) \in S$ with $s(q)cu^q \in I^{[q]}$. It follows that $cu^q \in I^{[q]} : s(q)$. With $s \in S$ as in 10.3.9(b) we have $s^{(n+1)q}cu^q = c(s^{n+1}u)^q \in I^{[q]}$. This implies $s^{n+1}u \in I^*$, and so $u/1 \in I^*R_S$. The other inclusion is trivial. $\square$

Now we can show

Proposition 10.3.10. *Let R be an F-rational ring that is a homomorphic image of a Cohen–Macaulay ring, and S a multiplicatively closed set in R. Then R_S is F-rational.*

PROOF. By 10.3.6 it suffices to show that $R_\mathfrak{p}$ is F-rational for every prime ideal $\mathfrak{p}$ of R. Since R is Cohen–Macaulay, we have height $\mathfrak{p} = \operatorname{grade}(\mathfrak{p}, R)$ (see 2.1.4). Therefore there exists an R-sequence $x_1, \ldots, x_d \in \mathfrak{p}$ of length $d = $ height $\mathfrak{p}$. In $R_\mathfrak{p}$ this sequence forms a system of parameters. By 10.3.5 we only need that $(x_1, \ldots, x_d)R_\mathfrak{p}$ is tightly closed. But this results from 10.3.8. $\square$

Another easily proved permanence property is given by the following

Proposition 10.3.11. *Let $(R, \mathfrak{m})$ be a local ring, and let $x \in \mathfrak{m}$ be an R-regular element. Then R is F-rational, if R/xR is F-rational.*

PROOF. R/xR is Cohen–Macaulay by 10.3.4, and so is R. In particular R is equidimensional. We may extend x to a system of parameters $x, x_2, \ldots, x_d$ of R. According to 10.3.5 it suffices to show that $I = (x, x_2, \ldots, x_d)$ is tightly closed. Choose $u \in I^*$ and $c \in R^\circ$ such that $cu^q \in I^{[q]}$ for $q \gg 0$. We may write $c = dx^t$ where $d \notin x^t R$ for some t. Then $du^q \in (x^{q-t}, x_2^q, \ldots, x_d^q)$ for $q \gg 0$. Since $d \neq 0$ and since the F-rational ring R/xR is a domain, the image of u in R/xR is in the tight closure of $(x_2, \ldots, x_d)R/xR$. Since this ideal is tightly closed, $u \in I$ as desired. $\square$

At this point it is useful to resume the discussion of the examples 10.1.3.

Examples 10.3.12. (a) We have seen that $x \in (y, z)^*$ for $R_1 = k[X, Y, Z]/(X^2 - Y^3 - Z^7)$ where k is a field of positive characteristic. Therefore R_1 is not F-rational, independently of k. Moreover, no ideal I generated by a system of homogeneous parameters of R_1 is tightly closed. Otherwise $I_\mathfrak{m}$ would be tightly closed in the localization $(R_1)_\mathfrak{m}$ with respect to $\mathfrak{m} = (x, y, z)$, and it would follow that $(R_1)_\mathfrak{m}$ is F-rational. Since $(R_1)_\mathfrak{p}$ is regular for prime ideals $\mathfrak{p} \neq \mathfrak{m}$, the ring R_1 would have to be F-rational, too. That R_1 is not F-rational follows also from the fact that $a(R_1) \geq 0$; see Exercise 10.3.28.

(b) We have also seen that (y, z) is tightly closed in $R_2 = k[X, Y, Z]/(X^2 - Y^3 - Z^5)$ where k is a field of characteristic at least 7. Therefore $(R_2)_\mathfrak{m}$ is F-rational, and so is R_2, by the same localization argument as in (a). Since R_2 is Gorenstein, it is even F-regular.

Remarks 10.3.13. (a) Let R be a positively graded ring with *maximal ideal $\mathfrak{m}$. In analogy to the assertions relating the homological properties of R and $R_\mathfrak{m}$ (for example, see 2.1.27), one may ask whether the (weak) F-regularity or F-rationality of R and that of $R_\mathfrak{m}$ are equivalent. At least for F-rationality there is a satisfactory theorem: a *homogeneous* k-algebra R over a perfect field is F-rational if and only if $R_\mathfrak{m}$ is F-rational. For weak F-regularity there is only a weaker result. See Hochster and Huneke [196], (4.7) and (4.6).

(b) Again in analogy to the homologically defined ring-theoretic properties, one may ask how (weak) F-regularity and F-rationality behave under flat ring extensions with 'good' fibres. We refer the reader to Hochster and Huneke [195] and Velez [385] for theorems of this type; unfortunately they are much harder to prove than their homological counterparts.

Test elements. The proofs of the next results require test elements. We briefly discuss this notion.

Definition 10.3.14. An element $c \in R°$ is called a *test element* if for all ideals I and all $x \in I^*$ one has $cx^q \in I^{[q]}$ for all q.

The following is the most general existence theorem for test elements.

Theorem 10.3.15 (Hochster–Huneke). *Let R be a reduced algebra of finite type over an excellent local ring $(S, \mathfrak{n})$. Let $c \in R°$ be an element such that R_c is F-regular and Gorenstein. Then some power of c is a test element.*

The theorem implies in particular that test elements exist in reduced excellent local rings: choose an element $c \in I \cap R°$ where I is an ideal with $\operatorname{Sing} R = V(I)$.

For the proof of 10.3.15 the reader is referred to [195], (6.2). We will show the existence of test elements only in the important special case of reduced F-finite rings: one calls R *F-finite* if R, viewed as an R-module via F, is finite. For example, every ring which is a localization of an affine algebra over a perfect field and every complete local ring with perfect residue class field is F-finite. By a theorem of Kunz [248], F-finite rings are excellent.

Theorem 10.3.16. *Let R be an F-finite reduced ring, and $c \neq 0$ an element of R such that R_c is regular. Then some power of c is a test element.*

Let R be a domain of characteristic p with quotient field K; for each integer e one may then identify R, viewed as an R-module via F^e, with the ring $R^{1/q}$, $q = p^e$, of the q-th roots of the elements of R in some algebraic closure of K. The R-algebra structure of $R^{1/q}$ is of course given by the inclusion map $R \subset R^{1/q}$. The notation $R^{1/q}$ is convenient in the next lemma that will be needed for the proof of 10.3.16.

Lemma 10.3.17. *Let R be an F-finite regular domain, and $d \in R$. Then there exist a power q of p and an R-linear map $\varphi \colon R^{1/q} \to R$ such that $\varphi(d^{1/q}) = 1$.*

PROOF. Krull's intersection theorem implies that for each maximal ideal $\mathfrak{m}$ there is a power $q_{\mathfrak{m}}$ of p with $d \notin \mathfrak{m}^{[q_{\mathfrak{m}}]}$. By Kunz's theorem 8.2.8, $R_{\mathfrak{m}}^{1/q_{\mathfrak{m}}}$ is a free $R_{\mathfrak{m}}$-module. Since $d^{1/q_{\mathfrak{m}}}/1$ is part of a basis of $R_{\mathfrak{m}}^{1/q_{\mathfrak{m}}}$, there exists an $R_{\mathfrak{m}}$-homomorphism $\alpha_{\mathfrak{m}} \colon R_{\mathfrak{m}}^{1/q_{\mathfrak{m}}} \to R_{\mathfrak{m}}$ with $\alpha_{\mathfrak{m}}(d^{1/q_{\mathfrak{m}}}/1) = 1$. The map $\alpha_{\mathfrak{m}}$ is of the form $\beta_{\mathfrak{m}}/a_{\mathfrak{m}}$ where $\beta_{\mathfrak{m}} \colon R^{1/q_{\mathfrak{m}}} \to R$ is R-linear and $a_{\mathfrak{m}} \in R \setminus \mathfrak{m}$; in particular $\beta_{\mathfrak{m}}(d^{1/q_{\mathfrak{m}}}) = a_{\mathfrak{m}}$.

Since the ideal generated by the elements $a_{\mathfrak{m}}$ is not contained in a maximal ideal, it is the unit ideal, and hence 1 is a linear combination of some elements $a_1 = a_{\mathfrak{m}_1}, \ldots, a_r = a_{\mathfrak{m}_r}$, say $1 = \sum b_i a_i$. Set $\beta_i = \beta_{\mathfrak{m}_i}$, $q_i = q_{\mathfrak{m}_i}$, and $q = \max\{q_1, \ldots, q_r\}$. Since $d \notin \mathfrak{m}_i^{[q_i]}$, a fortiori $d \notin \mathfrak{m}_i^{[q]}$. Running through the argument above once more, we may in fact assume that $q_i = q$ for all i and $\beta_i(d^{1/q}) = a_i$. Now $\varphi = \sum b_i \beta_i$ has the desired property. $\qquad\square$

PROOF OF 10.3.16. As R_c is regular, the previous lemma implies that there exist a power q of p and an R_c-linear map $\psi \colon R_c^{1/q} \to R_c$ with $\psi(1) = 1$. One can write $\psi = \Psi/c^n$ where $\Psi \colon R^{1/q} \to R$ is R-linear. It follows that $\Psi(1) = c^n$ for some n, and replacing c by c^n we may as well assume that $\Psi(1) = c$. Restricting Ψ to $R^{1/p}$ (which is contained in $R^{1/q}$) yields an R-linear map $\varphi \colon R^{1/p} \to R$ with $\varphi(1) = c$.

We claim that c^2 is a test element if $\operatorname{char} R \neq 2$, and that c^3 is a test element if $\operatorname{char} R = 2$. In fact, let $I \subset R$ be an ideal of R, and pick $x \in I^*$. Then there exists an element $d \in R^{\circ}$ with $dx^q \in I^{[q]}$ for all q. As before, we find a power q' of p and an R-linear map $\alpha \colon R^{1/q'} \to R$ such that $\alpha(d^{1/q'}) = c^N$ for some N. Taking the q'-th root of the relation $dx^{qq'} \in I^{[qq']}$, one obtains $d^{1/q'} x^q \in I^{[q]}$ for all q. Now we apply α and get $c^N x^q \in I^{[q]}$ for all q.

Let N be the smallest integer with this property and write $N = mp + r$ with $0 \leq r < p$. Then $(c^r)^{1/p} c^m x^q \in I^{[q]} R^{1/p}$, and multiplication by $(c^{p-r})^{1/p}$ yields $c^{m+1} x^q \in I^{[q]} R^{1/p}$ for all q. Applying the linear map φ constructed in the first paragraph of the proof we obtain $c^{m+2} x^q \in I^{[q]}$ for all q. Since N was chosen minimal, $m + 2 \geq N$. This implies that $N \leq 2$, if $p > 2$, and $N \leq 3$, if $p = 2$. $\qquad\square$

Using test elements we can now prove a result about the behaviour of tight closure under completion.

Proposition 10.3.18. *Let $(R, \mathfrak{m})$ be an excellent local ring with $\mathfrak{m}$-adic completion $\hat{R}$, and let I be an $\mathfrak{m}$-primary ideal of R. Then $I^* \hat{R} = (I\hat{R})^*$.*

PROOF. We denote by R_{red} the residue class ring of R modulo its nilradical $N = \operatorname{Rad}(0)$. Choose an element $c \in R$ such that $(R_{\mathrm{red}})_c$ is regular (this is possible since Sing R_{red} is Zariski-closed). It follows from 10.3.15 that some power of c is a test element. Replacing c by this power we may assume that c itself is a test element. Let q' be such that $N^{q'} = 0$. Then for all ideals J in R, c has the property that $x \in J^*$ if and only if $cx^q \in J^{[q]}$ for all $q \geq q'$. One therefore says that c is a q'-*weak* test element for R.

Since R is excellent, the ring $(\hat{R}_{red})_c$ is also regular (this uses the regularity of the formal fibres of R_{red}), and hence we may assume c is a q'-weak test element for $\hat{R}$ as well.

The inclusion $I^*\hat{R} \subset (I\hat{R})^*$ is obvious since $R° \subset (\hat{R})°$. For the proof of the other inclusion we first notice that $(I\hat{R})^* \subset (I^*\hat{R})^*$, so that it suffices to show that $(I^*\hat{R})^* \subset I^*\hat{R}$. We may therefore assume that I is tightly closed. Since $(I\hat{R})^*$ is $\hat{m}$-primary, there is an ideal $J \subset R$ containing I with $J\hat{R} = (I\hat{R})^*$. Suppose $I\hat{R}$ is not tightly closed. Then the inclusion $I \subset J$ is proper. Hence there exists an element $x \in ((I\hat{R})^* \setminus I\hat{R}) \cap R$. For all $q \geq q'$ we then have $cx^q \in I^{[q]}\hat{R} \cap R = I^{[q]}$. This implies $x \in I^* = I$, a contradiction.   □

Corollary 10.3.19. *Let* (R, m) *be an excellent local ring. Then* R *is* F-*rational if and only if its* m-*adic completion* $\hat{R}$ *is* F-*rational.*

PROOF. Suppose R is F-rational, and let I be an ideal of $\hat{R}$ generated by a system of parameters. Then there exists an ideal J of R with $I = J\hat{R}$ such that J is also generated by a system of parameters. By 10.3.18 and our assumption, $I^* = (J\hat{R})^* = J^*\hat{R} = J\hat{R} = I$.

Conversely, assume that $\hat{R}$ is F-rational and I is an ideal of R generated by a system of parameters. Since $\hat{R}$ is a faithfully flat R-module, $I^* = (I^*\hat{R}) \cap R \subset (I\hat{R})^* \cap R = (I\hat{R}) \cap R = I$.   □

The Frobenius and local cohomology. We shall see that the Frobenius homomorphism $F : R \to R$ induces a natural action on local cohomology. This leads to an important characterization of F-rationality in terms of local cohomology discovered by Smith.

Let $x_1, \ldots, x_d$ be a system of parameters of R. We know from Section 3.5 that local cohomology may be computed as the homology of the modified Čech complex

$$C^\bullet : 0 \longrightarrow C^0 \longrightarrow C^1 \longrightarrow \cdots \longrightarrow C^n \longrightarrow 0,$$

$$C^t = \bigoplus_{1 \leq i_1 < i_2 < \cdots < i_t \leq n} R_{x_{i_1} x_{i_2} \cdots x_{i_t}}.$$

The Frobenius acts naturally on each C^i, and it is easy to see that it is compatible with the differentiation of $C^\bullet$. This shows that F induces an action

$$F : H^i_m(R) \to H^i_m(R) \qquad \text{for all } i.$$

(This map obviously coincides with that induced by the ring homomorphism $F: R \to R$.) We will describe the action of F explicitly on the highest non-vanishing local cohomology module $H_\mathfrak{m}^d(R)$. Notice that an element $c \in H_\mathfrak{m}^d(R)$ is the homology class $[\frac{a}{x^t}]$ of an element $a/x^t \in C^d = R_x$, where $x = x_1 \cdots x_d$.

Lemma 10.3.20. (a) $[\frac{a}{x^t}] = [\frac{ax^n}{x^{t+n}}]$ *for all integers* $n \geq 0$;

(b) $[\frac{a}{x^t}] = 0$ *if and only if there exists an integer* $n \geq 0$ *such that* $ax^n \in (x_1^{t+n}, \ldots, x_d^{t+n})$;

(c) *if* R *is Cohen–Macaulay, then* $[\frac{a}{x^t}] = 0$ *if and only if* $a \in (x_1^t, \ldots, x_d^t)$;

(d) $F([\frac{a}{x^t}]) = [\frac{a^p}{x^{tp}}]$.

PROOF. (a) and (d) are obvious, while (c) follows from (b). Finally, $[\frac{a}{x^t}] = 0$ if and only if a/x^t is a boundary in $C^\bullet$. This is the case exactly when there exist elements $c_i \in R$ such that

$$\sum_{i=1}^d (-1)^{i+1} c_i \frac{x_i^{s_i}}{x^{s_i}} = \frac{a}{x^t}$$

for some integers $s_i \geq 0$. We may assume that $s_i = s$ for all i. Then such an equation holds if and only if there exists an integer $m \geq 0$ such that

$$\sum_{i=1}^d (-1)^{i+1} c_i' x_i^{t+(s+m)} = ax^{s+m}, \qquad c_i' = c_i x^m \prod_{j \neq i} x_j^t.$$

Thus the assertion follows with $n = s + m$. $\qquad\square$

As a first application of these concepts we prove a criterion for F-rationality, due to Fedder and Watanabe [104], which can often be used in concrete situations.

A local ring $(R, \mathfrak{m})$ is called *F-injective*, if $F: H_\mathfrak{m}^i(R) \to H_\mathfrak{m}^i(R)$ is injective for all $i = 0, \ldots, d$, $d = \dim R$. If R is Cohen–Macaulay, this is only a requirement on $H_\mathfrak{m}^d(R)$ that, by 10.3.20(c), is equivalent to the following condition: $x^p \in I^{[p]}$ implies $x \in I$ for each ideal I in R generated by a system of parameters .

Proposition 10.3.21. *Let* $(R, \mathfrak{m})$ *be an excellent Cohen-Macaulay local ring, and let* $f \in \mathfrak{m}$ *be an R-regular element such that* (i) $R/(f)$ *is F-injective and* (ii) R_f *is an F-regular Gorenstein ring. Then* R *is F-rational.*

PROOF. Since F-regular rings are normal (see 10.1.12(d)) and hence reduced, assumption (ii) and 10.3.15 imply that some power of f, say f^t, is a test element for R. Since f is R-regular we can extend f to a system of parameters $f, x_2, \ldots, x_d$. In order to prove that R is F-rational we apply 10.3.5 and show that $I = (f, x_2, \ldots, x_d)$ is tightly closed. Indeed, let $x \in I^*$; then for all q there exist $a_{iq} \in R$ such that $f^t x^q = a_{1q} f^q + a_{2q} x_2^q + \cdots + a_{dq} x_d^q$. Thus, for $q > t$ we get $f^t(x^q - a_{1q} f^{q-t}) \in (x_1^q, \ldots, x_d^q)$. Since R is Cohen–Macaulay, the sequence $f, x_2, \ldots, x_d$ is R-regular. Hence $x^q \in (f^{q-t}, x_2^q, \ldots, x_d^q)$. Since $R/(f)$ is F-injective, this implies $\bar{x} \in (\bar{x}_2, \ldots, \bar{x}_d)$ where $^-$ denotes reduction modulo f. Thus $x \in (f, x_2, \ldots, x_d)$. In other words, I is tightly closed. $\quad\square$

A submodule $N \subset H_{\mathfrak{m}}^d(R)$ is called F-*stable* if $F(N) \subset N$. We have the following characterization of F-rationality.

Theorem 10.3.22 (Smith). *Let $(R, \mathfrak{m})$ be an excellent local ring. Then the following conditions are equivalent:*
(a) *R is F-rational;*
(b) *R is Cohen–Macaulay and $H_{\mathfrak{m}}^d(R)$ has no proper non-zero F-stable submodule.*

PROOF. (b) $\Rightarrow$ (a): Assume R is not F-rational. Choose a system of parameters $x_1, \ldots, x_d$ of R and set $I = (x_1, \ldots, x_d)$. Then, by 10.3.5, there exists an element $a \in I^* \setminus I$. The element $\eta = [\frac{a}{x}] \in H_{\mathfrak{m}}^d(R)$ is non-zero since R is Cohen–Macaulay; see 10.3.4 and 10.3.20(c). Consider the smallest F-stable submodule $N \subset H_{\mathfrak{m}}^d(R)$ containing η. The non-zero module N is obviously spanned by the elements $F^e(\eta) = [\frac{a^q}{x^q}]$, $q = p^e$, $e \geq 0$. Since $a \in I^*$, there exists an element $c \in R^\circ$ such that $ca^q \in I^{[q]}$ for all q (see 10.1.2). This implies $cN = 0$.

Suppose $N = H_{\mathfrak{m}}^d(R)$; then c annihilates $H_{\mathfrak{m}}^d(R)$ and consequently its Matlis dual, which is the canonical module $\omega_{\hat{R}}$ of the completion of R. This is a contradiction, since $\omega_{\hat{R}}$ is faithful according to 3.3.11.

(a) $\Rightarrow$ (b): As R is excellent, the $\mathfrak{m}$-adic completion of R is again F-rational. Since furthermore R and its completion have the same local cohomology, we may assume that R is complete.

Suppose there is a proper non-zero F-stable submodule $N \subset H_{\mathfrak{m}}^d(R)$. Taking Matlis duals yields an epimorphism $\omega_R \longrightarrow N^\vee$ with non-zero kernel U. Since R is a domain and ω_R is a module of rank 1 (see 3.3.18), $N^\vee$ is a torsion module. Hence there exists an element $c \in R^\circ$ such that $cN^\vee = 0$. Therefore $cN = 0$.

As $N \neq 0$, one finds a non-zero $\eta = [\frac{a}{x}]$ in N, where $x = x_1 \cdots x_d$ for a system of parameters $x_1, \ldots, x_d$ of R. Since N is F-stable, c annihilates all elements $F^e(\eta)$, that is, $[\frac{ca^q}{x^q}] = 0$ for all q. Because R is Cohen–Macaulay, this implies $ca^q \in (x_1^q, \ldots, x_d^q)$ for all q; see 10.3.20(c). In other words, $a \in (x_1, \ldots, x_d)^*$. Since R is F-rational, $a \in (x_1, \ldots, x_d)$, and so $\eta = 0$, a contradiction. $\quad\square$

Pseudo-rational and rational singularities. A point x on a normal variety X is said to be a *rational singularity*, if there exists a desingularization $f : W \to X$ such that $(R^i f_* \mathcal{O}_W)_x = 0$ for all $i \geq 1$. (Since this condition is local, it suffices to compute the higher direct image sheaves when X is affine, in which case $R^i f_* \mathcal{O}_W$ is the sheaf associated to the module $H^i(W, \mathcal{O}_W)$.)

The disadvantage of this definition is that X may have no desingularization. Therefore Lipman and Teissier [261] introduced the notion of pseudo-rationality. It coincides with rationality for rings that are localizations of affine domains over fields of characteristic zero; furthermore they showed that regular rings are pseudo-rational.

Definition 10.3.23. Let $(R, \mathfrak{m})$ be a d-dimensional normal Cohen-Macaulay local ring whose completion is reduced. Then R is *pseudo-rational* if for any proper birational map $\pi : W \to X = \operatorname{Spec} R$ with W normal and closed fibre $E = \pi^{-1}(\mathfrak{m})$, the canonical map

$$\delta_\pi : H^d_{\mathfrak{m}}(R) \longrightarrow H^d_E(W, \mathcal{O}_W)$$

is injective.

In the definition, $H^i_E(W, \mathcal{O}_W)$ denotes cohomology with supports in E; see Hartshorne [152], Exercise III.2.3. Cohomology with supports is related to ordinary cohomology via the long exact sequence

$$(1) \quad 0 \longrightarrow H^0_E(W, \mathcal{O}_W) \longrightarrow H^0(W, \mathcal{O}_W) \longrightarrow H^0(W \setminus E, \mathcal{O}_W) \longrightarrow \cdots$$
$$\longrightarrow H^i_E(W, \mathcal{O}_W) \longrightarrow H^i(W, \mathcal{O}_W) \longrightarrow H^i(W \setminus E, \mathcal{O}_W) \longrightarrow \cdots$$

If $X = \operatorname{Spec} R$ and $x = \{\mathfrak{m}\}$, then cohomology with supports is just local cohomology: $H^i_x(X, \mathcal{O}_X) \cong H^i_{\mathfrak{m}}(R)$ for all i. Furthermore, for affine X the above long exact sequence implies that $H^i_x(X, \mathcal{O}_X) \cong H^{i-1}(X \setminus x, \mathcal{O}_X)$ for $i \geq 2$ since $H^i(X, \mathcal{O}_X) = 0$ for $i > 0$; see [152], Theorem III.3.7.

The homomorphism δ_π is the composition of the maps

$$H^d_{\mathfrak{m}}(R) = H^{d-1}(X \setminus x, \mathcal{O}_X) \xrightarrow{\alpha} H^{d-1}(W \setminus E, \mathcal{O}_W) \xrightarrow{\beta} H^d_E(W, \mathcal{O}_W),$$

where α is the edge homomorphism $E_2^{d-1,0} \to E^{d-1}$ of the Leray spectral sequence (see Godement [128], II.4.17)

$$E_2^{p,q} = H^p(X \setminus x, R^q \pi_* \mathcal{O}_X) \Rightarrow E^{p,q} = H^{p+q}(W \setminus E, \mathcal{O}_W),$$

and β is a connecting homomorphism in the long exact sequence (1). Since δ_π is defined naturally, it has good functorial properties.

Suppose R is of characteristic p; then we have a morphism of schemes $F : W \to W$, the *absolute Frobenius morphism*. This map is the identity on the underlying topological space and the p-th power map locally on sections of $\mathcal{O}_W \to F_* \mathcal{O}_W = \mathcal{O}_W$. This morphism of schemes induces a natural map on cohomology with supports, compatible with δ_π. In other words, one has a commutative diagram

$$
\begin{array}{ccc}
H_{\mathfrak{m}}^d(R) & \longrightarrow & H_{\mathfrak{m}}^d(R) \\
\delta_\pi \downarrow & & \downarrow \delta_\pi \\
H_E^d(W, \mathcal{O}_W) & \longrightarrow & H_E^d(W, \mathcal{O}_W),
\end{array}
$$

where the top horizontal map is just the Frobenius action on $H_{\mathfrak{m}}^d(R)$ defined above. Consequently, the kernel of δ_π is an F-stable submodule of $H_{\mathfrak{m}}^d(R)$. This observation is part of the proof of

Corollary 10.3.24 (Smith). *Let $(R, \mathfrak{m})$ be an excellent local ring of characteristic p. If R is F-rational, then it is pseudo-rational.*

PROOF. By 10.3.22 it suffices to show that the kernel of δ_π is not all of $H_{\mathfrak{m}}^d(R)$.

We may assume that $d \geq 2$, and prove that codim Ker $\delta_\pi \geq 2$ (which of course implies Ker $\delta_\pi \neq H_{\mathfrak{m}}^d(R)$). From the exact sequence (1) we obtain that Ker β is the image of $\alpha : H^{d-1}(W, \mathcal{O}_W) \to H^{d-1}(W \setminus E, \mathcal{O}_W)$. Since R is normal and π is birational, π is an isomorphism at primes of height 1; hence $H^q(W, \mathcal{O}_W)_\mathfrak{p} = 0$ for $q > 0$ and all $\mathfrak{p} \in X$ with height $\mathfrak{p} \leq 1$. This implies codim Ker $\beta \geq 2$.

It remains to show that codim Ker $\alpha \geq 2$. Pick $\mathfrak{p} \in X$ with height $\mathfrak{p} = s$; then

$$
(R^q \pi_* \mathcal{O}_W)_\mathfrak{p} = H^q(W, \mathcal{O}_W)_\mathfrak{p} = H^q(\pi^{-1}(\operatorname{Spec} R_\mathfrak{p}), \mathcal{O}_W) = 0
$$

for $q > 0$, $q \geq s$. In fact, by Chow's Lemma ([152], Exercise III.4.10) we can assume that $\pi : \pi^{-1}(\operatorname{Spec} R_\mathfrak{p}) \to \operatorname{Spec} R_\mathfrak{p}$ is projective, and is therefore obtained by blowing up an ideal of $R_\mathfrak{p}$. So the maximal dimension of the closed fibre of π is bounded by $s - 1$, whence the assertion on the vanishing of $(R^q \pi_* \mathcal{O}_W)_\mathfrak{p}$ follows from the comparison theorem for projective morphisms ([152], Corollary III.11.2).

The vanishing of $(R^q \pi_* \mathcal{O}_W)_\mathfrak{p}$ for $q \geq s$ implies dim Supp$(R^q \pi_* \mathcal{O}_W) \leq d - q - 1$. Hence

$$
\dim \operatorname{Supp}(R^q \pi_* \mathcal{O}_W) \cap (X \setminus x) \leq d - q - 2,
$$

and so $H^p(X \setminus x, R^q \pi_* \mathcal{O}_W) = 0$ for $p + q > d - 2$, $q > 0$. The Leray spectral sequence now yields $H^{d-1}(W \setminus E, \mathcal{O}_W) = E_\infty^{d-1,0}$. In particular, α may be identified with the map $E_2^{d-1,0} \to E_\infty^{d-1,0}$ which is the composition of the surjective maps

$$
E_2^{d-1,0} \longrightarrow E_3^{d-1,0} \longrightarrow E_4^{d-1,0} \longrightarrow \cdots
$$

where, for each $r \geq 2$, the kernel of $d_r : E_r^{d-1,0} \to E_{r+1}^{d-1,0}$ is the image of $E_r^{d-1-r,r-1} \to E_r^{d-1,0}$. Since each $E_r^{d-1-r,r-1}$ is a subquotient of

$$
E_2^{d-1-r,r-1} = H^{d-1-r}(X \setminus x, R^{r-1} \pi_* \mathcal{O}_W),
$$

and since codim Supp$(R^{r-1}\pi_*\mathcal{O}_W) \geq 2$, as observed above, we conclude codim(Ker d_r) ≥ 2 for all $r \geq 2$. Therefore codim Ker $\alpha \geq 2$. $\square$

The following corollary is the characteristic p analogue of Boutot's theorem [50] that a direct summand of a rational singularity is a rational singularity.

Corollary 10.3.25. *Let $(R, \mathfrak{m})$ be an excellent local ring of characteristic p which is a direct summand of an F-regular overring. Then R is pseudo-rational.*

PROOF. We know from 10.1.13 that a direct summand of an F-regular ring is again F-regular, and hence F-rational. Now we apply 10.3.24. $\square$

For the sake of completeness we quote without proofs the extension of the theory to characteristic 0.

Definition 10.3.26. Let k is a field of characteristic 0, and R an affine k-algebra. The ring R is of *F-rational type* if there exists a finitely generated $\mathbb{Z}$-subalgebra A of k and a finitely generated A-algebra R_A, with flat structure map $A \to R_A$ such that
(a) $(A \to R_A) \otimes_A k$ is isomorphic to $k \to R$;
(b) the ring $R_A \otimes_A A/\mathfrak{m}$ is F-rational for all maximal ideals $\mathfrak{m}$ in a dense open subset of Spec A.

A typical situation described in the definition is the following: R is an affine k-algebra $k[X_1,\ldots,X_n]/(f_1,\ldots,f_m)$ where the polynomials f_i are defined over $\mathbb{Z}$, $\mathbb{Z}[X_1,\ldots,X_n]/(f_1,\ldots,f_m)$ is a free $\mathbb{Z}$-module, and $(\mathbb{Z}/p)[X_1,\ldots,X_n]/(\bar{f}_1,\ldots,\bar{f}_m)$ is F-rational for all but finitely many prime elements p.

Let X be a scheme of finite type over a field of characteristic zero. One says that a point $x \in X$ has *F-rational type* if x has an open affine neighbourhood defined by a ring R of F-rational type. The scheme X *has F-rational type* if every point $x \in X$ has F-rational type.

The following fundamental theorem relates rational singularities and F-rational rings:

Theorem 10.3.27 (Smith and Hara). *Let X be a scheme of finite type over a field of characteristic 0. Then X has F-rational type if and only if X has rational singularities.*

Exercises

10.3.28. Let R be a positively graded k-algebra where k is a field of positive characteristic. Prove that $a(R) < 0$ if R is F-rational. (Hint: $a(R) = \max\{i : {}^*H_\mathfrak{m}(R) \neq 0\}$. See [196], (7.12) and (7.13) for converse results.)

10.3.29. Show that F-rationality implies F-injectivity for Cohen–Macaulay local rings.

10.3.30. One says that R is *F-pure* if R is a pure extension of R via the Frobenius map F (see 6.5.3(b) for this notion). Show that $k[X_1, \ldots, X_n]/I$ is F-pure for every field k of positive characteristic and each ideal I generated by squarefree monomials; indeed, R is a direct summand of R under F.

10.3.31. (a) Let R be an arbitrary ring and S a pure extension of R. Show that for every complex $C_\bullet$ of R-modules the natural map $H_i(C_\bullet) \to H_i(C_\bullet \otimes S)$ is injective for all i.
(b) Prove that F-purity implies F-injectivity. (One can show that weak F-regularity implies F-purity; see Fedder and Watanabe [104], 1.6.)

10.4 Direct summands of regular rings

In this section we return to a subject that has been treated several times before, namely the Cohen–Macaulay property of direct summands of regular rings, which we will now prove for rings containing a field – the general case seems to be unknown. The next theorem generalizes 6.5.2, in which we have considered graded direct summands of polynomial rings $k[X_1, \ldots, X_n]$, and 10.1.14, which covers rings of characteristic p.

Theorem 10.4.1 (Hochster–Huneke). *Let R be a Noetherian ring containing a field and suppose R is a direct summand of a regular ring S. Then R is Cohen–Macaulay.*

PROOF. Already the proof of 6.5.2 depended on reduction to characteristic p and eventually used a tight closure argument. However, not even in the relatively 'harmless' setting of 6.5.2 could the direct summand property be pushed through the reduction. Therefore we will have to prove a general local analogue of 6.5.4 from which we now derive the theorem.

Being Cohen–Macaulay is a local property. Thus, let $\mathfrak{p}$ be a prime ideal of R. Then the hypotheses are inherited by the submodule $R_\mathfrak{p}$ of $S_\mathfrak{p}$ so that we may assume R is local with maximal ideal $\mathfrak{m}$. Next we pass to the $\mathfrak{m}$-adic completion $\hat{R}$ and the $\mathfrak{m}S$-adic completion $\hat{S}$ of S. It is clear that $\hat{R}$ is a direct summand of $\hat{S}$. Also regularity has survived. Indeed, one has $\hat{S}/\mathfrak{m}\hat{S} \cong S/\mathfrak{m}S$, and $\mathfrak{m}\hat{S}$ is contained in the Jacobson radical of $\hat{S}$ (see [270], §8). It results from this fact and Nakayama's lemma that every maximal ideal of $\hat{S}$ is of the form $\mathfrak{n}\hat{S}$ where $\mathfrak{n}$ is a maximal ideal of S. Now one uses the natural isomorphism between the $\mathfrak{n}$-adic completion S_1 of S and the $\mathfrak{n}\hat{S}$-adic completion S_2 of $\hat{S}$ to conclude that $\hat{S}_{\mathfrak{n}\hat{S}}$ is a regular local ring. The isomorphism of S_1 and S_2 is not hard to prove: choose systems of generators $a_1, \ldots, a_r$ and $b_1, \ldots, b_s$ of $\mathfrak{m}S$ and $\mathfrak{n}$ respectively; with $X = X_1, \ldots, X_r$ and $Y = Y_1, \ldots, Y_s$ one then has

$$S_1 \cong S[\![X, Y]\!]/(X_1 - a_1, \ldots, X_r - a_r, Y_1 - b_1, \ldots, Y_s - b_1),$$
$$S_2 \cong (S[\![X]\!]/(X_1 - a_1, \ldots, X_r - a_r))[\![Y]\!]/(Y_1 - b_1, \ldots, Y_s - b_s)$$

by [270], 8.12 (for the first isomorphism we use that $a_1, \ldots, a_r b_1, \ldots, b_s$ also generates $\mathfrak{n}$).

From now on we may assume that R is a residue class ring of a Cohen–Macaulay ring. Notice that S is the direct product $S_1 \times \cdots \times S_s$ of regular integral domains. Let $e_i \in S$ be the idempotent representing $1 \in S_i$. For the R-homomorphism $\sigma : S \to R$ splitting the inclusion $R \to S$ we have $1 = \sigma(e_1) + \cdots + \sigma(e_s)$. Since R is local, one of the $\sigma(e_i)$ is a unit in R. It follows easily that the induced map $R \to S_i$ is split. Hence we can replace S by the domain S_i; especially, R is a domain.

Now choose a system of parameters $x_1, \ldots, x_d$ of R. Then

$$((x_1, \ldots, x_{m-1}) :_R x_m)S = (x_1, \ldots, x_{m-1})S, \qquad m = 1, \ldots, d,$$

by 10.4.3 below. Moreover, $IS \cap R = I$ for all ideals I of R, and we conclude immediately that $x_1, \ldots, x_d$ is an R-sequence, as desired. $\qquad \square$

Remark 10.4.2. The theorem holds under the slightly weaker hypothesis that R is a pure subring of S (see 6.5.3(b) for this notion). In fact, purity implies that $IS \cap R = I$ for all ideals I of R; furthermore it is stable under the reduction in the proof of 10.4.1 (see Hochster and Roberts [201], Section 6). If one assumes directly that R and S are domains and R is a residue class ring of a Cohen–Macaulay ring, then it is sufficient that $IS \cap R = I$ for all ideals I of R (as was the case for 6.5.2 and 10.1.14).

We refer the reader to 6.5.3 for a discussion of the predecessors and variants of 10.4.1.

Theorem 10.4.3. *Let $(R, \mathfrak{m})$ be Noetherian local domain of dimension d that contains a field and is a homomorphic image of a Cohen–Macaulay local ring. Furthermore let S be a regular domain extending R. Then one has $((x_1, \ldots, x_{m-1}) :_R x_m)S = (x_1, \ldots, x_{m-1})S$ for every system of parameters $x_1, \ldots, x_d$ of R and $m = 1, \ldots, d$.*

PROOF. If the claim should fail, then there exists a maximal ideal $\mathfrak{n}$ of S such that $((x_1, \ldots, x_{m-1}) :_R x_m)S_\mathfrak{n} \neq (x_1, \ldots, x_{m-1})S_\mathfrak{n}$. Evidently $\mathfrak{p} = R \cap \mathfrak{n}$ must contain $x_1, \ldots, x_m$. In order to replace R by $R_\mathfrak{p}$ we must only show that $x_1, \ldots, x_m$ can be extended to a system of parameters of $R_\mathfrak{p}$. This holds if $\text{height}(x_1, \ldots, x_m) = m$. Indeed, by assumption we have $\text{codim}(x_1, \ldots, x_m) = m$, and R is a (universally) catenary local domain (see 2.1.12). In such a ring one has $\text{height } I = \text{codim } I$ for all ideals I.

After this first step we can assume S is a regular *local* domain extending R. The completion of S (with respect to its maximal ideal) is a regular local ring extending S, and since it is faithfully flat over S, there is no harm in supposing that S is even a complete regular local ring. The homomorphism $R \to S$ induces a map $\hat{R} \to \hat{S} = S$, which however need not be injective. At least, its kernel $\mathfrak{q}$ is a prime ideal of $\hat{R}$ with

$q \cap R = 0$. Since $\hat{R}$ is flat over R, q is a minimal prime ideal of $\hat{R}$ and $\dim \hat{R}/q = \dim \hat{R}$ by virtue of Theorem 2.1.15.

As in Section 8.1 we use the ideals $a_i = \operatorname{Ann} H_m^i(R)$. Set $a(R) = a_0 \cdots a_{d-1}$ and recall from 8.1.3 that

$$a(R) \cdot ((x_1, \ldots, x_{m-1}) :_R x_m) \subset (x_1, \ldots, x_{m-1})R;$$

furthermore, by 8.1.1, $a(R) \not\subset q$. Since the image of $a(R)$ under the map $R \to S$ is non-zero, we can invoke the following lemma and conclude the proof.   □

It is necessary to relax the condition that the homomorphism $R \to S$ be injective. Actually we will have to reduce the next lemma to the case where this map is surjective in order to prove it in characteristic 0. (The hypothesis 'complete' is only included to save us another reduction.)

Lemma 10.4.4. *Let (R, m) be a complete Noetherian local ring containing a field, (S, n) a complete regular local ring, and $\varphi : R \to S$ a ring homomorphism such that $\varphi(a(R)) \neq 0$. Then*

$$((x_1, \ldots, x_{m-1}) :_R x_m)S = (x_1, \ldots, x_{m-1})S$$

for every system of parameters $x_1, \ldots, x_d$ of R and $m = 1, \ldots, d$.

PROOF. Let us first prove the lemma in characteristic p. Pick $d \in a(R)$ such that $c = \varphi(d) \neq 0$. For $y \in (x_1, \ldots, x_{m-1}) :_R x_m$ and all $q = p^e$ one has $y^q \in (x_1^q, \ldots, x_{m-1}^q) :_R x_m^q$. Since $x_1^q, \ldots, x_d^q$ is also a system of parameters, $dy^q \in (x_1^q, \ldots, x_{m-1}^q)$. Applying φ, we immediately see that $\varphi(y) \in ((x_1, \ldots, x_{m-1})S)^*$, whence $\varphi(y) \in (x_1, \ldots, x_{m-1})S$ by 10.1.7.

The next step is the reduction to the case in which φ is surjective. By Cohen's structure theorem A.21 there are representations $R \cong K[[Y_1, \ldots, Y_r]]/I$ and $S \cong L[[Z_1, \ldots, Z_s]]$ where $K \cong R/m$ and $L \cong S/n$ are coefficient fields of R and S, respectively. The map φ induces an inclusion $K \to L$ so that we may view K as a subfield of L. We set $A = K[[Y_1, \ldots, Y_r]]$ and $A' = L[[Y_1, \ldots, Y_r]]$.

Evidently, φ can only be surjective if $K = L$, and therefore we must extend R such that the extension R' has residue class field L. Consider the homomorphism $A \to S$ induced by φ. It clearly factors through A'. Therefore φ factors through $R' = A'/IA'$. Note that R' is flat over R: first, it is easily proved that A' is a flat A-algebra (see Exercise 9.1.15), and, second, if $C_\bullet$ is an exact sequence of R-modules, then $C_\bullet \otimes_R R' \cong C_\bullet \otimes_A A'$. Moreover, mR' is the maximal m' ideal of R'. Especially $\dim R' = \dim R$, and so every system of parameters of R is also a system of parameters of R'.

We can replace R by R' if we have shown that $a(R)R' = a(R')$. We set $b_i = \operatorname{Ann}_A H_m^i(R)$, and define b_i' similarly. Then b_i is the preimage of $\operatorname{Ann}_R H_m^i(R)$, and the corresponding statement holds for b_i'. Therefore it is enough that $\mathfrak{b}_i A' = \mathfrak{b}_i'$. By 8.1.1 we have $\mathfrak{b}_i = \operatorname{Ann}_A M$ where $M = \operatorname{Ext}_A^{d-i}(R, A)$. The flatness of A' over A implies $M \otimes A' \cong \operatorname{Ext}_{A'}^{d-i}(R', A')$ and $\operatorname{Ann}_A(M)A' = \operatorname{Ann}_{A'}(M \otimes A')$ (see the proof of 10.1.7). Using 8.1.1 once more, we arrive at the desired equality.

From now on we may assume that $K = L$, $R = R'$, and $A = A'$. Next we extend φ to a surjection $\psi \colon \widetilde{R} \to S$ by choosing $\widetilde{R} = R[\![Z_1, \ldots, Z_s]\!]$ and setting $\psi(Z_j) = Z_j \in S$. The extension $R \to \widetilde{R}$ is faithfully flat, and every system of parameters of R can be extended by $Z_1, \ldots, Z_s$ to a system of parameters of $\widetilde{R}$. Before we can replace R by $\widetilde{R}$, we need only to prove that $\mathfrak{a}(R)\widetilde{R} = \mathfrak{a}(\widetilde{R})$. This however results again from 8.1.1; the reader can easily check that $\mathfrak{a}_i\widetilde{R} = \widetilde{\mathfrak{a}}_{i+s}$ for all $i = 0, \ldots, d$ and $\widetilde{\mathfrak{a}}_i = \widetilde{R}$ for $i = 0, \ldots, s-1$: set $\widetilde{A} = A[\![Z_1, \ldots, Z_s]\!]$ and use that $\widetilde{R} = R \otimes_A \widetilde{A}$.

We may now replace R by $\widetilde{R}$ and A by $\widetilde{A}$. After this change of notation the failure of the lemma can be described as follows: there exist

(i) a regular local ring A with a regular system of parameters $a_1, \ldots, a_n$, a residue class ring $R = A/(b_1, \ldots, b_u)$ of dimension d, and a residue class ring $S = A/(a_1, \ldots, a_v)$, $v \leq n$, such that $(b_1, \ldots, b_u) \subset (a_1, \ldots, a_v)$ (in fact, the kernel of the homomorphism $A \to S$ is generated by a subset of a regular system of parameters);

(ii) elements $c_0, \ldots, c_{d-1}$ with $c_i \in \operatorname{Ann}_A \operatorname{Ext}_A^{d-i}(R, A)$ and $c = c_0 \cdots c_{d-1} \notin (a_1, \ldots, a_v)$;

(iii) elements $x_1, \ldots, x_d \in A$ whose residue classes form a system of parameters, a number m, $1 \leq m \leq d$, and an element $w \in A$ such that $wx_m \in (x_1, \ldots, x_{m-1}) + (b_1, \ldots, b_u)$, but $x_m \notin (x_1, \ldots, x_{m-1}) + (a_1, \ldots, a_v)$.

We want to show that, given such data in characteristic 0, we can also find them in characteristic p. To this end we must show that the data above have a regular equational presentation. Theorem 8.4.1 then yields a characteristic p counterexample to our contention, the desired contradiction.

All the relations '$\in$' and '$\subset$' can of course be expressed by polynomial equations. This holds also for the fact that $b_1, \ldots, b_u$ and $y_1, \ldots, y_d$ generate an ideal whose radical contains $(a_1, \ldots, a_n)$. Furthermore we have already seen in 8.4.4 that the dimension condition in (i) and the non-membership relations in (ii) and (iii) can be captured by equations.

For given i we write $\operatorname{Ext}_A^{d-i}(R, A)$ as a residue class module A^s/W. Then the isomorphism $\operatorname{Ext}_A^{d-i}(R, A) \cong A^s/W$ admits a regular equational presentation by 8.4.4, as does the relation $c_i A^s \subset W$ for trivial reasons. This finally shows that all the data given in (i)–(iii) can be encoded in a system of polynomial equations over $\mathbb{Z}$. □

Notes

The fundamental paper for tight closure is Hochster and Huneke [192]. Essentially all of the material of Sections 10.1 and 10.2 has been taken from this source. A detailed discussion of the not yet solved localization and completion problems can be found in Huneke's lecture notes [215] which we have consulted extensively in writing Chapter 10. Much of the work that preceded tight closure theory and motivated its creation has been discussed Chapters 8 and 9.

The theorem of Briançon and Skoda was originally proved by analytic methods, and the lack of an algebraic proof had been 'for algebraists something of a scandal – perhaps even an insult – and certainly a challenge' (Lipman and Teissier [261]). As pointed out in Section 10.2 algebraic proofs of slightly different theorems were given by Lipman and Teissier and Lipman and Sathaye [260]; the latter work uses differential methods. The proof of the tight closure version by Hochster and Huneke is contained in their article [190], which is still very useful as a first overview of our subject. For variants and generalizations of the Briançon–Skoda theorem see Aberbach and Huneke [3], [4], and Swanson [368]. For the connection with reduction numbers and Rees algebras see Aberbach, Huneke, and Trung [6].

F-rational (local) rings appeared first in Fedder and Watanabe [104]. Our treatment of their basic properties essentially follows Huneke [215]. The connection with local cohomology and the Frobenius action on it goes back to the work [202] of Hochster and Roberts that introduced F-purity. Special cases of Smith's theorem [349] that F-rational type implies rational singularity and its converse by Hara [149], which we have quoted in 10.3.27, had been proved in special cases by Fedder [101], [102], [103] and Hochster and Huneke [190]. The proof of 10.3.24 was suggested to us by Watanabe.

We could only prove the easiest result on the existence of test elements that in its general version presents the perhaps most intricate aspect of tight closure; see Hochster and Huneke [192], [195]. The existence of test elements is closely related with the so-called persistence theorem that under suitable conditions guarantees the relation $\varphi(I^*) \subset (\varphi(I)S)^*$ for a ring homomorphism $\varphi : R \to S$; see [195], (6.24).

Some results about the hierarchy of 'F-properties' have been indicated in Section 10.3. For more information, especially for examples delimiting these properties from each other and for the relation to singularity theory, the reader is referred to Fedder and Watanabe [104], Watanabe [390], [391], and Hara and Watanabe [150].

Theorem 10.4.1 was stated by Hochster and Huneke [190], 3.5 without proof. A complete proof appeared in their paper [197]. It uses the functoriality of big Cohen–Macaulay algebras. Our derivation of the theorem is certainly a variant of the idea behind [190], 3.5.

The definition of tight closure can be extended from the situation $I \subset R$ to that in which U is a submodule of the R-module M. In particular, this leads one to the notion of phantom homology and phantom acyclicity; see Hochster and Huneke [192], [194]. For phantom acyclic complexes one has a vanishing theorem similar to 10.4.3 where the ideal quotient is replaced by the homology of a phantom acyclic complex. It seems however that the strongest such vanishing theorem [197], (4.1) needs big Cohen–Macaulay algebras. One can also derive a 'phantom' version of the 'improved new intersection theorem' 9.4.2; thus tight closure offers another approach to the homological theorems of Chapter 9. Aberbach has developed 'phantom' homological algebra that includes phantom projective dimension, phantom depth and an Auslander–Buchsbaum formula.

Tight closure can be also defined in characteristic 0; see [198] and the Appendix of [215] by Hochster. So far there seems to be no definition of tight closure in mixed characteristic. Hochster has developed a theory of solid closure that does not depend on characteristic [188]. For 'good' rings of characteristic $p > 0$ solid closure coincides with tight closure; however, there exist examples showing that ideals in a regular ring containing a field of characteristic 0 need not be solidly closed.

There are many more aspects and applications of tight closure. We content ourselves with a list of cues and references: tight closure in graded rings (Smith [350]), Hilbert–Kunz functions and multiplicities (Kunz [248], Monsky [277]), uniform Artin–Rees theorems (O'Carroll [292], Huneke [214]), arithmetic Macaulayfication (Huneke and Smith [5]), strongly F-regular rings (Hochster and Huneke [191], Glassbrenner [127]), differentially simple rings (Smith and Van den Bergh [352]), Kodaira vanishing and other vanishing theorems of algebraic geometry (Huneke and Smith [218], Smith [351]).

Appendix: A summary of dimension theory

Dimension theory is a cornerstone of commutative ring theory, and is covered by every serious introduction to the subject. For ease of reference we have collected its main theorems in this appendix, together with the structure theorems for complete local rings.

Most of the theorems below have the names of their creators associated with them and should be easily located in the literature. For some of the results we outline a proof.

Height and dimension. There exist two principal lines of development for general dimension theory. The first and 'classical' approach, to which we shall adhere, starts from the Krull principal ideal theorem ([47], [231], [284], [344], [397]) whereas the second brings the Hilbert–Samuel function into play at a very early stage ([15], [270]).

Let R be a commutative ring, and $\mathfrak{p} \in \operatorname{Spec} R$. The *height* of $\mathfrak{p}$ is the supremum of the lengths t of strictly descending chains

$$\mathfrak{p} = \mathfrak{p}_0 \supset \mathfrak{p}_1 \supset \cdots \supset \mathfrak{p}_t$$

of prime ideals. For an arbitrary ideal I one sets

$$\operatorname{height} I = \inf\{\operatorname{height} \mathfrak{p} : \mathfrak{p} \in \operatorname{Spec} R, \ \mathfrak{p} \supset I\}.$$

The fundamental theorem on height is Krull's principal ideal theorem:

Theorem A.1. *Let R be a Noetherian ring, and $I = (x_1, \ldots, x_n)$ a proper ideal. Then $\operatorname{height} \mathfrak{p} \leq n$ for every prime ideal $\mathfrak{p}$ which is minimal among the prime ideals containing I.*

In particular, every proper ideal in a Noetherian ring has finite height. In a sense, the following theorem is a converse of the principal ideal theorem.

Theorem A.2. *Let R be a Noetherian ring, and I a proper ideal of height n. Then there exist $x_1, \ldots, x_n \in I$ such that $\operatorname{height}(x_1, \ldots, x_i) = i$ for $i = 1, \ldots, n$.*

The elements x_i are chosen successively such that x_i is not contained in any minimal prime overideal of $(x_1, \ldots, x_{i-1})$; that such a choice is possible follows from 1.2.2.

412

The (*Krull*) *dimension* of a ring R is the supremum of the heights of its prime ideals,

$$\dim R = \sup\{\text{height } \mathfrak{p} : \mathfrak{p} \in \text{Spec } R\}.$$

Because of the correspondence between $\text{Spec } R_\mathfrak{p}$, $\mathfrak{p} \in \text{Spec } R$, and the set of prime ideals contained in $\mathfrak{p}$, one has

$$\dim R_\mathfrak{p} = \text{height } \mathfrak{p}.$$

A fundamental and very easily proved inequality is

$$\text{height } I + \dim R/I \le \dim R$$

for all proper ideals I of R.

The dimension of a Noetherian local ring can be characterized in several ways:

Theorem A.3. *Let* $(R, \mathfrak{m})$ *be a Noetherian local ring, and* $n \in \mathbb{N}$. *Then the following are equivalent:*
(a) $\dim R = n$;
(b) $\text{height } \mathfrak{m} = n$;
(c) *n is the infimum of all m for which there exist $x_1, \ldots, x_m \in \mathfrak{m}$ with* $\text{Rad}(x_1, \ldots, x_m) = \mathfrak{m}$.
(d) *n is the infimum of all m for which there exist $x_1, \ldots, x_m \in \mathfrak{m}$ such that* $R/(x_1, \ldots, x_m)$ *is Artinian.*

The equivalence of (a) and (b) is trivial. That of (b) and (c) results from A.1 and A.2, and for (c) $\Longleftrightarrow$ (d) one uses the fact that a Noetherian ring is Artinian if and only if all its prime ideals are maximal, in other words, if it has dimension 0. If $\dim(R/(x_1, \ldots, x_n)) = 0$ with $n = \dim R$, then $x_1, \ldots, x_n$ is a *system of parameters* of R.

Sometimes it is appropriate to use the *codimension* of an ideal in a ring R which is given by

$$\text{codim } I = \dim R - \dim R/I.$$

Dimension of modules. The notion of dimension can be transferred to modules. Let M be an R-module; then $\dim M$ is the supremum over the lengths t of strictly descending chains

$$\mathfrak{p} = \mathfrak{p}_0 \supset \mathfrak{p}_1 \supset \cdots \supset \mathfrak{p}_t \quad \text{with} \quad \mathfrak{p}_i \in \text{Supp } M.$$

In the case of main interest in which M is a finite module one has $\text{Supp } M = \{\mathfrak{p} \in \text{Spec } R : \mathfrak{p} \supset \text{Ann } M\}$ so that $\dim M = \dim(R/\text{Ann } M)$. If $(R, \mathfrak{m})$ is local, then a *system of parameters* for a non-zero finite R-module M is a sequence $x_1, \ldots, x_n \in \mathfrak{m}$, $n = \dim M$, such that $M/(x_1, \ldots, x_n)M$ is Artinian. The following inequality is often useful:

Proposition A.4. *Let $(R, \mathfrak{m})$ be a Noetherian local ring, M a finite R-module, and $x_1, \ldots, x_r \in \mathfrak{m}$. Then*

$$\dim(M/(x_1, \ldots, x_r)M) \geq \dim M - r,$$

equality holding if and only if $x_1, \ldots, x_r$ is part of a system of parameters of M.

This is easy. One first replaces M by $R/\operatorname{Ann} M$ so that it is harmless to assume $M = R$. Then one chooses $y_1, \ldots, y_m \in \mathfrak{m}$ such that their residue classes in $R/(x_1, \ldots, x_r)$ form a system of parameters. Finally one applies A.3.

An important datum of a homomorphism of local rings $(R, \mathfrak{m}) \to (S, \mathfrak{n})$ is its *fibre* $S/\mathfrak{m}S$. For example it relates the dimensions of R and S:

Theorem A.5. *Let $(R, \mathfrak{m}) \to (S, \mathfrak{n})$ be a homomorphism of Noetherian local rings.*
(a) *Then $\dim S \leq \dim R + \dim S/\mathfrak{m}S$;*
(b) *more generally, if M is a finite R-module and N is a finite S-module, then $\dim_S(M \otimes_R N) \leq \dim_R M + \dim_S N/\mathfrak{m}N$.*

For the proof of (b) set $I = \operatorname{Ann} M$ and $\bar{R} = R/I$. Then $U \otimes_R N \cong U \otimes_{\bar{R}} N/IN$ for every $\bar{R}$-module U. Thus we may replace R by $\bar{R}$, S by S/IS, and N by N/IN. That is, we may assume $\operatorname{Supp} M = \operatorname{Spec} R$. Next, replacing S by $S/(\operatorname{Ann} N)$, we may suppose $\operatorname{Supp} N = \operatorname{Spec} S$. Under these conditions the desired inequality is equivalent with (a). For (a) one chooses a system $x_1, \ldots, x_n$ of parameters of R, and uses that $\operatorname{Rad}(x_1, \ldots, x_n)S = \operatorname{Rad} \mathfrak{m}S$.

Integral extensions. Recall that an extension $R \subset S$ of commutative rings is *integral* if every element $x \in S$ satisfies an equation $x^n + a_{n-1}x^{n-1} + \cdots + a_0 = 0$ with coefficients $a_i \in R$. Very often one uses that S is a finite R-module if and only if it is an integral extension and finitely generated as an R-algebra.

Theorem A.6. *Let $R \subset S$ be an integral extension, and $\mathfrak{p} \in \operatorname{Spec} R$.*
(a) *There exists a prime ideal $\mathfrak{q} \in \operatorname{Spec} S$ with $\mathfrak{p} = \mathfrak{q} \cap R$ (one says $\mathfrak{q}$ lies over $\mathfrak{p}$);*
(b) *there are no inclusions between prime ideals lying over $\mathfrak{p}$;*
(c) *in particular, when $\mathfrak{q}$ lies over $\mathfrak{p}$, then $\mathfrak{p}$ is maximal if and only if $\mathfrak{q}$ is.*

The following theorem comprises the Cohen–Seidenberg going-up and going-down theorems.

Theorem A.7. *Let $R \subset S$ be an integral extension.*
(a) *If $\mathfrak{p}' \supset \mathfrak{p}$ are prime ideals of R and $\mathfrak{q} \in \operatorname{Spec} S$ lies over $\mathfrak{p}$, then there exists a prime ideal $\mathfrak{q}' \supset \mathfrak{q}$ in S lying over $\mathfrak{p}'$;*

(b) *if, in addition, S is an integral domain and R is integrally closed, then, given prime ideals* $\mathfrak{p}' \subset \mathfrak{p}$ *of R and* $\mathfrak{q}$ *of S,* $\mathfrak{q}$ *lying over* $\mathfrak{p}$, *there exists* $\mathfrak{q}' \in \operatorname{Spec} S$, $\mathfrak{q}' \subset \mathfrak{q}$, *which lies over* $\mathfrak{p}'$.

Corollary A.8. *Let* $R \subset S$ *be an integral extension of Noetherian rings, and I a proper ideal of S. Then* $\dim S/I = \dim R/(I \cap R)$.

The first step in proving the corollary is to replace S by S/I and R by $R/(I \cap R)$ so that one may assume $I = 0$. Then given a strictly descending chain $\mathfrak{q}_0 \supset \mathfrak{q}_1 \supset \cdots \supset \mathfrak{q}_t$ of prime ideals, the chain of prime ideals $R \cap \mathfrak{q}_i$ is also strictly descending by A.6, and conversely, given a chain in $\operatorname{Spec} R$, one constructs a chain of the same length in $\operatorname{Spec} S$ using A.7(a).

In general, one says that *going-up* or *going-down* holds for a ring homomorphism $R \to S$ if it satisfies *mutatis mutandis* the conclusions of A.7(a) or (b) respectively.

Flat extensions. It is an important fact that flatness implies going-down:

Lemma A.9. *Let* $R \to S$ *be a homomorphism of Noetherian rings, and suppose there exists an R-flat finite S-module N with* $\operatorname{Supp} N = \operatorname{Spec} S$. *Then going-down holds.*

Going-down can be reformulated as follows: for all prime ideals $\mathfrak{p} \in \operatorname{Spec} R$ and $\mathfrak{q} \in \operatorname{Spec} S$ lying over $\mathfrak{p}$ the natural map $\operatorname{Spec} S_{\mathfrak{q}} \to \operatorname{Spec} R_{\mathfrak{p}}$ is surjective. Now, given such prime ideals $\mathfrak{p}$ and $\mathfrak{q}$, $N_{\mathfrak{q}}$ is even a faithfully flat $R_{\mathfrak{p}}$-module, and the surjectivity of $\operatorname{Spec} S_{\mathfrak{q}} \to \operatorname{Spec} R_{\mathfrak{p}}$ follows from the next lemma.

Lemma A.10. *Let* $R \to S$ *be a ring homomorphism. If an S-module N is faithfully flat over R, then the associated map* $\operatorname{Supp} N \to \operatorname{Spec} R$ *is surjective.*

In fact, let $\mathfrak{p} \in \operatorname{Spec} R$; we set $k(\mathfrak{p}) = R_{\mathfrak{p}}/\mathfrak{p} R_{\mathfrak{p}}$. Then $k(\mathfrak{p}) \otimes_R N \neq 0$, and the support of the $k(\mathfrak{p}) \otimes_R S$-module $k(\mathfrak{p}) \otimes_R N$ contains a prime ideal $\mathfrak{Q}$. If we choose $\mathfrak{q} = S \cap \mathfrak{Q}$, then $\mathfrak{q} \in \operatorname{Supp} N$, and furthermore $\mathfrak{q} \cap R = \mathfrak{p}$: one has $\mathfrak{q} \cap R = \mathfrak{Q} \cap R$, and $\mathfrak{Q} \cap R = \mathfrak{p}$ since the map $\operatorname{Spec}(k(\mathfrak{p}) \otimes_R S) \to \operatorname{Spec} R$ factors through $\operatorname{Spec} k(\mathfrak{p})$.

For flat extensions the inequalities in A.5 become equations:

Theorem A.11. *Let* $(R, \mathfrak{m}) \to (S, \mathfrak{n})$ *be a homomorphism of Noetherian local rings.*
(a) *If S is a flat R-algebra, then* $\dim S = \dim R + \dim S/\mathfrak{m}S$;
(b) *more generally, if M is a finite R-module and N is an R-flat finite S-module, then* $\dim_S(M \otimes_R N) = \dim_R M + \dim_S N/\mathfrak{m}N$.

As we did for A.5, the theorem is easily reduced to the case in which $\operatorname{Supp} N = \operatorname{Spec} S$. By virtue of the previous lemma the homomorphism $R \to S$ then satisfies going-down. We choose a prime ideal $\mathfrak{q}$ of S which contains $\mathfrak{m}S$ and has the same height as $\mathfrak{m}S$. Then going-down immediately implies height $q \geq$ height $\mathfrak{m}$. Hence

$$\dim S \geq \text{height } \mathfrak{m}S + \dim S/\mathfrak{m}S \geq \text{height } \mathfrak{m} + \dim S/\mathfrak{m}S,$$

as desired. The converse inequality is part of A.5.

Polynomial and power series extensions. The dimension of a polynomial or power series extension is easily computed:

Theorem A.12. *Let R be a Noetherian ring. Then*

$$\dim R[X] = \dim R[[X]] = \dim R + 1.$$

Let $S = R[X]$ or $S = R[[X]]$. Then $R \cong S/(X)$, and since height$(X) = 1$ one has $\dim S \geq \dim R + 1$. For the converse we first consider the polynomial case. Let $\mathfrak{n}$ be a maximal ideal of $R[X]$, and set $\mathfrak{p} = R \cap \mathfrak{n}$. As $S = R[X]$ is R-flat, one may apply A.11, and only needs to show that $\dim(S_{\mathfrak{n}}/\mathfrak{p}S_{\mathfrak{n}}) = 1$. It is a routine matter to check that $S_{\mathfrak{n}}/\mathfrak{p}S_{\mathfrak{n}}$ is a localization of the polynomial ring $(R_{\mathfrak{p}}/\mathfrak{p}R_{\mathfrak{p}})[X]$ with respect to a maximal ideal. Since $R_{\mathfrak{p}}/\mathfrak{p}R_{\mathfrak{p}}$ is a field, $S_{\mathfrak{n}}/\mathfrak{p}S_{\mathfrak{n}}$ is a discrete valuation ring and therefore of dimension 1. In the power series case $\mathfrak{p}$ is always a maximal(!) ideal of R, and $S_{\mathfrak{n}}/\mathfrak{p}S_{\mathfrak{n}}$ is therefore the discrete valuation ring $(R/\mathfrak{p})[[X]]$.

Corollary A.13. *Let k be a field. Then*

$$\dim k[X_1, \ldots, X_n] = \dim k[[X_1, \ldots, X_n]] = n.$$

Affine algebras. Let k be a field. A finitely generated k-algebra R is called an *affine k-algebra.* Excellent sources for the theory of affine algebras are Kunz [249] and Sharp [344]. The key result is Noether's normalization theorem :

Theorem A.14. *Let R be an affine algebra over a field k, and let I be a proper ideal of R. Then there exist $y_1, \ldots, y_n \in R$ such that*
(a) $y_1, \ldots, y_n$ are algebraically independent over k;
(b) R is an integral extension of $k[y_1, \ldots, y_n]$ (and thus a finite $k[y_1, \ldots, y_n]$-module);
(c)

$$I \cap k[y_1, \ldots, y_n] = \sum_{i=d+1}^{n} y_i k[y_1, \ldots, y_n] = (y_{d+1}, \ldots, y_n)$$

for some d, $0 \leq d \leq n$.
Moreover, if $y_1, \ldots, y_n$ satisfy (a) and (b), then $n = \dim R$.

If $y_1, \ldots, y_n$ satisfy (a) and (b), then $k[y_1, \ldots, y_n]$ is called a *Noether normalization* of R. That necessarily $n = \dim R$ follows from A.8 and A.13. That condition (c) can be satisfied in addition to (a) and (b) is crucial for dimension theory. The graded variant of Noether normalization (due to Hilbert) is given in 1.5.17.

An important consequence of Noether normalization is (the abstract version of) Hilbert's Nullstellensatz :

Theorem A.15. *Let k be a field, and K an extension field of k which is a finitely generated k-algebra. Then K is a finite algebraic extension of k.*

In fact, if $k[y_1, \ldots, y_n]$ is a Noether normalization of K, then $n = \dim K = 0$ and K is an integral extension of k, from which one easily concludes that it is a finite algebraic extension.

The following theorem contains the main results of the dimension theory of affine algebras.

Theorem A.16. *Let R be an affine algebra over a field k. Suppose that R is an integral domain. Then*
(a) $\dim R = \operatorname{tr deg}_k Q(R)$ *where* $\operatorname{tr deg}_k Q(R)$ *is the transcendence degree of the field of fractions of R over k,*
(b) $\operatorname{height} \mathfrak{p} = \dim R - \dim R/\mathfrak{p}$ *for all prime ideals $\mathfrak{p}$ of R.*

For part (a) we choose a Noether normalization $k[y_1, \ldots, y_n]$. Then $Q(R)$ is algebraic over $Q(k[y_1, \ldots, y_n])$, and the latter has transcendence degree n over k. For part (b) we require in addition that $k[y_1, \ldots, y_n]$ satisfies A.14 for $I = \mathfrak{p}$. Then the image of $k[X_1, \ldots, X_d]$ in $R/\mathfrak{p}$ is a Noether normalization for that ring, whence $\dim R/\mathfrak{p} = d$. On the other hand, note that going-down holds according to A.7: being a factorial ring (a UFD in other terminology) $k[y_1, \ldots, y_n]$ is integrally closed. It follows that $\operatorname{height} \mathfrak{p} \geq \operatorname{height} \mathfrak{p} \cap k[y_1, \ldots, y_n] = n - d$. Summing up, we have $\operatorname{height} \mathfrak{p} + \dim R/\mathfrak{p} \geq n = \dim R$, and the converse inequality is automatic as noticed above.

Hilbert rings. It is a consequence of Hilbert's Nullstellensatz that a prime ideal in an affine algebra over a field is the intersection of the maximal ideals in which it is contained. Rings with this property are therefore called *Hilbert rings* (Bourbaki prefers the term *Jacobson rings*). The following is the main theorem on Hilbert rings:

Theorem A.17. *Let R be a Hilbert ring, and S a finitely generated R-algebra. Then*
(a) S *is a Hilbert ring,*
(b) $\mathfrak{m} \cap R$ *is a maximal ideal of R for every maximal ideal $\mathfrak{m}$ of S.*

Corollary A.18. *Let R be a finitely generated $\mathbb{Z}$-algebra, and $\mathfrak{m}$ a maximal ideal of R. Then $\mathfrak{m} \cap \mathbb{Z} = (p)$ for some prime number $p \in \mathbb{Z}$, and $R/\mathfrak{m}$ is a finite field.*

In fact, $\mathbb{Z}$ is a Hilbert ring, and $R/\mathfrak{m}$ is a finite algebraic extension of $\mathbb{Z}/(p)$ by A.15.

A dimension inequality. For the study of the dimension of Rees rings and associated graded rings the following theorem (due to Cohen) is important.

Theorem A.19. *Let $R \subset S$ be an extension of integral domains, and suppose R is Noetherian. Let $\mathfrak{P} \in \operatorname{Spec} S$ and $\mathfrak{p} = \mathfrak{P} \cap R$. Then*

$$\dim S_{\mathfrak{P}} + \operatorname{tr} \deg_{Q(R/\mathfrak{p})} Q(S/\mathfrak{P}) \leq \dim R_{\mathfrak{p}} + \operatorname{tr} \deg_{Q(R)} Q(S).$$

We reproduce the proof given in [270], §15. The first step is a reduction to the case in which S is a finitely generated R-algebra. There is nothing to prove if the right hand side is infinite. So suppose it is finite, and let m and t be integers with $0 \leq m \leq \dim S_{\mathfrak{P}}$, $0 \leq t \leq \operatorname{tr} \deg_{Q(R/\mathfrak{p})} Q(S/\mathfrak{P})$. Then there exists a strictly descending chain $\mathfrak{P} = \mathfrak{P}_0 \supset \cdots \supset \mathfrak{P}_m$ of prime ideals in S. We choose $a_{i+1} \in \mathfrak{P}_i \setminus \mathfrak{P}_{i+1}$, and furthermore elements $c_1, \ldots, c_t \in S$ whose residue classes in $S/\mathfrak{P}$ are algebraically independent over $Q(R/\mathfrak{p})$. Let $S' = R[a_1, \ldots, a_m, c_1, \ldots, c_t]$, and $\mathfrak{P}' = \mathfrak{P} \cap S'$; then $\dim S'_{\mathfrak{P}'} \geq m$ and $\operatorname{tr} \deg_{Q(R/\mathfrak{p})} Q(S/\mathfrak{P}') = t$. Thus it is enough to prove the claim for S' and C'.

In the case in which S is finitely generated, we use induction on the number of generators so that only the case $S = R[x]$ remains. Write $S = R[X]/\mathfrak{Q}$.

If $\mathfrak{Q} = 0$, then $S = R[X]$, and $\dim S_{\mathfrak{P}} = \dim R_{\mathfrak{p}} + \dim(S_{\mathfrak{P}}/\mathfrak{p} S_{\mathfrak{P}})$ by A.11. As $S_{\mathfrak{P}}/\mathfrak{p} S_{\mathfrak{P}}$ is a localization of $Q(R/\mathfrak{p})[X]$, we have $\dim(S_{\mathfrak{P}}/\mathfrak{p} S_{\mathfrak{P}}) = 1 - \operatorname{tr} \deg_{Q(R/\mathfrak{p})} Q(S/\mathfrak{P}) = \operatorname{tr} \deg_{Q(R)} Q(S) - \operatorname{tr} \deg_{Q(R/\mathfrak{p})} Q(S/\mathfrak{P})$.

In the case $\mathfrak{Q} \neq 0$ we have $\operatorname{tr} \deg_{Q(R)} Q(S) = 0$. Since R is a subring of S, $\mathfrak{Q} \cap R = 0$ so that $R[X]_{\mathfrak{Q}}$ is a localization of $Q(R)[X]$, and therefore has dimension 1, equivalently height $\mathfrak{Q} = 1$. Let $\mathfrak{P}'$ the inverse image of $\mathfrak{P}$ in $R[X]$, and note that $Q(R[X]/\mathfrak{P}') \cong Q(S/\mathfrak{P})$ in a natural way. Then

$$\dim S_{\mathfrak{P}} \leq \dim R[X]_{\mathfrak{P}'} - \text{height } \mathfrak{Q}$$

$$= \dim R_{\mathfrak{p}} + 1 - \operatorname{tr} \deg_{Q(R/\mathfrak{p})} Q(R[X]/\mathfrak{P}') - 1$$

$$= \dim R_{\mathfrak{p}} - \operatorname{tr} \deg_{Q(R/\mathfrak{p})} Q(S/\mathfrak{P}).$$

Complete local rings. The theory of Noetherian complete local rings, for which we recommend Matsumura [270] or Bourbaki [47] as a source, leads to similar results as that of affine algebras.

For the relation between the characteristic $\operatorname{char} R$ of a local ring $(R, \mathfrak{m})$ and that of its residue field $R/\mathfrak{m}$ one of the following cases holds true: (i) $\operatorname{char} R/\mathfrak{m} = 0$; then R contains the field $\mathbb{Q}$ of rational numbers, in particular $\operatorname{char} R = 0$; (ii) $\operatorname{char} R/\mathfrak{m} = p > 0$ and $\operatorname{char} R = p$ too; then

R contains the field $\mathbb{Z}/p\mathbb{Z}$; (iii) char $R/\mathfrak{m} = p > 0$ and char $R = 0$ (the typical case in number theory); (iv) char $R/\mathfrak{m} = p > 0$ and char $R = p^m$ for some $m > 1$. In cases (i) and (ii) one says that R is *equicharacteristic*. (Note that R does not contain a field in cases (iii) and (iv), and that (iv) is excluded for a reduced ring.)

Theorem A.20. *Let $(R, \mathfrak{m})$ be a Noetherian complete local ring.*
(a) *If R is equicharacteristic, then it contains a coefficient field, i.e. a field k which is mapped isomorphically onto $R/\mathfrak{m}$ by the natural homomorphism $R \to R/\mathfrak{m}$.*
(b) *Otherwise let $p = $ char $R/\mathfrak{m}$. Then there exists a discrete valuation ring (S, pS) and a homomorphism $\varphi : S \to R$ which induces an isomorphism $S/pS \to R/\mathfrak{m}$ and furthermore*
  (i) *is injective, if char $R = 0$,*
  (ii) *has kernel $p^m S$, if char $R = p^m$.*

It is a standard technique to pass from a Noetherian local ring $(R, \mathfrak{m})$ to its completion $\hat{R}$ (with respect to the $\mathfrak{m}$-adic topology). Then one is in a position to apply Cohen's structure theorem :

Theorem A.21. *Let $(R, \mathfrak{m})$ be a Noetherian complete local ring. Then there exists a ring R_0 which is a field or a discrete valuation ring such that R is a residue class ring of a formal power series ring $R_0[\![X_1, \ldots, X_n]\!]$.*

In fact, let $x_1, \ldots, x_n$ be a system of generators of $\mathfrak{m}$. Then there exists a uniquely determined homomorphism $\varphi : R_0[\![X_1, \ldots, X_n]\!] \to R$ with $\varphi(X_i) = x_i$ where R_0 is either a coefficient field of R or, in the case of unequal characteristic, a discrete valuation ring S according to A.20. In Section 2.2 it is shown that $R_0[\![X_1, \ldots, X_n]\!]$ is a regular local ring, and often one uses A.21 'only' to the extent that a complete local ring is a residue class ring of a regular local ring.

The analogue of Noether normalization is

Theorem A.22. *Let $(R, \mathfrak{m})$ be a Noetherian complete local ring, and suppose that R is equicharacteristic or a domain.*
(i) *In the equicharacteristic case let $R_0 \subset R$ be a coefficient field of R, and $y_1, \ldots, y_n$ a system of parameters;*
(ii) *otherwise, let $p = $ char $R/\mathfrak{m}$ and $R_0 \subset R$ be a discrete valuation ring according to A.20, and $y_1, \ldots, y_n$ be elements such that $p, y_1, \ldots, y_n$ is a system of parameters.*
Then R is a finite $R_0[\![y_1, \ldots, y_n]\!]$-module, and $R_0[\![y_1, \ldots, y_n]\!]$ is isomorphic to the formal power series ring $R_0[\![Y_1, \ldots, Y_n]\!]$.

One first shows that R is a finite $R_0[\![y_1, \ldots, y_n]\!]$-module, so that $\dim R = \dim R_0[\![y_1, \ldots, y_n]\!]$. The substitution $Y_i \mapsto y_i$ induces a surjective homomorphism $\varphi : R_0[\![Y_1, \ldots, Y_n]\!] \to R_0[\![y_1, \ldots, y_n]\!]$ which is also injective since $\dim R_0[\![Y_1, \ldots, Y_n]\!] = \dim R_0[\![y_1, \ldots, y_n]\!]$.

References

1. I. M. ABERBACH. Finite phantom projective dimension. *Amer. J. Math.* **116** (1994), 447–477.

2. I. M. ABERBACH, M. HOCHSTER, AND C. HUNEKE. Localization of tight closure and modules of finite phantom projective dimension. *J. Reine Angew. Math.* **434** (1993), 67–114.

3. I. M. ABERBACH AND C. HUNEKE. An improved Briançon-Skoda theorem with applications to the Cohen–Macaulayness of Rees algebras. *Math. Ann.* **297** (1993), 343–369.

4. I. M. ABERBACH AND C. HUNEKE. A theorem of Briançon-Skoda type for regular local rings containing a field. *Proc. Amer. Math. Soc.* **124** (1996), 707–713.

5. I. M. ABERBACH, C. HUNEKE, AND K. E. SMITH. A tight closure approach to arithmetic Macaulayfication. *Ill. J. Math.* **40** (1996), 310–329.

6. I. M. ABERBACH, C. HUNEKE, AND N. V. TRUNG. Reduction numbers, Briançon–Skoda theorems and the depth of Rees rings. *Compositio Math.* **97** (1995), 403–434.

7. S. S. ABHYANKAR. *Enumerative combinatorics of Young tableaux.* M. Dekker, 1988.

8. K. AKIN, D. A. BUCHSBAUM, AND J. WEYMAN. Resolutions of determinantal ideals: the submaximal minors. *Adv. in Math.* **39** (1981), 1–30.

9. D. ANICK. A counter-example to a conjecture of Serre. *Ann. of Math.* **115** (1982), 1–33.

10. Y. AOYAMA. Some basic results on canonical modules. *J. Math. Kyoto Univ.* **23** (1983), 85–94.

11. R. APÉRY. Sur les courbes de première espèce de l'espace à trois dimensions. *C. R. Acad. Sc. Paris* **220** (1945), 271–272.

12. A. ARAMOVA, J. HERZOG, AND T. HIBI. Gotzmann theorems for exterior algebras and combinatorics. *J. Algebra* **191** (1997), 174–211.

13. M. ARTIN. Algebraic approximation of structures over complete local rings. *Publ. Math. I.H.E.S.* **36** (1969), 23–58.

14. E. F. ASSMUS. On the homology of local rings. *Ill. J. Math.* **3** (1959), 187–199.

15. M. F. ATIYAH AND I. G. MACDONALD. *Introduction to commutative algebra.* Addison–Wesley, 1969.

16. M. AUSLANDER AND M. BRIDGER. *Stable module theory.* Mem. Amer. Math. Soc. **94**, 1969.

17. M. AUSLANDER AND D. A. BUCHSBAUM. Homological dimension in Noetherian rings. *Proc. Nat. Acad. Sci. U.S.A.* **42** (1956), 36–38.

18. M. AUSLANDER AND D. A. BUCHSBAUM. Homological dimension in local rings. *Trans. Amer. Math. Soc.* **85** (1957), 390–405.

19. M. AUSLANDER AND D. A. BUCHSBAUM. Codimension and multiplicity. *Ann. of Math.* **68** (1958), 625–657.

20. M. AUSLANDER AND D. A. BUCHSBAUM. Unique factorization in regular local rings. *Proc. Nat. Acad. Sci. U.S.A.* **45** (1959), 733–734.

21. M. AUSLANDER AND R.-O. BUCHWEITZ. The homological theory of Cohen–Macaulay approximations. *Mem. Soc. Math. de France* **38** (1989), 5–37.

22. L. L. AVRAMOV. Flat morphisms of complete intersections. *Soviet Math. Dokl.* **16** (1975), 1413–1417.

23. L. L. AVRAMOV. Homology of local flat extensions and complete intersection defects. *Math. Ann.* **228** (1977), 27–37.

24. L. L. AVRAMOV. A class of factorial domains. *Serdica* **5** (1979), 378–379.

25. L. L. AVRAMOV. Obstructions to the existence of multiplicative structures on minimal free resolutions. *Amer. J. Math.* **103** (1981), 1–31.

26. L. L. AVRAMOV, R.-O. BUCHWEITZ, AND J. D. SALLY. Laurent coefficients and Ext of finite graded modules. *Math. Ann.* **307** (1997), 401–415.

27. L. L. AVRAMOV AND H.-B. FOXBY. Homological dimensions of unbounded complexes. *J. Pure Applied Algebra* **71** (1991), 129–155.

28. L. L. AVRAMOV AND H.-B. FOXBY. Locally Gorenstein homomorphisms. *Amer. J. Math.* **114** (1992), 1007–1048.

29. L. L. AVRAMOV AND H.-B. FOXBY. Cohen–Macaulay properties of ring homomorphisms. *Adv. in Math.* **132** (1997).

30. L. L. AVRAMOV AND H.-B. FOXBY. Ring homomorphisms and finite Gorenstein dimension. *Proc. London Math. Soc.* **75** (1997), 241–270.

31. L. L. AVRAMOV, H.-B. FOXBY, AND S. HALPERIN. Descent and ascent of local properties along homomorphisms of finite flat dimension. *J. Pure Applied Algebra* **38** (1985), 167–185.

32. L. L. AVRAMOV AND E. GOLOD. On the homology algebra of the Koszul complex of a local Gorenstein ring. *Math. Notes* **9** (1971), 30–32.

33. L. L. AVRAMOV, A. R. KUSTIN, AND M. MILLER. Poincaré series of modules over local rings of small embedding codepth or small linking number. *J. Algebra* **118** (1988), 162–204.

34. J. BACKELIN. Les anneaux locaux à relations monomiales ont des séries de Poincaré–Betti rationelles. *C. R. Acad. Sc. Paris Sér. I* **295** (1982), 607–610.

35. K. BACLAWSKI. Canonical modules of partially ordered sets. *J. Algebra* **83** (1983), 1–5.

36. H. BASS. Injective dimension in Noetherian rings. *Trans. Amer. Math. Soc.* **102** (1962), 18–29.

37. H. BASS. On the ubiquity of Gorenstein rings. *Math. Z.* **82** (1963), 8–28.

38. D. BAYER AND D. MUMFORD. What can be computed in algebraic geometry? In D. EISENBUD ET AL. (ed.), *Computational algebraic geometry and commutative algebra.* Symp. Math. **34**, Cambridge Uniuversity Press, 1993, pp. 1–48.

39. M. M. BAYER AND C. W. LEE. Combinatorial aspects of convex polytopes. In P. M. GRUBER AND J. WILLS (eds.), *Handbook of convex geometry*. North-Holland, 1993.

40. D. J. BENSON. *Polynomial invariants of finite groups*. LMS Lect. Note Ser. 190, Cambridge University Press, 1993.

41. M.-J. BERTIN. Anneaux d'invariants d'anneaux de polynômes en caractéristique *p*. *C. R. Acad. Sc. Paris Sér. A* **264** (1967), 653–656.

42. A. M. BIGATTI. Upper bounds for the Betti numbers of a given Hilbert function. *Commun. Algebra* **21** (1993), 2317–2334.

43. L. J. BILLERA AND C. W. LEE. A proof of the sufficiency of McMullen's conditions for *f*-vectors of simplicial convex polytopes. *J. Combinatorial Theory Ser. A* **31** (1981), 237–255.

44. J. BINGENER AND U. STORCH. Zur Berechnung der Divisorenklassengruppen kompletter lokaler Ringe. *Nova Acta Leopoldina (N.F.) 52 Nr.* **240** (1981), 7–63.

45. A. BJÖRNER. Shellable and Cohen–Macaulay partially ordered sets. *Trans. Amer. Math. Soc.* **260** (1980), 159–183.

46. A. BJÖRNER, P. FRANKL, AND R. P. STANLEY. The number of faces of balanced Cohen–Macaulay complexes and a generalized Macaulay theorem. *Combinatorica* **7** (1987), 23–34.

47. N. BOURBAKI. *Algèbre commutative, Chap. I–IX*. Hermann, Masson, 1961–1983.

48. N. BOURBAKI. *Algèbre, Chap. I–X*. Hermann, Masson, 1970–1980.

49. N. BOURBAKI. *Groupes et algèbres de Lie, Ch. I–IX*. Hermann, 1971–1975.

50. J.-F. BOUTOT. Singularités rationelles et quotients par les groupes réductifs. *Invent. Math.* **88** (1987), 65–68.

51. M. BRODMANN. A 'macaulayfication' of unmixed domains. *J. Algebra* **44** (1977), 221–234.

52. W. BRUNS. Die Divisorenklassengruppe der Restklassenringe von Polynomringen nach Determinantenidealen. *Revue Roumaine Math. Pur. Appl.* **20** (1975), 1109–1111.

53. W. BRUNS. 'Jede' endliche freie Auflösung ist freie Auflösung eines von drei Elementen erzeugten Ideals. *J. Algebra* **39** (1976), 429–439.

54. W. BRUNS. The Eisenbud–Evans generalized principal ideal theorem and determinantal ideals. *Proc. Amer. Math. Soc.* **83** (1981), 19–24.

55. W. BRUNS. The canonical module of a determinantal ring. In R. Y. SHARP (ed.), *Commutative algebra, Durham 1981*. LMS Lect. Note Ser. 72, Cambridge University Press, 1982, pp. 109–120.

56. W. BRUNS. Algebras defined by powers of determinantal ideals. *J. Algebra* **142** (1991), 150–163.

57. W. BRUNS AND J. HERZOG. On the computation of *a*-invariants. *Manuscripta Math.* **77** (1992), 201–213.

58. W. BRUNS AND J. HERZOG. *Math. Proc. Cambridge Philos. Soc.* **118** (1995), 245–257.

59. W. BRUNS AND J. HERZOG. Semigroup rings and simplicial complexes. *J. Pure Applied Algebra* (to appear).

60. W. BRUNS, A. SIMIS, AND NGÔ VIÊT TRUNG. Blow-up of straightening closed ideals in ordinal Hodge algebras. *Trans. Amer. Math. Soc.* **326** (1991), 507–528.

61. W. BRUNS AND U. VETTER. *Determinantal rings.* LNM **1327**, Springer, 1988.

62. B. BUCHBERGER. Ein algorithmisches Kriterium für die Lösbarkeit eines algebraischen Gleichungssystems. *Aequationes Math.* **4** (1970), 374–383.

63. D. A. BUCHSBAUM AND D. EISENBUD. What makes a complex exact? *J. Algebra* **25** (1973), 259–268.

64. D. A. BUCHSBAUM AND D. EISENBUD. Some structure theorems for finite free resolutions. *Adv. in Math.* **12** (1974), 84–139.

65. D. A. BUCHSBAUM AND D. EISENBUD. Algebra structures for finite free resolutions, and some structure theorems for ideals of codimension 3. *Amer. J. Math.* **99** (1977), 447–485.

66. L. BURCH. On ideals of finite homological dimension in local rings. *Proc. Camb. Philos. Soc.* **64** (1968), 941–948.

67. H. CARTAN AND S. EILENBERG. *Homological algebra.* Princeton University Press, 1956.

68. C. CHEVALLEY. On the theory of local rings. *Ann. of Math.* **44** (1943), 690–708.

69. W.-L. CHOW. On unmixedness theorem. *Amer. J. Math.* **86** (1964), 799–822.

70. G. CLEMENTS AND B. LINDSTRÖM. A generalization of a combinatorial theorem of Macaulay. *J. Combinatorial Theory Ser. A* **7** (1969), 230–238.

71. I. S. COHEN. On the structure and ideal theory of complete local rings. *Trans. Amer. Math. Soc.* **59** (1946), 54–106.

72. C. DE CONCINI, D. EISENBUD, AND C. PROCESI. Young diagrams and determinantal varieties. *Invent. Math.* **56** (1980), 129–165.

73. C. DE CONCINI, D. EISENBUD, AND C. PROCESI. Hodge algebras. Asterisque **91**, Soc. Math. de France, 1982.

74. C. DE CONCINI AND D. PROCESI. A characteristic free approach to invariant theory. *Adv. in Math.* **21** (1976), 330–354.

75. C. DE CONCINI AND E. STRICKLAND. On the variety of complexes. *Adv. in Math.* **41** (1981), 57–77.

76. V. I. DANILOV. The geometry of toric varieties. *Russian Math. Surveys* **33** (1978), 97–154.

77. P. DOUBILET, G.-C. ROTA, AND J. STEIN. On the foundations of combinatorial theory: IX, Combinatorial methods in invariant theory. *Stud. Appl. Math.* **53** (1974), 185–216.

78. A. J. DUNCAN. Infinite coverings of ideals by cosets with applications to regular sequences and balanced big Cohen–Macaulay modules. *Math. Proc. Cambridge Philos. Soc.* **107** (1990), 443–460.

79. S. P. DUTTA. Frobenius and multiplicities. *J. Algebra* **85** (1983), 424–448.

80. S. P. DUTTA. Generalized intersection multiplicities of modules. *Trans. Amer. Math. Soc.* **276** (1983), 657–669.

81. S. P. DUTTA. Symbolic powers, intersection multiplicity, and asymptotic behaviour of Tor. *J. London Math. Soc.* **28** (1983), 261–281.

82. S. P. DUTTA. On the canonical element conjecture. *Trans. Amer. Math. Soc.* **299** (1987), 803–811.

83. S. P. DUTTA. Ext and Frobenius. *J. Algebra* **127** (1989), 163–177.

84. S. P. DUTTA. Dualizing complex and the canonical element conjecture. *J. London Math. Soc.* **50** (1994), 477–487.

85. S. P. DUTTA, M. HOCHSTER, AND J. E. MCLAUGHLIN. Modules of finite projective dimension with negative intersection multiplicities. *Invent. Math.* **79** (1985), 253–291.

86. J. A. EAGON AND D. G. NORTHCOTT. Ideals defined by matrices and a certain complex associated with them. *Proc. Roy. Soc. London Ser. A* **269** (1962), 188–204.

87. J. A. EAGON AND D. G. NORTHCOTT. Generically acyclic complexes and generically perfect ideals. *Proc. Roy. Soc. London Ser. A* **299** (1967), 147–172.

88. J. A. EAGON AND V. REINER. Resolutions of Stanley–Reisner rings and Alexander duality. *J. Pure Applied Algebra* (to appear).

89. E. EHRHART. *Polynômes arithméthiques et méthode des polyédres en combinatoire.* Birkhäuser, 1977.

90. D. EISENBUD. Introduction to algebras with straightening laws. In B. R. MCDONALD (ed.), *Ring theory and algebra III.* Lect. Notes in Pure and Appl. Math. **55**, M. Dekker, 1980, pp. 243–268.

91. D. EISENBUD. *Commutative algebra with a view toward algebraic geometry.* Springer, 1995.

92. D. EISENBUD AND E. G. EVANS. A generalized principal ideal theorem. *Nagoya Math. J.* **62** (1976), 41–53.

93. D. EISENBUD AND S. GOTO. Linear free resolutions and minimal multiplicity. *J. Algebra* **88** (1984), 89–133.

94. D. EISENBUD AND C. HUNEKE. Cohen–Macaulay Rees algebras and their specializations. *J. Algebra* **81** (1983), 202–224.

95. D. EISENBUD AND B. STURMFELS. Binomial ideals. *Duke Math. J.* **84** (1996), 1–45.

96. J. ELIAS AND A. IARROBINO. The Hilbert function of a Cohen–Macaulay local algebra: extremal Gorenstein algebras. *J. Algebra* **110** (1987), 344–356.

97. E. G. EVANS AND P. GRIFFITH. The syzygy problem. *Ann. of Math.* **114** (1981), 323–333.

98. E. G. EVANS AND P. GRIFFITH. *Syzygies.* LMS Lect. Note Ser. **106**, Cambridge University Press, 1985.

99. G. EWALD. *Combinatorial convexity and algebraic geometry.* Springer, 1996.

100. G. FALTINGS. Über Macaulayfizierung. *Math. Ann.* **238** (1978), 175–192.

101. R. FEDDER. *F*-purity and rational singularity. *Trans. Amer. Math. Soc.* **278** (1983), 461–480.

102. R. FEDDER. *F*-purity and rational singularity in graded complete intersection rings. *Trans. Amer. Math. Soc.* **301** (1987), 47–62.

103. R. FEDDER. A Frobenius characterization of rational singularity in 2-dimensional graded rings. *Trans. Amer. Math. Soc.* **340** (1993), 655–668.

104. R. FEDDER AND K. WATANABE. A characterization of F-regularity in terms of F-purity. In M. HOCHSTER, J. D. SALLY, AND C. HUNEKE (eds.), *Commutative algebra*. Math. Sc. Res. Inst. Publ. **15**, Springer, 1989, pp. 227–245.

105. D. FERRAND. Suite régulière et intersection complète. *C. R. Acad. Sc. Paris Sér. A* **264** (1967), 427–428.

106. D. FERRAND AND M. RAYNAUD. Fibres formelles d'un anneau local noethérien. *Ann. Sci. Ec. Norm. Sup.* (4) **3** (1970), 295–311.

107. H. FLENNER. Rationale quasi-homogene Singularitäten. *Arch. Math.* **36** (1981), 35–44.

108. R. FOSSUM. *The divisor class group of a Krull domain.* Springer, 1973.

109. R. FOSSUM AND H.-B. FOXBY. The category of graded modules. *Math. Scand.* **35** (1974), 288–300.

110. R. FOSSUM, H.-B. FOXBY, P. GRIFFITH, AND I. REITEN. Minimal injective resolutions with applications to dualizing modules and Gorenstein modules. *Publ. Math. I.H.E.S.* **45** (1976), 193–215.

111. R. FOSSUM AND P. GRIFFITH. Complete local factorial rings which are not Cohen–Macaulay in characteristic p. *Ann. Sci. Ec. Norm. Sup.* (4) **8** (1975), 189–199.

112. H.-B. FOXBY. On the μ^i in a minimal injective resolution. *Math. Scand.* **29** (1971), 175–186.

113. H.-B. FOXBY. Gorenstein modules and related modules. *Math. Scand.* **31** (1972), 267–284.

114. H.-B. FOXBY. Isomorphisms between complexes with applications to the homological theory of modules. *Math. Scand.* **40** (1977), 5–19.

115. H.-B. FOXBY. On the μ^i in a minimal injective resolution II. *Math. Scand.* **41** (1977), 19–44.

116. H.-B. FOXBY. Bounded complexes of flat modules. *J. Pure Applied Algebra* **15** (1979), 149–172.

117. H.-B. FOXBY. The MacRae invariant. In R. Y. SHARP (ed.), *Commutative algebra: Durham 1981.* LMS Lect. Note Ser. **72**, Cambridge University Press, 1982, pp. 121–128.

118. E. FREITAG AND R. KIEHL. Algebraische Eigenschaften der lokalen Ringe in den Spitzen der Hilbertschen Modulgruppen. *Invent. Math.* **24** (1974), 121–148.

119. R. FRÖBERG. Determination of a class of Poincaré series. *Math. Scand.* **37** (1975), 29–39.

120. R. FRÖBERG AND LÊ TUÂN HOA. Segre products and Rees algebras of face rings. *Commun. Algebra* **20** (1992), 3369–3380.

121. O. GABBER. Work in preparation. See also P. BERTHELOT. Alterations de variétés algébriques (d'apres A. J. de Jong). *Seminaire Bourbaki, Exp.* **815** (1996).

122. P. GABRIEL. Des catégories abéliennes. *Bull. Soc. Math. France* **90** (1962), 323–348.

123. F. GAETA. Quelques progrès récents dans la classification des variétés algébriques d'un espace projectif. In *Deuxième colloque de géométrie algébrique.* Liège, 1952, pp. 145–183.

124. A. M. GARSIA. Combinatorial methods in the theory of Cohen–Macaulay rings. *Adv. in Math.* **38** (1980), 229–266.

125. H. GILLET AND C. SOULÉ. K-théorie et nullité des multiplicités d'intersection. *C. R. Acad. Sc. Paris Sér. I* **300** (1985), 71–74.

126. R. GILMER. *Commutative semigroup rings.* University of Chicago Press, 1984.

127. D. J. GLASSBRENNER. Strong F-regularity in images of regular rings. *Proc. Amer. Math. Soc.* **124** (1996), 345–353.

128. R. GODEMENT. *Topologie algébrique et théorie des faisceaux.* Hermann, 1973.

129. N. L. GORDEEV. Finite groups whose algebras of invariants are complete intersections. *Math. USSR-Izv.* **28** (1987), 335–379.

130. S. GOTO. The divisor class group of a certain Krull domain. *J. Math. Kyoto Univ.* **17** (1977), 47–50.

131. S. GOTO. On the Gorensteinness of determinantal loci. *J. Math. Kyoto Univ.* **19** (1979), 371–374.

132. S. GOTO AND Y. SHIMODA. On the Rees algebras of Cohen–Macaulay local rings. In R. N. DRAPER (ed.), *Commutative algebra (analytical methods).* Lect. Notes in Pure and Appl. Math. **68**, M. Dekker, 1982, pp. 201–231.

133. S. GOTO, N. SUZUKI, AND K. WATANABE. On affine semigroup rings. *Japan J. Math.* **2** (1976), 1–12.

134. S. GOTO AND K. WATANABE. On graded rings, I. *J. Math. Soc. Japan* **30** (1978), 179–213.

135. S. GOTO AND K. WATANABE. On graded rings, II ($\mathbb{Z}^n$-graded rings). *Tokyo J. Math.* **1** (1978), 237–261.

136. G. GOTZMANN. Eine Bedingung für die Flachheit und das Hilbertpolynom eines graduierten Ringes. *Math. Z.* **158** (1978), 61–70.

137. H.-G. GRÄBE. The canonical module of a Stanley–Reisner ring. *J. Algebra* **86** (1984), 272–281.

138. M. GREEN. Restrictions of linear series to hyperplanes, and some results of Macaulay and Gotzmann. In E. BALLICO AND C. CILIBERTO (eds.), *Algebraic curves and projective geometry.* LNM **1389**, Springer, 1989, pp. 76–86.

139. P. GRIFFITH. A representation theorem for complete local rings. *J. Pure Applied Algebra* **7** (1976), 303–315.

140. P. GRIFFITH. Maximal Cohen–Macaulay modules and representation theory. *J. Pure Applied Algebra* **13** (1978), 321–334.

141. W. GRÖBNER. *Moderne algebraische Geometrie.* Springer (Wien), 1949.

142. A. GROTHENDIECK. *Eléments de géometrie algébrique, Chap. IV.* Publ. Math. I.H.E.S. **20, 24, 28, 32**, 1964, 1965, 1966, 1967.

143. A. GROTHENDIECK. *Local cohomology.* LNM **41**, Springer, 1967.

144. B. GRÜNBAUM. *Convex polytopes.* J. Wiley and Sons, 1967.

145. I. D. GUBELADZE. Anderson's conjecture and the maximal monoid class over which projective modules are free. *Math. USSR-Sb.* **63** (1989), 165–180.

146. T. H. GULLIKSEN. A proof of the existence of minimal R-algebra resolutions. *Acta Math.* **120** (1968), 53–58.

147. T. H. GULLIKSEN AND G. LEVIN. *Homology of local rings.* Queen's papers in pure and applied mathematics **20**, Queen's University, Kingston, Ont., 1969.

148. S. HALPERIN. The non-vanishing of the deviations of a local ring. *Comment. Math. Helv.* **62** (1987), 646–653.

149. N. HARA. A characterization of rational singularities in terms of injectivity of Frobenius maps. (Preprint).

150. N. HARA AND K. WATANABE. The injectivity of Frobenius acting on local cohomology and local cohomology modules. *Manuscripta Math.* **90** (1996), 301–315.

151. R. HARTSHORNE. *Residues and duality.* LNM **20**, Springer, 1966.

152. R. HARTSHORNE. *Algebraic geometry.* Springer, 1977.

153. M. HASHIMOTO. Determinantal ideals without minimal free resolutions. *Nagoya Math. J.* **118** (1990), 203–216.

154. M. HASHIMOTO. Resolutions of determinantal ideals: t-minors of $(t+2) \times n$ matrices. *J. Algebra* **142** (1991), 456–491.

155. M. HERRMANN, S. IKEDA, AND U. ORBANZ. *Equimultiplicity and blowing up.* Springer, 1988.

156. J. HERZOG. Certain complexes associated to a sequence and a matrix. *Manuscripta Math.* **12** (1974), 217–248.

157. J. HERZOG. Ringe der Charakteristik p und Frobeniusfunktoren. *Math. Z.* **140** (1974), 67–78.

158. J. HERZOG AND M. KÜHL. On the Betti numbers of finite pure and linear resolutions. *Commun. Algebra* **12** (1984), 1627–1646.

159. J. HERZOG AND E. KUNZ (eds.). *Der kanonische Modul eines Cohen–Macaulay–Rings.* LNM **238**, Springer, 1971.

160. J. HERZOG AND E. KUNZ. *Die Wertehalbgruppe eines lokalen Rings der Dimension 1.* S.-Ber. Heidelberger Akad. Wiss. II. Abh., 1971.

161. J. HERZOG, A. SIMIS, AND W. V. VASCONCELOS. Approximation complexes and blowing-up rings, I. *J. Algebra* **74** (1982), 466–493.

162. J. HERZOG, A. SIMIS, AND W. V. VASCONCELOS. Approximation complexes and blowing-up rings, II. *J. Algebra* **82** (1983), 53–83.

163. J. HERZOG AND NGÔ VIÊT TRUNG. Gröbner bases and multiplicity of determinantal and pfaffian ideals. *Adv. in Math.* **96** (1992), 1–37.

164. T. HIBI. Every affine graded algebra has a Hodge algebra structure. *Rend. Sem. Math. Univ. Politech. Torino* **44** (1986), 277–286.

165. T. HIBI. Distributive lattices, affine semigroup rings, and algebras with straightening laws. In M. NAGATA AND H. MATSUMURA (eds.), *Commutative algebra and combinatorics.* Advanced Studies in Pure Math. **11**, North–Holland, 1987, pp. 93–109.

166. T. HIBI. Level rings and algebras with straightening laws. *J. Algebra* **117** (1988), 343–362.

167. T. HIBI. Flawless O-sequences and Hilbert functions of Cohen–Macaulay integral domains. *J. Pure Applied Algebra* **60** (1989), 245–251.

168. T. HIBI. Ehrhart polynomials of convex polytopes, h-vectors of simplicial complexes, and nonsingular projective toric varieties. In J. E. GOODMAN, R. POLLACK, AND W. STEIGER (eds.), *Discrete and computational geometry.* DIMACS Series **6**, Amer. Math. Soc., 1991, pp. 165–177.

169. T. HIBI. *Algebraic combinatorics on convex polytopes.* Carslaw Publications, 1992.

170. T. HIBI. Face number inequalities for matroid complexes and Cohen–Macaulay types of Stanley–Reisner rings of distributive lattices. *Pacific J. Math.* **154** (1992), 253–264.

171. D. HILBERT. Über die Theorie der algebraischen Formen. *Math. Ann.* **36** (1890), 473–534.

172. V. A. HINIČ. On the Gorenstein property of the ring of invariants. *Math. USSR-Izv.* **10** (1976), 47–53.

173. LÊ TUÂN HOA. The Gorenstein property depends upon characteristic for affine semigroup rings. *Arch. Math.* **56** (1991), 228–235.

174. M. HOCHSTER. Rings of invariants of tori, Cohen–Macaulay rings generated by monomials, and polytopes. *Ann. of Math.* **96** (1972), 318–337.

175. M. HOCHSTER. Cohen–Macaulay modules. In J. W. BREWER AND E. A. RUTTER (eds.), *Conference on commutative algebra.* LNM **311**, Springer, 1973, pp. 120–157.

176. M. HOCHSTER. Contracted ideals from integral extensions of regular rings. *Nagoya Math. J.* **51** (1973), 25–43.

177. M. HOCHSTER. Grassmannians and their Schubert subvarieties are arithmetically Cohen–Macaulay. *J. Algebra* **25** (1973), 40–57.

178. M. HOCHSTER. *Deep local rings.* Preprint Series **8**, Matematisk Institut, Aarhus Universitet, 1973/74.

179. M. HOCHSTER. Grade-sensitive modules and perfect modules. *Proc. London Math. Soc.* **29** (1974), 55–76.

180. M. HOCHSTER. Big Cohen–Macaulay modules and algebras and embeddability in rings of Witt vectors. In *Proceedings of the conference on commutative algebra, Kingston 1975.* Queen's Papers in Pure and Applied Mathematics **42**, 1975, pp. 106–195.

181. M. HOCHSTER. *Topics in the homological theory of modules over commutative rings.* CBMS Regional conference series **24**, Amer. Math. Soc., 1975.

182. M. HOCHSTER. Cohen–Macaulay rings, combinatorics, and simplicial complexes. In B. R. MCDONALD AND R. A. MORRIS (eds.), *Ring theory II.* Lect. Notes in Pure and Appl. Math. **26**, M. Dekker, 1977, pp. 171–223.

183. M. HOCHSTER. Some applications of the Frobenius in characteristic 0. *Bull. Amer. Math. Soc.* **84** (1978), 886–912.

184. M. HOCHSTER. Canonical elements in local cohomology modules and the direct summand conjecture. *J. Algebra* **84** (1983), 503–553.

185. M. HOCHSTER. Invariant theory of commutative rings. In S. MONTGOMERY (ed.), *Group actions on rings.* Contemp. Math. **43**, Amer. Math. Soc., 1985, pp. 161–179.

186. M. HOCHSTER. Intersection problems and Cohen–Macaulay modules. In S. J. BLOCH (ed.), *Algebraic geometry, Bowdoin 1985.* Proc. Symp. Pure Math. **46**, Part II, Amer. Math. Soc., 1987, pp. 491–501.

187. M. HOCHSTER. The canonical module of a ring of invariants. In R. FOSSUM ET AL. (eds.), *Invariant theory, Denton 1986.* Contemp. Math. **88**, Amer. Math. Soc., 1989, pp. 43–83.

188. M. HOCHSTER. Solid closure. In W. HEINZER, J. D. SALLY, AND C. HUNEKE (eds.), *Commutative algebra: syzygies, multiplicities, and birational algebra.* Contemp Math. **159**, Amer. Math. Soc., 1994, pp. 103–172.

189. M. HOCHSTER AND J. A. EAGON. Cohen–Macaulay rings, invariant theory, and the generic perfection of determinantal loci. *Amer. J. Math.* **93** (1971), 1020–1058.

190. M. HOCHSTER AND C. HUNEKE. Tight closure. In M. HOCHSTER, C. HUNEKE, AND J. D. SALLY (eds.), *Commutative algebra.* Math. Sc. Res. Inst. Publ. **15**, Springer, 1989, pp. 305–324.

191. M. HOCHSTER AND C. HUNEKE. Tight closure and strong *F*-regularity. *Mém. Soc. Math. France (N. S.)* **38** (1989), 119–133.

192. M. HOCHSTER AND C. HUNEKE. Tight closure, invariant theory, and the Briançon–Skoda theorem. *J. Amer. Math. Soc.* **3** (1990), 31–116.

193. M. HOCHSTER AND C. HUNEKE. Infinite integral closures and big Cohen–Macaulay algebras. *Ann. of Math.* **135** (1992), 53–89.

194. M. HOCHSTER AND C. HUNEKE. Phantom homology. *Mem. Amer. Math. Soc.* **103** (1993), 1–91.

195. M. HOCHSTER AND C. HUNEKE. *F*-regularity, test elements, and smooth base change. *Trans. Amer. Math. Soc.* **346** (1994), 1–62.

196. M. HOCHSTER AND C. HUNEKE. Tight closures of parameter ideals and splitting in module-finite extensions. *J. Algebr. Geom.* **3** (1994), 599–670.

197. M. HOCHSTER AND C. HUNEKE. Applications of the existence of big Cohen–Macaulay algebras. *Adv. in Math.* **113** (1995), 45–117.

198. M. HOCHSTER AND C. HUNEKE. Tight closure in characteristic zero. (Preprint).

199. M. HOCHSTER AND J. E. MCLAUGHLIN. Splitting theorems for quadratic ring extensions. *Ill. J. Math.* **27** (1983), 94–103.

200. M. HOCHSTER AND L. J. RATLIFF, JR. Five theorems on Cohen–Macaulay rings. *Pacific J. Math.* **44** (1973), 147–172.

201. M. HOCHSTER AND J. L. ROBERTS. Rings of invariants of reductive groups acting on regular rings are Cohen–Macaulay. *Adv. in Math.* **13** (1974), 115–175.

202. M. HOCHSTER AND J. L. ROBERTS. The purity of the Frobenius and local cohomology. *Adv. in Math.* **21** (1976), 117–172.

203. W. V. D. HODGE. Some enumerative results in the theory of forms. *Proc. Camb. Philos. Soc.* **39** (1943), 22–30.

204. W. V. D. HODGE AND D. PEDOE. *Methods of algebraic geometry.* Cambridge University Press, 1952.

205. S. HUCKABA AND C. HUNEKE. Powers of ideals having small analytic deviation. *Amer. J. Math.* **114** (1992), 367–404.

206. S. HUCKABA AND C. HUNEKE. Rees algebras of ideals having small analytic deviation. *Trans. Amer. Math. Soc.* **339** (1993), 373–402.

207. H. A. HULETT. Maximum Betti numbers of homogeneous ideals with a given Hilbert function. *Commun. Algebra* **21** (1993), 2335–2350.

208. J. E. HUMPHREYS. *Linear algebraic groups.* Springer, 1975.

209. C. HUNEKE. On the symmetric and Rees algebra of an ideal generated by a *d*-sequence. *J. Algebra* **62** (1980), 268–275.

210. C. HUNEKE. Linkage and the Koszul homology of ideals. *Amer. J. Math.* **104** (1982), 1043–1062.

211. C. HUNEKE. On the associated graded ring of an ideal. *Ill. J. Math.* **26** (1982), 121–137.

212. C. HUNEKE. The theory of *d*-sequences and powers of ideals. *Adv. in Math.* **46** (1982), 249–279.

213. C. HUNEKE. Numerical invariants of liaison classes. *Invent. Math.* **75** (1984), 301–325.

214. C. HUNEKE. Uniform bounds in Noetherian rings. *Invent. Math.* **107** (1992), 203–223.

215. C. HUNEKE. *Tight closure and its applications*. Amer. Math. Soc., 1996.

216. C. HUNEKE AND J.-H. KOH. Some dimension 3 cases of the canonical element conjecture. *Proc. Amer. Math. Soc.* **98** (1986), 394–398.

217. C. HUNEKE AND M. MILLER. A note on the multiplicity of Cohen–Macaulay algebras with pure resolutions. *Canad. J. Math.* **37** (1985), 1149–1162.

218. C. HUNEKE AND K. E. SMITH. Tight closure and the Kodaira vanishing theorem. *J. Reine Angew. Math.* **484** (1997), 127–152.

219. C. HUNEKE AND B. ULRICH. Divisor class groups and deformations. *Amer. J. Math.* **107** (1985), 1265–1303.

220. C. HUNEKE AND B. ULRICH. The structure of linkage. *Ann. of Math.* **126** (1987), 277–334.

221. C. HUNEKE AND B. ULRICH. Algebraic linkage. *Duke Math. J.* **56** (1988), 415–429.

222. C. HUNEKE AND B. ULRICH. Minimal linkage and the Gorenstein locus of an ideal. *Nagoya Math. J.* **109** (1988), 159–167.

223. S. IKEDA. The Cohen–Macaulayness of the Rees algebras of local rings. *Nagoya Math. J.* **89** (1983), 47–63.

224. S. IKEDA. On the Gorensteinness of Rees algebras over local rings. *Nagoya Math. J.* **102** (1986), 135–154.

225. S. IKEDA AND NGÔ VIÊT TRUNG. When is the Rees algebra Cohen–Macaulay? *Commun. Algebra* **17** (1989), 2893–2922.

226. F. ISCHEBECK. Eine Dualität zwischen den Funktoren Ext und Tor. *J. Algebra* **11** (1969), 510–531.

227. M.-N. ISHIDA. The local cohomology groups of an affine semigroup ring. In H. HIJIKATA ET AL. (eds.), *Algebraic geometry and commutative algebra in honor of Masayoshi Nagata, Vol. I.* Konikuniya, 1987, pp. 141–153.

228. V. KAC AND K. WATANABE. Finite linear groups whose ring of invariants is a complete intersection. *Bull. Amer. Math. Soc.* **6** (1982), 221–223.

229. G. KALAI. Many triangulated spheres. *Discrete Comp. Geom.* **3** (1988), 1–14.

230. I. KAPLANSKY. Commutative rings. In J. W. BREWER AND E. A. RUTTER (eds.), *Conference on commutative algebra*. LNM **311**, Springer, 1973, pp. 153–166.

231. I. KAPLANSKY. *Commutative rings.* University of Chicago Press (revised edition), 1974.

232. G. KATONA. A theorem for finite sets. In P. ERDÖS AND G. KATONA (eds.), *Theory of graphs.* Academic Press, 1968, pp. 187–207.

233. T. KAWASAKI. Local rings of relative small type are Cohen–Macaulay. *Proc. Amer. Math. Soc.* **122** (1994), 703–709.

234. T. KAWASAKI. On Macaulayfication of Noetherian schemes. (Preprint).

235. G. KEMPF. The Hochster–Roberts theorem of invariant theory. *Michigan Math. J.* **26** (1979), 19–32.

236. G. KEMPF, F. KNUDSEN, D. MUMFORD, AND B. SAINT-DONAT. *Toroidal embeddings I.* LNM. **339**, Springer, 1973.

237. H. KLEPPE AND D. LAKSOV. The algebraic structure and deformation of Pfaffian schemes. *J. Algebra* **64** (1980), 167–189.

238. F. KNOP. Der kanonische Modul eines Invariantenringes. *J. Algebra* **127** (1989), 40–54.

239. F. KNOP. Die Cohen–Macaulay Eigenschaft von Invariantenringen. 1991. *Unpublished.*

240. J. L. KOSZUL. Sur un type d'algèbres différentielles en rapport avec la transgression. In *Colloque de topologie, Bruxelles 1950.* Paris, 1950, pp. 73–81.

241. H. KRAFT. *Geometrische Methoden in der Invariantentheorie.* Vieweg, 1984.

242. W. KRULL. Dimensionstheorie in Stellenringen. *J. Reine Angew. Math.* **179** (1938), 204–226.

243. J. KRUSKAL. The number of simplices in a complex. In R. BELLMAN (ed.), *Mathematical optimization techniques.* University of California Press, 1963, pp. 251–278.

244. E. KUNZ. Vollständige Durchschnitte und Differenten. *Arch. Math.* **19** (1968), 47–58.

245. E. KUNZ. Characterizations of regular local rings of characteristic p. *Amer. J. Math.* **91** (1969), 772–784.

246. E. KUNZ. Almost complete intersections are not Gorenstein rings. *J. Algebra* **28** (1974), 111–115.

247. E. KUNZ. Holomorphe Differentialformen auf algebraischen Varietäten mit Singularitäten I. *Manuscripta Math.* **15** (1975), 91–108.

248. E. KUNZ. On Noetherian rings of characteristic p. *Amer. J. Math.* **98** (1976), 999–1013.

249. E. KUNZ. *Introduction to commutative algebra and algebraic geometry.* Birkhäuser, 1985.

250. E. KUNZ AND R. WALDI. *Regular differential forms.* Contemp. Math. **79**, Amer. Math. Soc., 1988.

251. A. KUSTIN AND M. MILLER. Constructing big Gorenstein ideals from small ones. *J. Algebra* **85** (1983), 303–322.

252. A. KUSTIN AND M. MILLER. Deformation and linkage of Gorenstein algebras. *Trans. Amer. Math. Soc.* **284** (1984), 501–534.

253. A. KUSTIN AND M. MILLER. Classification of the Tor-algebras of codimension four Gorenstein local rings. *Math. Z.* **190** (1985), 341–355.

254. A. KUSTIN AND M. MILLER. Tight double linkage of Gorenstein algebras. *J. Algebra* **95** (1985), 384–397.

255. R. E. KUTZ. Cohen–Macaulay rings and ideal theory of invariants of algebraic groups. *Trans. Amer. Math. Soc.* **194** (1974), 115–129.

256. D. LAKSOV. The arithmetic Cohen–Macaulay character of Schubert schemes. *Acta Math.* **129** (1972), 1–9.

257. A. LASCOUX. Syzygies des variétés déterminantales. *Adv. in Math.* **30** (1978), 202–237.

258. S. LICHTENBAUM. On the vanishing of Tor in regular local rings. *Ill. J. Math.* **10** (1966), 220–226.

259. J. LIPMAN. *Dualizing sheaves, differentials, and residues on algebraic varieties.* Asterisque **117**, Soc. Math. de France, 1984.

260. J. LIPMAN AND A. SATHAYE. Jacobian ideals and a theorem of Briançon–Skoda. *Michigan Math. J.* **28** (1981), 199–222.

261. J. LIPMAN AND B. TEISSIER. Pseudo-rational local rings and a theorem of Briançon–Skoda about integral closures of ideals. *Michigan Math. J.* **28** (1981), 97–116.

262. F. S. MACAULAY. *The algebraic theory of modular systems.* Cambridge University Press, 1916.

263. F. S. MACAULAY. Some properties of enumeration in the theory of modular systems. *Proc. London Math. Soc.* **26** (1927), 531–555.

264. S. MACLANE. *Homology.* Springer, 1975.

265. R. MACRAE. On an application of the Fitting invariants. *J. Algebra* **2** (1965), 153–169.

266. W. S. MASSEY. *Singular homology theory.* Springer, 1980.

267. J. R. MATIJEVIC. *Some topics in graded rings.* PhD thesis, University of Chicago, 1973.

268. J. R. MATIJEVIC AND P. ROBERTS. A conjecture of Nagata on graded Cohen–Macaulay rings. *J. Math. Kyoto Univ.* **14** (1974), 125–128.

269. E. MATLIS. Injective modules over Noetherian rings. *Pacific J. Math.* **8** (1958), 511–528.

270. H. MATSUMURA. *Commutative ring theory.* Cambridge University Press, 1986.

271. P. MCMULLEN. The maximum numbers of faces of a convex polytope. *Mathematika* **17** (1970), 179–184.

272. P. MCMULLEN AND G. C. SHEPHARD. *Convex polytopes and the upper bound conjecture.* LMS Lect. Note Ser. 3, Cambridge University Press, 1971.

273. P. MCMULLEN AND D. W. WALKUP. A generalized lower-bound conjecture for simplicial polytopes. *Mathematika* **18** (1971), 264–273.

274. F. MEYER. Zur Theorie der reductibeln ganzen Functionen von *n* Variablen. *Math. Ann.* **30** (1887), 30–74.

275. M. MIYAZAKI. Level complexes and barycentric subdivisions. *J. Math. Kyoto Univ.* **30** (1990), 459–467.

276. M. MIYAZAKI. On 2-Buchsbaum complexes. *J. Math. Kyoto Univ.* **30** (1990), 367–392.

277. P. MONSKY. The Hilbert–Kunz function. *Math. Ann.* **263** (1983), 43–49.

278. T. S. MOTZKIN. Comonotone curves and polyhedra. *Bull. Amer. Math. Soc.* **63** (1957), 35.

279. D. MUMFORD AND J. FOGARTY. *Geometric invariant theory.* Springer, 1982.

280. J. R. MUNKRES. *Elements of algebraic topology.* Addison–Wesley, 1984.

281. M. P. MURTHY. A note on factorial rings. *Arch. Math.* **15** (1964), 418–420.

282. C. MUSILI. Postulation formula for Schubert varieties. *J. Ind. Math. Soc.* **36** (1972), 143–171.

283. M. NAGATA. The theory of multiplicity in general local rings. In *Proc. Intern. Symp. Tokyo–Nikko 1955.* Science Council of Japan, 1956, pp. 191–226.

284. M. NAGATA. *Local rings.* Interscience, 1962.

285. H. NAKAJIMA. Affine torus embeddings which are complete intersections. *Tôhoku Math. J.* **38** (1986), 85–98.

286. H. NAKAJIMA. Rings of invariants of finite groups which are hypersurfaces II. *Adv. in Math.* **65** (1987), 39–64.

287. H. NAKAJIMA AND K. WATANABE. The classification of quotient singularities which are complete intersections. In S. GRECO AND R. STRANO (eds.), *Complete intersections, Acireale 1983.* LNM **1092**, Springer, 1984, pp. 102–120.

288. D. G. NORTHCOTT. A homological investigation of a certain residual ideal. *Math. Ann.* **150** (1963), 99–110.

289. D. G. NORTHCOTT. *Lessons on rings, modules, and multiplicities.* Cambridge University Press, 1968.

290. D. G. NORTHCOTT. *Finite free resolutions.* Cambridge University Press, 1976.

291. D. G. NORTHCOTT AND D. REES. Reductions of ideals in local rings. *Proc. Camb. Philos. Soc.* **50** (1954), 145–158.

292. L. O'CARROLL. A uniform Artin–Rees theorem and Zariski's main lemma on holomorphic functions. *Invent. Math.* **90** (1987), 647–652.

293. T. ODA. *Convex bodies and algebraic geometry.* Springer, 1985.

294. T. OGOMA. Cohen–Macaulay factorial domain is not necessarily Gorenstein. *Mem. Fac. Sci. Kôchi Univ. Ser. A Math.* **3** (1982), 65–74.

295. T. OGOMA. A note on the syzygy problem. *Commun. Algebra* **17** (1989), 2061–2066.

296. K. PARDUE. Deformation classes of graded modules and maximal Betti numbers. *Ill. J. Math.* **40** (1996), 564–585.

297. C. PESKINE AND L. SZPIRO. Dimension projective finie et cohomologie locale. *Publ. Math. I.H.E.S.* **42** (1972), 47–119.

298. C. PESKINE AND L. SZPIRO. Liaison des variétés algébriques. *Invent. Math.* **26** (1974), 271–302.

299. C. PESKINE AND L. SZPIRO. Syzygies et multiplicités. *C. R. Acad. Sc. Paris Sér. A* **278** (1974), 1421–1424.

300. J. H. POINCARÉ. Analysis situs. *J. de l'Ecole Polytechnique* (1895), 1–121.

301. M. RAYNAUD. *Anneaux locaux henséliens.* LNM **169**, Springer, 1970.

302. D. REES. A theorem of homological algebra. *Proc. Camb. Philos. Soc.* **52** (1956), 605–610.

303. D. REES. The grade of an ideal or module. *Proc. Camb. Philos. Soc.* **53** (1957), 28–42.

304. D. REES. On a problem of Zariski. *Ill. J. Math.* **2** (1958), 145–149.

305. D. REES. *Lectures on the asymptotic theory of ideals.* LMS Lect. Note Ser. **113**, Cambridge University Press, 1988.

306. G. A. REISNER. Cohen–Macaulay quotients of polynomial rings. *Adv. in Math.* **21** (1976), 30–49.

307. I. REITEN. The converse to a theorem of Sharp on Gorenstein modules. *Proc. Amer. Math. Soc.* **32** (1972), 417–420.

308. L. ROBBIANO. Introduction to the theory of Gröbner bases. In A. V. GERAMITA (ed.), *The curves seminar at Queen's, Vol. V.* Queen's Papers in Pure and Applied Mathematics **80**, 1988.

309. P. ROBERTS. Two applications of dualizing complexes over local rings. *Ann. Sci. Ec. Norm. Sup. (4)* **9** (1976), 103–106.

310. P. ROBERTS. Cohen–Macaulay complexes and an analytic proof of the new intersection conjecture. *J. Algebra* **66** (1980), 220–225.

311. P. ROBERTS. *Homological invariants of modules over commutative rings.* Séminaire de Mathématiques Supérieures, Université de Montréal, 1980.

312. P. ROBERTS. Rings of type 1 are Gorenstein. *Bull. London Math. Soc.* **15** (1983), 48–50.

313. P. ROBERTS. The vanishing of intersection multiplicities of perfect complexes. *Bull. Amer. Math. Soc.* **13** (1985), 127–130.

314. P. ROBERTS. Le théorème d'intersection. *C. R. Acad. Sc. Paris Sér. I* **304** (1987), 177–180.

315. P. ROBERTS. Local Chern characters and intersection multiplicities. In S. J. BLOCH (ed.), *Algebraic geometry, Bowdoin 1985.* Proc. Symp. Pure Math. **46**, Part II, Amer. Math. Soc., 1987, pp. 389–400.

316. P. ROBERTS. Intersection theorems. In M. HOCHSTER, C. HUNEKE, AND J. D. SALLY (eds.), *Commutative algebra.* Math. Sc. Res. Inst. Publ. **15**, Springer, 1989, pp. 417–436.

317. P. ROBERTS. An infinitely generated symbolic blow-up in a power series ring and a new counterexample to Hilbert's fourteenth problem. *J. Algebra* **132** (1990), 461–473.

318. J. ROTMAN. *An introduction to homological algebra.* Academic Press, 1979.

319. J. D. SALLY. *Numbers of generators of ideals in local rings.* Lect. Notes in Pure and Appl. Math. **35**, M. Dekker, 1978.

320. P. SAMUEL. La notion de multiplicité en algèbre et en géométrie algébrique. *J. math. pure et appl.* **30** (1951), 159–274.

321. U. SCHÄFER AND P. SCHENZEL. Dualizing complexes and affine semigroup rings. *Trans. Amer. Math. Soc.* **322** (1990), 561–582.

322. G. SCHEJA. Über die Bettizahlen lokaler Ringe. *Math. Ann.* **155** (1964), 155–172.

323. G. SCHEJA AND U. STORCH. Differentielle Eigenschaften der Lokalisierungen analytischer Algebren. *Math. Ann.* **197** (1972), 137–170.

324. G. SCHEJA AND U. STORCH. Über Spurfunktionen bei vollständigen Durchschnitten. *J. Reine Angew. Math.* **278** (1975), 174–190.

325. G. SCHEJA AND U. STORCH. Residuen bei vollständigen Durchschnitten. *Math. Nachr.* **91** (1979), 157–170.

326. P. SCHENZEL. Dualizing complexes and systems of parameters. *J. Algebra* **58** (1979), 495–501.

327. P. SCHENZEL. Über die freien Auflösungen extremaler Cohen–Macaulay–Ringe. *J. Algebra* **64** (1980), 93–101.

328. P. SCHENZEL. Cohomological annihilators. *Math. Proc. Cambridge Philos. Soc.* **91** (1982), 345–350.

329. P. SCHENZEL. *Dualisierende Komplexe in der lokalen Algebra und Buchsbaum–Ringe.* LNM **907**, Springer, 1982.

330. C. SCHOELLER. Homologie des anneaux locaux noethériens. *C. R. Acad. Sc. Paris Sér. A* **265** (1967), 768–771.

331. G. SEIBERT. Complexes with homology of finite length and Frobenius functors. *J. Algebra* **125** (1989), 278–287.

332. J.-P. SERRE. Sur la dimension homologique des anneaux et des modules noethériens. In *Proc. Int. Symp. Tokyo–Nikko 1955.* Science Council of Japan, 1956, pp. 175–189.

333. J.-P. SERRE. *Groupes algébriques et corps de classes.* Hermann, 1959.

334. J.-P. SERRE. *Algèbre locale. Multiplicités.* LNM **11**, Springer, 1965.

335. J.-P. SERRE. Groupes finis d'automorphismes d'anneaux locaux réguliers. In *Colloque d'algèbre ENSJF, Paris 1967.* Secrétariat mathématique, 1968.

336. R. Y. SHARP. The Cousin complex for a module over a commutative Noetherian ring. *Math. Z.* **112** (1969), 340–356.

337. R. Y. SHARP. Gorenstein modules. *Math. Z.* **115** (1970), 117–139.

338. R. Y. SHARP. On Gorenstein modules over a complete Cohen–Macaulay local ring. *Quart. J. Math. Oxford Ser.* (2) **22** (1971), 425–434.

339. R. Y. SHARP. Finitely generated modules of finite injective dimension over certain Cohen–Macaulay rings. *Proc. London Math. Soc.* **25** (1972), 303–328.

340. R. Y. SHARP. Dualizing complexes for commutative Noetherian rings. *Math. Proc. Cambridge Philos. Soc.* **78** (1975), 369–386.

341. R. Y. SHARP. Local cohomology and the Cousin complex for a commutative Noetherian ring. *Math. Z.* **153** (1977), 19–22.

342. R. Y. SHARP. Cohen–Macaulay properties for balanced big Cohen–Macaulay modules. *Math. Proc. Cambridge Philos. Soc.* **90** (1981), 229–238.

343. R. Y. SHARP. Certain countably generated big Cohen–Macaulay modules are balanced. *Math. Proc. Cambridge Philos. Soc.* **105** (1989), 73–77.

344. R. Y. SHARP. *Steps in commutative algebra.* Cambridge University Press, 1990.

345. G. C. SHEPHARD AND J. A. TODD. Finite unitary reflection groups. *Canad. J. Math.* **6** (1954), 274–304.

346. A. SIMIS AND W. V. VASCONCELOS. Approximation complexes. In *Atas da 6. escola de álgebra.* Coleção Atas **14**, Soc. Bras. Mat., 1981, pp. 87–157.

347. H. SKODA AND J. BRIANÇON. Sur la clôture intégrale d'un idéal de germes de fonctions holomorphes en un point de $\mathbb{C}^n$. *C. R. Acad. Sc. Paris Sér. A* **278** (1974), 949–951.

348. K. E. SMITH. Tight closure of parameter ideals. *Invent. Math.* **115** (1994), 41–60.

349. K. E. SMITH. *F*-rational rings have rational singularities. *Amer. J. Math.* **119** (1997), 159–180.

350. K. E. SMITH. Tight closure in graded rings. *J. Math. Kyoto Univ.* **37** (1997), 35–53.

351. K. E. SMITH. Vanishing, singularities, and effective bounds via prime characteristic local algebra. In J. KOLLAR, D. MORRISON, AND R. LAZARSFELD (eds.), *Algebraic Geometry, Santa Cruz, 1995*. Proc. Symp. Pure Math., Amer. Math. Soc., 1997.

352. K. E. SMITH AND M. VAN DEN BERGH. Simplicity of rings of differential operators in prime characteristic. *Proc. London Math. Soc.* **75** (1997), 32–62.

353. L. SMITH. *Polynomial invariants of finite groups*. A. K. Peters, 1995.

354. E. SPERNER. Über einen kombinatorischen Satz von Macaulay und seine Anwendungen auf die Theorie der Polynomideale. *Abh. Math. Sem. Univ. Hamburg* **7** (1930), 149–163.

355. T. SPRINGER. *Invariant theory*. LNM **585**, Springer, 1977.

356. R. P. STANLEY. The upper bound conjecture and Cohen–Macaulay rings. *Stud. Appl. Math.* **54** (1975), 135–142.

357. R. P. STANLEY. Hilbert functions of graded algebras. *Adv. in Math.* **28** (1978), 57–83.

358. R. P. STANLEY. Invariants of finite groups and their applications to combinatorics. *Bull. Amer. Math. Soc.* **1** (1979), 475–511.

359. R. P. STANLEY. The number of faces of a simplicial convex polytope. *Adv. in Math.* **35** (1980), 236–238.

360. R. P. STANLEY. Linear diophantine equations and local cohomology. *Invent. Math.* **68** (1982), 175–193.

361. R. P. STANLEY. *Enumerative combinatorics, Vol. I*. Wadsworth & Brooks/Cole, 1986.

362. R. P. STANLEY. On the Hilbert function of a graded Cohen–Macaulay domain. *J. Pure Applied Algebra* **73** (1991), 307–314.

363. R. P. STANLEY. *Combinatorics and commutative algebra*. Birkhäuser, second edition 1996.

364. J. R. STROOKER. *Homological questions in local algebra*. LMS Lect. Note Ser. **145**, Cambridge University Press, 1990.

365. J. STÜCKRAD AND W. VOGEL. *Buchsbaum rings and applications*. Springer, 1986.

366. B. STURMFELS. *Gröbner bases and convex polytopes*. Amer. Math. Soc., 1996.

367. T. SVANES. Coherent cohomology on Schubert subschemes of flag schemes and applications. *Adv. in Math.* **14** (1974), 369–453.

368. I. SWANSON. Joint reductions, tight closure, and the Briançon–Skoda theorem. *J. Algebra* **147** (1992), 128–136.

369. J. TATE. Homology of Noetherian rings and local rings. *Ill. J. Math.* **1** (1957), 14–27.

370. NGÔ VIÊT TRUNG. On the presentation of Hodge algebras and the existence of Hodge algebra structures. *Commun. Algebra* **19** (1991), 1183–1195.

371. NGÔ VIÊT TRUNG AND LÊ TUÂN HOA. Affine semigroups and Cohen–Macaulay rings generated by monomials. *Trans. Amer. Math. Soc.* **298** (1986), 145–167.

372. B. ULRICH. Gorenstein rings as specializations of unique factorization domains. *J. Algebra* **86** (1984), 129–140.

373. B. ULRICH. On licci ideals. In R. FOSSUM ET AL. (eds.), *Invariant theory, Denton 1986.* Contemp. Math. **88**, Amer. Math. Soc., 1989, pp. 84–94.

374. B. ULRICH. Sums of linked ideals. *Trans. Amer. Math. Soc.* **318** (1990), 1–42.

375. G. VALLA. Certain graded algebras are always Cohen–Macaulay. *J. Algebra* **42** (1976), 537–548.

376. G. VALLA. On the symmetric and Rees algebra of an ideal. *Manuscripta Math.* **30** (1980), 239–255.

377. G. VALLA. On the Betti numbers of perfect ideals. *Compositio Math.* **91** (1994), 305–319.

378. M. VAN DEN BERGH. Cohen–Macaulayness of modules of covariants. *Invent. Math.* **106** (1991), 389–409.

379. W. V. VASCONCELOS. Ideals generated by R-sequences. *J. Algebra* **6** (1967), 309–316.

380. W. V. VASCONCELOS. *Divisor theory in module categories.* North–Holland, 1974.

381. W. V. VASCONCELOS. On the equations of Rees algebras. *J. Reine Angew. Math.* **418** (1991), 189–218.

382. W. V. VASCONCELOS. *Arithmetic of blowup algebras.* LMS Lect. Note Ser. **195**, Cambridge University Press, 1994.

383. W. V. VASCONCELOS. Hilbert functions, analytic spread, and Koszul homology. In W. HEINZER, C. HUNEKE, AND J. D. SALLY (eds.), *Syzygies, multiplicities, and birational algebra.* Contemp. Math. **159**, Amer. Math. Soc., 1994, pp. 401–422.

384. W. V. VASCONCELOS. *Computational methods in commutative algebra and algebraic geometry.* Springer, to appear.

385. J. VELEZ. Openness of the F-rational locus and smooth base change. *J. Algebra* **172** (1995), 425–453.

386. J. WATANABE. A note on Gorenstein rings of embedding codimension three. *Nagoya Math. J.* **50** (1973), 227–232.

387. K. WATANABE. Certain invariant subrings are Gorenstein. I. *Osaka J. Math.* **11** (1974), 1–8.

388. K. WATANABE. Certain invariant subrings are Gorenstein. II. *Osaka J. Math.* **11** (1974), 379–388.

389. K. WATANABE. Rational singularities with k^*-action. In S. GRECO (ed.), *Commutative algebra.* Lect. Notes in Pure and Appl. Math. **84**, M. Dekker, 1983, pp. 339–351.

390. K. WATANABE. *F*-regular and *F*-pure normal graded rings. *J. Pure Applied Algebra* **71** (1991), 341–350.

391. K. WATANABE. *F*-regular and *F*-pure rings vs. log-terminal and log-canonical singularities. (Preprint).

392. D. WESTON. On descent in dimension two and non-split Gorenstein modules. *J. Algebra* **118** (1988), 263–275.

393. J. WEYMAN. On the structure of free resolutions of length 3. *J. Algebra* **126** (1989), 1–33.

394. F. WHIPPLE. On a theorem due to F. S. Macaulay. *J. London Math. Soc.* **8** (1928), 431–437.

395. H. WIEBE. Über homologische Invarianten lokaler Ringe. *Math. Ann.* **179** (1969), 257–274.

396. O. ZARISKI. The concept of a simple point of an abstract algebraic variety. *Trans. Amer. Math. Soc.* **62** (1947), 1–52.

397. O. ZARISKI AND P. SAMUEL. *Commutative algebra, Vols. I and II.* Van Nostrand (new edn. Springer 1975), 1958, 1960.

398. S. ZARZUELA. Systems of parameters for nonfinitely generated modules and big Cohen–Macaulay modules. *Mathematika* **35** (1988), 207–215.

399. G. M. ZIEGLER. *Lectures on polytopes.* Springer, 1995.

Notation

439

dim R	Krull dimension of the ring R, 413
$E(M)$	injective hull of the R-module M, 98
$^*E(M)$	*injective hull of the graded module M, 137
emb dim R	embedding dimension of the local ring R, 74
$e(I, M)$	multiplicity of M with respect to the ideal I, 188
$e(M)$	multiplicity of the (graded) module M, 150, 188
$e(x, M)$	multiplicity symbol, 193
$\varepsilon_1(R)$	first deviation of the local ring R, 75
$\varepsilon_2(R)$	second deviation of the local ring R, 81
$^*\mathrm{Ext}_i(M, N)$	i-th graded extension module of M by N, 33
$\mathscr{F}$	Frobenius functor, 328
$\mathscr{F}(D)$	face lattice of the cone D, 262
$\mathscr{F}(P)$	face lattice of the polyhedron P, 224
$f(\Delta)$	f-vector of the simplicial complex Δ, 208
$\mathscr{F}_0(R)$	category of finite graded R-modules, 142
$\mathscr{F}(R)$	category of finite R-modules, 105
$\Gamma * \Delta$	join of the simplicial complexes Γ and Δ, 221
Γ_γ	312
$\Gamma_{\mathfrak{m}}(M)$	submodule of elements of M with support in $\{\mathfrak{m}\}$, 127
G_γ	residue class ring of $G(X)$, 312
GL (n, k)	group of invertible $n \times n$ matrices over k
GL (V)	group of automorphisms of the vector space V
grade I	grade of the ideal I, 12
grade(I, M)	grade of the ideal I with respect to the module M, 10
grade M	grade of the module M, 12
$\mathrm{gr}_F(R)$	associated graded ring with respect to filtration F, 182
$\mathrm{gr}_F(M)$	182
$\mathrm{gr}_I(M)$	associated graded module of M with respect to ideal I, 6
$\mathrm{gr}_I(R)$	associated graded ring of R with respect to ideal I, 6
g^*	initial form of g, 186
$G(X), G_B(X)$	B-algebra generated by the maximal minors of X, 306
H^+, H^-	closed half-spaces defined by the hyperplane H, 223
height I	height of the ideal I, 412
$H^\bullet(f)$	cohomology of the Koszul complex of f, 45
$H_\bullet(f)$	homology of the Koszul complex of f, 44
$H^\bullet(f, M)$	cohomology of $K^\bullet(f, M)$, 45
$H_\bullet(f, M)$	homology of $K_\bullet(f, M)$, 44
$\widetilde{H}^i(\Delta; G)$	i-th reduced simplicial cohomology of Δ with values in G, 230
$\widetilde{H}_i(\Delta; G)$	i-th reduced simplicial homology of Δ with values in G, 230
$H_i(M, n)$	i-th iterated Hilbert function of the module M, 150
$H(M, n)$	Hilbert function of the graded module M, 147
$H_M(t)$	Hilbert series of the graded module M, 147
$^*H_{\mathfrak{m}}^i(M)$	i-th *local cohomology module of the module M, 143
$H_{\mathfrak{m}}^i(M)$	i-th local cohomology module of the module M, 128
$H^\bullet(F^\bullet)$	cohomology of the complex $F^\bullet$
$H_\bullet(F_\bullet)$	homology of the complex $F_\bullet$
$H^i(F^\bullet)$	i-th cohomology of the complex $F^\bullet$

$H_i(F_\bullet)$	i-th homology of the complex $F_\bullet$
$^\bullet\mathrm{Hom}_R(M, N)$	group of homogeneous homomorphisms $\varphi: M \to N$, 33
$H_\bullet(R)$	Koszul algebra of the local ring R, 75
$H_i(R)^j$	80
$[I]$	divisor class of the ideal I
$\bar{I}$	integral closure of the ideal I, 388
I_δ	312
id_M	identity map on the set M
$^\bullet\mathrm{inj\,dim}\,M$	$^\bullet$injective dimension of the graded module M, 138
$\mathrm{inj\,dim}\,M$	injective dimension of the module M, 92
$I^{[q]}$	q-th Frobenius power of I, 328
$I^\bullet$	ideal generated by the homogeneous elements in I, 28
I^*	tight closure of the ideal I, 378
I^*	ideal generated by leading monomials of the elements of I, 157
$I_t(\varphi)$	ideal generated by the t-minors of (a matrix of) φ, 21
$I_t(U)$	ideal generated by the t-minors of the matrix U, 21
$k[C]$	affine semigroup ring over C, 256
$k[C]_F$	262
$k[\Delta]$	Stanley–Reisner ring of the simplicial complex Δ, 209
$K^\bullet(f)$	dual of the Koszul complex with respect to f, 45
$K_\bullet(f)$	Koszul complex of f, 44
$K^\bullet(f, M)$	dual of $K_\bullet(f)$ with respect to M, 45
$K_\bullet(f, M)$	Koszul complex of f with coefficients in M, 44
$K_\bullet(x)$	Koszul complex with respect to sequence x, 46
$k(\mathfrak{p})$	residue class field of localization with respect to $\mathfrak{p}$, 9
$\lambda(I)$	analytic spread of the ideal I, 190
$\lambda(I, M)$	analytic spread of the ideal I with respect to M, 190
$\ell(M)$	length of the module M
$L(f)$	leading monomial of f, 157
$L(I)$	set of the leading monomials of the elements of I, 157
$\varinjlim M_i$	direct limit of the modules M_i
$\varprojlim M_i$	inverse limit of the modules M_i
$\mathrm{lk}_\Delta F$	link of the face F with respect to Δ, 232
$\log$	257
$\mathscr{L}_u$	158
$\mathscr{M}_0(R)$	category of graded R-modules, 28
$M^\bullet, M^{\bullet\bullet}$	(bi)dual of the module M, 19
M_f	module of fractions with respect to the powers of f
$\hat{M}$	completion of the module M over a local ring
$M(i)$	module with shifted grading, 32
$M_\chi(t)$	Molien series of the character χ, 284
$M_G(t)$	Molien series of the group G, 283
$M_{(\mathfrak{p})}$	homogeneous localization of M with respect to $\mathfrak{p}$, 31
M_S	module of fractions of M with respect to the multiplicatively closed set S
$\mu_i(\mathfrak{p}, M)$	i-th Bass number of M with respect to $\mathfrak{p}$, 101
$\mu(M)$	minimal number of generators of M, 16

$\mathbb{N}$	set of non-negative integers
nat	natural map
ω_R	canonical module of the ring R, 110
$\mathcal{O}(x)$	order ideal of the element x of a module, 360
$\mathrm{Pf}(\varphi)$	ideal generated by the submaximal Pfaffians of φ, 121
$\mathrm{pf}(\varphi)$	Pfaffian of the matrix φ, 120
$P_M(n)$	Hilbert polynomial of the graded module M, 149
proj dim M	projective dimension of the module M, 16
$\mathbb{Q}$	field of rational numbers
$\mathbb{Q}_+$	set of non-negative rational numbers
$\mathbb{Q}C$	$\mathbb{Q}$-vector space generated by the affine semigroup C, 257
rank M	rank of the module M, 20
rank φ	rank of the homomorphism φ, 20
rank (φ, M)	rank of φ with respect to the module M, 351
rank v	rank of the poset element v, 215
R^χ	module of semi-invariants of the ring R with respect to the character χ, 279
R_δ	determinantal ring, 312
$\mathbb{R}$	field of real numbers
$\mathbb{R}_+$	set of non-negative real numbers
$\mathbb{R}_+ S$	cone generated by the set S, 258
$\mathbb{R}C$	$\mathbb{R}$-vector space generated by the affine semigroup C, 257
$\mathscr{R}_+(F)$	Rees ring with respect to the filtration F, 182
$\mathscr{R}_+(F, M)$	182
$\mathscr{R}(F)$	extended Rees ring with respect to the filtration F, 182
$\mathscr{R}(F, M)$	182
$\mathscr{R}(I, M)$	182
reg(M)	regularity of the graded module M, 168
relint C	relative interior of the affine semigroup C, 261
relint X	relative interior of the set X, 225
R_f	ring of fractions with respect to the set of powers of f
R^G	ring of invariants of R under the action of G, 278
$\hat{R}$	completion of the local ring R
$r_k(\Delta)$	type of the simplicial complex Δ over the field k, 240
$R * M$	trivial extension of the ring R by the module M, 111
$(R, \mathfrak{m})$	˙local ring R with ˙maximal ideal $\mathfrak{m}$, 35
$(R, \mathfrak{m})$	local ring R with maximal ideal $\mathfrak{m}$
$(R, \mathfrak{m}, k)$	local ring R with maximal ideal $\mathfrak{m}$ and residue class field $k = R/\mathfrak{m}$
$r(M)$	type of the module M, 13
(R_n)	Serre's condition (R_n), 71
$R_{(\mathfrak{p})}$	homogeneous localization of R with respect to $\mathfrak{p}$, 31
$R^{1/q}$	ring of q-th roots of the elements of R, 398
$R_{r+1}, R_{r+1}(X)$	determinantal ring, 306
R_S	ring of fractions of R with respect to the multiplicatively closed set S
$\mathscr{R}_u$	158

$R[[X_1, \ldots, X_n]]$	formal power series ring over R
R°	set of elements of R not contained in a minimal prime ideal, 378
Sing R	singular locus of R, 382
SL (n, k)	group of $n \times n$ matrices over k with determinant 1
SL (V)	group of automorphisms of V with determinant 1
$\Sigma_M^I(n)$	Hilbert–Samuel polynomial of M with respect to I, 188
(S_n)	Serre's condition (S_n), 63
Soc M	socle of the module M, 15
*Soc M	homogeneous socle of the graded module M, 142
$\mathrm{st}_\Delta F$	star of the face F with respect to Δ, 232
supp a	support of the element $a \in \mathbb{Z}^n$, 211
supp u	support of the monomial u, 300
supp x^a	support of the monomial x^a, 211
$S(V)$	symmetric algebra of V, 278
$\mathrm{tr\,deg}_k K$	transcendence degree of K over k
Tr ρ	trace of the linear map ρ, 284
$u \prec v$	v covers u, 215
$V(I)$	set of prime ideals containing I
vol P	volume of the polytope P, 277
$\bigwedge^i M$	i-th exterior power of the module M, 41
$\bigwedge M$	exterior algebra of the module M, 40
$\bigwedge \varphi$	extension of φ to exterior algebra, 41
x^F	monomial corresponding to the face F, 220
$x \wedge y$	product of x and y in exterior algebra, 40
$\langle x, y \rangle$	scalar product of x and y, 223
$\mathbb{Z}$	ring of integers
$\mathbb{Z}C$	group generated by the affine semigroup C, 256

Index